Connective Tissue Disease

THE BIOCHEMISTRY OF DISEASE

A Molecular Approach to Cell Pathology

A Series of Monographs

SERIES EDITORS

Emmanuel Farber
Department of Pathology
University of Toronto
Toronto, Ontario, Canada

Henry C. Pitot
McArdle Laboratory for Cancer Research
University of Wisconsin
Madison, Wisconsin

Volume 10 Nutritional Pathology: Pathobiochemistry of Dietary Imbalances
Edited by Herschel Sidransky

Volume 11 Biochemistry and Biology of Plasma Lipoproteins
Edited by Angelo M. Scanu and Arthur A. Spector

Volume 12 Connective Tissue Disease: Molecular Pathology of the Extracellular Matrix
Edited by Jouni Uitto and Andrea J. Perejda

Additional Volumes in Preparation

Connective Tissue Disease

Molecular Pathology of the Extracellular Matrix

edited by

JOUNI UITTO
ANDREA J. PEREJDA

UCLA School of Medicine and
Harbor – UCLA Medical Center
Torrance, California

MARCEL DEKKER, INC. New York and Basel

Library of Congress Cataloging-in-Publication Data

Connective tissue disease.

(Biochemistry of disease ; v. 12)
Includes index.
1. Connective tissues--Diseases. 2. Extracellular matrix--Diseases. 3. Pathology, Molecular. I. Uitto, Jouni, 1943- . II. Perejda, Andrea J., 1952- III. Series. [DNLM: 1. Connective Tissue Diseases. 2. Extracellular Matrix. W1 BI64F v.12 / WD 375 C752]
RC924.C553 1986 616.7'7 86-16665
ISBN 0-8247-7533-3

MARCEL DEKKER, INC.
270 Madison Avenue, New York, New York 10016

Current printing (last digit):
10 9 8 7 6 5 4 3 2 1

PRINTED IN THE UNITED STATES OF AMERICA

Preface

Biomedical research pertaining to the structure and metabolism of connective tissues has recently reached a new level of sophistication, leading to tremendous progress in our understanding of various aspects of basic biochemistry and applied biology of the extracellular matrix. For example, elucidation of the gene structures of several connective tissue proteins is currently underway, and application of recombinant DNA technology to connective tissue research promises to further our understanding of the regulatory mechanisms of gene expression. Progress made in understanding the basic biology of the extracellular matrix has provided us with novel means by which to explore the connective tissue defects in a variety of heritable and acquired diseases. Thus, we are now able to link specific molecular genetic defects to phenotypic expression of diseases more precisely than has been possible before. Better understanding of the etiologies and pathogeneses of these diseases has allowed us to initiate studies to formulate rational treatment modalities to counteract the pathological processes.

This book is a collection of works that canvass many of the recent developments in various areas of connective tissue research. However, it does not attempt to be a comprehensive presentation of all aspects of connective tissue biology and biochemistry. Rather, the individual chapters provide up-to-date reviews of the subject matter, corresponding to the expertise of the individual contributors. The text has been designed with three primary goals in mind: First, we have attempted to present various aspects of the basic biology and biochemistry of connective tissue macromolecules. Second, chapters that emphasize pertinent state-of-the-art research accomplishments by contributors in the forefront of their respective research fields have been included. Third, discussions of selected connective tissue disorders serve to demonstrate the merger of basic science with clinical research and the enhancement of our understanding of human diseases. We believe this text provides a summary of the highlights of progress in connective tissue research to date. In addition to documentation of specific scientific achievements, many of the contributors have presented their views of future scientific directions. It is hoped that these speculations will serve to stimulate our imagination and creativity for the advancement of research of the extracellular matrix in coming years.

We would like to extend our sincere appreciation to each author for his or her superb contribution. We are particularly indebted to Paul Dolgert, Elaine Grohman, and Daniel Gonneau at Marcel Dekker, Inc., whose expert editorial assistance has been invaluable. We would also like to extend special thanks to Dr. Maurits Dekker for his graciousness. The excellent secretarial assistance of Charlene D. Aranda has been greatly appreciated.

Jouni Uitto
Andrea J. Perejda

Contributors

R. Patrick Abergel, M.D. Chief Resident and Fellow, Division of Dermatology, Department of Medicine, Harbor-UCLA Medical Center, Torrance, California

Oscar M. Alvarez, Ph.D. President, Bioderm Medical Inc., Princeton, New Jersey

Donald Bartel, M.D. Associate Professor, School of Mechanical and Aerospace Engineering, Cornell University, Ithaca, New York

Eugene A. Bauer, M.D. Professor of Medicine, Division of Dermatology, Department of Internal Medicine, Washington University School of Medicine, St. Louis, Missouri

Richard A. Berg, Ph.D. Associate Professor, Department of Biochemistry, University of Medicine and Dentistry of New Jersey-Rutgers Medical School, Piscataway, New Jersey

Merton Bernfield, M.D. Professor of Human Biology and Pediatrics, Department of Pediatrics, Stanford University Medical Center, Stanford, California

Robert E. Burgeson, Ph.D. Associate Research Director, Shriners Hospital for Crippled Children and Associate Professor, Department of Biochemistry, Oregon Health Sciences University, Portland, Oregon

Peter H. Byers, M.D. Professor, Departments of Pathology and Medicine, University of Washington School of Medicine, Seattle, Washington

Joan G. Clark, M.D.* Assistant Professor of Medicine, Pulmonary Division, Department of Internal Medicine, Washington University School of Medicine, St. Louis, Missouri

Edward G. Cleary, M.D. Reader of Pathology, Department of Pathology, University of Adelaide, Adelaide, South Australia

Lloyd A. Culp, Ph.D. Professor, Department of Molecular Biology and Microbiology, Case Western Reserve University, School of Medicine, Cleveland, Ohio

Jeffrey M. Davidson, Ph.D.† Associate Research Career Scientist, Medical Research Service, Veterans Administration, and Associate Professor, Department of Pathology, University of Utah School of Medicine, Salt Lake City, Utah

Arthur Z. Eisen, M.D. Professor and Head, Division of Dermatology, Department of Medicine, Washington University School of Medicine, St. Louis, Missouri

Arvan L. Fluharty, Ph.D. Professor, MRRC-UCLA Research Group at Lanterman Developmental Center, University of California, Los Angeles, Los Angeles, California

Karen A. Holbrook, Ph.D. Professor, Department of Biological Structure; Adjunct Professor, Department of Medicine (Dermatology); and Associate Dean for Scientific Affairs, University of Washington School of Medicine, Seattle, Washington

Kari I. Kivirikko, M.D., Ph.D. Professor, Collagen Research Unit and Department of Medical Biochemistry, University of Oulu, Oulu, Finland

Joy E. Koda, Ph.D.‡ Postdoctoral Fellow, Department of Pediatrics, Stanford University Medical Center, Stanford, California

Helena Kuivaniemi, M.D., Ph.D.§ Teaching Assistant, Collagen Research Unit and Department of Medical Biochemistry, University of Oulu, Oulu, Finland

Michael W. Lark, Ph.D.|| Research Associate, Department of Molecular Biology and Microbiology, Case Western Reserve University, School of Medicine, Cleveland, Ohio

---

Present affiliations:

*Assistant Member, Division of Oncology, The Fred Hutchinson Cancer Research Center, and Assistant Professor of Medicine, Division of Respiratory Diseases, Department of Medicine, University of Washington, Seattle, Washington

†Associate Professor of Pathology, Vanderbilt University School of Medicine, and Associate Research Career Scientist, Veterans Administration Medical Center, Nashville, Tennessee

‡Research Scientist, Department of Cell Biology, Hybritech Incorporated, San Diego, California

§Jefferson Institute of Molecular Medicine and Department of Biochemistry, Thomas Jefferson University, Philadelphia, Pennsylvania

||Senior Research Immunologist, Department of Experimental Pathology, Merck and Co., Inc., Rahway, New Jersey

Carlo L. Mainardi, M.D.* Associate Professor of Medicine and Chief, Division of Rheumatology, University of Medicine and Dentristy of New Jersey-Rutgers Medical School, Piscataway, New Jersey

Richard Mayne, D.Phil. Professor, Department of Cell Biology and Anatomy, University of Alabama at Birmingham, Birmingham, Alabama

Ronald R. Minor, V.M.D., Ph.D. Associate Professor, Department of Pathology, New York State College of Veterinary Medicine, Cornell University, Ithaca, New York

Nicholas P. Morris, Ph.D. Shriners Hospital for Crippled Children and Assistant Professor, Oregon Health Sciences University, Portland, Oregon

Deane F. Mosher, M.D. Professor, Departments of Physiological Chemistry and Medicine, University of Wisconsin, Madison, Wisconsin

Joanne E. Murphy-Ullrich, Ph.D. Postdoctoral Fellow, Departments of Physiological Chemistry and Medicine, University of Wisconsin, Madison, Wisconsin

Donald S. Neblock, Ph.D.† Research/Teaching Specialist, Department of Biochemistry, University of Medicine and Dentistry of New Jersey-Rutgers Medical School, Piscataway, New Jersey

Aarne I. Oikarinen, M.D., Ph.D.‡ Visiting Assistant Professor, Division of Dermatology, Department of Medicine, Harbor-UCLA Medical Center, Torrance, California

Donald F. Patterson, D.V.M., D.Sc. Professor of Veterinary Medicine and Chief, Section of Medical Genetics, School of Veterinary Medicine, University of Pennsylvania, Philadelphia, Pennsylvania

Andrea J. Perejda, Ph.D. Assistant Research Professor, UCLA School of Medicine; Division of Dermatology, Department of Medicine, Harbor-UCLA Medical Center, Torrance, California

Alan Rapraeger, Ph.D.§ Postdoctoral Fellow, Department of Pediatrics, Stanford University Medical Center, Stanford, California

Lasse J. Ryhänen, M.D., Ph.D. Research Associate, Division of Dermatology, Department of Medicine, Harbor-UCLA Medical Center, Torrance, California

---

Present affiliations:
*Associate Professor of Medicine, Department of Medicine, University of Tennessee Center for the Health Sciences, and Chief, Medical Service, Veterans Administration Medical Center, Memphis, Tennessee
†Senior Scientist, Cistron Technology, Inc., Pine Brook, New Jersey
‡Assistant Professor, Department of Dermatology, Oulu University Central Hospital, Oulu, Finland
§Assistant Professor, Department of Pathology and Laboratory Medicine, University of Wisconsin, Madison, Wisconsin

Lawrence B. Sandberg, M.D., Ph.D.* Professor, Department of Pathology, University of Utah Medical Center, Salt Lake City, Utah

Daniel J. Santa Cruz, M.D. Associate Professor of Pathology, Division of Surgical Pathology, Department of Pathology, Washington University School of Medicine, St. Louis, Missouri

Jeremiah E. Silbert, M.D. Medical Investigator, Veterans Administration Outpatient Clinic, and Professor of Medicine, Department of Immunology/Rheumatology, Harvard University, and Brigham and Women's Hospital, Boston, Massachusetts

Norman T. Soskel, M.D. Assistant Professor of Medicine, Department of Internal Medicine, Veterans Administration Medical Center and University of Tennessee, Memphis, Tennessee

Jouni Uitto, M.D., Ph.D. Professor of Medicine, UCLA School of Medicine; Associate Chief and Director of Research, Division of Dermatology, Department of Medicine, Harbor-UCLA Medical Center, Torrance, California

David T. Woodley, M.D. Associate Professor, Department of Dermatology, University of North Carolina School of Medicine, Chapel Hill, North Carolina

Joyce A. M. Wootton, Ph.D. Research Associate, Department of Pathology, New York State College of Veterinary Medicine, Cornell University, Ithaca, New York

Present affiliation:

*Loma Linda Veterans Administration Medical Center, Loma Linda, California

Contents

I

STRUCTURE AND BIOLOGY OF THE COMPONENTS OF THE EXTRACELLULAR MATRIX

1
The Collagen Family of Proteins

Robert E. Burgeson and Nicholas P. Morris

Shriners Hospital for Crippled Children and
Oregon Health Sciences University,
Portland, Oregon

I. INTRODUCTION

The study of the extracellular matrix is a rapidly expanding field that includes the isolation and characterization of multiple connective tissue macromolecules, the molecular interactions between these connective tissues components, between these components and adjacent cell surfaces, and the effects of these components on the determination and regulation of cellular behavior. Substantial progress has been made in enumerating a variety of glycoprotein and proteoglycan components, some of which are unique to specific types of matrix, and others of which are common to all extracellular matrices. Perhaps because they are the most abundant of these matrix components, the collagens have received a great deal of attention during the past decade (for review, see Refs. 1,2). It is now clear that the term "collagen" represents not a single protein species, but a family of closely related glycoproteins, each of which is genetically distinct, and which has unique structural features and, presumably, a unique function. While considerable direct information has been accumulated regarding the structures of these individual molecules, only indirect information presently exists for their functions.

Since this area of research is in rapid transition, there is considerable confusion concerning identification of these macromolecules, as well as their structure-function relationships in the extracellular matrix. Therefore, it is the purpose of this chapter to review the generally accepted information and to place less emphasis on details presently being disputed. We provide a current catalogue of the various forms of collagen, pointing out similarities and differences, and relate these properties to the ability of these molecules to form specialized fiber systems that have unique functions in the connective tissue matrix. In addition, we briefly summarize the complex intracellular biosynthetic pathways and extracellular processing that leads to the formation of stable fiber forms. It is hoped that this general review will provide a framework for more specific discussion of these events in subsequent chapters.

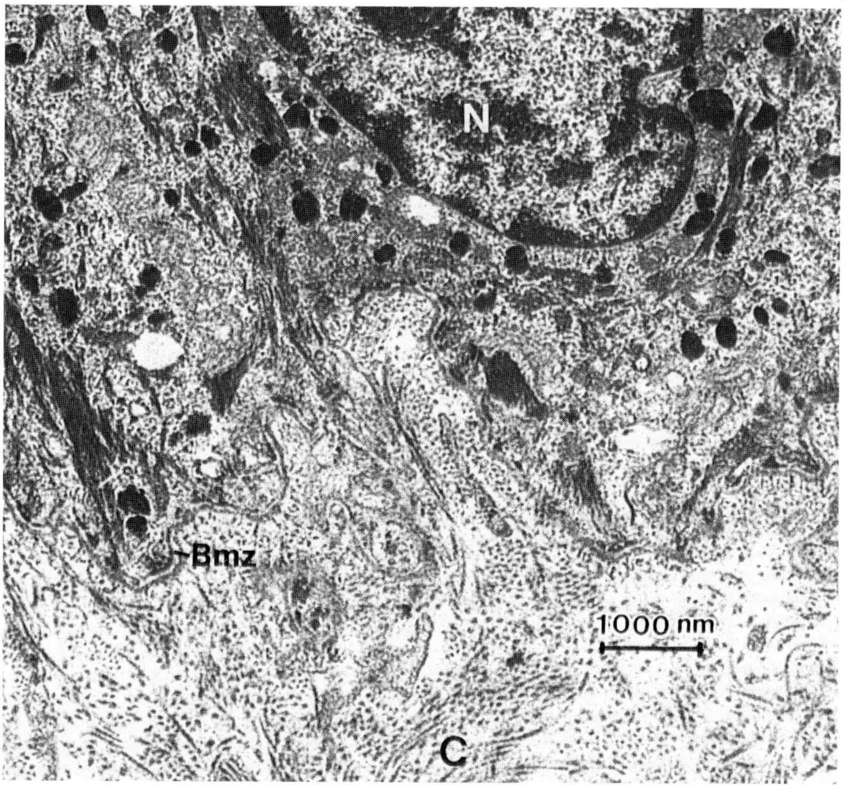

Figure 1 Transmission electron micrographs of the dermal-epidermal junction at two magnifications. A variety of fibrous structure are identified: collagen fibers (C); the basement membrane zone (Bmz), which includes the lamina lucida (L1) and the lamina densa (Ld); anchoring fibrils (AF); anchoring filaments (af); and the reticular network (Rn). The nucleus (N) and hemidesmosomes (H) are also identified.

When the ultrastructure of any connective tissue is examined, a large variety of extracellular fibrous structures can be identified. For example, Figure 1 illustrates the ultrastructure of human skin dermal-epidermal junction. The cytoplasm and plasma membranes of the basal cells are adjacent to the connective tissue matrix. In this zone, the basal cells are separated from subjacent matrix by a dense, amorphous structure known as the basement membrane or basal lamina. Subjacent to the basement membrane is the connective tissue matrix. This matrix contains a variety of fibrous structures, the most obvious of which are the large, cross-striated fibers showing a 67-nm periodicity. These are the structures first identified as collagens, and they represent the most common fiber form of these molecules. But, in addition to the banded collagen fibers, a fine, reticular network is also visible. Also, in this tissue, banded fibers with a centrosymmetric cross-striation can be seen that appear to originate in the lamina densa of the basement membrane and run perpendicularly from the basement membrane into the upper layers of the

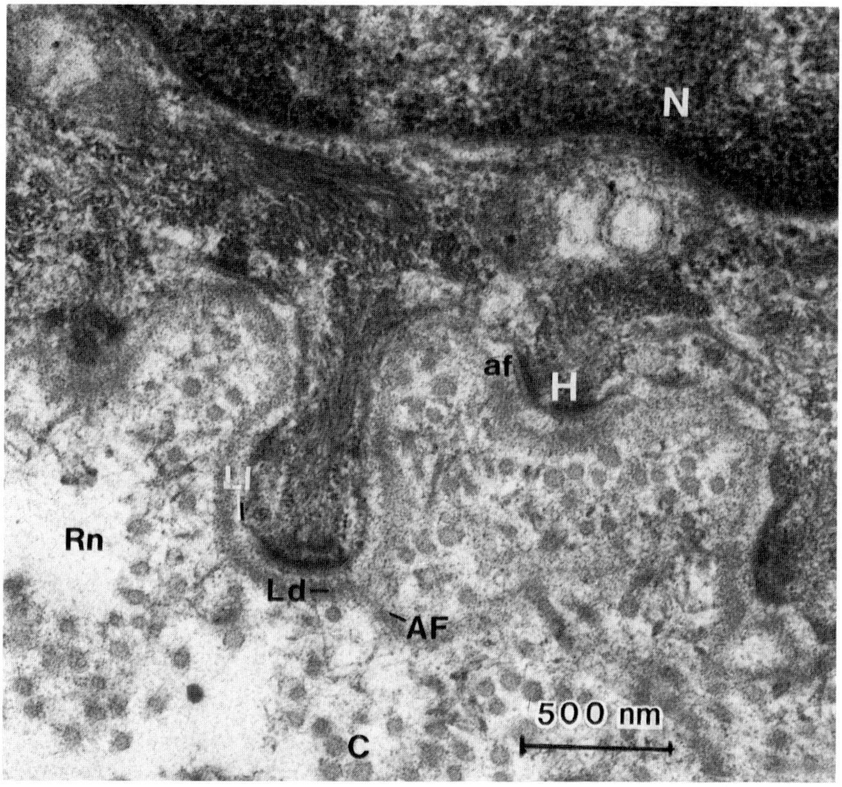

Figure 1 (Continued)

dermis. These structures, which are common to basement membrane regions of striated epithelial origin, are known as anchoring fibrils (3). Another fibrous structure can be identified between the hemidesmosomes in the plasma membrane of the basal cells and the lamina densa. These fine filamentous structures have been termed anchoring filaments (4). Many of these fibrous elements are believed to be composed of different collagen types. The lamina densa of the basement membrane is clearly collagenous. The fine filamentous feltwork is an aggregate of at least one other collagen type. There is some evidence that the anchoring fibril is a specialized condensation of yet another collagen type (5). The anchoring filament appears not to be collagenous.

Figure 1 illustrates several important points. (a) The extracellular matrix is made from a variety of collagen types. (b) The fibrous elements in connective tissues have different fiber forms, and these fiber forms can be attributed to different organizational patterns of the various collagen types. (c) Many of these fiber forms have restricted distributions, such as the basal lamina and the anchoring fibrils.

All these observations suggest that the various members of the collagen family have structural similarities that allow them to form insoluble fiber systems, and each has specific properties that determine unique organizational patterns to serve specific functions which determine and stabilize tissue architecture. This chapter outlines the information contributing to these concepts.

II. THE COLLAGENS

Presently, at least 18 gene products have been identified as subunits of at least 15 collagen molecules. Because of the rapid accumulation of data in this area, the nomenclature of these proteins is confusing and in transition. The most commonly used system for naming these molecules is the use of roman numerals to designate the type of molecule. Therefore, the different collagen molecules are individually indicated as type I, type II, etc. Each collagen molecule contains three polypeptide subunit chains held in a specialized triple-helical array known as the collagen helix. The polypeptide subunits are individually termed α chains. In a given collagen molecule, the α chains may be identical or nonidentical. For example, type I collagen is a heterotrimer of two nonidentical chains. Therefore, these chains are denoted as $\alpha1(I)$ and $\alpha2(I)$. The molecular composition of type I collagen is noted as follows: $[\alpha1(I)]_2\alpha2(I)$. In contrast, type II collagen is a homotrimer of three identical chains known as $\alpha1(II)$ chains which form molecules with the composition $[\alpha1(II)]_3$. While this system of nomenclature is extremely useful and specific, it has the inherent difficulty that a newly discovered collagen, which is often initially observed as an individual α chain, cannot be correctly identified as a collagen type and given a roman numeral until the α-chain composition of the parent molecule has been elucidated. For instance, when type III collagen was first identified, the nondenatured molecule (i.e., the native molecule still containing all three polypeptide chains held in triple-helical array) was isolated as a single molecular species and was subsequently denatured into a single type of α chain. In this case, the designation of these α chains as a specific collagen type could be made with relative certainty. In contrast, preparations of native type V collagen isolated from human chorioamniotic membrane released two chains upon denaturation. Therefore, in that tissue several possibilities existed for the molecules containing these chains. They could both be derived from a single heterotrimeric molecule, or each could arise from an independent homotrimeric molecule where the mixture could not be adequately separated under experimental conditions. Therefore, the description of these chains as $\alpha1(V)$ and $\alpha2(V)$ chains could not be made until a molecule with the composition $[\alpha1(V)]_2\alpha2(V)$ was identified. To add to the confusion, yet another distinct α chain similar to both the type V chains was identified in other tissues, and the suggestion has been made that it arises from a parent heterotrimeric molecule with the chain composition $\alpha1(V)\alpha2(V)\alpha3(V)$. All these observations suggest that the individual α chains can associate with other chains in a variety of ways to produce different molecular species. This has been strongly suggested for the three chains of type V collagen and for the $\alpha1(I)$ chain, which can associate in the more common heterotrimeric type I molecule or infrequently, into a homotrimeric molecule known as the type I "trimer." Another limitation of the nomenclature is that in the above example it implies that the $\alpha1(I)$ chain in the type I molecule and in the type I "trimer" are genetically identical. This is a commonly held supposition, but has not yet been directly demonstrated. The same objections can be raised to the nomenclature of the type V α chains.

As a result of the difficulties mentioned above, the current nomenclature system contains chains identified by the roman numeral system as well as by the use of other means of identification. In general, the term $\alpha(X)$ is reserved for chains whose molecular stoichiometry has been clearly identified, while newly identified chains usually receive a capital letter or letters.

There are a variety of ways to classify the various collagen types. Ultimately, the most useful classification system would be based on the function of the collagen form. However, this information is currently unavailable for the majority of the different collagen types, and other, more arbitrary, systems are used. For the purposes of this chapter, the collagens are grouped into two main categories: those that form periodically banded fibers and those that do not. The second category contains a variety of collagens, some of which have been identified as specific ultrastructural fiber forms, and several others whose fiber structures are not known but which appear to be unable to form banded structures in vitro. This classification is presented in Table 1. In addition to the collagen type, the α-chain composition and presently known tissue locations are also indicated.

A. Collagens That Form Fibers with Uniform Periodic Cross-Striations

This group of collagens are capable of forming large fibers, which, under appropriate conditions, bind stain in a cross-striated pattern with a uniform 67-nm interband distance (6). The molecules align to form elongated and broad structures capable of sustaining the two-dimensional forces applied to connective tissues. These fibers represent the most common fiber form in connective tissue and are largely responsible for the rigidity of connective tissue matrices. These banded fibers can be formed both in vivo and in vitro from three major types of collagen, types I, II, and III. This group is often referred to as the interstitial collagens.

1. *Type I Collagen*

Type I collagen was the first characterized member of this protein family. As indicated in Table 1, it is a heterotrimer with the chain composition $[\alpha1(\mathrm{I})]_2\alpha(\mathrm{I})$ (7). It is broadly distributed throughout connective tissues and is the major collagen found in bone, tendon, skin, and dentin (8). Type I "trimer" collagen is believed to be a subtype of type I collagen (9,10). This collagen is a homotrimer with the chain composition $[\alpha1(\mathrm{I})]_3$. This collagen type has been observed as a minor component of skin and is a biosynthetic product of a variety of embryonic cells and tumor cell lines. The functional difference between type I and type I "trimer" collagens has not been elucidated.

2. *Type II Collagen*

Type II collagen is a homotrimeric species that has been identified in two forms. The predominant form is usually referred to only as type II collagen, occasionally as type IIM collagen to indicate that it is the major species (11,12). A minor form of this collagen type, type IIm, has been identified by amino acid sequencing studies (13). It differs from the major form by several single amino acid substitutions. It is presently unclear whether these reflect allelic differences in the same gene or whether the two forms represent products of different genes. Another collagen chain, commonly referred to as the 3α chain, has been partially characterized and is strongly reminiscent of type II collagen chains by amino acid composition and peptide maps (14). The 3α chain is a minor component of tissues containing type II collagen. This chain appears to be more highly glycosylated than the type II collagen chains, and there is indirect evidence that it is functionally different from type II collagen. The relationship between the 3α chain and the type IIm chain is presently unknown.

Table 1 Genetically Distinct Collagens

A. The interstitial collagens–Collagens that form broad-banded fibers		
Type I	$[\alpha 1(I)]_2\alpha 2(I)$	Found in most tissues except cartilage; major component of bone, tendon, skin, and dentin
Type I "trimer"	$[\alpha 1(I)]_3$	Some fetal tissues; product of certain malignant and normal cell lines
Type IIM	$[\alpha 1(IIM)]_3$	Cartilaginous tissues
Type IIm	$[\alpha 1(IIm)]_3$	Cartilaginous tissues
Type III	$[\alpha 1(III)]_3$	Found together with type I; in relatively high concentrations in extensible tissues such as blood vessels, skin, and gut
B. The minor collagens–Less abundant collagens that do not form broad-banded fibers		
Type IV (basement membrane collagen)	$[\alpha 1(IV)]_2\alpha 2(IV)$	Major component of all basal laminae
Type V	$[\alpha 1(V)]_2\alpha 2(V)$ $[\alpha 1(V)]_3$ $[\alpha 3(V)]_3$ $\alpha 1(V)\alpha 2(V)\alpha 3(V)$	Minor components of most tissues except cartilage; fiber forms unknown
Type VI	$[\alpha 1(VI)]_2\alpha 2(VI)$	First identified in aortic intima, but now thought to have a broad, but yet undefined, tissue distribution
Type VII ("long-chain collagen")	$[\alpha 1(VII)]_3$	Identified in amniotic membrane and skin; believed to be associated with all stratified epithelia; may be anchoring fibril protein
Type VIII ("endothelial cell collagen")	$[\alpha 1(VIII)]_3$	Identified as product of a variety of cell types
Type IX ("HMW-LMW")	$\alpha 1(IX)\ \alpha 2(IX)\ \alpha 3(IX)$	Cartilaginous tissues
Type X ("G collagen")		Product of cartilage hypertrophic cells
C. Minor cartilage collagens with unknown subunit compositions		
1α 2α		Compositionally similar to type V; structures, fiber forms, and functions unknown
3α		Compositionally similar to type II.

Type II collagens are restricted in their distribution to cartilagenous tissues. They are the major structural macromolecules of hyaline cartilage and are also found in elastic and fibrous cartilage as well as in intervertebral disk, where they occur predominantly in the nucleus pulposus and in combination with type I collagen in the annulus fibrosus (15). The morphological, ultrastructural and biomechanical differences between cartilage and other connective tissues correlate with the increased content of proteoglycan aggregates in these highly specialized tissues. This observation strongly suggests that type II collagen, but not type I collagen, is capable of performing these associations between cartilage collagens and cartilage proteoglycans, but these specific associations have not been unequivocably established.

3. Type III Collagen

The final member of the interstitial collagens is the type III collagen molecule. Like type II collagen, this molecule is a homotrimeric aggregate of $\alpha 1$(III) chains (16,17). With the exceptions of tendon and bone, type III collagen occurs in all connective tissues in which type I collagen is found. The function of type III collagen in these tissues has not been directly demonstrated, but there appears to be a direct correlation between the relative proportion of type III to type I collagen and the extensibility of the connective tissue. For example, relatively high concentrations of type III collagen are found in highly extensible tissues such as skin, blood vessels, gut, and chorioamniotic membranes. No type III collagen was found in inextensible tissues such as bone, tendon, and dentin. Furthermore, fetal skin contains a high percentage of type III collagen relative to type I collagen, which decreases with increasing fetal and newborn age, paralleling the decrease in extensibility of skin (18). Finally, patients having the inherited condition of Ehlers-Danlos type IV are unable to accumulate substantial amounts of type III collagen in their tissues (19). These patients have thin, inextensible skin and vascular tissue irregularities leading to frequent and abnormal scarring and, in most cases, to premature death due to dissecting aneurysms. These observations strongly suggest that the presence of type III collagen influences tissue architecture in such a way as to ensure tissue extensibility. The molecular mechanisms underlying this defect are not yet understood.

The structures of all the collagens in this major group are similar. All are initially synthesized in high-molecular-weight forms known as procollagens (for review, see Ref. 20). A model of procollagen structures is depicted in Figure 2. The interstitial procollagens contain two kinds of structural domains, the collagen triple helix and the globular domains. The terminal globular regions are often referred to as the amino or carboxyl propeptides. In fact, all collagens thus far described contain variations of these two major types of structural domains. The sizes and distributions of these domains in the interstitial procollagens are quite similar. All these collagens contain a major triple helix approximately 300 nm in length, containing nearly 1000 amino acids in each of the three component α chains. The formation of a thermally stable triple helix requires that each of the α chains individually maintain a helical conformation (for review, see Ref. 2). In turn, this requirement imposes a specific restriction on the amino acid sequence of these chains. The amino acid glycine must occur in every third position along the entire sequence. The α-chain amino acid sequence can be described as a repeating triplet with the structure $[\text{gly-X-Y}]_n$. At appropriately low temperatures, peptides with this repeating structure are capable of spontaneously forming a helical conformation. The resulting helix contains 3.27 residues per turn of the helix (Figure 3) (21). This noninteger value of residues per turn indicates that the glycine residues are slightly staggered as they align along one face of the resulting helical cylinder. The fact that glycine does not contain a bulky side group protruding from its α carbon allows this face of the helical cylinder to be flattened relative to the faces of the cylinder occupied by the amino acids in X and Y positions. The thermal stability of this α-chain helix is dependent on posttranslational hydroxylation of a minimum number of proline residues along the sequence. The prolylhydroxylation allows for increased hydrogen bonding between the turns of the helix, thus stabilizing this unique conformation.

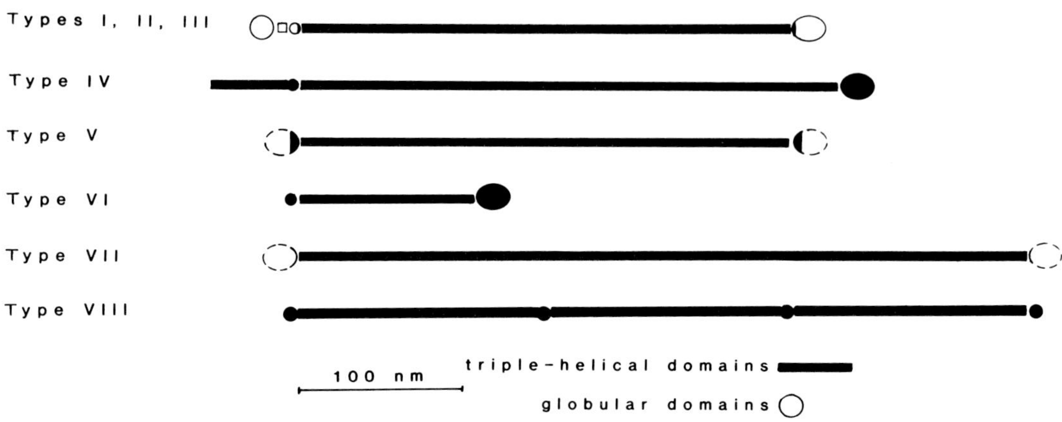

Figure 2 The triple-helical and globular domains of types I-VII procollagen molecules are represented in approximate proportion. Dashed lines represent some of the proposed globular domains or those with unknown dimensions. The globular domains of several of these collagens are partially removed prior to fiber formation. The darkened regions represent parts of the globular and helical domains believed to be retained in the fiber form of the molecules. In the case of type III collagen, the amino terminal globular domain persists in the tissue of an extended period.

The collagen triple helix forms spontaneously from these helical α chains by lateral alignment that occurs between the flattened faces of the α-chain cylinders occupied by the glycine residues. Because the glycine residues are slightly staggered, and the flattened face of the cylinder spirals along the length of the helix, the resulting triple-helical structure must also be spiraled to attain the tightest fit between the three α chains that is most thermodynamically stable. Therefore, the collagen triple-helical conformation is intimately dependent on the presence of glycine in every third position in the α-chain amino acid sequence, and its thermal stability is dependent on adequate prolyl-hydroxylation.

This specific structural domain has a variety of unique features, including a resistance to proteolytic degradation, as long as the conformation is maintained as an extended, semirigid, ropelike structure. Any disruption in the gly-X-Y sequence will sufficiently perturb the structure to impart protease sensitivity and decreased linear rigidity at the site of the disruption (22). The interstitial collagens have major triple-helical domains where the helicity is maintained along the entire length of the domain. Type III collagen is a minor exception, in that it contains a single trypsin cleavage site in the triple helix very near the collagenase binding region (23). Type III collagen is also unique in that it contains disulfide-bonded cysteinyl residues within the triple helix near the carboxyl terminus (24).

In addition to the major triple-helical domain, the procollagens of this first class of structural macromolecules contain globular domains at both the amino and carboxyl termini (for review see Ref. 20). The three α chains are covalently cross-linked by interchain disulfide bonds within the carboxyl terminal globular domain (25). The amino terminal globular domain is much more complex. At the extreme amino end is a large globular region that also contains disulfide bonds, but unlike the carboxyl terminal domain, the amino terminus has intrachain bonds (that is, they occur between cystine residues located in the same chain and not between residues in different chains) (26). These

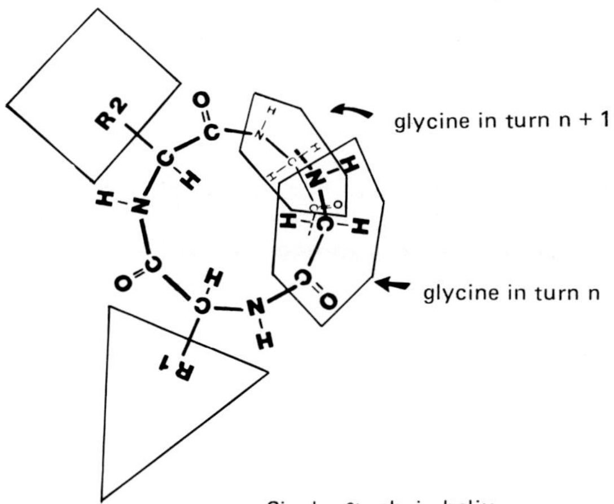

Single α - chain helix

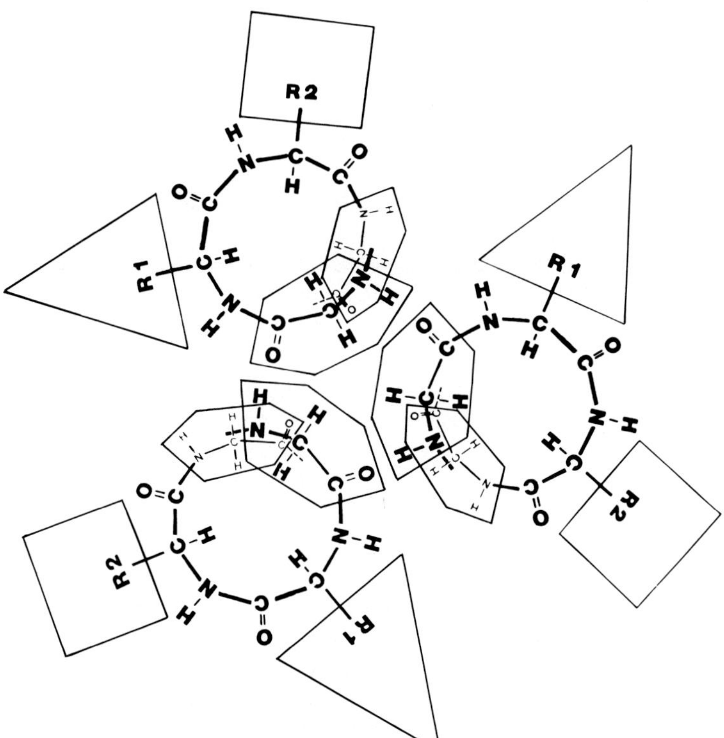

Collagen triple helix

Figure 3 Representation of an end-on view of the peptide backbone of the collagen helix of a single α chain and of a collagen triple helix. One turn of the helix contains 3.27 amino acids, and every third amino acid is glycine, which has no bulky side group (R1 and R2). The α chains come in contact with each other at glycine residues to form the triple helix. Since the glycyl residues are staggered along the helix, to attain the closest approach the chains must twist, forming a triple-helical structure.

intrachain disulfides stabilize the conformation of this region. This large globular region is followed by a short triple-helical domain (27), which is then followed by another, but much smaller, globular domain (28). This more internal globular domain contains the cleavage site for procollagen amino-propeptidase, the enzyme responsible for the proteolytic excision of the amino terminal globular domain prior to fiber formation (29-31). An analogous site occurs near the end of the major triple helix at its carboxyl terminus for another enzyme known as procollagen carboxyl-propeptidase, required for the proteolytic removal of that globular domain (29-31). Unlike many of the other collagens, the interstitial collagens require removal of these globular domains from both ends of the molecules to allow formation and stabilization of the characteristic large-banded fiber systems. This has been clearly demonstrated in the disease known as dermatosparaxis in cattle and sheep (32) and in the inherited human condition known as Ehlers-Danlos type VII in which the procollagen amino-propeptidase is deficient or inactive (33). In this disease, only thin, wispy fibers can be formed, resulting in a fiber system that is incapable of adequate stabilization, and as a consequence these individuals demonstrate extreme skin and tendon laxity.

The enzymatic sites for removal of the amino and carboxy terminal globular domains do not occur exactly at the transition between the triple helix and the propeptide, but rather these sites are several amino acid residues within the globular domains. Cleavage at these sites leaves the short non-triple-helical regions known as telopeptides. These telopeptides are major sites for cross-links between molecules in fibers, probably because of the increased flexibility of these non-triple-helical regions.

B. Collagens That Do Not Form Uniformly Banded Fiber Systems

Over the past 10 years a large number of collagens have been described with properties quite dissimilar to those reviewed above. The most consistent feature distinguishing these collagens from the interstitial collagen is their inability to form broad-banded collagen fibers either in vivo or in vitro. Like all collagens, these macromolecules contain both globular and triple-helical structural domains. They differ by the relative size and distribution of these domains, the extent of posttranslational modification of these structures, the extent to which they are proteolytically processed prior to fiber formation, the integrity of their triple-helical domains, and the manner in which they aggregate to form new fiber systems specialized for quite different physiological functions.

1. Type IV Collagen

The best studied molecule in this class is type IV collagen (for review see Refs. 34,35). This collagen form contains two α chains, $\alpha 1$(IV) and $\alpha 2$(IV). It is most generally thought to be a heterotrimer with the structure $[\alpha 1(\text{IV})]_2\alpha 2(\text{VI})$, but the possibility of two independent homotrimeric molecules has not been ruled out. Immunolocalization of type IV collagen by monospecific antibodies indicates that these molecules are limited in their distribution to the basal laminae of basement membranes (36). All ultrastructurally defined basement membranes are composed of this collagen type (37). Unlike the interstitial collagenous matrices, and in addition to providing tissues with tensile strength, the basal lamina is also involved in membrane ultrafiltration, serves as specific anchoring sites for overlying and underlying cell types, and is involved in the development and maintenance of a variety of tissues (36).

Like the interstitial procollagens, type IV collagen contains both globular and triple-helical domains, though the size and orientation of these domains differ. The major type IV triple-helical domain is longer than that of the insterstitial collagens, being approximately 330 nm in length (38). In addition, as can be seen from its amino acid composition in Table 2, the type IV molecule contains cysteinyl residues that form disulfide bonds within its triple helix. Also, the type IV triple helix is imperfect, containing several regions in which the sequence gly-X-Y is disrupted, allowing this domain to be susceptible to a large number of proteolytic enzymes (8). These disruptions also increase the flexibility of the triple-helical domain at these sites. The amino and carboxyl terminal domains of type IV collagen are also different. The amino terminal domain is largely triple-helical, and this minor triple helix is separated from the major triple helix by only a small globular domain, frequently referred to as NC-2 (39). (This terminology is used to indicate that this region is not susceptible to a bacterial collagenase enzyme that is specific for triple-helical structures.) The minor triple helix is a site for interaction between individual type IV molecules to stabilize the specialized fiber form present in basement membranes (40). The carboxyl terminal globular domain is similar in size to that of the interstitial collagens.

2. *Type V Collagen*

Type V collagen contains several molecules composed of genetically distinct chains with similar structural and compositional features (41-45). The exact number of molecules containing these chains and their molecular stoichiometry are not yet unambiguously known. Thus far, there is evidence for perhaps as many as six distinct chains in this group. The best studied are the chains known as $\alpha1(V)$, $\alpha2(V)$, and $\alpha3(V)$. For these three chains, the molecules $[\alpha1(V)]_2\alpha2(V)$ (41,45) and $[\alpha1(V)]_3$ (46,47) have been established. The $\alpha3(V)$ chain may occur as a separately homotrimeric structure, although there is evidence of a molecule with the structure $\alpha1(V)\alpha2(V)\alpha3(V)$ (48). Many tissues such as chorioamniotic membranes appear not to contain the $\alpha3(V)$ chain, while others such as placenta contain all three type V chains. Recent evidence suggests that the $\alpha1(V)$ chain forming homotrimeric molecules may be a different gene product from that employed in the heterotrimers.

In addition to these three, or possibly four, type V chains, two additional α chains with similar structural and compositional properties have been identified in cartilaginous tissues (14). These are commonly referred to as the 1α and 2α chains, as their molecular stoichiometry has not yet been established. These chains have amino acid compositions strikingly similar to those of the other type V chains, but distinct differences in peptide maps generated by a variety of fragmentation procedures indicate that they are the products of distinct genes (49). It appears that the 1α and 2α chains are the cartilage analogs of the other type V chains, just as type II collagen is structurally similar, and in many ways analogous, to type I collagen.

The function of type V collagen is not yet understood. Current immunolocalization studies provide conflicting evidence regarding the distribution of these materials in tissues. Initial observations indicated that type V collagen may be a component of basement membranes (50), and it has been seen associated with cell surfaces leading to its description as a cytoskeletal collagen (51). More recent evidence employing monoclonal antibodies suggests that type V collagen is distributed uniformly throughout the connective tissue stroma (52,53). This distribution suggests that type V collagens may be related to the fine reticular network often seen underlying the dense interstitial collagen fibers in most connective tissues.

Although distinguishable in detail, the amino acid compositions of the type V collagen chains are very similar to those of the interstitial collagens. These similarities extend

Table 2 Amino Acid and Neutral Hexose Composition of the Helical Domains of Several Collagens (Residues/1000 Residues)

Amino acid	α1(I)	α2(I)	α1(II)	α1(III)	α1(IV)[a]	α2(IV)[a]	α1(V)	α2(V)	α3(V)	1α	2α	α1(VII)
3-OH-proline	1.1	1.2	1.1	0	1	1	2.9	2.5	2.2	2.3	1.1	0.1
4-OH-proline	114	105	96	126	118	107	109	109	92	91	95	84
Aspartic acid	46	45	36	42	45	50	50	51	42	50	46	51
Threonine	18	18	22	13	20	28	19	26	19	24	17	26
Serine	35	30	27	38	36	30	36	31	34	28	24	49
Glutamic acid	77	70	95	71	88	64	91	84	98	97	107	105
Proline	118	114	106	107	88	73	118	97	99	118	109	81
Glycine	330	331	334	351	324	328	344	341	332	322	330	307
Alanine	119	105	108	95	30	47	46	52	49	49	54	64
Cystine/2	0	0	0	2	2	3	0	0	1.3	0	0	6.7
Valine	19	35	19	15	32	25	25	24	29	18	27	35
Methionine	7	8	11	8	14	14	8	11	8.1	9	10	7.6
Isoleucine	8	16	11	13	32	39	19	16	20	16	15	10.5
Leucine	20	33	27	23	50	56	39	35	56	39	35	49
Tyrosine	2	3	1	3	8	6	2	2	2.4	3	2	28
Phenylalanine	13	11	13	8	28	36	12	14	9.2	11	11	9.8
Hydroxylysine	10	12	18	6	49	39	35	24	43	40	37	41
Lysine	27	20	20	29	7	5	20	18	15	15	18	18
Histidine	4	10	2	6	6	7	8	11	14	11	6	3.1
Arginine	49	51	52	48	22	42[b]	45	50	42	47	42	48
Glucose	1	1	5	1	43	n.r.	18	10	17	36	29	n.r.
Galactose	1	2	10	2	45	n.r.	30	14	24	34	28	n.r.

[a]T. F. Kresina and E. J. Miller, *Biochemistry, 18*:3089, 1979.
[b]n.r., not reported.

to the structural analyses as well. Like the interstitial collagens, type V collagen is synthesized with extensive amino and carboxyl terminal globular domains (54). Although the details of the structures of these domains remain to be determined, the present data indicate a grossly similar distribution of globular and triple-helical domains in the type V procollagens. The triple-helical domain is approximately the same length as that of the interstitial collagens and contains no disruptions in the gly-X-Y sequence throughout the length of this structural domain. The most striking difference between the type V collagens and the interstitial collagens is observed during extracellular processing of these molecules to form fibers. Unlike the interstitial collagens in which the entire amino and carboxyl terminal globular domains are removed during processing, only a portion of these domains is excised during processing of the type V collagens (55). In addition, there is evidence for the covalent attachment of a yet unknown peptide to one of the residual globular domains (56). The collagen fiber system arising from these monomeric subunits has not yet been described, but it is likely that the presence of remaining globular portions at both ends of the molecule influences its ability to form fibers, and one might expect to ultimately find a unique fiber system composed of type V molecules.

3. *Type VI Collagen*

Type VI collagen, or intima collagen, has recently been isolated from pepsin digests of aortic intima (42,51), human kidney (58), human and bovine placenta (59), and bovine skin (60). Current evidence suggests that it contains two nonidentical α chains present in a single heteropolymer. This molecule is unique from most other collagens in that it contains a triple-helical domain that is considerably shorter than that of the interstitial collagens. This triple helix appears to be uninterrupted and has a molecular length of approximately 103 nm, where each α chain has a relative molecular weight of approximately 56,000 (61). The triple-helical domain contains a small globular domain at one end and a larger domain at the other. The molecule is stabilized by disulfide bonds within both the helical and globular domains. To date, the amino acid composition of the triple-helical domain alone has not yet been determined for comparison with other collagens and is not contained in Table 2.

The type VI collagen molecules associate into an elaborate, disulfide-bonded aggregate (62). Its unique structure suggests that it originates from a microfibrillar component of connective tissues similar to those found at the interface between elastic and collagenous fibrils, but thus far no function has been unambiguously assigned to this collagen type.

4. *Type VII Collagen*

While type VI has the shortest length of any known collagen, type VII is distinguished by having the longest triple-helical length yet observed without large intervening globular domains. The major triple-helical domain of type VII collagen is approximately 450 nm in length, made up of three identical α chains, each with a relative molecular weight (M_r) of 170,000 (64). Type VII collagen has only been isolated following pepsin extraction of human amniotic membrane and skin, and therefore the extent to which this molecule contains globular domains is not yet known. Preliminary indications suggest that the fiber form of this collagen type contains at least small regions of globular domain at one end of the molecule. The triple-helical domain appears to contain only a single discontinuity that can serve as a substrate site for a variety of proteolytic enzymes. This site occurs about halfway along the molecular length. The individual $\alpha 1$(VII) chains are covalently cross-linked by disulfide bonds, some of which may be within the helical domain (63). The amino acid composition of type VII collagen indicates that the protein obtained following pepsin digestion is approximately 90% triple-helical and therefore may contain up to 10% of its mass as one or more globular domains. Its amino acid composition indicates a high degree of lysyl-hydroxylation and a relatively high content of cysteinyl residues. The composition is otherwise unremarkable.

The fiber form of type VII collagen has not yet been directly demonstrated. It has been suggested that type VII collagen may be closely related to the anchoring fibril protein, based on comparisons of the banding patterns of SLS-crystallites formed from this collagen with the centrosymmetric banding patterns observed in toadskin (64). (SLS-crystallites are artificially created aggregates induced in vitro under appropriate ionic strength conditions, in which the molecules undergo lateral association with their amino and carboxyl termini aligned.) If this correlation proves to be correct, it suggests that type VII collagen molecules form fibers in a nonstaggered array by lateral accretion.

5. *Type VIII Collagen*

One of the more unusual collagens thus far described has recently been designated as type VIII collagen (65-67). Like other collagens, this molecule is a combination of triple-helical and globular domains. Structural studies suggest a model for this molecule in which three regions of collagen triple helix, each approximately M_r 50,000, are linked to one another by small globular domains of approximately M_r 10,000. Biosynthetic studies suggest that extended globular domains do not occur at the amino and carboxyl terminal regions. Unlike types IV, V, VI, and VII, each of these three triple-helical domains contains specific sites for the action of vertebrate collagenase. This molecule was first detected as a biosynthetic product of aortic endothelial cells and was therefore initially described as EC collagen. It has since been shown to be synthesized by a large variety of normal cells and tumor cell lines suggesting a broad distribution in vertebrate tissues. The fiber form of this molecule and its function are presently undetermined.

The molecular length of type VIII collagen in its procollagen form or as it occurs in tissues as fibrous structures is not yet clear. Biosynthetic studies in vitro show that the molecule is synthesized in two forms. The first contains three α chains of M_r 177,000 and the second with chains M_r 125,000. Peptide mapping studies suggest that both forms contain the same triple-helical structural domains. The lower-molecular-weight form may be a proteolytic product of the higher, which occurs very rapidly following synthesis, or conversely, the lower form may be converted to the higher form by the covalent attachment of a M_r-50,000 peptide to each chain of the lower-molecular-weight form. In pulse-chase studies, a third form, M_r 100,000 is frequently seen. This is believed to be a proteolytic product of the higher-molecular-weight forms.

6. *Type IX Collagen*

During the past several years, several α-chain pepsin fragments, termed HMW (68), LMW (68), M, (69-71), J, or K (72), have been isolated from cartilage. Recent recombinant DNA techniques have allowed the characterization of a new collagen gene that contains the coding sequences for the HMW and LMW chains (73). M, J, and K fragments appear to be fragments of HMW (74). Tryptic peptide mapping studies of the LMW fragment of the molecule indicate that this new collagen is composed of three nonidentical α chains. Similar to type VIII collagen, the molecule appears not to contain large globular domains at either the amino or carboxyl terminus. The type IX molecule has been obtained as the biosynthetic product of an in vitro cartilage organ culture system (74). This molecule has a relative molecular weight of approximately 300,000, but it is not yet known whether this is the initial gene transcript, or whether this molecule has undergone proteolytic processing. Two of the possible three chains can be separated electrophoretically. The molecule contains two domains of a pepsin-stable triple-helix, which accounts for 60% of the molecular mass of the molecule. The structural characteristics of the pepsin-sensitive portions of the molecule are presently unknown. These molecules are associated into high-molecular-weight aggregates by disulfide bonding, but the details of their fiber forms are unknown.

C. Other Collagenous Proteins

Cartilagenous tissues have proven to be a rich source for the isolation of unique proteins with regions of collagen triple helix (for review, see Ref. 75). Because of the extreme insolubility of this tissue, these peptides have been isolated only after pepsin solubilization of the matrix. Collagenous protein fragments have been characterized from this digest, but it has been difficult to determine the structure of their parent molecules. As a result, several collagenous peptides have been characterized and are identified by a temporary nomenclature.

The 1α, 2α, and 3α chains were the first of these minor cartilage collagens to be described (14). Each of these has been discussed in previous sections. The 1α and 2α chains appear to be closely related to type V collagen, while the 3α chain is similar to type II (49).

The last collagenous fragment to be considered here has also been isolated from cartilage (76,77). This molecule, termed G collagen, is synthesized by cells in the hypertrophic zone of the cartilage growth plate. This collagen, observed after metabolic labeling of organ-cultured hypertrophic zone or of cultured chondrocytes derived from the hypertrophic zone contains three presumably identical chains, each M_r 59,000. A higher-molecular-weight precursor to this molecule has not yet been identified. Proteolytic treatment of G collagen results in removal of approximately 20% of the molecular mass of the molecule, leaving a triple-helical domain of approximately M_r 40,000. This domain is insensitive to further pepsin digestion, but contains a site for vertebrate collagenase cleavage. The fiber form and function of G collagen are unknown, but its synthesis by cells of the hypertrophic zone of cartilage suggests that this molecule is a marker for the latest stages in chondrocyte differentiation.

The above information indicates a broad structural heterogeneity for members of the collagen family. It suggests that evolution has utilized the structural domain known as the collagen helix in a large variety of proteins that require a semirigid and extended conformation. As expected, this structure occurs frequently in the proteins of connective tissues. The collagen triple helix also occurs in nonstructural proteins. The enzyme acetylcholinesterase and the clotting factor C1q also contain a short region of this structural domain. In each case, triple-helical domains of various sizes are interspersed with globular domains to generate proteins with diverse functions and distributions. Within the next decade we expect to have a clearer understanding of the functions of these molecules in connective tissues.

III. BIOSYNTHESIS OF PROCOLLAGEN

From the preceding discussion we observed that the structure of procollagen is very complex. It is not surprising, then, that the biosynthesis and assembly of such a structure is also an involved and complex process. We shall use type I procollagen as an example for biosynthesis since it was the first to be described, is still the most completely characterized, and incorporates features in common with all the other procollagen types (see Refs. 8,20,78 for recent review).

The biosynthesis of procollagen begins with the selection of the correct genes for translation. The collagen genes are among the most complex yet described (78,79). The complete gene for the proα2(I) chain has been isolated and characterized. The genetic information is divided into about 50 exons by large intervening sequences. It is proposed that the entire gene, introns and exons, is transcribed into heterogeneous nuclear RNA from which the introns are removed and the exons spliced together to form the final messenger RNA (78,79). This 5000-base messenger RNA, containing about 150 nucleotides of poly-A (20), is then transported from the nucleus into the cytoplasm of the cell, where translation of procollagen begins.

There follow concurrent and sequential events including attachment of the messenger RNA and ribosomes to the rough endoplasmic reticulum, a host of posttranslational enzymatic modifications, assembly of proα chains into a trimeric procollagen molecule, and finally secretion into the extracellular space. The secretory pathway is dictated by the initial translation of a highly hydrophobic series of amino acids anywhere from 5000 to 20,000 molecular weight (80-82), which results in the attachment of the messenger RNA ribosome complex to the endoplasmic reticulum. This "signal peptide" or prepeptide causes the nascent chain to be fed through the wall of the endoplasmic reticulum into the lumen (83,84). The signal is removed before synthesis of the nascent chain is complete. An intact preproα chain can be obtained only when translation is carried out in a cell-free system where the enzymes responsible for cleaving the signal peptide are absent (80-82).

Concurrent with translation of messenger RNA, as the nascent chain appears in the lumen of the ER, modification of certain proline and lysine residues occurs. Two separate enzymes convert some of the prolyl residues to either 4-hydroxyproline or 3-hydroxyproline and a third enzyme converts lysine to hydroxylysine (85). Hydroxylation of proline in the 4 position allows for the contribution of an additional hydrogen bond to the conformation of the triple helix and thus increases the stability and the melting temperature of collagen (86). The function of 3-hydroxyproline is not known. Hydroxylation of lysine residues is necessary for the glycosylation of collagen in the helical region and for the formation of certain covalent intermolecular cross-links. The three enzymes are related mechanistically, all requiring iron, ascorbic acid, α-ketoglutarate and molecular oxygen as cofactors (85). Cell fractionation studies have indicated that all three enzymes are found in the rough endoplasmic reticulum in either soluble form [prolyl hydroxylase (87,88)] or attached to the membranes [lysyl hydroxylase (89)]. The minimum sequence for a substrate to be recognized by prolyl-4-hydroxylase is an -X-pro-gly triplet, and for the lysyl hydroxylase, an -X-lys-gly triplet, and hence both enzymes are specific for the helical domain of the collagen molecule (85,90,91). However, both enzymes require a nascent unfolded polypeptide chain as substrate and neither is active on triple-helically folded molecules (85). The extent of prolyl and lysyl hydroxylation varies between the different types of collagen (see Table 2), and it also varies between different tissues for the same type of collagen and between the same tissue at different ages. Hence, the sequence alone cannot determine whether or not a given site will be hydroxylated. Other factors that may be involved in control of hydroxylation are poorly understood. In general, there is coincidence between the extent of hydroxylation and glycosylation and the length of time it takes to synthesize and fold a procollagen molecule (20). However, no causal relationship has yet been established.

While the nascent chain is being translated and hydroxylated, certain of the hydroxylysyl residues are selected for glycosylation. Galactose or glucosyl galactose is found incorporated into the helical region of interstitial collagens, attached in O-glycosidic linkage to hydroxylysine (85). The addition of galactose to hydroxylysine is catalyzed by hydroxylysyl-galactosyl transferase. This enzyme recognizes the triplet X-hyl-gly, which occurs in the nascent polypeptide chain. Glucose is added to some, but not all, of the galactosyl hydroxylysine residues by the enzyme galactosyl-hydroxylysyl-glucosyl transferase. Both these enzymes are found distributed in the lumen of the rough and smooth endoplasmic reticulum, and 80% of the activity is membrane associated. As with the hydroxylases, glycosyl transferases are also inhibited by the formation of triple helix (85). All these enzymes are active on the elongating or completed polypeptide chains, and it is proposed that lysyl hydroxylase and both glycosyl transferases belong to a multienzyme complex so that residues of lysine are glycosylated concomitantly with hydroxylation, and activity ceases only when prolyl hydroxylation is complete and triple-helical conformation is attained (92).

In addition to the glucosyl galactose occurring on hydroxylysine residues within the triple-helical domain of the procollagen, asparagine-linked carbohydrate is attached via the dolichol phosphate oligosaccharide mechanism to the carboxyl globular domains of type I procollagen (25,93) and to both the amino and carboxyl globular domains of type II procollagen (94).

Once the proα chains have been synthesized and posttranslationally modified, these proα chains must assemble into trimers and fold. The formation of trimeric molecules is a highly specific event. Since procollagen type I contains two different types of proα chains, proα1 and proα2 in a 2:1 ratio, the association is not random. In addition, immunofluorescent studies have shown that a single cell can make type I and type III at the same time (95). Hence, the endoplasmic reticulum and the Golgi must contain at least three different proα chains, which can form only 1 out of 10 possible types of trimeric molecules. The cell must provide a mechanism for the rapid association of three very long polypeptide chains in register and in parallel so that triple helix formation can occur. Helix formation in vitro with α chains is slow, and specificity is poor (96). The globular domains of the amino propeptides of proα1 have been shown to have very little affinity for each other (97). The interchain disulfide bonds in the type III amino propeptides are dependent on triple helix formation (98,99), while the interchain disulfide bonds between the carboxyl propeptides form independently. It is proposed that the carboxyl globular domains of proα1 and proα2 provide for the alignment and chain selection necessary for rapid and specific association of three proα chains into the trimeric procollagen molecule. The mechanism appears to involve disulfide-bond-exchange enzymes in the rough endoplasmic reticulum that lead to formation of a covalently stabilized association of the carboxyl-terminal globular domains. Such enzymatic activity correlates well with the ability of cells to synthesize collagen (100).

The total time necessary to synthesize one proα chain of type I collagen is approximately 5-6 min (29), and about 20-30 min (101,102) is required to both synthesize and secrete a procollagen molecule. Collagen molecules follow the classical secretory pathway, moving from the endoplasmic reticulum to the Golgi apparatus, where they are packaged into vesicles and then secreted into the extracellular space by fusion of the vesicle with the plasma membrane (78).

The pathway for type I collagen fibril synthesis has been elucidated in detail, but how does it compare to that of the other types? The other interstitial collagens, types II and III, have a very similar structure to type I procollagen, and the rate of synthesis and secretion are approximately the same. Type III procollagen contains interchain disulfide bonds between the globular domains of the amino propeptide and near the telopeptide region at the carboxyl end of the molecule in addition to those in the carboxyl propeptides (98,99,103). The total biosynthesis and secretion time for types IV, V and VIII procollagen is considerably longer than that for the interstitial collagens. For type V procollagen, the delay may be due to an extended length of time required to attain a triple-helical conformation (104). For these procollagens, secretion does not occur until after 60-75 min (8,78,104). This prolonged period of time inside the cell correlates well with increased levels of hydroxylation and glycosylation observed for these molecules (see Table 2). There is little information yet available for the biosynthesis of types VI, VII, and IX, although type IX has been recovered from the medium of chick sternal cartilage in organ culture labeled with [^{14}C-] proline (74).

IV. PROCESSING AND FIBER FORMATION

Following the biosynthesis of type I procollagen, the propeptides are excised from molecules, leaving the helix and the telopeptide region intact (8,20,78) (Figure 3). As the procollagen molecules are being secreted from the cell, the amino propeptide is cleaved enzymatically. It is not known whether this cleavage occurs intracellularly or extracellularly, although the enzymatic activity has been shown to be associated with the cell layer of cells in culture (30). The enzyme makes a single cut across all three polypeptide chains of the procollagen molecule, liberating an intact propeptide. This cleavage product is now called pC collagen because it contains only the carboxyl propeptide. Starting at about 40 min after synthesis, the carboxyl propeptides are cleaved, and this is clearly an extracellular event (29). The enzyme responsible for this cleavage is found in the medium of cells in culture (30) and converts pC collagen to collagen by a stepwise scission of each of the propeptides from its corresponding α chain. Intermediates in this process can be observed that have had only one or two of the three chains cleaved (31,105). Again, an intact or nearly intact propeptide is released. The fate and function of these propeptides once released from the procollagen molecules is unclear.

In the extracellular space, interstitial collagen molecules associate by means of staggered lateral interactions into the classic banded collagen fiber seen ultrastructurally. The banded pattern arises from alternating regions of higher and lower electron density. Such a result can be explained by using a two-dimensional representation (Figure 4). Each collagen molecule of 300 nm is staggered with respect to its neighbor by 67 nm. A gap exists between one molecule and the next one in line because the length of the molecule is not an even multiple of the stagger distance. Each 67-nm length or "D" unit contains a gap region and a region where there is an overlap. The gap regions then will contain only four-fifths the mass of the overlap regions, creating the alternating dark and light banding pattern (106).

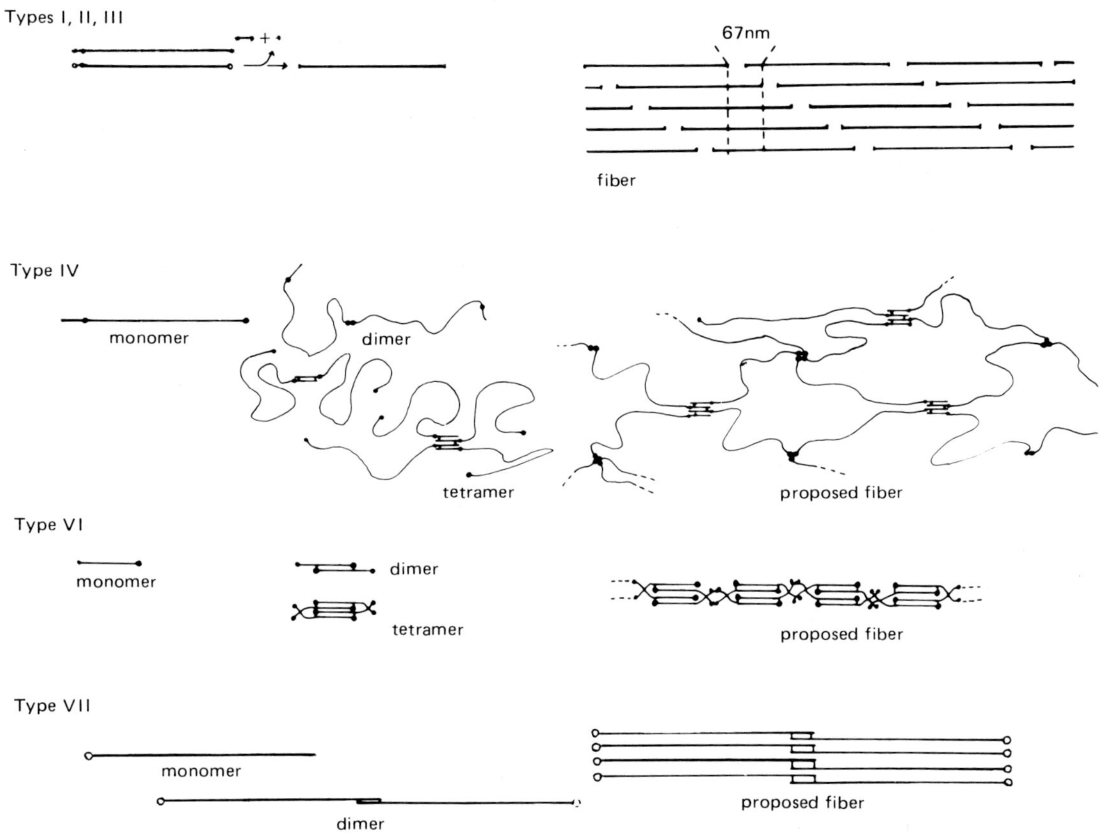

Figure 4 Models for the known and proposed fiber forms of collagen types I, II, III, IV, VI, and VII. The 67-nm periodic staining of the interstitial collagens results from increased stain accumulation in the "hole" regions of these fibers as indicated.

Interstitial collagen molecules will form fibers spontaneously from solution upon warming to 37°C. This suggests that exclusion of water and the resulting increase of the entropy of the solution drive the assembly of molecules into fibers. Interactions between the positively and negatively charged amino acids and clusters of hydrophobic amino acids distributed along the length of the molecule contribute to the precise alignment of each molecule relative to its neighbor (8,107).

Once the fiber is formed, the introduction of covalent bonds within and between molecules provides the fiber with tremendous tensile strength and insolubility. Certain lysyl or hydroxylysyl residues, particularly in the nonhelical telopeptide region, are converted to the corresponding aldehyde, allysine or hydroxyallysine, by oxidative deamination. The enzyme responsible for this conversion is lysyl oxidase. It is an extracellular enzyme with a strong preference for collagen molecules in fiber form as substrate (108). These aldehydes can condense with other lysine or hydroxylysine residues or other allysines to form a covalent bond (109). Such cross-links are recovered from tissue as aldimines after reduction (109,110). Those condensation reactions involving hydroxylysine or hydroxylallysine can undergo an Amadori rearrangement to yield a stable cross-link which does not require reduction for isolation (110). These types of cross-links disappear as the tissue ages and may serve as intermediates in the formation of mature cross-links (109,110). The identity and mechanism of synthesis of mature cross-links are unresolved. One possibility is a trifunctional cross-link involving histidine, known as histidino-hydroxymerodesmosine (109-111). A second is a continued reaction of the stable hydroxyallysine containing a bifunctional cross-link to form polyfunctional cross-links (110, 112). A third possible candidate, one for which substantial data have been accumulated and for which some consensus has been achieved, is the trifunctional cross-link, pyridinoline (113,114). This compound is derived from two hydroxyallysines and one hydroxylysine (110) and has been found in tendon, bone (113,114), dentin and cartilage (115). A mechanism for the formation of pyridinoline from two bifunctional cross-links has been proposed (115).

In our discussion, we have separated the process of the removal of the propeptide regions of procollagen from the assembly of the collagen molecules into fibers. For type I procollagen, these may not necessarily be separate processes, and as we shall see for other types of collagens, the propeptides may be directly involved in fiber formation. The form of collagen that first actually contacts the fiber is probably pC collagen, the collagen molecules containing the carboxyl propeptide. Such molecules have a substantial half-life in the extracellular space and may be involved in the alignment of collagen molecules and the rate and extent of fibril formation (20).

How do other collagen types vary from this pattern of complete processing and fiber formation by means of lateral association in a quarter-stagger fashion? Type II collagen processing and fiber formation is very similar to that of type I. Both the carboxyl and amino propeptides are completely removed during the process of fiber formation (116). However, there seems to be no preference for the removal of one amino propeptide prior to the other. The collagen fibers of type II tend to be of smaller diameter compared to type I, and this may be related to the increased content of neutral hexose in type II collagen (117).

Type III collagen is again closely related to type I. In this case, the carboxyl propeptide is removed prior to removal of the amino propeptide (99). In addition, biosynthetic studies have indicated that the amino propeptide may never be removed or removed very slowly (99) such that it is incorporated into the fiber structure of the type III collagen. pN collagen type III has been extracted from tissue in significant amounts, indicating that this form of type III persists and probably contributes to the fibrils of type III collagen (97,118). In vitro, type III decreases the diameter of the fibers made by the mixture of type I and type III, and this effect is proprotional to the percent of type III (119). While this suggests that, in vivo, type III and type I can interact to modulate fiber diameters, the ultrastructural element to which type III belongs is still unresolved. In addition to the collagen cross-links, type III may also contain covalent disulfide bridges between molecules in the collagen fiber (120).

Type IV collagen appears not to be processed at all after secretion from the cell (8) (Figure 2). Type IV does not form banded fibers, and the details of the fibrous network formed by type IV are unresolved. The intermolecular interactions that have been described for type IV collagen appear to take place at the ends of the molecule (see Figure 4). At the carboxyl end of the procollagen molecule, the propeptide seems to interact with one other carboxyl propeptide to form a dimer (40,104,121,122). The interactions at the amino end are more complex and involve separate collagenous domains of about 30 nm in length at the extreme amino terminus. At least four molecules overlap throughout this region, having both parallel and antiparallel associations (40,123). This overlap region is stabilized by disulfide as well as lysyl-derived cross-links and has been isolated after extensive pepsin digestion as a molecule with a sedimentation coefficient of 7S (58). The combination of the tetrameric association at the amino end and the dimeric association at the carboxyl end results in the chicken-wire-type model for type IV collagen fiber (2,40,58,122,123). It is not understood how this chicken-wire network fits into the basement membrane.

The association of type IV molecules with each other may not be limited to interactions at the ends. It has recently been reported that type IV molecules can interact through their central helical regions as well (124). Dimers of type IV molecules (linked through their carboxyl propeptides) associated side by side with a stagger of 170 nm. A model is proposed whereby the carboxyl globular domains are associated with a flexible region of the helix forming the vertices of a hexagonal lattice. In combination with the previous model, the amino terminal ends would interact to form the 7S structure, which would be perpendicular to and connect two planes of hexagonal lattice. Further work is needed to clearly define the nature and geometry of the type IV fiber system.

The biosynthesis and structure of type V procollagen has been studied extensively since it was first reported in 1976 (47,104,125-128). These studies have been complicated by the fact that type V collagen appears to contain alternative molecular forms for the $\alpha 1$ and $\alpha 2$ type V chains, and in some tissues the presence of a third type V chain, $\alpha 3$, further complicates the issue (see above). Type V procollagen is synthesized with globular domains at either end of the molecule analogous to type I, although these domains are larger than those of type I (104,125) (Figure 2). Only some of the folded trimeric procollagen molecules contain interchain disulfide bonds. Of those, some contain two disulfide-bonded chains and some three, but all disulfide bonding appears to occur via the proα2(V) chain (104,127). Processing of type V procollagen is similar to processing of type I in that two proteolytic stages are involved, proα(V) $\rightarrow$ pα(V) $\rightarrow$ α(V). But, this processing involves an attachment of mass to pα2, and fully processed type V retains substantial portions of the globular propeptide domains (104,127,128) (Figure 2). How the presence of these nonhelical regions affects the fiber form of type V is unknown since no fiber structure either in vivo or in vitro has been identified for type V.

Type VI collagen contains two globular domains (61,129), one at each end of the molecule which persist even after extraction from the tissue with pepsin (62) (Figure 1). Based on transmission electron microscopy following rotary shadowing, a model for the structure of type VI collagen aggregates has been proposed that involves interactions of both the helical domain and the globular domains (62) (see Figure 4). Two monomers associate side by side in antiparallel orientation with an overlap encompassing about 70% of the molecular length. Most of this overlap involves the helical region of the molecule, and there appears to be a weak interaction between the two helices. The dimer is stabilized by disulfide bonds between the globular regions within the overlap and the corresponding helical portion of the other molecule. Tetramers are formed from dimers by the nonstaggered alignment of two dimers side by side. The nonoverlapped arms of the dimer cross over forming a scissorlike region. This crossover region can be stabilized by disulfide bonds. It is hypothesized that thc scissorlike regions of the tetramer interact with each other to form filamentous polymers. There are no processing data available so we do not know whether additional peptide domains contribute to the multiple interactions.

At this time only preliminary data are available for the fiber structure of type VII collagen (5,63,64) (Figure 4). Pepsin-extracted type VII collagen appears to be predominantly triple-helical; electron microscopy of rotary-shadowed molecules reveals no obvious globular domain. However, amino acid analysis (see Table 2) indicates that there may be some nonhelical regions (see Figure 2). Two molecules of type VII interact at "like" ends with an antiparallel orientation. In this interaction, the ends of each molecule overlap each other by about 60 nm out of a total dimeric length of 840 nm. It is not known whether there is any interaction between the helical regions within the overlap zone. However, the overlap is stabilized by intermolecular disulfide bonds, which are probably located in the nonhelical region. This orientation and alignment is reminiscent of the model for the structure of type IV collagen fiber. However, there is no evidence for interactions at the opposite ends of the molecule to generate polymers or fibers of greater length. In the absence of biosynthetic data, it is not known whether type VII contains additional peptide regions that survive processing and contribute to fiber formation. As discussed earlier, it has been proposed that type VII may be the anchoring fibril protein. These structures in toadskin have the same length as the type VII dimer; however, they are much broader. One would expect aggregation of type VII molecules to follow a process of side-by-side association until a thickness approaching an anchoring fiber was reached. Such a structure can be generated from type VII molecules in vitro; however, this occurs under nonphysiological conditions.

At this time we have no information on the fiber structure of type VIII or type IX collagen. Also, we have no information on the intermolecular interaction of the various unique α chains mentioned above.

V. SUMMARY

In this brief overview of collagen, we have attempted to illustrate that each of the different types or classes of collagen molecules has unique features. This specialization results from differences in primary structure, posttranslational modification, and extracellular processing. Considerable progress has been made in characterizing new collagen molecules. Additional study is required to determine the geometry and other fiber assembly characteristics of these new collagen molecules. Only the fiber form of the interstitial collagens is known in some detail, and only for types I, II, and IV are the corresponding ultrastructural elements known without question. Such knowledge must be extended to the other members of the collagen family of proteins if we are to understand the structure and function of the extracellular matrix, the interactions of collagens with cells and other matrix components, the regulatory effects of the matrix on collagen synthesis, remodeling of the matrix, and, finally, the host of pathological conditions that can arise from aberrations in the normal course of these complex events.

REFERENCES

1. L. W. Cunningham and D. W. Fredericksen, in *Methods in Enzymology* (S. P. Colowick and N. O. Kaplan, eds.), Academic Press, New York, 1982, p. 3.
2. R. E. Burgeson, *J. Invest. Dermatol., 79*(suppl. 1):25s (1982).
3. R. R. Bruns, *J. Cell Biol., 42*:418 (1969).
4. G. E. Palade and M. G. Farquhar, *J. Cell Biol., 27*:215 (1965).
5. H. Bentz, N. P. Morris, L. W. Murray, L. Y. Sakai, D. W. Hollister, and R. E. Burgeson, *Proc. Natl. Acad. Sci. USA, 80*:3168 (1983).
6. A. Miller, in *Biochemistry of Collagen* (G. N. Ramachandran and A. H. Reddi, eds.), Plenum Press, New York, 1976, p. 85.

7. K. A. Piez, M. S. Lewis, G. R. Martin, and J. Gross, *Biochim. Biophys. Acta, 53*: 596 (1961).
8. P. Bornstein and H. Sage, *Ann. Rev. Biochem., 49*:957 (1980).
9. C. D. Little, R. L. Church, R. A. Miller, and F. H. Ruddle, *Cell, 10*:287 (1977).
10. C. D. Little and R. L. Church, *Arch. Biochem. Biophys., 190*: 632 (1978).
11. R. L. Trelstad, A. H. Kang, S. Igarashi, and J. Gross, *Biochemistry, 9*:4993 (1970).
12. E. J. Miller, *Biochemistry, 10*:1652 (1971).
13. W. T. Butler, J. E. Finch, Jr., and E. J. Miller, *J. Biol. Chem., 252*:639 (1977).
14. R. E. Burgeson and D. W. Hollister, *Biochem. Biophys. Res. Commun., 87*:1124 (1979).
15. D. R. Eyre, in *International Review of Connective Tissue Research* (D. A. Hall and D. S. Jackson, eds.), Vol. 8, Academic Press, New York, 1979, p. 227.
16. E. J. Miller, E. H. Epstein, Jr., and K. A. Piez, *Biochem. Biophys. Res. Commun., 42*:1024 (1971).
17. E. Chung and E. J. Miller, *Science, 183*:1200 (1974).
18. E. H. Epstein, Jr., *J. Biol. Chem., 249*:3225 (1974).
19. F. M. Pope, G. R. Martin, J. R. Lichtenstein, R. P. Penttinen, B. Gerson, D. W. Rowe, and V. A. McKusick, *Proc. Natl. Acad. Sci. USA, 72*:1314 (1975).
20. J. H. Fessler and L. I. Fessler, *Ann. Rev. Biochem., 47*:129 (1978).
21. G. N. Ramachandran and C. Ramakrishnan, in *Biochemistry of Collagen*, Plenum Press, New York, 1976, p. 45.
22. R. W. Glanville, T. Voss, and K. Kuhn, in *New Trends in Basement Membrane Research* (K. Kuehn, H. H. Schoene and R. Timpl, eds.), Raven Press, New York, 1981, p. 69.
23. E. D. Harris, Jr., and C. A. Vater, in *Methods in Enzymology* (L. W. Cunningham and D. W. Fredericksen, eds.), Vol. 8, Academic Press, New York, 1982, p. 423.
24. E. H. Epstein, Jr., and N. H. Munderloh, *J. Biol. Chem., 250*:9304 (1975).
25. B. R. Olsen, N. A. Guzman, J. Engel, C. Condit, and S. Aase, *Biochemistry, 16*: 3030 (1977).
26. U. Becker, R. Timpl, O. Helle, and D. J. Prockop, *Biochemistry, 15*:2853 (1976).
27. D. Horlein, P. P. Fietzek, E. Wachter, C. M. Lapiere, and K. Kuhn, *Eur. J. Biochem., 99*:31 (1979).
28. D. Horlein, P. P. Fietzek, and K. Kuhn, *FEBS Lett., 89*:279 (1978).
29. L. I. Fessler, N. P. Morris, and J. H. Fessler, *Proc. Natl. Acad. Sci. USA, 72*:4905 (1975).
30. M. K. K. Leung, L. I. Fessler, D. B. Greenberg, and J. H. Fessler, *J. Biol. Chem., 254*:224 (1979).
31. N. P. Morris, L. I. Fessler, and J. H. Fessler, *J. Biol. Chem., 254*:11024 (1979).
32. C. M. Lapiere and B. Nusgens, *Bilchim. Biophys. Acta., 342*:237 (1974).
33. B. Steinmann, L. Tuderman, L. Peltonen, G. R. Martin, V. A. McKusick, and D. J. Prockop, *J. Biol. Chem., 255*:8887 (1980).
34. N. A. Kefalides, *Renal Physiol., 4*:57 (1981).
35. J. R. Stanley, D. T. Woodley, S. I. Katz, and G. R. Martin, *J. Invest. Dermatol., 79*(suppl. 1):69s (1982).
36. A. Martinez-Hernandez and D. S. Amenta, *Lab Invest., 48*:656 (1983).
37. L. Y. Sakai, E. Engvall, D. W. Hollister, and R. E. Burgeson, *Am. J. Pathol., 108*: 310 (1982).
38. K. Kuhn, H. Wiedemann, R. Timpl, J. Risteli, H. Dieringer, T. Voss, and R. W. Glanville, *FEBS Lett., 125*:123 (1981).
39. R. Timpl, J. Risteli, and H. P. Bachinger, *FEBS Lett., 101*:265 (1979).
40. R. Timpl., H. Widemann, V. VanDelden, H. Furthmayr, and K. Kuhn, *Eur. J. Biochem., 120*:203 (1981).

41. R. E. Burgeson, F. A. El Adli, I. I. Kaitila, and D. W. Hollister, *Proc. Natl. Acad. Sci. USA, 73*:2579 (1976).
42. E. Chung, R. K. Rhodes, and E. J. Miller, *Biochem. Biophys. Res. Commun., 71*: 1167 (1976).
43. R. A. Brown, C. A. Shuttleworth, and J. B. Weiss, *Biochem. Biophys. Res. Commun., 80*:866 (1978).
44. H. Sage and P. Bornstein, *Biochemistry, 18*:3815 (1979).
45. H. Bentz, H. P. Bachinger, R. Glanville, and K. Kuhn, *Eur. J. Biochem., 92*:563 (1978).
46. R. K. Rhodes and E. J. Miller, *Biochemistry, 17*:3442 (1978).
47. M. A. Haralson, W. M. Mitchell, R. K. Rhodes, T. F. Kresina, R. Gay, and E. J. Miller, *Proc. Natl. Acad. Sci. USA, 77*:5206 (1980).
48. R. K. Rhodes and E. J. Miller, *Collagen Rel. Res., 1*:337 (1981).
49. R. E. Burgeson, P. A. Hebda, N. P. Morris, and D. W. Hollister, *J. Biol. Chem., 257*:7852 (1982).
50. K. S. Stenn, J. A. Madri, and F. S. Roll, *Nature (London), 277*: 229 (1979).
51. S. Gay, R. K. Rhodes, R. E. Gay, and E. J. Miller, *Collagen Rel. Res., 1*:53 (1981).
52. A. Poschl and K. VonDerMark, *FEBS Lett., 115*:100 (1980).
53. T. F. Linsenmayer, J. M. Fitch, T. M. Schmid, N. B. Zar, E. Gibney, R. D. Sanderson, and R. Mayne, *J. Cell Biol., 96*:124 (1983).
54. C. A. Kumamoto and J. H. Fessler, *Proc. Natl. Acad. Sci. USA, 77*:6434 (1980).
55. C. A. Kumamoto and J. H. Fessler, *J. Biol. Chem., 256*:7053 (1981).
56. L. I. Fessler, W. J. Robinson, and J. H. Fessler, *J. Biol. Chem., 256*:9646 (1981).
57. D. K. Furuto and E. J. Miller, *J. Biol. Chem., 255*:290 (1980).
58. J. Risteli, H. P. Bachinger, J. Engel, H. Furthmayr, and R. Timpl, *Eur. J. Biochem., 108*:239 (1980).
59. R. Jander, J. Rauterberg, B. Voss, and D. B. von Bassewitz, *Eur. J. Biochem., 114*: 17 (1981).
60. G. Laurain, T. Delvincourt, and A. G. Szymanovicz, *FEBS Lett., 120*:44 (1980).
61. E. Odermatt, J. Risteli, V. VanDelden, and R. Timpl, *Biochem. J., 211*:295 (1983).
62. H. Furthmayr, H. Widemann, R. Timpl, E. Odermatt, and J. Engel, *Biochem. J., 211*:303 (1983).
63. H. Bentz, N. P. Morris, L. W. Murray, L. Y. Sakai, D. W. Hollister, and R. E. Burgeson, *Proc. Natl. Acad. Sci. USA 80*:3168 (1983).
64. R. E. Burgeson, N. P. Morris, L. W. Murray, K. G. Duncan, D. R. Keene, and L. Y. Sakai, *Ann. N. Y. Acad. Sci.* (in press).
65. H. Sage, P. Pritzl and P. Bornstein, *Biochemistry, 19*:5747 (1980).
66. P. D. Benya, *Renal Physiol., 3*:30 (1980).
67. H. Sage, B. Trueb and P. Bornstein, *J. Biol. Chem., 258*:13391 (1983).
68. C. A. Reese and R. Mayne, *Biochemistry, 20*:5443 (1981).
69. M. Shimokomaki, V. C. Duance, and A. J. Bailey, *FEBS Lett., 121*:51 (1980).
70. S. Ayad, M. Z. Abedin, S. M. Grundy, and J. B. Weiss, *FEBS Lett., 123*:195 (1981).
71. K. VonDerMark, M. vanMenxel, and H. Wiedemann, *Eur. J. Biochem., 124*:57 (1982).
72. G. J. Gibson, C. M. Kieltiz, C. Garner, S. L. Schor, and M. E. Grant, *Biochem. J., 211*:417 (1983).
73. Y. Ninomiya and B. R. Olsen, *Proc. Natl. Acad. Sci. USA, 81*:3014 (1984).
74. P. Bruckner, R. Mayne, and L. Tuderman, *Eur. J. Biochem., 136*:333 (1983).
75. R. Mayne and K. VonDerMark, in *Cartilage* (B. K. Hall, ed.), Vol. 1, Academic Press, New York, 1983, p. 181.
76. C. S. Gibson, S. L. Schor, and M. E. Grant, *J. Cell Biol., 93*:767 (1982).

77. T. M. Schmid and E. Conrad, *J. Biol. Chem., 257*:12444 (1982).
78. B. R. Olsen, in *Cell Biology of the Extracellular Matrix* (E. D. Hay, ed.), Plenum Press, New York, 1981, p. 139.
79. H. Boedtker, F. Fuller, and V. Tate, *Int. R. Conn. 10*:1 (1983).
80. R. D. Palmiter, J. M. Davidson, J. Gagnon, D. W. Rowe, and P. Bornstein, *J. Biol. Chem., 254*:1433 (1979).
81. L. Sandell and A. Veis, *Biochem. Biophys. Res. Commun., 92*:554 (1980).
82. P. Graves, B. R. Olsen, P. P. Fietzek, D. J. Prockop, and J. M. Monson, *Eur. J. Biochem., 118*:363-369 (1981).
83. P. N. Campbell and G. Blobel, *FEBS Lett., 72*:215 (1976).
84. G. Blobel, *Proc. Natl. Acad. Sci. USA, 77*:1496 (1980).
85. K. I. Kivirikko and R. Myllyla, in *Methods In Enzymology* (L. W. Cunningham and D. W. Frederiksen, eds.), Vol. 8, Academic Press, New York, 1982, p. 245.
86. D. J. Prockop, R. A. Berg, K. I. Kivirikko, and J. Uitto, in *Biochemistry of Collagen* (G. N. Ramachandran and A. H. Reddi, eds.), Plenum Press, New York, 1976, p. 163.
87. B. Peterkofsky and R. Assad, *J. Biol. Chem., 251*:4770 (1976).
88. B. R. Olsen, R. A. Berg, Y. Kishida, and D. J. Prockop, *J. Biol. Chem., 64*:340 (1975).
89. B. Peterkofsky and R. Assad, *J. Biol. Chem., 254*:4714 (1979).
90. K. I. Kivirikko and R. Myllyla, in *Collagen In Health and Disease* (M. I. Jayson and J. B. Weiss, eds.), Churchill-Livingston, Edinburgh, 1982, pp. 101-120.
91. K. I. Kivirikko and R. Myllyla, in *The Enzymology of Posttranslational Modification of Proteins* (R. B. Friedman and H. Hawkins, eds.), Academic Press, New York, 1980, p. 53.
92. N. Blumenkrantz, R. Assad, and B. Peterkofsky, *J. Biol. Chem., 259*:854 (1984).
93. C. C. Clark, *J. Biol. Chem., 254*:10798 (1979).
94. N. A. Guzman, P. N. Graves, and D. J. Prockop, *Biochem. Biophys. Res. Commun., 84*:691 (1978).
95. S. Gay, G. R. Martin, P. K. Miller, R. Timpl, and K. Kuhn, *Proc. Natl. Acad. Sci. USA, 73*:4037 (1976).
96. J. H. Fessler, N. P. Morris, G. M. Greenberg, and L. I. Fessler, in *Extracellular Matrix Influences on Gene Expression* (H. A. Slavkin, ed.), Academic Press, New York, 1975, p. 101.
97. J. Engel, P. Bruckner, U. Becker, R. Timpl, and B. Rutschmann, *Biochemistry, 16*: 4026 (1977).
98. L. I. Fessler, and J. H. Fessler, *J. Biol. Chem., 254*:223 (1979).
99. L. I. Fessler, R. Timpl, and J. H. Fessler, *J. Biol. Chem., 256*:2531 (1981).
100. R. Myllyla, J. Koivu, T. Pihlajaniemi, and K. I. Kivirikko, *Eur. J. Biochem., 134*:7 (1983).
101. N. P. Morris, L. I. Fessler, A. Weinstock, and J. H. Fessler, *J. Biol. Chem., 250*:5719 (1975).
102. C. P. Leblond, G. W. Laurie, and G. Wright, in *New Trends In Basement Membrane Research* (K. Kuhn, H. H. Schoene, and R. Timpl, eds.), Raven Press, New York, 1982, p. 163.
103. P. Bruckner, H. P. Bachinger, R. Timpl, and J. Engel, *Eur. J. Biochem., 90*:595 (1978).
104. J. H. Fessler, H. P. Bachinger, G. Lunstrum, and L. I. Fessler, in *New Trends In Basement Membrane Research* (K. Kuhn, H. H. Schoene, and R. Timpl, eds.), Raven Press, New York, 1982, p. 145.
105. J. M. Davidson, L. S. G. McEneany, and P. Bornstein, *Biochemistry 14*:5188 (1975).
106. B. Brodsky and E. F. Eikenberry, in *Methods in Enzymology* (L. W. Cunningham and D. W. Fredericksen, eds.), Vol. 8, Academic Press, New York, 1982, p. 127.

107. R. L. Trelstad and F. H. Silver, in *Cell Biology of Extracellular Matrix* (E. D. Hay, ed.), Plenum Press, New York, 1981, p. 179.
108. R. C. Siegel, *Proc. Natl. Acad. Sci. USA, 71*:4826 (1974).
109. M. L. Tanzer, in *Biochemistry of Collagen* (G. N. Ramachandran and A. H. Reddi, eds.), Plenum Press, New York, 1976, p. 137.
110. S. P. Robbins, *Meth. Biochem. Anal., 28*:329 (1982).
111. M. L. Tanzer, R. Fairweather, C. Franzblau, and P. M. Gallop, *J. Biol. Chem., 248*: 393 (1973).
112. A. J. Bailey, N. D. Light, and E. D. T. Atkins, *Nature, 228*:408 (1980).
113. D. Fugimoto, T. Moriguchi, T. Ishida, and H. Hayashi, *Biochem. Biophys. Res. Commun., 84*:52 (1978).
114. D. Fugimoto, *Biochem. Biophys. Res. Commun., 93*:948 (1980).
115. D. R. Eyre and H. Oguchi, *Biochem. Biophys. Res. Commun., 92*:403 (1980).
116. J. Uitto, *Biochemistry, 16*:3421 (1977).
117. P. J. Morgan, H. G. Jacobs, J. P. Segrest, and L. W. Cunningham, *J. Biol. Chem., 245*:5042 (1970).
118. H. Nowak, B. R. Olsen, and R. Timpl, *Eur. J. Biochem., 70*:205 (1976).
119. C. M. Lapiere, B. Nusgens, and G. E. Pierard, *Connective Tiss. Res., 5*:21 (1977).
120. D. T. Cheung, P. Dicesare, P. D. Benya, E. Libau, and M. E. Nimni, *J. Biol. Chem., 258*:7774 (1983).
121. R. Timpl, I. Oberbaumer, H. Furthmayr, and K. Kuhn, in *New Trends in Basement Membrane Research* (K. Kuhn, H. H. Schoene, and R. Timpl., eds.), Raven Press, New York, 1982, p. 57.
122. L. I. Fessler and J. H. Fessler, *J. Biol. Chem., 257*:9804 (1982).
123. K. G. Duncan, L. I. Fessler, H. P. Bachinger, and J. H. Fessler, *J. Biol. Chem., 258*: 5869 (1983).
124. P. D. Yurchenko and H. Furthmayr, *Biochemistry, 23*:1839 (1984).
125. H. P. Bachinger, K. J. Doege, J. P. Petschek, L. I. Fessler, and J. H. Fessler, *J. Biol. Chem., 257*:14590 (1982).
126. J. M. Foidart, K. Tryggvason, P. G. Robey, L. A. Liotta, and G. R. Martin, *Coll. Rel. Res., 1*:137 (1981).
127. K. Alitalo, R. Myllyla, H. Sage, P. Pritzl, A. Vaheri, and P. Bornstein, *J. Biol. Chem., 257*:9016 (1982).
128. A. S. Narayanan and R. C. Page, *J. Biol. Chem., 258*:11694 (1983).
129. J. Rauterberg, R. Jander, B. Voss, H. Furthmayr, and R. Glanville, in *New Trends In Basement Membrane Research* (K. Kuhn, H. H, Schoene, and R. Timpl, eds.), Raven Press, New York, 1982, p. 107.

2

Elastin

Structure and biology

Jeffrey M. Davidson*
Medical Research Service, Veterans Administration, and
University of Utah School of Medicine,
Salt Lake City, Utah

1. INTRODUCTION

A. Elastic Tissue

Connective tissues provide multicellular animals with the basic fabric for stabilizing the three-dimensional organization of cells. In addition to the important structural parameters of strength and rigidity, higher vertebrates, including man, depend on the resiliency of certain organs and tissues. These elastic tissues are composed, in part, of a unique structural entity, the elastic fiber, whose physical properties approach those of an ideal rubber. The loss or dysfunction of elastic tissue can have severe pathological consequences.

B. The Elastic Fiber

The elastic fiber has been recognized as a histological entity for at least a century (1). This material, whose principal component is the structural protein elastin (2), is a prominent constituent of diverse connective tissues where resiliency is an important functional parameter: blood vessels, skin, and lung. Elastic fibers are recognized in these and other tissues at the light microscopic level by their wavy appearance, frequent codistribution with interstitial collagen fibers, and characteristic staining properties: At the ultrastructural level, elastic fibers are conspicuously lacking in obvious substructure and bind conventional stains poorly in the absence of mordants, but are typically bounded by associated microfibrillar components. As discussed in the chapter by Cleary (3, this volume), the microfibrils of elastic fibers are probably a chemically distinct entity. Microfibrils can be found in tissues in the absence of amorphous elastin and can be selectively removed from elastic fibers by denaturing solvents and reduction.

**Present affiliation*: Vanderbilt University School of Medicine and Veterans Administration Medical Center, Nashville, Tennessee.

The purpose of this chapter is to provide an overview of the molecular and cellular biology of the polypeptide elastin in the context of the known or suspected involvement of elastin in human disease states. Several reviews have recently been published on the topics of elastin structure and biology (4-7). Consequently, this review will not deal with those topics already comprehensively discussed in previous publications, but rather will focus on what may be the more current questions to be addressed at the biochemical and molecular level. The reader is referred to Chapter 3 for a detailed description of the nonelastic components of elastic fibers, the microfibrillar component. In addition, the reader is referred to Chapters 18 and 19 for a detailed discussion of specific disease states that involve the altered morphology or function of elastic fibers.

II. ELASTIN CHEMISTRY

A. Isolation of Elastin

The structural protein elastin is the most insoluble protein present in the human body. This feature, as well as elastin's unusual structure, has certainly contributed to the moderate pace at which our understanding of the biology of elastin has proceeded.

The early chemical isolation procedures devised by Lansing et al. (8) and Partridge et al. (9) involved vigorous extraction and denaturation of other connective tissue components by hot alkali or autoclaving. The insoluble residue had most of the compositional and physical properties of other types of elastin preparations (10). Subsequently, numerous studies have attempted to isolate a more "pure" form of the protein as judged by the criterion of amino acid composition (11).

Only with the discovery of the soluble precursor tropoelastin (12) was it possible to define the precise chemical nature of elastin. The monomeric, secreted form of elastin has a molecular weight of about 70,000 when extracted from elastin-synthesizing tissues such as aorta. This hydrophobic polypeptide has the unusual property of being soluble in alcohols and of undergoing spontaneous precipitation (coacervation) as it is warmed in aqueous solutions above 15-20°C. Tropoelastin is rapidly cross-linked into an insoluble polymer.

B. Amino Acid Composition and Sequence

The compositional hallmarks of elastin and its soluble precursor tropoelastin are extremely high contents of the small, hydrophobic amino acids: glycine, alanine, and valine (Table 1). Another important feature of purified elastin is the absence of methionine, tryptophan, and histidine. The presence of any of these three amino acids is, therefore, a good indication of the extent of contamination of the elastic protein; however, as discussed below, several authors have suggested that some forms of elastin may have compositional differences (13-15). Virtually unique to elastin are the desmosines, which are formed by the covalent cross-linking of four lysyl residues. Desmosines have been reported in eggshell proteins as well (16). The amino acid composition of elastin is highly conserved among higher vertebrate species. While some elastin-like proteins have been described in lower vertebrates, the protein seems to have evolved in recent evolutionaly history, being present only in the vertebrate classes (17).

While collagen and elastin fall under the same generic classification of connective tissue proteins, there is very little known to support a structural or genetic relationship between these two classes of molecules. However, elastin and collagen can be synthesized by the same cells (18) and found in close anatomical association. The posttranslational modifi-

Table 1 Amino Acid Composition of Porcine and Human Elastin[a]

Amino acid	Porcine Soluble	Porcine Insoluble	Human[b] (Insoluble)
Glycine	245	256	237
Alanine	187	181	182.8
Valine	103	92	114.8
Proline	92	90	103.5
Leucine	41	41	44.9
Isoleucine	14	14	18.8
Glutamic and glutamine	12	16	15.6
Phenylalanine	14	12	16.4
Threonine	10	11	11.0
Hydroxyproline	7	8	5.2
Serine	8	11	7.4
Tyrosine	14	12	18.4
Aspartic acid and asparagine	3	5	5.0
Arginine	4	5	5.8
Lysine[c]	37	5	3.1
Desmosines[d]	–	3	3.4
Histidine	0	0	0
Cysteine	<1	<1	–
Methionine	0	0	0
Tryptophan	0	0	–

[a]Expressed as residues per 800 residues.
[b]Taken from Ref. 13 (Table 4). Human thoracic aorta.
[c]Seven lysines are unaccounted for in mature elastin.
[d]Equivalent of 12 lysyl residues; other intermediates account for a total of 25 lysines.

cations that are shared by elastin and collagen may be, to some extent, a consequence of cosynthesis of these two structural elements, as discussed below.

The reported amino acid compositions of purified elastin have frequently been deduced after exhaustive extraction of connective tissue structures, such as the artery wall, with collagen-degrading enzymes (19), strong denaturing solvents (20), and reducing agents (21). However, reasonably pure preparations of elastin, from a compositional point of view, have been prepared by much cruder techniques such as refluxing with hot sodium hydroxide (8) or simply by extensive autoclaving (11). All these procedures have been designed to denature and solubilize collagen, associated glycoproteins, and proteoglycans in the connective tissue matrix. There is some disagreement among investigators in the field as to which type of extraction leads to the material of highest "purity." As discussed below, however, many of these questions have been resolved by a more detailed examination of the soluble elastin precursor tropoelastin.

C. Primary and Higher-Order Structure

Elastin is highly insoluble in denaturing solvents owing to the extensive network of cross-linked polypeptide chains within the protein. This feature of the molecule, which had hampered structural analysis for years, was finally overcome by the discovery that cross-link formation was a copper-dependent process mediated by the enzyme lysyl oxidase (22-25). These findings led to isolation of the soluble elastin precursor tropoelastin and studies of its primary amino acid sequence (26). Amino acid sequence analysis to date has been derived from tryptic peptides of porcine (and some chick) tropoelastin (4). These structural studies, while not quite complete, have revealed some surprising features of the protein. Two general classes of tryptic peptides have been produced: large (10-80 residues) hydrophobic peptides, and small lysine- and alanine-rich peptides. Sequence analysis of the large peptides has revealed numerous polypeptide repeat units within these fragments. Most frequent among these repeat sequences are the pentapeptide repeat (PGVGV) and the hexapeptide repeat structure (PGVGVA). The smaller peptides consist of polyalanyl tracts ending with lysyl residues (AAKAAK), which are likely to represent the regions of the molecule with potential for cross-link formation.

The relationship of the hydrophobic polypeptide repeat units to the rubbery properties of the elastin molecule is not fully understood. Gray, Sandberg, and Foster have proposed that they participate in an "oiled-coil" structure (27). X-ray crystallographic and nuclear magnetic resonance studies of elastic fibers suggest that the bulk protein exists largely in a random coil conformation (28). Nevertheless, optical diffraction studies on isolated elastic fiber material (29), as well as NMR and diffraction studies on the synthetic polymers of the pentapeptide repeat (30), have led to proposals for a more ordered structure in these regions of the elastomer. It has been suggested that under certain circumstances the polypentapeptide repeat, as represented by the synthetic oligopolypentapeptide $(PGVGV)_n$, can assume a "β spiral" conformation in which a ribbon of residues, stabilized by β turns, is wrapped around a core of bound water (31). It must be emphasized, however, that the thermodynamic characteristics of elastin are entirely consistent with its role as a biological rubber in an anisotropic system (32). The extent to which the amino acid sequences within the molecule assume an ordered conformation in the relaxed state, or the extent to which this conformation is favored in biological systems, is incompletely understood.

The elastin component of elastic fibers is amorphous at the ultrastructural level; however, some investigators have reported the appearance, largely in vitro, of subassemblies (aggregates) of elastin or tropoelastin exhibited in a fibrillar conformation, suggestive of an ordered array of molecules within the elastic fiber (33). It is conceivable that ordered arrays of tropoelastin molecules occur as intermediates in the assembly of elastic fibers and later become disordered upon the completion of fibrillogenesis.

D. Soluble Forms of Elastin

Because of the insoluble nature of elastin, numerous methods have been devised to obtain solubilized forms of the protein for a variety of biological and immunological experiments. Insoluble elastin is hydrolyzed by a variety of elastolytic proteinases, the elastases, particularly the serine proteinases of pancreatic and leukocyte origin (34). Insoluble elastin can also be nonspecifically hydrolyzed by refluxing repeatedly in the presence of 0.25 M oxalic acid (9). This preparation results in a mixture of various sized peptides that can be fractionated by gel filtration or dialysis to produce two fractions: α- and β-elastin. α-Elastin can also be fractioned from β-elastin because it shares with tropoelastin the chemical property of spontaneously precipitating from acidic or neutral solutions upon warming to physiological temperatures, a process known as coacervation. The amino acid compositions of α- and β-elastin are quite similar except for the lower content of desmosine cross-links in β-elastin. α-Elastin, however, has an average molecular weight of around 50,000 daltons, whereas the dialyzable β-elastin peptides fall in the range of 10,000 daltons or less. An additional form of soluble elastin, referred to as kappa elastin (35), is created by the alcoholic hydrolysis of insoluble elastin with potassium hydroxide.

Since each of these chemical cleavage methods results in a more or less random assortment of peptide sequences, the major purpose of producing these soluble forms of elastin is the production of a more antigenic material for the development of antielastin antisera and elastin immunoassay. Although it has been used to produce antibodies (36), insoluble elastin is a very poor immunogen (37). In contrast, solubilized elastin peptides, either directly injected into animals or previously conjugated to keyhole limpet hemocyanin, have been used very successfully to produce high-titer antisera against the polypeptide (38,39,60). Other forms of elastin are the synthetic polypentapeptide and polyhexapeptide repeat polymers, which have been used for extensive studies of secondary structure of the molecule (31). These structures are based on known repeating sequences in elastin, and they share with elastin the property of coacervation. They also are capable of forming fibrillar structures reminiscent of some of the proposed intermediates in elastic fiber assembly.

E. Cross-Links of Elastin

The characteristic intermolecular covalent cross-links of elastin are the desmosines: desmosine and isodesmosine. These entities are tetrafunctional adducts of four adjacent, but not contiguous, lysyl residues on two tropoelastin chains. The existence and structure of desmosinelike cross-links were first proposed by Partridge et al. in the early 1950s (9); however, it was not until later in the decade that the biochemical mechanism for generating these cross-links became apparent. It was shown that cross-links in collagen and elastin were generated by the oxidative deamination of residues by the enzyme lysyl oxidase (26,40). It became apparent that lathyrism, a toxic condition that prevents cross-linking of collagen and elastin, was the result of the direct inhibition of lysyl oxidase by the naturally occurring compound β-aminopropionitrile.

Studies by other investigators had shown that copper deficiency affected the strength of connective tissue and suggested that the covalent cross-linking of collagen and elastin was severely impaired (41). Lysyl oxidase was indeed found to be a copper-dependent enzyme (24), and many of the abnormalities of connective tissue in copper-deficient animals could be ascribed to defective cross-linking of the molecules (42).

While the desmosines are the diagnostic cross-links of elastin, they are the end result of a maturation process that can take several days to occur (2). The initial steps in cross-linking, which are described in Section III, produce three deaminated lysyl side chains terminated by reactive aldehyde groups (allysine). It is likely that desmosines result from the successive condensation of three reactive allysine side chains with a fourth lysyl residue to condense ultimately into the desmosine structure. Intermediates in the process, which are also stable to acid hydrolysis, include lysinonorleucine and the aldol condensation product. Reduction with sodium borotritide labels many of these cross-link species, which have been subsequently identified by hydrolysis and ion-exchange chromatography (11). Since the aldol condensation product and allysine are structures that are shared with collagen, they are less diagnostic of elastin than are the desmosines and lysinonorleucine. For detailed discussion of the cross-linking process in elastin, the reader is referred to reviews by Ryhänen and Uitto (5) and Sandberg (10).

F. Elastin Isotypes

A significant number of studies have suggested the existence of variant forms of elastin, either chemically or functionally related to the elastin molecule described above. As a general caveat, the reader should be aware that the major criterion by which homogeneity of an elastin preparation is judged must be, at present, its amino acid composition. Therefore, any deviation from a conventional composition may be the result of contamination by other proteins associated with elastin, possibly in covalent linkage with that molecule. The very low turnover rate of elastin (see below) may permit a variety of covalent modifications to the molecule during the course of a normal (human) lifetime.

Three major variant forms of elastin have been described. In the first instance, certain authors (14,43) have suggested that the elastin of elastic (ear) cartilage had a distinct amino acid composition and primary structure. Morphological studies have suggested that elastin in this tissue is assembled in a very different way from that in other elastic tissues (44), perhaps because of the overriding influence of the abundant proteoglycan molecules present under these circumstances. Nevertheless, these studies remain somewhat controversial for the following reasons: (a) the tropoelastin secreted by rabbit auricular chondrocytes in cell culture appears to be immunologically identical to conventional tropoelastin (37); (b) studies by two groups have shown that cartilage slices incubated in vitro produced a tropoelastin molecule that is indistinguishable from tropoelastin synthesized by aorta (45,46); and (c) the compositional analyses of cartilage elastin are equally consistent with the presence of contaminating proteins (11).

The second unusual elastin reported, a molecule described by Foster et al. (47,48), has been referred to as "tropoelastin A." It was initially identified as an elastinlike molecule that migrated more slowly on SDS polyacrylamide gels. This material has been identified both as a product of the cell-free translation of messenger RNA from chick embryonic tissues and in urea extracts of chick aorta (47). In addition, it has been reported that developmentally regulated, differential expression of tropoelastin A and B occurred in chick embryonic aorta (49). Tropoelastin A has been reported to have a similar but distinct primary structure as compared with conventional tropoelastin ("B"). The biological significance of this material is not known. The existence of these alternate forms has not been confirmed by other laboratories to date, although tropoelastin frequently appears as a doublet on SDS polyacrylamide gels.

Based on currently available information, it would appear that the chemical relationship between elastin and collagen is, to some extent, convergent. Although elastin and collagen can: (a) be secreted by the same cell (18), (b) be hydroxylated at certain prolyl residues (2), and (c) act as substrates for lysyl oxidase, current evidence would suggest that the collagen-synthesizing machinery found in all metazoans was in place long before elastin evolved in the higher chordates (17). Elastin synthesis appears to take advantage of some of the collagen posttranslational machinery. Nevertheless, hydroxyprolyl residues in elastin are not an essential feature of its structure, and work by Urry and others (50) has suggested that hydroxyproline in elastin may be deleterious to its integrity and function. When cells that produce collagen and elastin were administered the iron chelator α,α'-dipyridyl, collagen synthesis was largely arrested owing to the impaired hydroxylation of proline, whereas elastin synthesis and secretion were unimpaired (51). The same has been found true for proline analogs (52), suggesting that prolyl and hydroxyprolyl residues play very different roles in determining the secondary structure of these two classes of proteins. Interestingly, certain forms of tropoelastin have been reported to be sensitive to bacterial collagenase digestion (53). Smith ct al. have recently described an amino acid sequence in chick tropoelastin that does contain 13 gly-X-Y repeats, which could act as a substrate for the bacterial enzyme (53). In general, however, elastin in mammals is entirely refractory to collagenolytic enzymes, and susceptibility of an elastin or tropoelastin molecule to collagenase should caution the investigator to be aware of nonspecific proteolysis.

III. ELASTIN BIOSYNTHESIS

Prior to about 1960, elastin was described only in terms of partial or complete hydrolysis products. It was fairly certain, however, that insoluble elastin resulted from the polymerization of a soluble precursor molecule at some point during synthesis or secretion. This putative precursor of elastin was referred to as either "proelastin" (41) or tropoelastin (12). Subsequently, "proelastin" was also used to describe high-molecular-weight proteins thought to be precursors of tropoelastin (54,55). The initial isolation of tropoelastin was facilitated by use of the copper-deficient (and aneurysm-prone) pig. Elastin-rich tissues such as the aorta yielded a soluble, hydrophobic protein that shared many of the molecular properties of elastin, except for a higher lysine content and the conspicuous absence of the characteristic desmosine cross-links. By analogy with the contemporaneous term "tropocollagen," which was used to describe the soluble precursor form of that matrix macromolecule, this monomeric material was termed "tropoelastin." The sections below describe the synthetic pathway of tropoelastin as presently understood (see Table 2).

A. Elastin DNA

Current concepts of tropoelastin biosynthesis have come from studies of cells or tissues highly differentiated for production of elastin: the aorta of chick (51,56,57) and pig (58), and the nuchal ligament of cow and sheep (59,60). Expression of elastin commences by transcription of its messenger RNA from the elastin gene. Our laboratory has recently described (61) genomic clones derived from the sheep that contain 30,000 nucleotides of elastin-coding sequences, but not the entire gene. The size of the segments of DNA and the preliminary R-loop analysis of one gene fragment indicate that elastin mRNA, like most other eukaryotic mRNAs, is transcribed from a gene that contains large stretches of noncoding or intervening sequence. Similar observations have recently been made by Rosenbloom and his co-workers (62). Both R-loop analysis (61) and DNA sequence studies (62) suggest that the elastin gene is "dilute," containing a high proportion of noncoding DNA. Studies by Olliver et al. (63) have suggested that this gene probably exists as a single copy within the sheep genome. Subclones of the sheep gene that are homologous with human DNA have also been used to examine the human genome for restriction fragment length polymorphisms, and unique restriction fragments have been detected (64), again suggesting that elastin exists as a single-copy gene in the human species as well.

In recent studies, the cloned sheep and human DNA sequences have been used to establish the chromosomal localization of the human elastin gene. This work relies on detection of elastin sequences in hamster-human hybrid cells by Southern blot hybridization or by in situ hybridization to human metaphase chromosomes. The human elastin gene has been localized to the $\alpha^{31} \rightarrow \alpha$ter region of chromosome 2 (64).

Little is known at this time about the pretranslational events that may be involved in regulation of elastin gene expression. As discussed below, most of the systems that have

Table 2 Steps in Elastin Biosynthesis

Site	Step	Target	Mediators	Function
Nucleus	Activation of gene	Chromatin; promoter	Methylase/ demethylase?	Allows expression of elastin gene
	Transcription of $mRNA_E$	Promoter; enhancer	RNA polymerase II glucocorticoids	Modulates expression of elastin mRNA
	Processing of $mRNA_E$	Intervening sequences	U1 RNA? Splicing enzymes?	Produce functional RNA transcript
	Capping	Capping site	Methylase	Retard mRNA degradation
	Polyadenylation	Adenylation site	Terminal transferase	Transport? Stability?
	Packaging into mRNP	Poly[A]-mRNA	Ribonucleoprotein	Transport to cytoplasm; stabilization
Cytoplasm	Initiation of translation	$mRNA_E$	Ribosomes; initiation factors; $tRNA^f{}_{met}$	Binding of mRNA to translational machinery
	Chain elongation	Polysomal $mRNA_E$	Ribosomes + RNAs	Continued synthesis of protein
	Extrusion into RER[a]	Nascent elastin chains	Signal peptide recognition complex; docking protein	Vectorial transport into RER
	Signal peptide cleavage	Signal peptide	Signal peptidase	Removal of signal peptide
	Prolyl hydroxylation	X-Pro-Gly sequence	Prolyl hydroxylase; ascorbate	?
	Translocation to Golgi	Completed elastin chains	Microfilaments	Formation of secretory vesicles
	Translocation to extracellular space	Secretory vesicles	Microfilaments/ microtubules	Secretion of tropoelastin
Extracellular space	Lysyl oxidation	Lysyl residues	Lysyl oxidase; Cu^{2+}	Formation of reactive aldehydes
	Chain association	Tropoelastin monomers	Hydrophobicity? Chain recognition	Association of monomers
	Association with microfibrillar component(s)	Soluble elastin	?	Formation of elastic fiber
	Cross-linking	Deaminated lysyl residues	?	Formation of cross-links

[a]RER, rough endoplasmic reticulum.

been described to date indicate that elastin expression is under transcriptional control (65); i.e., the rate of synthesis is largely regulated by the amount of elastin mRNA. It will be important in future studies to focus on understanding how elastin gene transcription is influenced by the cellular and/or nuclear environment.

B. Elastin mRNA

Elastin messenger RNA (mRNA) is processed from a large transcript into a mature cytoplasmic message with a size ranging between 3500 and 4000 nucleotides (66,67). The cell-free translation product of elastin mRNA is a polypeptide of approximately 70,000 molecular weight; therefore, elastin mRNA must contain a rather high content of untranslated nucleotide sequences. The nucleotide sequence of a partial cDNA clone of the sheep elastin mRNA molecule has been reported (68,69). This clone, which extends from the polyadenylation site at the 3′ terminus of elastin mRNA into the known carboxy-terminal sequence of (porcine) tropoelastin, confirms that elastin mRNA from the sheep contains at least 974 bases of untranslated nucleotide sequence (61). The functional significance of these untranslated sequences is unknown, but it is quite interesting that the corresponding exon segment of the sheep elastin gene, which has been subcloned into a plasmid vector and which contains a 663-base-pair homology with the cDNA clone (69), is capable of hybridizing to a wide range of elastin mRNA's from vertebrate species: human, bovine, swine, hamster, rat, and chick. Studies of other 3′ untranslated sequences frequently show high divergence, even between protein isotypes within the same species. Conservation of this region has been reported in the fibronectin gene, among others (70). It is difficult to imagine that such conservation through the course of vertebrate evolution would persist without selective, evolutionary pressure to retain such a structure.

C. Translation of Elastin mRNA

Studies by Ryhänen et al. (71) showed that tropoelastin mRNA was translated on membrane-bound polysomes, as is the case with most secreted proteins. This and subsequent studies by others using partially purified or purified elastin mRNA have consistently shown that elastin mRNA is translated into a polypeptide whose molecular weight ranges between 70,000 and 74,000 daltons (47,60,66,72). This size, relative to secreted tropoelastin, is consistent with the presence of a signal peptide sequence. Studies with chick aorta (73) and with sheep nuchal ligament (74) have confirmed the presence of a 24-26-residue leader sequence that is quite homologous with other signal peptides described from eukaryotes (75). Although numerous studies have shown that the cell-free translation product of elastin messenger RNA is frequently resolved into a doublet by SDS polyacrylamide gels, the molecular basis for this electrophoretic difference is not known.

1. The Proelastin Question

The elastin literature contains reports of other types of elastin precursor molecules, some of considerably larger molecular mass than that described for the elastin cell-free product (54,55,76). At this time, the identity of the isolated product is not precisely clear in these studies. Many laboratories have now confirmed that the translation product of the elastin message is a molecule that is only 2000 daltons larger than the secreted tropoelastin molecule (47,60,66,72). In any case, the term "proelastin," which was used to describe these entities, is currently considered to be an inappropriate term. It is conceivable that the molecule isolated by these investigators may have been an adduct of tropoelastin with itself or with some other molecule. A dimeric form of tropoelastin that may be an assembly intermediate for the elastic fiber has been described (77); however, the "proelastin" entities previously described did not have amino acid compositions consistent with a tropoelastin dimer.

D. Hydroxylation

During the process of translation, certain prolyl residues of tropoelastin are hydroxylated by the membrane-associated enzyme prolyl hydroxylase (78). The presence of hydroxyproline within tropoelastin may, however, be an accident of its translation coincident with collagen in many tissues. This point is amplified by a number of studies that show that while proline analogs, or inhibitors of prolyl hydroxylase, impair collagen triple-helix stability, and in fact promote intracellular degradation of collagen (79), these inhibitors have little or no effect on the production of elastin (51,52). Studies of synthetic elastin polypeptides based on porcine sequences have suggested that hydroxylation of prolines in elastin may destabilize secondary structure (50). Smooth muscle cells grown in the presence of high levels of ascorbic acid, a known cofactor for prolyl hydroxylation, accumulate less elastin (80). It is not clear whether this reduction arises from decreased production of the protein or increased proteolysis of elastin as a result of overhydroxylation. The effect may be unrelated to prolyl hydroxylation.

E. Secretion of Tropoelastin

After vectorial translocation of the nascent elastin polypeptide chains into the endoplasmic reticulum, the leader sequence of the molecule is cleaved off (73), and transport proceeds through the Golgi apparatus and into secretory vesicles (81). Despite the clear immunochemical identification of tropoelastin in the trans cisternae of the Golgi (82), tropoelastin is not known to be glycosylated. In the bovine nuchal ligament, a specialized secretory granule, the "acanthosome" has been described as a secretory vesicle for soluble elastin material (81). Elastin secretion has been shown to be impaired by colchicine (52). Interestingly, the secretion of tropoelastin is also impaired in rat smooth muscle cells by other conditions that alter the shape of the cells, such as cytochalasin B treatment and/or plating on poly-hydroxyethylmethacrylate-coated dishes (83). Recent studies have shown that the sodium ionophore monensin, which has been shown to impair the secretion of other extracellular matrix molecules (84), also inhibited tropoelastin secretion. More important, these studies indicated that when tropoelastin secretion was impaired, expression of the gene was also down-regulated, suggesting the presence of a negative feedback mechanism to regulate intracellular protein levels (85).

Tropoelastin is released at the cell surface into the extracellular space. The fate of the molecule from this point onward seems to vary, depending on the biological system analyzed. In intact tissues, or in culture systems where cells are either embedded within or grown upon an extracellular matrix, the accretion or cross-linking of tropoelastin into insoluble elastin is considerably facilitated (86). In contrast, when most elastin-synthesizing cells are grown on plastic, tropoelastin accumulates in the culture medium and is cross-linked only slowly and inefficiently (23,80,86). This observation is quite similar to that seen with production of interstitial procollagen molecules by cultured cells (87). These secretory products are released into the artificial compartment (medium), very slowly processed from precursor into product, and inefficiently incorporated into the extracellular matrix.

1. Microfibrillar Interactions—Fiber Assembly

The chapter in this volume by Cleary (Chapter 3) discusses in great detail the chemistry and biology of the putative microfibrillar component(s) (21,88,143). Accordingly, this chapter will only make brief mention of the fact that a number of detailed morphological studies have suggested that the assembly of elastin into elastic fibers proceeds by a mechanism wherein a microfibrillar structural glycoprotein is laid down at or near the surface of elastin-producing cells and that this structure may serve as a nexus for condensation of tropoelastin molecules into the mature elastic fiber (21). It is further known from these studies that the staining properties, density, and immunological reactivity of elastic fiber material indeed changes with time (age), with development, and in certain pathological conditions (89,90). The chemical basis for these differences in reactivity of elastic fibers during their assembly, maturation, and degeneration is incompletely understood. The basis of the noncovalent interaction between tropoelastin and microfibrillar elements is not known.

In one study (91), elastin-producing cells were cultivated under conditions where coacervation did not occur (15°C). Under these circumstances, tropoelastin cross-linking was not initiated until the cultures were incubated above 25°C. These results suggested that the sequence of events that initiate the conversion of tropoelastin into insoluble elastin can be divided into two phases. First, soluble tropoelastin molecules condense (or coacervate) into an aggregated form; subsequently, the cross-linking process is initiated (91). It is not known whether microfibrillar component(s) facilitate the latter process.

It is possible that lysyl oxidase and the tropoelastin monomer(s) are compartmentalized and are secreted by somewhat different pathways, since only soluble elastin molecules have been found within cells. It is likely, however, that lysyl oxidase rapidly associates with its substrate after secretion and thereafter remains very tightly associated with its substrate, as has been shown in many studies (92,93). One such study has suggested that lysyl oxidase itself may become incorporated into the elastic fiber (94).

F. Cross-Linking Processes

Under physiological conditions, elastin is rapidly covalently cross-linked by the oxidative deamination of lysyl residues. It is not known whether these cross-links occur randomly or between specific sites (as in collagen molecules) on adjacent tropoelastin chains that have formed fibrils (95). The time course of the cross-linking process in vitro is variable (96). In several cell culture systems, the appearance of the final desmosine product has been shown to take place over the course of several days or even weeks in culture (23,59, 80,97). Nevertheless, the initial deamination is a rapid process, and the initial association of lysyl residues probably occurs within several hours of secretion of the molecule, since desmosine precursors can already be identified at that time.

The cross-linking process is blocked by addition of the lathyrogens β-aminopropionitrile or α-amino-acetonitrile (98). It is also possible to block the aldehydes per se by addition of D-penicillamine. Copper depletion in animals or in vitro will also inhibit the action of lysyl oxidase resulting in the production of elastic fibers with abnormal morphology (41) and an excess of soluble elastin (42). It is likely that impaired cross-linking in vivo increases elastin degradation, since tropoelastin is much more susceptible to nonspecific proteolysis than elastin; it has also been demonstrated that tropoelastin from chick aorta copurifies with a serine proteinase capable of specifically degrading the protein (99). Smith et al. (100) had reported decreased synthesis in copper-deficient aorta; however, recent results from our laboratory (101) suggest that aortic elastin (and collagen) synthesis is actually increased in copper-deficient animals relative to total protein synthesis.

G. Elastin Turnover

A number of studies have addressed the question of biological turnover of elastin, all of which infer that elastin is principally synthesized during development or growth of tissues. During adult life, this component of connective tissue appears to turn over at a very low rate. For example, the half-life of elastin in rat lung has been estimated to be on the order of six years (102). This is in marked contrast to the well-known remodeling of collagen in development, growth, and wound repair (103), as well as its constitutive turnover. Low elastin turnover is also consistent with the progressive appearance of damaged elastic fibers in aging skin, where there is extensive fragmentation of the fibers owing to the cumulative effects of inflammatory episodes, mechanical injury, or ultraviolet radiation damage. Little is known about the involvement of elastin in tissue repair processes. As reviewed in Chapter 19 by Soskel and Sandberg, artificially induced proteolytic lung injury can stimulate the elastin production (34); however, deposition seems to occur in a nonfunctional pattern. Thus, it is likely that the quality of elastin that is synthesized early in life, during fetal development and juvenile growth, is critical to the future health of the organism. Although it is known that elastin synthesis can be stimulated in adult organisms by injury and in fibroblasts derived from adult tissues (113), it appears that elastic tissues are generally not well equipped to undergo a regenerative process.

IV. BIOLOGICAL SYSTEMS THAT DEPEND ON ELASTIN

A. Lung

A detailed discussion of lung pathology as it relates to elastin and pulmonary emphysema is presented elsewhere in this volume (34). Therefore, this description will be confined to certain other aspects of lung elastin metabolism. The presence of elastin in lung has been clearly documented by numerous morphological and ultrastructural studies (104). Elastin is distributed not only in the blood vessel wall and pleura, but in the alveolar interstitium. Presumably, it is the progressive loss of elastin from interstitial alveolar walls that leads to the degenerative condition of emphysema (34,104). Elastin biosynthesis or accumulation has been studied in the lung of several species (47,72,104,105). Studies in the rat lung (106,107) have shown that elastin accumulation begins late in development, principally during neonatal life, since the rat lung is in an immature, glandular stage at the time of birth. There is a concomitant accumulation and organization of elastin and elastic fibers in the rat lung as morphogenesis of the tissue proceeds over the next three weeks of postnatal development.

In the chick lung, the expression of elastin messenger RNA was followed from approximately the 11th day of development onward (49). Two putative forms of tropoelastin were present at a constant ratio in the cell-free translation product of lung elastin mRNA. In addition, total elastin mRNA activity during this period increased in parallel with elastin synthesis.

Detailed studies have been made of the developmental regulation of elastin accumulation in the sheep lung (72,108). Using a combination of an in vitro explant culture system, cell-free translation, and molecular hybridization of elastin messenger RNA, it was demonstrated that elastin biosynthesis increased in the developing lung, from about 0.5% to 1.5% of total protein synthesis between the 60th and 147th day of gestation (parturition). This period of sheep development corresponds to the progressive differentiation of the lung from a poorly organized epithelial structure into a mature gas-exchange organ. Recent studies by Fukuda et al. using light and electron immunohistochemistry (109), have confirmed the accumulation of elastin, as well as microfibrillar components, in the lungs of sheep during this same interval. Translation and hybridization studies have shown that, in this system, elastin expression is under transcriptional regulation (67,72); that is, the amount of elastin made seems to be largely regulated by the abundance of elastin messenger RNA.

B. Skin

Elastin fibers are a prominent component of the dermis. Histologically, some of the fibers of the upper dermal layers have altered morphology or stain affinity giving rise to the terms "oxytalan" and "elaunin" fibrils (110). There is, however, no evidence to date that the elastic protein of these different fiber forms is chemically different.

Little is known about the mechanism of elastin accumulation in skin. It has been presumed from numerous morphological studies that skin fibroblasts are involved in the elaboration of this protein (111,112). Recently, our laboratory has established that human skin fibroblasts in culture can indeed produce elastin at significant levels, up to 5×10^4 molecules/cell/hr (113). Elastin production is dependent on cell density and serum concentration. In addition, we have established that human skin fibroblasts also contain a significant amount of elastin messenger RNA, as detected by a molecular hybridization probe. Future studies should be able to capitalize on this observation, since the numerous genetic and acquired disorders that involve elastic fibers (see Chapter 18) may be far more accessible to study in cell culture systems than in patients. This was demonstrated by a recent examination of one elastic fiber derangement, atrophoderma of Pasini and Pierini (113).

Among the characteristics of elastin in human skin are its progressive destruction, fragmentation, and disappearance during the course of senescence. While there is little evidence that elastin is newly synthesized under most circumstances in the adult organism, an important exception may be the process of dermal wound healing. There is no current biosynthetic evidence available to show that new elastic fibers are laid down during normal dermal wound repair, though there are several reports of increased accumulations of elastin in breast tumors (114) and liver fibrosis (115,116).

In other disorders of elastic tissue, elastin may be diminished, increased, or altered (calcified, fragmented) (117-119). Many of these disorders are not obvious at birth, suggesting that the initial synthesis and deposition of elastin during growth and development is normal. Defective elastin may, however, have reduced resistance to proteases or mechanical stress. Constant or intermittent inflammatory episodes occurring over the life-span of the human may, therefore, cause gradual deterioration of defective elastic fibers.

C. Blood Vessels

Elastin structure and function are an intricate part of the physiology of the blood vessel wall. Here the elastic fiber network, which extends in concentric sheets around the lumina of all major blood vessels, is critical for providing the elastic recoil of these structures. Vessel wall resiliency smooths pulsatile blood flow from the ventricles and accommodates changing pressure. Studies using the copper-deficiency model (120) or experimental lathyrism (121) have implicated elastin in maintaining the integrity and function of these vessels; however, the cross-linking of both collagen and elastin is affected under these experimental circumstances. Nevertheless, at least three human disease conditions involving elastic tissue (pseudoxanthoma elasticum, cutis laxa, and Marfan syndrome) are known to affect blood vessel integrity (6, Chapter 18).

Rapid accumulation of elastin occurs during the development of the aorta. The biosynthetic process has been studied in detail in fetal or neonatal aortas from both avian (23,56,57) and mammalian (58,122) sources. In general, there is a developmental shift from a collagen-synthesizing phenotype to one in which, for a period of development, elastin synthesis and accumulation predominate. In the aorta of chick (56,66) and of pig (58), this has been shown to be controlled, in large part, by the abundance of elastin messenger RNA.

Recently, we have begun to examine in greater detail the segmental variation in elastin accumulation in the aorta, as well as broader developmental aspects (58,123). These studies have shown that the thoracic aorta, as previously indicated by its histological appearance, was differentiated toward elastin production earlier and to a greater extent than were the corresponding medial layers of the abdominal aorta. This situation persisted throughout the neonatal growth of the organ; however, the abdominal aorta accumulated elastin more slowly, primarily during the postnatal period of development. A similar study has been made of the developing rat aorta by Leslie (124), although the differential expression of collagen and elastin in this vessel is not as marked as in the pig.

The principal site of accumulation of aortic elastin is the tunica media, where this protein is produced in significant quantities by the vascular smooth muscle cells. Vascular endothelium may make some contribution to the subintimal elastin and the internal elastic lamina in particular (125,126). Elastic fibers are also found in smaller arterioles and venules and are presumably synthesized by the same cellular sources. There is evidence to suggest that accumulation of elastin in the blood vessel wall is, to some extent, regulated by the amount of stress the vessel wall experiences (127); for example, the pulmonary artery contains fewer elastic lamellae per cross-sectional area and lower relative elastin content as compared with the thoracic aorta.

One critical aspect of elastin structure and metabolism pertains to the role of elastin in the onset and progression of cardiovascular disease. Breakage of elastic fibers within blood vessel walls has been noted under a variety of conditions. In the genetic diseases cutis laxa and Marfan syndrome, abnormal elastic fibers have been reported (see Chapter 18); cystic medial necrosis also involves degeneration of elastic fibers (128). In atherosclerosis, the intimal or subintimal proliferation of vascular smooth muscle cells is accompanied by an accumulation of some elastic fibers in the atheromatous lesion. It is not known to what extent the accumulation of elastin in the lesional site contributes to the progression of this disease. It is most likely, however, that the accumulation of elastin in this lesion is secondary to the ingression and proliferation of smooth muscle cells (129), which may be a variant subpopulation (130).

As is the case with the other elastic organs, no clear-cut example is known of defective elastin biosynthesis leading to a vascular disease state. The principal pathological mechanisms involve breakage or dissolution of preformed elastic fibers. It may be that in many of these disorders, elastin accumulation during development and growth proceeds normally; however, subsequent injury or inflammation leads to the release of elastolytic proteases by inflammatory cells, fibroblasts (119) or smooth muscle cells (131). The apparently low rate of renewal of elastin during adult life suggests that elastic fibers may gradually accumulate proteolytic "nicks." Mechanical damage may also lead to a gradual deterioration of this important connective tissue structure. However, it is not clear from the literature precisely how much elastin repair can or might occur after injury to the vessel wall.

V. THE CELL BIOLOGY OF ELASTIN

This section will briefly touch on the cellular systems that have been used to evaluate the regulation of elastin synthesis or as potential models of elastin pathobiology. The major cell types that have been used are vascular smooth muscle cells, nuchal ligament fibroblasts, skin fibroblasts, and, to a limited extent, vascular endothelial cells.

A. Smooth Muscle Cells

The vascular smooth muscle cell has been the most common cell type used for the study of elastin synthesis. Studies by a variety of investigators (39,80,132,133), have shown that smooth muscle cells derived from aortic explants of rabbit, pig, calf, and monkey are capable of synthesizing tropoelastin, insoluble elastin, or desmosine cross-links. Although elastin is produced initially by smooth muscle cells in continuous culture, there is frequently a rapid decline in the production of elastin by the cells, presumably owing to the lack of an appropriate stimulus. A contrast to this is seen in studies of neonatal rat smooth muscle cells, which continue to produce significant amounts of elastin and collagen over numerous population doublings and several serial passages (134,135). Rat smooth muscle cells tend to grow in multilayered arrays very reminiscent of the arrangement of medial smooth muscle cells in the intact aorta, while most other mammalian smooth muscle cells grow in a "hill-and-valley" mixture of multilayered and monolayered cells (136). After several weeks in culture, elastic fibers can be visualized in cultures of bovine or porcine smooth muscle cells.

The phenotype of the vascular smooth muscle cell is also characterized by a relatively high level of production of type III collagen (about 40%) and type V collagen (about 5%). Type III collagen is thought to be more abundant in extensible tissues (137). We have recently examined the stability of the elastin phenotype in porcine vascular smooth muscle cells, its relationship to collagen phenotype, and the dependency of elastin production on several tissue culture parameters (39,138). These data have shown that elastin production by porcine vascular smooth muscle cells in culture is: (a) a relatively unstable phenotypic condition, having substantially declined by the time of a third serial passage; (b) a serum- and density-dependent phenomenon; and (c) diminished in vascular smooth muscle cell cultures, which at postconfluency assume a nodular, or hill-and-valley morphology. These studies have not distinguished between decreased elastin synthesis and increased elastin degradation as the mechanism by which these phenomena arise.

The cells of the aortic wall of the developing chick have also been used extensively for studies of production of elastin (56,139,140). While it is not clear whether all of the cells in the chick vascular wall are in fact smooth muscle cells, it has been demonstrated by immunological methods that some of the cells of this tissue simultaneously express collagen and elastin (18). Matrix-free chick aorta cells have also been used to demonstrate the differential susceptibility of collagen and elastin production to proline analogs (51) and inhibitors of prolyl hydroxylase (52). Studies with these cells have also provided the best estimates of the secretion time for newly synthesized tropoelastin molecules, which is approximately 1 hr (141).

B. Fibroblasts

Two types of fibroblasts have now been examined in detail for production of elastin. The first is the fibroblast derived from the bovine or ovine nuchal ligament. The nuchal ligament rapidly develops as an elastin-rich organ during the last trimester of fetal life and continues its growth during the neonatal period. Chemical studies by Cleary et al. (142), cell biology studies by Mecham et al. (59) and Jones et al. (97), and molecular biology studies by Davidson et al. (60,67), have shown that elastin gene expression is highly differentiated in the tissues of this organ. Elastin synthesis in the newborn sheep nuchal ligament approaches 40% of total protein synthesis.

Interestingly, when cells from an undifferentiated ligament have been grown on the (dead) matrix derived from an older (240-day-gestational) ligament, the matrix was capable of "inducing" higher levels of elastin production in undifferentiated cells (86). These results suggest that some type of positive feedback loop exists between the nuchal ligament fibroblast and its extracellular matrix to regulate the extent of elastin gene expression. The precise molecular mechanism causing increased production is unclear, but induction is blocked by bromodeoxyuridine, suggesting control is at the level of gene activation. Collagen, elastin, and fibronectin had no inductive effect (86).

Additional studies have shown that the differentiated nuchal ligament fibroblast is chemotactic to certain elastin peptides: i.e., the elastase-derived soluble fragments of insoluble elastin, tropoelastin, and, more important, the synthetic polyhexapeptide produced by Urry (144). Chemotactic responsiveness is elicited in undifferentiated nuchal ligament fibroblasts by contact with older nuchal ligament. The biological significance of fibroblast chemotaxis to elastin fragments remains somewhat uncertain, but it is interesting to note that (a) this is the only known biological activity described for elastin, and (b) the appearance of the chemotactic response to elastin peptides is also correlated with differentiation of the nuchal ligament fibroblast. Chemotaxis to elastin peptides may conceivably be involved in a wound repair response. In this context, it should be pointed out that elastin peptides are also chemotactic for circulating macrophage precursors (145, 146), which has led to the proposal that breakdown of elastin in lung and other connective tissues may be an important mechanism for recruiting inflammatory cells to sites of tissue destruction. Since tropoelastin per se has been shown to have chemotactic properties for monocytes (144,145), it is likely that the desmosine cross-link structure is not a critical part of the recognition mechanism. The fibroblast chemotactic activity of the hydrophobic sequences in the synthetic polyhexapeptide reinforces this hypothesis.

The human skin fibroblast has recently been examined for its capacity to produce elastin. Work from this laboratory has demonstrated that tropoelastin molecules are synthesized and secreted by cultured human skin fibroblasts (113). A large number of human cell lines, including fetal foreskin fibroblasts, and skin fibroblasts from neonate through adult have now been examined. All these fibroblasts produce immunologically detectable tropoelastin; in each case, the principal product of the cells is identical in electrophoretic mobility to the tropoelastin monomer produced by other animal tissues such as the aorta and nuchal ligament.

Two important opportunities stem from these observations. One is the possibility of understanding the regulation of human elastin metabolism in the dermis, and the other is the potential for examining genetic diseases involving elastic tissue. A preliminary characterization of elastin production by human skin fibroblasts indicates that such production, like that in the porcine smooth muscle cell, is dependent on cell density, serum concentration, and cell population doublings (103). Further studies are now being aimed at delineating those factors present in serum that may be directly involved in stimulating (or repressing) elastin production.

C. Endothelial Cells

Elastin production is not unique to parenchymal cells. Recent reports have demonstrated that endothelial cells in culture are capable of producing an elastinlike protein (125,126). In one study (147), the cells were derived from a continuously growing line of rabbit vascular endothelial cells, while the other two studies were made with subcultivated cells probably having a limited life-span in culture. Elastin production was demonstrated by techniques including the synthesis of desmosines, production of immunoreactive elastin, and immunofluorescent staining of the cells in culture. Thus, it appears that the endothelial cell of the vascular wall is also capable, at least under the conditions of cell culture, of producing elastin that is indistinguishable from the interstitial protein. This observation may have significance with regard to the repair of vascular wall injury, the origin of the internal elastic lamina, and the maintenance of vascular integrity.

D. Cells That Degrade Elastin

A thorough discussion of the elastolytic capacity of cells is considerably beyond the scope of this chapter. Other chapters in this volume address the question in the context of several disease states (Mainardi, Chapter 23; Soskel and Sandberg, Chapter 19). Cells involved in inflammatory responses and wound repair are known to contain elastolytic proteinases. Principal among these cells is the polymorphonuclear granulocyte, whose specific granules contain a potent serine proteinase, known as neutrophil elastase or leukocyte elastase. This enzyme is structurally and functionally very similar to the pancreatic elastolytic enzyme. In humans (148) and rodents (149), it has been reported that macrophages also contain elastolytic enzymes; however, the biological relevance remains controversial, since the biochemical nature of the enzyme appears to be different in different reports and among different species. Other investigators have shown that human macrophages are capable of taking up and incorporating neutrophil serine proteinase elastase into intracellular granules (150). It has been suggested that the immature human monocyte is also capable of producing a neutral metalloproteinase, i.e., an enzyme that is inhibited by metal chelators and functions at neutral pH to degrade elastin. In the mouse peritoneal macrophage, it is clear that a metalloproteinase with elastolytic activity (149) is endogenously produced and secreted.

Many other cell types have been reported to express elastolytic activity, either as a secreted enzyme or associated with the cell surface. Included are smooth muscle cells, platelets, and fibroblasts (151). The relative contribution of the proteinases present in these cells compared with the total burden of elastin degradation is not clear at this time. Furthermore, as pointed out early in this chapter, "elastase" is a generic term. There are no known enzymes whose exclusive property it is to degrade elastin; rather, these are enzymes that *can* degrade elastin or its synthetic analogs.

VI. THE MOLECULAR BIOLOGY OF ELASTIN

The broad advances in molecular biology have begun to affect the elastin field. Because the complete characterization of the elastin polypeptide has proven to be a difficult task, a full description of the structure of elastin messenger RNA and protein-coding regions of the elastin gene has been begun by capitalizing on the progress made in nucleic acid technology over the last several years. These new approaches have also had a major influence on our concepts of elastin gene expression.

A. The Elastin Gene

Our current knowledge of the elastin gene is confined to recombinant molecules that have been isolated from a sheep genomic library constructed in the bacteriophage Charon 4A (61,65), and a set of bovine genomic clones (62). The sheep molecules were isolated by screening the phage library with a highly enriched [^{32}P]-labeled elastin messenger RNA. The identity of these clones was established by the ability of cloned DNA to hybridize specifically to elastin mRNA both in Northern blotting experiments and by hybrid-selected translation. Further confirmation has come from collaborative studies between Rosenbloom and Davidson, which have shown that at least a 663-nucleotide sequence homology exists between a complementary DNA recombinant constructed from sheep elastin mRNA (68) and a 1.3-kb region of the sheep elastin gene (69). The bovine genes were isolated by screening with the sheep elastin cDNA (68).

1. Structural Organization

The sheep elastin gene is partly represented in two nonoverlapping clones containing 9.9 kb (SE1) and 20.1 kb (SE2) of genomic DNA. It is not known, at this time, how much further the elastin gene extends 5′ (N-terminal); however, mRNA-homologous sequences have been identified further downstream of the 3′ terminus of SE1 by Cicila (62), although the precise site of transcription termination has not been isolated. The annealing of elastin messenger RNA (R-loop analysis) to one of the sheep elastin subclones (SE1), which is about 10 kb long, has indicated that seven or eight very large intervening sequences exist within this region of the molecule, and that the average size of the expressed sequences, that is, those regions of the gene that directly hybridize with the RNA molecules, is rather short (less than 150 nucleotides). Like the collagen gene, the elastin gene appears to be constructed from small building blocks.

Recent RNA sequence information on bovine- and sheep-cloned fragments has refined these observations (62,69). SE1 and its bovine homolog consists of at least 8 small exons, ranging in size from 50 to 120 nucleotides. Large intervening sequences are present, and cross-linking and hydrophobic domains appear to be in different exons (62,152). Interestingly, there is very high homology between the bovine and ovine gene at the nucleotide levels.

2. Number and Location of Elastin Genes

A second aspect of the organization of the elastin gene is its copy number. In the sheep, hybridization of genomic DNA to cloned cDNA and genomic DNA has suggested a unique gene. Hybridization of cloned sheep elastin gene sequences to Southern blots of restriction enzyme-digested human and sheep genomic DNA has revealed the presence of a low number of restriction enzyme fragments, consistent with the concept that the sheep and human elastin genes are present in low copy number. It is not clear whether the proposed tropoelastin A molecule is a product of a second elastin gene in the chicken. This protein has not been described in mammalian systems. Chromosomal assignment of the human elastin gene is tentative at this writing. Different locations have been determined by Southern blotting and in situ hybridization.

3. *Origin of the Elastin Gene*

The elastin molecule is apparently built up from several types of repeating hydrophobic peptides, which are interspersed by cross-linking domains. It thus seems likely that gene duplication has played a role in the evolution of this polypeptide. Current concepts suggest that the elastin molecule is a uniquely chordate protein with obscure origins. By using appropriate cloned segments of the elastin gene, it should be possible to address the question of the evolutionary origins of this molecule (17) at the more basic level of nucleic acid homology. These studies would depend on the conservation of nucleotide sequence over the course of vertebate evolution. Present limited information on nucleotide sequence homologies among mammals suggests high conservation, which is reflected in the similarities at the level of amino acid composition. The unusual amino acid sequences and repeat units of elastin may have a unique heritage that can be examined by analyzing primitive chordate genomes.

4. *Gene Regulation*

In general, elastin gene transcription has been shown to be critical to developmental regulation (65). Future studies will have to identify the putative regulatory elements, which probably lie adjacent to or within the elastin gene. Based on general models of eukaryotic gene organization (153), it is very likely that once the 5′ terminus of elastin gene has been identified, several elements of promoter sequences will be found immediately upstream of the transcribed gene. Elucidating the interactions of these regulatory sequences with effector molecules in the cytoplasm and nucleus will be of critical importance for understanding the mechanisms of transcriptional control of elastin. Recombinant DNA molecules that fuse putative promoter or enhancer sequences to expressed bacterial proteins when transfected into eukaryotic cells will be a useful bioassay for control regions of the gene. Similar types of transfections using strong promoters may also provide insight into some of the steps in processing and utilizing of elastin gene transcripts.

5. *Derangements of the Elastin Gene*

A number of genetic disorders of elastic tissue are known to exist (see Chapter 18, this volume). It is likely that some of these disorders involve alterations in the elastin structural gene or its controlling elements. The techniques that have been developed to isolate and characterize the normal elastin gene can probably help to determine the site and extent of these lesions within or near the elastin gene locus. Restriction fragment length polymorphisms, which use cloned (elastin) sequences to assess the rearrangements of a gene locus (154), may prove a useful tool for detecting defective elastin genes in the human population. Elastin mRNA levels can be readily determined by hybridization to cloned DNA probes (58,67). Human elastin recombinant DNAs will undoubtedly become important diagnostic tools in the several cutaneous disorders that involve the elastin gene.

B. Elastin Messenger RNA

The messenger RNA for elastin has been purified from a number of systems. The neonatal sheep nuchal ligament is an excellent source of abundant elastin mRNA, since the tissue expresses about 40% of its total protein synthesis as elastin at the time of birth. It has been a fairly straightforward task to extract and purify a poly[A]-containing messenger RNA fraction and to isolate microgram quantities of elastin messenger RNA with about 80% purity by sucrose density gradient centrifugation. Similar studies have been accomplished using the embryonic (14-18-day) chick (155) and pig (58) aortas as starting material.

1. Physical Properties

As discussed above, the sedimentation characteristics of this mRNA from sheep (as well as cow, pig, chick, rat, and human) indicate a molecule of about 25S, which is unexpectedly large for the size of its translation product. Studies on the electrophoretic mobility of elastin mRNA under denaturing conditions show that this molecule contains at least 3400-3500 nucleotides, whereas the polypeptide would require only 2400 nucleotides to code for its entire structure, including a signal peptide. It is highly likely that there are untranslated sequences present at the 3′ and 5′ terminis of the messenger molecule. DNA sequence analysis of both chick and sheep cDNA clones derived from elastin mRNA molecules (68,156) have demonstrated that the 3′-terminal untranslated region of elastin mRNA (those sequences which would be adjacent to the carboxyterminal coding regions) is as large as 974 nucleotides. The precise carboxyterminus of the protein has now been identified in the sheep cDNA clone (69). Studies of cDNA or genomic clones to date have not revealed the structure of the 5′ region of the gene, which may also contain untranslated sequences.

In the rat smooth muscle cell, elastin messenger RNA appears to have an even more extensive untranslated region, since it migrates on denaturing gels as a molecule containing about 4000 nucleotides, yet in vitro translation produces the typcial, $M_R = 72{,}000$ tropoelastin molecule (85). It is not known whether alternative processing of elastin mRNA may occur during some biological situations. Multiple forms of tropoelastin might arise by such a mechanism. For example, alternate forms of certain other eukaryotic messages, such as those of invertase (157), immunoglobulin M (158), and fibronectin (70), have been shown to arise from differential splicing of pre-mRNA. It is not known at this time how processing of elastin messenger RNA from its precursor is carried out.

DNA sequence analysis has been made of a 1.3-kb sheep cDNA clone (68,69), and these studies, as mentioned above, confirm the presence of a large 3′ untranslated region in elastin mRNA, and this cDNA was large enough (about 1200 nucleotides) to establish homology at the amino acid level with a known porcine peptide of tropoelastin (4) and also to extend the COOH-terminal sequence by more than 30 amino acid residues, previously unidentified. Progress in constructing longer elastin cDNA clones for sequence analysis has been somewhat hampered by the expected presence of a very high content of guanine and cytosine residues, which results in RNA molecules having a high degree of secondary structure. This secondary structure tends to make both the copying of mRNA molecules into cDNA by reverse transcriptase and the actual sequence analysis a more difficult task. We can hope that in the near future, however, more complete copies of elastin mRNA will be constructed, which will eventually allow the completion of the amino acid sequence and peptide ordering of the protein.

2. Dynamics in Development

The abundance of elastin messenger RNA has been shown to vary dramatically in several different systems. These studies have used either cell-free translation or molecular hybridization to detect and quantitate elastin sequences. In each case, it has been demonstrated that the production of elastin is largely regulated by the availability of its mRNA. These systems include the sheep nuchal ligament (60,67), the pig aorta (58), the chick aorta (66,106), and the sheep lung (67,72). Therefore, it seems appropriate to address mechanisms of elastin regulation at the level of expression of the elastin gene.

VII. SUMMARY AND PERSPECTIVES

Elastin is an unusually insoluble protein with an exceptionally long life-span. The presence of unique covalent cross-links, the desmosines, interspersed between repeating hydrophobic domains, results in a durable, elastic meshwork. Elastin shares certain post-translational modifications with collagen, but it is biochemically and genetically distinct. Elastic fibers consist of elastin plus other partly described structural proteins, which may be important in fiber assembly.

The elastin gene is highly dispersed, consisting of small exons and large introns. The exons appear to correspond to the two types of domains seen in the polypeptide. Sequence conservation among mammals may be unusually high, and elastin mRNA contains a large 3′-untranslated region with considerable interspecies sequence homology. Synthesis of elastin varies markedly during development, largely as a function of cellular elastin mRNA content.

The molecular pathology of elastic tissue is in its infancy. Techniques have been developed to quantify elastin synthesis, mRNA levels, and gene alterations, but few applications have yet been made. The cultured human skin fibroblast can act as an important diagnostic tool in the dissection of acquired and genetically determined disease states, since it appears to reflect elastin metabolism of the donor. As the complete structure of the protein and its gene are unraveled, we can hope to create more specific probes for analyzing the control of production and the diseases that involve this biological elastomer.

ACKNOWLEDGMENTS

I am most grateful to G. Sephel and M. Mason for their critical reading of the manuscript and especially to M. Mason for editorial assistance. The manuscript preparation is a credit to G. Shaw. Original contributions from this laboratory were supported in part by the Veterans Administration, NIH Grant GM 32480, a grant from the Dean's Research Committee, University of Utah School of Medicine, the Montana and Utah Heart Associations, and by a gift from R. J. Reynolds, Inc.

REFERENCES

1. J. P. Ayer, *Int. Rev. Connect. Tiss. Res., 2*: 33 (1964).
2. C. Franzblau, in *Comprehensive Biochemistry* (M. Florkin and E. H. Stotz, eds.), Vol. 26C, Elsevier, New York, 1971, p. 65.
3. E. G. Cleary, Chapter 3, this volume.

4. L. B. Sandberg and J. M. Davidson, *Peptide and Protein Reviews* (M. Hearn, ed.), Vol. 3, 1984, p. 169.
5. L. Ryhänen and J. Uitto, in *The Biochemistry and Physiology of the Skin* (L. Goldsmith, ed.), Oxford Press, Fairview, NJ, 1982.
6. L. B. Sandberg, N. T. Soskel, and J. G. Leslie, *N. Engl. J. Med., 304*:566 (1981).
7. J. A. Foster, C. B. Rich, and S. R. Karr, *Int. Rev. Connect. Tiss. Res., 10*:65 (1984).
8. A. L. Lansing, T. B. Rosenthal, M. Alex, and B. W. Dempsey, *Anat. Rec., 114*:555 (1952).
9. S. M. Partridge, H. F. Davis, and G. S. Adair, *Biochem. J., 61*:11 (1955).
10. L. B. Sandberg, *Int. Rev. Connect. Tiss. Res., 7*:159 (1976).
11. M. A. Paz, D. A. Keith, and P. M. Gallop, *Meth. Enzymol., 82*:571 (1982).
12. D. W. Smith, N. Weissman, and W. H. Carnes, *Biochem. Biophys. Res. Commun., 31*:309 (1968).
13. M. Spina, S. Garbisa, J. Hinnie, J. C. Hunter, and A. Serafini-Fracassini, *Arteriosclerosis, 3*:64 (1983).
14. D. A. Keith, M. A. Paz, P. M. Gallop, and M. J. Glimcher, *J. Histochem. Cytochem., 25*:1154 (1977).
15. J. A. Foster, C. B. Rich, S. Fletcher, S. R. Karr, and A. Przybyla, *Biochemistry, 19*:857 (1980).
16. R. M. Leach and R. B. Rucker, *Poult. Sci., 57*:1151 (1978).
17. H. Sage, *J. Invest. Dermatol. , 79*:146S (1982).
18. W. W.-Y. Kao, S. Gay, G. M. Bressan, and D. J. Prockop, *Artery, 7*:176 (1980).
19. A. Serafini-Fracassini, J. M. Field, G. W. Rodger, and M. Spina, *Biochim. Biophys. Acta, 386*:80 (1975).
20. V. Richmond, *Biochim. Biophys. Acta, 351*:173 (1974).
21. R. Ross and P. Bornstein, *J. Cell Biol., 40*:366 (1969).
22. S. R. Pinnell and G. R. Martin, *Proc. Natl. Acad. Sci. USA, 6*:708 (1968).
23. A. S. Naranayan, R. C. Siegel, and G. R. Martin, *Biochem. Biophys. Res. Commun., 46*:745 (1972).
24. R. C. Siegel, *Int. Rev. Connect. Tiss. Res., 8*:73 (1979).
25. E. D. Harris, W. A. Gonnerman, J. E. Savage, and B. L. O'Dell, *Biochim. Biophys. Acta, 341*:332 (1974).
26. J. A. Foster, E. Bruenger, W. R. Gray, and L. B. Sandberg, *J. Biol. Chem., 248*:2876 (1973).
27. W. R. Gray, L. B. Sandberg, and J. A. Foster, *Nature, 246*:461 (1973).
28. D. A. Torchia and K. A. Piez, *J. Mol. Biol., 76*:419 (1973).
29. L. Gotte, D. Volpin, R. W. Horne, and M. Mammi, *Micron, 7*:95 (1976).
30. D. W. Urry, *Ultra. Path., 4*:227 (1983).
31. D. W. Urry, C. M. Venkatachalam, M. M. Long, and K. U. Prasad, in *Conformation in Biology* (R. Srinivasan and R. H. Sarma, eds.), Adenine Press, New York, 1982, p. 11.
32. B. B. Aaron and J. M. Gosline, *Biopolymers, 20*:1247 (1981).
33. G. M. Bressan, I. Castellani, M. G. Giro, D. Volpin, C. Fornieri, and I. Pasquali-Ronchetti, *J. Ultrastruct. Res., 82*:335 (1983).
34. N. T. Soskel and L. B. Sandberg, Chapter 19, this volume.
35. N. Kornfeld-Poullain and L. Robert, *Bull. Soc. Chim. Biol. (Paris), 50*:759 (1968).
36. B. C. Sykes and J. W. Chidlow, *Febs. Lett., 47*:222 (1974).
37. R. P. Mecham and G. Lange, *Meth. Enzymol., 82*:744 (1982).
38. R. A. Daynes, M. Thomas, V. L. Alvarez, and L. B. Sandberg, *Conn. Tiss. Res., 5*:75 (1977).
39. M. G. Giro, K. E. Hill, L. B. Sandberg, and J. M. Davidson, *Collagen Rel. Res., 4*:21 (1984).

40. R. C. Siegel, S. R. Pinnell, and G. R. Martin, *Biochemistry, 9*:4486 (1970).
41. G. S. Shields, W. F. Coulson, D. A. Kimball, W. H. Carnes, G. E. Cartwright, and M. M. Wintrobe, *Am. J. Pathol., 41*:603 (1962).
42. L. B. Sandberg, T. N. Hackett, and W. H. Carnes, *Biochim. Biophys. Acta, 181*: 201 (1969).
43. J. M. Field, G. W. Rodger, J. C. Hunter, A. Serafini-Fracassini, and M. Spina, *Arch. Biochem. Biophys, 191*:705 (1978).
44. G. Quintarelli, B. C. Starcher, A. Vocaturo, F. Ginanfilippo, L. Gotte, and R. P. Mecham, *Connect. Tiss. Res., 4*:1 (1979).
45. P. Heeger and J. Rosenbloom, *Connect. Tiss. Res. 8*:21 (1980).
46. J. A. Foster, C. B. Rich, and M. D. DeSa, *Biochim. Biophys. Acta, 626*:383 (1980).
47. J. A. Foster, C. B. Rich, S. Fletcher, S. R. Karr, M. D. DeSa, T. Oliver, and A. Pryzybyla, *Biochemistry, 20*:3528 (1981).
48. S. R. Karr and J. A. Foster, *Collagen Rel. Res., 3*:459 (1984).
49. L. L. Barrineau, C. B. Rich, and J. A. Foster, *Develop. Biol., 87*:46 (1981).
50. R. S. Bhatnagar, R. S. Rapaka, and D. W. Urry, *FEBS Lett., 95*:61 (1978).
51. J. Rosenbloom and A. Cywinsky, *FEBS Lett., 65*:246 (1976).
52. J. Uitto, H-P. Hoffmann, and D. J. Prockop, *Arch. Biochem. Biophys., 173*:187 (1976).
53. D. W. Smith, L. B. Sandberg, B. H. Leslie, T. B. Wolt, S. T. Minton, B. Myers, and R. B. Rucker, *Biochem. Biophys. Res. Commun., 103*:880 (1981).
54. J. A. Foster, R. P. Mecham, C. B. Rich, M. F. Cronin, A. Levine, M. Imberman, and L. L. Salcedo, *J. Biol. Chem., 253*:2797 (1978).
55. R. B. Rucker, C. S. Heng-Khoo, M. Dubick, M. Lefevre, and C. E. Cross, *Biochemistry, 18*:3854 (19790.
56. J. A. Foster, R. Shapiro, P. Voynow, G. Crombie, B. Faris, and C. Franzblau, *Biochemistry, 14*:5343 (1975).
57. R. Eichner and J. Rosenbloom, *Arch. Biochem. Biophys., 198*:414 (1979).
58. J. M. Davidson, K. E. Hill, M. L. Mason, and M. G. Giro, *J. Biol. Chem., 260*:1901 (1985).
59. R. P. Mecham, G. Lange, J. Madaras, and B. J. Starcher, *Cell Biol., 90*:322 (1981).
60. J. M. Davidson, K. Smith, and S. Shibahara, *J. Biol. Chem., 257*:747 (1982).
61. J. M. Davidson, S. Shibahara, M. P. Schafer, M. Harrison, C. Leach, P. Tostoshev, and R. G. Crystal, *Biochem. J., 220*:643 (1984).
62. G. Cicila, M. May, N. Ornstein-Goldstein, Z. Indik, S. Morrow, H. S. Yeh, J. Rosenbloom, and K. Yoou, *Biochemistry, 24*:3075 (1985).
63. L. Olliver, J. M. Davidson, J. Rosenbloom, and C. Boyd (in preparation).
64. B. S. Emanuel, L. Cannizzaro, N. Ornstein-Goldstein, Z. K. Indik, K. Yoou, M. May, L. Olliver, C. Boyd, and J. Rosenbloom, *Am. J. Hum. Genet., 37*:873 (1985).
65. J. M. Davidson and R. G. Crystal, *J. Invest. Dermatol., 79*:133S (1982).
66. W. Burnett, R. Eichner, and J. Rosenbloom, *Biochemistry, 19*:1106 (1980).
67. J. M. Davidson, S. Shibahara, C. Boyd, M. L. Mason, P. Tolstoshev, and R. G. Crystal, *Biochem. J., 220*:653 (1984).
68. K. Yoon, M. May, N. Goldstein, Z. K. Indik, L. Oliver, C. Boyd, and J. Rosenbloom, *Biochem. Biophys. Res. Commun., 118*:261 (1984).
69. K. Yoou, J. M. Davidson, C. Boyd, M. May, P. LuValle, N. Ornstein-Goldstein, J. Smith, Z. Indik, A. Ross, E. Golub, and J. Rosenbloom, *Arch. Biochem. Biophys., 241*:684 (1985).
70. J. E. Schwarzbauer, J. W. Tamkun, I. R. Lemischka, and R. O. Hynes, *Cell, 35*:421 (1983).
71. L. Ryhänen, P. N. Graves, G. M. Bressan, and D. J. Prockop, *Arch. Biochem. Biophys., 185*:344 (1978).

72. S. Shibahara, J. M. Davidson, K. S. Smith, and R. G. Crystal, *Biochemistry, 20*: 6755 (1981).
73. S. R. Karr and J. A. Foster, *J. Biol. Chem., 256*:5946 (1981).
74. J. M. Davidson, B. Leslie, T. Wolt, R. G. Crystal, and L. B. Sandberg, *Arch. Biochem. Biophys., 218*:31 (1982).
75. S. N. Thibodeau, D. C. Lee, and R. D. Palmiter, *J. Biol. Chem., 253*:3771 (1978).
76. J. A. Foster, R. P. Mecham, and C. Franzblau, *Biochem. Biophys. Res. Commun., 72*:1399 (1976).
77. P. A. Abraham and W. H. Carnes, *J. Biol. Chem., 253*:1993 (1979).
78. K. I. Kivirikko and R. Myllyla, in *The Enzymology of Post-Translational Modifications* (R. B. Freedman and H. Hawkins, eds.), Academic Press, New York, 1980, p. 53.
79. R. A. Berg, M. L. Schwartz, and R. G. Crystal, *Proc. Natl. Acad. Sci. USA, 77*: 4746 (1980).
80. D. M. Dunn and C. Franzblau, *Biochemistry, 21*:4195 (1982).
81. W. H. Fahrenbach, L. B. Sandberg, and E. G. Cleary, *Anat. Rec., 155*:563 (1966).
82. V. Damiano, A. L. Tsang, G. Weinbaum, P. Christner, and J. Rosenbloom, *Collagen Rel. Res., 4*:153 (1984).
83. S. M. Frisch and Z. Werb, *Fed. Proc., 42*:1982 (1983).
84. N. Uchida, H. Smilowitz, and M. L. Tanzer, *Proc. Natl. Acad. Sci. USA, 76*:1868 (1979).
85. S. M. Frisch, J. M. Davidson, and Z. Werb, *Molec. Cell. Biol., 5*:253 (1985).
86. R. P. Mecham, J. G. Madaras, and R. M. Senior, *J. Cell Biol., 98*:1804 (1984).
87. J. M. Davidson and R. A. Berg, *Meth. Cell. Biol., 23*:119 (1981).
88. E. G. Cleary and M. A. Gibson, *Int. Rev. Connect. Tiss. Res., 10*:97 (1983).
89. S. S. Spicer, R. M. Brissie, and N. T. Thompson, *Am. J. Pathol., 79*:481 (1975).
90. P. Pizzolato and R. D. Lillie, *Virchows Arch. A. Path. Anat. Histol., 380*:299 (1978).
91. A. S. Naranayan, R. C. Page, F. Kuzan, and C. G. Cooper, *Biochem. J., 173*:857 (1978).
92. A. S. Naranayan and R. C. Page, *J. Biol. Chem., 251*:1125 (1976).
93. R. Siegel, *Proc. Natl. Acad. Sci. USA, 71*:4826 (1974).
94. A. Serafini-Fracassini, G. Ventrella, M. J. Field, J. Hinnie, N. I. Onyezili, and R. Griffiths, *Biochemistry, 20*:5424 (1981).
95. G. Bressan, D. Volpin, and I. Pasquali-Ronchetti, personal communication.
96. R. P. Mecham and J. A. Foster, *Biochem. J., 173*:617 (1978).
97. C. J. P. Jones, C. H. J. Sear, and M. E. Grant, *J. Pathol., 131*:35 (1980).
98. C. B. Rich and J. A. Foster, *Meth. Enzymol., 82*:665 (1982).
99. R. P. Mecham and J. A. Foster, *Biochemistry, 16*:3825 (1977).
100. D. W. Smith, P. A. Abraham, and W. H. Carnes, *Adv. Exp. Med. Biol., 79*:385 (1977).
101. K. E. Hill and J. M. Davidson, *Arteriosclerosis*, 1986 (in press).
102. H. G. B. Slack, *Nature, 174*:512 (1954).
103. J. Gross, in *Biochemistry of Collagen* (G. N. Ramachandran and A. H. Reddi, eds.), Plenum Press, New York, 1976, p. 275.
104. A. L. Horowitz, N. A. Elson, and R. G. Crystal, in *The Biochemical Basis of Pulmonary Function* (R. G. Crystal, ed.), Marcel Dekker, New York, 1976, p. 273.
105. B. C. Starcher, C. Kuhn, and J. E. Overton, *Am. Rev. Resp. Dis., 117*:299 (1978).
106. J. S. Brody and C. Vaccaro, *Fed. Proc., 38*:315 (1979).
107. J. G. Clark, C. Kuhn, J. A. McDonald, and R. P. Mecham, *Int. Rev. Connect. Tiss. Res., 10*:249 (1983).
108. J. M. Davidson, S. Shibahara, K. Smith, and R. G. Crystal, *Connect. Tiss. Res., 8*: 209 (1981).

109. Y. Fukuda, A. J. Ferrans, and R. G. Crystal, *Am. J. Anat., 170*:597 (1984).
110. G. Cotta-Pereira, R. Guerra, and S. Bittencourt-Sampaio, *Invest. Dermatol., 66*:143 (1976).
111. I. M. Braverman and E. Fonferko, *J. Invest. Dermatol., 78*:434 (1982).
112. J. Uitto, in *Dermatology in General Medicine* (T. B. Fitzpatrick et al., eds.) McGraw-Hill, New York, 1979, pp. 182-188.
113. M. G. Giro, A. I. Oikarinen, H. Oikarinen, G. Sephel, J. Uitto, and J. M. Davidson, *J. Clin. Invest., 75*:672 (1985).
114. M. G. Reyes, D. B. Bazile, T. Tosch, and A. I. Rubenstone, *Arch. Pathol. Lab. Med., 106*:610 (1982).
115. S. N. Thung and M. A. Gerber, *Arch. Path. Lab. Med., 106*:468 (1982).
116. K. Tanaka, W. Mori, and K. Suwa, *Acta Path. Jpn., 31*:93 (1981).
117. P. H. Byers, R. C. Siegel, K. A. Holbrook, A. S. Naranayan, P. Bornstein, and J. G. Hall, *N. Engl. J. Med., 303*:61 (1980).
118. L. Danielsen and T. Kobayasi, *Acta Derm. Venereol., 52*:1 (1972).
119. M. C. Bourdillon, D. Brechmeier, N. Blaes, J. C. Derouette, W. Hornebeck, and L. Robert, *Cell Biol. Int. Rep., 4*:313 (1980).
120. W. H. Carnes, *Int. Rev. Conn. Tiss. Res., 4*:197 (1968).
121. R. C. Page and E. P. Benditt, *Lab. Invest., 26*:22 (1972).
122. L. B. Sandberg, J. G. Leslie, and B. W. Oakes, *J. Connect. Tiss. Res., 8*:219 (1981).
123. J. M. Davidson, K. E. Hill, J. L. Alford, and M. G. Giro, 1985 (submitted for publication).
124. J. G. Leslie, Doctoral Dissertation, University of Utah, 1980.
125. J. O. Cantor, S. Keller, M. S. Parshley, T. V. Darnule, A. T. Darnule, J. M. Cerreta, G. M. Turino, and I. Mandl, *Biochem. Biophys. Res. Commun., 95*:1381 (1980).
126. R. P. Mecham, J. G. Madaras, J. A. McDonald, and U. Ryan, *J. Cell. Physiol., 116*: 282 (1983).
127. H. Wolinsky, *Circ. Res., 26*:507 (1970).
128. T. J. M. Schlatmann and A. E. Becker, *Am. J. Cardiol., 39*:13 (1977).
129. R. Ross, *Arteriosclerosis, 1*:293 (1981).
130. J. Chamley-Campbell, G. Campbell, and R. Ross, *J. Cell Biol., 89*:378 (1981).
131. W. Hornebeck, G. Bellon, D. Brechemier, G. Godeau, and L. Robert, in *Connective Tissue Research: Chemistry, Biology, and Physiology*, Alan R. Liss, New York, 1981, p. 233.
132. B. Faris, L. L. Salcedo, V. Cook, L. Johnson, J. A. Foster, and C. Franzblau, *Biochim. Biophys. Acta, 418*:93 (1976).
133. A. S. Naranayan, L. B. Sandberg, R. Ross, and D. Layman, *J. Cell Biol., 68*:411 (1976).
134. B. W. Oakes, A. C. Batty, C. J. Handley, and L. B. Sandberg, *Eur. J. Cell Biol., 27*:34 (1982).
135. P. A. Jones, T. Scott-Burden, and W. Gevers, *Proc. Natl. Acad. Sci. USA, 76*:353 (1979).
136. J. Chamley-Campbell, and R. Ross, *Physiol. Rev., 59*:1 (1979).
137. S. Gay and E. J. Miller, *Collagen in the Physiology and Pathology of Connective Tissue.* Gustav Fischer Verlag, Stuttgart, New York, 1978.
138. J. L. Alford, P. A. LuValle, K. E. Hill, M. G. Giro, and J. M. Davidson (submitted for publication).
139. J. Rosenbloom and A. Cywinski, *Biochem. Biophys. Res. Commun., 69*:613 (1976).
140. G. M. Bressan and D. J. Prockop, *Biochemistry, 16*:1406 (1977).
141. W. W-Y. Kao, G. M. Bressan, and D. J. Prockop, *Connect. Tiss. Res., 10*:263 (1982).
142. E. G. Cleary, L. B. Sandberg, and D. S. Jackson, *J. Cell Biol., 33*:469 (1967).

143. C. H. J. Sear, M. A. Kewley, M. E. Grant, and D. S. Jackson, *Biochem. J., 170*:715 (1978).
144. R. M. Senior, G. L. Griffin, R. P. Mecham, D. S. Wrenn, K. U. Prasad, and D. W. Urry, *J. Cell Biol., 99*:870 (1984).
145. R. M. Senior, G. L. Griffin, and R. P. Mecham, *J. Clin. Invest., 66*:859 (1980).
146. G. W. Hunninghake, J. M. Davidson, S. Rennard, S. Szapiel, J. Gadek, and R. G. Crystal, *Science, 212*:925 (1981).
147. W. H. Carnes, P. A. Abraham, and V. Buonassissi, *Biochem. Biophys. Res. Commun., 90*:1393 (1979).
148. C. Kuhn, R. M. Senior, and J. Pierce, in *Mechanisms in Respiratory Toxicology* (H. P. Witschi and P. Nettesheim, eds.), CRC Press, Boca Raton, FL, 1982, p. 156.
149. M. J. Banda and Z. Werb, *Biochem. J., 193*:589 (1980).
150. E. J. Campbell, R. B. White, R. M. Senior, R. J. Rodriguez, and C. Kuhn, *J. Clin. Invest., 64*:824 (1979).
151. P. Codogno, W. Hornebeck, Y. Gossault, L. Robert, and M. Aubery, *Cell Biol. Int. Rep., 7*:433 (1983).
152. J. Rosenbloom, K. Yoon, L. Oliver, and C. Boyd, *J. Cell Biochem., 8b*:270 (1984).
153. C. Benoist and P. Chambon, *Nature, 290*:304 (1981).
154. D. Botstein, R. L. White, M. Skolnick, and R. W. Davis, *Am. J. Hum. Genet., 32*:314 (1980).
155. W. Burnett, K. Yoon, A. Finnigan-Bunick, and J. Rosenbloom, *J. Invest. Dermatol., 79*:138S (1982).
156. W. Burnett, K. Yoon, and J. Rosenbloom, *Biochem. Biophys. Res. Commun., 99*:364 (1981).
157. D. Perlman and H. O. Halvorson, *Cell, 25*:525 (1981).
158. J. Rogers, P. Early, C. Carter, K. Calame, M. Bond, L. Hood, and R. Wall, *Cell, 20*:303 (1980).

3

The Microfibrillar Component of the Elastic Fibers
Morphology and biochemistry

Edward G. Cleary
University of Adelaide,
Adelaide, South Australia

I. INTRODUCTION

Elastic tissue is a fibrous, rubberlike structure composed predominantly of the protein elastin. It exhibits characteristic staining reactions in histological sections, in which it is seen in a variety of arrangements. These include thick, nonbranching, rodlike structures, as in the nuchal ligament of the ungulates; sheets or lamellae, as in the walls of blood vessels; and reticular meshworks in elastic cartilage and skin. Mixed or intermediate forms of these arrangements are found in many other tissues.

At the ultrastructural level, elastic fibers from all these forms are recognized in conventionally stained sections in the electron microscope as consisting of two components: a major amorphous, poorly staining component surrounded by a lesser and variable rim of more densely staining, tubular-appearing structures, about 10-14 nm in diameter. The latter, called *elastin-associated microfibrils*, are the subject of this chapter. Their composition, function, and relationship to amorphous elastic tissue is only now, after more than 20 years of research effort, beginning to be understood. A detailed review of the background and history of research in this field has been undertaken recently (1), and that work should be consulted for a comprehensive treatment of those aspects.

II. DEFINITIONS

A. Microfibrils

The extracellular matrix contains many filamentous structures ranging in diameter from 3-4 nm up to more than 20 nm. Those filaments in this size range which do not exhibit the 64-nm band pattern typical of collagen were called *microfibrils* by Low in 1961 (2).

Since this original definition it has been shown that most of the matrix filaments less than 5 nm in diameter can be related to specific matrix components including proteoglycans (3,4), fibronectin (4), or type VI collagen (5). Within the microfibrillar population between 5 and 20 nm in diameter, some were recognized from the beginning to have a specific association with amorphous elastic tissue (2) and some have since been identified as nonstriated interstitial collagen fibrils (6). After all these exclusions, there are still many such microfibrils occurring in a wide variety of tissues, alone or in groups, whose identity has not been established. Apparently typical microfibrils have also been identified as essential components of amyloid deposits in which they are known to be γ-globulin molecules arranged in β-pleated sheets (7).

It is thus clear that the term "microfibril" is descriptive of a particular morphology that may be assumed by a variety of very different proteins in the extracellular matrix. For the present, the microfibrillar component of the elastic fiber can be distinguished from these other microfibrils only by its association with amorphous elastic tissue, as determined electron microscopically. To acknowledge that the relationship of the microfibrillar component to the amorphous component of elastic tissue is not yet determined, it was suggested that they be described as *elastin-associated microfibrils* (1) in preference to "elastic fiber microfibrils" (8).

B. Elastin-Associated Microfibrils

Elastin-associated microfibrils are thus the microfibrillar structures directly associated with amorphous elastic tissue. They exhibit different staining patterns (9) and different susceptibilities to enzyme digestion (10) from those of the amorphous elastic tissue and are resistant to digestion by highly purified collagenase (10,11). Their diameter ranges from 8 to 16 nm, but is most commonly 11-14 nm. In longitudinal section they exhibit a periodic staining pattern in which 13-nm segments alternate with 8-nm segments. In cross-section they appear as complex structures made up of four to six densely staining tubular structures, each about 3 nm in diameter, arranged around the perimeter of a less electron-dense central region (12). These features are demonstrated in Figure 1. The elastin-associated microfibrils have an affinity for cationic stains and have been shown to bind lectins and to exhibit specific staining reactions indicative of the presence of glycoprotein material associated with them (13). An example of the periodic acid-bismuth staining of elastic tissue is shown in Figure 2. Because the microfibrillar components are intimately associated with elastogenesis in all tissues examined to date, it has been suggested that they are either precursors of amorphous elastic tissue (14,15) or structural organizers of the amorphous component (10). They exhibit variations in diameter between species, during development in a single species, and in response to copper deficiency in sheep and pigs (12).

Chemical characterization has not as yet been achieved, despite claims for many proteins and glycoproteins. Virtually all these claims have been based on identification of microfibrillar protein in one of two ways, neither of which is adequate for the purpose and both of which may be incorrect.

There has been an almost universally held belief that microfibrillar protein may be identified by a unique amino acid composition characterized by relatively high contents of aspartic and glutamic acids and a small but significant content of cysteine. Alternatively, it has been concluded that *everything* that can be extracted from tissues with highly concentrated solutions of guanidinium chloride (GuHCl) under reducing conditions (reductive GuHCl), following a variety of preliminary nonreductive extractions, including GuHCl, is microfibrillar in origin.

Both these conclusions derive from the experimental work of Ross and Bornstein (10,16), which has been the catalyst for most of the biochemical research on elastin-

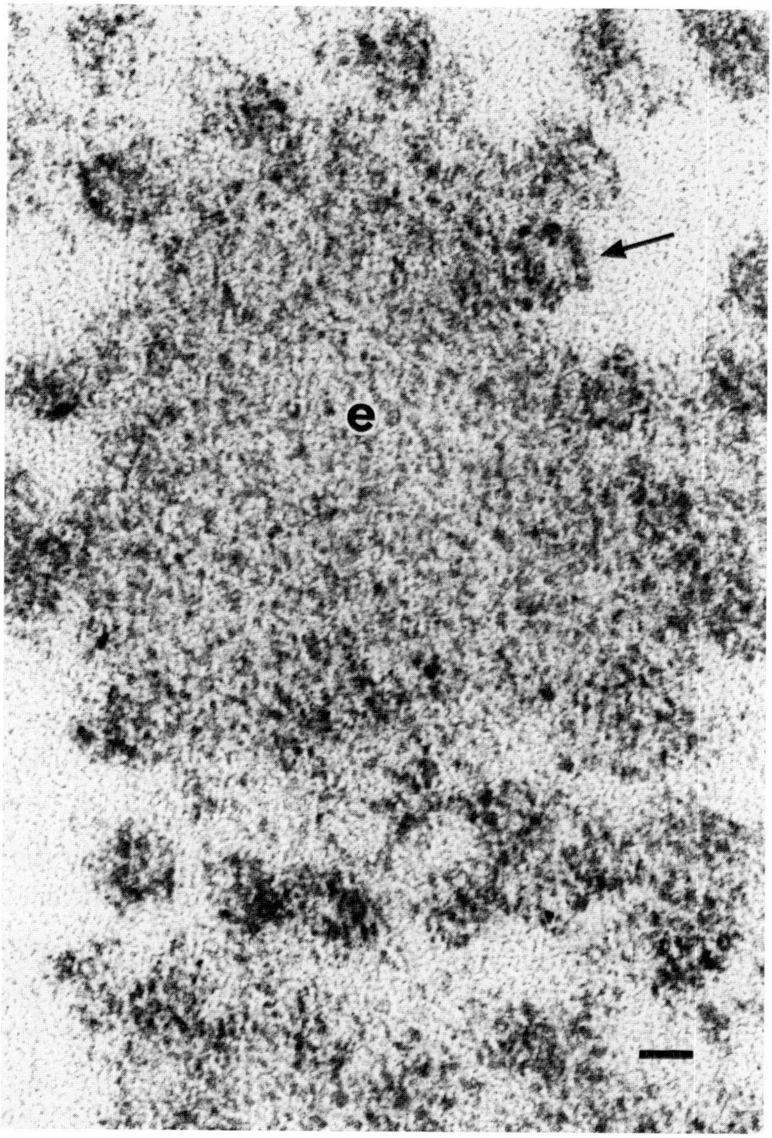

Figure 1 High-resolution electron micrograph of transverse section of elastic lamella from thoracic aorta of a 6-month-old pig. Microfibrils are seen around the periphery of the amorphous elastic tissue (e). The microfibrils have an average diameter of 14 nm and are seen to be made up of 4 to 6 more densely staining tubular structures (arrow) surrounding a less electron-dense central region. Bar, 10 nm. (Courtesy of Dr. J. C. Fanning.)

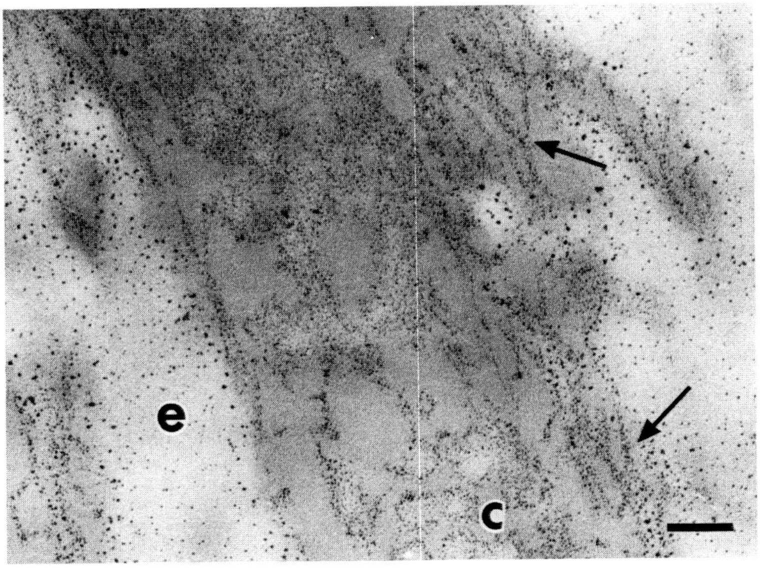

Figure 2 Section of elastic lamella from the thoracic aortic media of a 210-day-old bovine fetus. Alkaline bismuth stain after periodic acid oxidation of the section. Fine electron-dense bismuth granules are seen in linear arrays along microfibrils (arrows) at the periphery of the amorphous elastic tissue (e). Similar granules are seen around the periphery of collagen fibers (c) in cross-section. Bar, 100 nm. (Courtesy of Dr. J. C. Fanning.)

associated microfibrils. However, these authors subsequently acknowledged that reductive GuHC1 extracts of nuchal ligament under these conditions were quite heterogeneous in protein composition (17). As Bornstein (18) has since drawn attention to the inadequacy of such nonspecific amino acid analyses for identifying proteins, it seems clear that neither of the original proponents of these criteria for identifying microfibrillar constituents would now accept them as valid identification. Thus, it must be acknowledged that the chemical composition of elastin-associated microfibrils, and of other morphologically similar microfibrils, is unknown.

C. Oxytalan and Elaunin Fibers

Many connective tissues contain fibers which, in the electron microscope, consist of collections of microfibrils, which are morphologically similar to elastin-associated microfibrils but which lack an associated amorphous elastic tissue component.

These structures were first identified, by light microscopy, in periodontal ligament as fibers exhibiting the staining characteristics of elastic tissue, but only after treatment with strong oxidizing agents. They are called *oxytalan fibers* and occur widely in skin, tendons, fasciae, the adventitial layer of blood vessels, and other sites (19).

Another group of fibers, containing morphologically similar microfibrils and exhibiting histological staining properties intermediate between those of elastic and oxytalan fibers (20) has been recognized, especially in skin. These fibers have been aptly described as "partially elastinized bundles of microfibrils" (21) and are called *elaunin fibers* (20). In skin, the oxytalan fibers of the dermal-epidermal junction region have been shown to be connected to the elastic fibers of the reticular dermis by elaunin fibers (22). It is claimed that these three groups of fibers belong to a single system of microfibrillar fibers and that they correspond to different stages in the development of elastic fibers (20,22). The differentiation between the subgroups in skin is postulated to be due to the presence, in the region of the dermal-epidermal junction, of some factor capable of interrupting the elastogenic process, so that amorphous elastic tissue is not secreted or deposited among the microfibrils in this zone (23).

The histological staining reactions appear to be in some way related to the microfibrillar components, as microfibril-rich, elastic fibers in developing tissues exhibit oxytalan staining reactions (20,24). For our purpose it is important to recognize that the microfibrils of structures exhibiting oxytalan and elauninlike features have not been shown to be identical to elastin-associated microfibrils and thus that they may have a different composition.

III. HISTORICAL OVERVIEW

One can identify three phases in research in elastin-associated microfibrils. The initial phase was concerned with the ultrastructural identification of these microfibrils. The second phase began with the recognition that they could be extracted by chaotropic agents in the presence of reducing agents. It was predominantly biochemical and has been characterized by attempts to identify microfibrillar proteins on the basis of conformity to a particular amino acid profile, the validity of which has not been established. The third phase is at present concerned with the immunological characterization of microfibrillar constituents and immunoelectron microscopic proof of identity between elastin-associated microfibrils and putative microfibrillar constituents.

A. Phase 1. Ultrastructural Identification of Microfibrils

Microfibrils were first demonstrated in association with elastic fibers in tissue sections of trachea in 1958 (25). Subsequently, Low, in making a detailed study of the extensive population of filamentous structures in the extracellular tissue spaces, observed that collagen fibrils tended to be greater than 20 nm in diameter while those smaller than this were seldom striated like collagen fibrils. He proposed that structures less than 20 nm in diameter be referred to as microfibrils and went on to note that a group of such microfibrils, about 8-10 nm in diameter, were commonly seen around the periphery of elastic fibers (2).

There followed a steady stream of reports of microfibrils associated with vascular and nonvascular elastic tissue in many tissues, in many mammals and even in birds (reviewed in Ref. 1). Haust (26) also drew attention to the presence of apparently identical microfibrils without an association with elastic tissue in tissues such as skin.

In the same year Cleary and Jackson and their colleagues published the first of a series of systematic studies on elastogenesis in bovine nuchal ligament during fetal and postnatal development (14,27-31).

As part of this study, Fahrenbach et al. (14) showed that collections of microfibrils were usually secreted into the extracellular matrix by the ligament fibroblasts prior to the deposition of amorphous elastic tissue among the microfibrils. These microfibrils stained positively with the cationic stain ruthenium red.

Greenlee and his colleagues (9) confirmed the relationship of microfibrillar materials to developing elastic tissue and drew attention to the major differences in staining behavior of the two components. These were interpreted as indicating that the two components had either different chemical compositions or different organizational structures. Subsequently, Ross and Bornstein (10) noted further differences between the two components when they showed that they had different susceptibilities to digestion with enzymes. They concluded that the many differences between the two components of elastic tissue indicated that microfibrils were very likely to consist of proteins different from elastin, and they went on to show that the microfibrils appeared to be resistant to digestion with collagenase, even after extraction of the tissue with GuHCl (10,16).

B. Phase 2. Extraction and Biochemical Characterization of Microfibrillar Proteins

Ross and Bornstein (10,16) undertook concurrent biochemical and electron microscopic studies on the effects of various extractants on fetal bovine nuchal ligament. Their data were interpreted as indicating that elastin-associated microfibrils were selectively extracted with 5 M GuHCl in the presence of a reducing agent (to cleave disulfide bonds). The material thus extracted was therefore claimed to be entirely microfibrillar in origin. Unfortunately, the extent of the heterogeneity in this preparation was not documented, although it was shown to consist of at least two components with different solubilities and amino acid compositions. Both contained sugar residues, which was interpreted as confirming the glycoprotein nature of the elastic tissue microfibrils. This had been previously suspected from their staining behavior (9). Both fractions were relatively rich in acidic amino acids, as would be expected (we now recognize) of a protein mixture (32). However, as the heterogeneity and its consequences were overlooked, this amino acid pattern came to be regarded, almost universally, as that typical of microfibrillar glycoproteins. Claims for a relationship between a purified matrix protein or a tissue extract and the elastin-associated microfibrils, based primarily on these amino acid criteria, have continued to be made even after further work from Ross and Bornstein's laboratories showed that microfibrillar protein isolated from fetal nuchal ligament by the Ross and Bornstein extraction procedure "revealed great molecular weight heterogeneity," in spite of additional purification by ion-exchange chromatography (17).

Attention has been directed to the presence in tissue matrices of glycoproteins other than those derived from serum (33). These, however, proved to be difficult to extract and, because of their insolubility, difficult to purify and characterize. It is now recognized that the harsh extraction methods caused significant degradation of these structural glycoproteins (34,35) and that these preparations also were heterogeneous (35,36). It is thus not surprising that all these preparations were rich in acidic amino acids and contained cysteine, nor that they were therefore claimed to be "microfibrillar" constituents when obtained from elastin-rich tissues. Electron microscopic studies attempting to relate these preparations to elastin-associated microfibrils were, however, unconvincing (37, 38).

The inclusion of inhibitors of protease activity in the extracting buffers has allowed immunological identification of actin, myosin, and fibronectin as major constituents of structural glycoprotein preparations from tissues (35) and cell cultures (39).

Serafini-Fracassini and his colleagues have isolated, in pure form, two polypeptides from reductive GuHCl extracts of adult bovine nuchal ligament. One of these, with M_r 34,000, exhibited amine oxidase activity (40). Although a relationship between this glycoprotein and elastin-associated microfibrils has not been established, this material has not been excluded as a potential microfibrillar constituent, despite the fact that adult bovine nuchal ligament has only a minor microfibrillar content.

A noncollagenous matrix glycoprotein, M_r 115,000, isolated from reductive GuHCl extracts of chick aorta has been shown in immunofluorescence studies to form an extracellular fibrillar network in the aortic wall, but this was clearly not related to elastin-associated microfibrils (41).

In recent years, attention has turned to tissue cultures as a potential source of more soluble forms of the constituents of elastin-associated, or morphologically similar, microfibrils (17,42,43). The most favored materials identified in these experiments to date have proven to be fibronectin (17,44) and type VI collagen (45), neither of which is distributed in tissues in association with elastin-associated microfibrils (46).

C. Phase 3. Immunological Characterization of Microfibrillar Constituents

The latest and most promising phase in the identification of microfibrillar constituents began in 1976 with the development by Kewley and Steven of an antibody to components of the reductive GuHCl extract from fetal bovine nuchal ligament and the use of immunoelectron microscopy to establish localization of the antibody to elastin-associated microfibrils in embryonic chick aorta (47). This observation established that the reductive GuHCl extract of fetal bovine nuchal ligament, prepared by the method of Ross and Bornstein (10), contained an antigenically active component, the antibody against which localized in tissue sections to elastin-associated microfibrils. Thus, although the extraction procedure was such as to lead to the denaturation of the extracted protein(s), they retained antigenic determinants that were recognizable in undenatured tissue sections. This study demonstrated the possibility of developing immunological probes for identification of the constituents of elastin-associated microfibrils. At the same time, it has demonstrated the need for extreme care in establishing the specificity of an immunological probe used for this purpose. In immunofluorescence studies it was found that the antiserum also bound to a variety of basement membranes and collagenous tissues (48). This was interpreted as indicating that elastin-associated microfibrils shared antigenic determinants with components of these other tissues. Thus, when the antibody was used to immunoprecipitate radioactive glycoprotein material from conditioned medium from nuchal ligament fibroblast cultures, both the glycoproteins so identified, viz., MFPI and MFPII, were named microfibrillar proteins. These glycoproteins had quite different compositions, and one, MFPI, which contained collagenous amino acids, was claimed to be the putative microfibrillar protein (42). Subsequently, a very similar protein, CL glycoprotein, has been isolated from reductive GuHCl extracts of fetal bovine nuchal ligament and aorta (49). A polyclonal antibody against the purified, isolated protein has been shown to localize in the nonelastic tissue distribution of the Kewley antiserum (46). It is now claimed that MFPI is related to type VI collagen (45), and it has been shown that CL glycoprotein is a native form of type VI collagen that can be extracted from tissues as well as from tissue culture medium (Gibson and Cleary, unpublished data). Neither CL glycoprotein (and thus type VI collagen) nor MFPI is microfibrillar in origin (46).

Cleary and his colleagues (50) modified the Kewley extraction procedure and have developed polyclonal antibodies in rabbits against the antigenic mixture. These have been purified by extensive immunoadsorption with bovine serum and tissue components to yield a much more specific antiserum that localized almost exclusively to elastin-associ-

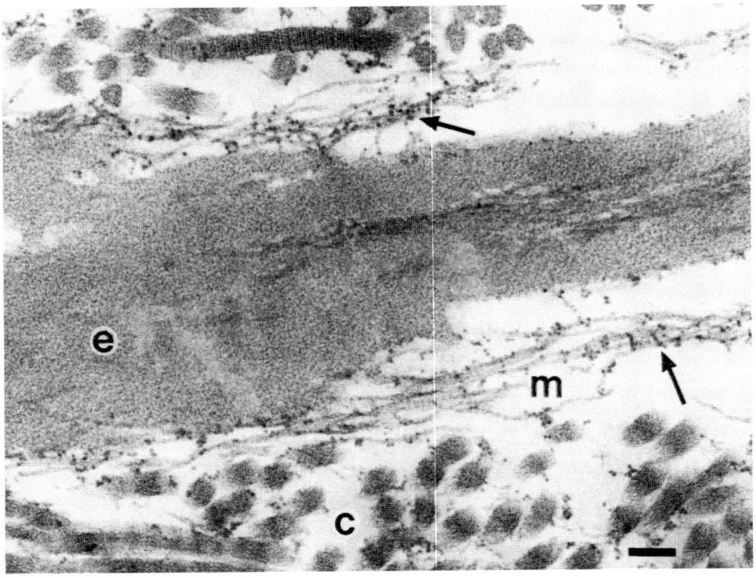

Figure 3 Electron micrograph of longitudinal section of fetal bovine nuchal ligament showing elastin-associated microfibrils (m), amorphous elastic tissue (e), and collagen (c). The tissue has been treated with ferritin-conjugated anti-(microfibrillar preparation) antibody. Ferritin particles are predominantly associated with the microfibrils (arrows). Small amounts of ferritin are also bound in the intercollagenous spaces but were absent when the section was blocked beforehand with anti-(CL glycoprotein) antibody. Bar, 100 nm. (From Ref. 11, by permission of the editors of the *Australian Journal of Experimental Biology and Medical Science*.)

ated microfibrils, as shown in Figure 3 (11). This antibody therefore constitutes an extremely useful probe for monitoring extraction steps (by immunofluorescence studies on the tissues) and fractionation procedures (by ELISA and immunoblotting) (Gibson and Cleary, unpublished data) and if soluble forms of microfibrillar constituents are identified, the antibody could be useful for immunoprecipitation- or immunoadsorption-type preparative methods.

D. Summary

It is possible, from reviewing these mainstream findings to reach a number of conclusions that must be taken into account when examining research data relating to elastin-associated microfibrils. Thus:

1. Elastin-associated microfibrils share a similar morphology with other microfibrillar structures. The relationship, if any, with these other microfibrils is unknown.
2. High-resolution ultrastructural studies have shown that elastin-associated microfibrils are complex structures (12,14), which may therefore be composed of more than one constituent.
3. Elastin-associated microfibrils are associated with glycoprotein material (13).
4. Elastin-associated microfibrils in tissues are resistant to extraction in usual solvents, requiring high concentrations of denaturing chaotropic agents for their extraction(10).

5. The constituents of elastin-associated microfibrils appear to have limited solubility (10,11).
6. Constituent(s) of Ross and Bornstein-type extracts of elastin-rich tissues contain antigenically active sites, which are represented on elastin-associated microfibrils in tissues (11,47,50).
7. The amino acid composition of the constituent protein(s) of elastin-associated microfibrils is unknown.
8. Immunoelectron microscopic identification of putative microfibrillar components with elastin-associated microfibrils is essential to establishing their relationship with those microfibrils (11,47).
9. The latter methodology is subject to error, and rigorous controls are essential to the interpretation of the findings (11,47,50).

IV. ELASTOGENESIS AND ELASTIN-ASSOCIATED MICROFIBRILS IN VIVO

A. In Elastin-Rich Soft Tissues

Fibroblasts, vascular smooth muscle cells, and endothelial cells are all able to synthesize elastic tissue. The process has been documented best in the fetal bovine nuchal ligament (14), but it follows a similar course in other tissues. The earliest recognizable feature of elastogenesis is deposition within the extracellular space, close to the cell surface, of collections of microfibrils. These stain positively with cationic stains, such as ruthenium red, and are usually seen within electron-dense embayments in the cell membrane. Adjacent to these embayments are collections of acanthosomes, or spiny vesicles, containing ruthenium red-staining material. These membrane-bound vesicles appear to be budded off from the Golgi, to travel through the cell cytoplasm and to discharge their contents to the extracellular space by fusing with the cell membrane. Within the embayments microfibrils polymerize first, and subsequently, amorphous elastic tissue is seen to be deposited within the collections of microfibrils. Further growth of the elastic fibers occurs by accretion of additional amorphous material, but always in such a manner that the growing amorphous masses are continually surrounded by a rim of microfibrillar material (14).

The peripheral microfibrils stain with cationic uranyl groups, while the amorphous elastic tissue has an affinity for the anionic phosphotungstate groups (9). In conventionally stained sections (uranyl acetate/lead citrate stain) newly formed amorphous elastic components of vascular elastic lamellae retain a greater affinity for electron-dense uranyl groups than do older elastic deposits (12).

Elastogenesis follows this same pattern, during fetal and postnatal development, in the nuchal ligament and in both the thoracic and the abdominal segments of the aorta. However, the process begins earlier and at a given age is more advanced in thoracic aorta than in abdominal aorta, which in turn is in advance of the nuchal ligament (Fanning and Cleary, unpublished data).

It has been demonstrated that the mean diameter of the elastin-associated microfibrils in thoracic aortic sections increases systematically with age, during fetal and postnatal development. The mechanism of this alteration was not clear but was typical of each species, highly significant in extent, and the increase in microfibril diameter was most rapid during the period of maximal elastin deposition (12).

The alignment of the microfibrils during fetal elastogenesis is predominantly in the longitudinal axis of the nuchal ligament so that the microfibrils parallel the principal direction of the elastic fibers. This arrangement created the impression that the microfibrils were acting as tissue organizers for the deposits of amorphous elastic tissue (10). In sections of nuchal ligament fixed in a stretched state, the microfibrils align in the long axis of the elastic fibers as regular parallel arrays, which can be traced for considerable distances into the amorphous elastic tissue (Figure 4; Fanning and Cleary, unpublished data). These appearances are considered to favor the view that elastin-associated microfibrils become incorporated into the amorphous elastic tissue and retain a fibrillar structure within the elastic fiber (12). The apparent incorporation, by loss of stainability, of the microfibrils into the amorphous elastic tissue is sometimes observed in elastic fibers cut in cross-section (12). Such appearances are visible in Figure 1. According to this interpretation, microfibrils are considered to be made up of tropoelastin molecules with a coating of glycoprotein material, which is removed (possibly by proteolytic digestion) as the microfibrils are incorporated into the elastic fiber. The temporary persistence of available anionic groups during this process would account for the greater electron density in stained sections of newly forming elastic fibers compared with older fibers.

The morphology of vascular elastic tissue, and the changes in this tissue during development, conform to these descriptions in a wide range of vessels down to the small arterioles in all mammals and birds and in virtually all vertebrates with the exception of some fishes (1). The exceptions are the aortas of salmon and trout (51) and of the octopus (52).

In the salmon and the trout aorta, the elastic tissue consisted entirely of collections of fibrils, 25 nm in diameter, that were interwoven into a three-dimensional network, within which were embedded smooth muscle cells. No amorphous elastic tissue was present in the sections. However, there were areas in which the fibrils tended to coalesce to form bundles, and these exhibited reduced stainability with uranyl acetate/lead citrate. Typical elastin-associated microfibrils were not seen (51).

Biochemical characterization of the elastins isolated from these species (53) revealed that although they contained appreciable amounts of desmosine and isodesmosine, their amino acid compositions were different from those of mammalian and chick elastins (54). On electron microscopic examination, the purified elastin was found to consist of fibrils similar to those seen in the tissue sections (51).

These findings together provide clear evidence that "elastic tissue," even though containing lysine-derived covalent cross-links, can exist in smaller fibrillar form. This favors the possibility that conventional elastin-associated microfibrils could contain tropoelastin complexes. It would be of considerable interest to determine the regulatory mechanisms involved in elastin fibrillogenesis in the fish aorta.

B. In Soft Tissues with Lesser Elastin Contents

1. Lung

Lung is a complex tissue in which there is a relatively high content (20-30% of the crude connective tissue dry weight) of elastin in the respiratory parenchyma, but much lower concentrations in the other portions of the tissue (55). In each location, however, the elastic tissue exhibits classical ultrastructural features, with amorphous elastic material surrounded by a rim of microfibrils (56). There have been conflicting reports as to the diameter of microfibrils in lung elastic tissue and as to the susceptibility of these microfibrils to digestion with proteolytic enzymes, but it seems probable that differences in processing of the tissue could account for these results (1).

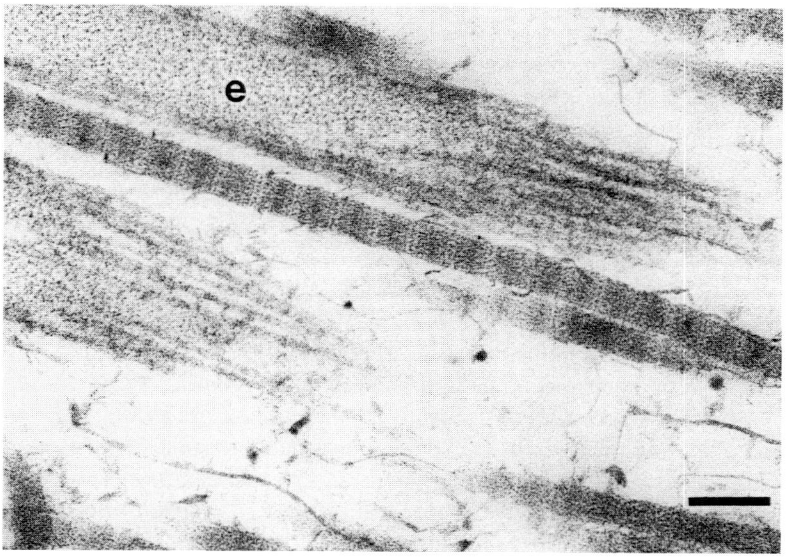

Figure 4 Longitudinal section of 210-day-old bovine fetal nuchal ligament that has been held in a stretched state during fixation and embedding. The elastin-associated microfibrils are seen running into the amorphous elastic tissue (e). Collagen fibers and proteoglycan filaments are also visible. Bar, 100 nm. (Courtesy of Dr. J. C. Fanning.)

Elastic tissue is claimed to exercise a critical role in regulating alveolar septum formation (evidence reviewed in Refs. 1,57). It is surprising, therefore, to note that it is not yet known which cell(s) is/are responsible for elastin synthesis in the lung (57). However, the development of elastic fibers in the alveolar septa follows the classical sequence observed in other soft tissues, involving microfibrils and thin amorphous elastic tissue (1, 56).

When rat pups were fed the lathyrogen, β-aminopropionitrile (BAPN), during the phase of rapid lung development, the extracellular bays of the septal myofibroblasts were observed to be filled with what was interpreted as "microfibrillar material," but little amorphous elastic tissue formed (58). It thus seems that it may be possible, in lung tissue, to modify the usual relationships between microfibrils and amorphous elastic tissue. This could thereby provide a useful model for further examination of their interrelationship.

2. *Skin*

As pointed out in Section II.C, skin has a complex arrangement of microfibrillar constituents. In human skin, the distribution of these components varies with age and location and is also affected by exposure to sunlight (21-23). Furthermore, almost all the information about these constituents and the changes with aging and disease are based on morphological and empirical histochemical observations, which in turn have been founded on assumptions, some of which are now recognized to be inadequate. In this tissue, perhaps more than any other, the need for methods of analyzing and identifying possible differences between groups of microfibrillar constituents is critical to an understanding of normal and abnormal processes.

In skin protected from actinic damage in young human adults, elastic fibers with typical staining reactions and electron microscopic appearances are seen forming a network in the deeper portion of the reticular dermis. Similar fibers ascend toward the papillary dermis. In the upper part of the reticular dermis and in the adjacent papillary dermis, the fibers assume the staining properties and electron microcopic appearances of elaunin fibers. In areas of thin skin the elaunin fibers form a plexus of fine fibers immediately below and parallel to the dermal-epidermal junction. From this plexus even finer filaments extend across the lamina lucida in brushlike configuration to be inserted directly into the basal lamina of the dermal-epidermal junction (23) as "anchoring" fibrils (59). The superficial fine filaments have the staining properties of oxytalan fibers and in the electron microscope are seen to be made up entirely of microfibrillar constituents (23). They are continuous with the anchoring fibrils above (59) and with the elaunin fibers below (22).

The oxytalan-elaunin-elastic fibers thus form a continuous system of microfibrillar structures in which there is an increasing content of amorphous components (23). These fibers are claimed to serve to anchor the epidermis to the underlying dermis (59).

In regions of thick skin, in normal young adults, although the oxytalan-elaunin-elastic fiber sequence is maintained, the fiber pattern is different in that the plexus of elaunin fibers below the dermal-epidermal junction is not formed. Instead, oxytalan fibers arise directly from ascending elaunin fibers (23).

During fetal development in humans, microfibrillike structures were recognized in skin, during the second month of development. At first the microfibrils were unoriented, but by the third uterine month small bundles of microfibrils were identified. In the fourth to sixth uterine months small amounts of amorphous elastic tissue had been deposited in the microfibril bundles in the reticular dermis. These were thus classified as elaunin fibers. By the eighth month, elastic fibers could be recognized forming a network in the reticular dermis, and by birth, the elastic tissue framework had assumed the adult pattern of distribution. There was continued development of dermal elastic tissue elements postnatally (60).

In the electron microscope, dermal elastic fibers differ in appearance from those seen in aorta or nuchal ligament, in that they exhibit many small, round, or irregular electron-dense zones within which are collections of microfibrillar material. These may be seen in the earliest electron micrographs of skin (61). They were named microfibrillar dense zones by Varadi (62) and have been emphasized to be normal constituents of dermal elastic fibers by Braverman and Fonferko (21), who stressed the importance of recognizing this difference when interpreting dermal elastic tissue changes in disease.

It has been claimed that dermal elastic tissue from normal sun-protected areas of adult skin lacks microfibrils around the periphery of the amorphous elastic tissue masses (21,62) but peripheral microfibrils are clearly recognizable in dermal elastic tissue in several published electron micrographs (21,60,63,64). The reasons for these discrepancies are not clear and will obviously require, for their reconciliation, detailed examination of the role of variables such as location, sampling, age, and staining.

Braverman and Fonferko (21) have reported a detailed study of the changes in dermal elastic fibers with aging. They found that after the age of 30, in skin protected from actinic damage, elastic fibers in the papillary dermis (more than in the reticular dermis) showed an increase in diameter of the microfibrillar dense zones, and adjacent to these small cystic spaces developed. Beyond the age of 50, these cystic spaces became larger and were often filled with a finely fibrillar or granular material. Some elastic fibers exhibited a porous look, owing to apparent disaggregation of electron-dense elastic fibers (with their associated microfibrils). Sometimes this change affected whole fiber masses while in others only the peripheral parts were affected. These peripheral porosities stained with ruthenium red. Beyond the age of 70 years, additional elastic tissue changes were seen, including granular and fibrillar degeneration of the peripheral portions of elastic fibers; the appearance of aggregates of short fibrils, 23.5-33.5 nm wide, around the periphery of the amorphous elastic tissue; and the appearance of collections of haphazardly arranged microfibrils, often separate from amorphous elastic material. These changes were distinguished from the effects of actinic damage, which itself produced major changes in elastic tissue (21).

The anatomical continuity between the microfibrillar components of the dermal "elastic system" has been interpreted as indicating that they are identical throughout (23), but there is evidence for differences in behavior of the microfibrils in different parts of the system. While dermal oxytalan fibers, in normal adult skin, were unaffected by autoclaving (65), the peripheral microfibrils associated with dermal elastic tissue were completely removed by this treatment (64) although the electron-dense inclusions within the elastic tissue masses remained (64). Clearly, failure to test such fundamental assumptions could easily lead to confusion.

There have been few biochemical studies on elastic tissue constituents of skin. Hot alkali-insoluble elastin has been prepared from human skin and has been shown to have an amino acid composition similar to that of elastin from other locations including the presence of the lysine-derived cross-links desmosine and isodesmosine, which in mammals are peculiar to elastin (62). Attempts to study microfibrillar constituents have involved reductive GuHCl extractions (which will obviously have yielded very heterogeneous protein mixtures) and amino acid analysis. As we have seen, this approach can provide no information that can be specifically related to microfibrillar constituents.

3. *The Eye*

Elastic tissue with typical histochemical staining properties and ultrastructural appearances is a component of many of the structures in the eye. These include the sclera (66), the glassy membrane (of Bruch) within the choroid—in which there is a plexus of very fine elastic fibers (67)—and the several components that make up the ciliary apparatus—including the scleral spur and trabecular meshwork of the ciliary tendon (68), the ciliary muscles, and the pectinate ligament.

Raviola (69) showed that the ciliary zonule exhibited the staining characteristics of oxytalan fibers and that it consisted of bundles of microfibrils. These were indistinguishable in appearance from elastin-associated microfibrils and were similarly susceptible to digestion with α-chymotrypsin and elastase and resistant to digestion by collagenase.

These findings have been confirmed in a recent, detailed study of the ocular "elastic fiber system" by Streeten and her colleagues (70,71). The relationship between the microfibrillar structures in this system and those in skin and blood vessels is of particular interest as several diseases associated with obvious defects in vascular connective tissues, e.g., Marfan's syndrome and homocystinuria, are associated with dislocation of the ocular lens (70).

While the zonule consisted of bundles of microfibrils, similar to those in elastic tissues, the elastic fibers of the bovine ciliary body exhibited the staining characteristics and ultrastructural features of elaunin fibers (70). In young adult human eyes the elastic fibers had only a sparse microfibrillar content. This difference was attributed to the greater age of the human eyes, but the effects of this variable in a single species were not examined (70).

The zonule is a circular ligamentous structure attached to the edge of the ocular lens and separating the posterior chamber of the eye from the vitreous body. It can thus be readily dissected out and analyzed. Biochemical studies (71) on whole zonules showed that it was proteinaceous, had a relatively high carbohydrate content, and was associated with a small amount of proteoglycan. Extraction studies, under a variety of conditions, yielded a complex mixture of soluble and relatively insoluble components, some of which were glycoproteins, and whose apparent molecular weights ranged from 24 kDa to very large molecules excluded on polyacrylamide gel electrophoresis. As would be predicted, amino acid analysis of zonules showed a relatively high content of acidic amino acids and a small amount of cysteine. Similar biochemical observations had been reported previously (72).

The distribution of antibodies raised in rabbits, to a sonicated mixture of bovine zonules was studied in ocular sections, in bovine nuchal ligament, and in chick aorta using an indirect immunoperoxidase technique (70,73). The antiserum reacted with the zonular fibers but not with the lens capsule or fibrils of the vitreous body, thus confirming that the antigens had not been derived from contamination with these structures. The antibodies also reacted with other elastic tissue fibers in the eye and with those in the calf nuchal ligament and the chick aorta (70).

In the latter study (70) the distribution of this obviously very heterogeneous antiserum was shown to be similar to that obtained with antiserum produced in animals immunized, by a single-stage injection schedule, with the "microfibrillar antigen" used by Kewley et al. (47). It seems inevitable that this antiserum, obtained with this antigenic mixture in this manner, would be even more heterogeneous than that used by Kewley et al. (47), which was obtained by a secondary immunization regime.

In spite of the obvious interest in comparing zonular microfibrils with elastin-associted microfibrils, considerable constraint must be exercised in interpreting the findings. Similar, or even greater, reservations must be held about the specificity of the zonular antiserum to those expressed above concerning the Kewley antiserum (see Section III.C). These findings are not proof of identity between these two microfibrillar structures, nor do they even necessarily establish that zonular microfibrils are part of the elastic tissue fiber system. However, further refinement of the immunological studies of the zonular constituents will be of considerable interest, particularly as to their relationship to vascular elastin-associated microfibrils.

The biochemical studies (71) again highlight the problems that arise when one attempts to use amino acid analysis of a heterogeneous mixture of proteins as an index of identity, in the absence of a unique marker amino acid.

4. *Other Soft Tissue Structures*

A wide variety of soft tissues have been shown to contain small amounts of typical elastic tissue. In some of these tissues significant changes have been observed in the proportions of amorphous and microfibrillar constituents, apparently in relation to location, function, and hormonal effects (see Ref. 1).

Small amounts of normal elastic tissue are seen in the tunica propria of the seminiferous tubule of the human testis. In Klinefelter's syndrome, this tissue exhibited a massive increase in microfibrils in the absence of amorphous elastic tissue, while in hypogonadotropic hypogonadism there was a reduced number of microfibrils in this region (74).

The ratio of amorphous elastic tissue to microfibrils in the small amounts of elastic tissue in the hip joint of the rat has been shown to be responsive to hormonal influences.

When female rats were treated with estrogens the ratio increased, whereas testosterone treatment or ovariectomy was associated with a reduction in this ratio (75).

The literature relating to the oxytalan fiber-containing periodontal ligament has recently been reviewed in detail (76).

C. In Elastic Cartilage

The cartilages of the external ear and the epiglottis differ from other forms of cartilage in having relatively large amounts of elastic tissue in the extracellular matrix. The evidence detailing the many apparently conflicting reports as to the presence of microfibrils in association with this elastic tissue has been reviewed in detail recently (1). In summary, one might conclude that typical microfibrils had been identified unequivocally in elastic cartilage only in the mouse pinna and only during the early postnatal period of development (77).

In the light of this evidence, and noting that ear cartilage elastin had been claimed to be genetically distinct from other types of elastin (78), Quintarelli and his colleagues undertook a study of elastogenesis in the pinna of young rabbits, aged 2 days and 16 days, both in vivo and in chondrocyte cultures (79). In a careful and detailed electron microscopic study, elastin-associated microfibrils could not be demonstrated in the elastic tissue that was laid down in both systems.

If confirmed, this finding would indicate that elastogenesis did not invariably follow the pattern described in soft tissues (9,14,26) and thus would cast doubt on the postulated integral roles of microfibrils in elastogenesis (9,10,14-16,50). An alternative explanation was that elastogenesis in the mouse pinna was exceptional (and thus of additional interest). However, this seemed less likely when later evidence established that the tropoelastin produced by auricular chondrocytes from hamsters (80) and pigs (81) was identical to that produced in aorta and nuchal ligament.

In view of the importance of these conclusions for the study of elastin-associated microfibrils, a series of experiments has recently been completed in this laboratory (Cleary, Fanning, and Prosser, unpublished data). These include studies of elastogenesis in the pinna of young rabbits and bovine fetuses. The results indicate that:

1. There are significant differences in the stage of development of ear cartilage at the time of birth in different species; e.g., bovine ear cartilage is much more highly developed at birth than is that of the rabbit.
2. In each species, the early stages of ear cartilage development (prenatal in bovines) involve the deposition of collections of typical elastin-associated microfibrils within which amorphous elastic material subsequently develops.
3. With further development and with the deposition of proteoglycan material in the matrix in close proximity to the elastic fibers the microfibrils become inaccessible to staining, so that they are no longer seen in the electron microscope.
4. Treatment of tissue sections with testicular hyaluronidase or with chondroitinase ABC removes the masking proteoglycan material and partially reveals elastin-associated microfibrils at the periphery of masses of developing amorphous elastic tissue in younger animals, but fails to do so in older animals.
5. Similar observations were made on elastogenesis in auricular chondrocyte cultures (see Section V.A.4).

It can therefore be concluded that elastogenesis in ear cartilage follows a pattern similar to that in soft tissues. It appears justified to conclude that elastogenesis in mammals and birds invariably involves elastin-associated microfibrils. No information is available concerning the chemical composition of elastin-associated microfibrils in ear cartilage elastin.

D. In Hyaline and Fibrous Cartilage

Gawlik (20) reported demonstrating oxytalan fibers in human meniscus fibrocartilage, and typical elastic fibers, with microfibrils, have been demonstrated electron microscopically in the fibrocartilage of human intervertebral disk (82). An interesting aspect of this study was the demonstration that, during postnatal growth, changes in the relative amounts of microfibrils and amorphous elastic tissue were observed in only one of the two regions of the disk (82).

Gawlik (20) also reported demonstrating elaunin fibers in hyaline cartilage, but later workers have reported an inability to identify elastic tissue or elastinlike microfibrils in this type of cartilage (reviewed in Ref. 83). Reconciliation of these findings has been achieved in recent studies on hyaline cartilage of rats. Oxytalan fibers were demonstrated in the extracellular matrix around the chondrocytes, and elaunin fibers were found in the transition layer of chondrocytes adjacent to the perichondrium (83). Oxytalan and elaunin fibers were also demonstrated along the periphery of collagen bundles in the fibrocartilaginous intervertebral disk. These studies, done on adult rats, used both light and electron microscopy after the tissues had been digested with testicular hyaluronidase. The authors were of the opinion that the removal of the proteoglycan from the matrix had permitted the identification of these microfibrillar elements. Typical elastic fibers were present in the perichondrium but were not seen in the cartilages (83).

V. ELASTOGENESIS AND ELASTIN-ASSOCIATED MICROFIBRILS IN VITRO

Under favorable conditions, elastic tissue is produced in vitro by a variety of cell types, including vascular smooth muscle cells, vascular endothelial cells, nuchal ligament fibroblasts, chick wing ligament fibroblasts, and auricular chondrocytes. In the absence of specific biochemical or immunological means of identifying microfibril production, it is necessary to undertake concurrent, high-resolution electron microscopy on these cultures if one is to comment on the production of microfibrils and on their relationship to the formation of amorphous elastic tissue. This has been done relatively infrequently.

Even so, a wealth of data is accumulating as different workers develop a multitude of culture systems and assay procedures, for their own diverse purposes. As these studies, of themselves, call for reproducible and stable culture conditions, and thus for adequate characterization of the culture systems, we are beginning to see reports of systematic studies of the effects of variations in a single parameter on the synthetic activity of stable, and predictable, cell culture systems. Although these findings are not necessarily applicable to other cell types or to other species (and in some instances we already know they are not), some general rules that describe factors governing the biosynthesis of extracellular matrix components are beginning to emerge, and notable exceptions to these rules are being recognized. Investigation of these exceptions often provides additional insights into regulatory mechanisms. From one such diversion we are beginning to recognize that physical contact with constituents of the extracellular matrix itself is capable of influencing the elastogenic activity of the matrix-producing cells.

There have been recent reviews of the extensive literature in this field (1,84). They should be consulted for more details. I propose to examine a selection of the data potentially relevant to production of elastin-associated microfibrils. I shall then attempt to derive some generalizations from these data.

A. Data from Individual Experiments

1. Vascular Smooth Muscle Cells

Factors Determined by the Donor Cells: Smooth muscle cells (SMCs) from many species, including guinea pigs (85), grow well as cell cultures in media supplemented with fetal calf serum, in the absence of additional supplements of ascorbic acid. In these systems they produce microfibrils first and later amorphous elastic tissue (by electron microscopic observation) (17,85-89) or elastin (by chemical or immunological assay) (39,90-92).

The production of elastic fibers, including amorphous elastic material, by cells in these systems is enhanced by

1. growing the cells at high density ($>0.5 \times 10^6$ cells/25 cm^2) (39,87-89,92-95);
2. gently aggregating the cells into masses or conglomerates (93);
3. growing the cells as tissue transplants, again in three-dimensional association (96);
4. using primary cultures (88,93,95);
5. using cells from young animals in whom active elastogenesis was occurring in vivo (85,87-89,93,95,96).

The Role of Tissue Sampling Within a Single Organ: Smooth muscle cells from the "aortic arch" region of bovine calves (of unspecified age) appear to differ from aortic SMCs of other species in that, as first or second subcultures in a careful study, they could not be maintained in cell cultures for longer than 21 days in the absence of ascorbate (94). Apparently conflicting results were obtained with thoracic aortic SMCs from "young male calves" that were maintained at the same cell density under similar culture conditions, with weekly subculture through 3-18 generations without ascorbate supplements (39). The latter cells were shown to be unable to form insoluble elastin.

These two studies draw attention to the importance of defining adequately the population of cells being sampled. The arch and upper thoracic aorta of the herbivores has a musculoelastic morphology that is very different from the thoracic aortic wall anatomy seen in man and laboratory animals (97). Attention is directed to the nonuniform chemical composition of the aortic wall in different animal species (98) and to the likelihood that smooth muscle cells at different locations have different synthetic capabilities in vivo. For in vitro studies this constitutes an additional variable which, in general, is ignored in tissue sampling and in interpreting the results.

The Response to Ascorbic Acid: There is additional evidence suggesting there may be differences between species in the effects of ascorbate on elastogenesis by aortic SMCs. While supplements of 50 μg of ascorbate/ml of medium have been found to switch most vascular SMCs in culture to predominantly collagen production at the expense of elastogenesis (94,99,100), two studies have suggested that SMCs from the thoracic aorta of young rats (one to five days old) may respond differently to ascorbate supplements at this dose (87,95). Both studies used primary cultures of aortic SMCs grown at high density in the presence of 20% FCS. Both reported the formation of significant amounts of amorphous elastic tissue. Elastin-associated microfibrils were present in both studies but in lesser amounts than usually seen in vivo or in cultures grown without ascorbate supplementation.

The effects of ascorbate supplements on elastogenesis have been elucidated to some extent by a recent systematic study of the response of rabbit aortic SMCs to varying doses of ascorbate. In this system, relatively small supplements of ascorbate (0.5 μg/ml), on alternate days, promoted maximal elastogenesis, while the larger doses ($\geq$ 2 μg/ml) promoted collagen production and reduced the formation of amorphous elastic tissue (89).

The Role of Serum Factors: While supplements of fetal calf serum (FCS) appear to promote elastogenesis in most SMC culture systems, supplementation with adult homologous serum has been shown to inhibit elastin cross-link formation (90). Although electron microscopy was not done in this study, the results may indicate that adult homologous serum inhibits formation of amorphous elastic tissue.

Effects of Repeated Subculture: Repeated subcultures of aortic SMCs from young guinea pigs (85) and monkeys (17) have been shown to produce cultures in which there is plentiful production of apparently normal elastin-associated microfibrils with only minor amounts of amorphous elastic tissue. Unfortunately, these cultures were plated at unspecified cell densities, so that population doublings cannot be estimated. Attempts to use such cultures as a source of microfibrils for biochemical studies have indicated the degree of complexity of this system (17).

Rat Heart SMCs: Rat heart SMCs have been grown in standard culture conditions and have been shown to produce collagen, elastin, glycoproteins, and proteoglycans (101). Scorbutic conditions were shown to favor deposition of cross-linked elastin (102,103). Electron microscopy was not undertaken.

2. *Vascular Endothelial Cells*

The Role of Cell Source and of Ascorbic Acid: Endothelial cells from vascular sources in several species have been shown to be capable of producing, in cell cultures, either cross-linked elastin or typical elastin with amorphous and microfibrillar components. These include human umbilical vein endothelium (104) and bovine calf pulmonary artery endothelium (105). Their behavior was, in general, similar to that described for vascular smooth muscle cells in culture. Rabbit aortic endothelial cells in ascorbate-supplemented cultures produced tropoelastin but did not produce cross-linked elastin (106). On the other hand, a rat lung endothelial cell line grown in cultures not supplemented with ascorbate produced a small amount of cross-linked elastin (107).

The Role of "Factors" in "Conditioned" Media: There have been reports that endothelial cells grown in monolayer culture synthesize and release one or more factors that inhibit the proliferation of SMCs in culture (108,109). They may also produce factors that stimulate SMCs (108,110). The influence of such activities on elastogenesis by SMCs does not appear to have been studied.

The reverse relationship has been examined recently with interesting results. Calf pulmonary artery endothelial cells grown in the presence of ascorbic acid supplements, at 50 μg/ml concentration, were unable to produce extracellular amorphous elastic tissue or cross-linked elastin in cultures supplemented with 10% FCS. However, when Fetal Bovine Serum Replacement (a commercial product derived from calf serum that had been "conditioned" by growth factors produced by aortic smooth muscle cell cultures) was substituted for the FCS, the endothelial cells produced both elastic tissue and cross-linked elastin. Electron microscopic observations, which would allow evaluation of effects on elastin-associated microfibrils, were, unfortunately, not made (105).

Endothelial Cell/Smooth Muscle Cell Interactions: The relationship between endothelial cells and vascular smooth muscle cells has been the subject of several investigations in tissue culture and has produced some interesting observations. Cloned bovine pulmonary artery endothelial cells have been grown on preformed layers of rat heart smooth muscle cells. These cells were supplemented daily with ascorbic acid and multiplied rapidly to reach a higher saturation density than similar cells grown on plastic. In this configuration, the endothelial cells attached to the underlying smooth muscle cells, and both cells were stable in cultures for up to 6 months. Under these conditions, the endothelial cells were not observed to form amorphous elastic tissue although significant amounts of amorphous elastic material were observed in the smooth muscle cell layer (111). These findings are not directly comparable to other observations concerning the interrelationship of these cell types, but they offer a novel approach to further investigations in an important area.

3. Nuchal Ligament Fibroblasts

Dissociation of Production of Microfibrils and Amorphous Elastic Tissue: Nuchal ligament fibroblasts in culture produce plentiful elastin-associated microfibrils (42,112,113). Such cultures have been used as a source of microfibrils for biochemical characterization. Whole microfibrils were partially purified by removing cell-related contaminants. The product was examined in the electron microscope, by enzyme digestion and by amino acid analysis (113). This could provide a useful source of material for more definitive experiments.

Fetal bovine nuchal ligament fibroblasts maintained in culture synthesize soluble elastin (114) but under usual culture conditions are unable to form elastic fibers or amorphous elastic tissue (42,112,114).

Matrix Factors in Elastogenesis: Nuchal ligament fibroblasts tend to detach from the surface of the plastic culture dish 5-15 days after confluence. If these detached aggregates or "cell knots" were maintained in culture, further cellular proliferation occurred, and after several weeks amorphous elastic tissue was observed, surrounded by beaded 11-nm-diameter microfibrils. When the cultures were supplemented with ascorbic acid, detachment, cellular knotting and further matrix production occurred, but microfibril production was reduced and amorphous elastic tissue was never observed (112).

These findings are reminiscent of those in transplanted SMCs (87) and suggest that a three-dimensional cellular arrangement and contact with the matrix could be important factors in promoting elastogenesis. Evidence that the matrix has such a role has been provided by an elegant series of experiments with bovine nuchal ligament cells cultured in tissue minces and on dead ligament tissue (114-116). These studies confirm that nuchal ligament fibroblasts can produce elastic fibers (identified by immunofluorescence) and insoluble elastin, but as electron microscopy was not undertaken, no comment can be made about changes in microfibrils in these conditions.

Other important relevant observations were made in these studies. Immature nuchal ligaments (from fetuses less than 140 days gestation) are not competent to produce elastin, but elastin gene expression can be activated in these young cells by growing them in physical contact with acellular ligament matrix from fetuses older than 180 days (116). These observations will need to be taken into account in planning experiments on the relationship of microfibrils to elastic fiber formation. These observations reinforce doubts that this writer has about the reported age of bovine fetus ("45 days") used by Sear et al. (42) in their original nuchal ligament tissue culture studies. At 45 days, the fetus is only 3.1 cm in crown-rump length, and the nuchal ligament is extremely difficult to identify, let alone dissect out. I suspect that the fetus used was either 145 days old or 45 cm crown-rump (or about 160 days gestation).

Immunochemical Attempts to Identify Microfibrillar Components in Cultures: Sear and his colleagues have made a detailed study of the glycoproteins synthesized by nuchal ligament fibroblasts in cultures. Using the "antimicrofibrillar" antiserum produced by Kewley et al. (47), they immunoprecipitated two glycoproteins of different sizes (M_r 150K and 300K) and chemical composition (42,43). These were called MFPI and MFPII. Subsequent studies have shown that MFPI is related to type VI collagen (45) and thus to CL glycoprotein (49,117,118), which has been shown not to be microfibrillar in location (46).

These studies highlight the need for electron microscopic monitoring of experiments concerned with microfibril production. It is clear that immunoelectron microscopic monitoring will be necessary in order to establish the chemical identify of microfibrils. The experience of Sear and his colleagues emphasizes the need to establish, rigorously, the monospecificity of antibodies (including monoclonal antibodies) used for this purpose. Even so, the work of Kewley and Sear and their colleagues has clearly indicated useful directions for further study.

4. Auricular Chondrocytes

The Role of Cellular Variables and of Ascorbic Acid: Auricular chondrocytes in monolayer cell cultures produce elastic tissue by a process that is not obviously different from that observed in other tissues. Thus, apparently typical elastin-associated microfibrils are deposited in the extracellular tissues in collections, and later amorphous elastic tissue is deposited among these microfibrils (119). As in other elastogenic cells in culture, addition of ascorbic acid (at 50 μg/ml concentration) stimulates collagen production and reduces elastogenesis with concomitant reduction in the numbers of 11-nm microfibrils seen in the cultures (120).

Elastic fiber formation is maximal in chondrocyte cultures from newborn and early neonatal rabbits (up to seven days of age). It declines with increasing donor age (119) and with increasing population doublings in culture (121,122). Cells from adult rabbit ear cartilage produce very little elastic tissue in culture (122).

Biochemical characterization of the insoluble residue (prepared by the rigorous autoclave/hot alkaline extraction method) from cultures of young rabbit ear chondrocytes has confirmed that they are forming significant amounts of elastin. Amino acid analysis of this material is similar to that of elastin prepared from ear cartilage and from other tissue sources, but there is a significant residual contamination with nonelastic protein. Formation of the elastin-specific cross-links isodesmosine and desmosine was confirmed by studies of [^{14}C-]lysine incorporation into the elastin. Electron microscopic examination of the elastic tissue forming in the cultures was not carried out (79). These findings have been confirmed in a more recent systematic biochemical study on rabbit ear chondrocytes in culture (123).

Growth of Chondrocytes as Intramuscular Transplants: Chondrocytes have been grown in a novel manner as transplants by intramuscular injection into rabbits both as autotransplants and as allogeneic transplants (120,122,124,125). In these circumstances they are capable of reconstituting the elastic cartilage. Chondrocytes from younger rabbits produced more ordered elastic cartilage than did those from older animals (125), and freshly isolated chondrocytes were more effective than chondrocytes that had undergone more than eight population doublings in culture (121).

In the transplants, the intercellular matrix did not differ from that in the intact ear. The pattern of elastogenesis was similar to that observed in soft tissues. However the microfibrillar components of the developing elastic fibers formed within one week of transplantation and were more prominent than in the intact ear cartilage (120).

Fetal Bovine Ear Cartilage Cells in Culture: Virtually all the reported experiments have been done on rabbit ear cartilage. We have made a detailed study of fetal bovine ear cartilage (Prosser and Cleary, unpublished data). It has been shown that fetal bovine chondrocytes can be maintained in cultures without ascorbic acid supplements, for prolonged periods (up to 56 days). They are capable of producing typical amorphous elastic fibers with typical peripheral elastin-associated microfibrils (Fig. 5). When ascorbic acid was added to the cultures, there was much greater collagen production, but small amounts of amorphous elastic tissue with microfibrillar components continued to be formed.

The ear cartilage seems to be a promising system for the study of elastogenesis, particularly in relation to the interaction between the cells and the more simple matrix of this structure.

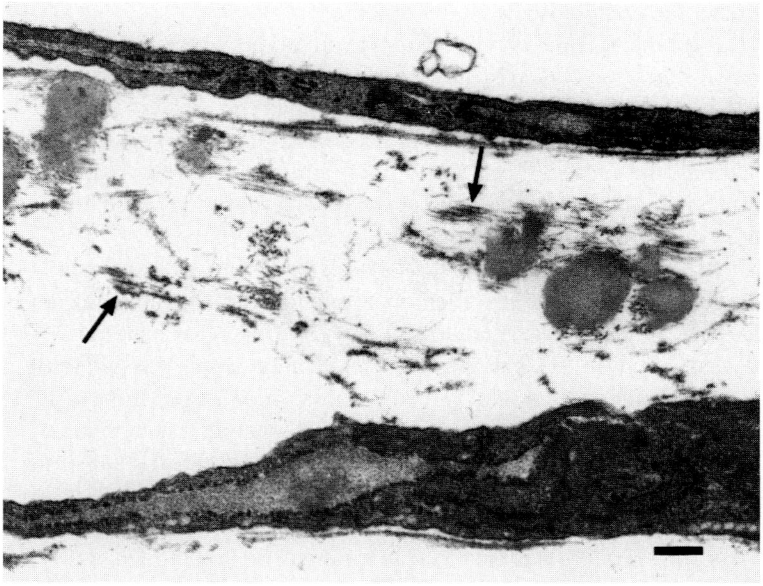

Figure 5 Electron micrograph of auricular chondrocytes from a 210-day-old fetal calf, maintained in culture, under scorbutic conditions, for 35 days after confluence. The cells are attentuated and relatively widely separated. A number of small masses of amorphous elastic tissue are seen within the intercellular matrix. Collections of microfibrils (arrows) are prominent and related to amorphous elastic tissue. Occasional striated small collagen fibers are present. Bar, 250 nm. (Courtesy of Dr. I. W. Prosser.)

VI. CONCLUDING REMARKS

It is becoming increasingly clear that the extracellular matrix has a major influence on cellular biosynthetic activity and thus on tissue organization. As attention focuses on the matrix components and as techniques evolve that will facilitate analysis of its constituents, it is apparent that we have stumbled into a whole new world of microconstituents and microorganization. Prominent within this environment are microfibrillar structures of a wide range of diameters, within which can be identified a subpopulation(s) of tubular-appearing, glycoprotein-containing microfibrils, ranging from 5 to 18 nm in diameter and with a periodic staining pattern in conventional electron micrographs. These constituents appear to contain covalent cross-links and are relatively insoluble, but they are readily susceptible to protease digestion. Although these properties may be well suited to their biological roles, they make the microfibrils difficult to extract and to characterize.

We have seen that many different proteins can assume a similar microfibrillar morphology. We must therefore resist the temptation to classify and identify microfibril types on the basis of this morphology alone. From this review we have identified a group of microfibrils that are distinguished by a regular intimate physical relationship with the protein elastin in elastic fibers. These microfibrils appear to have a central role in the process of elastogenesis, but the nature of this role and the exact relationships are, as yet, unknown. Microfibrils that are at present indistinguishable from these elastin-associated microfibrils are seen to occur in many other tissues, sometimes in apparent physical continuity with those associated with elastic fibers. Even here we must acknowledge the possibility, indeed the likelihood, that these microfibrils may be chemically, and/or physically, different from the elastin-associated microfibrils.

We have noted that these microfibrils appear to form in situations where the tissues are subjected to repeated stresses in several planes, which could point to a common functional role or response, including anchoring epithelia to the underlying tissues. While it is possible, even probable, that these responses will involve constituents that are common to many of these situations, the variety of cell types that produce microfibrils and the presence of microfibrillar constituents in many diseases affecting the matrix should caution against assuming that they are all chemically identical, or even related.

We have seen that these microfibrils are intimately associated with a variety of constituents derived from the serum, with cellular membrane and basement membrane components, with collagens and proteoglycans, and doubtless with many other materials. We must therefore acknowledge that, until a unique amino acid marker is specifically identified with some or all microfibrillar groups, amino acid analysis will not serve to identify microfibrillar proteins. Furthermore, the complexity of the microfibrillar environment imposes obligations to establish rigorously the purity of putative microfibrillar constituents rather than acting as an excuse for not establishing their homogeneity.

There is an increasing body of evidence that many cells will produce typical microfibrillar constituents in vitro. Already it is established that many factors influence the production, in these conditions, of microfibrillar components as well as of other matrix materials. Although it is tempting to assume that these are simplified circumstances, that the range of contaminating matrix components is somehow reduced, and that the microfibrillar output is probably homogeneous, it must be acknowledged that all these are assumptions, that some or all of them are patently false, and that each of them must eventually be tested and established or discarded.

There is already sufficient high-resolution microscopic evidence to indicate that elastin-associated microfibrils are themselves complex structures with internal substructure. Critical examination of the similar-appearing microfibrils of skin, the periodontal ligament, and the ocular zonule suggest (Fanning and Cleary, unpublished data) that there are subtle differences in the organization and substructure of these microfibrils.

The way forward appears to be through the antigenicity of microfibrillar constituents. We have seen that it is difficult to achieve, and to prove, monospecificity of polyclonal antibodies raised against putative microfibrillar constituents. However, provided the possible limitations of this tool are taken into account, these antibodies offer the prospect of progress by their use as a screening method that will enable progressive constriction of the focus to the microfibrillar substituents. While the use of monoclonal antibodies is intuitively attractive, they too have their limitations for immunolocalization, and these must be kept constantly in mind. Ultimately, it is necessary to establish rigorous immunological identity between the putative microfibrillar substituents and a particular subpopulation of microfibrils. This involves high-resolution immunoelectron microscopy. Fortunately, there have been new technological developments in this field involving improved tissue preservation (with low-temperature embedding media) and new antibody-labeling procedures. In addition, there may be some advantages to microfibrillar research from the development of specific lectins (with electron-dense markers). It is true that these methods are being shown to involve further technical problems and limitations on interpretation. Even so, there is promise that the next decade will witness great advances in our understanding of the composition and function of elastin-associated microfibrils, of those microfibrils of similar morphology that have been considered in this chapter, and of the role that these structures play in a very extensive range of common and important diseases that affect the connective tissue matrix.

ACKNOWLEDGMENTS

Original contributions from this laboratory were supported by grants from the Australian Research Grants Committee, The National Health and Medical Research Council, and The National Heart Foundation of Australia.

REFERENCES

1. E. G. Cleary and M. A. Gibson, *Int. Rev. Connect. Tiss. Res., 10*:97-211 (1983).
2. F. N. Low, *Anat. Rec., 142*:131-137 (1962).
3. R. G. Frederickson and F. N. Low, *Am. J. Anat., 130*:347-376 (1971).
4. B. W. Mayer, E. D. Hay, and R. O. Hynes, *Dev. Biol., 82*:267-286 (1981).
5. J. C. Fanning, M. A. Gibson, and E. G. Cleary (submitted for publication).
6. R. G. Frederickson, D. E. Morse, and F. N. Low, *Am. J.Anat., 150*:1-34 (1977).
7. E. S. Cohen and A. Calkins, *Nature (London), 183*:1202-1203 (1959).
8. R. Ross, *J. Histochem. Cytochem., 21*:199-208 (1973).
9. T. K. Greenlee, Jr., R. Ross, and J. L. Hartman, *J. Cell Biol., 30*:59-71 (1966).
10. R. Ross and P. Bornstein, *J. Cell Biol., 40*:366-381 (1969).
11. I. W. Prosser, M. A. Gibson, and E. G. Cleary, *Aust. J. Exp. Biol. Med. Sci., 62*: 485-505 (1984).
12. J. C. Fanning, N. G. Yates, and E. G. Cleary, *Micron, 12*:339-348 (1981).
13. J. C. Fanning and E. G. Cleary, *J. Histochem. Cytochem., 33*:287-294 (1985).

14. W. H. Fahrenbach, L. B. Sandberg, and E. G. Cleary, *Anat. Rec., 155*:563-576 (1966).
15. J. Waisman and W. H. Carnes, *Am. J. Pathol., 51*:117-135 (1967).
16. R. Ross and P. Bornstein, in *Chemistry and Molecular Biology of the Intercellular Matrix* (E. A. Balazs, ed.), Vol. 1, Academic Press, New York, 1970, pp. 641-655.
17. L. W. Muir, P. Bornstein, and R. Ross, *Eur. J. Biochem., 64*:105-114 (1976).
18. P. Bornstein, in *The Comparative Molecular Biology of the Extracellular Matrices* (H. C. Slavkin, ed.), Academic Press, New York, 1972, pp. 355-356.
19. H. M. Fullmer and R. D. Little, *J. Histochem. Cytochem., 6*:425-430 (1958).
20. Z. Gawlik, *Folia Histochem. Cytochem., 3*:233-251 (1965).
21. I. M. Braverman and E. Fonferko, *J. Invest. Dermatol., 78*:434-443 (1982).
22. G. Cotta-Pereira, F. G. Rodrigo, and S. Bittencourt-Sampaio, *J. Invest. Dermatol., 66*:143-148 (1976).
23. G. Cotta-Pereira, F. G. Rodrigo, and J. F. David-Ferreira, *Biol. Cellulaire, 31*:297-302 (1978).
24. H. M. Fullmer, *J. Histochem. Cytochem. 8*:290-295 (1960).
25. H. E. Karrer, *J. Ultrastruct. Res., 2*:96-121 (1958).
26. M. D. Haust, *Amer. J. Pathol., 47*:1113-1137 (1965).
27. D. S. Jackson, L. B. Sandberg, and E. G. Cleary, *Biochem. J., 96*:813-817 (1965).
28. E. G. Cleary, D. S. Jackson, and L. B. Sandberg, in *Biochemistry and Physiology of Connective Tissue* (P. Comte, ed.), Soc. Ormeco et Imprimerie du Sud Est, Lyon, 1966, pp. 167-172.
29. D. S. Jackson, L. B. Sandberg, and E. G. Cleary, *Nature (London), 210*:195-196 (1966).
30. E. G. Cleary, L. B. Sandberg, and D. S. Jackson, *J. Cell Biol., 33*:469-479 (1967).
31. Z. T. Wirtschafter, E. G. Cleary, D. S. Jackson, and L. B. Sandberg, *J. Cell Biol., 33*:481-488 (1967).
32. G. R. Reeck and L. Fisher, *Int. J. Peptide Protein Res., 5*:109-117 (1973).
33. L. Robert, J. Parbelas, P. Oudea, A. Zweimbaum, and B. Robert, in *Structure and Function of Connective and Skeletal Tissue* (S. Fitton Jackson, R. D. Harkness, S. M. Partridge, and G. R. Tristram, eds.), Butterworth, London, 1965, pp. 406-412.
34. M. Moczar, E. Moczar, and L. Robert, in *Méthodologie de la structure et du métabolisme des conjuguées* (A. M. Robert and L. Robert, eds.), CNRS, Paris, 1974, pp. 445-456.
35. R. Bach and J. P. Bentley, *Connect. Tiss. Res. 7*:185-196 (1980).
36. I. Wolff, W. Fuchswans, M. Weiser, H. Furthmayr, and R. Timpl, *Eur. J. Biochem., 20*:426-431 (1971).
37. A. Kadar, B. Robert, and L. Robert, *Path-Biol., 21*(suppl):80-88 (1973).
38. K. G. McCullagh, S. Derouette, and L. Robert, *Exp. Mol. Pathol., 18*:190-201 (1973).
39. E. Schwartz, A. M. Adamany, and O. Blumenfeld, *Biochim. Biophys. Acta, 624*:531-544 (1980).
40. A. Serafini-Fracassini, G. Ventrella, J. M. Field, N. I. Onyezili, and R. Griffiths, *Biochemistry, 20*:5424-5429 (1981).
41. G. M. Bressan, I. Castellani, A. Colombatti, and D. Volpin, *J. Biol. Chem., 258*:13262-13267 (1983).
42. C. H. J. Sear, M. A. Kewley, C. J. P. Jones, M. E. Grant, and D. S. Jackson, *Biochem. J., 170*:715-718 (1978).
43. C. H. J. Sear, M. E. Grant, and D. S. Jackson, *Biochem. J. 194*:587-598 (1981).
44. J. M. Burke and R. Ross, *Int. Rev. Connect. Tiss. Res., 8*:119-157 (1979).
45. K. R. Knight, S. Ayad, C. A. Shuttleworth, and M. E. Grant, *Biochem. J., 220*:395-403 (1984).

46. M. A. Gibson and E. G. Cleary, *Collagen Rel. Res., 3*:469-488 (1983).
47. M. A. Kewley, F. S. Steven, and G. Williams, *Immunology, 32*:483-489 (1977).
48. M. A. Kewley, F. S. Steven and G. Williams, *Immunology, 33*:381-386 (1977).
49. M. A. Gibson and E. G. Cleary, *Biochem. Biophys. Res. Comm., 105*:1288-1295 (1982).
50. E. G. Cleary, J. C. Fanning, and I. W. Prosser, *Connect. Tiss. Res., 8*:161-166 (1981).
51. A. Serafini-Fracassini, J. M. Field, M. Spina, S. Garbisa, and R. J. Stuart, *J. Ultrastruct. Res., 65*:1-12 (1978).
52. R. E. Shadwick and J. M. Gosline, *Science, 213*:759-761 (1981).
53. M. Spina, S. Garbisa, J. M. Field, and A. Serafini-Fracassini, *Arch. Biochem. Biophys., 192*:430-437 (1979).
54. H. Sage and W. R. Gray, *Comp. Biochem. Physiol., 64B*:313-327 (1979).
55. J. A. Pierce and R. V. Ebert, *Thorax, 20*:469-476 (1965).
56. J. S. Brody and C. Vaccaro, *Fed. Proc., 38*:215-223 (1979).
57. J. G. Clark, C. Kuhn III, J. A. McDonald, and R. P. Mecham, *Int. Rev. Connect. Tiss. Res., 10*:259-331 (1983).
58. J. S. Brody, H. Kagan, and A. Manalo, *Am. Rev. Resp. Dis., 120*:1289-1295 (1979).
59. T. Kobayasi, *J. Invest. Dermatol., 68*:389-390 (1977).
60. L. T. Smith, K. A. Holbrook, and P. H. Byers, *J. Invest. Dermatol., 79*:93s-104s (1982).
61. A. Charles, *Br. J. Dermatol., 73*:57-60 (1961).
62. D. P. Varadi, *J. Invest. Dermatol., 59*:238-246 (1972).
63. T. Kobayasi, *Acta derm.-venereol., 48*:303-312 (1968).
64. T. Tsuji, *Br. J. Dermatol., 102*:413-417 (1980).
65. T. Tsuji, *J. Invest. Dermatol., 77*:452-457 (1981).
66. A. Kanai and H. E. Kaufman, *Invest. Ophthalmol., 11*:816-821 (1972).
67. M. J. Hogan, J. A. Alvarado, and J. E. Weddell, *Histology of the Human Eye. An Atlas and Textbook*, W. B. Saunders, Philadelphia, 1971, p. 303.
68. T. Iwamoto, *Invest. Ophthalmol., 3*:144-156 (1964).
69. G. Raviola, *Invest. Ophthalmol., 10*:851-869 (1971).
70. B. W. Streeten and P. A. Licari, *Invest. Ophthalmol. Vis. Sci., 24*:667-681 (1983).
71. B. W. Streeten, D. A. Swann, P. A. Licari, M. R. Robinson, S. A. Gibson, N. J. Marsh, J-P. Vergnes, and I. L. Freeman, *Invest. Ophthalmol. Vis. Sci. 24*:119-123 (1983).
72. E. Buddecke and J. Wollensak, *Z. Naturforschg., 21*:337-341 (1966).
73. B. W. Streeten, P. A. Licari, A. A. Marucci, and R. M. Dougherty, *Invest. Ophthalmol. Vis. Sci., 21*:130-135 (1981).
74. J. Wiedmar-Bridel, A. Vogel, and C. Hedinger, *Virchows Arch. B Cell Path., 27*:267-277 (1978).
75. G. D. Rooker, G. N. Steins, L. J. Tate, and F. W. Keeley, *Clin. Orthop. Rel. Dis., 152*:292-295 (1980).
76. C. A. Shuttleworth and J. W. Smalley, *Int. Rev. Connect. Tiss. Res., 10*:211-247 (1983).
77. C. F. Sanzone and E. J. Reith, *Am. J. Anat., 146*:31-72 (1976).
78. D. A. Keith, M. A. Paz, P. M. Gallop, and M. J. Glimcher, *J. Histochem. Cytochem., 25*:1154-1162 (1977).
79. G. Quintarelli, B. C. Starcher, A. Vocaturo, F. D. Gianfilippo, L. Gotte, and R. P. Mecham, *Connect. Tiss. Res., 7*:1-19 (1979).
80. P. Heeger and J. Rosenbloom, *Connect. Tiss. Res., 8*:21-25 (1980).
81. J. A. Foster, C. B. Rich, and M. D. DeSa, *Biochim. Biophys. Acta, 626*:383-389 (1980).

82. J. A. Buchwalter, R. P. Cooper, and J. A. Maynard, *J. Bone Jt. Surg., 58A*:73-76 (1976).
83. G. Cotta-Pereira, L. M. Del-Caro, and G. S. Montes, *Acta anat., 119*:80-85 (1984).
84. J. A. Foster, C. B. Rich, and S. R. Carr, *Int. Rev. Connect. Tiss. Res., 10*:65-95 (1983).
85. R. Ross, *J. Cell Biol., 50*:172-186 (1971).
86. A. S. Daoud, K. E. Fritz, J. Singh, J. M. Augustyn, and J. Jarmolych, *Adv. Exp. Med. Biol., 43*:281-298 (1974).
87. A. Hinek and J. Thyberg, *J. Ultrastruct. Res., 60*:12-20 (1977).
88. P. Toselli, L. L. Salcedo, P. Oliver, and C. Franzblau, *Connect. Tiss. Res., 8*:231-239 (1981).
89. B. Faris, R. Ferrera, P. Toselli, J. Nambu, W. A. Gonnerman, and C. Franzblau, *Biochim. Biophys. Acta, 797*:71-75 (1984).
90. A. S. Narayanan, L. B. Sandberg, R. Ross, and D. L. Layman, *J. Cell Biol., 68*:411-419 (1976).
91. B. Faris, L. L. Salcedo, V. Cook, L. Johnson, J. A. Foster, and C. Franzblau, *Biochim. Biophys. Acta, 418*:93-103 (1976).
92. R. Snider, B. Faris, V. Verbitzki, R. Moscaritolo, L. L. Salcedo, and C. Franzblau, *Biochemistry, 20*:2614-2618 (1981).
93. S. Moskalewski, M. Konwinski, and A. Hinek, *Anat. Embryol., 150*:113-122 (1976).
94. J. Mazurkowitz, D. W. Vaughan, and C. Franzblau, *In Vitro, 16*:337-345 (1980).
95. B. W. Oakes, A. C. Batty, C. J. Handley, and L. B. Sandberg, *Eur. J. Cell Biol., 27*:34-46 (1982).
96. A. Hinek, J. Thyberg, and U. Friberg, *Cell Tiss. Res., 172*:59-79 (1976).
97. E. G. Cleary, Thesis for M. D., University of Sydney, 1963.
98. E. G. Cleary, *Adv. Exp. Med. Biol., 79*:577-588 (1977).
99. D. M. Dunn and C. Franzblau, *Biochemistry, 21*:4195-4202 (1982).
100. L. L. Salcedo and C. Franzblau, *In Vitro, 17*:114-120 (1981).
101. P. A. Jones, T. Scott-Burden, and W. Gevers, *Proc. Natl. Acad. Sci. USA, 76*:353-357 (1979).
102. T. Scott-Burden, P. J. Davies, and W. Gevers, *Biochem. Biophys. Res. Commun., 91*:739-746 (1979).
103. Y. A. De Clerck and P. A. Jones, *Biochem. J., 186*:217-225 (1980).
104. E. A. Jaffe, R. Minick, B. Adelman, C. G. Becker, and R. Nachman, *J. Exp. Med., 144*:209-225 (1976).
105. R. P. Mecham, J. Madaras, J. A. McDonald, and U. Ryan, *J. Cell Physiol., 116*:282-288 (1983).
106. W. H. Carnes, P. A. Abraham, and V. Buonassisi, *Biochem. Biophys. Res. Commun., 90*:1393-1399 (1979).
107. J. O. Cantor, S. Keller, M. S. Parshley, T. V. Darnule, A. T. Darnule, J. M. Cerreta, G. M. Turino, and I. Mandl, *Biochem. Biophys. Res. Commun., 95*:1381-1386 (1980).
108. J. J. Castellot, Jr., M. L. Addonizio, R. Rosenberg, and M. I. Karnovsky, *J. Cell Biol., 90*:372-379 (1981).
109. Ch. Willems, G. C. B. Astaldi, Ph. G. de Groot, M. C. Janssen, M. D. Gonsalvez, W. P. Zeijlemaker, J. A. van Mourik, and W. G. van Aken, *Exp. Cell Res., 139*:191-197 (1982).
110. C. Gajdusek, P. DiCorleto, R. Ross, and S. M. Schwartz, *J. Cell Biol., 85*:467-472 (1980).
111. P. A. Jones, *Proc. Natl. Acad. Sci. USA, 76*:1882-1886 (1979).
112. C. J. P. Jones, C. H. J. Sear, and M. E. Grant, *J. Pathol., 131*:35-53 (1980).

113. S. I. Lamberg, D. C. Poppke, and B. R. Williams, *Connect. Tiss. Res., 8*:1-8 (1980).
114. R. P. Mecham, G. Lange, J. Madaras, and B. Starcher, *J. Cell Biol., 90*:332-338 (1981).
115. R. P. Mecham, *Connect. Tiss. Res., 8*:155-160 (1981).
116. R. P. Mecham, J. G. Madaras, and R. M. Senior, *J. Cell Biol., 98*:1804-1812 (1984).
117. M. A. Gibson and E. G. Cleary, *J. Biol. Chem., 260*:11149-11159 (1985).
118. R. Jander, D. Troyer, and J. Rauterberg, *Biochemistry, 23*:3675-3681 (1984).
119. S. Moskalewski, *Am. J. Anat., 146*:443-448 (1976).
120. J. Thyberg and A. Hinek, *Cell Tiss. Res., 180*:341-356 (1977).
121. S. Moskalewski, I. Adamiec, and A. Golaszewska, *Amer. J. Anat., 155*:339-348 (1979).
122. S. Moskalewski, *Connect. Tiss. Res., 8*:171-174 (1981).
123. K. Madsen, S. Moskalewski, K. von der Mark, and U. Friberg, *Dev. Biol., 96*:63-73 (1983).
124. J. Kawiak, S. Moskalewski, and A. Hinek, *Acta anat., 76*:530-544 (1970).
125. S. Moskalewski and E. Rybicka, *Acta anat., 97*:231-240 (1977).

4

Advances in the Biochemistry of Proteoglycans

Jeremiah E. Silbert
Veterans Administration Outpatient Clinic,
Harvard University, and Brigham and Women's Hospital,
Boston, Massachusetts

The proteoglycans and glycosaminoglycans represent a group of compounds that are related to one another by common structural characteristics. Their trivial names are: hyaluronic acid, chondroitin, chondroitin 4-sulfate (formerly called chondroitin sulfate A), chondroitin 6-sulfate (formerly called chondroitin sulfate C), dermatan sulfate (formerly called chondroitin sulfate B), heparin, heparan sulfate (or heparitin sulfate), and keratan sulfate.

These substances play an important role as part of the supporting matrix of connective tissue. They function in maintaining salt and water distribution and may control a "domain," so that a volume of water as much as 1000 times the volume of the proteoglycan can be contained within the molecule. In addition, these molecules are present on cell surfaces where they appear to play a role in cell-cell and cell-matrix interactions.

I. TERMINOLOGY

Historically, the term "mucopolysaccharide" was applied to these substances in order to describe a polysaccharide material that was viscous or mucinous. Initially, only the structure of the polysaccharide (glycosaminoglycan) portions was recognized, but subsequently, it became clear that the polysaccharides are usually linked covalently to protein. This complication in structure led to further difficulties in nomenclature. Because of historic and widespread use, the term mucopolysaccharide continues to be used on occasion as an all-inclusive name. Thus, mucopolysaccharide is sometimes used when referring to the polysaccharide portion alone, or when referring to the entire polysaccharide-protein (proteoglycan) combination. The term is not rigorously defined.

The term "glycosaminoglycan" was introduced in order to clarify some of the confusion over the use of "mucopolysaccharide." This term is descriptive of the polysaccharide (glycan) structure containing hexosamines (glycosamino) and does not include the protein moiety. The compounds listed above by trivial name are the only substances in vertebrates that satisfy these structural criteria. However, some substances in lower animals and bacteria fulfill the requirements of being a glycosaminoglycan. These compounds are, in general, clearly separate from the vertebrate compounds by virtue of totally different function, location, etc.

Since hyaluronic acid may not be linked covalently to protein, the term "proteoglycan" should not be applied except in those instances where there is an unequivocal protein component.

II. STRUCTURE

A. Glycosaminoglycan

Glycosaminoglycans consist of hexosamine (either glucosamine or galactosamine), alternating with another sugar (glucuronic acid, iduronic acid, or galactose). Although an individual chain may have a mixture of uronic acids (i.e., some glucuronic acid and some iduronic acid), the hexosamine will always be all glucosamine or all galactosamine, and there are no descriptions of glycosaminoglycan chains with mixtures of galactose and uronic acid. Because the hexosamine occurs as every other sugar, it is most convenient to think of the glycosaminoglycans as compounds consisting of multiple disaccharide repeating units, as shown in Figure 1. However, heparin and heparan sulfate have a more complicated structure and cannot be adequately represented in this fashion. A representative section of these compounds is shown in Figure 2. Variations in sulfate content and the uronic acid occur throughout the chain. In addition, heparan sulfate has variations in the amount of *N*-sulfate and *N*-acetyl.

The glycosaminoglycan chains of the various compounds range in size from approximately 15 disaccharide units (several thousand daltons) to 5000 disaccharide units (several million daltons) and are linear (unbranched). All hexosamines are substituted on the amino group with either acetyl or sulfate. In addition, most glycosaminoglycans have ester *O*-sulfate on one or both sugars of a repeating unit. All the glycosaminoglycans are highly anionic by virtue of their sulfate groups (as many as three per disaccharide unit) and/or their uronic acids. The sugars are linked by α or β glycosidic bonds and are 1,3 or 1,4, as detailed in Table 1.

B. Oligosaccharide Linkage

The glycosaminoglycan portions of chondroitin 4-sulfate and 6-sulfate, dermatan sulfate, heparin, and heparan sulfate are each linked at their reducing end to a trisaccharide consisting of galactosyl-galactosyl-xylose (1). Xylose is glycosidically linked to the hydroxyl of serine in the protein chain; the first sugar of the glycosaminoglycan chain (linked to the galactose) is always glucuronic acid. This first glucuronic acid could be considered to be a part of the linkage oligosaccharide since, biosynthetically, it is added by a specific enzyme different from the enzyme that is involved in the incorporation of glucuronic acid into the polysaccharide chain (2).

The linkage of keratan sulfate to core protein is different from that of the other glycosaminoglycans. There are two forms: Corneal keratan sulfate is of the mucin type, linked by *N*-acetylglucosamine to asparagine residues in the core protein, while skeletal keratan sulfate is linked by *N*-acetylgalactosamine to serine or threonine residues in the core protein. In addition, there is galactose attached to the *N*-acetylgalactosamine, and sialic acid attached to the galactose.

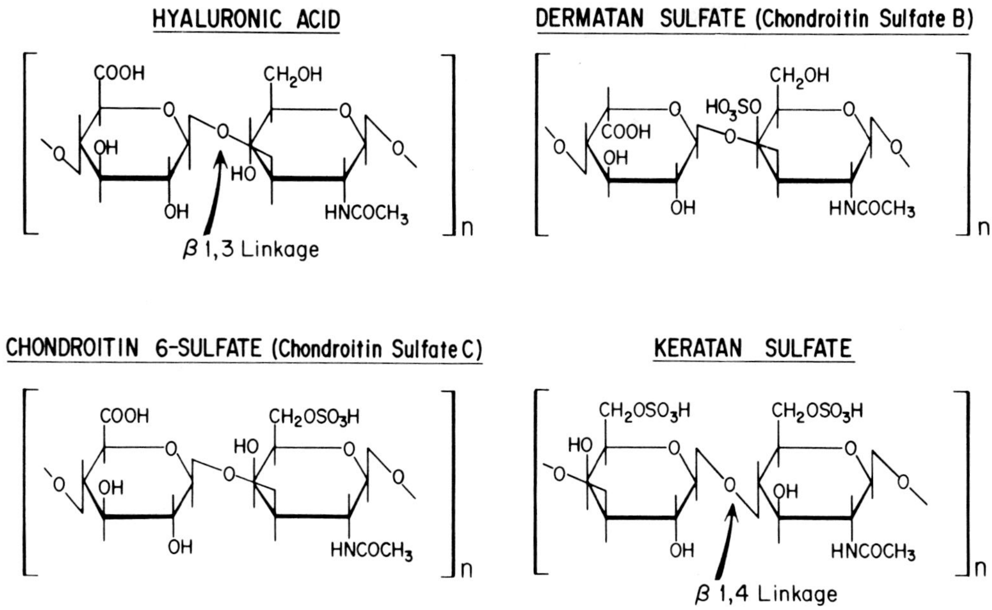

Figure 1 Disaccharide repeating units of glycosaminoglycans.

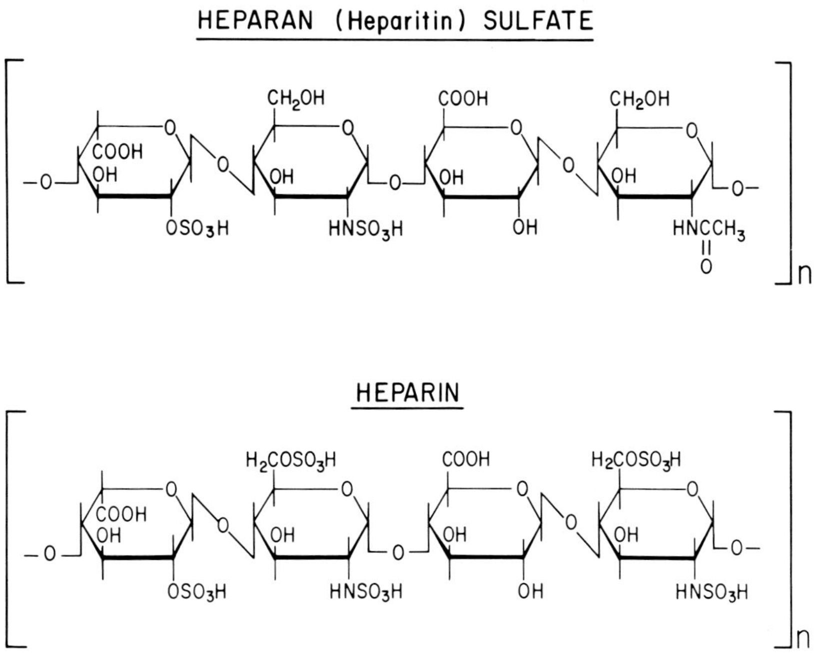

Figure 2 Typical oligosaccharide structures of heparin and heparan sulfate.

Table 1 Components of the Glycosaminoglycan Portion of Proteoglycans

	Approximate molecular weight of glycosamino-glycan chains	Major component sugars	Location of sulfate	Linkage
Hyaluronic acid	5-50 × 10^5	*N*-acetylglucosamine,	–	β-1,4
		glucuronic acid		β-1,3
Chondroitin 4-sulfate	2-6 × 10^4	*N*-acetylgalactosamine,	4	β-1,4
(chondroitin sulfate A)		glucuronic acid		β-1,3
Chondroitin 6-sulfate	2-6 × 10^4	*N*-acetylgalactosamine,	6	β-1,4
(chondroitin sulfate C)		glucuronic acid		β-1,3
Dermatan sulfate	2-6 × 10^4	*N*-acetylgalactosamine,	4	β-1,4
(chondroitin sulfate B)		iduronic acid,	2	α-1,3[a]
		glucuronic acid (trace)		β-1,3
Heparin	0.5-4 × 10^4	Glucosamine,	3,6,N	α-1,4
		glucuronic acid,		β-1,4
		iduronic acid	2	α-1,4[a]
Heparan sulfate	1-5 × 10^4	Glucosamine,	6,N	α-1,4
		N-acetylglucosamine,		α-1,4
		glucuronic acid,		β-1,4
		iduronic acid	2	α-1,4[a]
Keratan sulfate	0.5-2 × 10^4	*N*-acetylglucosamine,	6	β-1,3
		galactose	6	β-1,4

[a]These linkages are identical to the β-1,3 or β-1,4 linkages found in hyaluronic acid and chondroitin sulfates or heparin. However, iduronic acid has an L rather than a D configuration, which results in these bonds being designated as α rather than β.

C. Proteoglycan

The proteoglycan structure has mainly been described in detail for cartilage (3); however, proteoglycans from other tissues have considerable similarities to this structure. In cartilage, both chondroitin sulfate and keratan sulfate glycosaminoglycan chains of 1-5 × 10^4 mol. wt. are connected through the above-described oligosaccharide linkages to the same protein core molecule (2-3 × 10^5 mol. wt.), forming a proteoglycan unit, or monomer. Each proteoglycan may have 50 or more glycosaminoglycan chains attached to this core in a "bottle brush" configuration (Figure 3) so that the total molecular weight may be more than 10^6. The chains may vary in size and sulfate content within a single proteoglycan unit, resulting in a high degree of heterogeneity. In addition, there may be varying amounts of *O*-linked oligosaccharides and *N*-linked mannose-rich oligosaccharides. The structures of the various core proteins have not been defined, even to the extent of whether or not there are repeating sequences.

As many as 100 individual proteoglycan monomers in turn can be attached noncovalently around a core of hyaluronic acid to form aggregates of 100-150 × 10^6 mol. wt. (3,4). The proteoglycans are spaced along the chain at intervals of 10 disaccharide units, and a separate protein has been shown to be a stabilizer of the noncovalent link. The aggregate can be dissociated to proteoglycan monomer by high salt concentration, and at least part can be reaggregated by lowering the salt concentration.

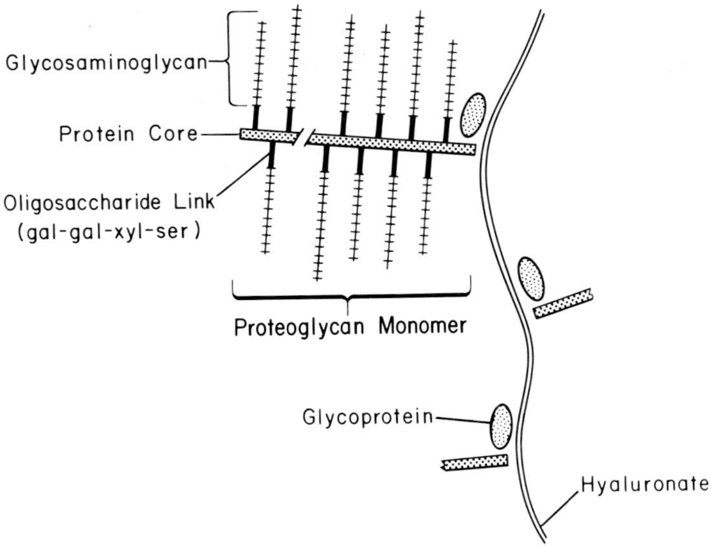

Figure 3 Proteoglycan monomer and aggregate.

III. BIOSYNTHESIS

A. Precursor Activation

The synthesis of the glycosaminoglycan portion of proteoglycans proceeds through uridine sugar nucleotide precursors. Sugar nucleotides are formed by a pyrophosphorylase reaction, such as:

UTP + *N*-acetylglucosamine-1-P

$\rightleftharpoons$

UDP-*N*-acetylglucosamine + pyrophosphate

Thus, as shown in Figure 4, the key intermediate compounds related to glycosaminoglycans are UDP-glucose and UDP-*N*-acetylglucosamine. Both can be formed by reaction of UTP with a sugar-1-P (either glucose-1-P or *N*-acetylglucosamine-1-P) in the presence of homogenates from proteoglycan-producing tissues. UDP-glucose can be oxidized to form UDP-glucuronic acid (5), which can in turn be decarboxylated to form UDP-xylose (6). Both UDP-glucose and UDP-*N*-acetylglucosamine can undergo epimerization at the C-4 position to form UDP-galactose and UDP-*N*-acetylgalactosamine (7), respectively. Thus, the synthetic steps in the formation of all the sugar nucleotides necessary for synthesis are known to occur in proteoglycan-producing tissues. It is possible that sugar nucleotides with bases other than uridine might be involved in proteoglycan synthesis, but there is no evidence for this at present.

The sugar moieties of the glycosaminoglycans have been shown by use of labeled sugars to arise from glucose without any rearrangement of the label (8). The pathways are shown in Figure 4. In addition, alternate pathways have been found in liver involving the direct phosphorylation of galactose and phosphorylation and acetylation of glucosamine with subsequent conversion to the appropriate sugar nucleotides (9,10). These pathways might also be involved in other tissues but have not as yet been demonstrated.

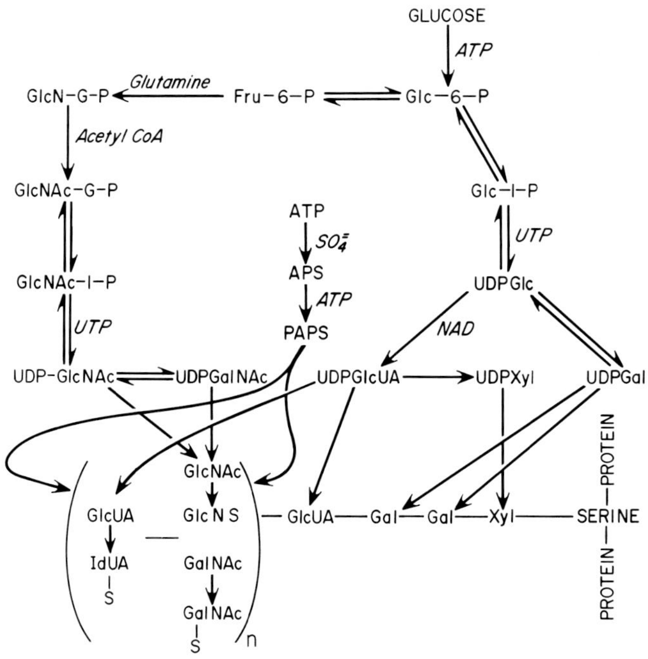

Figure 4 Biosynthesis of the carbohydrate portion of proteoglycan.

Other pyrophosphorylase reactions (such as UTP with glucuronic acid-1-P and xylose-1-P) have not been observed in mammalian systems, although some of these reactions occur in plants.

The intermediate for sulfation of glycosaminoglycans is 3′-phosphoadenosine 5′-phosphosulfate (PAPS) (11). The reactions involved in synthesis of this compound are shown in Figure 4, and have been described in many tissues.

Two of the sugar nucleotides have been shown to act as feedback inhibitors of the key enzymes in the pathway of sugar nucleotide formation in liver. It is likely that they function similarly in other glycosaminoglycan-producing tissues. UDP-*N*-acetylglucosamine has been shown to inhibit the formation of glucosamine (L-glutamine-D-fructose-6-phosphate aminotransferase) (12). Thus an excess of UDP-*N*-acetylglucosamine would turn off its own synthesis and control the amount of hexosamine available for glycosaminoglycan formation. UDP-xylose has been shown to inhibit the oxidation of UDP-glucose to UDP-glucuronic acid (UDP-glucose dehydrogenase) (13). This turns off the synthesis of the uronic acid nucleotide, thus affecting the availability of uronic acid for synthesis of glycosaminoglycans. In addition, the formation of UDP-xylose would be limited by the supply of its precursor, UDP-glucuronic acid. Therefore, the formation of the oligosaccharide that links protein to glycosaminoglycan would be controlled.

It is assumed that all cells capable of synthesizing proteoglycans contain the machinery for synthesizing all the necessary precursors, since phosphorylated sugars and sugar nucleotides do not readily enter the cell. The presence of all the necessary enzymes has been documented in one or another of the proteoglycan-synthesizing tissues, and it can be assumed that these enzymes are present in all tissues that synthesize proteoglycans.

B. Protein Core

It has been established that synthesis of proteins is necessary before formation of the glycosaminoglycan can take place. This has been demonstrated by pulse-chase experiments (14) and by experiments showing that puromycin inhibits synthesis of proteoglycans, but does not inhibit the transfer of sugars to an endogenous proteoglycan acceptor (15). However, it is likely that some modifications of the protein core take place before the proteoglycans leave the cell.

C. Oligosaccharide Linkage

Biosynthesis of the galactosyl-galactosyl-xylosyl oligosaccharide that links protein to glycosaminoglycans other than keratan sulfate appears to take place by sequential addition of sugars to a protein acceptor. The linkage oligosaccharide appears to be identical for these glycosaminoglycans, so that synthesis is probably similar in each case. Formation of the keratan sulfate linkage region has not been examined, but would be expected to be identical to the formation of the oligosaccharide portions of the glycoproteins that have similar structure.

Transfer of [^{14}C]xylose from UDP-[^{14}C]xylose to the hydroxyl of a serine on endogenous protein acceptors has been demonstrated with enzyme from cartilage and mast cell tumors (16,17). The specificity of this reaction is not clear, but it is likely that the activity is related to the synthesis of chondroitin sulfate and heparin. The reaction appears to be directly from UDP-xylose to acceptor, and xylosylphosphoryldolichol does not act as an intermediate (18).

The incorporation of [^{14}C]galactose from UDP-[^{14}C]galactose onto an endogenous protein acceptor has been demonstrated with preparations from cartilage (19). Treatment of the product with alkaline borohydride resulted in [^{14}C]galactosyl-xylitol, thus demonstrating that the galactose was transferred to a xylose attached to a protein. However, no stimulation of galactose incorporation was achieved when UDP-xylose was present. Thus, the sequential addition could not be demonstrated directly but remained only presumptive. Various xylosides have been used in intact cells where they lead to extensive glycosaminoglycan polymerization (20-26). Nevertheless, these same xylosides are poor galactose acceptors in cell-free systems. This suggests that there may be additional steps in formation of the gal-gal-xyl linkage region and in initiation of polymerization.

The amounts of incorporation into endogenous acceptors have been relatively low. Furthermore, it is difficult to distinguish between the amount of enzymatic activity and the amount of acceptor as the limiting factors in the activities of these preparations. For these reasons synthetic exogenous acceptors have been utilized to test the relative activities of the enzymes and to help in characterization of these enzymes. Various compounds such as xylose, xylosyl-serine, galactosyl-xylose, and galactosyl-xylosyl-serine have been utilized with UDP-[^{14}C]galactose and particulate preparations from chick cartilage. However, the amounts of galactose incorporated have still been small, suggesting that some other undefined factors are involved. By use of experiments with mixed substrates, it has been demonstrated that there are two separate galactosyl transferases (19). The same particulate preparation described above was shown to catalyze the addition of glucuronic acid onto galactosyl-galactosyl-xylose when this trisaccharide was added to the incubation mixture (2). Galactosyl-galactose and galactosyl-galactosyl-xylosyl-serine also acted as acceptors. By use of mixed substrate experiments (galactosyl-galactose and oligosaccharides obtained after degradation of chondroitin-6-sulfate) it was shown that this enzyme was not the same as the one involved in addition of glucuronic acid for polymerization. Similar experiments with similar conclusions have been performed with enzyme preparations from chick brain (27).

D. Glycosaminoglycan Polymerization

As with the formation of the oligosaccharide link, the glycosaminoglycan chains are synthesized from sugar nucleotide precursors. Thus, UDP-glucuronic acid and UDP-*N*-acetylglucosamine are precursors for hyaluronic acid (28), heparin (29), and heparan sulfate (30,31); and UDP-glucuronic acid and UDP-*N*-acetylgalactosamine are precursors for chondroitin sulfate (32,33).

In chondroitin synthesis, the sugars are added alternately, one at a time (34,35), and some glycosaminoglycan chains terminate in hexosamine, while others terminate in uronic acid (36). It is likely that other glycosaminoglycans are formed similarly, although this has not been demonstrated directly.

The nature of the endogenous primers for these polymerizations has not been completely defined, although the reactions, in part, represent elongation of glycosaminoglycan chains that are already of considerable size. The primer is bound to protein that may be larger than the final protein of proteoglycan (37), suggesting some postsynthetic modification. The whole complex is part of a particulate fraction of cells and may be attached in turn to the active enzyme. When oligosaccharide primers are utilized instead of the endogenous primers, the polymerization only proceeds to the addition of a few sugars, rather than the polymerization of a long chain (35). This would suggest that the natural primer and enzyme may be in juxtaposition in order to facilitate optimal biosynthesis.

Sulfation apparently occurs by direct transfer of a sulfate group from PAPS to the appropriate sites of the glycosaminoglycan. Even though a sulfated sugar nucleotide (UDP-*N*-acetylgalactosamine-4-S) has been described, it does not appear to be an intermediate (38). The possibility of other intermediates does exist, however, since the enzymes accomplishing sulfation have not been purified and various cofactors might be present in the crude preparations utilized in most experiments.

Incorporation of sulfate into proteoglycans has been demonstrated most extensively with preparations from cartilage (39) and mast cell tumors (40), although other tissues have been used as sources for the enzymatic activity. There appears to be a wide range of glycosaminoglycans that will accept sulfate with enzymes from a single source, so it has not been clear how specific the sulfotransferases are. The synthesis of highly sulfated chondroitin sulfate and heparin glycosaminoglycan was accomplished with cell-free systems. In both cases, the particulate enzyme preparations that catalyzed glycosaminoglycan polymerization also catalyzed the incorporation of sulfate onto the newly formed glycosaminoglycan. In this instance, as in the case of polymerization itself, it seems that the optimal transfer of sulfate is dependent on a degree of organization of the glycosaminoglycan acceptor with protein and perhaps with the enzyme. With these systems, sulfation can be completed on an already completely preformed glycosaminoglycan or may take place at the same time as polymerization. It occurs as an "all-or-nothing" phenomenon (39), with some molecules being essentially completely sulfated and others not sulfated at all. Only small amounts of partly sulfated chains are found.

An unusual sulfation reaction has been found to take place in the course of heparin synthesis and also applies to heparan sulfate. Heparin contains *N*-sulfate in contrast to the *N*-acetyl found in other glycosaminoglycans. However, the synthesis proceeds through the formation of a glycosaminoglycan containing *N*-acetylglucosamine (29-31), which is then deacetylated with replacement of the acetyl by sulfate (30,31,41,42). The formation of the *O*-sulfate on heparin and heparan sulfate appears to be similar to that of chondroitin sulfate.

The formation of the iduronic acid moiety of heparin, heparan sulfate, and dermatan sulfate takes place by epimerization of glucuronic acid to iduronic acid after polymerization is complete (43,44). In the case of heparin this epimerization may be linked to sulfation of the iduronic acid.

E. Cellular Biosynthetic Organization

There is a distinct separation between the enzymes of precursor activation and those of macromolecular assembly. Most of the enzymes involving formation of sugar nucleotides and PAPS are found in a high-speed supernatant fraction from tissue homogenates, implying that they are cytoplasmic in origin. The activities involved in glycosaminoglycan assembly, on the other hand, are particulate, being found in a crude "microsomal" fraction. This implies that these enzymes are either attached to, or part of, membranous structures in the cell. The enzymes involved in sulfate transfer from PAPS are readily found in the supernatant fraction, but the activity there appears to be relatively low. The maximal transfer of sulfate is found with the same microsomal fractions that promote polysaccharide polymerization (39,40). It seems reasonable to suppose that the state of the membrane-bound proteoglycan substrate contributes greatly to the overall efficiency of sulfation.

Radioautographic studies to locate the sites of proteoglycan biosynthesis have indicated that sulfation and probably glycosaminoglycan polymerization take place in the Golgi apparatus (45,46), while the synthesis of protein occurs in the rough endoplasmic reticulum. This is in agreement with work on glycoprotein biosynthesis with isolated Golgi preparations (47). Isolated Golgi preparations from mast cells have been shown to contain essentially all of the glycosaminoglycan-synthesizing activity (48).

Work on biosynthesis of proteoglycans has provided sufficient information to suggest the following overall pattern. Proteoglycan biosynthesis takes place so that all the glycosaminoglycan chains on a single proteoglycan core protein are polymerized rapidly, with sulfation taking place during chain elongation. Exogenous acceptors not fixed at the site of synthesis are therefore not readily maintained adjacent to the synthesizing enzymes. The glycosaminoglycan chains do not have an absolute molecular size, nor do they end in one specific sugar. This suggests that a model can be explained by a membrane-related synthesis, as shown in Figure 5. The protein core of the proteoglycan could be extruded into the Golgi site of glycosaminoglycan synthesis where it would be firmly retained, attached to the membrane. Enzymes arrayed on the membrane or adjacent to the membrane would remain in juxtaposition to the protein core and to the oligosaccharide linkage region. In this fashion polymerization would take place rapidly on all of the chains followed directly with sulfation. If a proteoglycan were removed from this site of synthesis, it would not readily reattach to the synthetic site so that effective synthesis would not readily take place.

There still remain major questions concerning important aspects of biosynthesis, such as processing of the protein portion and the organization of synthesis within the cell. Of particular importance would be the way that a specific substructure of glycosaminoglycan might be produced. It had long been considered that the heterogeneity of glycosaminoglycans and proteoglycans was due to a certain degree of random synthesis with no speci-

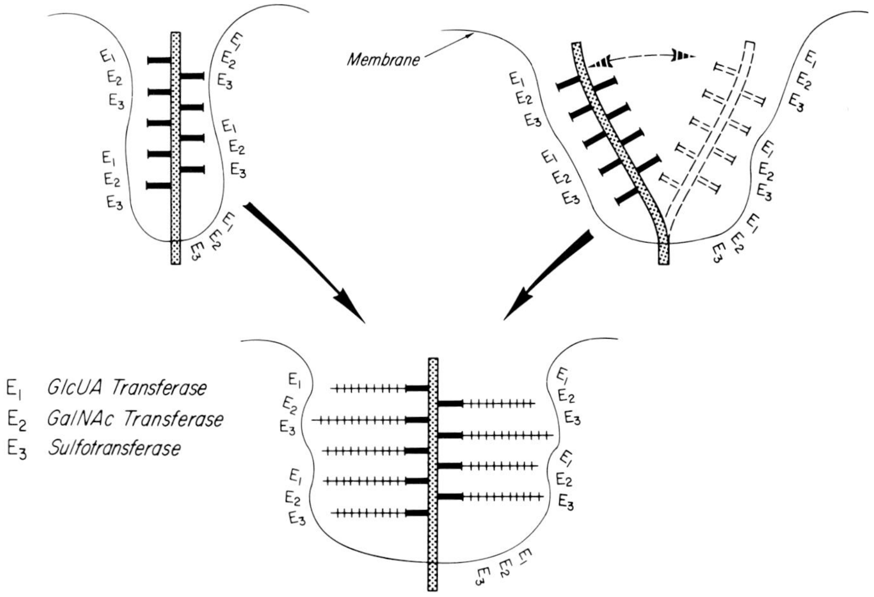

Figure 5 Model of membrane-bound glycosaminoglycan synthesis.

ficity of sulfate placement, chain size, etc. This may in part be true, but it is hard to believe that variations in iduronic acid/glucuronic acid, sulfation degree and position, or degree of acetylation, etc. are not controlled to produce specific substructures with specific functions. Such functional specificity has now been demonstrated clearly by the definition of a specific heparin oligosaccharide sequence that is active in binding to and activating antithrombin III (49,50). Molecules of heparin that do not contain this sequence have no anticoagulant activity. In addition, heparin may have other functions that require other small oligosaccharide sequences.

Control of the synthetic processes is largely unsolved. This may prove to be of great importance, since it has become increasingly apparent that pathological processes are frequently related to alterations of aspects of control and metabolic organization rather than to alterations in individual steps in the metabolic pathways. This would be expressed on a cellular basis resulting in multiple changes, rather than as a single, easily described enzyme effect. It is probable that most effects on proteoglycan or glycosaminoglycan formation are secondary to more widespread changes in the cell.

IV. DEGRADATION

A. Turnover

The half-life in vivo of connective tissue hyaluronic acid, chondroitin sulfate, and dermatan sulfate has been estimated to be from two days to two weeks depending on the tissues and animals examined (51-53). Thus, there must be an active turnover with continual synthesis taking place. The nature of the degradation products is such, however, that it is unlikely that there is much of any recycling of constituents.

Incorporation of radioactively labeled precursors and turnover of these precursors have been studied in synovial cells (54) and in cartilage. Simultaneous incorporation of [^{35}S] sulfate, [^{14}C] proline, and [^{14}C] phenylalanine into proteochondroitin sulfate of chondrocytes has been shown (55) with both the ^{14}C-labeled and ^{35}S-labeled material being turned over at the same rate. This has also been found with [^{14}C] lysine and [^{35}S] sulfate (56), indicating a concomitant turnover of both the protein and the glycosaminoglycan moieties of the proteoglycan. Turnover of heparin and heparan sulfate is less well understood.

Radioautography of rat mast cells has been utilized to show that sulfate is incorporated into heparin and then slowly lost. Approximately 75% of the heparin was found to be present after 18 days (57). When [^{14}C] hyaluronic acid was injected into guinea pigs (58), most of the label was found to be excreted as hyaluronic acid with less being metabolized. Rats injected with chondroitin [^{35}S] sulfate were found to excrete free [^{35}S] sulfate in the urine (59), indicating that a sulfatase was present in tissues of the rats. When chondroitin 4-sulfate, chondroitin 6-sulfate, dermatan sulfate, and heparan sulfate were injected into dogs, as much as 60% of the injected glycosaminoglycan could not be accounted for (60). It was suggested that this material might have been degraded by the tissues. When small amounts of [^{35}S] heparin were injected into dogs, radioactivity was obtained quantitatively in the urine (61) as inorganic [^{35}S] sulfate. When larger doses of [^{35}S] heparin were utilized, there was some [^{35}S] heparin as well as [^{35}S] sulfate found in the urine.

It can readily be seen by these studies that the mechanism of the proteoglycan turnover is not clear. It seems that there is probable degradation of the glycosaminoglycan moiety, but the degree to which this takes place is not known. The studies indicate that at least some of the glycosaminoglycans can be excreted intact in the urine. Therefore, it is possible that turnover is both by degradation and by excretion of intact or partially intact macromolecules.

B. Endoglycosidases

Enzymatic degradation of glycosaminoglycans or proteoglycans was first noted as "spreading factor" (62). When India ink or similar material was injected subcutaneously, an increased diffusibility of the dye was noted if a "spreading factor" was injected at the same time. Identification of these factors as glycosaminoglycan-degrading enzymes helped to stimulate interest in, and led to studies (mainly in the laboratory of Karl Meyer) defining the structures of the glycosaminoglycans. It was assumed that enzymes similar to the "spreading factors" were probably involved in glycosaminoglycan breakdown and turnover in vivo.

The best defined glycosaminoglycan-lyase is the hyaluronidase that is found in testicular tissue. This enzyme degrades hyaluronic acid to a family of oligosaccharides ranging in size from tetrasaccharide to larger even-numbered oligosaccharides (63). When hyaluronic acid is incubated with the enzyme, the primary product is the tetrasaccharide with a small amount of larger oligosaccharides and a small amount of disaccharide. The disaccharide has been identified as *N*-acetylhyalobiuronic acid, and the tetrasaccharide consists of two *N*-acetylhyalobiuronic acid units linked in the same fashion as they are linked in the intact hyaluronic acid molecule. The larger oligosaccharides are multiples of this.

The enzyme has been found to be an endo-β-*N*-acetylhexosaminidase, attacking β-linked *N*-acetylglucosaminyl groups at random within the chain, with a preferential reaction on the larger molecules rather than on smaller oligosaccharides (64). Thus, the hyaluronic acid molecule initially can be depolymerized rapidly, with consequent loss of viscosity. Further degradation results in release of measurable amounts of reducing sugars, but completion of degradation requires a longer time. The enzyme has been shown to be a transglycosylase, capable of transferring disaccharide or larger oligosaccharide units from one molecule to another (64,65). As a result of this activity, when hexasaccharide is incubated with testicular hyaluronidase, some octasaccharide can be found as well as tetrasaccharide.

Desulfated chondroitin, chondroitin 4-sulfate, and chondroitin 6-sulfate have also been shown to be degraded by the testicular enzyme, resulting in oligosaccharide products similar in size to those from hyaluronic acid (65,66). Sulfate groups are not removed during this reaction. Degradation of chondroitin sulfate is not as rapid as degradation of hyaluronic acid (64), while nonsulfated chondroitin reacts similarly to hyaluronic acid in rapidity of degradation.

A hyaluronidase from rat skin has also been described (67). This enzyme acts in a fashion similar to that of testicular hyaluronidase as a *β-N*-acetylhexosaminidase with activity as a transglycosylase. However, it differs somewhat from testicular hyaluronidase, yielding mainly hexasaccharide and larger products with no disaccharide and only a small amount of tetrasaccharide. It has been concluded that the amount of hyaluronidase obtained in fractions from skin is sufficient to account for all the necessary degradation in the known turnover of hyaluronic acid in skin tissue. The enzyme has been noted to have activity in degrading chondroitin 4-sulfate and chondroitin 6-sulfate. Hyaluronidase has also been found in embryonic chick skin and in chick cell cultures, but not in human skin fibroblast cultures (68).

Actively differentiating tissues are also sources of hyaluronidase. This enzyme has been found in the back skin of tadpoles (69) and has been shown to be active throughout all of the stages of tadpole metamorphosis, but not to be present in the adult bullfrog (70). Hyaluronidase also appears before or during the development of the regenerating newt limb (71) and chick cornea (72). Glycosaminoglycan-degrading enzymes apparently have an important active part in tissue remodeling and growth (73). Hyaluronate on cell surfaces may be of particular importance.

Since dermatan sulfate contains a small amount of glucuronic acid in addition to iduronic acid (74), it has been suggested that hyaluronidase can cause some depolymerization by acting at these points. In addition, there have been reports of enzymes from spleen (75), liver (76), and platelets (77) capable of degrading heparin or heparan sulfate to large oligosaccharides. In general, these do not degrade extensively but only cleave at a few places in the glycosaminoglycan chains (78-80). Some of these enzymes may be involved in processing proteoglycans during synthesis rather than in true degradation. However, there are increasing indications that these enzymes, particularly those that degrade heparan sulfate, may play key roles in controlling cellular interactions. For example, metastatic tumor cells have been reported to be capable of degrading vascular subendothelial matrix (81). A smaller heparan sulfate was found, suggesting that these tumor cells might be secreting an endoglycosidase. It should be noted that connective tissue endoenzymes capable of degrading dematan sulfate or heparan sulfate extensively have not been found.

C. Exoglycosidases and Sulfatases

The products of degradation of glycosaminoglycans by most glycosaminoglycan-lyases are oligosaccharides containing a glucuronic acid moiety at their nonreducing end (82).

β-Glucuronidase will attack these ends of nonsulfated oligosaccharide, yielding free glucuronic acid and a terminal *N*-acetylhexosamine on the remaining oligosaccharide (83). This moiety can in turn be attacked by *N*-acetylhexosaminidases (83,84) if it is nonsulfated. Xylosidases (84) and galactosidases (85) have also been found in mammalian tissue, suggesting that they might be involved in degradation of the oligosaccharide linkage between protein and glycosaminoglycan. To date, these glycosidases have been found to act only as exoglycosidases; that is, they attack only a terminal nonreducing sugar and are not capable of cleaving within an intact glycosaminoglycan or oligosaccharide. Sulfatases also generally have no action on intact glycosaminoglycan, only cleaving sulfate from a nonreducing terminal end (86). Therefore, these enzymes would have to act sequentially from the nonreducing end of a glycosaminoglycan. Furthermore, removal of sulfate from a terminal nonreducing sugar appears to be necessary before an exoglycosidase will remove the sugar.

During the last decade, a series of degradative enzymes has been examined in cultures of skin fibroblasts. An extensive literature exists concerning the presence or absence of these various enzymes and their relationship to the genetic mucopolysaccharidoses (87). It has been shown that intracellular oligosaccharides or small glycosaminoglycans can be degraded from their nonreducing ends by sequential removal of terminal glycosyl groups and/or sulfate. Absence of one or more of the degradative enzymes will result in a glycosaminoglycan storage disease owing to inability to degrade the remainder of the molecule. Many of these diseases have been described, with a single missing degradative enzyme for each. Information has been obtained regarding these missing enzymes and the manner in which cells can incorporate these enzymes (88).

D. Proteases

One possible method of turning over proteoglycans in tissues is by degradation of the protein moiety. This has been graphically demonstrated by experiments involving the injection of the protelytic enzyme papain into rabbit ears (89). This procedure results in a rapid loss of cartilage rigidity accompanied by mobilization and excretion of glycosaminoglycans in the urine. Thus, it has been conclusively demonstrated that proteolytic enzymes can mobilize large amounts of glycosaminoglycans from connective tissue in vivo.

Acid proteases (cathepsins) of bone and cartilage have been shown to be capable of degrading the protein portion of proteoglycans (90), followed by solubilization of the glycosaminoglycans. In contrast, the protein core of heparin appears to be highly resistant to proteases (91).

REFERENCES

1. U. Lindahl and L. Rodén, in *Glycoproteins* (A. Gottschalk, ed.), Elsevier, Amsterdam, 1972, pp. 491-517.
2. T. Helting and L. Rodén, *J. Biol. Chem., 244*:2799-2805 (1969).
3. V. C. Hascall, *J. Supramol. Struct., 7*:101-120 (1977).
4. T. E. Hardingham and H. Muir, *Biochim. Biophys. Acta, 279*:401-405 (1972).
5. B. Jacobson and E. A. Davidson, *J. Biol. Chem., 237*:635-637 (1962).
6. J. E. Silbert and S. DeLuca, *Biochim. Biophys. Acta, 141*:193-196 (1967).
7. B. Jacobson and E. A. Davidson, *Biochim. Biophys. Acta, 73*:145-151 (1963).
8. L. Rodén and A. Dorfman, *J. Biol. Chem., 233*:1030-1033 (1958).
9. K. J. Isselbacher, *J. Biol. Chem., 232*:429444 (1959).

10. R. DelGiacco and F. Maley, *J. Biol. Chem., 239*:2400-2402 (1964).
11. P. W. Robbins and F. Lipmann, *J. Biol. Chem., 229*:837-851 (1957).
12. S. Kornfeld, R. Kornfeld, E. F. Neufeld, and P. J. O'Brien, *Proc. Natl. Acad. Sci. USA, 52*:371-379 (1964).
13. E. F. Neufeld and C. W. Hall, *Biochem. Biophys. Res. Commun., 19*:456-461 (1965).
14. S. DeLuca, A. I. Caplan, and V. C. Hascall, *J. Biol. Chem., 253*:4713-4720 (1978).
15. A. Telser, H. C. Robinson, and A. Dorfman, *Proc. Natl. Acad. Sci. USA, 54*:912-919, (1965).
16. H. C. Robinson, A. Telser, and A. Dorfman, *Proc. Natl. Acad. Sci. USA, 56*:1859-1866 (1966).
17. E. Grebner, C. W. Hall, and E. F. Neufeld, *Arch. Biochem. Biophys., 116*:391-398 (1966).
18. C. R. Faltynek, J. E. Silbert, and L. Hof, *J. Biol. Chem., 257*:5490-5495 (1982).
19. T. Helting and L. Rodén, *J. Biol. Chem., 244*:2790-2798 (1969).
20. N. Schwartz, L. Galligani, P. Ho, and A. Dorfman, *Proc. Natl. Acad. Sci. USA, 71*: 4047-4051 (1974).
21. C. Robinson, M. Brett, P. Tralaggan, D. Lowther, and M. Okayama, *Biochem. J., 148*:24-34 (1975).
22. K. Gibson and B. Segen, *Biochem. J., 168*:65-79 (1977).
23. Y. Kato, K. Kimata, K. Karasawa, and S. Suzuki, *J. Biol. Chem., 253*:2784-2789 (1978).
24. S. Lohmander, K. Madsen, and A. Hinek, *Arch. Biochem. Biophys., 192*:148-157 (1979).
25. K. Sandy, H. Brown, and D. Lowther, *Biochem. J., 188*:119-130 (1980).
26. R. Stevens and K. Austen, *J. Biol. Chem., 257*:253-259 (1982).
27. A. E. Brandt, J. Distler, and G. W. Jourdian, *Proc. Natl. Acad. Sci. USA, 64*:374-380 (1969).
28. S. Schiller, G. A. Slover, and A. Dorfman, *Biochem. Biophys. Res. Commun., 5*:344-348 (1961).
29. J. E. Silbert, *J. Biol. Chem., 238*:3542-3546 (1963).
30. J. Riesenfeld, M. Höök, and U. Lindahl, *J. Biol. Chem., 257*:7050-7055 (1982).
31. J. E. Silbert and C. T. Baldwin, *Glycoconjugate J., 1*:63-71 (1984).
32. J. E. Silbert, *J. Biol. Chem., 239*:1310-1315 (1964).
33. R. L. Perlman, A. Telser, and A. Dorfman, *J. Biol. Chem., 239*:3623-3629 (1964).
34. A. Telser, H. C. Robinson, and A. Dorfman, *Arch. Biochem. Biophys., 116*:458-465 (1966).
35. J. E. Silbert and A. C. Reppucci, Jr., *J. Biol. Chem., 251*:3942-3947 (1976).
36. J. E. Silbert, *J. Biol. Chem., 253*:6888-6892 (1978).
37. C. R. Faltynek and J. E. Silbert, *J. Biol. Chem., 256*:7202-7206 (1981).
38. S. Suzuki and J. L. Strominger, *J. Biol. Chem., 235*:257-266 (1960).
39. S. DeLuca, M. E. Richmond, and J. E. Silbert, *Biochemistry, 12*:3911-3915 (1973).
40. J. E. Silbert, *J. Biol. Chem., 242*:5146-5152 (1967).
41. J. E. Silbert, *J. Biol. Chem., 242*:5153-5157 (1967).
42. U. Lindahl, G. Bäckström, L. Jansson, and A. Hallén, *J. Biol. Chem., 248*:7234-7241 (1973).
43. M. Höök, U. Lindahl, G. Bäckström, A. Malmström, and L-A. Fransson, *J. Biol. Chem., 249*:3908-3915 (1974).
44. A. Malmström, L-A. Fransson, M. Höök, and U. Lindahl, *J. Biol. Chem., 250*:3419-3425 (1975).
45. M. Peterson and C. P. Leblond, *J. Cell Biol., 21*:143-148 (1964).
46. G. C. Godman and N. Lane, *J. Cell Biol., 21*:353-356 (1964).
47. H. Schachter and S. Roseman, in *The Biochemistry of Glycoproteins and Proteoglycans* (W. J. Lennarz, ed.), Plenum Press, New York, 1980, pp. 85-160.

48. J. E. Silbert and L. S. Freilich, *Biochem. J., 190*:307-313 (1980).
49. U. Lindahl, G. Bäckström, and L. Thunberg, *J. Biol. Chem., 258*:9826-9830 (1983).
50. D. H. Atha, A. W. Stephens, and R. D. Rosenberg, *Proc Natl. Acad. Sci. USA, 81*: 1030-1034 (1984).
51. S. Schiller, M. B. Mathews, J. A. Cifonelli, and A. Dorfman, *J. Biol. Chem., 218*:139-145 (1956).
52. E. A. Davidson and W. Small, *Biochim. Biophys. Acta., 69*:453-458 (1963).
53. H. Boström, *J. Biol. Chem., 196*:477-481 (1952).
54. P. Barland, C. Smith, and D. Hamerman, *J. Cell Biol., 37*:13-26 (1968).
55. R. D. Campo and D. D. Dziewiatkowski, *J. Biol. Chem., 237*:3729-2735 (1962).
56. J. I. Gross, M. B. Mathews, and A. Dorfman, *J. Biol. Chem., 235*:2889-2892 (1960).
57. E. Jorpes, E. Odeblad, and H. Boström, *Acta Haematol., 9*:273-276 (1953).
58. R. S. Jones and E. V. Howell, *Proc. Soc. Exp. Biol. Med., 110*:547-552 (1962).
59. D. D. Dziewiatkowski, *J. Biol. Chem., 223*:239-249 (1956).
60. D. Kaplan and K. Meyer, *J. Clin. Invest., 41*:743-749 (1962).
61. H. Danishefsky and H. B. Eiber, *Arch. Biochem. Biophys., 85*:53-61 (1959).
62. D. C. Hoffman and F. Duran-Reynals, *J. Exp. Med., 53*:387-398 (1931).
63. B. Weissmann, K. Meyer, P. Sampson, and A. Linker, *J. Biol. Chem., 208*:417-429 (1954).
64. B. Weissmann, *J. Biol. Chem., 216*:783-794 (1955).
65. P. Hoffman, K. Meyer, and A. Linker, *J. Biol. Chem., 219*:653-663 (1950).
66. K. Meyer and M. M. Rapport, *Arch. Biochem., 27*:287-293 (1950).
67. D. C. Cashman, J. U. Laryea, and B. Weissmann, *Arch. Biochem. Biophys., 135*: 387-395 (1969).
68. R. W. Orkin and B. P. Toole, *J. Biol. Chem., 255*:1036-1042 (1980).
69. J. E. Silbert, Y. Nagai, and J. Gross, *J. Biol. Chem., 240*:1509-1511 (1965).
70. M. J. Lipson, R. Cerskus, and J. E. Silbert, *Dev. Biol., 25*:198-208 (1971).
71. B. P. Toole and J. Gross, *Dev. Biol., 25*:57-77 (1971).
72. R. L. Trelstad, K. Hayashi, and B. P. Toole, *J. Cell Biol., 62*:815-830 (1974).
73. B. P. Toole, in *Cell Biology of Extracellular Matrix* (E. D. Hay, ed.), Plenum Press, New York, 1981, pp. 259-294.
74. L-A. Fransson and L. Rodén, *J. Biol. Chem., 242*:4164-4169 (1967).
75. M. Höök, I. Pettersson, and S. Ögren, *Thromb. Res., 10*:857-861 (1975).
76. M. Höök, Å. Wasteson, and Å. Oldberg, *Biochem. Biophys. Res. Commun., 67*: 1422-1427 (1975).
77. Å. Wasteson, B. Glimelius, C. Busch, B. Westermark, C. H. Heldin, and B. Norling, *Thromb. Res., 11*:309-321 (1977).
78. S. Ögren and U. Lindahl, *J. Biol. Chem., 250*:2690-2697 (1975).
79. B. Arbogast, J. J. Hopwood, and A. Dorfman, *Biochem. Biophys. Res. Commun., 75*:610-617 (1977).
80. U. Klein, H. Kress, and K. von Figura, *Biochem. Biophys. Res. Commun., 69*:158-166 (1976).
81. R. H. Kramer, K. G. Vogel, and G. L. Nicolson, *J. Biol. Chem., 257*:2678-2686 (1982).
82. B. Weissmann, S. Hadjiioannou, and J. Tornheim, *J. Biol. Chem., 239*:59-63 (1964).
83. A. Linker, K. Meyer, and B. Weissmann, *J. Biol. Chem., 213*:237-248 (1955).
84. V. Patel and A. L. Tappel, *Biochim. Biophys. Acta, 191*:86-94 (1969).
85. G. Vaes and P. Jacques, *Biochem. J., 97*:380-388 (1965).
86. J. H. Glaser and H. E. Conrad, *J. Biol. Chem., 254*:2316-2325 (1979).
87. C. W. Hall, I. Liebaers, P. M. DiNatale, and E. F. Neufeld, *Meth. Enzymol., 50*: 439-456 (1978).
88. E. F. Neufeld and G. Ashwell, in *The Biochemistry of Glycoproteins and Proteoglycans* (W. J. Lennarz, ed.), Plenum Press, New York, 1980, pp. 241-266.

89. L. Thomas, *Biophys. J., 4*(suppl.):207-213 (1964).
90. J. F. Woessner, Jr., *Fed. Proc., 32*:1485-1488 (1973).
91. R. W. Yurt, R. W. Leid, K. F. Austen, and J. E. Silbert, *J. Biol. Chem., 252*:518-521 (1977).

II

MOLECULAR ORGANIZATION OF THE EXTRACELLULAR MATRIX

5

Diseases of the Extracellular Matrix
Structural alterations of collagen fibrils in skin

Karen A. Holbrook and Peter H. Byers
University of Washington School of Medicine,
Seattle, Washington

I. INTRODUCTION

During the past several years our understanding of the protein, proteoglycan, and glycoprotein composition of the dermis has increased enormously. We now have detailed structural information of the constituent macromolecules, their genes and gene structure, and insight into the complexity of their biosynthetic, processing, and degradative pathways. In the last few years, there have been many studies of the biochemistry of abnormal molecules synthesized by cells isolated from patients with heritable dermal and other connective tissue disorders that affect the collagens of skin. Yet despite this expansion of understanding, the complex nature of the extracellular matrix remains poorly understood, partly because the intermolecular interactions are just now being explored.

For the last 10 years we have systematically studied dermal tissue from patients with a variety of single gene heritable disorders including many forms of the Ehlers-Danlos syndrome and osteogenesis imperfecta, the Marfan syndrome, cutis laxa, and several other conditions more commonly thought of as disorders that do not affect either collagen or elastin. Our studies have included morphological examination of dermal tissue and characterization of the abnormal proteins and abnormal gene structure. From these, we have developed a general picture of the types of alterations that can occur in the dermis of individuals with inherited connective tissue disorders, the extent to which such alterations are disease-specific, and how or whether they may be used in diagnosis. In many ways, the analysis of alterations in dermis has increased our understanding of the nature and types of interactions among molecules of the extracellular matrix.

A. Current Morphological Approaches to the Study of Altered Connective Tissue

Current morphological analysis of the extracellular matrix in normal and diseased tissues can provide information about the structure of the tissue, fibers, fibrils, and molecules, their biochemical identity, and, in some instances, the manner in which the molecules interact. A number of technological advances have made such studies possible.

Histo(cyto)chemical and immunohisto(cyto)chemical staining procedures are now used routinely to localize proteins [e.g., the site of deposition of proα1(I) in the dermis (1)] or to identify the biochemical composition of a specific structural entity [e.g., a collagen fibril as type I or III (2)]. Autoradiography is used to trace the biosynthetic pathway of a radiolabeled molecule (e.g., [^{3}H] proline-labeled collagens) through the cell into the extracellular matrix (3). The elemental composition of a matrix component [e.g., an atypical association of Ca^{2+} with collagen fibrils in soft tissues (4)] can be determined qualitatively and quantitatively by x-ray microanalysis or backscattered electron imaging combined with scanning or transmission electron microscopy. Other types of morphological information [e.g., density of elastic fibers (5)], can be quantitated through the use of stereologic or morphometric techniques. All these methods can be applied at the light and ultrastructural levels. Thus, the modern techniques of tissue analysis are bridging the gap that once existed between morphology and biochemistry. We now know a great deal about the biochemistry of the inherited connective tissue diseases and the biochemical identity of the components in tissues, but we have relatively little understanding of how the abnormalities in structure and quantity of matrix molecules lead to the phenotypic expression of the disease.

B. The Problem of Matrix Interactions in Analyzing Patterns of Molecular-Structural Alteration

In those instances where studies of the connective tissue in inherited disease have been pursued at multiple levels (e.g., the gene-molecule-fibril-fiber), one would expect that we might be closer to understanding the relationship between a molecular defect, the altered ultrastructure of the fibril and the histology of the tissue, the compromised functional properties of the tissue or organ, and the clinical presentation of the patient. Unfortunately, this is not the case; the primary molecular defect appears to have little predictable relationship to the structural changes. This suggests that, in addition to structural or quantitative alterations in a connective tissue component, the interactions among elements in the matrix are important in the connective tissue architecture. Thus, understanding the molecular basis of a disease, although fundamental to revealing defects at gene and protein levels, is only a beginning and is the simplest level to investigate. The tissue alterations and phenotypic expression of a molecular disorder are the result of disordered interactions among the matrix molecules; this is another order of complexity.

In skin, the structurally altered collagen molecules have to be considered in their environment of a connective tissue milieu, associated with an epithelium (epidermis) and epithelial derivatives (hair follicles and sweat glands), and in contact with vascular and nervous networks and a variety of cells that produce and release chemical mediators into the extracellular matrix. Thus, even though we can obtain morphological and biochemical information, the structure of the matrix and the nature of interactions are so complex that correlation between the structure (histology and ultrastructure) and biochemistry of matrix components is difficult. Although this may seem pessimistic, we view it as a challenge to continue work to understand the complexity of matrix interactions in tissues; the study of single gene disorders provides an important vehicle for this purpose.

C. Systematic Analysis of Skin Biopsy Samples for Structural Alterations in the Connective Tissue

Our own studies have focused on the dermis because of its easy accessibility and because it contains all the molecules of interest to us organized in a predictable manner. In some connective tissue disorders, the primary manifestations of a disease are not in the skin (e.g., bone in osteogenesis imperfecta, vascular tissue in the Marfan syndrome), but because the same collagens and most other macromolecules that are present in bone and vascular tissue are present in skin, the dermis serves as a window to effects of the primary defect.

Our approach is to systematically examine the structure of full-thickness skin, obtained by punch biopsy, using light microscopy (LM) and scanning (SEM) and transmission (TEM) electron microscopy. Each biopsy is divided lengthwise. One-half is embedded in plastic and sectioned in semithin and thin sections for light and electron microscopic study. We use semithin sections and light microscopy to determine the thickness and organization of the dermis; the extent of the papillary and reticular regions; the size, integration, and density of collagen fiber bundles; the amount, structure, and distribution of elastic fibers; and the relative numbers of fibroblastic cells. Other information about sites of inflammation, abnormal patterns of vessels, or alterations in epidermis, epidermal appendages, and subcutaneous tissue also can be obtained. Scanning electron microscopy is useful for the same kind of evaluation but has the added advantage of accommodating bulk specimens (compared with a section), which can be rotated and tilted and examined at potentially higher magnifications (10-50,000X) and with substantially improved resolution (20-500 Å). At low magnifications, the SEM is useful in studying the three-dimensional relationships among matrix components. High-resolution SEM allows one to see altered structure of individual fibrils or labeled antibodies attached to a fibril as identifying markers.

Screening by LM and SEM provides direction for examination of thin sections by TEM, allowing us to focus on a particular fibrous component or one region of the dermis. However, it is important to examine all portions of the dermis for other evidence of pathological change that might not have been apparent with lower resolution.

With TEM, the papillary and reticular dermis are surveyed for changes in the diameter and shape of collagen fibrils (the individual banded elements), packing of fibrils into fibers (collections of fibrils spaced equidistantly and aligned in a similar direction) and fibers into fiber bundles, and unexpected relationships between collagen and other matrix components (e.g., association with excess proteoglycans, altered elastic fibers). Alterations in any of these features are more likely to occur in the reticular dermis; in the papillary dermis they are usually limited to changes in the density and packing of fibrils and only rarely involve the structure of the individual fibrils (see below). Elastic fibers are also examined for quantity, their position relative to collagen bundles, proportion of microfibrils and elastin, and structure (i.e., size, fragmentation, calcification, granular degeneration). Fibroblastic cells can provide evidence for abnormal storage of matrix molecules (dilated rough endoplastic reticulum) or enhanced collagen degradation (fibrils within lysosomes); the relationship of these cells to components of the fibrous matrix (whether in interstitial spaces, associated with elastic fibers, or insinuated into collagen bundles) provides additional clues to the tissue pathology. The connective tissue between fiber bundles often includes fibrils with atypical structure or reduced diameter, unusual matrix components, such as fibrous long spacing collagen (FLS) and altered elastic fibers.

A concern in examining the tissue for structural alterations is the ability to define the range of normal and to sort out pathological changes from local perturbations, e.g., those caused by degranulation of a mast cell, or an artifact such as might be found at the edges of the biopsy sample. It is difficult to set guidelines for making these distinctions, but as a result of studies of the dermis from many patients with a variety of inherited connective tissue diseases, we feel we have gained enough experience to allow us to overlook minor alterations such as a sporadic composite collagen fibril, minor alterations in diameter, small areas of collagen that appear tangled, elastic fibers that have an irregular surface, or fibroblasts that have some dilated rough endoplasmic reticulum, from major changes that involve large numbers of individual fibrils, a significant alteration of an entire region or fiber bundle, or a modified overall pattern of dermal organization.

II. STRUCTURE AND COMPOSITION OF THE NORMAL DERMIS

In order to discuss alterations in the normal pattern of dermal architecture at the tissue and ultrastructural levels, it is necessary first to identify the regions of the dermis and their morphological organization and biochemical composition.

A. General Dermal Organization

The organization of the dermis, the amount of collagen, and the size and density of the collagen bundles are related to the age and sex of the individual (6,7). In general, the dermis can be divided into papillary and reticular regions (Figure 1). Each of these can be subdivided on the basis of its morphology and biochemical composition. The papillary dermis lies immediately beneath the epidermis and is approximately twice its thickness. It is equivalent structurally and biochemically to the connective tissue that surrounds the epidermal appendages and vessels; the two zones together are called the adventitial dermis (1). The papillary dermis is set apart from the reticular dermis by a horizontal vascular plexus and is distinguished from it by differences in the size and organization of collagen fibrils and fiber bundles (Figure 2a). The reticular dermis forms the bulk of the skin and is responsible for most of its mechanical properties (Figure 2b). A second interface separates the deep reticular dermis from the subcutaneous tissue or hypodermis (Figure 1).

B. Connective Tissue Components of the Dermis

Approximately 70% of the dermis is collagen, elastic fibers, proteoglycans, and glycoproteins. The collagens in dermis include types I, $\alpha 1$(I) trimer, III, IV, and V and types I and III procollagen (8). Type IV is restricted to basal laminae found at the dermal-epidermal junction and surrounding vessels, nerves, and epidermal appendages. The interstitial collagens (I, III, and V) are distributed throughout the dermis and in the subcutaneous tissue. Overall, type I collagen accounts for approximately 80-85% of the collagen in the dermis and appears as large-diameter fibrils that organize into large bundles in the reticular dermis (2). Type III collagen makes up only 10-15% of the dermal collagen (9); it forms small-diameter fibrils that are assembled into small-sized fiber bundles. There are regional differences in the localization of type III collagen in skin. Type V collagen accounts for about 4-5% (10) of the collagen in the skin. It is present uniformly throughout the dermis (8) in a form that has yet to be determined. Type I trimer is present as 1-2% of the collagenous protein in skin (10,11); its location and structure are unknown. Type I procollagen is found in the upper papillary dermis (1) and

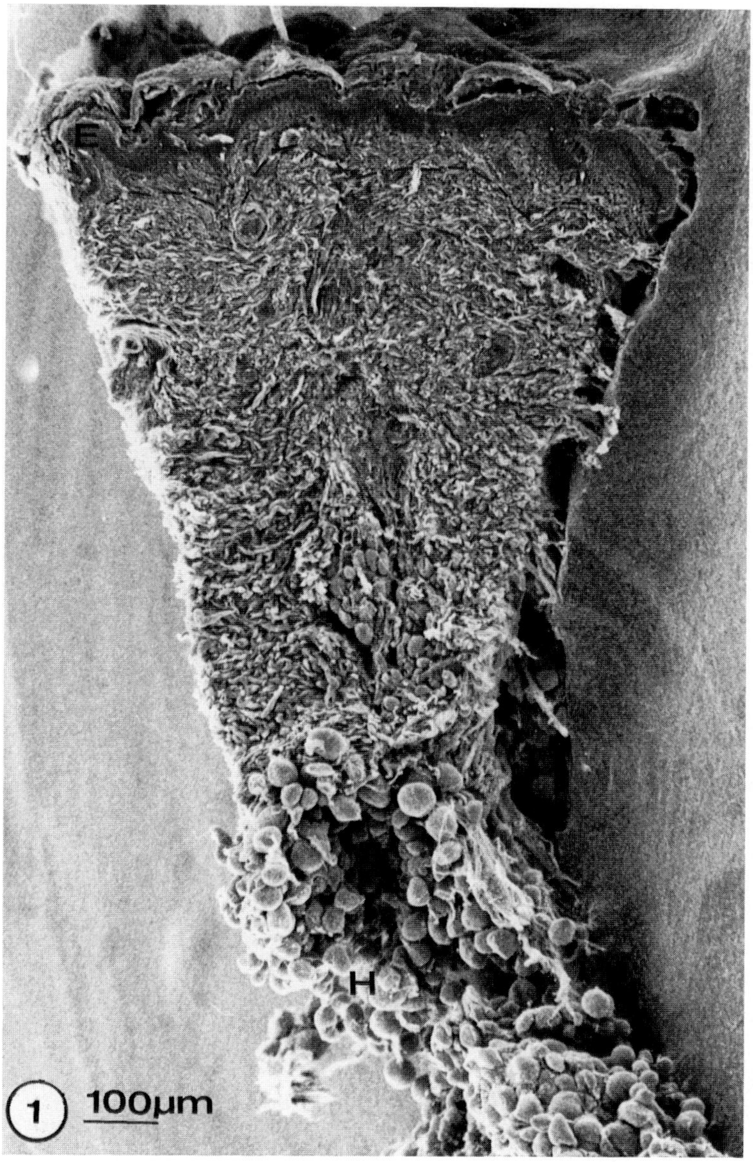

Figure 1 Full-thickness skin of a newborn infant showing the relative thickness of the epidermis (E) and the papillary and reticular regions of the dermis (separated by a broken line), the progressive increase in size of collagen bundles, and the hypodermis (H). X90.

around cells (12); type III procollagen is dispersed throughout the dermis and subcutaneous tissue in association with type III collagen fibrils (8,12).

Collagen and elastic fibers are integrated structurally and interact functionally to maintain the normal mechanical properties of the tissue; elastic fibers are organized in a network that extends from the dermal-epidermal junction to the dermal-subcutaneous interface. The elastic fibers closest to the epidermis, called oxytalan fibers (13), are bundles of microfibrils, with little or no elastic matrix. They course through the papillary dermis

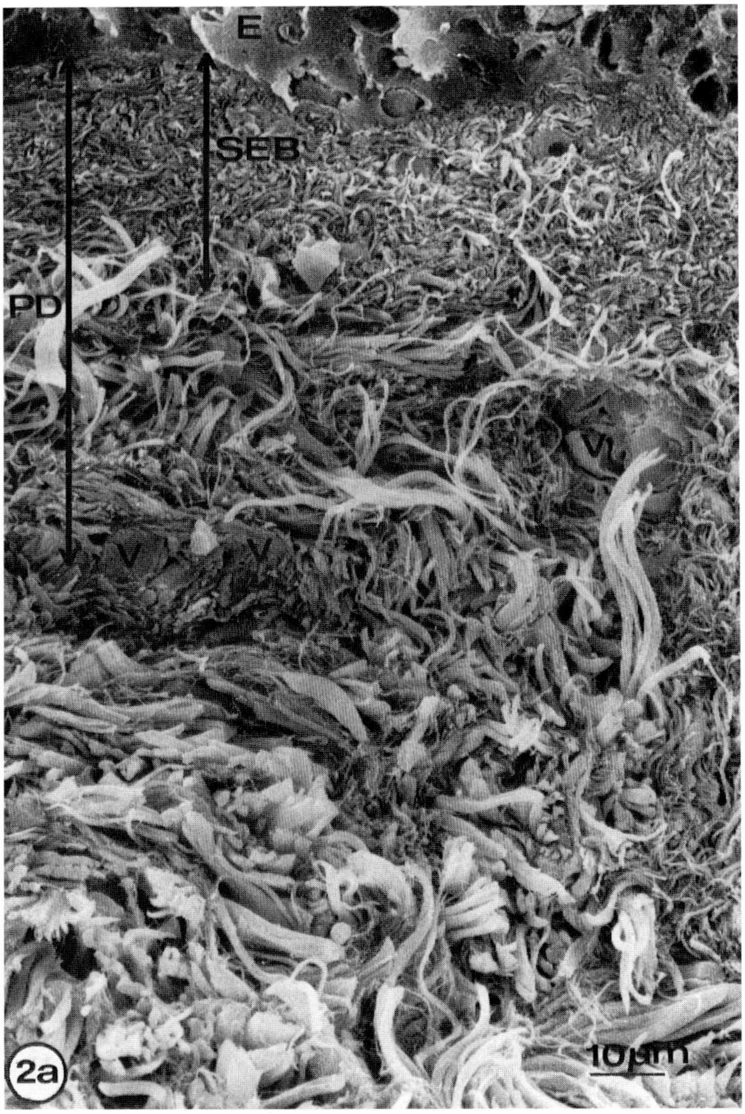

Figure 2 (a) Papillary dermis (PD) and upper part of the reticular dermis (RD_u). The thin collagen fibers seen beneath the epidermis (E) are within the subepidermal band (SEB). The vessels (V) of the subpapillary plexus form an approximate boundary between papillary and reticular regions of the dermis. X950. (b) Reticular dermis of the same individual and at the same magnification as Figure 2(a). Note the large size and the interwoven nature of the collagen fiber bundles. X950.

perpendicular to the dermal-epidermal junction and join with a horizontal network of elaunin fibers, which contain a small amount of elastin added to the microfibrillar bundles (14), in the upper reticular dermis. The largest, most complex elastic fibers are composed mainly of elastin and are found in the deep reticular dermis. Elastic fibers are components of vessel walls and the connective tissue sheaths surrounding epidermal appendages.

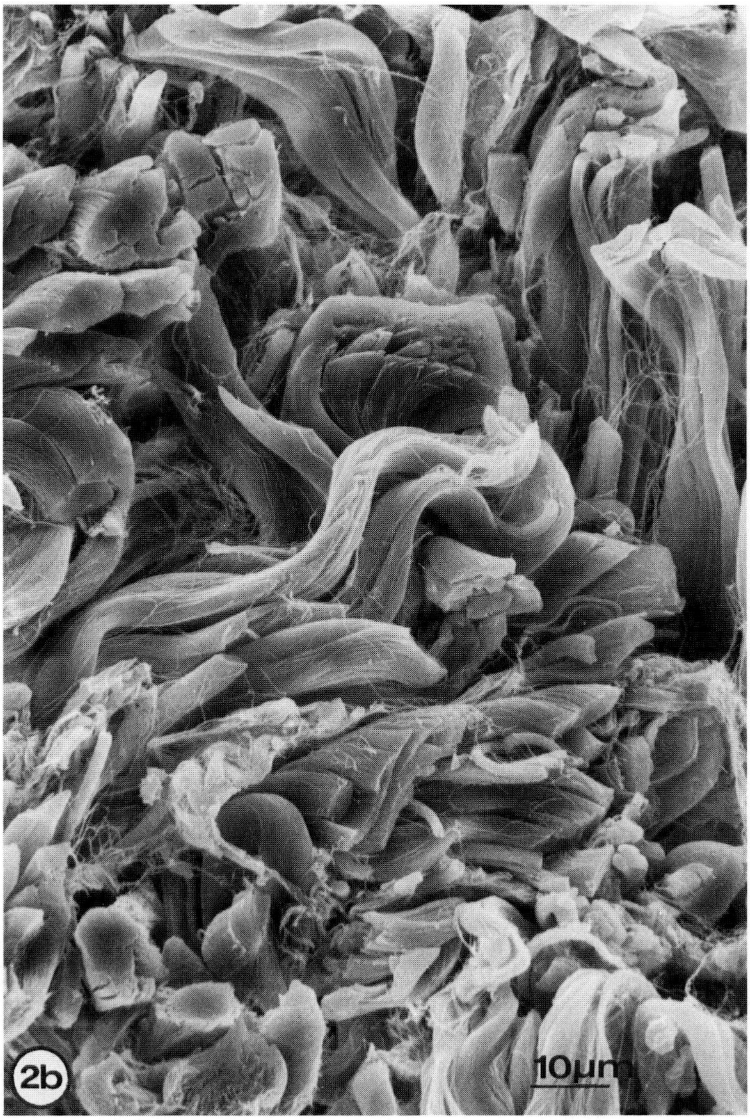

Figure 2 (Continued)

The proteoglycans and structural glycoproteins are the continuous phase of the matrix that enmeshes the fibrous components (Figure 3). They are difficult to study by light or electron microscopy because they are easily lost during tissue processing unless special ligands and stains such as ruthenium red, bismuth nitrate, or alcian blue are used. Proteoglycans bind to collagen fibrils and influence fibrillogenesis and the deposition of mineral in hard connective tissues (15). Certain glycoproteins form fine fibrillar elements in the dermis and can interact specifically with collagen. Fibronectin, for example, plays a role in organizing the extracellular matrix. It interacts with glycosaminoglycans (GAGs), binds to a specific region of collagen, mediates platelet binding to collagen fibrils and adhesion of cells to the extracellular matrix, and is found in fibrinogen-fibrin

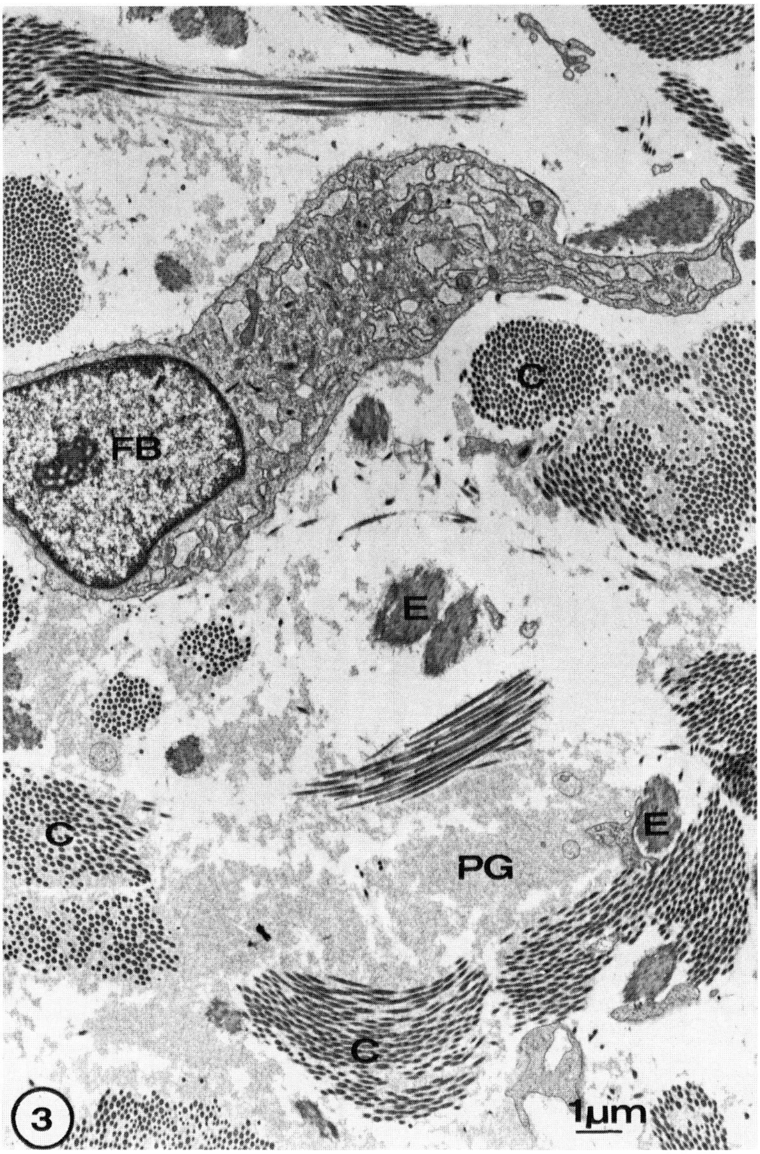

Figure 3 Dermis of a patient showing the diffuse (proteoglycan-PG) phase of the matrix. Collagen (C) and elastic fibers (E) are seen in cross-section near a fibroblast (FB). X5630.

complexes (16). Hyaluronate microfibrils have also been identified in the dermis in association with collagen and elastic fibers (17).

C. Regions of the Dermis

1. Papillary Dermis

The papillary dermis consists of two regions—a subepidermal band and the papillary dermis proper. The papillary dermis proper contains fine-diameter collagen fibrils that are organized into loosely woven bundles that are typically smaller than those of the reticular dermis (Figure 2). The fine-diameter fibrils and the small size of the fiber bundles are presumed to be a consequence of the higher ratio of type III:I collagen in this region. The diameter of type III collagen fibrils is thought to be limited by the retention of procollagen amino terminal propeptides on the collagen molecules as they polymerize (18). Enrichment for type III collagen in the papillary dermis has been demonstrated by studies with fluorescein-labeled anticollagen antibodies (1) and by biochemical analysis of papillary and reticular regions of skin that has been maintained in organ culture (10). These observations have been difficult to confirm, however, by biochemical analysis of collagens extracted directly from papillary and reticular regions of skin (10) or of collagens synthesized by fibroblasts from each region cultured separately (10,19,20). Compared with the reticular dermis, the papillary dermis has a higher density of cells, a greater proportion of vascular tissue, different glycosaminoglycans (20,21), and the highest level of collagenase activity in the skin (22). Signals from the epidermis are probably important in regulating the dimensions of the papillary dermis and the organization of the matrix contained within it.

The subepidermal band is a narrow zone of papillary dermis found immediately beneath the basal lamina (Figure 2a). It consists of a diffuse network of fine collagen fibrils that interlace with anchoring fibrils from the basal lamina and oxytalan fibers of the elastic connective tissue network. Type I procollagen and fibronectin are concentrated in this zone (1). The subepidermal band may be highly variable in thickness among individuals, although the papillary dermis varies little.

The papillary dermis appears to be a structurally stable region of the skin, even in cases where the deeper dermis is altered substantially. It is rarely affected in patients with genetic diseases, and even then it is more common for elastic fibers in this zone to be altered than collagen. In EDS type IV, for example, the thickness and organization of collagen in the papillary dermis is normal, but mature elastic fibers are found in place of oxytalan fibers that are normally present. Since these patients lack or have decreased amounts of type III collagen, one might expect the amount of collagen in the papillary dermis to be reduced and the overall dimensions of the region decreased. Substantial changes in the structure of the papillary dermis are noted, however, in induced conditions, such as actinic damage (23), and with aging (24). As in the genetic diseases, the major structural changes involve elastic fibers that undergo hypertrophy and fragmentation. A decrease in the amount of collagen has also been noted (25,26).

2. *Reticular Dermis*

Coarse, densely interwoven bundles of collagen, described as a "wickerwork" (27), characterize the reticular dermis (Figure 2b). Two regions, the uppermost and deep regions, can be identified. Types I, III, and V collagen occur in the reticular dermis. Type I collagen is presumed to form the large bundles with type III fibrils distributed at their peripheries in the loose connective tissue between them and around vessels and the dermal portions of epidermal appendages. The uppermost middermis or intermediate-dermis contains elaunin elastic fibers and intermediate-sized collagen bundles (Figure 2a). Individual collagen fibrils are intermediate in diameter between those of the papillary dermis and deep reticular dermis and appear particularly prone to alteration in diameter in the skin of individuals with inherited connective tissue disorders. There is no indication, however, that this collagen is biochemically different from that of the deeper dermis, nor is there a boundary between intermediate and deep regions; an increase in the size of both the collagen fibrils and fiber bundles occurs progressively toward the deeper levels of the tissue. The deep reticular dermis contains the largest-diameter fibrils and largest fiber bundles of any dermal region. At this level, elastic fibers form a coarse network in the loose connective tissue that surrounds the collagen bundles. Fibronectin and type III collagen have also been localized in these regions by immunostaining (28). Elastic fibers in normal skin are not incorporated into collagen fiber bundles. The intermediate and deep regions of the reticular dermis can be altered independently, or one of them can be altered exclusively, in certain connective tissue diseases. For example, the middermis is the primary location of the histopathology in the skin of patients with pseudoxanthoma elasticum (PXE) (29), type I osteogenesis imperfecta (30), and the Buschke-Ollendorff syndrome (30,31).

3. *Hypodermis*

The hypodermis or subcutaneous tissue is continuous with the reticular dermis through vascular and nerve connections, by accommodating the terminal portions of epidermal appendages (follicles and sweat glands), and through the integration of connective tissue networks of the two regions (Figure 1). Collagen in the hypodermis is organized into trabeculae, which support the adipocytes of the subcutaneous fat. Fine collagenous networks wrap around each fat cell giving them the surface appearance of randomly wound balls of yarn. Types I, III, and V collagen are all present in the hypodermis; however, like the papillary dermis, this region has a high ratio of type III:I collagen (10). The hypodermis is rarely involved in the pathological alterations of skin of patients with inherited connective tissue disorders, but it is the primary location of enhanced collagen deposition in scleroderma (12).

III. ALTERATIONS IN COLLAGENOUS CONNECTIVE TISSUE AT THE LEVEL OF THE FIBRIL

A. Overview

We have found that collagen fibril structure in single gene disorders can be altered in three major ways: by (a) an increase or decrease in diameter, (b) an unusually loose assembly in which the fibril varies along its length and is irregular in cross-section, and (c) unraveling into filamentous subunits of varying width. As we indicated above, these structural alterations do not bear a predictable relationship to the observed molecular defects, and they may be seen in skin and other tissues in both genetic and acquired diseases. Nonetheless, their occurrence in various single gene disorders provides a basis for determining different mechanisms that seem to be involved in the perturbation of fibril structure and tissue organization.

B. Altered Fibril Diameters

Fibrils within collagen bundles of the reticular dermis normally have a mean diameter of approximately 100 nm with a relatively small deviation. By contrast, collagen fibrils within the dermis of patients with several inherited connective tissue disorders can be larger or smaller in diameter than normal, or there may be variation among fibrils in diameter with the extent of variability markedly different from normal. Variability in fibril diameter can be restricted to a few collagen bundles in a region or may occur in all bundles of the reticular dermis.

1. Small-Diameter Fibrils

We have consistently found alterations in the diameter of collagen fibrils in the dermis of patients with type IV Ehlers-Danlos syndrome. The biochemical defects in this disorder include reduced or absent synthesis or secretion of type III collagen (32). EDS IV is heterogeneous in the mode of inheritance (autosomal dominant or recessive), the specific biochemical defect that leads to diminished type III collagen, and the structure of the collagen fibrils and organization of the dermis. We began our correlated structural/biochemical studies of patients with EDS type IV by examining the skin of an individual whose disorder was apparently due to a new mutation. Electron microscopic studies revealed that, compared with normal age- and sex-matched individuals, the dermis was only 25% of the normal thickness, the fiber bundles were small, and collagen fibrils were decreased in diameter by 35-40%. An apparent increase in elastic fibers was thought to reflect decreased total collagen (33) (Figure 4a). Biochemical studies of collagen from her skin revealed that type III was virtually absent (34). Fibroblasts throughout the dermis had extensively dilated rough endoplasmic reticulum (RER) suggesting storage of a protein within the cells (33) (Figure 4b). Dermal fibroblasts from this patient were grown in vitro and the collagens present in the cells and those secreted into the medium were analyzed. Her cells synthesized a form of proα1(III) that was overmodified and had altered electrophoretic mobility (34) and was apparently unable to be released from the cells. Immunostaining of permeabilized cells in culture revealed large, intracellular pools of type III collagen compared with the normal controls. The small-diameter fibrils in the skin must be type I collagen that for some reason is limited in fibril growth. These features characterize a group of patients with EDS type IV studied by us and others (35).

There are other patients with EDS type IV who are indistinguishable clinically from those above and have reduced or absent type III collagen in the skin but, instead of having all small-diameter collagen fibrils in the dermis, have a population of different-sized fibrils including those which are unusually small and large (36-38) (see below).

Dermis from patients in each group is thinner than normal, the fiber bundles are small, fibrils are small or of variable diameter, and there is an excessive amount of elastic fibers. This pattern occurs with enough consistency to suggest that these features are related to a change in the normal ratio of types I and III collagen in the dermis. We do not know enough, however, about the interactions of types I and III collagen in tissues to understand how the absence or near absence of type III collagen might lead to reduction or variability in diameter of the type I collagen fibrils in the dermis. The opposite effect might have been predicted because in vitro studies have shown that when type I collagen polymerizes in the absence of type III collagen, large-diameter fibrils and fiber bundles form. The addition of type III collagen in this system reduces the growth of type I fibrils and modifies the size of the fiber bundles depending on the ratio of the two collagen types. Type III collagen alone will form thin fibrils and assemble into small-diameter bundles (39).

Unusually thin fibrils have also been observed in the dermis of patients with OI type I (Figure 5a). In one individual we studied, the fibril diameter was 50% smaller than normal (40). Fibroblasts from that patient synthesized an abnormal α2(I) chain of type I collagen, which resulted in a decrease in the overall amount of type I collagen produced and thus decreased the ratio of type I:type III in the matrix. As in type IV EDS, an alteration at the fibril level seemed to influence the organization of the matrix at higher levels; fiber bundles were smaller than normal and widely separated, and the dermis was unusually thin (Figure 5). Small-diameter fibers have also been observed in the

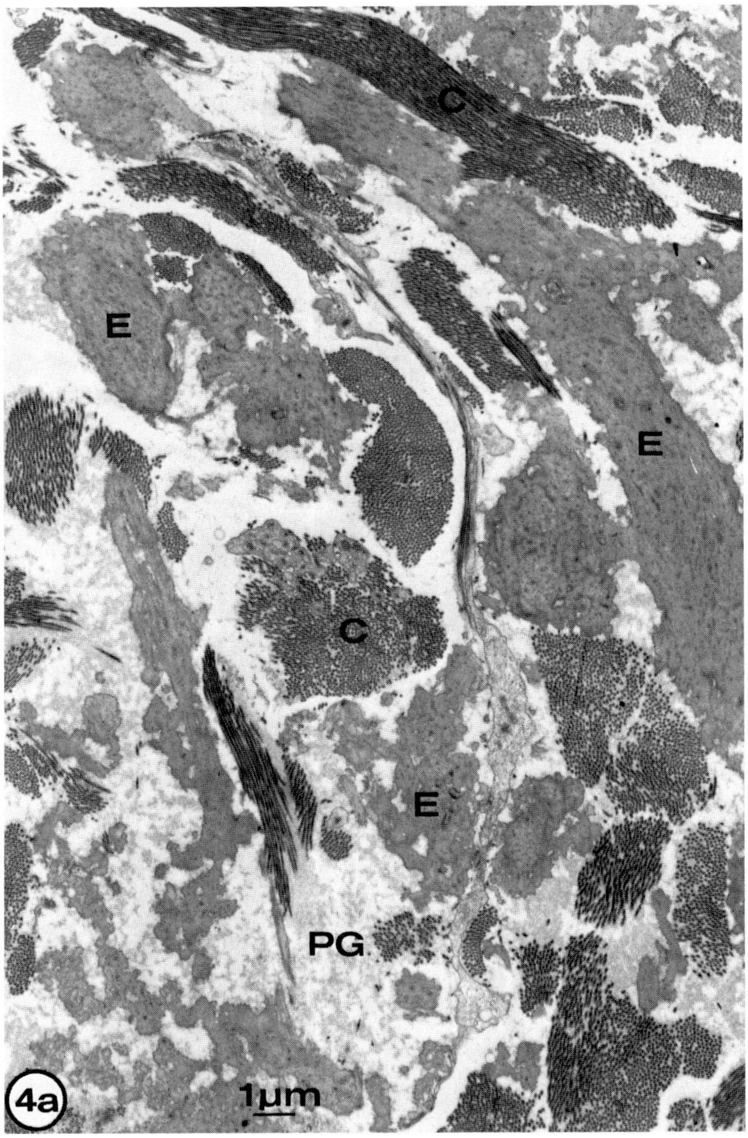

Figure 4 (a) Reticular dermis of a patient with EDS type IV. Large, branched, and irregularly shaped elastic fibers (E) occupy much of the dermis. Small collagen bundles (C) and proteoglycans (PG) are also seen. X5700. (b) Fibroblast in the dermis of a patient with EDS type IV showing the characteristically dilated rough endoplasmic reticulum (arrows). X8600.

dermis of patients with the mild form of 0I (type I) by others who have correlated their appearance with a decrease in the total amount of dermal collagen and a relative increase in the amount of type III collagen (41).

Fine-diameter fibrils are also found in the developing and remodeling dermis (healing wounds) and in the deep subcutaneous tissue of patients with scleroderma (2). In the

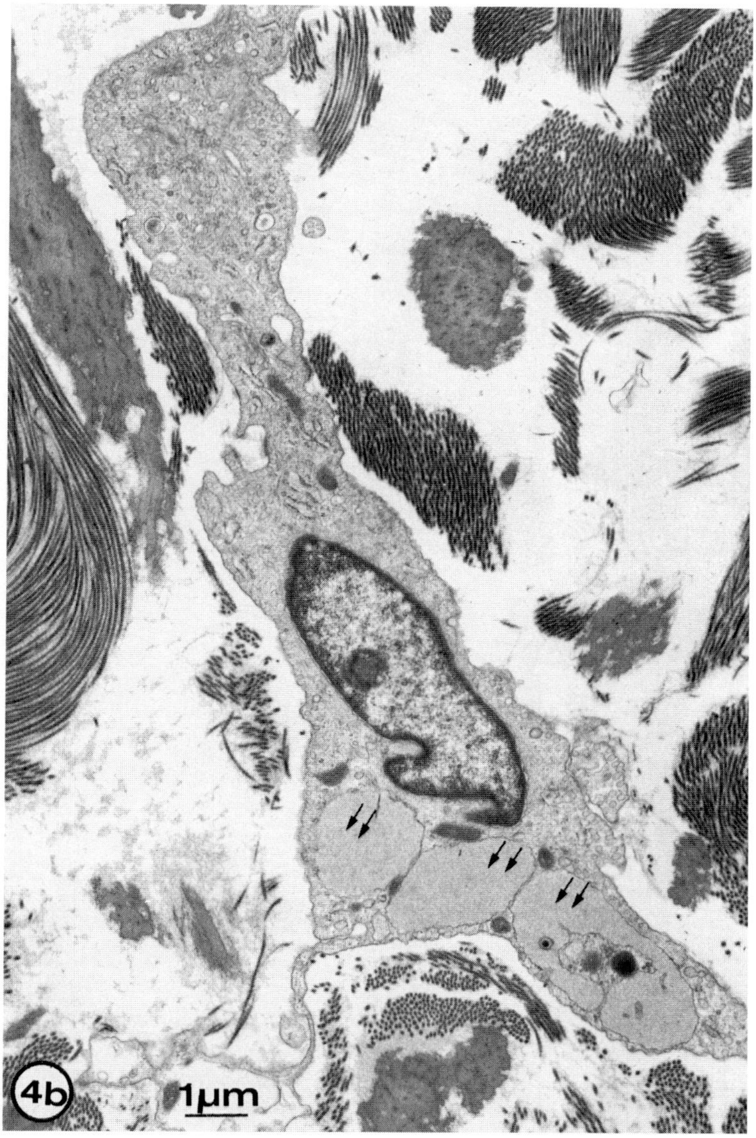

Figure 4 (Continued)

first two situations, type III collagen is present in greater abundance than in normal adult dermis (9,42,43). The unusually small-diameter fibrils characteristic of the subcutaneous connective tissue in scleroderma stain with antibodies to type III collagen. Thus, there are a number of examples of genetic diseases and developmental processes in which the presence of fine-diameter fibrils appears to correlate with an increased amount of type III collagen. The observation of fine fibrils in an environment of elevated type III collagen is more consistent with the effects of type III collagen on the polymerization and organization of type III collagen in vitro and with the observation that type III collagen in tissue forms finer-diameter fibrils than type I (2).

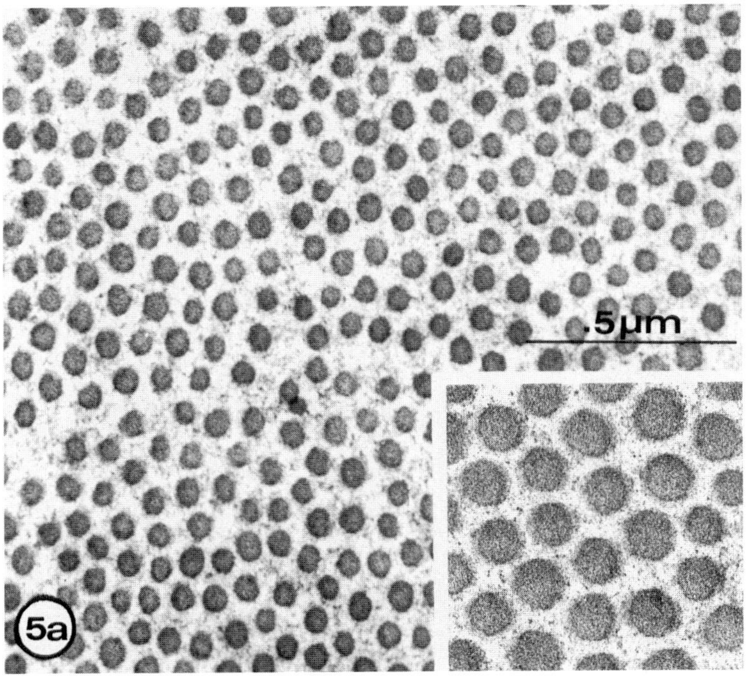

Figure 5 (a) Small-diameter, widely spaced collagen fibrils in the dermis of a patient with OI type I compared with fibrils from a normal control individual (inset, X57,000). (b) Full-thickness dermis of the patient with OI type I (Fig. 5a). The fiber bundles are very small and widely spaced. Note the clear distinction between the papillary (PD) and reticular (RD) regions of the dermis. X685.

2. Variable-Diameter Fibrils

The concurrent distribution of fibrils with large and small diameters (Figure 6) has been noted in the reticular dermis of individuals with EDS types II, III (44), IV (36,44), VII (45), and VIII (44). They have also been observed in the dermis of certain patients with cutis laxa (44) or multicentric reticulohistiocytosis (46) and as an apparently normal occurrence at early stages of dermal development in fetal skin (42), in normal tendon, and in the papillary and intermediate levels of the dermis of some normal adult individuals (Holbrook, unpublished observations). They have been seen in regions of the vessel wall where a high rate of synthesis of new fibrils was demonstrated (47).

A biochemical defect in collagen has not been found in EDS types II and III. In type VII EDS there is a defect in the collagen molecule at the site where the amino terminal propeptide is normally cleaved from the procollagen molecule (48), and in EDS type VIII, like type IV, a reduced proportion of type III collagen in the skin has been demonstrated (49). Retention of the amino terminal propeptide on types I and III collagens has been shown to regulate the diameters of the fibrils. Fibrils of type III collagen contain molecules that retain their propeptides; the extensions are believed to restrict fibril growth to a diameter of 40 nm. Type I fibrils retain the aminopropeptide only during an early phase of fibril growth when the fibril diameter is less than 40 nm. Once the propeptide is cleaved, the fibrils increase in size to approximately 100 nm (18). It is not known, however, what finally limits the diameter of type I fibrils at that size in the tissue

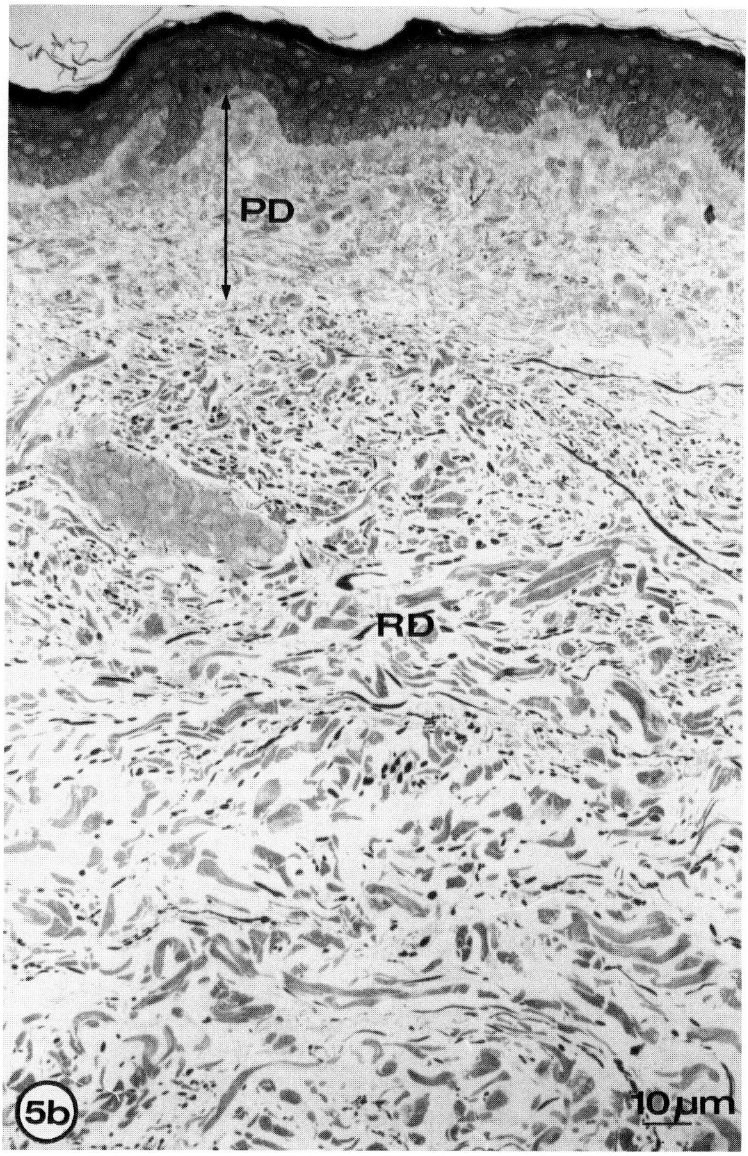

Figure 5 (Continued)

or why the fibril diameter is not better regulated in the dermis of patients with some heritable connective tissue diseases.

3. *Large-Diameter Fibrils*

Large-diameter fibrils, 40-45% greater than normal, were observed in the reticular dermis of patients with X-linked cutis laxa (decreased lysyl oxidase activity) (50) (Figure 7a). Fibrils that measured 13-40% greater than normal were found in patients with types I, II, III, V, and VI Ehlers-Danlos syndrome (EDS) (51-53). Fibrils with exceptionally large diameters are often closely packed (Figure 7b) and can become distorted in shape by the

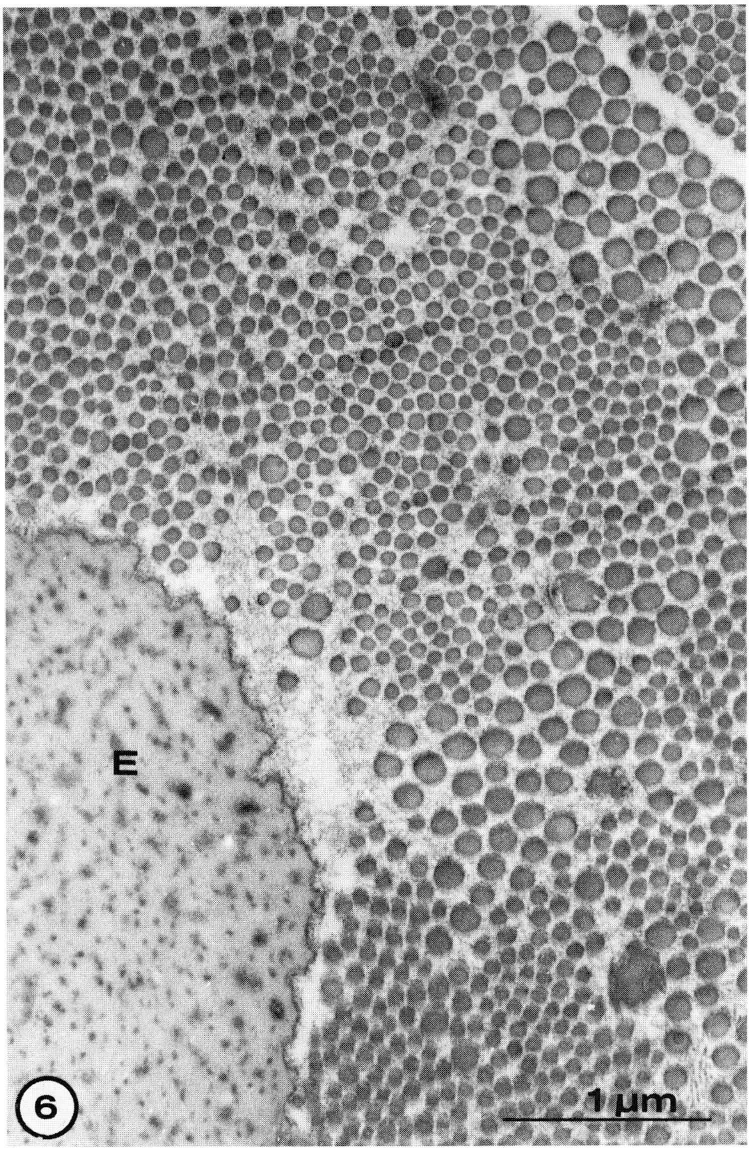

Figure 6 Collagen fibrils with a range of diameters seen in the dermis of a patient with EDS type IV. Note that there are also some fibrils with an irregular contour. A portion of an elastic fiber (E) is also shown. X27,000.

apparent pressure of contact (Figure 7a). Since fiber bundles lack rigid boundaries, it is not apparent why the close packing and distortion occur, unless the amount of proteoglycan associated with the fibril is diminished or the total amount of collagen in the dermis is excessive.

4. Summary

From the examples discussed, it appears that several factors may regulate fibril diameter during fibrillogenesis: the ratio of types I and III collagen in the tissue (see discussion of EDS types IV and VIII, OI type I (9,54), the retention of the amino terminal propeptides on collagen chains (EDS type VII) (18), the developmental status of the tissue (e.g., fetal dermis and healing wounds) and the rate of collagen synthesis (vessel wall). Thus far, only the first two factors have been shown to correlate with altered fibril diameters in diseases.

A relationship may exist between fibril diameter and the amount and type of carbohydrate that associates with the collagen fibrils (through glycosylation or interaction with proteoglycans and glycoproteins in the extracellular matrix). Collagens and collagenous structures (e.g., in cornea, heart valve, wound tissue) that have a high sugar content seem to contain or form fine-diameter fibrils (55,56). Proteoglycans added to collagen solutions during polymerization affected both the rate of polymerization and the diameter of the resultant fibril. When they were added at early stages of polymerization, the assembly of the fibril was delayed and its diameter was large (57,58). On the other hand, fibrils that polymerized in the presence of glycoproteins were thin (59). In the disorders we have mentioned, it is not known whether or how proteoglycans might be involved; however, with another type of fibril alteration, the composite fibril (see below), there is morphological evidence, at least, that they may play a role in the assembly of this structure.

C. Composite Collagen Fibrils/Collagen Flowers

Composite fibrils are large in diameter and lobulated (Figure 8a). They have been seen in several heritable connective tissue disorders (EDS types I, II, III, VI, X, type I OI, the Marfan syndrome, PXE, etc.) and have been given a number of names that describe their appearance in longitudinal and cross-section. Lengthwise, they look like loosely aggregated bundles of collagen subunits wound in a shallow spiral (Figure 8b). From this impression they have been called collagen ropes or wires. The names given to their appearance in cross-section reflect the extent to which the subunits coalesce; those which are near normal in diameter and are more tightly associated are called serrated fibrils (53) (Figure 8c). Other larger-diameter, loosely associated fibrils have been described as lobulated, fused, composite, dysplastic, and aggregate, and as collagen flowers (Figure 8a). Collagen pinwheels have been seen in the tissue of some patients with OI type I (Figure 8d). Some of the fibrils in this category are so irregular (e.g., in the reticular dermis of very young children with EDS type I) that they are impossible to call by any descriptive term (Holbrook and Byers, unpublished observations) (Figure 8e). The term "composite" (60) appears to be the best general term to describe such fibrils, recognizing that the pathogenesis may be heterogeneous.

1. Composite Fibrils in Inherited Disorders with an Unknown Biochemical Defect

In our experience, composite fibrils are invariably present in the dermis of patients with the dominant forms of EDS (I, II, and III), particularly type I, and thus can be listed among the diagnostic criteria. They occur among fibrils that have a normal, round cross-sectional shape, but typically a larger-than-normal diameter (Figure 8a). The numbers of these fibrils and the severity of their altered structure vary and seem to influence the structure of the dermis at higher levels of organization. In most adult patients with EDS type I, the percentage of composite fibrils is usually around 5%; the fiber bundles are

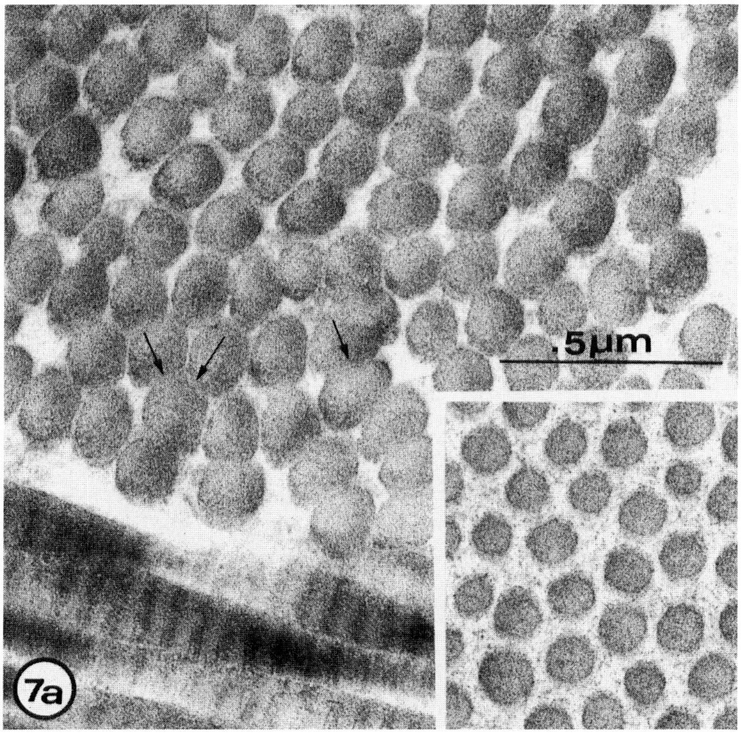

Figure 7 (a) Large-diameter collagen fibrils in the dermis of a patient with EDS type V compared with those from normal control individual (inset). Note the distorted fiber shape apparently caused by pressure from adjacent fibrils (arrows). X57,000. (b) Densely packed collagen fiber bundles in the dermis of a patient with EDS type V. Note that an elastic fiber (E) is embedded between the fiber bundles. X12,150.

often smaller than expected but can be packed into the dermis in a seemingly normal manner. At the light microscopic level the only atypical finding is the presence of small fiber bundles. Nonetheless, these patients exhibit the classic characteristics of hyperextensible, fragile skin, hypermobile joints, and easy bruisability and laceration. Wounds gape and heal leaving atrophic, shiny scars. Jansen (27) describes the "insufficient twining of the collagen elements" in the dermis, hypodermis, and joints as a cause of unrestricted sliding of bundles across one another, hence failure to resist extension. This is not evident at the microscopic level in most adult patients with EDS type I, but could be a factor in the altered mechanical properties of the skin and joints in small children who are affected.

We have examined the skin from a newborn infant and two young children who have EDS type I. In all three, the only normal-appearing fibrils are found at the periphery of the fiber bundles in the reticular dermis (Figure 9). The majority of the fibrils are randomly oriented, composite fibrils that are more abnormal appearing in structure than the more symmetrical, flowerlike fibrils seen in the dermis of affected adult members of their families (Figure 83). The organization of these fibrils in the bundles gives them a "mushy" appearance at the histological level. The bundles are not only disorganized internally, but they fail to intercalate with others in a normal, woven pattern or "wicker-

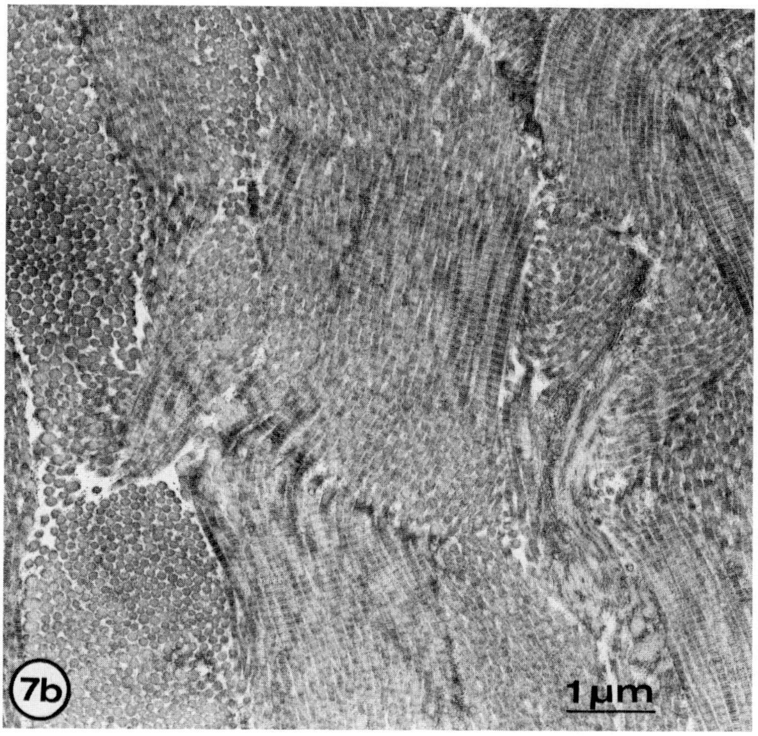

Figure 7 (Continued)

work"; they appear instead as oval units, isolated by exaggerated amounts of loose connective tissue in the interstitial spaces. This organization is similar to Jansen's observations and could account for the cutaneous hyperextensibility. We and others have observed altered packing of fibrils into bundles in the dermis of patients with several types of EDS (44,53,61-63).

The reason for the formation of composite fibrils in the dermis of patients with type I EDS is not clear. Biochemical studies of collagens in the skin from these patients and of their cultured fibroblasts have not revealed alterations in the collagen molecule. However, Shinkai et al. (64) found that the fibroblasts from one individual with type I EDS synthesized more proteoglycans in culture than control cells. The influence of proteoglycans and glycoproteins on collagen fibrillogenesis and fibril diameter has already been discussed (p. 117). It is particularly noteworthy, with regard to the composite fibril, that structures with this morphology were formed in vitro when collagen molecules were allowed to polymerize in an environment enriched in proteoglycan monomer (58).

2. Composite Fibrils in Disorders with Recognized Collagen Defects

Composite fibrils occur in single gene disorders (other than EDS I, II, and III) in which there is a primary defect in the collagen molecule (EDS VI, OI type I, and the Marfan syndrome) (44,53,65-67). In EDS type VI, there is a deficiency in the enzyme lysyl hydroxylase (68) in the connective tissues of certain, but not all, organs. Types I and III collagen are affected; types IV and V are not, suggesting different isoenzymes of lysyl hydroxylase (69). There is a complete absence of the enzyme in skin of most of these

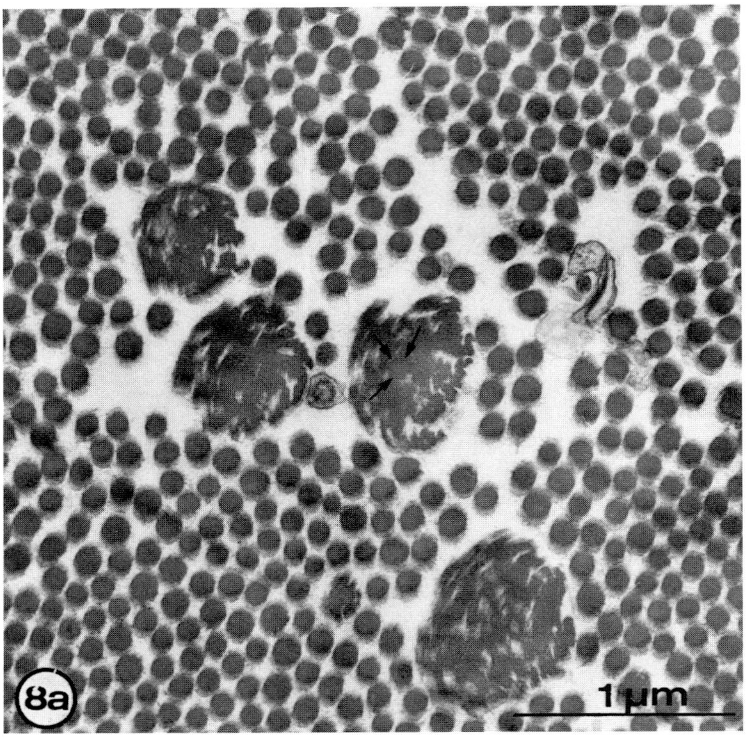

Figure 8 (a) Composite fibrils typical of those found in the dermis of patients with EDS types I, II, III, VI, and X viewed in cross-section. Note that a fibril with a diameter similar to that of normal-appearing fibrils can be seen in the center of some of the composite fibrils (arrows). X27,150. (b) Composite fibrils seen in longitudinal section. Note that the banding pattern is carried across the loosely integrated fibril. X15,150. (c) Composite fibrils with a serrated appearance found in the dermis of a patient with the Buschke-Ollendorff syndrome. Note that the peripheral material of the abnormal fibrils appears to be added onto a fibril of normal diameter and contour. X43,350. (d) Collagen "pinwheels" in the dermis of a patient with OI type I. Note the variation in diameters of the more normally shaped fibrils. X45,300. (e) Collagen fibrils from the dermis of an infant with EDS type I.

patients, and as a result, types I and III collagen fibrils in the dermis are underhydroxylated. Because some hydroxylysyl residues are sites of glycosylation, the collagen molecules are likely to be underglycosylated as well (70). Underhydroxylation and altered glycosylation may each interfere with the formation of intermolecular cross-links, decrease the stability of the collagen fibril, and perhaps account for the fragility and hyperextensibility of the skin that are among the clinical features of the disease. Although lysyl residues are substrates for lysyl oxidase, the enzymatic component of the cross-link pathway, hydroxylysyl-derived cross-links are more stable. Decreased glycosylation of the molecule can alter its surface charge and modify the interaction of collagen with other components of the extracellular matrix. Thus, the molecular defect in collagen modification in EDS VI probably affects fibril structure through aberrant interactions of collagen molecules with other components of the matrix.

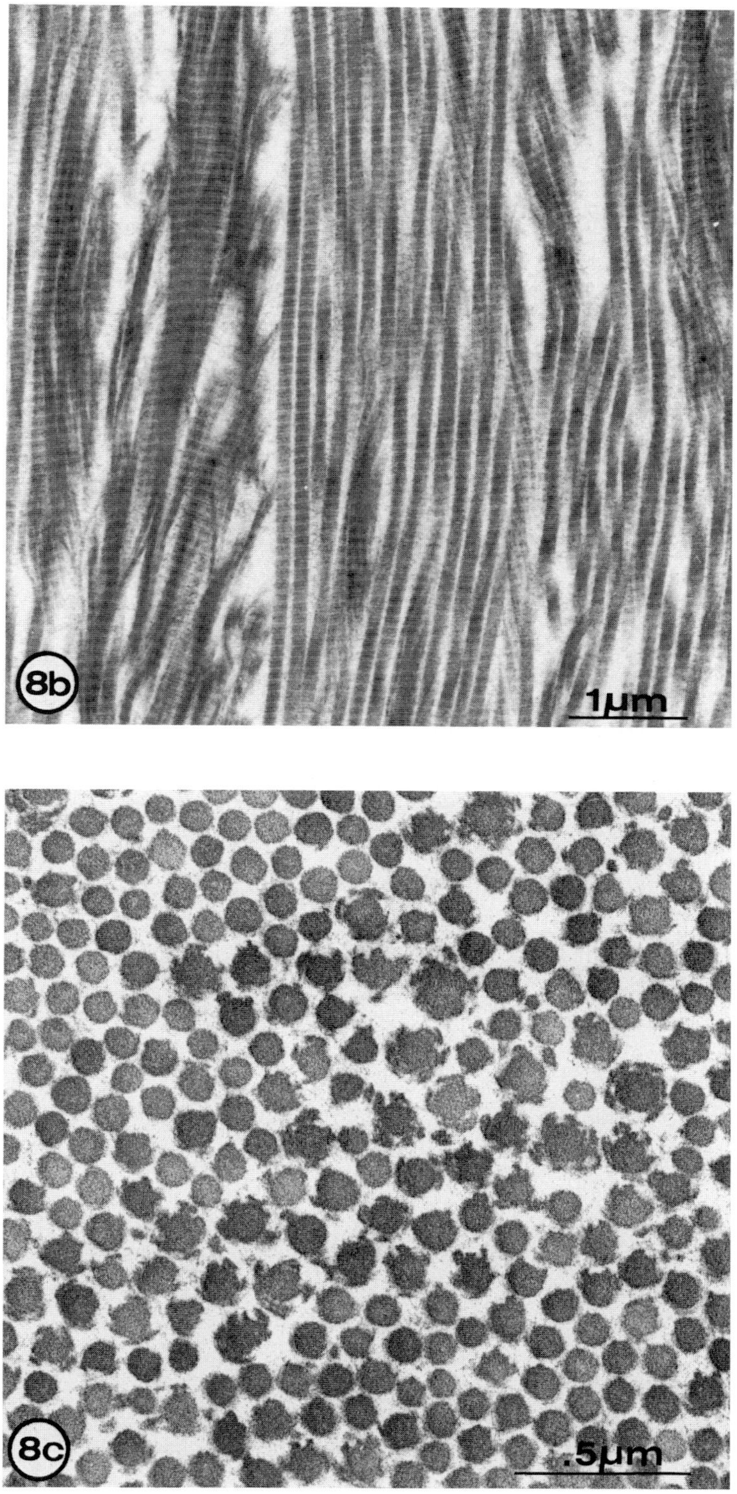

Figure 8 (Continued)

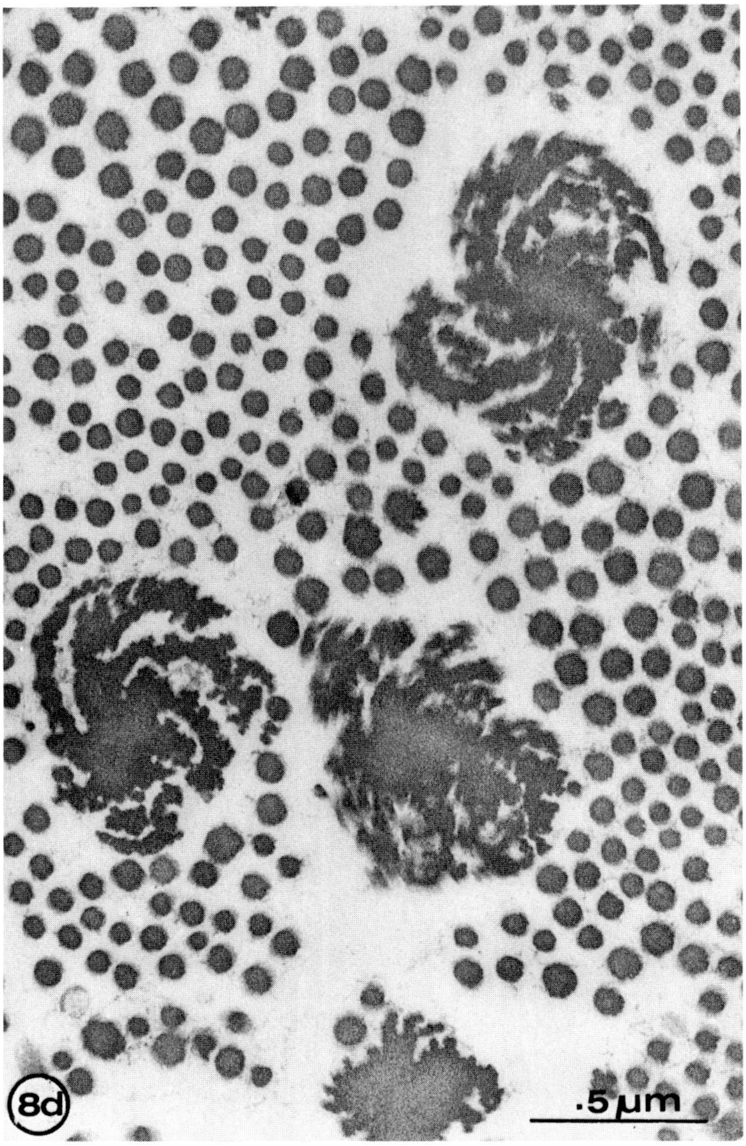

Figure 8 (Continued)

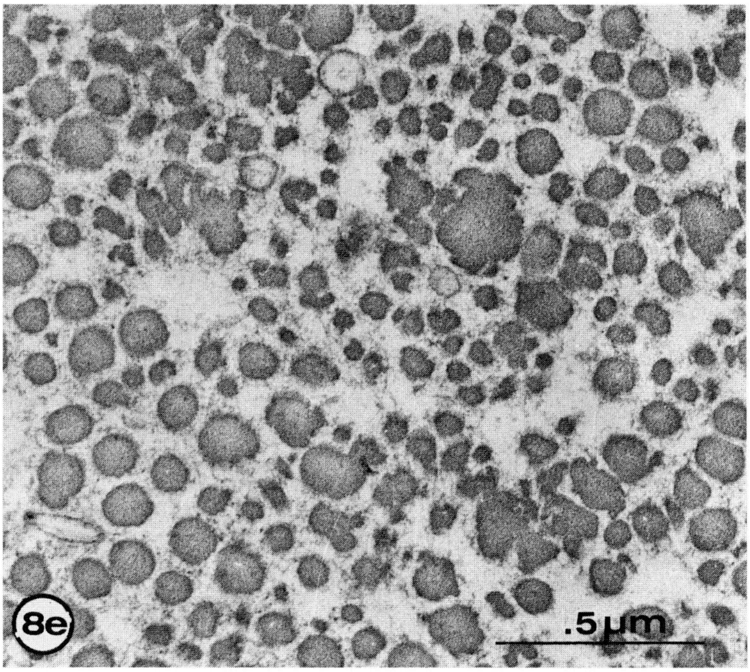

Figure 8 (Continued)

Composite fibrils are also prominent in the dermis of the affected members of two families with 0I type I that we have studied. In members of one family, the altered fibrils are extremely large in diameter and have a lobulated "flower" morphology (Figure 8b). They are present only in bundles of the reticular dermis, which, like those of patients with EDS type I, are quite small. The fibroblasts from the patients with 01 type I secrete about half the normal level of type I procollagen because of a defect in the synthesis of proα1(I) chain (71). The overall effect of the reduced proα1(I) synthesis is a decrease in type I collagen in the matrix and thus an altered ratio of types I and III collagen in the dermis (54). We have already indicated that the collagen fibrils are small in diameter and the fiber bundles are smaller than normal in patients with 0I where there is a relative increase of type III collagen; composite fibrils can also be found in an environment of altered collagen ratios. It should be pointed out that this family serves as an example of the many instances in which there may be differences in the morphology of collagen fibrils in patients with the same disorder, even in the same family.

The biochemical defect is the same in the second family, decreased production of type I collagen, but the clinical disease is variable among the family members. The alterations in fibril structure are correspondingly variable; the individuals with the most severe clinical disease have large numbers of composite fibrils that are more extreme in their structural alteration than any others we have seen (Figure 8d). All the altered fibrils have long, ragged-appearing arms that seem to radiate spirally from a central axis. Those of less severely affected members have the most typical EDS I-type composite fibrils (Figure 8a).

Composite fibrils have been seen in the aorta and valves of an individual with the Marfan syndrome (67), a disorder in which vascular fragility is a prominent feature. In this

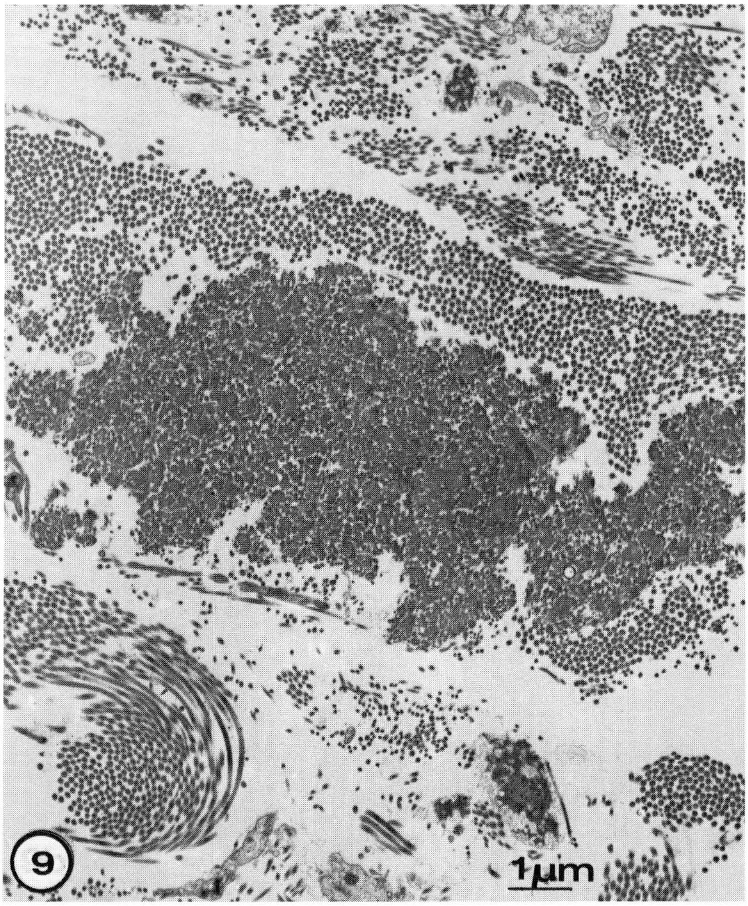

Figure 9 Altered collagen fiber bundle in the dermis of an infant with EDS type I. Note that the fibrils within the center of the bundle are abnormal; those at the periphery are apparently normal in diameter and contour. X7400.

particular individual about half the α2(I) chains contained an insertion of approximately 20 amino acids in the central region of the triple helical domain (72). Although this insertion appeared to interfere with the normal cross-linking (the solubility of collagen in the skin was increased), it did not result in fibril alterations in the dermis. It is possible that the increased mechanical stress on the vascular tissue, compared with the skin, was a factor. Increased solubility of collagen (73) or elastin (74) and abnormal metabolism of GAGs (75) with accumulation in the wall of the aorta (76) have been suggested as basic molecular defects in the Marfan syndrome. In all instances where collagen fibrils in vessels have been examined, except the case of the individual with the insertion in the α2(I) chain (67,72,77), the structure of collagen fibrils appears to be normal. Dermal collagen fibrils in all instances have been normal.

3. *Composite Fibrils in Disorders Where Defects Occur in Noncollagenous Matrix Components*

Composite fibrils can also occur in genetic diseases where the biochemical basis of the disease is unknown but structural alterations in elastic fibers are prominent and exist concurrently with structural alterations in collagen. The skin of patients with the Buschke-Ollendorff syndrome illustrates this point. This disease is inherited as an autosomal dominant disorder involving a bone dysplasia (osteopoikilosis) and characteristic cutaneous papules or nodules (dermatofibrosis lenticularis disseminata). Elastic fibers are large, branched, and fragmented and in some areas are assembled as irregular globular units of elastin (31). Elastic fibers can also insinuate into collagen bundles (Holbrook and Pinnell, unpublished observations) (Figure 10). Composite fibrils are found in portions of fiber bundles adjacent to the altered elastic fibers (31) or throughout the entire fiber bundle in large areas of the dermis. The extent of alteration of the collagen fibril structure can also vary; they can resemble the composite fibrils found in the dermis of patients with EDS I, or they can have more nearly circular outlines that appear serrated

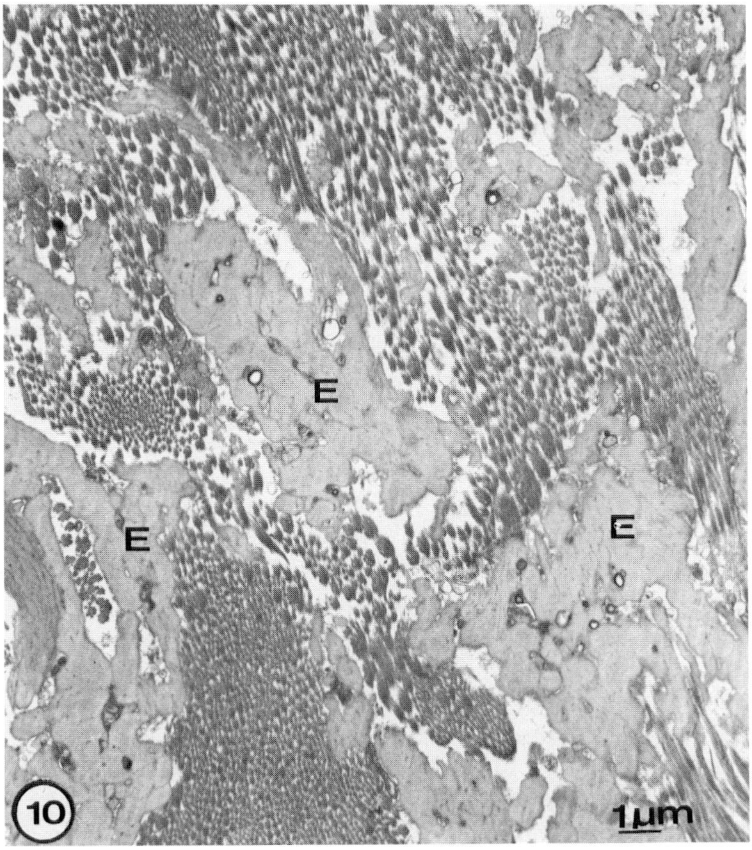

Figure 10 Dermis of a patient with the Buschke-Ollendorff syndrome. Note the structurally abnormal elastic fibers (E), which insinuate among collagen fiber bundles, and the abnormal structure and diameter of the collagen fibrils in this region. (Patient of Dr. S. R. Pinnell.) X4850.

(Holbrook and Pinnell, unpublished observations) (Figure 84c). In both instances the abnormal connective tissue was localized in the middermis and created a disorganized appearance in this region (30). It is not known whether or how the structure of collagen is altered in bone.

Lesional tissue from the skin of several patients with the Buschke-Ollendorff syndrome revealed a marked increase in the amount of desmosine, in accord with the morphological observations of increased elastic fibers in the tissue (31). The levels of collagen were normal. It is not known, however, how the increase in amount of elastin relates to the underlying defect in this disorder. Because bone, skin, collagen, and elastic fibers are affected, it is likely that the defect is not one of elastin primarily, but instead an alteration that is common to the metabolic pathways or physical aggregation of collagen and elastin.

Abnormal relative amounts of elastic and collagenous connective tissue, associated with composite fibrils, are also seen in other genetic disorders: cutis laxa, PXE, and α1-antitrypsin deficiency, and in the elastosis perforans serpiginosa (EPS) lesions often noted in patients with EDS IV, the Marfan Syndrome, OI, and PXE (78). In the autosomal dominant or recessive forms of cutis laxa, the amount of elastic connective tissue is decreased; the fibrils present in skin are fragmented and/or have an abnormal association of the elastin and microfibrillar components. Composite fibrils were observed in low numbers in the fiber bundles in the reticular dermis (79). Dermis from a patient with α1-antitrypsin deficiency had a decreased amount of elastic connective tissue, elastic fibers that were ragged and deficient in the amount of elastin, and collagen fibrils with the composite morphology (80).

An increased amount of elastic connective tissue in the middermis is also a hallmark of the tissue pathology in PXE (81-83). Composite collagen fibrils occur in PXE as a conspicuous and consistent ultrastructural feature (82,84,85). The basic biochemical defect in PXE is unknown, although there are biochemical data which show that collagen, elastin, and GAGs in the skin can be abnormal. Biochemical studies of the collagen from the skin of affected individuals have revealed an increase in the amount of soluble collagen (86), an apparently normal rate of collagen synthesis, but a decrease in total collagen—interpreted as an abnormality in collagen degradation (87)—elevated lysine and hydroxylysine residues in lesional tissue compared with uninvolved sites (88), or collagen that is modified neither in amount nor in composition (84). A total increase in elastin in the skin of affected patients has been measured, and its amino acid composition has been found to deviate from that of normal elastin (84). Increased levels of proteoglycans and glycoproteins have been determined biochemically (84,86), observed in the tissue histochemically (86,89), and visualized ultrastructurally as abundant masses of ruthenium red-positive granulofilamentous material (84). These may be present as a consequence of the activity of an elevated proteoglycan protease in the tissue (90). Unusually high levels of fibronectin, associated with the microfibrils of elastic fibers and types I and III collagen in lesional sites, have been found in the dermis by immunoelectron microscopy (91). Thus, even though the primary defect in PXE is still unclear, much of the evidence for a biochemical defect could support a hypothesis that the extracellular matrix environment (proteoglycans, glycoproteins, GAGs) in which collagen fibrils and elastic fibers polymerize interferes with their normal assembly.

EDS X was recently described as a mild, recessively inherited form of EDS in which the joints are hypermobile and the skin is hyperextensible. A defect in the ability of the platelets to bind to collagen was found to be correctable in vitro by the addition of normal human plasma fibronectin (92). Large composite fibrils were observed in collagen bundles of the reticular dermis (44) (Figure 8a), thus suggesting that an abnormality in the glycoprotein fibronectin that is regularly present in the dermis may be important in the normal assembly of collagen fibrils. Fibers were also poorly integrated into fiber bundles (30).

4. Composite Fibrils in Nongenetic Disorders of the Skin

From the results of the correlated structural and biochemical studies of the dermis and cultured fibroblasts from patients with inherited connective tissue disorders, it appears that composite fibrils can form in conditions where there are primary defects in the collagen molecule, when the ratios of types I and III collagen are altered in the tissue, or in disorders where the primary alteration appears to be in another component of the extracellular matrix. The variety of circumstances under which composite fibrils can occur may explain why similar appearing composite fibrils can be observed in other skin diseases, e.g., amyloidosis (93), Shagreen patch in tuberous sclerosis (94), and hyalinosis cutis et mucosae (95) and in a variety of nongenetic conditions of the skin and other tissues. They have been seen in chronically sun-exposed skin (96), in PXE-like plaques caused by contact of the skin with the fertilizer, Norwegian Hydrous Salpeter (97), in D-penicillamine-induced elastosis perforans serpiginosa lesions in patients with Wilson's disease (98) and in cutis laxa-like lesions in patients infected with *Onchocerca volvulus* (60). They have been recognized in the fibrotic liver (99), fibrolamellar carcinoma of the liver (Hanna, personal communication), the synovium in rheumatoid arthritis, palmar aponeurosis (Dupuytren's) (100), flexor retinaculum (carpal tunnel syndrome), vessels of patients with temporal arteritis, diabetic angiopathy and tumors of the brain (101,102), varicose veins (60,102), and in distressed respiratory tissue (60). They are also common in the walls of arteries of the spontaneously hypertensive rat (101).

It is not known when composite fibrils form, whether they reflect a defect in fibrillogenesis or are a consequence of degradative phenomena or are a permanent or transient phenomenon in a tissue. Several factors support the origin of composite fibrils during fibrillogenesis, although the biochemical evidence for this is scant. It has been shown that fibrils with the composite morphology polymerize in vitro if proteoglycan monomers are added to the mixture (58). These conditions might occur in the tissue in a number of disorders where elevated proteoglycans and GAGs have been measured or in instances where lysosomal enzymes in the tissue could hydrolyze proteoglycans and in turn provide an environment that would promote "intrafibrillar dysplasia" of the collagen (101,103). In vessels, lysosomes were abundant in areas of the tissue where progressive development of composite fibril structure was observed. No one has specifically looked for the lysosomes or tested if hydrolytic enzymes are elevated at the sites where composite fibrils occur in the dermis, but because proteoglycans can also influence the formation of elastic fibers (104), a hypothesis that recognizes an abnormality in this component of the matrix would be an attractive explanation of the many instances where there are simultaneous alterations in collagen fibrils and elastic fibers (44,60). The composite fibrils often contain a "core," which is approximately the same in density and diameter as a typical collagen fibril (Figures 8a and c). The extra peripheral material appears to be added secondarily, in excess of what would be incorporated normally into the fibril.

Some degradative events might also be involved in the formation of composite fibrils, perhaps directly through the action of hydrolytic enzymes on the collagen. Extracellular lysosomes have been seen in association with fragmented elastic fibers in PXE (105), and the release of elastase into the tissue of a patient with 1α-antitrypsin deficiency resulted in elastic fiber degradation, in both instances composite collagen fibrils were found adjacent to these areas of elastic fiber damage. There are many other potential sources of connective tissue proteases in the dermis (e.g., mast cells, fibroblasts, polymorphonuclear leukocytes, macrophages, etc.), but a direct correlation between release of enzymes from these cells and the structural alterations in patients with inherited connective tissue diseases has not been made.

D. Hieroglyphic Fibrils

The hieroglyphic fibrils (Figure 11) found in the skin and tendon of sheep (106), cattle (107,108), cats (109, 110), and dogs (44) with dermatosparaxis could be considered a specific type of composite collagen fibril (Figure 11) because like them, they are loosely aggregated and composed of individual, spiral subunits. They differ from composite fibrils, however, in that they are small, twisted ribbons or sheets that in cross-section bear no resemblance to the normal or deviated, round fibril contour. This is one of the few examples (if not the only one) of an alteration in collagen fibril structure that correlates uniquely with a molecular defect. Animals with dermatosparaxis are deficient (to varying degrees) in the enzyme that cleaves the amino terminal propeptide of type I procollagen (109,111-113); the persistent propeptide hinders normal fibril assembly and functionally impairs the connective tissue (114). The extent of fibrillar disarray is correlated with the degree of enzyme inhibition. Thus, in the cat (109,110), in which only a small portion of the fibrils retain the propeptide extensions, the fibrils are less severely altered than in other dermatosparactic animals. An effect of the fibril alteration on dermal architecture has also been observed in all the dermatosparactic animals. The fiber bundles do not have the normal interwoven structure but are organized in a laminar pattern, in some cases parallel to the plane of the tissue (115); in the cat, the bundles lie perpendicular to the skin surface (110). This aberrant architecture may account for the extreme fragility of the skin, which tears during birth leaving denuded areas of the body. As a consequence many animals survive for only a short period.

The molecular alteration-structural alteration relationship seen in dermatosparaxis does not carry over to the analogous disorder in humans in which the collagen fibrils also retain the amino terminal propeptides of type I collagen fibrils. Patients with EDS type VII have a defect in about half of the proα2(I) chains that prevents normal cleavage of the amino terminal propeptide (48) but do not have collagen fibrils in the skin or other tissues with the same structure as seen in dermatosparaxis, laminar organization of the dermis, or marked skin fragility.

E. Unraveled Collagen Fibrils

Collagen fibrils with an unraveled appearance often occur in tissues in which fibrils with an altered diameter and/or the composite morphology are also found. Two types of "unraveled" fibrils can be distinguished by the size of the components into which they dissociate. In one type, the fibril separates into parallel, banded subunits (often referred to as microfibrils) approximately 200 Å in diameter (Figure 12a). It is not known whether this alteration occurs only at the ends of fibrils or whether the fibrils can dissociate into these elements at intervals along their length. In the other type, the fibril dissociates into fine filaments (often called microfilaments) that retain a banded pattern only at the point

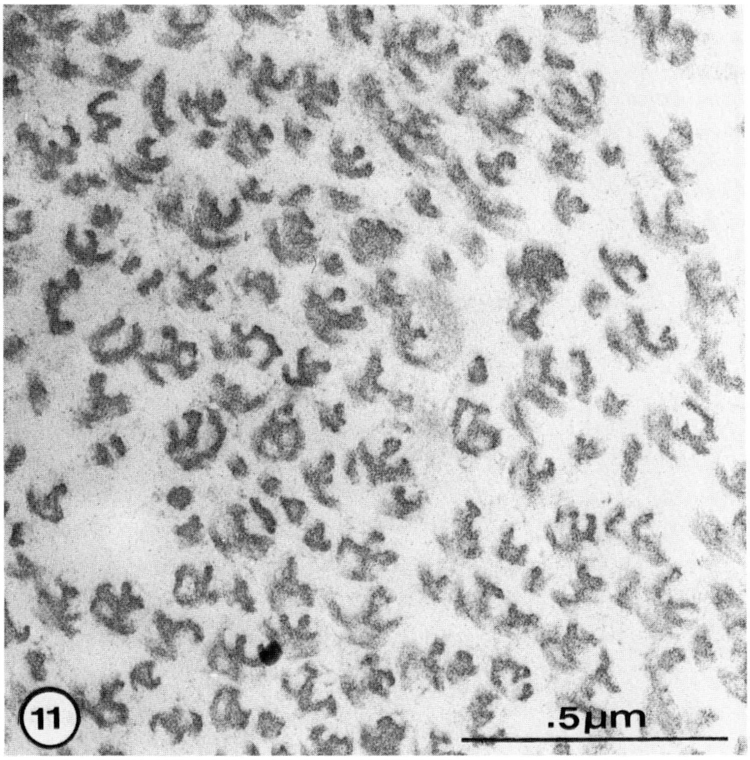

Figure 11 Hieroglyphic fibrils in the dermis of a dog affected with dermatosparaxis. X70,300.

where they begin to splay out from the fibril (Figure 12b). The altered regions can occur at the ends of the fibril and at intervals along its length. It is possible that these are not separate types of alterations but that microfibrils are formed as a stage in the progression to the microfilaments. This sequence is observed in experimental conditions where collagen is denatured by treatment with urea or guanidine (116).

There are examples where the microfibrils are clearly continuous with the collagen fibril (117) (Figure 12a). In other instances they seem to intermingle with (and either coexist with or replace) collagen fibrils within bundles (Figure 12c); thus, they may be newly synthesized matrix fibrils or degraded products of the old. They have been recognized in the dermis of individuals with the Marfan syndrome (118), spondyloepiphyseal dysplasia (SED) (117), hyalinosis cutis et mucosae (95), amyloidosis (93), and tuberous sclerosis (Shagreen patch) (94). They are also observed in damaged tissue (e.g., solar elastosis; 119).

1. Unraveled Fibrils in Heritable Disorders

An example of a situation where the more completely unraveled collagen fibrils have been observed is the dermis of a patient who had corneal dystrophy and short stature with SED (117). His skin was soft and velvety, and when it was studied by light microscopy, large areas of disorganized collagen bundles were seen to be scattered in the reticular dermis (Figure 13a). At the ultrastructural level, these were found to correspond to collagen

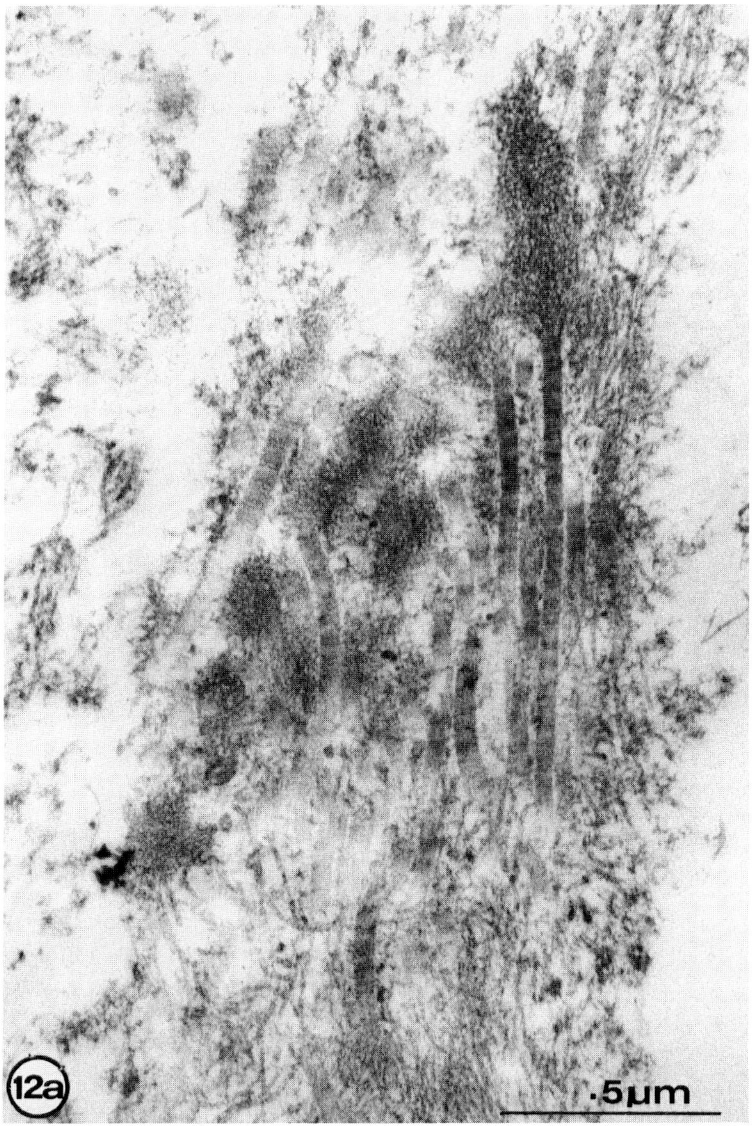

Figure 12 (a) Microfibrils interwoven with an unraveled collagen fibril. X57,000. (b) Collagen fibrils in the dermis of a patient with a form of SED. The fibrils appear to be unwound into filaments. X38,000. (c) Collagen fibrils intermingled with microfibrils in the dermis of a pateint with SED. X38,000.

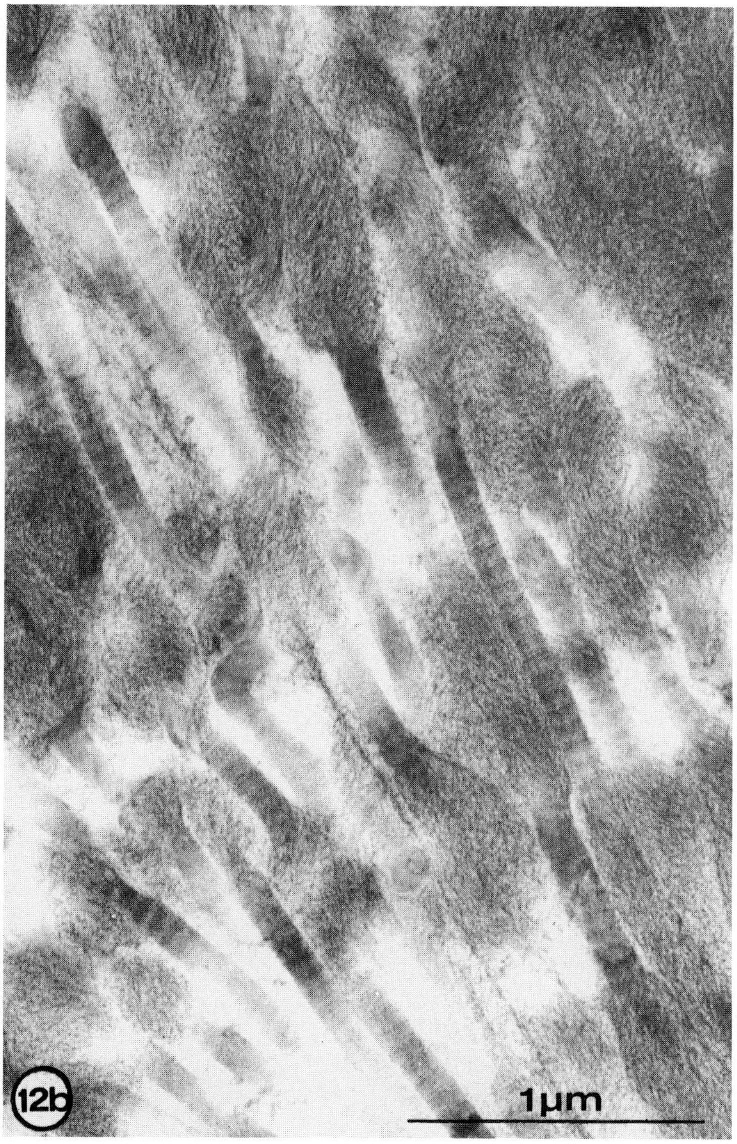

Figure 12 (Continued)

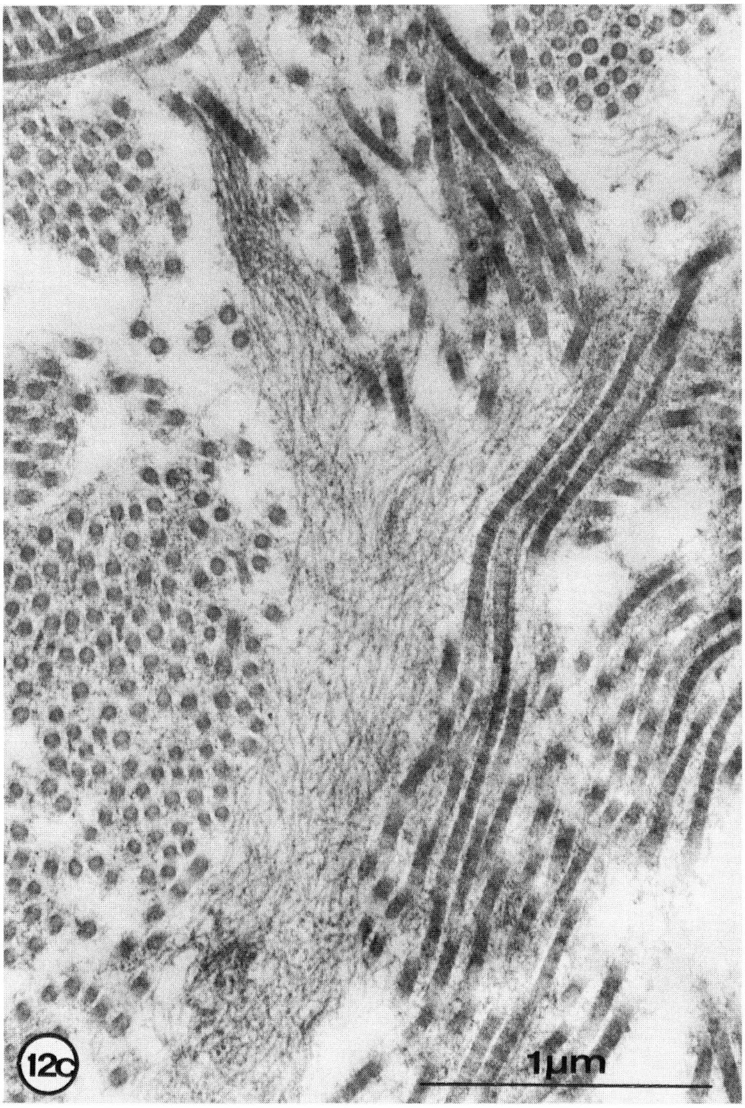

Figure 12 (Continued)

bundles in which all or a portion of the fibrils were unraveled. The unraveled fibrils were also present among structurally normal fibrils in areas of loose connective tissue between the fiber bundles (Figure 13b). Although the collagen fibrils were the major connective tissue components altered, the variety of tissues involved in this disorder (cornea, cartilage, skin) suggests that it is unlikely that the primary defect was in collagen structure. The collagens in the skin appeared to be normal biochemically, but the cultured dermal fibroblasts accumulated increased amounts of chondroitin sulfate in the extracellular matrix. Thus, this alteration, as in the other examples where composite fibrils or fibrils with altered diameters have been shown to occur, may result from an improper interaction of normal collagen molecules with other matrix components that are abnormal in either composition or amount.

The opposite situation, decreased total amount of proteoglycans, correlated with the formation of unraveled fibrils in a number of experimental studies. For example, segments of aorta treated with guanidine HCl lost about 85% of their proteoglycans and the collagen fibrils unraveled into their component microfilaments (118). Similarly, urea injected into the stroma of the human cervix (120) softened the tissue, solubilized the GAGs (removed dermatan sulfate), and caused an unraveling of collagen fibrils identical to the experimental changes in dermis (116) and to the in vivo alterations seen in the skin of our patient with SED (117).

Fibrils with this morphology have also been recognized in the dermis of a patient with EDS type I (30), but no biochemical studies have been carried out using her fibroblasts on culture. These fibrils were present in small, knotlike bundles of collagen in the middermis and in larger tangled bundles in the deep dermis.

2. *Unraveled Fibrils in Focal Accumulations in Loose Connective Tissue*

Small, focal areas of loose connective tissue, fibroblasts, and macrophages, found between bundles of collagen in the reticular dermis, are other common sites for fibrils with an unraveled appearance. In these areas individual collagen fibrils are oriented randomly, bent, frayed, and mixed with small, twisted elastic fibers (Figure 14). They appear to be areas of localized tissue destruction, possibly through the action of proteolytic enzymes. Fibroblasts and macrophages are known to synthesize and release proteases, including collagenase and elastase, but what creates these circumscribed accumulations and what is the eventual disposition of the seemingly degraded matrix components are not known. Such accumulations are characteristic of a variety of inherited diseases in addition to other changes in the collagen fibrils in larger fiber bundles. Unraveled fibrils are also found in hyalin (95) and amyloid (93,121), and in degradative vacuoles of cells that have phagocytized collagen and are presumably in the process of degrading it (46,122).

IV. CALCIFICATION OF COLLAGEN IN SOFT TISSUES

There are only a few examples where collagen calcifies in soft tissues (e.g., dermis) of patients with inherited connective tissue diseases, and in those instances, calcification of another component (notably elastin) is usually more prominent. Mineralization of collagen fibrils has been seen in hyalin and amyloid accumulations in the skin and in both collagen fibrils and elastic fibers in the dermis of patients with PXE. Degenerating collagen fibrils in hyalin mineralizes (95); in cutaneous amyloid, calcium apatite crystals are aligned with the collagen fibrils in the amyloid and with those fiber bundles which border the islands of amyloid in the deep dermal tissue (93). Sclerosis of both elastic and collagenous connective tissue in the deep dermis has also been observed in PXE (123). In one study of patients with PXE, it was shown that calcium could associate in dotlike accumulations with both normal and abnormal-appearing collagen fibrils located in the middermis (29). Calcification of the collagen was suggested to be an early phase of the tissue damage, which resolves as the diseases progresses.

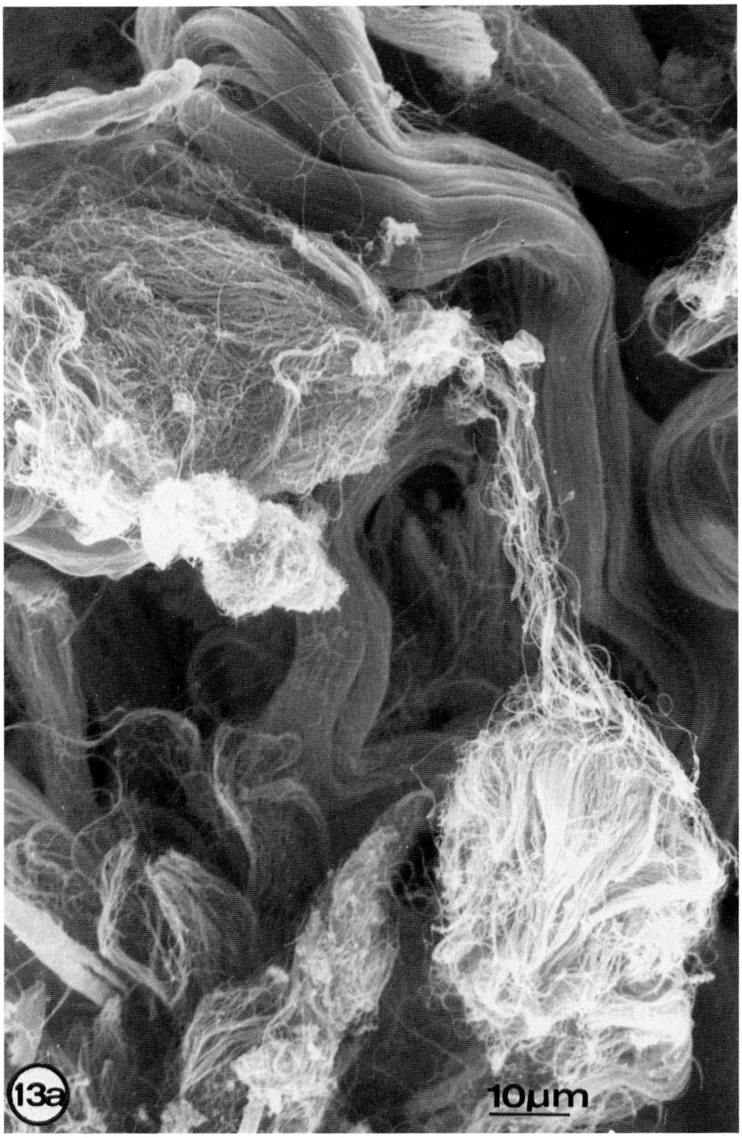

Figure 13 (a) Bundles of disorganized collagen in ther dermis of a patient with SED. These areas correspond to sites where collagen fibrils are unraveled (see Figs. 12a-c). X1120. (b) Unraveled collagen fibrils among structurally normal fibrils in the dermis of the patient with SED. X10,300.

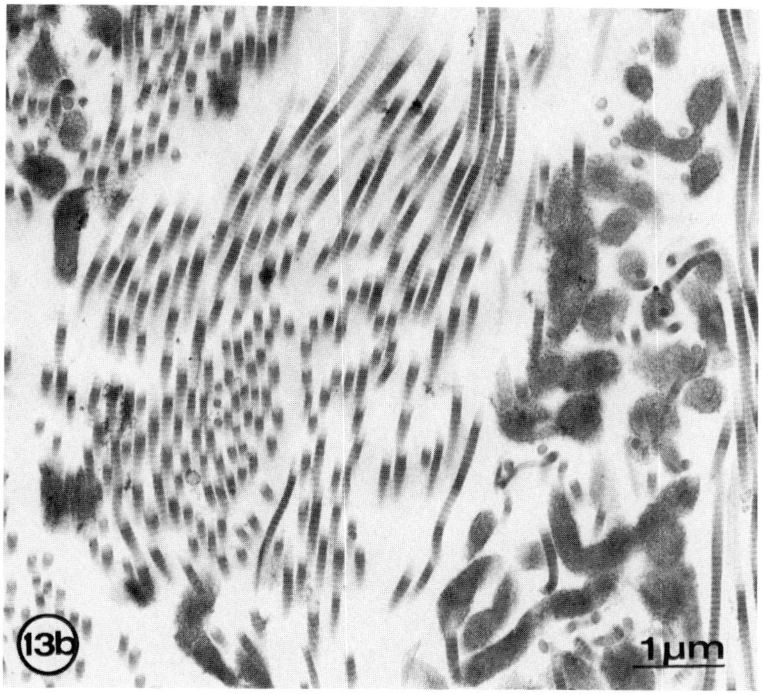

Figure 13 (Continued)

V. CONCLUSIONS

Substantial progress has been made in identifying the molecular basis for genetic diseases of collagen, but at this stage, we are uncertain how specific alterations in collagen structure produce the structural abnormalities that are so readily detectable. We have yet to understand all changes in the collagen molecules themselves or in the interactions of matrix molecules that provoke the various altered fibril, fiber, and fiber bundle morphologies. Thus, the best way to summarize or integrate the information reviewed in this chapter is to provide a series of generalizations, some of which have been mentioned at various points in the discussion.

1. Collagen fibrils undergo limited repertoire of changes in the skin of individuals with genetic or induced abnormalities of the connective tissue. The alterations may occur during fibril assembly or as a consequence of degradation and can modify their diameter (resulting in fibrils of smaller- or larger-than-normal diameters or mixed populations of the two) or integrity (composite and unraveled fibrils).
2. Structural alterations at one level of organization (e.g., the fibril) can affect the organization of the matrix at other levels (e.g., the fiber, fiber bundle, and dermis as a whole).
3. Structural alterations in two or more matrix components can occur in the dermis concurrently.
4. A molecular defect in one connective tissue molecule (either collagen, elastin, a glycoprotein, or proteoglycan) can have consequences for the structural properties of other components of the connective tissue. This is well illustrated by the many

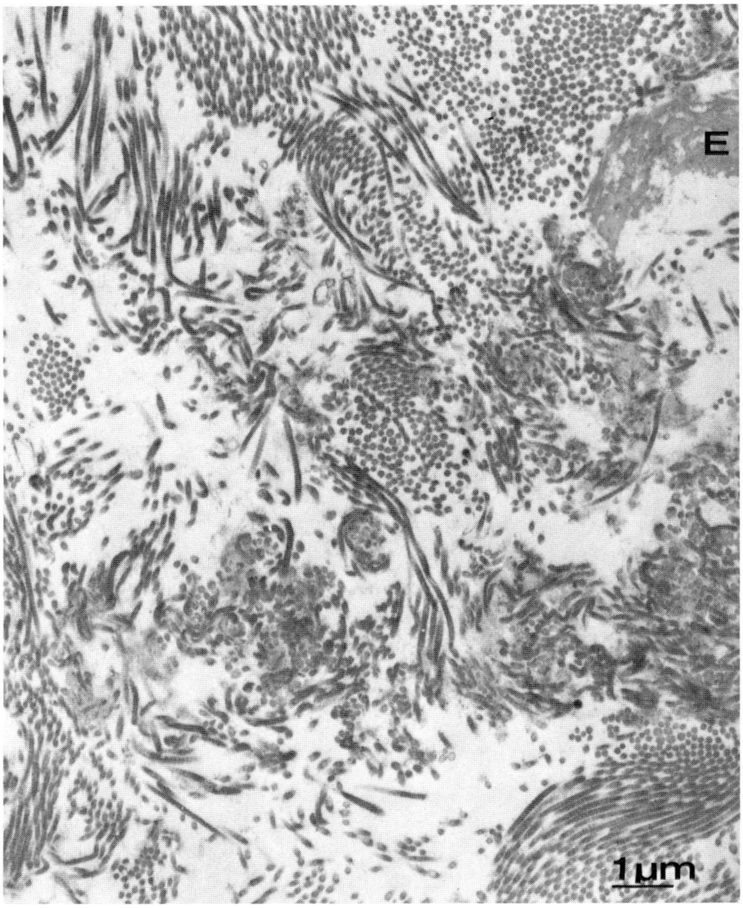

Figure 14 Focal area of the dermis where the normal structure and relationships of collagen and elastic fibers (E) are disrupted. X7440.

examples where the structure of collagen fibrils and elastin fibers are both altered even when the molecular basis for the disorder is known to occur in only one of the proteins.

5. A structural defect in a fibrous component is not necessarily explained by the molecular defect in the molecule; i.e., different molecular defects may be manifest in the dermis by the same structural defect. This is well illustrated by surveying the different disorders in which composite fibrils are part of the histopathology.
6. Structural alterations in connective tissue fibers are not specific for a given disease; nonetheless, such alterations (or the composite of several alterations) are usually constant for a disease and as such may be useful in the diagnosis. Patterns of structural change in the matrix as a whole are characteristics, in particular, for some of the inherited connective tissue diseases (e.g., EDS type IV).
7. Certain regions of the dermis are affected more prominently by genetic diseases of the connective tissue. The papillary dermis is usually spared, and in the reticular dermis there is often differential expression of the alterations in the upper and deep levels of the tissue.

8. The relationship between the molecular biology, structural characteristics in the dermis, and clinical expressions of the disease may be summarized as follows:

Gene defect → Abnormal protein structure → Abnormal interactions of matrix and nonmatrix molecules; abnormal rates of synthesis and/or degradation → Structural alterations in fibrils/fibers or matrix organization → Phenotypic expression of the disease

ACKNOWLEDGMENT

The authors gratefully acknowledge the expert preparation of biopsy specimens by Mrs. Mary Hoff. The work was supported in part from grants from the National Institutes of Health AM-21557, HD-17664, GM-07266, GM-15253, a Clinical Research Grant from the March of Dimes (6-298), and an award from the Poncin Scholarship Fund. PHB is an Established Investigator of the American Heart Association.

REFERENCES

1. W. N. Meigel, S. Gay, and L. Weber, *Arch. Derm. Res., 259*:1 (1977).
2. R. H. Fleischmajer, S. Gay, J. S. Perlish, and J. P. Cesarini, *J. Invest. Dermatol., 75*: 189 (1980).
3. M. Weinstock and C. P. Leblond, *J. Cell Biol., 60*:92 (1974).
4. R. G. Frederickson, E. R. Walker, and M. D. Mayers, *Anat. Rec., 208*:58A (1984).
5. J. Uitto, J. L. Paul, K. Brockley, R. H. Pearce, and J. G. Clark, *Lab. Invest., 49*:499 (1983).
6. S. Schuster, M. M. Black, and E. McVitie, *Br. J. Dermatol., 93*:639 (1975).
7. E. Maurel, B. Bouissou, M. T. Pieraggi, and M. Julian, *Conn. Tiss. Res., 8*:33 (1980).
8. L. Weber, in *Connective Tissue Research: Chemistry, Biology, and Physiology*, Alan R. Liss, Inc., New York, 1981, p. 131.
9. E. H. Epstein, Jr., *J. Biol. Chem., 249*:3231 (1974).
10. L. Weber, E. Kirsch, P. K. Müller, and T. Krieg, *J. Invest. Dermatol., 82*:156 (1984).
11. J. Uitto, *Arch. Biochem. Biophys., 192*:371 (1979).
12. R. H. Fleischmajer, W. Dessau, R. Timpl, T. Krieg, C. Luderschmidt, and M. Wiestner, *J. Invest. Dermatol., 75*:270 (1980).
13. H. M. Fullmer and R. D. Lillie, *J. Histochem. Cytochem., 6*:425 (1958).
14. A. Gawlik, *Folia Histochem. Cytochem., 8*:233 (1965).
15. J. E. Scott and C. R. Orford, *Biochem. J., 197*:213 (1979).
16. D. F. Mosher and L. T. Furcht, *J. Invest. Dermatol., 77*:175 (1981).
17. T. Kobayasi, L. Danielsen, and G. Asboe-Hansen, *Acta Dermatovener., 51*:27 (1971).
18. R. H. Fleischmajer, R. Timpl, L. Tuderman, L. Raischer, M. Wiestner, J. S. Perlish, and P. N. Graves, *Proc. Natl. Acad. Sci. USA, 78*:7360 (1981).
19. S. Tajimi and S. R. Pinnell, *J. Invest. Dermatol., 77*:410 (1981).
20. S. Tajimi, T. Nishikawa, I. Hatano, and Y. Nagai, *Arch. Derm. Res., 273*:115 (1982).
21. L. C. U. Junquiera, O. M. S. Toleda, and G. S. Montes, *Cell Tiss. Res., 217*:171 (1981).
22. A. Z. Eisen, *J. Invest. Dermatol., 52*:442 (1969).
23. R. E. Mitchell, *J. Invest. Dermatol., 48*:203 (1967).
24. I. M. Braverman and E. Fonferko, *J. Invest. Dermatol., 78*:434 (1982).
25. W. G. Banfield and D. C. Brindley, *J. Invest. Dermatol., 41*:9 (1963).
26. R. E. Mitchell, *Aust. NZ J. Surg., 36*:310 (1967).
27. L. H. Jansen, *Dermatologica, 110*:108 (1955).

28. Ö. Fyrand, *Br. J. Derm., 101*:263 (1979).
29. A. Eng and J. Bryant, *Int. J. Derm., 14*:586 (1975).
30. K. A. Holbrook, P. H. Byers, and S. R. Pinnell, *Scan Electron Micros./1978, IV*:1731 (1982).
31. J. Uitto, D. J. Santa-Cruz, B. Starcher, M. Whyte, and W. A. Murphy, *J. Invest. Dermatol., 76*:284 (1981).
32. F. M. Pope, G. R. Martin, J. R. Lichtenstein, R. P. Penttinen, B. Gerson, D. Rowe, and V. A. McKusick, *Proc. Natl. Acad. Sci. USA, 72*:1314 (1975).
33. K. A. Holbrook and P. H. Byers, *Lab. Invest., 44*:342 (1981).
34. P. H. Byers, K. A. Holbrook, G. S. Barsh, L. T. Smith, and P. Bornstein, *Lab Invest., 44*:336 (1980).
35. R. Laurent and P. Agache, *Dermatologica, 148*:28 (1974).
36. P. H. Byers, K. A. Holbrook, B. McGillivray, and P. M. MacLeod, *Hum. Genet. 47*: 141 (1979).
37. V. I. Sulica, P. H. Cooper, F. M. Pope, G. W. Hambrick, B. M. Gerson, and V. A. McKusick, *Arch Dermatol., 115*:40 (1979).
38. J. Dorling, F. M. Pope, G. Skavin, and T. Sowter, *J. Invest. Dermatol., 76*:433 (1981).
39. Ch. M Lapière, B. Nusgens, and G.-E. Pièrard, *Connect. Tiss. Res., 5*:21 (1977).
40. P. H. Byers, J. R. Shapiro, D. W. Rowe, K. E. David, and K. A. Holbrook, *J. Clin. Invest., 71*:689 (1983).
41. R. Smith, M. J. O. Francis, and R. J. Bauze, *Q. J. Med. 44, 176*:555 (1975).
42. L. T. Smith, K. A. Holbrook, and P. H. Byers, *J. Invest. Dermatol., 79*:93s (1982).
43. R. Ross and E. Benditt, *J. Biochem. Biophys. Cytol., 11*:67 (1971).
44. K. A. Holbrook and P. H. Byers, *J. Invest. Dermatol., 79*:7s (1982).
45. B. Williams, R. Cranley, S. Doty, and J. Lichtenstein, *Am. J. Hum. Genet., 25*: 86A (1973).
46. R. Caputo, E. Alessi, and E. Berti, *J. Invest. Dermatol., 76*:342 (1981).
47. U. N. Reide, F. Steimer, and J. Staubesand, *Atherogenese, 1*:101 (1976).
48. B. Steinmann, L. Tuderman, L. Peltonen, G. R. Martin, V. A. McKusick, and D. J. Prockop, *J. Biol. Chem., 255*:8887 (1980).
49. Ch. M. Lapière and B. Nusgens, *J. Invest. Dermatol., 76*:422 (1981).
50. P. H. Byers, R. C. Siegel, K. A. Holbrook, A. S. Narayanan, P. Bornstein, and J. G. Hall, *N. Engl. J. Med., 303*:61 (1980).
51. A. Vogel, K. A. Holbrook, B. Steinmann, R. Gitzelmann, and P. H. Byers, *Lab. Invest., 40*:201 (1979).
52. K. A. Holbrook and P. H. Byers, *Clin. Res., 27*:570 (1978).
53. M. Sevenich, U. Schultz-Ehrenburg, and C. E. Orfanos, *Arch. Dermatol. Res., 267*: 237 (1980).
54. B. Sykes, M. J. O. Francis, and R. Smith, *N. Engl. J. Med., 296*:1200 (1977).
55. L. Forrest, J. Dixon, and D. S. Jackson, *Conn. Tiss. Res., 1*:243 (1972).
56. J. D. Schofield, I. L. Freeman, and D. S. Jackson, *Biochem. J., 124*:467 (1971).
57. G. C. Wood, *Biochem. J., 75*:605 (1960).
58. T. R. Oegma, J. Laidlaw, R. C. Hascall, and D. D. Dziewiatkowski, *Arch. Biochem. Biophys., 170*:698 (1975).
59. J. C. Anderson, R. I. Labedz, and M. A. Kewley, *Biochem. J., 168*:345 (1977).
60. H. Mensing and G. Schaeg, *Dermatologica, 168*:1 (1984).
61. C. M. Black, L. J. Gathercole, A. J. Bailey, and P. Beighton, *Br. J. Dermatol., 102*: 85 (1980).
62. H. Julkunen, P. Rokkanen, and H. Inoue, *Ann. Med. Exp. Fenn., 84*:201 (1970).
63. B. Steinmann, R. Gitzelman, A. Vogel, M. E. Grant, R. Harwood, and C. H. J. Sear, *Helv. Pediatr. Acta, 30*:255 (1975).
64. H. Shinkai, O. Hirabayashi, A. Tamaki, S. Matsubayashi, and S. Sano, *Arch. Derm. Res., 257*:113 (1976).

65. A. Vogel and B. Steinmann, *J. Cut. Pathol., 6*:177 (1980).
66. P. H. Byers, G. S. Barsh, D. W. Rowe, K. E. Peterson, K. A. Holbrook, and J. Shapiro, *Am. J. Hum. Genet., 32*:37A (1980).
67. M. Scheck, R. C. Siegel, J. Parker, Y-H. Chang, and J. C. C. Fu, *J. Anat., 219*:645 (1979).
68. S. R. Pinnell, S. M. Krane, J. E. Kenzora, and M. J. Glimcher, *N. Engl. J. Med., 286*: 1013 (1972).
69. A. Ihme, T. Krieg, A. Nerlich, U. Feldmann, J. Rauterberg, R. W. Glanville, G. Edel, and P. K. Müller, *J. Invest. Dermatol., 83*:161 (1984).
70. D. R. Eyre and M. J. Glimcher, *Proc. Natl. Acad. Sci. USA, 69*:2594 (1972).
71. G. Barsh, K. E. David, and P. H. Byers, *Proc. Natl. Acad. Sci. USA, 79*:3838 (1982).
72. P. H. Byers, R. C. Siegel, K. E. Peterson, K. A. Holbrook, D. W. Rowe, L. T. Smith, Y. H. Chang, and J. C. Fu, *Proc. Natl. Acad. Sci. USA, 78*:7745 (1981).
73. R. E. Priest, J. F. Moinuddin, and J. H. Priest, *Nature(London), 245*:264 (1972).
74. P. Abraham, A. J. Perejda, W. H. Carnes, and J. Uitto, *J. Clin. Invest., 70*:1245 (1982).
75. S. I. Lamberg and A. Dorfman, *J. Clin. Invest., 52*:2428 (1973).
76. S. Takebayashi, I. Kubota, and T. Takagi, *Acta Path. Jap., 23*:847 (1973).
77. R. C. Siegel and Y. H. Chang, *Clin. Res., 26*:510A (1978).
78. D. Volpin, I. Pasquali-Ronchetti, M. G. Castellani, A. Peserico, and G. Mori, *Dermatologica, 156*:209 (1978).
79. P. Marchase, K. A. Holbrook, and S. R. Pinnell, *J. Invest. Dermatol., 75*:399 (1980).
80. M. Ledoux-Corbousier and G. Achten, *J. Cutan. Pathol., 2*:25 (1975).
81. R. Ross, P. J. Fialkow, and L. K. Altman, *Clin. Genet., 13*:213 (1978).
82. L. Danielsen, *Int. J. Derm., 21*:449 (1982).
83. P. H. McKee, C. H. S. Cameron, D. B. Archer, and W. C. Logan, *J. Cut. Pathol., 4*: 146 (1977).
84. I. Pasquali-Ronchetti, D. Volpin, M. Baccarani-Contrii, I. Castellani, and A. Peserico, *Dermatologica, 163*:307 (1981).
85. S. Saito and G. Klingmüller, *Arch. Derm. Res., 260*:179 (1977).
86. J. G. Smith, Jr., W. M. Sams, Jr., E. A. Davidson, and R. D. Clark, *Arch. Derm., 86*: 741 (1962).
87. J. Uitto, S. Lindy, H. Turto, and L. Danielsen, *J. Invest. Dermatol., 52*:44 (1971).
88. N. Blumenkrantz, L. Danielsen, and G. Asboe-Hansen, *Acta Dermatovener., 53*: 429 (1973).
89. A. Martinez-Hernandez and W. F. Huffer, *Lab Invest., 31*:181 (1974).
90. S. G. Gordon, M. Overland, and J. Foley, *Conn. Tiss. Res., 6*:61 (1978).
91. M. Lebwohl, R. S. Fleischmajer, O. Lovelace, and R. Timpl, *J. Invest. Dermatol., 82*:433 (1984).
92. M. A. Arneson, D. E. Hammerschmidt, L. T. Furcht, and R. A. King, *JAMA, 244*: 144 (1980).
93. L. Danielsen and T. Kobayasi, *Acta Dermatovener., 53*:13 (1973).
94. T. Kobayasi, P. Wolf-Jurgensen, and L. Danielsen, *Acta Dermatovener., 53*:275 (1973).
95. K. Hashimoto, G. Klingmüller, and O-E. Rodermund, *Acta Dermatovener., 52*: 179 (1972).
96. L. Danielsen and T. Kobayasi, *Acta Dermatovener., 52*:1 (1972).
97. A. Otkjaer-Nielsen, O. B. Christensen, B. Hentzer, E. Johnson, and T. Kobayasi, *Acta Dermatovener., 58*:323 (1978).
98. J. L. Reymond, P. Stoebner, P. Zambelli, J. C. Beani, and P. Amblard, *Cut. Pathol., 9*:352 (1982).
99. S. Peyrol, J. A. Grimaud, and R. Borojevic, *C. R. Acad. Sci. (Paris), 282*:333 (1976).

100. Th. Nemetschek, A. Meinel, H. Nemetschek-Gansler, P. Reill, and H. Riedl, *Virchows Arch. A. Path. Anat. Histol., 372*:57 (1976).
101. J. Staubesand and N. Fischer, *Path. Res. Pract., 165*:374 (1979).
102. J. Staubesand, *Med. Welt., Bd. 28/Heft, 48*:1943 (1977).
103. J. Staubesand and N. Fischer, *Conn. Tiss. Res., 7*:213 (1980).
104. U. Lindahl and M. Höök, *Ann. Rev. Biochem., 47*:385 (1978).
105. J. J. Adnet, B. Kalis, C. Hopfner, M. Pluot, L. Harmel, P. Mallet, and T. Caulet, *Ann. Derm. Syph., 6*:530 (1973).
106. M. Fjølstad and O. Helle, *J. Pathol., 112*:183 (1974).
107. R. Hanset and M. Ansay, *Ann. Med. Vet., 7*:451 (1967).
108. P. J. O'Hara, W. K. Read, W. M. Romane, and C. H. Bridges, *Lab. Invest., 23*:307 (1970).
109. D. F. Counts, P. H. Byers, K. A. Holbrook, and G. A. Hegreberg, *J. Invest. Dermatol., 74*:96 (1980).
110. K. A. Holbrook, P. H. Byers, D. F. Counts, and G. A. Hegreberg, *J. Invest. Dermatol., 74*:100 (1980).
111. A. Lenaers, M. Ansay, B. V. Nusgens, and Ch. M. Lapière, *Eur. J. Biochem., 23*:533 (1971).
112. U. Becker, U. Timpl, O. Helle, and D. Prockop, *Biochemistry, 15*:3853 (1976).
113. A. J. Bailey and Ch. M. Lapière, *Eur. J. Biochem., 34*:91 (1973).
114. R. Jonak, Ch. M. Lapière, A. Meinel, H. Nemetschek-Gansler, T. Nemetschek, and H. Riedl, *Z. Naturforsch., 32C*:743 (1977).
115. G. E. Pierard and Ch. M. Lapière, *J. Invest. Dermatol., 66*:2 (1976).
116. J. H. Lillie, D. K. MacCallum, L. J. Scarletta, and J. C. Occhino, *J. Ultrastr. Res., 58*:134 (1977).
117. P. H. Byers, K. A. Holbrook, J.W. Chandler, P. Bornstein, and J. G. Hall, *Birth Defects Original Article Series XIV.*, (*6B*):221 (1978).
118. K. Saruk and R. Eisenstein, *Arch. Pathol. Lab. Med., 101*:74 (1977).
119. M. Ledoux-Corbousier and G. Achten, *Br. J. Dermatol., 91*:287 (1974).
120. C. W. Kischer, W. Droegemueller, M. Shetlar, M. Shvapil, and J. Vining, *Am. J. Pathol., 99*:525 (1980).
121. K. Hashimoto, B. G. Gross, and W. F. Lever, *J. Invest. Dermatol., 45*:204 (1965).
122. A. R. Ten Cate, *J. Anat., 112*:401 (1972).
123. G.-E. Piérard, *Ann. Dermatol. Venereol., 111*:111 (1984).

6

The Molecular Organization of Basement Membrane

David T. Woodley
University of North Carolina School of Medicine,
Chapel Hill, North Carolina

I. INTRODUCTION

In the process of organogenesis and tissue organization, endothelia, parenchyma, and epithelia are separated from their underlying connective tissue (stroma) by thin extracellular matrices called basement membranes. Basement membranes frequently serve as an interface between histologically dissimilar tissues that originate from different primary germ layers. For example, in skin there is a basement membrane between the epidermis, which is of ectodermal origin, and the dermis, which is derived from the mesoderm. In gestational development, routine tissue remodeling, and wound healing, interactions between tissues of different primordial origins are mediated at the basement membrane interface. Whole basement membranes or individual components may influence the behavior (proliferation, spreading, migration, differentiation) of cells by so-called *cell-matrix* interactions. Basement membranes may also serve a structural role in the support and organization of tissues. The adhesiveness between dissimilar tissues forming an organ may be determined by specific basement membrane compounds. Soluble factors, trace elements, and ions transported between tissues may be filtered selectively by basement membrane molecules that serve as permeability barriers. While an individual isolated component of the basement membrane may be responsible for one or more of these functions, it is likely that in vivo these components act in concert and that their tertiary structures and alignment with each other ultimately determine the functional capacities of each component and the whole integrated basement membrane. Therefore, how the individual components are organized and interact on a molecular level is relevant to the understanding of basement membranes.

II. ULTRASTRUCTURAL ORGANIZATION

Basement membranes from various organs differ somewhat in ultrastructural appearance. However, the major ultrastructural compartments of the basement membrane can usually be identified and analogous structures discerned when comparing different tissues and species. The basement membrane that lies between the epidermis and dermis of skin (the dermal-epidermal junction) is a complex structure that has been well studied and clearly described (1-4). Therefore, for the purposes of discussion in this chapter, it will be used as a general model for all basement membranes.

Based on electron microscopic studies, the basement membrane zone of the dermal-epidermal junction has been divided into four major compartments (Figure 1): (a) the *plasma membrane* of the epidermal cells apposed to the basement membrane, the basal cell keratinocytes, (b) an electron lucent space of 30-40 nm beneath the plasma membrane of the epidermal cells that is termed the *lamina lucida* (*lamina rara* in the kidney), (c) an amorphous or finely fibrillar electron dense layer lying beneath the lamina lucida and of approximate equal size, called the *lamina densa*, and (d) a fibrillar area below the lamina densa containing anchoring fibrils, bundles of microfibrillike elements and single collagen fibers (1). In the basement membrane of the kidney glomerulus, there is no correlate to the skin's sub-lamina densa fibrillar zone. Further, glomerular basement

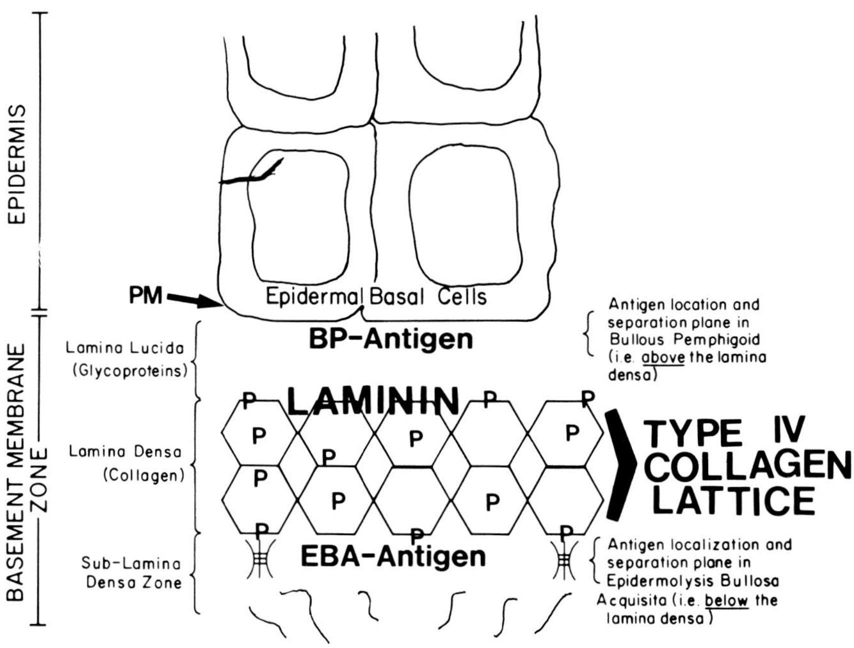

Figure 1 Schematic representation of the basement membrane zone at the dermal-epidermal junction of human skin. The four major structural compartments are the *plasma membrane* (PM) of the epidermal basal cells, the *lamina lucida* space containing the glycoproteins laminin and the bullous pemphigoid antigen, the *lamina densa* containing type IV collagen, and the *sub-lamina densa zone* comprised of anchoring fibrils, microfibrils, and the epidermolysis bullosa acquisita antigen. Bullous pemphigoid antigen (BP antigen) is localized to the *lamina lucida* and shares this space with laminin. Laminin may also be found partially within the *lamina densa*, which is comprised of type IV collagen in a latticelike arrangement. Heparan sulfate proteoglycan (P) is localized in the lamina densa but is also found somewhat in the lamina lucida and sub-lamina densa space. The epidermolysis bullosa acquisita (EBA) antigen is localized within and immediately below the lamina densa.

membrane has a lamina rara interna and a lamina rara externa on each side of the lamina densa. These two lamina rarae are analogous to the lamina lucida of skin since they contain similar macromolecules (see below).

The basement membrane within the dermal-epidermal junction has certain unique structures. For example, along the plasma membranes of the epidermal cells there are electron dense foci called hemidesmosomes (Figure 2). Tonofilaments within the cells converge on the attachment plaques within the hemidesmosomes. The lamina lucida is more complex immediately beneath the hemidesmosomes than in areas between these structures (Figure 2). A horizontal dense line, the *subbasal dense plaque* or *H line*, is seen in the lamina lucida beneath the hemidesmosomes parallel to the plasma membrane. Perpendicularly oriented filaments, termed *anchoring filaments*, extend from the hemidesmosomes to the lamina densa. At one time it was thought that anchoring filaments were connected to the attachment plaque and tonofilaments of the hemidesmosome and to the elastin-reticulin system in the upper dermis, and so-called oxytalan-elaunin system

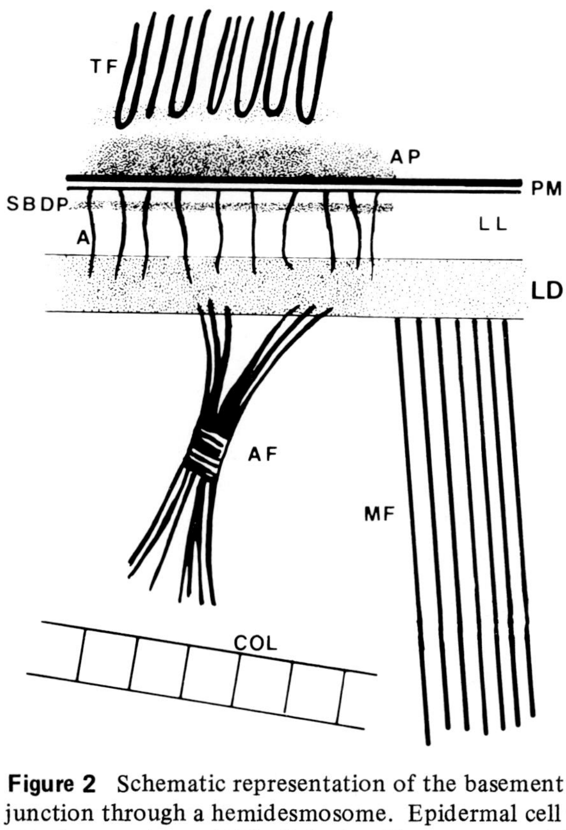

Figure 2 Schematic representation of the basement membrane of the dermal-epidermal junction through a hemidesmosome. Epidermal cell tonofilaments (TF) converge on the attachment plaque (AP) of the hemidesmosome, which is apposed to the plasma membrane (PM) of the basal cell. The lamina lucida space (LL) is bisected by an electron-dense line called the subbasal dense plaque (SBDP). Vertical electron dense filaments called anchoring filaments (A) traverse the lamina lucida perpendicular to the attachment plaque. Beneath the lamina densa collagen fibers (Col), anchoring fibrils (AF), and microfibrils (MF) can be seen. (Reprinted with permission from Ref. 1.)

(7-10). The elaunin fibers are arranged horizontally to the basement membrane, while oxytalan fibers branch off of the elaunin meshwork at 90° angles and ascend vertically toward the lamina densa. It was thought that perhaps they connected to the tonofilaments via anchoring filaments within the hemidesmosomes, and that this linkage was responsible for epidermal-dermal adherence. This theory has now been generally disregarded. However, the functions of these structures remain a mystery.

Anchoring fibrils are structures shaped similar to miniature "wheat stacks" that appear to be dangling beneath the lamina densa. These structures are absent or greatly reduced in certain scarring blistering diseases of the skin such as hereditary epidermolysis bullosa dystrophica and epidermolysis bullosa acquisita. Although the function of anchoring fibrils is unknown, in vitro recombinant studies have shown that they are of dermal origin (11) and are capable of inducing hemidesmosome formation directly adjacent to themselves when apposed to epithelial sheets (12). This is a clear example of an epidermal-mesenchymal interaction in which an epidermal structure is induced by a morphological dermal component.

The weakest area of the dermal-epidermal junction appears to be the lamina lucida. That is, when epidermis and dermis are separated by various means (induction of a suction blister, salt extraction, trypsinization, etc.), the cleavage plane occurs through the lamina lucida (1,13). The exceptions to this rule are certain immune-initiated blistering diseases of the skin such as epidermolysis bullosa acquisita and bullous systemic lupus erythematosus, which are characterized by immunoglobulin and complement deposits beneath the lamina densa. In these diseases, blisters form and their cleavage plane occurs beneath the lamina densa at the site of the immune deposits.

III. COMPONENTS OF BASEMENT MEMBRANE

A. Overview

It is impossible to address the subject of the molecular organization of basement membranes without first presenting a general overview of the known molecular components of the basement membrane. Detailed biochemical descriptions of the components have been reviewed (3,4,14-16) recently and are beyond the scope of this chapter. A brief summary of the main components will be presented here (Table 1).

Basement membranes are known to contain collagen (17), proteoglycans (18,19), and glycoproteins (20-22). To date, the major components that have been identified and reasonably well characterized are a basement membrane specific collagen termed type IV collagen (23,24); laminin, a large glycoprotein (21); and a heparan sulfate-rich proteoglycan termed BM-1 (18). These basement membrane molecules have been isolated from a transplantable murine tumor (Engelbreth-Holme-Swarm tumor) that is known to produce large quantities of extracellular matrix (24). In addition to production of prodigious amounts of basement membrane, this tumor allows the components of the basement membrane to be isolated in a relatively native form. For example, type IV collagen can be isolated without the use of pepsin digestion of the tissue, which is so often used when isolating collagens from other sources. In fact, while pepsin does not cleave the helical segments of most collagens, it will make multiple cleavages in type IV collagen because there are nonhelical, protease-sensitive regions throughout the molecule. Therefore, proteolytic treatment during tissue extraction might produce matrix molecules that are missing various domains that could be crucial for determining the functional aspects of the component.

Table 1 Properties of Basement Membrane Components

Macromolecule	Native molecular weight (daltons)	Chain structure	Chain size (daltons)	Role in BMZ	Molecular shape
Bullous pemphigoid	220,000–240,000	A_n	200,000–240,000	Unknown	Unknown
Laminin	1,000,000	AB_3	A = 200,000 B = 400,000	Cell substrate attachment	
Type IV collagen (BMZ specific collagen)	540,000	Pro $\alpha 1$ $(IV)_3$ Pro $\alpha 2$ $(IV)_3$ or heteropolymer	185,000 170,000	Tissue support	
Heparan sulfate proteoglycan	750,000	Central protein core with heparan sulfate glycosamino-glycan (GAG) side chains	GAG side chains of 70,000	Permeability barrier	[a]
Fibronectin	450,000	Disulfide-linked dimers	200,000–225,000	Cell attachment and matrix attachment	
Epidermolysis bullosa acquisita antigen	Over 800,000	An	290,000	Unknown	Unknown
Entactin	Unknown	?	158,000	Unknown	Unknown

[a]Theoretical shape from known molecular properties.

Type IV collagen and laminin have also been isolated from other tissues such as placenta (25-27) and tissue culture (28-30). Heparan sulfate proteoglycan has been isolated from kidney and other basement membranes (31,32). Antibodies raised to basement membrane components isolated from the EHS tumor stain basement membranes from a variety of tissues and across species. Antibodies to laminin, heparan sulfate proteoglycan, and collagens IV and V stain the dermal-epidermal junction and dermal vessels in adult human skin (13).

B. Type IV Collagen

Type IV collagen is a basement-membrane-specific collagen found in the lamina densa of virtually all basement membranes (33). Affinity-purified antibodies to the type IV collagen isolated from the EHS tumor bind to the type IV collagen in all basement membranes. A monoclonal antibody has been developed against type IV collagen (34). Compared to interstitial collagens, type IV collagen has several special characteristics: (a) a high carbohydrate content rich in hexose sugars, (b) absence of the banded fibrillar appearance of interstitial collagen, (c) less dependency on ascorbic acid for its cellular synthesis, (d) nonhelical segments interspersed within the helical domain of the molecule that represent pepsin sensitive sites and flexible areas, (e) a "procollagen" biosynthetic form that has a molecular size close to the "end product" collagen, and (f) an unusually high content of hydroxylysine and hydroxyproline (16,35).

Type IV collagen was originally isolated from the EHS tumor and later from a variety of cells in culture (23,36-39). In the initial studies, type IV collagen, when electrophoresed with reduction on SDS-polyacrylamide gels, appeared as two bands with molecular weights of 160,000 [alpha 1 (IV)] and 140,000 [alpha 2 (IV)] daltons. However, this collagen species was missing a piece of the molecule known as the 7S region that was degraded by proteolysis during the extraction with acetic acid (4). Type IV collagen with an intact 7S region and proalpha 1 (IV) and proalpha 2 (IV) chains of 185,000 and 170,000 daltons, respectively, have recently been isolated from the EHS tumor (15) and cells in culture (38,39). These procollagen chains are distinct gene products (38,39). Yaoita and colleagues (33) have found by immunoelectron microscopy that type IV collagen is localized within the lamina densa compartment of the basement membrane. Briggaman, Dalldorf, and Wheeler (11) have shown that stationary living epidermis can reconstitute the lamina densa when recombined with nonliving dermis. Others have suggested that when epidermal cells are in a migratory state, such as occurs in wound healing, there is a delay or absence in the reconstitution of the lamina densa and type IV collagen, especially when the dermal substrate is nonviable (40-42). In a wound-healing model, it was found that type IV collagen formation beneath the epithelium occurred only after the epithelium had stopped migrating (43). One explanation for the differences between migratory and stationary epidermis in basement membrane reconstitution studies is the fact that migratory epidermal cells produce collagenolytic enzymes against both interstitial and basement membrane collagens (44). Nevertheless, cultured endothelial cells, epithelial cells, and keratinocytes are capable of synthesizing type IV collagen (28,45-47). The molecular shape and interactions of type IV collagen will be discussed later.

C. Laminin

Laminin is a large glycoprotein found within all basement membranes (21). The molecule is rich in cysteine and highly sulfated. It has a native molecular weight of 1×10^6

daltons. When reduced and electrophoresed on a SDS-polyacrylamide gel, two bands appear (M_r 400,000 and M_r 200,000) (4). The laminin molecule has a cruciate shape by rotary shadowing and is composed of three 35-nm short arms (B chains) each arm corresponding to the 200-kd band seen on gel, and one long arm (A chain) of 75 nm consisting of the 400-kd band (48-52). At the ends of each chain are two closely set globular domains (Figure 3). It has been suggested that laminin, like fibronectin, has specific functional domains for binding cells, collagen, and proteoglycan (48,51,53-58). Laminin is thought to be an attachment factor for epithelial cells (53), and a laminin cell receptor, "connectin," has been isolated (54,55). Laminin has affinity for type IV collagen and basement membrane proteoglycan (51,54,56-58). The 400-kd long arm of laminin binds to proteoglycans (48,51) while the type IV collagen binding domain remains on one of the globular areas at the ends of one of the short arms (48,51,54). The cell-binding domain is the central cross of the molecule where there is a concentration of sulfated residues and bonds (48,50). Couchman and colleagues (59) found that laminin is an adhesion molecule for fibroblasts with activity equal to that of fibronectin. Engvall and Ruoslahti (60) were unable to confirm that laminin is a specific attachment factor for epithelial cells, as previously reported (53). However, the nature of the laminin used in these studies is unclear since it was not characterized by gel electrophoresis, amino acid analysis, or rotary shadowing. Since laminin is particularly protease sensitive, it is conceivable that the biological functions of individual laminin preparations are not uniform.

Antibodies to laminin react with all basement membranes. The ultrastructural location within the basement membrane is controversial (61,62) but may be explained by tissue and species differences in the published studies. In skin basement membrane, laminin was found to be localized to the lamina lucida (61) while in kidney basement membrane it was limited to the lamina densa (62). Laminin can be synthesized in culture by a variety of cell types including endothelial and epithelial cells (28,30,45,53,54,63,64).

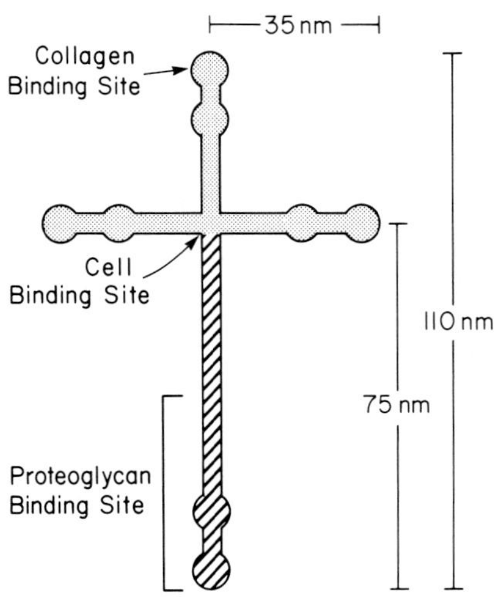

Figure 3 Schematic representation of laminin, a major glycoprotein of the basement membrane. The cell-binding site is located at the central cross while the collagen-binding site is one of the globular regions on the small arms (35 nm) of the cross. The heparan sulfate proteoglycan binding region is on the long arm (75 nm). Modifed from Refs. 48, 53, and 51.)

D. Proteoglycan

Basement membrane heparan sulfate proteoglycan was first recognized by Kanwar and Farquhar (19) in glomerular basement membrane. This highly sulfated and negatively charged molecule probably explains the presence of anionic domains within basement membrane that bind ruthenium red dye (65). These sulfated domains may be responsible for the ready incorporation in culture of [^{35}S] sulfate into epithelial cell-derived basement membrane (66). Hassell et al. (18) and later Kanwar and colleagues (67) isolated basement-membrane-specific heparan sulfate proteoglycan from the EHS tumor and glomerular basement membrane, respectively. The proteoglycan from the EHS tumor, a so-called BM1, has a molecular weight of 750,000 and is composed of a protein core (50% of the weight) and chains of repeating dissacharide units characteristic of heparan sulfate of approximately 70,000 daltons each. Heparan sulfate proteoglycans isolated from kidney or teratocarcinoma cell lines have different molecular sizes than that isolated from the EHS tumor (67,68). However, antibodies to the protein core of basement membrane heparan sulfate proteoglycans bind only to basement membrane and not to other tissues (18,67,68) where cell surface or connective tissue proteoglycans may occur. Laurie, LeBlond, and Martin (62) found that antibodies to the protein core of the heparan sulfate proteoglycan isolated from the EHS tumor bound the lamina densa compartment of kidney basement membrane. Thus, the heparan sulfate proteoglycan in basement membrane appeared to codistribute with type IV collagen.

Heparan sulfate proteoglycan is a highly charged molecule and is thought to function as a selective molecular sieve within the basement membrane. When heparan sulfate proteoglycan within basement membrane is disturbed by treatment with heparitinase, the basement membrane becomes permeable to molecules that are usually excluded, such as albumin and serum proteins (69). Recently, the glomerular basement membranes in nephrotic syndrome have been found to have altered proteoglycan and anionic domains (70).

E. Bullous Pemphigoid Antigen

The bullous pemphigoid antigen is a normal constituent of basement membranes beneath stratified squamous epithelium such as that found in skin, esophagus, bronchi, and bladder. It is not found in renal or vascular basement membranes. It is produced within human keratinocytes without mesenchymal influences (71-73) and deposited at the cell-matrix interface (71,73,74). It has a native molecular weight between 220,000 and 240,000 daltons (75,76) and appears as a closely set dimer when electrophoresed on a 5% sodium dodecyl-sulfate polyacrylamide gel. Recently, it has been determined that the antigen produced by epidermal cell culture is similar to that directly extracted from adult human skin (76) (Figure 4). Bullous pemphigoid antigen is localized to the lamina lucida compartment of human skin (77,78). Of all the components of the basement membranes, including those of the lamina lucida, the bullous pemphigoid antigen is the most closely associated with the epidermal cell (13) (Figures 5 and 6). When the epidermis is separated from the dermis by the induction of a suction blister, trypsinization, intrinsic protease activation, or salt extraction, the bullous pemphigoid antigen remains with the epidermis (Figure 6) while all the other known components of the basement membranes remain with the dermis (Figure 7) (13). It is not clear whether bullous pemphigoid antigen is in some way partially integrated into the plasma membrane of the epidermal cells. Stanley et al. (75) were able to extract the antigen from epidermal tissue culture with physiological

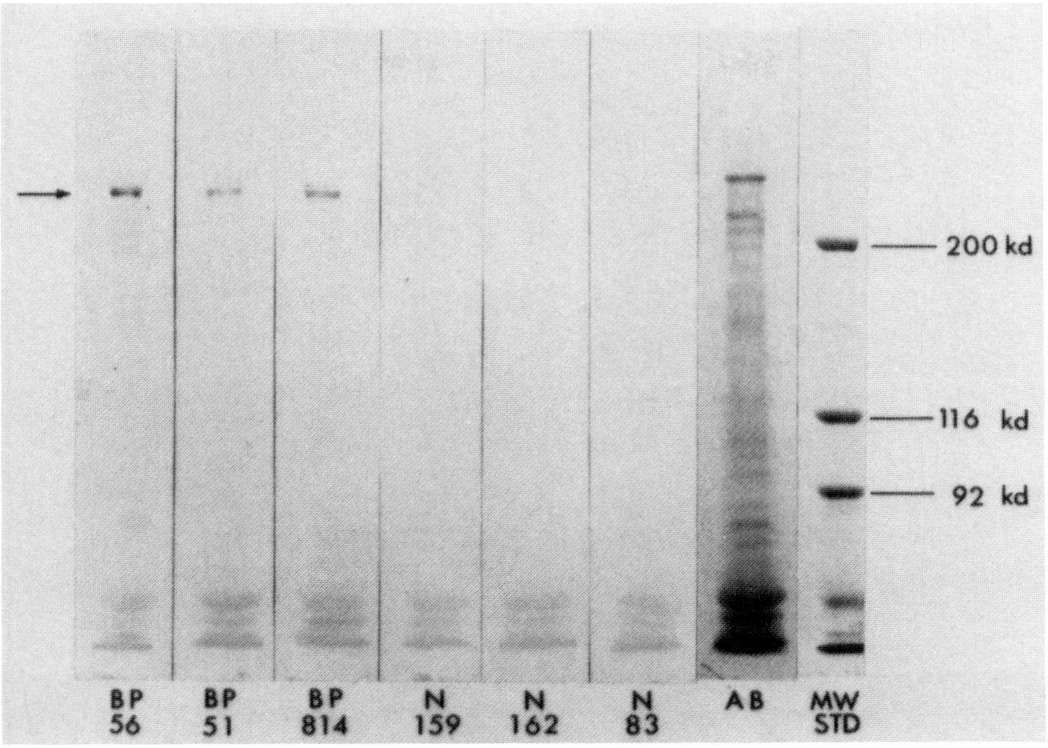

Figure 4 Proteins directly extracted from suction-blister-derived epidermis are separated on gels, electrophoretically transferred to nitrocellulose strips, and reacted in a Western immune blot assay with three different bullous pemphigoid sera (BP) or three different normal human sera (N). An amido-black-stained strip (AB) shows the total epidermal cell proteins transferred to nitrocellulose. From the myriad of protein bands, the BP sera specifically bind to a closely set dimer of approximately 225 kd (arrow). None of the normal sera react with this dimer. MW STD is a molecular weight standard stained with amido black. (Reprinted with permission from Ref. 76.)

saline. In contract, Grekin et al. (79) could only remove the antigen from dispersed human keratinocytes with harsh denaturing agents (79).

Very recent evidence suggests that at least part of the antigen is within the epidermal basal keratinocyte (80,81). The function of bullous pemphigoid antigen is unknown. However, its close apposition to the dividing pool of keratinocytes and the fact that it is implicated in a blistering disease suggest that it may be involved in cell-matrix interactions, such as cell proliferation and cell-to-substrate adhesion. The biochemistry and functional aspects of the bullous pemphigoid antigen await the isolation and purification of the molecule.

F. Other Basement Membrane Components

In addition to these four known components of basement membranes, there are other extracellular matrix molecules that may be components of some basement membranes, but their actual tissue localizations are disputed or unclear. This includes fibronectin, type V collagen, and entactin.

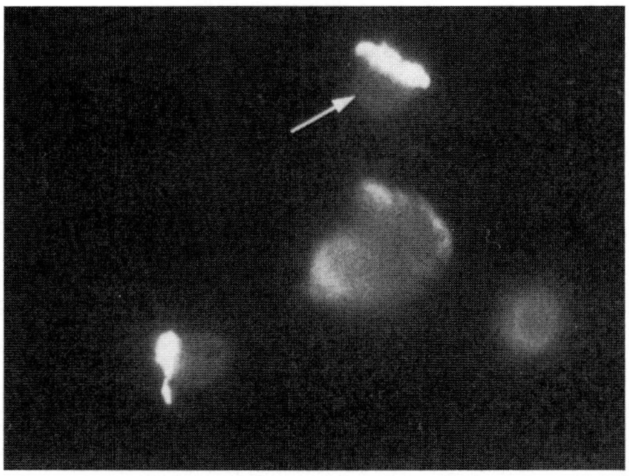

Figure 5 A human adult epidermal cell suspension was prepared by cold trypsinization of split-thickness skin. The suspension was air-dried and stained by indirect immunofluorescence with sera from a bullous pemphigoid patient diluted 1:40. The bullous pemphigoid antigen is seen localized to the end of cells (arrow) that were previously in contact with the lamina lucida of whole skin.

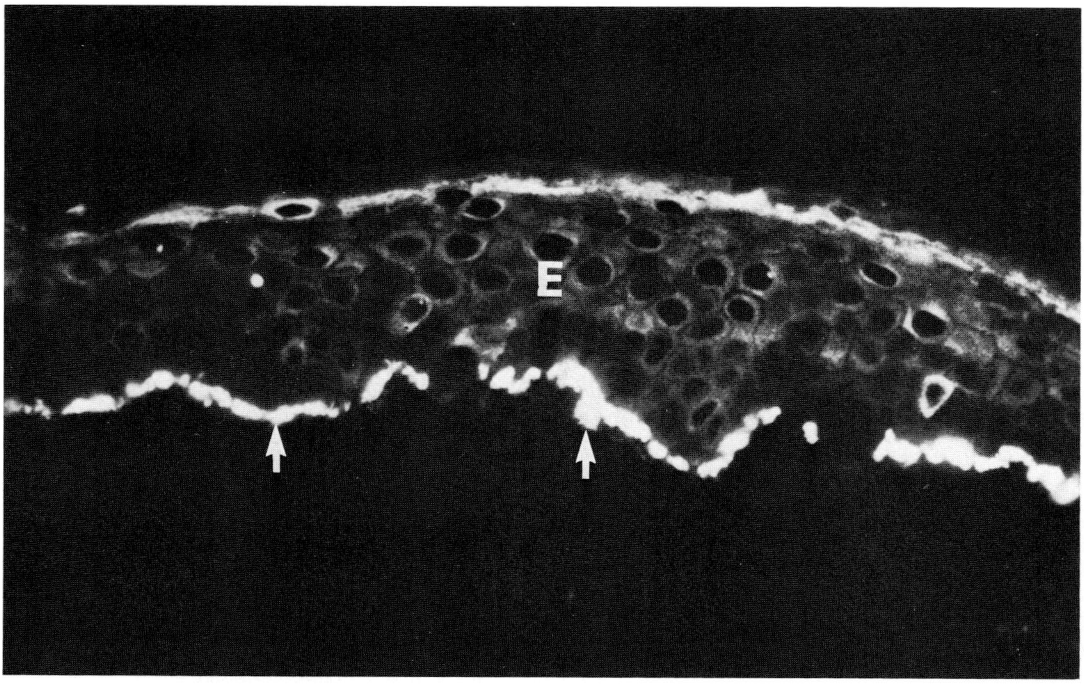

Figure 6 A sheet of human adult epidermis (E) was prepared by incubating split-thickness skin in 1 M saline for 72 hr and then stained by indirect immunofluorescence with sera from a patient with bullous pemphigoid diluted 1:40. The fluorescent band along the lower poles of the basal cells of the epidermal sheet is bullous pemphigoid antigen (arrows).

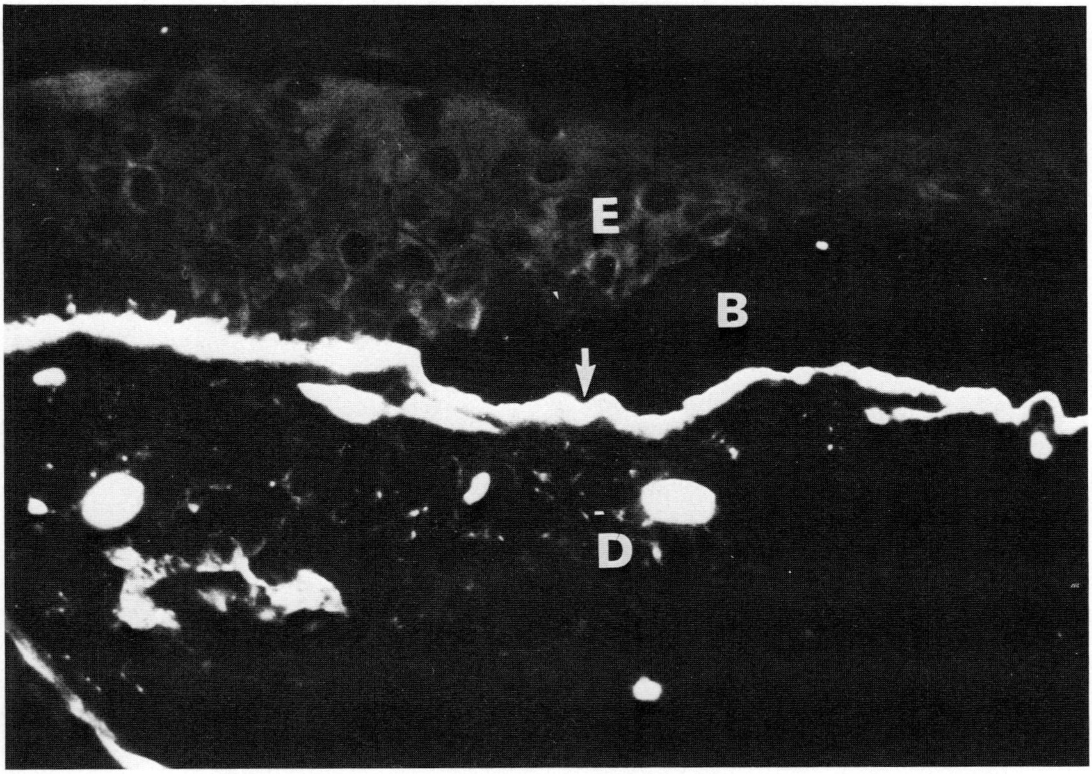

Figure 7 A suction blister was raised on normal human skin, biopsied, sectioned with a cryostat cryotome, and stained by indirect immunofluorescence with purified antibody to type V collagen. Type V collagen remains with the dermal side (D) of this separation through the lamina lucida and is not seen on the epidermis (E). Similar results are obtained with antibodies to laminin, type IV collagen, and heparan sulfate proteoglycan (see Ref. 13). B = blister space.

1. Fibronectin

Fibronectin is a major extracellular matrix glycoprotein with a native molecular weight of approximately 440 kd (43,74). On sodium dodecyl-sulfate polyacrylamide gels it appears as a single band or two closely set dimers of approximately 220 kd. The molecule has been well studied and is known to consist of biologically active segments that provide the molecule with a host of biological functions, as outlined in Figure 9 (48,82-85). Fibronectin may function at the cell surface to maintain a nontransformed phenotype since fibronectin has been found to be lost or altered during transformation (86-88). Furthermore, adding fibronectin to transformed cells partially restores the nontransformed phenotype (88). The molecular shape of fibronectin has been determined by rotary shadowing (52) and is known to consist of two long chains of approximately equal length that are joined at one end by a disulfide bond. This gives the molecule a V-shaped appearance (52). Fibronectin binds collagens (89,90), attaches cells to collagen substrate (91-95) and induces cell spreading (14,93,95-97). Grinnell and associates (96,97) have shown that the tertiary structure of fibronectin is critical in expressing its biological activity. Fibronectin is produced locally within connective tissue by fibroblasts. The

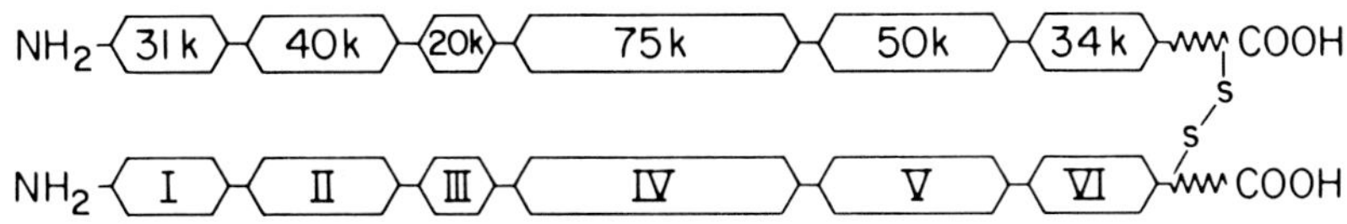

DOMAIN	BINDING SITE FOR
I	Fibrin, bacteria, heparin, actin, Factor XIII a, transglutaminase substrates
II	Collagen, gelatin
III	Unknown
IV	Cell surface
V	Heparin
VI	Fibrin

Note : The molecule has heparin and fibrin binding sites.

Figure 8 Schematic representation of molecular binding domains of fibronectin and a summary list of its biologically active sites. (Modified from Ref. 85.)

precise binding sites on fibronectin for cell surfaces and other matrix molecules are known (48,85). There is also abundant (300-600 mg/ml) fibronectin in serum and tissue fluids (14). When whole adult human skin is stained with antibody to fibronectin, the entire dermis is stained while the epidermis is not. Nevertheless, O'Keefe et al. (98) have shown that human keratinocytes cultured without the presence of a fibroblast feeder layer in fibronectin-free medium are capable of synthesizing and depositing extracellular fibronectin. Human keratinocytes migrating across the floor of a Petri dish leave a fibronectin "track" (Figure 9).

If keratinocytes can produce fibronectin in culture, the questions that arise are: does this occur in vivo and, if so, where does the epidermal cell-derived fibronectin go? Couchman et al. (99) have found fibronectin within the lamina lucida space of rat skin when examined by immunoelectron microscopy with antifibronectin antibody. However, this finding awaits confirmation. Saksela et al. (100) found fibronectin at the basement membrane zone in skin that had been separated at the lamina lucida space by the induction of a suction blister. Fibronectin was found on the dermal side of the separated skin. However, during wounding in this manner, the entire skin is literally swamped with fibronectin from plasma. Since fibronectin has avidity for connective tissue molecules (collagen, proteoglycans), it is difficult to determine whether or not it is a normal constituent of the basement membrane when examining wounded skin flooded by plasma-derived fibronectin. It may be that fibronectin is a constituent of fetal basement membranes or plays a role in newly synthesized basement membranes at gestation or after wound healing (43,101).

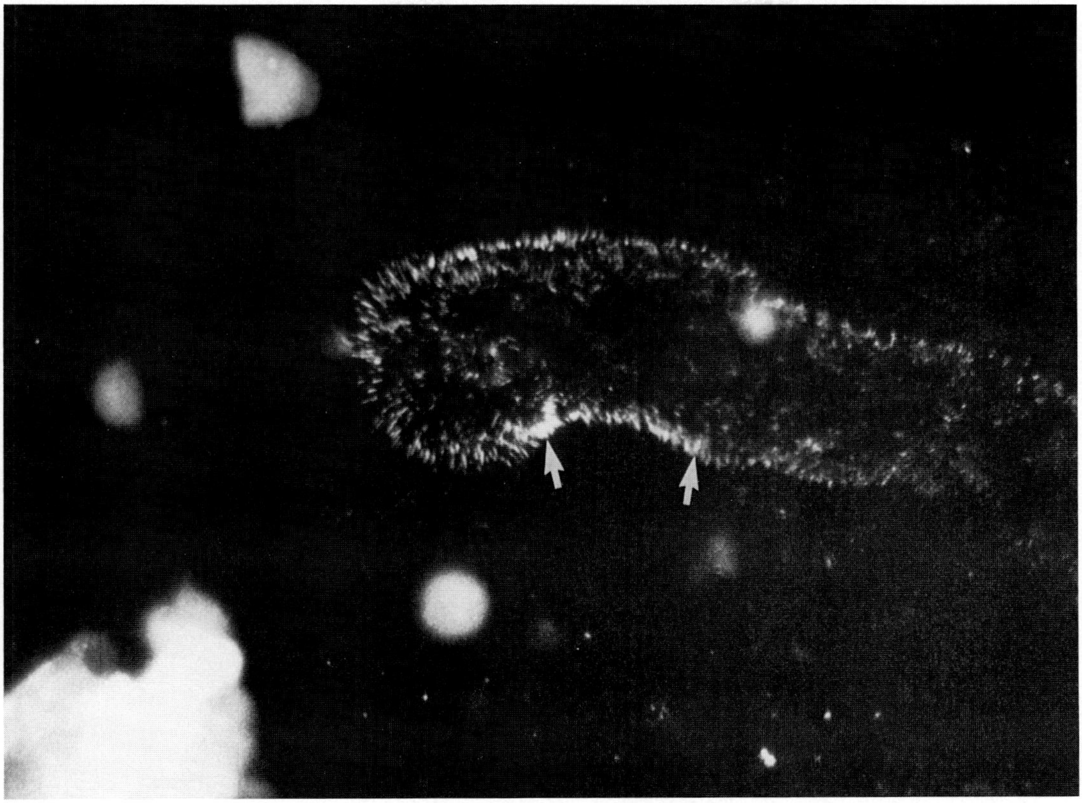

Figure 9 Human keratinocytes grown without a fibroblast feeder layer and in fibronectin-free media are capable of synthesizing fibronectin themselves and depositing it as "tracks" (arrows) on the Petri dish floor. (Reprinted with permission from Ref. 98.)

2. Type V Collagen

Type V collagen (also known as AB_2 collagen) may also be a component of basement membranes (13,102). The collagen was first isolated from human placenta by Burgeson et al. (27) and later by others (25,26,28,103). It has a molecular weight of approximately 330,000 daltons. Type V collagen is codistributed with type IV collagen in mesangial basement membrane of the kidney (102). Affinity-purified antibodies to type V collagen stain the dermal-epidermal junction and around the dermal vessels of human skin. When the skin is separated at the lamina lucida by various techniques, type V collagen remains with the dermis (Figure 7) (13). When monkey esophagus is stained with antibodies to type V collagen, type V collagen is found at the junction between the epithelium and lamina propria and around the blood vessels within the lamina propria. It is also found around the muscle fascicle bundles throughout the external muscular coat of the esophagus (13). However, its precise ultrastructural localization in human skin or in monkey esophagus has not been determined.

The immunoelectron microscopic studies of Gay et al. (104) have shown that type V collagen is related to the surface of smooth muscle cells. Similar studies have shown type V collagen at the luminal surface of endothelial cells lining blood vessels. It was suggested that type V collagen at this location served to maintain blood vessels in a nonthrombogenic state (105). When the endothelial surface and its type V collagen coat were disturbed during injury, the coagulation cascade would be initiated (35,105). Stenn and colleagues (106) provided evidence that the synthesis of type V collagen is related to keratinocyte migration. They showed that the migration of mouse keratinocytes across an insoluble matrix was halted when the synthesis of type V collagen was interrupted. This suggests that type V collagen may be synthesized by keratinocytes, deposited extracellularly, and used as an attachment factor to drive cellular migration. Grotendorst and colleagues (107) have shown that smooth muscle cells are capable of directional migration through a type V collagen-coated filter toward a chemotactic agent, platelet-derived growth factor. In this study, the smooth muscle cells were found to attach to the type V collagen via a carbohydrate moiety within the cell membrane. These studies, like those of Stenn and colleagues (106), suggest that some cells are capable of attaching and migrating on a type V collagen matrix.

3. *Entactin*

Entactin is another new component of basement membrane that is found localized to the foot processes of the renal epithelial cell within the lamina rara and somewhat to the lamina densa proper (22,108). It is thought to act as a bridge between the cell surface and the lamina densa. The molecule has a molecular weight of 158,000 and is highly sulfated. However, its responses to proteolytic enzymes and amino acid composition indicate that it is not a proteoglycan or related to laminin (109). Its function is undetermined.

G. Epidermolysis Bullosa Acquisita Antigen

A new component of basement membranes beneath stratified squamous epithelium has been noted recently (110). This basement membrane component binds to antibodies in the serum of patients who have a primary blistering disease of the skin called epidermolysis bullosa acquisita (EBA). This component is located within and immediately below the lamina densa (Figure 1) (110,111). The molecule has a native molecular weight above 800,000 daltons, but when reduced with dithiothreitol and subjected to gel electrophoresis, it appears as two protein bands: a major band of 290,000 daltons and a minor band of 145,000 daltons (Figure 10). How these two chains are related to one another is unclear. These bands stain with both Coomassie blue and fuchsin-sulfite stain, suggesting that they are composed of both protein and carbohydrate. Since this component is recognized by the sera of EBA patients, it has been termed the "EBA antigen." However, a monoclonal antibody has been raised recently to the EBA antigen (112), and the molecule is a normal constituent of cutaneous basement membrane. It is clear that the EBA antigen is a cutaneous basement membrane molecule that is distinct from fibronectin, laminin, elastin, anchoring fibrils, bullous pemphigoid antigen, and collagen types I, IV, and V (110). The function of this component is unclear, but the fact that it is implicated in a blistering disease of the skin suggests that it may have a role in epidermal-dermal adherence.

H. A Possible Model of the Molecular Organization of Basement Membrane

The precise molecular organization of the basement membrane zone has not been elucidated. However, the ultrastructural tissue localizations of the components, in vitro studies of the affinities between matrix molecules, and the technique of visualizing the

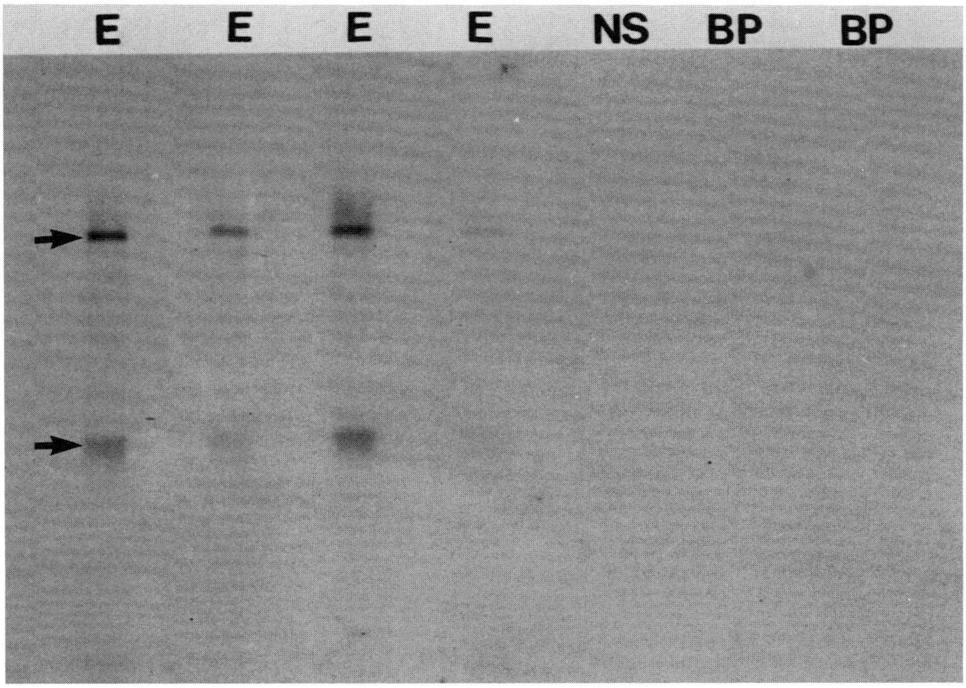

Figure 10 Protein extracts of human skin basement membrane are separated by SDS-PAGE, electrophoretically transferred to nitrocellulose paper strips, and reacted in a Western immune blot assay with sera from patients with epidermolysis bullosa acquisita (EBA) (lanes E). The sera only bind to two protein bands in the extract: a major band of 290 kd and a minor band of 145 kd (arrows). Normal human sera (lane NS), sera from patients with other blistering diseases (lanes BP = stained with bullous pemphigoid sera), and antibodies to known matrix components do not bind to these bands. (Reprinted with permission from Ref. 110.)

shapes of individual components of the basement membranes by rotary shadowing have provided some information about the possible molecular arrangements within basement membranes.

Studies of the ultrastructural tissue location and molecular shape of type IV collagen have provided a working model of the molecular organization of the collagenous lamina densa. All the immunoelectron microscopic studies have indicated that type IV collagen is localized to the lamina densa (33,62,113). Electron microscopists have recognized for a long time that the lamina densa does not have a banded appearance typical of interstitial collagens. Rather the lamina densa has a granular, amorphous, "felt-tip-pen-like" appearance in Epon-embedded sections (1,3,4,114). The studies of Timpl et al. (115) have shown that type IV collagen molecules are arranged in a latticelike or honeycomb pattern within the lamina densa. These observations were made during experiments in which native type IV collagen was subjected to various degrees of degradation with collagenase. Under maximal enzymatic treatment a 30-nm, bar-shaped piece of the type IV collagen molecule remains that was originally termed 7S collagen because of its sedimentation coefficient (116,117). Rotary shadowing experiments with pure type IV collagen have shown that this 7S piece is actually a convergence point where four type

IV collagen molecules come together at their N termini (like-end-to-like-end) and are tightly cross-linked by disulfide bonds. The C-terminal ends of these chains are nonhelical, globular domains and also connect to the globular ends of other type IV collagen chains (Figure 11). Thus, it appears that native type IV collagen molecules are shaped like hockey sticks with a ball at the end of the handle and the blade representing the 7S domain. The 7S domain at the N-terminal ends (hockey stick blade) is about 60 nm in length while the rodlike, discontinuous, helical portion (the hockey stick handle) is 330 nm long (115). Four such molecules align by their 7S domains and form a dense, overlapped imbrication that is resistant to collagenase extensive disulfide bonding. This unit of four molecules joined together at their 7S domains is called a "spider." Spider units connect to other similar units by the globular domains of the free ends of the collagen molecules emanating out from the 7S convergence. By this arrangement of spider units, a lattice network or "honeycomb" structure is formed (Figure 11). The tightly bonded 7S regions that are collagenase resistant would provide very stable regions to the lattice work. Although the tensile strength of this lattice structure or basement membrane collagens is unknown, this model of the collagenous lamina densa is consistent with the notion that the lamina densa functions as a primary structural component to all basement membranes (4).

Laminin, a major glycoprotein in basement membranes, has a cruciate molecular shape (48-52). The long axis of this cross is approximately 110 nm, and the width through the short arms of the cross near 70 nm. The three equal-sized (35 nm) short arms and the one long arm (75 nm) all have globular ends of 5-7 nm (48,51). The long arm has only one globular domain at its extremity while the short arms each have two closely set globular domains (Figure 3).

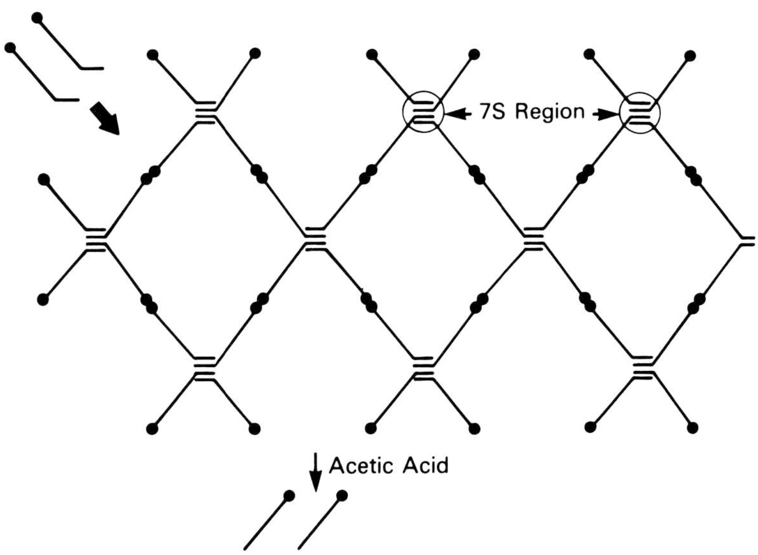

Figure 11 Type IV collagen molecules are shaped like hockey sticks with a ball on the end of the handle. Four such molecules converge to form a 7S collagenase-resistant piece. This piece is lost during an acid extraction of the EHS tumor. The arrangement of type IV collagen molecules aligned like-end-to-like-end forms a latticework lamina densa. (Reprinted with permission from Ref. 4.)

Laminin binds to heparan sulfate, heparin, and the heparan sulfate proteoglycan within basement membranes (48,51,56-58). The binding site on the laminin molecule for heparin and heparan sulfate proteoglycan is located within the long arm, the 400,000-dalton A chain (48,51).

Laminin also has an affinity for native basement membrane collagen (56). The affinity for interstitial collagens or denatured type IV collagen is considerably less. In contrast, fibronectin binds better to denatured collagen than native collagen and will bind to virtually all classes of collagen except the cuticle collagen of *Ascaris* (56). These observations suggest that the affinity of laminin for basement membrane collagen depends on the secondary or tertiary structure of the molecule, whereas fibronectin binds to a linear sequence of amino acids (56) within the collagen molecule independent of conformational considerations. In contrast, conformational considerations are extremely important for the cell-binding capabilities and biological activities of fibronectin (96,97). The binding site on the laminin molecule for type IV collagen is located within one or more of the short arms (54). When laminin is treated with alpha-thrombin, the long arm is degraded leaving a fragment consisting of three short arms (M_r = 600,000) (50). This α_3 fragment binds to type IV collagen as well as or better than the native molecule (56). When the globular domains were cleaved from the terminal ends of the α_3 fragment, laminin no longer had the capacity to bind to type IV collagen (54). This suggests that the type IV collagen-binding site is on a short-arm (B-chain) terminus of laminin and that the binding region is protease sensitive (54). This may be important when considering that the basement membrane zone of whole skin, when subjected to proteases, separates through the lamina lucida space where laminin has been found to be localized (1,13,61). It may be that proteases such as trypsin induce this separation in part by the degradation of the globular ends of laminin B chains and inhibit the molecule from binding with type IV collagen. Somewhat against this possibility is the finding that when the basement membrane zone of the skin is separated through the lamina lucida by a variety of techniques involving protease activation, laminin predominantly remains with the dermal side of the separation along with the basement membrane collagens types IV and V (13,100). However, a small amount of laminin can be detected on the epidermal side of the separation. The antibody-binding site on the laminin molecule is unknown and may vary with different polyclonal sera. Therefore, it is possible that the antibody-binding site and the type IV collagen-binding site are on opposite ends of the molecule such that the cell and collagen-binding regions are localized within the B chains and the antibody-binding site within the A chain.

Laminin's long arm has affinity for basement membrane proteoglycan (48,51). This proteoglycan has been found in both the lamina rarae interna and externa of the glomerular basement membrane (19,62,63,67,69) and within the lamina densa of the skin basement membrane (62). Therefore, the heparan sulfate proteoglycan appears to straddle and be within the lamina densa. Considering the "lattice-shaped" molecular organization of type IV collagen discussed previously, an appropriate hypothetical scheme for how proteoglycan and the lamina densa are associated might be by imagining that the collagenous lamina densa is a Nabisco Triscuit wheat wafer that has been dipped and saturated in a honey of heparan sulfate proteoglycan.

The basement membrane proteoglycan has affinity for type IV collagen and laminin (56), and the interactions between these three matrix components appear independent of each other. For example, mixing type IV collagen with laminin does not enhance laminin binding to proteoglycan (56). Since the proteoglycan-binding site is on the laminin A

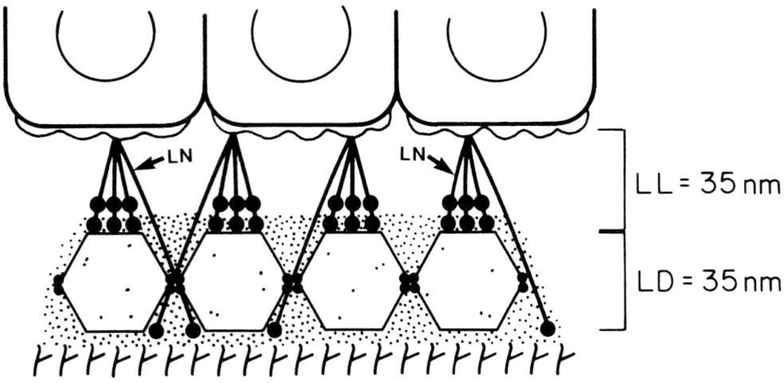

Figure 12 A possible model of the basement membrane zone. The lamina densa is formed by a lattice work of type IV collagen (represented as hexagons), which is saturated with heparan sulfate proteoglycan (represented as small dots) that "overflows" into both the lamina lucida and sub-lamina densa spaces. Bullous pemphigoid antigen (represented as scalloping along the cell plasma membrane) is localized high in the lamina lucida and is closely related to the plasma membrane of the basal keratinocytes. Laminin (LN) transverses both the lamina densa and lamina lucida with its long (75 nm) arm (400 kd, A chain) and binds the cells at the junction of the A and B chains (the center of the cross). It then flexes back across the lamina lucida, and the globular ends of one or more short (35 nm) arms (200 kd, B chain) bind directly to type IV collagen in the lamina densa. This "tenting" of laminin across the lamina lucida and attaching to the lamina densa at two ends of the molecule brings the sizes of the molecular domains and available space into accordance. The epidermolysis bullosa acquisita antigen (110) is somewhat within and immediately below the lamina densa.

chain, it is possible that the A chain, by its affinity for proteoglycan is embedded in the lamina densa and secondarily bound to type IV collagen via a proteoglycan bridge (Figure 12) (56). This 75-nm A chain would then ascend across the lamina lucida at a slight angle, and at the cross of the molecule (the junction of the A and B chains and the cell binding site), the molecule would attach to the plasma membrane (or a closely associated molecule such as the bullous pemphigoid antigen) and then flex sharply back toward the lamina densa. The three short B chains then span the lamina lucida space and attach via their globular termini (the type IV binding domains) to type IV collagen in the lamina densa. This spatial configuration would solve the problem of how a large 110-nm molecule such as laminin could fit into a 35-nm lamina lucida space. This tenting of the molecule across both the lamina densa and the lamina lucida could also explain why the ultrastructural location of laminin has been contradictory (61,62). It is interesting that the thickness of the lamina densa and lamina lucida together is approximately 70 nm and the long arm A chain of laminin is 75 nm. Thus, at a slight angle, this long arm could span both compartments. Furthermore, the short arms have a length of 35 nm, which is almost exactly the width of the lamina lucida. The molecular affinities between laminin, heparan sulfate proteoglycan, and basement membrane collagen and the size constraints of the basement membrane compartments in relation to the size of laminin indicate that this model of molecular organization is possible. Obviously, this is a tentative hypothetical model that will be modified and refined by new knowledge within the field.

ACKNOWLEDGMENTS

This work was supported by Grant Number 1 RO1 AM33625=01 from the National Institutes of Health.

REFERENCES

1. R. A. Briggaman, C. E. Wheeler, *J. Invest. Dermatol., 65*:71-84 (1975).
2. M. Prunieras, M. Regnior, S. Fougere, and D. Woodley, *J. Invest. Dermatol., 81*:74s-80s (1983).
3. R. A. Briggaman, *J. Invest. Dermatol., 78*:1-6 (1982).
4. J. R. Stanley, D. T. Woodley, S. I. Katz and G. R. Martin, *J. Invest. Dermatol., 79*: 69s-72s (1982).
5. A. S. Breathnach and J. Robins, *Br. J. Dermatol., 81*:504-510 (1969).
6. K. A. Holbrook, in *Biochemistry and Physiology of the Skin* (L. A. Goldsmith, ed.), Oxford University Press, Oxford, 1983, pp. 64-101.
7. Fullmer, H., *J. Histochem. Cytochem., 8*:290-295 (1960).
8. J. H. Ccoper, *J. Invest. Dermatol., 52*:169-176 (1969).
9. H. Fullmer and R. Lillie, *J. Histochem. Cytochem., 6*:425-430 (1958).
10. G. G. Carmichael and H. Fullmer, *J. Cell Biol., 28*:33-36 (1966).
11. R. A. Briggaman, F. Dalldorf and C. E. Wheeler, *J. Cell Biol., 51*:384-395 (1971).
12. I. K. Gipson, S. M. Grill, S. J. Spurr and S. J. Brennan, *J. Cell Biol., 97*:849-857 (1983).
13. D. T. Woodley, D. Sauder, M-J. Talley, M. Silver, G. Grotendorst and E. Qwarnstrom, *J. Invest. Dermatol., 81*:149-153 (1983).
14. H. K. Kleinman, R. J. Klebe and G. R. Martin, *J. Cell Biol., 88*:473-485 (1981).
15. H. K. Kleinman, M. L. McGarvey, L. A. Liotta, P. Gehron-Robey, K. Tryggvason and G. R. Martin, *Biochemistry, 21*:6188-6193 (1983).
16. H. Sage, *J. Invest. Dermatol., 79*:51s-59s (1982).
17. N. A. Kefalides, *Int. Rev. Conn. Tiss. Res., 6*:63-104 (1973).
18. J. R. Hassell, P. Gehron-Robey, H. J. Barrach, J. Wilczek, S. I. Rennard and G. R. Martin, *Proc. Natl. Acad Sci. USA, 77*:4494-4498 (1980).
19. Y. S. Kanwar and M. G. Farquhar, *Proc. Natl. Acad. Sci. USA, 76*:4493-4497 (1979).
20. R. G. Spiro, *J. Biol. Chem., 242*:1923-1932 (1967).
21. R. Timpl, H. Rohde, R. P. Gehron, S. I. Rennard, J-M. Foidart and G. R. Martin, *J. Biol. Chem., 254*:9933-9937 (1979).
22. B. Carlin, R. Jaffe, B. Bender and A. E. Chung, *J. Biol. Chem., 256*:5209-5214 (1981).
23. R. Timpl, G. R. Martin, P. Bruckner, G. Wick and H. Wiedemann, *Eur. J. Biochem., 120*:203-211 (1981).
24. R. W. Orkin, P. Gehron, E. B. McGoodwin, G. R. Martin, T. Valentine and R. Swarm, *J. Exp. Med., 145*:204-220 (1977).
25. H. Sage and P. Bornstein, *Biochemistry, 18*:3815-3822 (1979).
26. R. W. Glanville and K. Kuhn, *Front. Matrix Biol., 7*:19-26 (1979).
27. R. E. Burgeson, F. A. El Adli, I. J. Kaitila and D. W. Hollister, *Proc. Natl. Acad. Sci. USA, 73*:2579-2583 (1976).
28. K. Alitalo, E. Kuismanen, R. Myllyla, U. Kiistala, S. Asko-Seljavaara and A. Vaheri, *J. Cell Biol., 94*:497-505 (1982).
29. D. Gospodarowicz, D. Delgado and I. Vlodavsky, *Proc. Natl. Acad. Sci. USA, 77*: 4094-4098 (1980).
30. J. R. Stanley, P. Hawley-Nelson, M. Yaar, G. R. Martin and S. I. Katz, *J. Invest. Dermatol., 78*:456-459 (1982).
31. R. L. Trelstad, K. Hayashi and B. P. Toole, *J. Cell Biol., 62*:815-830 (1974).
32. A. Oldberg, C. Schwartz and E. Ruoslahti, *Arch. Biochem. Biophys., 216*:400-406 (1982).
33. M. Yaoita, J. M. Foidart and S. I. Katz, *J. Invest. Dermatol., 70*:191-193 (1978).
34. L. Sakai, E. Engvall, D. W. Hollister and R. E. Burgeson, *Am. J. Pathol., 108*:310-318 (1982).

35. P. Bornstein and H. Sage, *Annu. Rev. Biochem., 49*:957-1003 (1980).
36. R. Timpl, P. Bruckner and P. Fietzek, *Eur. J. Biochem., 95*:255-263 (1979).
37. K. Tryggvason, P. Gehron, P. Robey and G. Martin, *Biochemistry, 19*:1284-1389 (1980).
38. K. Alitalo, A. Vaheri, T. Creig and R. Timpl, *Eur. J. Biochem., 109*:247-255 (1980).
39. E. Crouch, H. Sage and P. Bornstein, *Proc. Natl. Acad. Sci. USA, 77*:745-749 (1980).
40. D. Woodley, M. Regnier and M. Prunieras, *Br. J. Dermatol., 103*:397-404 (1980).
41. J. R. Stanley, O. M. Alvarez, E. W. Bere, W. H. Eaglstein and S. I. Katz, *J. Invest. Dermatol., 77*:240-243 (1981).
42. H. Hintner, P. O. Fritsch, J-M. Foidart, G. Stingl, G. Schuler and S. I. Katz, *J. Invest Dermatol., 74*:200-204 (1980).
43. R. A. F. Clark, J. M. Lanigan, P. Dellapelle, E. Manseau, H. F. Dvorak and R. B. Colvin, *J. Invest. Dermatol., 79*:264-269 (1982).
44. D. T. Woodley, L. A. Liotta and R. Brundage, *J. Invest. Dermatol., 78*:238 (1982).
45. J-M. Foidart, J. J. Berman, L. Paglia, S. I. Rennard, S. Abe, A. Perantoni and G. R. Martin, *Lab. Invest., 42*:525-532 (1980).
46. B. V. Howard, E. J. Macarak, D. Gunson and N. A. Kefalides, *Proc. Natl. Acad. Sci. USA, 73*:2361-2364 (1976).
47. T. Hirone and S. Taniguchi, *J. Invest. Dermatol., 80*:341 (1983).
48. K. M. Yamada, in *Cell Biology of Extracellular Matrix* (E. D. Hay, ed.), Plenum Press, New York, 1983, pp. 95-110.
49. C. N. Rao, I. M. K. Margulies, R. H. Goldfarb, J. A. Madri, D. T. Woodley and L. A. Liotta, *Arch. Biochem. Biophys., 219*:65-70 (1982).
50. C. N. Rao, I. M. K. Margulies, T. S. Tralka, V. P. Terranova, J. A. Madri and L. A. Liotta, *J. Biol. Chem., 257*:9740-9744 (1982).
51. R. Timpl and G. R. Martin, in *Immunochemistry of the Extracellular Matrix*, Vol. II (H. Furthmayr, ed.), CRC Press, Boca Raton, FL, 1982, pp. 119-150.
52. J. Engel, E. Odermatt, A. Engel, T. A. Madri, H. Furthmayr, H. Rohde and R. Timpl, *J. Mol. Biol., 150*:97-120 (1981).
53. V. P. Terranova, D. H. Rohrbach and G. R. Martin, *Cell, 22*:719-729 (1980).
54. V. P. Terranova, C. N. Rao, T. Kalebic, I. M. Marguilies and L. A. Liotta, *Proc. Natl. Acad. Sci. USA, 80*:444-448 (1983).
55. H. Malinoff and M. S. Wicha, *J. Cell Biol., 96*:1475-1479 (1983).
56. D. T. Woodley, C. N. Rao, J. R. Hassell, L. A. Liotta, G. R. Martin and H. K. Kleinman, *Biochim. Biophys. Acta, 761*:278-283 (1984).
57. S. Sakashita, E. Engvall and E. Ruoslahti, *FEBS Lett., 116*:243-246 (1980).
58. M. Del Rosso, R. Capelletti, M. Viti, S. Vannucchi and V. Chiarugi, *Biochem. J., 199*: 699-704 (1981).
59. J. R. Couchman, M. Hook, D. A. Rees and R. Timpl, *J. Cell Biol., 96*:177-183 (1983).
60. E. Engvall and E. Ruoslahti, *Coll. Rel. Res., 3*:359-369 (1983).
61. J. M. Foidart, E. W. Bere, M. Yaar, S. I. Rennard, M. Guillino, G. R. Martin and S. I. Katz, *Lab. Invest., 42*:336-342 (1980).
62. G. W. Laurie, C. P. LeBlond and G. R. Martin, *J. Cell Biol., 95*:340-344 (1982).
63. A. E. Chung, R. Jaffe, I. L. Freeman, J-P. Vergnes, J. E. Braguiski and B. Carlin, *Cell, 16*:277-287 (1979).
64. I. Vlodavsky and D. Gospodarowicz, *Nature (London), 289*:304-306 (1980).
65. R. L. Trelstadt, K. Hayashi and B. P. Toole, *J. Cell Biol., 62*:815-830 (1979).
66. E. D. Hay and S. Meier, *J. Cell Biol., 62*:889-898 (1974).
67. Y. S. Kanwar, V. C. Hascall and M. G. Farquhar, *J. Cell Biol., 90*:527-533 (1981).
68. A. Oohira, T. Wright, J. McPherson and P. Bornstein, *J. Cell Biol., 92*:357-367 (1982).

69. Y. S. Kanwar, A. Linker and M. G. Farquhar, *J. Cell Biol., 86*:688-693 (1980).
70. R. L. Vernier, D. J. Klein, S. P. Sisson, J. D. Mahan, T. R. Oegema and D. M. Brown, *N. Engl. J. Med., 309*:1001-1009 (1983).
71. D. T. Woodley, L. Didierjean, M. Regnier, J. Saurat and M. Prunieras, *J. Invest. Dermatol., 75*:148-151 (1980).
72. J. R. Stanley, P. Hawley-Nelson, M. Poirier, E. M. Shevach, S. I. Katz and S. H. Yuspa, *J. Invest. Dermatol., 75*:183-186 (1980).
73. D. Woodley, J. H. Saurat, M. Prunieras and M. Regnier, *J. Invest. Dermatol., 79*: 23-29 (1982).
74. D. Woodley and M. Regnier, *Arch. Dermatol. Res., 266*:319-322 (1979).
75. J. R. Stanley, P. Hawley-Nelson, S. H. Yuspa, E. M. Shevach and S. I. Katz, *Cell, 24*:897-904 (1981).
76. J. R. Stanley, D. T. Woodley and S. I. Katz, *J. Invest. Dermatol., 82*:108-111 (1984).
77. K. Holubar, K. Wolff, K. Konrad and E. H. Beutner, *J. Invest. Dermatol., 64*: 220-227 (1975).
78. G. Schaumberg-Lever, A. Rule, B. Schmidt-Ullrich and W. F. Lever, *J. Invest. Dermatol., 64*:47-49 (1975).
79. P. M. Grekin, G. N. Levy, A. J. King and L. A. Diaz, *J. Invest. Dermatol., 76*: 190-192 (1981).
80. D. F. Mutasim, Y. Takahashi, R. S. Labib, G. J. Anhalt, H. P. Patel and L. A. Diaz, *Clin. Res., 32*:482A (1984).
81. Y. Yamasaki and T. Nishikawa, *Acta Derm. Venerol. (Stockholm), 63*:501-506 (1983).
82. K. M. Yamada and K. Olden, *Nature (London), 275*:179-184 (1978).
83. E. Ruoslahti and E. Engvall, *Biochim. Biophys. Acta, 631*:350–358 (1980).
84. R. O. Hynes and K. M. Yamada, *J. Cell Biol., 95*:369-377 (1982).
85. M. Hayashi and K. M. Yamada, *J. Biol. Chem., 258*:3332-3340 (1983).
86. R. O. Hynes, I. U. Ali, A. T. Destree, V. M. Mautner, D. R. Perkins, D. Senger, D. Wagner and K. K. Smith, *Ann. NY Acad. Sci., 312*:317-342 (1978).
87. R. O. Hynes, *Proc. Natl. Acad. Sci. USA, 70*:3170–3173 (1973).
88. K. M. Yamada, S. S. Yamada and I. Pastan, *Proc. Natl. Acad. Sci. USA, 73*:1211-1221 (1976).
89. G. Balian, E. M. Glick, E. Crouch, J. M. Davidson and P. Bornstein, *J. Biol. Chem., 254*:1429-1432 (1979).
90. L. E. Hahn and K. M. Yamada, *Proc. Natl. Acad. Sci. USA, 76*:1160-1163 (1979).
91. R. J. Klebe, *Nature (London), 250*:248-251 (1974).
92. E. Pearstein, *Nature (London), 262*:497-500 (1976).
93. F. Grinnell, *Int. Rev. Cytol., 53*:65-144 (1978).
94. M. Chicquet, E. C. Puri and D. C. Turner, *J. Biol. Chem., 254*:5475-5482 (1979).
95. F. Grinnell and M. K. Field, *Cell, 10*:117-128 (1979).
96. F. Grinnell, in *Biocompatible Polymers, Metals and Composites* (M. Szycher, ed.) Technomic Publishing Co., Lancaster, PA, 1983, pp. 673-699.
97. F. Grinnell and M. K. Feld, *J. Biomed. Materials Res., 15*:363-381 (1981).
98. E. J. O'Keefe, D. T. Woodley, G. Castillo, N. Russell and R. E. Payne, *J. Invest. Dermatol., 82*:150-155 (1984).
99. J. R. Couchman, W. T. Gibson, D. Thom, A. C. Weaver, D. A. Rees and W. E. Parrish, *Arch. Dermatol. Res., 266*:295-310 (1979).
100. O. Saksela, K. Alitalo, U. Kiistala and A. Vaheri, *J. Invest. Dermatol., 77*:283-286 (1981).
101. I. Thesleff, H. J. Barrach, J-M. Foidart, A. Vaheri, R. M. Pratt and G. R. Martin, *Dev. Biol., 81*:182-192 (1981).

102. F. J. Roll, J. A. Madri, J. Albert and H. Furthmayr, *J. Cell Biol., 85*:597-616 (1980).
103. C. A. Kumamoto and J. H. Fessler, *Proc. Natl. Acad. Sci. USA, 77*:6434-6438 (1980).
104. S. Gay, A. Martinez-Hernandez, R. K. Rhodes and E. J. Miller, *Coll. Rel. Res., 1*: 377-384 (1981).
105. J. A. Madri, B. Dreyer, F. A. Pitlick and H. Furthmayr, *Lab. Invest., 43*:303-315 (1980).
106. K. S. Stenn, J. A. Madri and F. J. Roll, *Nature, 277*:229-232 (1979).
107. G. R. Grotendorst, H. Seppa, H. K. Kleinman and G. R. Martin, *Proc. Natl. Acad. Sci. USA, 78*:3669-3772 (1981).
108. A. Chung, R. Jaffe, I. Freeman, J. Vergnes, J. Bragenski and B. Carlin, *Cell, 22*: 177-287 (1980).
109. B. L. Bender, B. Carlin, R. Jaffe, T. Temple, and A. E. Chung, *Exp. Cell Res., 137*:415-425 (1982).
110. D. T. Woodley, R. A. Briggaman, E. J. O'Keefe, A. O. Inman, L. L. Queen and W. R. Gammon, *N. Engl. J. Med., 310*:1007-1013 (1984).
111. H. Yaoita, R. A. Briggaman, T. J. Lawley, T. T. Provost and S. I. Katz, *J. Invest. Dermatol., 76*:288-292 (1981).
112. A. Paller, L. Queen, D. T. Woodley, W. R. Gammon, E. J. O'Keefe and R. A. Briggaman, *J. Invest. Dermatol., 82*:207 (1984).
113. J. M. Foidart and M. Yaar, *Front. Matrix Biol., 9*:175-188 (1981).
114. J. Daroczy, J. Feldman and K. Kiraly, *Front. Matrix Biol., 7*:208-232 (1979).
115. R. Timpl., H. Weidemann, V. Van Delden, H. Furthmayr and K. Kuhn, *Eur. J. Biochem., 120*:203-211 (1981).
116. R. Timpl, J. Risteli and H. Bachinger, *FEBS Lett., 101*:265-268 (1979).
117. J. Risteli, H. Bachinger, J. Engel, H. Furthmayr and R. Timpl, *Eur. J. Biochem., 108*:239-250 (1980).

7

Vascular Connective Tissue

Normal biology and derangement in human diseases

Richard Mayne
University of Alabama at Birmingham,
Birmingham, Alabama

I. INTRODUCTION

In all blood vessels the components of connective tissue possess important structural and biomechanical functions. The lamellae of elastin are especially important in providing the elastomeric properties of the larger arteries, and in the human aorta elastin represents 20% of the total dry weight. However, collagen is also present in a similar proportion in the human aorta, and provides the tensile strength throughout the vascular system. Several different types of collagen have now been isolated from blood vessels, and these will be discussed in detail later in this chapter. The proportion of proteoglycans in the aorta is relatively smaller (approximately 0.5-1% of the dry weight), but even at this concentration the extended structures of these molecules are sufficient to completely fill the aqueous spaces of the tissues. It is probable that the proteoglycans serve similar functions to those of hyaline cartilage, both resisting compression and also providing a gellike medium through which diffusion of nutrients to the smooth muscle cells can easily occur. Thin basement membranes are closely associated with both the endothelial and smooth muscle cells. For the endothelium the basement membrane is the site of cell attachment and also provides an attachment site to the connective tissue of the subendothelium, or to the elastin of the internal elastic lamina. It may also act as a barrier to the passage of large macromolecules, such as lipoproteins to the subendothelium. In blood vessels, all smooth muscle cells are surrounded by a continuous basement membrane, and in atherosclerotic lesions, the presence of a basement membrane is an important criterion in identification of this cell type. The smooth muscle basement membrane probably acts as a permeability barrier, but fusion of basement membranes between smooth muscle cells is often observed, and this may be important in cell-to-cell attachments. The smooth muscle basement membrane probably also acts as a site of attachment of the collagen fibrils, which often appear to interconnect adjacent smooth muscle cells.

II. HISTOLOGY OF BLOOD VESSELS IN RELATION TO THE CONNECTIVE TISSUE COMPONENTS

For a detailed review of the ultrastructure of blood vessels, including the cytoplasm of both the endothelial and smooth muscle cells, readers should consult Ref. 1.

A. Arteries

Three different size classes of arteries are normally distinguished, although no sharp morphological transitions exist between each class.

1. Elastic or Conducting Arteries

These vessels have a large content of elastin and expand during systole with a subsequent contraction during diastole, thus serving to maintain the blood pressure. In man, lining the lumen of the aorta and its larger branches, the tunica intima consists of a single layer of endothelial cells beneath which there is a prominent subendothelium containing connective tissue and smooth muscle cells. The tunica media consists of 40-60 fenestrated lamellae of elastin between which are spirals of smooth muscle cells organized at highly ordered angles (2). External to the tunica media, the tunica adventitia consists of fibroblasts, connective tissue, and an occasional smooth muscle cell. Within the tunica adventitia numerous small blood vessels or vasa vasorum are present that also penetrate the outer half of the tunica media.

Tunica Intima: The endothelial cells form a tight pavement of cells that are interconnected to each other by both occluding (tight) and communicating (gap) junctions. In addition, numerous plasmalemmal vesicles are present, which are probably involved in transendothelial transport of small and large molecules as occurs in capillaries (1). Beneath the endothelium there is a thin basement membrane to which the endothelial cells are attached. Desquamation of the endothelium allows circulating platelets to interact with the basement membrane and subendothelial connective tissue, and this may have important consequences regarding the "response to injury" theory for the origins of atherosclerosis (see Section V). In man, but not normally in other species, the subendothelium steadily thickens with increasing age, a process known as diffuse intimal thickening. Several electron microscopic analyses have been made of diffuse intimal thickening in an attempt to understand its potential relationship to the origins of both the fatty streaks and atherosclerotic lesions to which the human aorta is especially prone (3-5). In the early human fetus the endothelial cells rest directly on an internal elastic lamina, but beginning at birth intimal thickening of the subendothelium occurs by the deposition first of proteoglycans and a few collagen fibrils, followed later by smooth muscle cells and cells with the characteristics of modified (5) or proliferative (6) smooth muscle cells. The latter cells possess a more extensive rough endoplasmic reticulum, fewer myofilaments, and an incomplete basement membrane. They probably arise by migration of smooth muscle cells from the tunica media. Other wandering cells are sometimes observed that are probably macrophages and are derived from circulating monocytes. Fibers of elastin and collagen also appear in increasing amounts, especially in the region closest to the internal elastic lamina. Intimal thickening occurs most extensively in regions of the aorta that are prone to atherosclerosis, such as the lower aorta (3), and is accelerated in the presence of hypertension (3). Intimal thickening is probably part of the normal aging process, but accelerated intimal thickening gives rise to a predisposition for the development of atherosclerotic lesions.

Tunica Media: In mammals, only smooth muscle cells are present between the lamellae of elastin of the tunica media. In the cytoplasm of each cell there are characteristic myofilaments, dense bodies, and 10-nm filaments. In addition, sarcolemmal vesicles are often present in considerable numbers on discrete areas of the cell surface. (For a detailed description of the ultrastructure of the arterial smooth muscle cell see Ref. 7.) Around each smooth muscle cell smaller elastin fibers are present that interconnect the larger elastin lamellae. Considerable numbers of small collagen fibrils are also present, all of which are of the same diameter and appear to interconnect the smooth muscle cells both to each other and to the elastic fibers. From electron microscopic and earlier histochemical analyses (8), a separate system of reticular fibers cannot be recognized in blood vessels. No nerves are present in the tunica media, and conduction of impulses for muscular contraction may occur through communicating (gap) junctions between the smooth muscle cells.

Tunica Adventitia: This layer is not highly organized and consists predominantly of fibroblasts and collagen fibers. It tends to merge with the surrounding connective tissue.

2. Muscular or Distributing Arteries

The muscular arteries have an extensive smooth muscle component, and the degree of contraction of these cells determines the distribution of blood to the various organs and tissues.

Tunica Intima: In most muscular arteries the endothelium is directly attached by a basement membrane to the internal elastic lamina. However, in the human coronary arteries, longitudinally oriented smooth muscle cells are always present in the tunica intima, and at arterial branch points there are extensive numbers of these cells, called intimal cushions (9). Such branches are a common site for the development of atherosclerosis, although extensive intimal thickening of the coronary arteries occurs with increasing age regardless of the development of fatty streaks or atherosclerotic lesions (10). In the left anterior descending branch of the coronary artery the intima can become twice as thick as the media (11), and this is the most common site of atherosclerosis. Figure 1 shows an electron micrograph of a muscular artery from a normal monkey in which smooth muscle cells are present in the tunica intima.

Tunica Media: The smooth muscle cells are the only cells present in the tunica media and are arranged in a spiral manner (2). Between the cells both elastin and collagen fibers are present, but the connective tissue is less extensive than in the elastic arteries. The internal elastic lamina has many fenestrations and with age undergoes extensive splitting.

Tunica Adventitia: In muscular arteries the tunica adventitia is prominent, and in addition to fibroblasts and collagen fibers, longitudinally arranged elastin fibers are also present.

3. Arterioles

These are the smallest vessels of the arterial system, the tunica media consisting of one to three rings of smooth muscle cells (12). The tonus of the smooth muscle cells is controlled by the autonomic system and by the level of the hormone angiotensin II, and is the principal determinant of resistance to flow and therefore of blood pressure. The clinical condition of hypertension exists if the tonus remains continuously elevated.

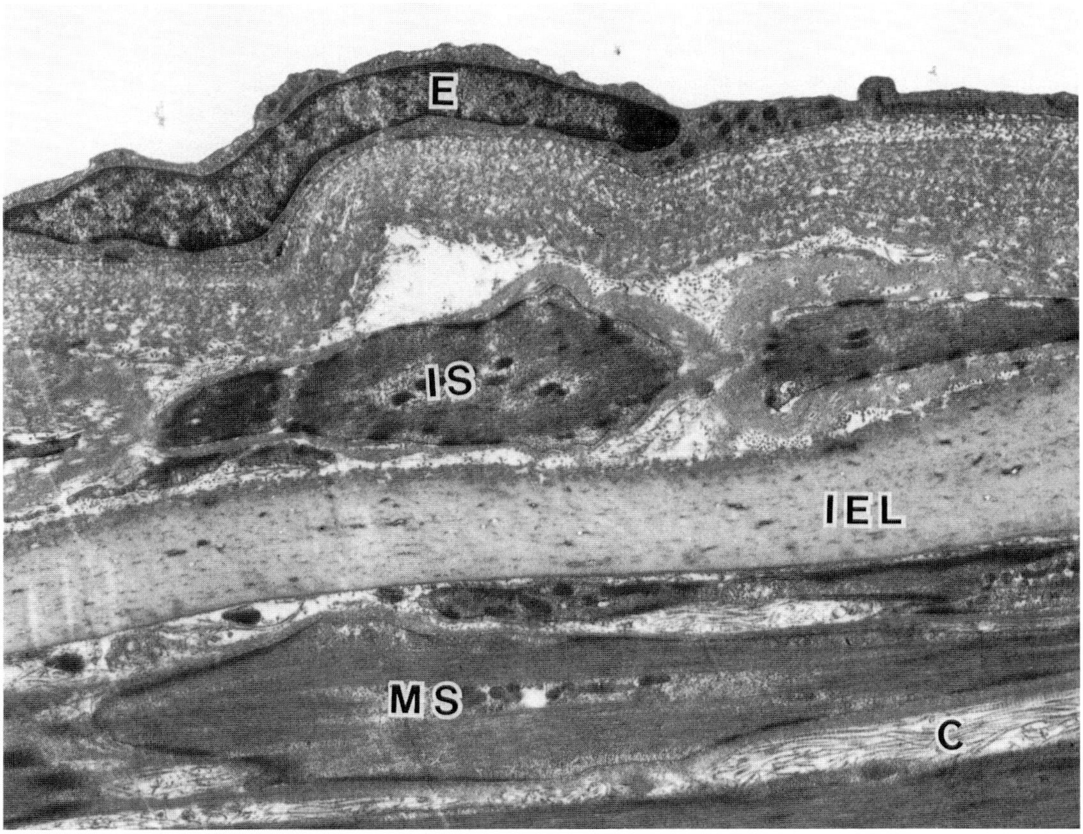

Figure 1 Electron micrograph of the basilar artery of a normal monkey. Note the extensive basement membrane-like matrix underneath the endothelial cells. In the tunica intima smooth muscle cells are present and are surrounded by an extensive basement membrane. In the tunica media, large numbers of collagen fibrils are present between the smooth muscle cells and, in areas, appear to interconnect the cells. E = endothelium, IS = intimal smooth muscle cell, IEL = internal elastic lamina, MS = medial smooth muscle cell, C = collagen. (Electron micrograph kindly provided by Dr. Jack Geer, Department of Pathology, University of Alabama at Birmingham.) ×10,000.

Tunica Intima: The endothelial cells lie on a thin internal elastic lamina, which contains extensive fenestrations through which cell-to-cell contacts are made with the smooth muscle cells of the tunica media.

Tunica Media: Each smooth muscle cell is surrounded by a basement membrane, and between the smooth muscle cells a few collagen and elastin fibers are present.

Tunica Adventitia: If present, it consists of a thin layer of collagen fibers and an occasional fibroblast.

B. Microvasculature

1. Precapillary Sphincter, Metarteriole, and Capillary

The first extensive exchange of nutrients and gases begins at the level of the capillaries, which consist of a single layer of endothelial cells surrounded by a continuous basement membrane. The pericyte is another cell type that is always found closely associated with capillaries. It is surrounded by a basement membrane that is continuous with the basement membrane of the endothelium. The functional importance of the pericyte is still not established, and it may represent a relatively undifferentiated form of the smooth muscle cell. External to the pericytes and endothelial cells, there is an extensive array of fine collagen fibrils many of which appear to be closely associated with the basement membrane. In tissues, the entry of blood to a capillary bed is controlled by precapillary sphincters that consist of a ring of a few smooth muscle cells surrounding the endothelium. The degree of contractility of these smooth muscle cells is controlled by the autonomic nervous system. Metarterioles are also present in capillary beds, and consist of a slightly larger vessel with a discontinuous layer of smooth muscle cells. Metarterioles serve as a shunt between the arterial and venous systems.

The continuous type of capillary is found in all muscles, the dermis, and the central nervous system. Considerable controversy exists with regard to the passageway of molecules across the continuous type of endothelium. Still unclear are the relative roles of the invaginations of the cell membrane called plasmalemmal vesicles in forming transient channels across the endothelium, or of the permeability of the intercellular junctions between the endothelial cells (compare Refs. 1 and 13). However, in tissues such as the endocrine organs, the intestines, or kidney glomeruli, small fenestrations are present in the endothelium, and it is in these regions that rapid exchange probably occurs. Except in the glomeruli, a very thin membrane is always observed across the fenestrations. In the liver, lymph nodes, and bone marrow the sinusoid endothelial cells contain large perforations through which, in the liver, the plasma but not the cells can freely pass to the space of Disse and the hepatocyte cell surface.

2. *Venules*

The venous system initially begins by several capillaries coming together to form a postcapillary venule. These venules represent a major site of fluid exchange and are extensively involved in inflammatory processes. The intercellular junctions between the endothelial cells do not appear to be as extensively developed as those present in the capillaries, and during inflammation they tend to widen and allow the passage of large macromolecules. The endothelial cells are attached to a thin basement membrane, which is continuous with the basement membrane of the pericytes. Thin collagen fibrils are present between endothelial cells and the pericytes. In the muscular venules the pericytes are replaced by smooth muscle cells, and a more extensive adventitia is present containing fibroblasts and collagen fibers.

C. Veins

In the small veins, the media is composed of two to four rings of smooth muscle cells between which elastic and collagen fibers are located. Considerable variation in arrangement of the cells is encountered for the media of large veins dependent on their location in the body; for a detailed description of these variations Ref. 2 should be consulted. Generally, in medium veins the tunica media is not extensive and consists of a few bundles of smooth muscle cells. However, the tunica adventitia is often extensive and contains bundles of smooth muscle cells arranged in a longitudinal manner. In the large veins these longitudinal bundles are highly developed and, by contracting, are in part responsible for the return of blood to the heart. All veins contain extensive amounts of collagen and elastic fibers, but no elastic lamellae are present.

III. ISOLATION AND LOCALIZATION OF CONNECTIVE TISSUE MACROMOLECULES FROM NORMAL BLOOD VESSELS

The structure and biology of the elastic fibers and lamellae with their associated microfibrils are described in detail in Chapters 2 and 3 of this volume and will not be described again in detail in this chapter.

A. Collagen

1. Isolation and Characterization

Type I and Type III Collagens: The collagen of blood vessels is largely insoluble in neutral salt or acidic solutions, and limited pepsin digestion is therefore usually used to solubilize a proportion of the total collagen. By this procedure, several groups have reported the presence of type I and type III collagens in human (14-17), pig (18), or calf (19,20) aorta and from human common iliac arteries (21). However, even after extensive pepsin digestion, only limited amounts of collagen are solubilized. In one study, it was reported that human aortas consists of 70% type III collagen and 30% type I collagen (10). However, these relative proportions of type I and type III collagens in the pepsin extract are probably not representative of the total collagen present, the type III collagen probably being preferentially solubilized during pepsin digestion (21).

Another approach to determining the relative proportions of type I and type III collagens is to fractionate and quantitate selected cyanogen bromide peptides of the α1(I) and α1(III) chains (17,22,23). In this method, after extraction of noncollagenous proteins, the samples are lyophilized and subsequently cleaved directly with cyanogen bromide. Estimates are then made of the relative proportions of CNBr peptides derived from the α1(I) and α1(III) chains. In one study, the relative proportion of type III collagen to type I collagen from human aorta was calculated to be 33%, 44%, and 38%, respectively, for the intima, media, and adventitia (17). In another study (23), very similar values were found for the proportion of type III collagen to type I collagen in the human intima plus media of the aortic arch (37% type III) and the abdominal aorta (42% type III). Interestingly, a somewhat lower proportion of type III collagen (21%) was found in the vena cava.

Type IV Collagen: In preliminary experiments, a high-molecular-weight fraction was isolated from human aorta which by amino acid analysis was suggested to consist predominantly of type IV collagen (15). Subsequently, using a method involving partial reduction of disulfide bonds with dithiothreitol prior to pepsin digestion (24), the yield of all collagens from bovine aorta was markedly increased (20). After differential salt precipitation, two α-sized chains were isolated with the same properties as the pepsin-resistant fragments of the α1(IV) and α2(IV) chains isolated from other tissues (25-30, see Chapter 1 for a discussion of the structure of type IV collagen). Approximately 5% of the total solubilized collagen was type IV, probably derived from the basement membrane that surrounds each smooth muscle cell.

Type V Collagen: Small amounts of type V collagen can be solubilized from human aorta after limited pepsin digestion (17,31,32). From bovine aorta the yield was markedly increased after partial reduction of the tissue with dithiothreitol (20), and type V collagen represented approximately 5% of the total solubilized collagen. Both the α1(V) and α2(V) chains were isolated from human (17) and bovine (20) aorta, but were not present in a consistent 2:1 ratio as might be expected for a molecule of chain composition

$[\alpha 1(V)]_2\ \alpha 2(V)$. The $\alpha 3(V)$ chain was not detected in human or bovine aorta (17,20), but may be present in a small proportion in the pig aorta (17).

Type VI Collagen: Type VI collagen was first isolated from human tunica intima (31) and is therefore sometimes called intimal collagen. However, it has subsequently been isolated from placenta (33-35) and from skin (36) and may have a wider distribution in connective tissues. Type VI collagen is isolated after pepsin digestion as aggregates containing both collagenous and noncollagenous domains that are stabilized by disulfide bonds (33,34,37). Three genetically distinct chains are present in the native molecule (38). Recently, a model was presented for the structure of a fibrillar form of the molecule based on both biochemical data and rotary shadowing observations (39,40; for details of the model see Chapter 1). It was suggested that type VI collagen may form some of the microfibrils found between the elastic and collagenous fibrils.

2. Localization in Blood Vessels

Type I and Type III Collagens: Using specific antibodies against type I and type III collagens, very different patterns of immunofluorescent staining were observed for human aorta (41) and calf common carotid artery (42). Type I collagen tended to localize in close proximity to the smooth muscle cells of the tunica media and gave extensive immunofluorescent staining of the tunica adventitia. In contrast, immunofluorescent staining for type III collagen was prominent in the subendothelium and close to the elastic lamellae of the tunica media. Separate studies confirmed the localization of type III collagen in close proximity to the elastic lamellae of the tunica media (43,44), and it was suggested that type III collagen may form a network of fine fibrils interconnecting the elastic lamellae to the smooth muscle cells and the larger bundles of type I collagen. For the tunica intima of both fetal and adult bovine aorta extensive immunofluorescent staining for type III collagen was observed throughout the layer and not just in the immediate subendothelium (44,45). In another study with bovine aorta, no staining for either type I or type III collagen was observed in the subendothelium (46).

Type IV Collagen: Using a polyclonal antibody to mouse EHS tumor type IV collagen, immunofluorescent staining of the basement membrane surrounding the smooth muscle cells of the bovine aortic media was demonstrated (47). Similar results were obtained using a monoclonal antibody to chicken type IV collagen in which immunofluorescent staining was observed around the smooth muscle cells, but not the fibroblasts, of the larger chicken arteries (48). Type IV collagen was also localized in the subendothelium of the calf aorta by immunofluorescent staining with antibodies to mouse tumor type IV collagen (45,46). In capillaries, a strong immunofluorescent staining of the basement membrane was always observed with antibodies to type IV collagen, and in skeletal muscle using either polyclonal antibodies (49) or a monoclonal antibody (48) the staining of the capillary basement membrane was always far stronger than the basement membrane of the muscle fibers.

Type V Collagen: One group, using both an immunofluorescent staining technique and immunoelectron miscroscopy (50,51), has reported that type V collagen is localized in the basement membranes of the kidney glomerulus and the lung. In a separate study, using different antibody preparations, these results could not be confirmed (52), and instead type V collagen appeared to be present as a pericellular matrix surrounding the smooth muscle cells and underlying capillary endothelial cells (53). Some type V collagen reactivity also appeared to be closely associated with the interstitial collagen fibers

(52). Other groups have reported a diffuse location of type V collagen throughout the tunica media (43,44,46), and intense immunofluorescent staining of the subendothelium of calf aorta was also reported (46). In all these analyses, polyclonal antibodies to type V collagen were used. Recently, two monoclonal antibodies were prepared against chicken type V collagen and used for immunofluorescent staining (54). Staining of blood vessels and other tissues was not observed unless the sections were first treated with dilute acetic acid, suggesting that this treatment may unmask the epitopes. After acetic acid treatment, type V collagen was localized principally in the fibroblastic lamellae of the tunica media and associated bundles of interstitial collagen fibers. It was proposed that acetic acid swells the fibers of type I and type III collagen and this unmasks the epitopes. This conclusion is supported by recent experiments with chick cornea in which the epitopes for the monoclonal antibodies to type V collagen were exposed by first digesting the sections with corneal collagenase, which cleaves type I but not type V collagen (55). It appears from these experiments that type V collagen is closely associated with the interstitial collagens and may be involved in the organization or formation of interstitial collagen fibrils. How type V collagen is able to perform this function in tissues is still unclear, although it is known that the pepsin-digested form of the molecule will form fibrils in vitro (56). An understanding of the role of type V collagen in blood vessel structure may be of great significance since in atherosclerotic lesions, the type V collagen content appears to be increased (see Section V), and the aortic smooth muscle cell may possess a specific receptor for type V collagen (57).

Type VI Collagen: In one study, immunofluorescent staining of blood vessels with antibodies prepared against type VI collagen suggested a general location throughout the matrix without any specific localization to basement membranes (34).

B. Proteoglycans

1. *Biochemical Characterization*

Until recently, no attempts were made to extract intact proteoglycans from blood vessels, and instead glycosaminoglycans were isolated after being liberated by protease digestion of core proteins.

Glycosaminoglycans: Most studies of the glycosaminoglycans demonstrated the presence of chondroitin 4- and 6-sulfates, dermatan sulfate, heparan sulfates, and hyaluronic acid in blood vessels, with the highest concentration being located in the intima and then a successive decrease in concentration toward the media and adventitia (58-60). The chondroitin 4-sulfate and dermatan sulfate were shown to exist as hybrid molecules, whereas the chondroitin 6-sulfate appeared to exist only as a polymer containing glucuronic acid and *N*-acetylgalactosamine-6-sulfate (59).

Proteoglycans: Several groups have extracted aortic proteoglycans in the presence of 4 M guanidine HCl and subsequently achieved partial fractionation (61-66). In general, it appears that aggregates of proteoglycans can be formed with hyaluronic acid, as occurs in cartilage, and which may be stabilized by link protein (67). In some experiments, two proteoglycans were fractionated, one of which consisted almost entirely of chondroitin sulfate and the other of a mixed polymer of chondroitin sulfate and dermatan sulfate chains (63,64,67).

2. *Tissue Localization*

To localize proteoglycans in monkey aorta, sections were stained with the cationic dye alcian blue and most of the marked staining reaction was observed in the intimal layer (68). More specific localization was achieved with the hexavalent cationic dye ruthenium red, which results in the collapse of proteoglycans to electron-dense particles (68,69). Networks of these particles were observed between the collagen and elastin fibers. These networks were interconnected by thin filaments, and a 64-nm periodic association with the collagen fibrils was often present. The glycosaminoglycan content of these granules was demonstrated by their susceptibility to testicular hyaluronidase and chondroitinase ABC. Smaller granules were also observed in close association with the basement membrane of the endothelial and smooth muscle cells, and considerable staining with ruthenium red was present on the luminal surface of the endothelial cell. To what extent the staining reactions of the endothelial cell surface can be attributed to proteoglycans, and more especially heparan sulfate, is at present unclear. However, the anionic staining properties of the endothelial surface were partially removed by heparitinase digestion, especially in the regions of the diaphragms of fenestrated endothelium (1).

With an antibody to mouse tumor heparan sulfate proteoglycan, immunofluorescent staining of the subendothelium was reported (45), which presumably represents staining of the endothelial basement membrane. Intense staining was observed in the tunica intima with lesser staining in the inner media (70), using antibodies to a dermatan sulfate/chondroitin sulfate-containing bovine aortic proteoglycan. In a separate study, staining of bovine aortic intima and media for type III collagen, but not type I collagen, was increased after treatment of sections with chondroitinase ABC, suggesting that type III collagen is closely associated with proteoglycan (44). The increased staining was predominantly located in proximity to the elastic laminae, a result that is also suggested by differential staining of glycosaminoglycans using alcian blue staining at different electrolyte concentrations (71).

C. Laminin

Laminin is a large glycoprotein isolated from a transplantable mouse tumor, which from immunological evidence appears to be present in all basement membranes (reviewed in Ref. 72). Present biochemical evidence suggests that it is assembled from one component of molecular weight 440,000 and two components of molecular weight 220,000, which together form a crosslike structure as viewed in the electron microscope after rotary shadowing (73). Although laminin has not been isolated from blood vessels, intense immunofluorescent staining with antibodies to laminin was reported around smooth muscle cells of the aortic media (74), and in the subendothelium of arteries and capillaries (45,75).

D. Fibronectin

Immunofluorescent staining shows that fibronectin is located in the subendothelium of both capillaries and arteries, around the smooth muscle cells of the media especially of muscular arteries, and throughout the adventitia (45,76,77). It appears to be specifically concentrated in the basement membrane of the endothelium (77).

IV. BLOOD VESSEL CELLS IN CULTURE

A. Smooth Muscle Cells

Several different procedures have been described to establish smooth muscle cells in culture (reviewed in Ref. 78). In general, either smooth muscle cells are liberated by digestion of the aortic media with combinations of proteases such as trypsin, elastase, and collagenase (79), or the cells are obtained as outgrowths from small explants of aortic media (80). With a method initially involving enzymatic dissociation, the cells, on plating, initially contract in response to angiotensin II and the cytoplasm remains filled with contractile filaments (81). The cells do not divide, and division cannot be stimulated by serum growth factors. After a few days in culture, the contractile filaments decrease and an extensive rough endoplasmic reticulum develops in the cytoplasm. The cell begins to respond to a growth factor derived from platelets. It is argued that the smooth muscle cell can modulate between a "contractile" and "synthetic" state, the contractile state to some extent returning when the culture reaches a high density (6). However, after passaging, the cells always remain in the synthetic state. It is suggested that this modulation of smooth muscle cell phenotype also occurs during the development of atherosclerotic lesions, so that the intimal smooth muscle cells enter into the synthetic state, become responsive to mitogens, and form an extensive extracellular matrix (82). The properties of the smooth muscle cell in culture may therefore be more representative of the cells in atherosclerotic lesions than in normal media. Most analyses of smooth muscle cells in culture, especially if the cells have been passaged a few times, are undoubtedly performed on cells in the synthetic state and therefore probably cannot be directly compared to the properties of the normal smooth muscle cells of the media.

1. Collagen

Morphological observations have shown that collagen is synthesized by smooth muscle cells in culture (83,84), and several attempts have been made to analyze the different proportions of collagen types (18,85-89). For both monkey (87,89) and guinea pig (88) smooth muscle cells, an excess of type III collagen over type I collagen was observed in comparison to fibroblast populations, for which type I collagen always predominated. However, other analyses of pig (18,85) or human (86) smooth muscle cells showed that the synthesis of type I collagen predominated over type III collagen. These differences may reflect the synthetic versus contractile state of smooth muscle cells at the time of analysis or may reflect species differences. Synthesis of appreciable amounts of the $\alpha1(V)$ and $\alpha2(V)$ chains of type V collagen was also observed for monkey aortic smooth muscle cells (89), and after pepsin digestion of both monkey and guinea pig aortic smooth muscle cells, a collagenous peptide was isolated of apparent molecular weight 45,000 (88,89). The origin of this peptide (called CP45) is at present unknown.

In one analysis of collagen synthesis by monkey smooth muscle cells grown at different densities, no difference was found in the rate of collagen synthesis per cell during rapid growth and at higher cell densities (90). However, in other studies with calf aortic smooth muscle cells, differences were found in the relative proportions of type I and type III collagen at differing cell densities, so that with time in culture the ratio of procollagen type I to type III decreased from 2.7 to 1.0 (91). In the presence of 17β-estradiol, for reasons that are unclear, the synthesis of type III collagen was preferentially suppressed. In another study rabbit smooth muscle cells were grown on purified elastin membranes and the membranes were either stretched or merely agitated (92). Stretching resulted in a 2-4 times increase in collagen synthesis as well as an increase in proteoglycan synthesis.

During stretching no difference was observed in the relative proportions of synthesis of type I and type III collagens. It was suggested that this system may be a model for hypertension research, since similar increases in collagen and proteoglycan are known to occur during experimentally induced hypertension in laboratory animals (see below).

2. *Elastin*

Morphological evidence has been presented for the formation of elastin by smooth muscle cells (80,84,93), and in one study with primary cultures of neonatal rat aortic smooth muscle grown to high density, large aggregates of elastin were observed in the electron microscope (94). It also proved possible to isolate a hot alkali-insoluble residue from these cultures whose amino acid analysis was the same as that of authentic elastin. In other studies a protein similar in properties to tropoelastin was isolated from monkey (95) and pig (96) smooth muscle cell cultures, and the cross-link amino acids desmosine and isodesmosine, which are specific to elastin, were identified in several smooth muscle cell cultures (94,95,97).

3. *Glycosaminoglycans and Proteoglycans*

In general, analyses of the glycosaminoglycans synthesized by smooth muscle have demonstrated chondroitin 4- and 6-sulfate, dermatan sulfate, heparan sulfate, and small amounts of keratan sulfate and hyaluronic acid (94,98-101). In a study of monkey smooth muscle cells an excess of dermatan sulfate was observed with somewhat less chondroitin 4- and 6-sulfates and very little hyaluronic acid, this result contrasting with dermal fibroblasts, which synthesize predominantly hyaluronic acid (102). A subsequent analysis of the proteoglycans synthesized by these cells resulted in the fractionation of two classes, the larger containing predominantly chondroitin sulfate and the smaller containing dermatan sulfate (103). Only the larger proteoglycan fraction would form higher-molecular-weight aggregates in associative conditions. In another study, pig smooth muscle cells were subjected to a centrifugal force, and a marked increase in glycosaminoglycan synthesis was observed. This result suggests that a response of these cells to stress, as may occur in hypertension, will result in an increased synthesis of proteoglycans (100).

B. Endothelial Cells

The development of techniques to grow endothelial cells successfully in culture with the retention of their phenotypic properties has led to many important insights into the behavior of these cells (reviewed in Refs. 104 and 105). Endothelial cells grow in culture as a contact-inhibited monolayer, but also sometimes lose contact inhibition by a process known as "sprouting" in which a few cells begin to undergrow the contact-inhibited layer (106). In certain culture conditions endothelial cells form vessellike structures with an enclosed lumen (107-109), this phenomenon being stimulated by the inclusion of capillary endothelial cells within a native collagen gel (110). In the latter experiments, it was possible to demonstrate by electron microscopy the formation of a basement membrane around the ring of endothelial cells. Migrating endothelial cells also have the ability to degrade basement membrane (type IV) collagen, and this may be a key event in the initiation of neovascularization (111).

1. *Collagen*

Preliminary studies have suggested that aortic endothelial cells can synthesize type IV collagen (112), and other analyses have clearly demonstrated the synthesis of type I and type III collagens by these cells (113,114). In an extensive series of experiments, Helene Sage, Paul Bornstein, and their colleagues have analyzed a variety of endothelial cell cultures for different collagen types. Initial analyses of bovine aortic endothelial cells demonstrated the secretion predominantly of type III collagen into the culture medium (115). However, a novel collagen called EC collagen (116), or now type VIII collagen (117), was also isolated from the culture medium and partially characterized. EC collagen appears to consist of non-disulfide-bonded chains of M_r 177,000 (called EC1), which are very sensitive to protease digestion and give rise to products of M_r 125,000 (EC2) and 100,000 (EC3). Analyses of the cell-matrix-associated collagens from bovine aortic endothelial cells showed the presence of type III collagen and the $\alpha1(V)$ and $\alpha2(V)$ chains of type V collagen (118). However, when bovine aortic endothelial cells were grown until the secondary growth pattern called "sprouting" occurred, then the synthesis of type I collagen predominated in both the cell layer and medium (106), suggesting a possible phenotypic change in these cells. Comparisons were also made between the collagens synthesized by bovine endothelial cells derived from aorta, superior vena cava, cornea, and an established line of capillary endothelial cells (119). The results for aorta, vena cava, and corneal endothelium were all similar, with the predominant secretion of type III procollagen and EC collagen in the medium, whereas type IV and type V collagens were associated with the cell layer. Collagen synthesis for capillary endothelial cells was much higher and consisted of a mixture of type I and type III collagens, without type IV or type V collagens. The significance of this result is still unclear and may relate either to a loss in phenotypic characteristics by the capillary endothelial cells in culture or to an intrinsic difference in the behavior of these cells compared to other endothelial cells. In another study (120), analyses were made of endothelial cell cultures derived from human umbilical vein and from a murine hemangioendothelioma. Type IV collagen was now the predominant collagen secreted into the culture medium, and for the human umbilical vein endothelial cells, EC collagen was also secreted. The production of type IV collagen by human umbilical vein endothelial cells was confirmed in a separate study (46). However, these authors also reported the presence of type V collagen in the cell layer, where it appeared to be associated with the cell surface. In the same study cultured bovine aortic endothelial cells also showed predominant synthesis of type IV and type V collagens without the synthesis of interstitial collagens, especially type III collagen.

It appears that the synthesis of interstitial collagens by endothelial cells may be very dependent on the conditions of culture and the phenotypic properties of the cells. Considerable interest exists in the possibility that endothelial cells have the potential to lay down fibrils of type I and type III collagen in the subendothelium. These collagens are known to cause platelet aggregation in vitro whereas type IV and type V collagens do not (reviewed in Ref. 56), and this could form the basis of platelet interaction with the subendothelium after endothelial denudation.

2. *Elastin*

An established line of rabbit aortic endothelial cells was shown to synthesize a similar protein to soluble elastin, most of this protein being retained in the cell layer (121). Also, the elastin-specific cross-linking amino acids desmosine and isodesmosine were identified in a hydrolysate derived from a clone of rat lung endothelial cells (122). These cells also gave positive immunofluorescent staining with an antibody raised against rat lung elastin peptides. It therefore appears that endothelial cells in vivo may also be able to synthesize elastin and could contribute to the elastic fibers of the subendothelium.

3. *Glycosaminoglycans and Proteoglycans*

Endothelial cells in culture synthesize a variety of glycosaminoglycans, the proportion of heparan sulfate always being higher than for smooth muscle cells (101,123). Some of this heparan sulfate appears to be associated with the cell surface where it is more highly sulfated (101,124), and from its heparinlike properties could contribute in vivo to the antithrombogenic properties of the endothelium. Cultured bovine aortic endothelial cells synthesize heparan-containing proteoglycans (125,126). In one study (126), two proteoglycans were identified in the culture medium, one of which was a large heparan-sulfate-containing proteoglycan of low buoyant density and the other was smaller and contained chondroitin sulfate with a lesser amount of dermatan sulfate. In the cell layer, the large heparan-sulfate-containing proteoglycan was present, and a smaller heparan sulfate proteoglycan of high buoyant density was also identified. Immunofluorescent staining of the cells was also observed with an antibody raised against the heparan sulfate proteoglycan of a transplantable mouse basement membrane tumor, suggesting that the cell-associated proteoglycans probably represent, in part, a basement membrane being laid down by the endothelial cells.

4. *Fibronectin*

Both bovine aortic endothelial cells and human umbilical vein endothelial cells were shown to synthesize and secrete fibronectin (127-129). However, it is less certain whether endothelial cells in vivo represent the major source of the plasma form of fibronectin.

5. *Laminin*

Synthesis and secretion of laminin by bovine aortic endothelial cells was reported, and a comparison made with the distribution of fibronectin (130). Secretion of laminin decreased markedly at higher cell densities whereas secretion of fibronection was unaffected by cell density.

V. DISEASES OF THE VASCULAR SYSTEM

The three major diseases that affect the blood vessels of man, and for which something is known of the involvement of connective tissue components, are diabetes, hypertension, and atherosclerosis.

A. Diabetes

In diabetes, the major vascular changes are at the level of the microcirculation. These are discussed in detail in Chapter 21. However, the diabetic condition also predisposes the larger arteries to atherosclerosis (131), and this aspect of the disease will form part of the general discussion of the connective tissue changes during the development of atherosclerotic lesions.

B. Hypertension

In hypertension, many of the pathological changes are observed at the level of the small arteries and arterioles, leading to the hyalinization or deposition of connective tissue in the subendothelial space. These changes appear to be associated with an increase in permeability of the endothelial layer, possibly associated with the disruption of the integrity of the junctions between the endothelial cells, or with the increased contractility of the endothelial cells (reviewed in Ref. 132). In the larger arteries, with hypertension, intimal thickening is accelerated and leads to the progressive development of localized atherosclerotic lesions.

1. Changes in Collagen Synthesis

In rats with experimentally induced hypertension a marked increase was observed within eight weeks in the diameter of the aortic media. This was accompanied by an increase in total amount of collagen and elastin, but without a change in their relative proportions to total dry weight (133). Subsequent studies showed increased collagen biosynthesis in the aorta and mesenteric arteries of experimentally induced and spontaneously hypertensive rats. This increase did not occur if the antihypertensive substance chlorothiazide or reserpine was given (134). Subsequently, it was found that β-aminopropionitrile, a specific inhibitor of the formation of collagen and elastin cross-links, would reduce blood pressure in hypertensive rats (135). This suggests that the increase in the amount of connective tissue may in part directly contribute to the increased blood pressure. However, subsequent studies have shown that the presence of β-aminopropionitrile does not affect the hypertension-induced hypertrophy of the aorta (136), and that during development of the spontaneously hypertensive rat, the increase in blood pressure precedes the increase in collagen biosynthesis (137). No change in collagen biosynthesis occurred in the veins of hypertensive rats, suggesting that a circulating humoral factor is probably not responsible for increased collagen synthesis in the arteries (138). The increased synthesis of collagen in arteries is also accompanied by increased turnover, so that the half-life of aortic collagen is reduced from 60-70 days to 17 days (139).

2. Changes in Glycosaminoglycans and Proteoglycans

In general, during hypertension, the synthesis of glycosaminoglycans appears to be increased. For example, in the human aorta the ratio of chondroitin sulfate to hyaluronic acid in the intima-media layer was increased in hypertension (140). In dogs, in which hypertension was induced in the thoracic aorta by surgical coarctation, the incorporation of radiosulfate into glycosaminoglycans was markedly increased in the hypertensive part of the aorta (141). More recently, an analysis was made of the proteoglycans synthesized by cultured smooth muscle cells derived from normotensive and hypertensive rats (142). Consistently, the synthesis of proteoglycans by hypertensive cells was greater than by normal cells provided the results were expressed on a per cell basis.

C. Atherosclerosis

Atherosclerotic lesions often occur at discrete locations in the arterial system and involve the proliferation of smooth muscle cells together with the formation of extensive amounts of connective tissue and the deposition of lipid (5). Many, but not all, lesions appear to arise from the progressive development of fatty streaks. Macrophages derived from circulating monocytes are commonly found in atherosclerotic lesions and, together with some of the smooth muscle cells, become distended with lipid to form foam cells. A recent experimental approach toward understanding the origins of atherosclerosis has focused on discovering factors that influence the proliferation of smooth muscle cells (143). The first of these to be identified was a platelet-derived growth factor that is released during platelet degranulation. This factor was shown to stimulate the growth of smooth muscle cells established in tissue culture (144). However, it does not stimulate the proliferation of freshly isolated aortic smooth muscle cells until the cells have modulated from a "contractile" to a "synthetic" state (81). Platelet-derived growth factor is the only growth factor that is chemotactic for smooth muscle cells (57) and could potentially stimulate the migration of smooth muscle cells from the media to the intima. However, for the platelet-derived growth factor to be active in promoting atherosclerosis in man, there must be a loss of endothelium and exposure of the subendothelial connective tissue to circulating platelets. Although endothelial loss and intimal proliferation can easily be achieved experimentally with a balloon catheter, whether exposure of the subendothelium normally occurs in blood vessel has not been established. Thus, there still remains a lack of direct evidence for the "response-to-injury hypothesis" involving the effects of platelet-derived growth factor on arterial smooth muscle cells as proposed by Ross (143). More recently, it was shown that macrophages also synthesize a growth factor for smooth muscle cells that is unrelated to platelet-derived growth factor (145). This factor is still poorly characterized, but could profoundly influence the growth of smooth muscle cells as macrophages are often found closely associated with fatty streaks. Endothelial cells also produce a growth factor that may influence the growth of smooth muscle cells (146).

Other results, which must be considered in any attempt to understand the pathogenesis of atherosclerosis, are the demonstration of the monoclonal or monotypic nature of many lesions (147,148). Whether such lesions arise from the transformation of smooth muscle cells by chemical agents, viruses, etc. (149) or simply represent the result of the selective survival of the progeny of the most actively proliferating cells after a series of waves of cell growth and degeneration (150) is not established.

To investigate a possible modulation in phenotypic properties of the smooth muscle cells present after intimal thickening and in atherosclerotic lesions, analyses were made of the proteins vimentin and desmin present in intermediate filaments (151) and of the different isoforms of actin (152). In normal rat aortic media, most smooth muscle cells were found to contain vimentin and some also contained desmin. However, in thickened intima after balloon catheterization, the cells were found to contain only vimentin, which was present in increased amounts (151). It was suggested that the migration and proliferation of medial smooth muscle cells occurs only with the vimentin containing cells, or that only vimentin can be expressed by proliferating aortic smooth muscle cells. During rat aortic intimal thickening or in human atherosclerotic lesions, instead of the α form of actin of smooth muscle, the β form predominated with detectable amounts of the γ form also being present (152). Similar changes in myosin isoforms were observed for smooth muscle cells in cull culture (152), again suggesting that the study of cultured smooth muscle cells may be directly relevant to the changes that occur in these cells during the development of atherosclerosis (6).

1. Morphology of Atherosclerotic Lesions

Electron microscopic analyses of human atherosclerotic lesions have shown that the fibrous cap consists of highly elongated smooth muscle cells between which are extensive layers of basement membrane-like material, dense bands of collagen fibrils, and proteoglycan (153-155). Below the cap, most of the cells are smooth muscle cells, but some macrophages are also present. Often the smooth muscle cells are distended by lipid droplets and are not easily distinguished from the macrophages.

2. Collagen

Several studies have shown that the collagen content of atherosclerotic lesions is increased when compared to normal intima (156-158). In one study, the collagen content of normal intima was 25 mg/100 mg protein, whereas in fibrous plaques obtained from the same aortas, the content was increased to 41 mg/100 mg protein (157). Increases in collagen biosynthesis were also observed in experimentally induced atherosclerosis in dogs (159) and rabbits (160-163). Cultured smooth muscle cells from atherosclerotic rabbit aorta were also found to have increased collagen biosynthesis when compared with cultured smooth muscle cells obtained from normal aortas (164). Initial estimates of the relative proportions of type I and type III collagen in aortas suggested that the normal aorta consists of 70% type III collagen, whereas in atherosclerotic lesions only 35% is type III collagen (16). From this result, it was argued that the smooth muscle cell has undergone a phenotypic change toward a more fibroblastic behavior. These analyses were based on pepsin-solubilized collagen, which subsequently was shown to be a relatively small proportion of the total collagen (17,163). Analyses of peptides derived from type I and type III collagen after cyanogen bromide cleavage of aortic fragments gave somewhat different results (17,23). In normal aortic intima or media, type I collagen was always in excess of type III collagen, and in atherosclerotic plaques there was a small change in the proportions toward type I collagen. Interestingly, a very high proportion of type V collagen was isolated from pepsin-digested atherosclerotic lesions (17,32), although this increase could not be detected by immunofluorescent staining with an antibody directed against type V collagen (43). The significance of this result is still uncertain and the function of type V collagen still remains poorly understood, even in normal tissues, where it may be closely associated with dense connective tissue (54).

3. *Glycosaminoglycans*

In general, the content of glycosaminoglycans appears to increase slightly during intimal thickening but then is relatively unchanged in the atherosclerotic intima (157,158). However, in one study, a marked decrease in total glycosaminoglycan content was noted in atherosclerotic lesions, the decrease occurring in chondroitin 6-sulfate and heparan sulfate but not in hyaluronic acid (59). The dermatan sulfate content in the lesions increased, which may reflect its close association with collagen fibrils, as occurs in tendon (165). Of considerable interest is the presence in atherosclerotic lesions of lipoprotein complexed with proteoglycan (166). These complexes consist primarily of chondroitin sulfate proteoglycan and low-density lipoprotein or very low-density lipoprotein. Similar interactions have also been shown to occur in vitro (167). Recently, it was shown that after balloon catheterization of rabbits, all of the different types of aortic glycosaminoglycans (chondroitin sulfates, dermatan sulfate, and heparan sulfate) accumulated in the reendothelialized areas but not in those that remained denuded of endothelium (168). It was speculated that this may relate to earlier observations that lipid deposition also selectively occurs in the reendothelialized areas (169) and suggests that the proteoglycans may entrap the lipid. In all these studies, the glycosaminoglycans were isolated after protease digestion, and nothing is known at present of the structure of the proteoglycans from which these chains were derived.

VI. CONCLUSIONS

Considerable progress has been made in the isolation and characterization of the different types of collagens present in arteries that can be solubilized by pepsin digestion. However, less is known of the function of some of these collagens. Type I and type III collagens form the fibrils of the extracellular matrix of the tunica media, but appear to have different locations relative to the elastic lamellae and the smooth muscle cells (41-44).

It is not known at present whether there are completely separate fibrils of type I and type III collagen, or whether mixed fibrils can also exist. Type IV collagen appears to be present normally only in the basement membranes of the endothelial and smooth muscle cells. Whether it is also present in the large amounts of amorphous basement-membrane-like material often described in atherosclerotic lesions is not known. This material may also contain type V collagen which is known to be present in increased proportions in atherosclerotic lesions and for which a function remains to be assigned.

The composition of the glycosaminoglycans present in arteries has been known for some time, and only recently have the aortic proteoglycans begun to be investigated. It will be especially interesting to understand the nature of the population of proteoglycans that can undergo aggregation, and to show that link proteins are involved in the stabilization of these aggregates, as occurs for cartilage proteoglycans. In general, atherosclerotic lesions appear to possess many of the properties of scar tissue containing a high proportion of type I collagen and dermatan sulfate. This appears to be associated with a shift in properties of the smooth muscle cell from a "contractile" to a "synthetic" state. However, most of the cells present in atherosclerotic lesions are, by morphological criteria alone, still smooth muscle cells. It may be that they are similar in properties and behavior to smooth muscle cells growing in cell culture.

ACKNOWLEDGMENTS

Original studies were supported by National Institutes of Health Grants AM 30481 and AM 31394 and by the Muscular Dystrophy Association. The author would like to thank Kathy Neighbors for typing the manuscript.

REFERENCES

1. N. Simionescu and M. Simionescu, in *Histology; Cell and Tissue Biology* (L. Weiss, ed.), Elsevier Biomedical, New York, 1983, p. 371.
2. J. A. G. Rhodin, in *Handbook of Physiology;* Section 2, *The Cardiovascular System* (D. F. Bohn, A. P. Somlyo, and H. V. Sparks, eds.) Waverly Press, Baltimore, 1980, p. 1.
3. S. L. Wilens, *Am. J. Pathol., 27*:825 (1951).
4. H. Z. Movat, R. M. More, and M. D. Haust, *Am. J. Pathol., 34*:1023 (1958).
5. J. C. Geer and W. S. Webster, *Adv. Exp. Med. Biol., 43*:9 (1973).
6. G. R. Campbell and J. H. Chamley-Campbell, *Pathology, 13*:423 (1981).
7. A. V. Somlyo, in *Handbook of Physiology;* Section 2, *The Cardiovascular System* (D. F. Bohn, A. P. Somlyo, and H. V. Sparks, eds.), Waverly Press, Baltimore, 1980, p. 33.
8. D. C. Pease, in *Blood Vessels and Lymphatics* (D. I. Abramson, ed.), Academic Press, New York, 1962, p. 12.
9. D. Jaffé, S. Hartroft, M. Manning, and G. Eleta, *Acta Paediatr. Scand., Suppl. 219*:1 (1971).
10. J. C. Geer, H. C. McGill, W. B. Robertson, and J. P. Strong, *Lab. Invest., 18*:565 (1968).
11. L. Gross, E. Z. Epstein, and M. A. Kugel, *Am. J. Pathol., 10*:253 (1934).
12. J. A. G. Rhodin, *J. Ultrastruct. Res., 18*:181 (1967).
13. M. J. Karnovsky, *J. Cell. Biol., 35*:213 (1967).
14. E. Chung and E. J. Miller, *Science, 183*:1200 (1974).
15. R. L. Trelstad, *Biochem. Biophys. Res. Commun., 57*:717 (1974).

16. K. A. McCullagh and G. Balian, *Nature, 258*:73 (1975).
17. L. F. Morton and M. J. Barnes, *Atherosclerosis, 42*:41 (1982).
18. D. M. Scott, R. Harwood, M. E. Grant, and D. S. Jackson, *Connect. Tiss. Res., 5*:7 (1977).
19. J. Rauterberg and D. B. von Bassewitz, *Hoppe-Seyler's Z. Physiol. Chem., 356*:95 (1975).
20. R. Mayne, J. G. Zettergren, P. M. Mayne, and N. W. Bedwell, *Artery, 7*:262 (1980).
21. A. G. Szymanowicz, G. Bellou, G. Laurain-Guillaume, T. Delvincourt, A. Randoux, T. Caulet, and J. P. Borel, *Artery, 10*:250 (1982).
22. E. H. Epstein and N. H. Munderloh, *J. Biol. Chem., 250*:9304 (1975).
23. A. N. Hanson and J. P. Bentley, *Anal. Biochem., 130*:32 (1983).
24. S. A. Jimenez and R. I. Bashey, *Biochem. J., 173*:337 (1978).
25. R. W. Glanville, A. Rauter, and P. P. Fietzek, *Eur. J. Biochem., 95*:383 (1979).
26. S. N. Dixit, *FEBS Lett., 106*:379 (1979).
27. A. J. Bailey, T. J. Sims, V. C. Duance, and N. D. Light, *FEBS Lett., 991*:361 (1979).
28. T. F. Kresina and E. J. Miller, *Biochemistry, 18*:3089 (1979).
29. H. Sage, R. G. Woodbury, and P. Bornstein, *J. Biol. Chem., 254*:9893 (1979).
30. S. Gay and E. J. Miller, *Arch. Biochem. Biophys., 198*:370 (1979).
31. E. Chung, R. K. Rhodes, and E. J. Miller, *Biochem. Biophys. Res. Commun., 71*: 1167 (1976).
32. A. Ooshima, *Science, 213*:666 (1981).
33. D. K. Furuto and E. J. Miller, *J. Biol. Chem., 255*:290 (1980).
34. R. Jander, J. Rauterberg, B. Voss, and D. B. von Bassewitz, *Eur. J. Biochem., 114*: 17 (1981).
35. M. Z. Abedin, S. Ayad, and J. B. Weiss, *Biosci. Rep., 2*:493 (1982).
36. G. Laurain, T. Delvincourt, and A. G. Szymanowicz, *FEBS Lett., 120*:44 (1980).
37. D. K. Furuto and E. J. Miller, *Biochemistry, 20*:1635 (1981).
38. R. Jander, J. Rauterberg, and R. W. Glanville, *Eur. J. Biochem., 133*:39 (1983).
39. E. Odermatt, J. Risteli, V. van Delden, and R. Timpl, *Biochem. J., 211*:295 (1983).
40. H. Furthmayr, H. Wiedemann, R. Timpl, E. Odermatt, and J. Engel, *Biochem. J., 211*:303 (1983).
41. S. Gay, L. Balleisen, K. Remberger, P. P. Fietzek, B. C. Adelmann, and K. Kühn, *Klin. Wochenschr., 53*:899 (1975).
42. S. Gay, P. Walter, and K. Kühn, *Klin. Wochenschr., 54*:889 (1976).
43. K. G. McCullagh, V. C. Duance, and K. A. Bishop, *J. Pathol., 130*:45 (1980).
44. J. S. Bartholomew and J. C. Anderson, *Histochem. J., 15*:1177 (1983).
45. A. Palotie, K. Tryggvason, L. Peltonen, and H. Seppä, *Lab. Invest., 49*:362 (1983).
46. J. A. Madri, B. Dreyer, F. A. Pittick, and H. Furthmayr, *Lab. Invest., 43*:303 (1980).
47. G. Wick, R. W. Glanville, and R. Timpl, *Immunobiology, 156*:372 (1979).
48. J. M. Fitch, E. Gibney, R. D. Sanderson, R. Mayne, and T. F. Linsenmayer, *J. Cell. Biol., 95*:641 (1982).
49. J. R. Sanes, *J. Cell. Biol., 93*:442 (1982).
50. F. J. Roll, J. A. Madri, J. Albert, and H. Furthmayr, *J. Cell. Biol., 85*:597 (1980).
51. J. A. Madri and H. Furthmayr, *Am. J. Pathol., 94*:323 (1979).
52. A. Martinez-Hernandez, S. Gay, and E. J. Miller, *J. Cell Biol., 92*:343 (1982).
53. S. Gay, A. Martinez-Hernandez, R. K. Rhodes, and E. J. Miller, *Collagen Rel. Res., 1*:377 (1981).
54. T. F. Linsenmayer, J. M. Fitch, T. M. Schmid, N. B. Zak, E. Gibney, R. D. Sanderson, and R. Mayne, *J. Cell Biol., 96*:124 (1983).
55. J. M. Fitch, J. Gross, R. Mayne, B. Johnson-Wint, and T. F. Linsenmayer, *Proc. Natl. Acad. Sci. USA, 81*:2791 (1984).
56. M. J. Barnes and D. M. Scott, in *Biochemical Interactions at the Endothelium* (A. Cryer, ed.), Elsevier Science Publishers, Amsterdam, 1983, p. 111.

57. G. Grotendorst, H. E. J. Seppä, H. K. Kleinman, and G. R. Martin, *Proc. Natl. Acad. Sci. USA, 78*:3669 (1981).
58. K. Murata, K. Nakazawa, and A. Hamai, *Atherosclerosis, 21*:93 (1975).
59. R. L. Stevens, M. Colombo, J. J. Gonzales, W. Hollander, and K. Schmid, *J. Clin. Invest., 58*:470 (1976).
60. T. A. Massaro, C. E. Glatz, N. A. Peppas, G. M. Chisolm, and C. K. Colton, *Artery, 5*:1 (1979).
61. T. R. Oegema, V. C. Hascall, and R. Eisenstein, *J. Biol. Chem., 254*:1312 (1979).
62. J. McMurtrey, B. Radhakrishnamurthy, E. R. Dalferes, Jr., G. S. Berenson, and J. D. Gregory, *J. Biol. Chem., 254*:1621 (1979).
63. B. G. J. Salisbury and W. D. Wagner, *J. Biol. Chem., 256*:8050 (1981).
64. R. Kappor, C. F. Phelps, L. Cöster, and L-A. Fransson, *Biochem. J., 197*:259 (1981).
65. M. Breton, J. Picard, and E. Berrou, *Biochimie, 63*:515 (1981).
66. A. Schmidt, M. Prager, P. Selmke, and E. Buddecke, *Eur. J. Biochem., 125*:95 (1982).
67. S. Gardell, J. Baker, B. Caterson, D. Heinegard, and L. Rodén, *Biochem. Biophys. Res. Commun., 95*:1823 (1980).
68. T. N. Wight and R. Ross, *J. Cell Biol., 67*:660 (1975).
69. R. Eisenstein and K. Kuettner, *Atherosclerosis, 27*:37 (1976).
70. M. Mangkornkanok-Mark, R. Eisenstein, and R. M. Bahu, *J. Histochem. Cytochem., 29*:547 (1981).
71. J. S. Bartholomew and J. C. Anderson, *Histochem. J., 15*:941 (1983).
72. R. Timpl, J. Engel, and G. R. Martin, *Trends Biochem. Sci., 8*:207 (1983).
73. J. Engel, E. Odermatt, A. Engel, J. Madri, H. Furthmayr, H. Rohde, and R. Timpl, *J. Mol. Biol., 150*:97 (1981).
74. H. Rohde, G. Wick, and R. Timpl, *Eur. J. Biochem., 102*:195 (1979).
75. J. M. Foidart, E. W. Bere, M. Yaar, S. I. Rennard, M. Gullino, G. R. Martin, and S. I. Katz, *Lab. Invest., 42*:336 (1980).
76. E. Linder, A. Vaheri, E. Ruoslahti, and J. Wartiovaara, *J. Exp. Med., 142*:41 (1975).
77. S. Stenman and A. Vaheri, *J. Exp. Med., 147*:1054 (1978).
78. J. Chamley-Campbell, G. R. Campbell, and R. Ross, *Physiol. Rev., 59*:1 (1979).
79. J. H. Chamley, G. R. Campbell, J. D. McConnell, and U. Gröschel-Stewart, *Cell Tissue Res., 177*:503 (1977).
80. R. Ross, *J. Cell. Biol., 50*:172 (1971).
81. J. H. Chamley-Campbell, G. R. Campbell, and R. Ross, *J. Cell. Biol., 89*:379 (1981).
82. G. R. Campbell and J. H. Chamley-Campbell, *Med. Hypoth., 7*:729 (1981).
83. J. Jarmolych, A. S. Daoued, J. Landau, K. E. Fritz, and E. McElvene, *Exp. Mol. Pathol., 9*:171 (1968).
84. J. P. Boisel, M. C. Boirdillon, R. Loire, and B. Crozet, *Atherosclerosis, 25*:107 (1976).
85. M. J. Barnes, L. F. Morton, and C. I. Levene, *Biochem. Biophys. Res. Commun., 70*:339 (1976).
86. D. L. Layman, E. H. Epstein, R. F. Dodson, and J. L. Titus, *Proc. Natl. Acad. Sci. USA, 74*:671 (1977).
87. J. M. Burke, G. Balian, R. Ross, and P. Bornstein, *Biochemistry, 16*:3243 (1977).
88. R. Mayne, M. S. Vail, E. J. Miller, S. H. Blose, and S. Chacko, *Arch. Biochem. Biophys., 181*:462 (1977).
89. R. Mayne, M. S. Vail, and E. J. Miller, *Biochemistry, 17*:446 (1978).
90. J. M. Burke and R. Ross, *Exp. Cell Res., 107*:387 (1977).
91. J. C. Beldekas, L. Gerstenfeld, G. E. Sonenshein, and C. Franzblau, *J. Biol. Chem., 257*:12252 (1982).
92. D. Y. M. Leung, S. Glagov, and M. B. Mathews, *Science, 191*:475 (1976).
93. A. Hinek and J. Thyberg, *J. Ultrastruct. Res., 60*:12 (1977).
94. B. W. Oakes, A. C. Batty, C. J. Handley, and L. B. Sandberg, *Eur. J. Cell Biol., 27*:34 (1982).

95. A. S. Narayanan, L. B. Sandberg, R. Ross, and D. L. Layman, *J. Cell Biol., 68*: 411 (1976).
96. P. A. Abraham, D. W. Smith, and W. H. Carnes, *Biochem. Biophys. Res. Commun., 58*:597 (1974).
97. P. A. Abraham, D. W. Smith, and W. H. Carnes, *Biochem. Biophys. Res. Commun., 67*:723 (1975).
98. K. Kresse, K. von Figura, E. Buddecke, and H. G. Fromme, *Hoppe Seyler's Z. Physiol. Chem., 356*:929 (1975).
99. O. Namiki, B. Faris, F. Tschopp, P. Füglistaller, W. Hollander, C. Franzblau, and K. Schmid, *Biochemistry, 19*:1900 (1980).
100. M. J. Merrilees, M. A. Merrilees, P. S. Birnbaum, P. J. Scott, and M. H. Flint, *Atherosclerosis, 27*:259 (1977).
101. G. Gamse, H. G. Fromme, and H. Kresse, *Biochim. Biophys. Acta, 544*:514 (1978).
102. T. N. Wight and R. Ross, *J. Cell Biol., 67*:675 (1975).
103. T. N. Wight and V. C. Hascall, *J. Cell Biol., 96*:167 (1983).
104. S. M. Schwarz, *In Vitro, 14*:966 (1978).
105. M. A. Gimbrone, in *Progress in Hemostasis and Thrombosis* (T. H. Spaet, ed.), Grune & Stratton, New York, 1976.
106. G. Cotta-Pereira, H. Sage, P. Bornstein, R. Ross, and S. Schwartz, *J. Cell. Physiol., 102*:183 (1980).
107. J. Folkman and C. Haudenschild, *Nature, 288*:551 (1980).
108. T. Maciag, J. Kadish, L. Wilkins, M. B. Stemerman, and R. Weinstein, *J. Cell. Biol., 94*:511 (1982).
109. J. Feder, J. C. Marasa, and J. V. Olander, *J. Cell. Physiol., 116*:1 (1983).
110. R. Montesano, L. Orci, and P. Vassalli, *J. Cell Biol., 97*:1648 (1983).
111. T. Kalebic, S. Garbisa, B. Glaser, and L. A. Liotta, *Science, 221*:281 (1983).
112. B. V. Howard, E. J. Macarek, D. Gunson, and N. A. Kefalides, *Proc. Natl. Acad. Sci. USA, 73*:2361 (1976).
113. M. J. Barnes, L. F. Morton, and C. I. Levene, *Biochem. Biophys. Res. Commun., 84*:646 (1978).
114. R. Mayne and P. M. Mayne, *J. Cell Biol., 79*:147a (1978).
115. H. Sage, E. Crouch, and P. Bornstein, *Biochemistry, 18*:5433 (1979).
116. H. Sage, P. Pritzl, and P. Bornstein, *Biochemistry, 19*:5747 (1980).
117. H. Sage, B. Trüeb, and P. Bornstein, *J. Biol. Chem., 258*:13391 (1983).
118. H. Sage, P. Pritzl, and P. Bornstein, *Biochemistry, 20*:436 (1981).
119. H. Sage, P. Pritzl, and P. Bornstein, *Arteriosclerosis, 1*:427 (1981).
120. H. Sage and P. Bornstein, *Arteriosclerosis, 2*:27 (1982).
121. W. H. Carnes, P. A. Abraham, and V. Buonassisi, *Biochem. Biophys. Res. Commun., 90*:1393 (1979).
122. J. O. Cantor, S. Keller, M. S. Parshley, T. V. Darnule, A. T. Darnule, J. M. Cerreta, G. M. Turino, and I. Mandl, *Biochem. Biophys. Res. Commun., 95*:1381 (1980).
123. V. Buonassisi, *Exp. Cell Res., 76*:363 (1973).
124. V. Buonassisi and M. Root, *Biochim. Biophys. Acta, 385*:1 (1975).
125. R. H. Kramer, K. G. Vogel, and G. L. Nicolson, *J. Biol. Chem., 257*:2678 (1982).
126. A. Oohira, T. N. Wight, and P. Bornstein, *J. Biol. Chem., 258*:2014 (1983).
127. E. J. Macarak, E. Kirby, T. Kirk, and N. A. Kefalides, *Proc. Natl. Acad. Sci. USA, 75*:2621 (1978).
128. C. R. Birdwell, D. Gospodarowicz, and G. L. Nicolson, *Proc. Natl. Acad. Sci. USA, 75*:3273 (1978).
129. E. A. Jaffe and D. F. Mosher, *J. Exp. Med., 147*:1779 (1978).
130. D. Gospodarowicz, G. Greenburg, J. M. Foidart, and N. Savion, *J. Cell. Physiol., 107*:171 (1981).
131. W. B. Robertson and J. P. Strong, *Lab. Invest., 18*:78 (1968).

132. A. V. Chobanian, *Prog. Cardiovasc. Dis., 26*:177 (1983).
133. H. Wolinsky, *Circ. Res., 26*:507 (1970).
134. A. Ooshima, G. C. Fuller, G. J. Cardinale, S. Spector, and S. Udenfriend, *Proc. Natl. Acad. Sci. USA, 71*:3019 (1974).
135. K. Iwatsuki, G. J. Cardinale, S. Spector, and S. Udenfriend, *Proc. Natl. Acad. Sci. USA, 74*:360 (1977).
136. T. J. Franklin, W. P. Morris, and B. E. Loveday, *Blood Vessels, 19*:217 (1982).
137. R. A. Newman and R. O. Langner, *Exp. Gerontol., 13*:83 (1978).
138. K. Iwatsuki, G. J. Cardinale, S. Spector, and S. Udenfriend, *Science, 198*:403 (1977).
139. R. Nissen, G. J. Cardinale, and S. Udenfriend, *Proc. Natl. Acad. Sci. USA, 75*:451 (1978).
140. G. Manley, *Br. J. Exp. Pathol., 46*:125 (1965).
141. W. Hollander, D. M. Kramsh, M. Farmelant, and I. M. Madoff, *J. Clin. Invest., 47*: 1221 (1968).
142. A. Schmidt, J. Grünwald, and E. Buddecke, *Atherosclerosis, 45*:299 (1982).
143. R. Ross, *Arteriosclerosis 1*:293 (1981).
144. R. Ross, J. A. Glomset, B. Kariya, and L. Harker, *Proc. Natl. Acad. Sci. USA, 71*: 1207 (1974).
145. S. J. Liebovich and R. Ross, *Am. J. Pathol., 84*:501 (1976).
146. C. Gajdusek, P. DiCorleto, R. Ross, and S. Schwartz, *J. Cell Biol., 85*:467 (1980).
147. E. P. Benditt and J. M. Benditt, *Proc. Natl. Acad. Sci. USA, 70*:1753 (1973).
148. T. A. Pearson, A. Wang, K. Solez, and R. H. Heptinstall, *Am. J. Pathol., 81*:379 (1975).
149. E. P. Benditt, T. Barrett, and J. K. McDougall, *Proc. Natl. Acad. Sci. USA, 80*: 6386 (1983).
150. W. A. Thomas and D. N. Kim, *Lab. Invest., 48*:245 (1983).
151. G. Gabbiani, E. Rungger-Brändle, C. de Chastonay, and W. W. Franke, *Lab. Invest., 47*:265 (1982).
152. G. Gabbiani, O. Kocher, W. S. Bloom, J. Vandekerckhove, and K. Weber, *J. Clin. Invest., 73*:148 (1984).
153. M. D. Haust, R. H. More, and H. Z. Movat, *Am. J. Pathol., 37*:377 (1960).
154. J. C. Geer, H. C. McGill, and J. P. Strong, *Am. J. Pathol., 38*:263 (1961).
155. R. Ross, T. N. Wight, E. Strandness, and B. Thiele, *Am. J. Pathol., 114*:79 (1984).
156. C. I. Levene and J. C. F. Poole, *Br. J. Exp. Pathol., 43*:469 (1962).
157. E. B. Smith, *J. Atheroscler. Res., 5*:224 (1965).
158. T. Anastassiades, P. A. Anastassiades, and O. F. Denstedt, *Biochim. Biophys. Acta, 261*:418 (1972).
159. K. G. McCullagh and L. A. Ehrhart, *Atherosclerosis, 19*:13 (1974).
160. R. O. Langner and J. B. Modrak, *Blood Vessels, 13*:257 (1976).
161. G. M. Fischer, M. L. Swain, and K. Cherian, *Atherosclerosis, 35*:11 (1980).
162. L. A. Ehrhart and D. Holderbaum, *Atherosclerosis, 37*:423 (1980).
163. M. Aumailley and H. Bricaud, *Atherosclerosis, 39*:1 (1981).
164. K. Pietilä and T. Nikkari, *Atherosclerosis 37*:11 (1980).
165. J. E. Scott, C. R. Orford, and E. W. Hughes, *Biochem. J., 195*:573 (1981).
166. S. R. Srinivasan, P. Dolan, B. Radhakrishnamurthy, P. S. Pargaonkar, and G. Berenson, *Biochim. Biophys. Acta, 388*:58 (1975).
167. P-H. Iverius, *J. Biol. Chem., 247*:2607 (1972).
168. T. N. Wight, K. D. Curwen, M. M. Litrenta, D. R. Alonso, and C. R. Minick, *Am. J. Pathol., 113*:156 (1983).
169. C. R. Minick, M. B. Stemerman, and W. Insull, Jr., *Proc. Natl. Acad. Sci. USA, 74*: 1724 (1977).

8

Cell-Matrix Interactions: Biochemistry of Close and Tight-Focal Adhesive Contacts of Fibroblasts to Extracellular Matrices

Michael W. Lark* and Lloyd A. Culp
Case Western Reserve University, School of Medicine,
Cleveland, Ohio

I. RATIONALE AND EXPERIMENTAL APPROACHES—SIMPLIFYING THE MODEL SYSTEM

A. Morphogenesis of Substratum-Attached Material from Close and Tight-Focal Adhesive Contacts

Extracellular matrices (ECM) form the biochemically precise foundation on which many animal cells adhere and conduct their specialized functions. Yet, an understanding of the molecular mechanisms of these processes has been slow to evolve because of the biochemical complexity of the surface of cells, as well as the multiplicity and complexity of ECMs. However, in the past 10 years, new experimental approaches have been developed to identify macromolecules that are critically important in these processes (reviewed in Refs. 1-3). The approach taken in our laboratory has been to study the simplest model system from the perspective of both the cell and the substratum with which it interacts. Mechanistic considerations from the model system can then be applied to cells that require a much more complex ECM for cell division and/or maintenance of differentiated functions.

The model cell we chose was the Balb/c 3T3 fibroblastlike cell line, which is a relatively poor collagen producer and which fails to deposit collagen onto the tissue culture substratum (4); therefore, these cells interact with the substratum by binding one or more cell surface receptors with adhesion-promoting molecules in the serum layer adsorbed to the substratum. From its initial isolation in the early 1970s (5), it has been a cell line studied in many laboratories, is anchorage-dependent, and is relatively nontumorigenic in

**Present affiliation*: Merck and Co., Inc., Rahway, New Jersey.

nude mice, except for the long-term generation of spontaneous transformants that are highly tumorigenic and anchorage-independent (6,7). Also available are oncogenic viral transformants of these cells using SV40 (the SVT2 cell line) and the Kirsten murine sarcoma virus (the KiMSV cell line) (7).

The substratum originally chosen for study was the serum-coated tissue culture plastic surface (8). However, extraction of these surfaces with SDS and subsequent analyses of bound serum proteins by SDS-PAGE revealed *at least* 12 different serum proteins adsorbed to the surface that could potentially be adhesion-promoting factors (9,10). This complexity was overcome by the discovery in a number of laboratories (11-14) that the serum protein plasma fibronectin (pFN) is the major adhesion-promoting activity in the substratum-adsorbed serum layer for mesenchymal cells. Our studies and this review will focus on the mechanisms by which fibronectin adsorbed to the artificial tissue culture ECM mediates adhesion of Balb/c3T3 cells and its derivatives.

The basic approach that this laboratory has taken to study the molecular mechanisms of fibroblast adhesion can be broken down into three steps. (a) The first step required detaching cells from the tissue culture substratum in such a gentle way that these cells leave their adhesive footpad structures as substratum-attached material (SAM) for further biochemical characterization. SAM from 3T3 cells is approximately a 100-fold enrichment in cell surface molecules participating in adhesive events (8,10,15). By studying SAM composition under varying conditions of cell attachment, movement, and confluence, preliminary assignment of functionally important molecules could be made. (b) The second step involved the probing of adhesive structures with enzymes, antibodies, and selective extraction conditions in order to learn something about the organization of supramolecular complexes in these sites. (c) This was complemented with the third approach, which analyzes the adhesive responses of cells to functional domains of the fibronectin molecule, as well as to binding proteins that have one of the particular binding activities of the fibronectin molecule. These three approaches will be reviewed here.

Scanning electron microscopy demonstrated the effects that the Ca^{2+}-specific chelator EGTA has on the morphological contours of the 3T3 cell (16). It disorganizes the cytoskeleton, causing the surface of the cell to actively bleb and retract from the footpad adhesion sites; these latter structures appear completely resistant to the chelation treatment, indicating that the major effects of chelation inhibition of adhesion are due to perturbation of cytoskeletal reorganization, rather than inhibition of the cell surface binding reactions (this is confirmed by fibronectin binding studies, discussed later). The retracted and rounded cells under the influence of EGTA generate very fragile retraction fibers of plasma membrane that break easily upon culture agitation. This liberates the cell into suspension and leaves membranous adhesive footpads as SAM (16).

At approximately the same time that these studies were undertaken, another light microscopic technique was utilized to measure the contact distances between the 3T3 cell surface and the serum-coated substratum—interference reflection microscopy (IRM) (17-19). Two different adhesive regions were identified with living cells. *Close* contacts with a spacing distance of 25-30 nm were prominent at the leading edge of moving cells and at other regions of the periphery. *Tight-focal* contacts were small regions of intimate contact (12-14 nm) where microfilament stress fibers condensed. The latter contacts were more prominent in density-inhibited nontransformed cells, whereas the former were more common in cells moving over the substratum (18,19).

In a collaboration between our laboratory and that of Colin Izzard, the origin of EGTA-generated SAM was investigated with IRM (20). Figure 1 illustrates the retracting

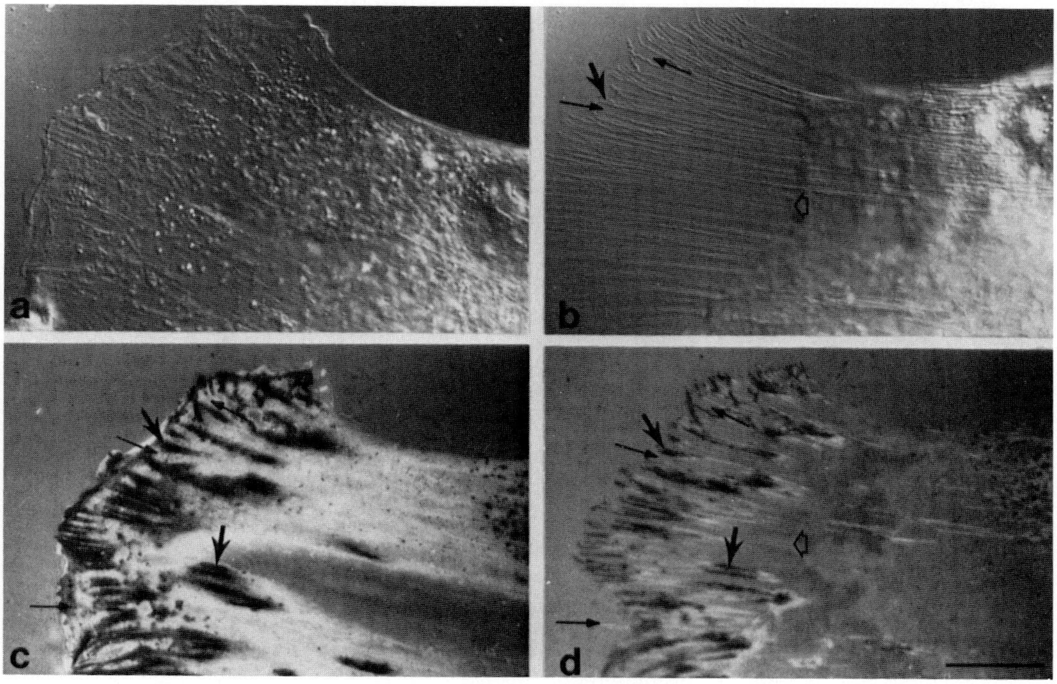

Figure 1 Origin of Balb/c 3T3 substratum-attached material from close and tight-focal adhesive contacts. BALB/c 3T3 cells were detached from stock cultures with 0.5 mM EGTA in PBS, rinsed well, and allowed to spread on serum-coated coverslips in serum-containing medium for 17 hr, after which the coverslips were mounted and the live cells observed with (c, d) interference reflection and (a, b) differential interference contrast optics, as previously described (20). While under observation, the same cell was photographed prior to EGTA treatment (a and c) or after 20 min of treatment with 0.5 mM EGTA in PBS to initiate cell retraction (b and d). In the IRM image (c and d), tight-focal contacts appear as black streaks, close contacts as broader gray zones, and cell-to-substratum contacts of greater than 30 nm as white zones. Large arrows indicate two of the many well-developed tight-focal contacts. Small arrows indicate regions of more distant contact between the focal contacts. The persistence of EGTA-resistant adhesions (large arrows in b and d) results in the formation of numerous fine linear retraction fibers. Small arrows mark areas void of retraction fibers in b and d. Large arrows mark positions of two tight-focal contacts present in c; these correspond to EGTA-resistant adhesions of b. The majority of the black streaks in d are derived from the tight-focal contacts in c and correspond to the EGTA-resistant adhesions seen under differential contrast optics in b. Small arrows (d) correspond to EGTA-sensitive areas of b and to regions lacking tight-focal or close contacts marked by small arrows in c. Such regions often contained close contacts (not shown). Outline arrows (b,d) indicate the point to which the cytoplasm has retracted. Careful examination of close contacts demonstrates that some of these regions are sensitive to EGTA treatment while others are resistant (20). Therefore, SAM is composed of tight-focal contacts, as well as *some* close contact material. Bar = 10 μm. (Taken from Ref. 20 with permission.)

regions of a 3T3 cell while photographed prior to (a and c) or during (b and d) the EGTA treatment with either differential contrast microscopy (a and b) or IRM (c and d) on a pFN-coated substratum. Tight-focal contacts (heavy black arrows) are highly resistant to the EGTA treatment and persist as SAM during the treatment. Similar analyses of close contacts (20) indicate that there are two subsets of these contacting regions—some areas are sensitive to the EGTA treatment and retract away from the substratum, while other regions are resistant and persist as SAM. Biochemically, at least three distinguishable adhesive contacts are found in these cells.

B. Biochemical Composition of SAM: Changes with Cellular Physiological Responses

SAM from Balb/c 3T3 cells comprises approximately 1-1.5% of the cell's total protein and phospholipid contents and 12-15% of its glucosamine-radiolabeled polysaccharide—data consistent with its origin as carbohydrate-rich cell surface material (8,10,15). Both SV40- and Kirsten murine sarcoma virus-transformed 3T3 cells leave only one-third to one-half of these amounts of SAM, whereas revertant variants deposit almost as much as 3T3 cells (8). Since revertant variant cells have regained the ability to reorganize large bundles of microfilaments at their undersurfaces when compared to bundle-poor transformed cells (21,22) and since SAM contains a sizable amount of membrane-associated cytoskeletal protein (10,23,24), the quantities of SAM derived from these three cell types correlate with the extent of microfilament bundle reorganization at the undersurface, which is a function of transmembrane linkage to certain proteoglycans and other "receptor" glycoproteins at the outer cell surface, which are in turn bound to pFN on the substratum (see section II.C).

The origin of the proteins and polysaccharides in SAM was investigated carefully. It was demonstrated that these result from cellular contact with the substratum and not from artifactual rebinding to the substratum of medium-secreted components (reviewed in Ref. 23). Similarly, SDS-PAGE analyses after various radiolabeling protocols revealed that SAM represents a specialized portion of the surface of the cell enriched in glycosaminoglycan (GAG)-containing proteoglycans (PG); cell surface fibronectin (cFN); vimentin from intermediate filaments; actin, myosin, and a very small amount of α-actin from microfilaments; and a number of other proteins that appear in constant proportion to known cytoskeletal proteins and are probably unidentified cytoskeletal elements (10,15,23,25). The constancy of these components led us to propose that there is an "adhesion complex" containing these components linked together with a well-defined stoichiometry (15). The specialization of SAM regions of surface membrane was particularly evident when lactoperoxidase-radioiodinated cells deposited footpads containing only radiolabeled cFN and proteoglycan (26).

Since SAM protein and polysaccharide could be quantitatively solubilized from the substratum with 0.2% SDS (26), methods were developed to analyze the GAG in this portion of the cell (27). Again, specialization was evident since the ratios of heparan sulfate (HS)/chondroitin sulfates (CS)/hyaluronic acid (HA) were very different for SAM and the rest of the cell (27) or whole cell surface membrane preparations (15). The EGTA-detached cell contained significant amounts of dermatan sulfate (DS) but SAM virtually none. As cells attach, spread, and become confluent, the GAG distribution in SAM changes dramatically while the distribution in the cell fraction remains relatively constant (27). The newly attaching cell deposits considerable HS into SAM and very little CS and HA. These latter components accumulate in SAM only after the cell initiates spreading and movement and become very prominent as 3T3 cells become confluent (in contrast, confluence has little effect on these distributions in SVT2 and KiMSV cells) (28). KiMSV cells, which are very weakly adherent at all times to the substratum, generate SAMs that always have a highly elevated content of HA (28). Changes also occur in protein distribution from the time of initial attachment of cells into their spreading-movement phase—vimentin is much more prominent in newly formed SAM, whereas microfilament proteins become predominant in the SAM of spreading-moving cells (10,15). These studies provided early insight into the dynamics of the biochemistry of these adhesion processes and the possible functional importance of some of these components (e.g., see Refs. 15 and 23).

C. Initial Identification of Functional Roles for Fibronectin and Proteoglycans

A number of experimental observations indicated that the glycoprotein fibronectin and the GAG-containing proteoglycans in SAM played critical roles in adhesion of these cell types to the tissue culture substratum. First, cFN from these fibroblastlike cells is the principal lactoperoxidase-iodinatable *and* metabolically radiolabeled glycoprotein enriched in these cell surface regions (26). Second, substrata adsorbed with pure pFN permitted cells to undertake all the attachment, spreading, and movement functions that they perform on serum-coated substrata (11-14). Third, since [^{3}H]glucosamine is incorporated into cellular GAGs at a very low specific activity, the amount of radiolabeled GAG recovered from SAM (27) indicated that these sites contained a very large amount of GAG which would undoubtedly influence adhesive processes (23-25). Fourth, these components resisted extraction off the substratum with chaotropic agents, salt, and nonionic detergents (26), revealing their tenacity of binding to the substratum (see Section III for further analyses along these lines). Fifth, the enrichment for heparan sulfate in newly formed sites indicated a binding reaction that is critical in the formation of new sites (27). Sixth, treatment of SAM with low concentrations of trypsin or hyaluronidase indicated that heparan sulfate, cFN, and the cytoskeletal proteins coordinately resisted solubilization as if they were linked in an "adhesion complex" (29). Furthermore, pulse-chase analyses indicated that the newly synthesized cFN *and a 49K glycoprotein* deposited into SAM were rapidly and coordinately turned over (15) and could be easily solubilized with hyaluronidase digestion, suggesting an association of this labile cFN with hyaluronic acid and/or chondroitin sulfate entities (29). For all these reasons, we were led into investigating the possibility that cFN and/or pFN can and do bind to certain GAGs.

Taking essentially different approaches, four laboratories demonstrated that FN binds to certain GAGs with considerable specificity (30-33). When GAGs purified from EGTA-detached cells or SAM were passed through a pFN-Sepharose affinity column, only a subset of the polysaccharide bound to the pFN—a reaction that was independent of divalent cations (33). SAM was significantly enriched in sequences that bound to pFN, particularly newly formed SAM where as much as 40-45% of the sequences bound. Virtually all of the HS in these samples bound to the affinity column, whereas no binding of HA or CS could be detected (33). And, as Stathakis and Mosses-son (30) had found earlier, HS binding to pFN was as effective at 4°C as at 37°C (33).

At the same time, Yamada et al. (31), using a filter-trapping assay, had detected significant binding of both HA and HS to cFN at different sites on the molecule, leading us to compare the binding properties of cFN and pFN (34). HS from 3T3 cells bound equiva-

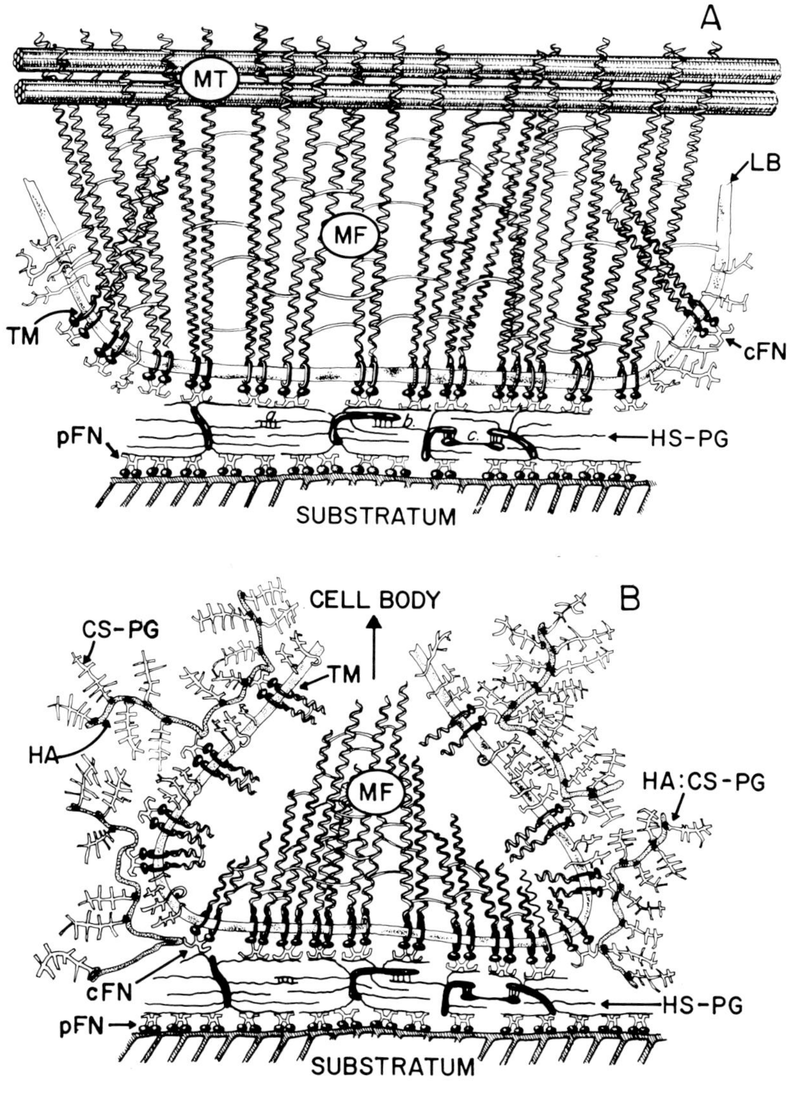

Figure 2 Prospective models for the initial adhesion and eventual detachment of cells. The initial events in formation of an adhesive bond between the cell surface and the plasma fibronectin-coated substratum are depicted in A. This model would apply principally to the formation of close contacts (see Section II). Dimeric plasma fibronectin (pFN) on the substratum probably binds to aggregated heparan sulfate proteoglycans (HS-PG) on the cell surface; HS-PG aggregation may occur by unknown mechanisms involving polysaccharide chain:chain interaction (a), core protein:GAG chain binding (b), and/or the involvement of a "link" protein that cross-links GAG chains (c). Cell surface fibronectin (cFN) may act as a multivalent "link" glycoprotein for HS-PG's. HS-PG would then bind to cell surface fibronectin forming an ordered array of these latter molecules. cFN would bind to some unidentified transmembrane protein (TM) that also binds to microfilaments on the interior of the lipid bilayer (LB). Microfilaments (MF) would be cross-linked into bundled arrays and may interact with the microtubules (MT) of the

lently to pFN or cFN when three different assays were used—FN-Sepharose affinity columns, gelatin-Sepharose columns to which the particular FN was bound, or aggregates of the particular FN that could be trapped on filters (34). In contrast, HA binding to pFN could not be detected in any of these assays. Nor could HA binding to cFN be detected on either type of affinity column. When cFN was permitted to aggregate and then tested for HA binding, results identical with those of Yamada et al. (31) were observed with considerable binding of HA (34). These GAG binding studies taken together indicate that pFN or cFN are functionally different in their abilities to bind to HA and may reflect an important mechanism that was revealed in previous studies (23-25,35)—namely, that HA accumulation in SAM is an important parameter in making cells more facile in their detachment from the substratum.

These data were then assembled into a primitive model of the initial adherence and eventual detachment of the cell from the substratum (Figure 2) (24,25). Initial adherence (Figure 2A) would be mediated by the multiple bindings of pFN on the substratum to patching networks of HS-PG on the surface of the cell, generating a highly multivalent and extremely tight adhesion complex. The HS-PGs in turn would be transmembrane-linked to the cell's microfilament networks, possibly via the participation of cFN (as shown) or by the HS-PG core protein spanning the phospholipid bilayer itself (not shown) with a more direct linkage to a cytoskeletal element (see below). There are three possible mechanisms for association of HS-PGs into supramolecular aggregates on the surface of the cell (at a, b, and c in Figure 2A). Since cFN has multiple binding sites for HS sequences and is dimeric, cFN could act as a "link" glycoprotein as in c of Figure 2A. Evidence supporting this is the solubilization of sizeable amounts of cFN off the surface of the cell upon treatment with *Flavobacter* heparanase under conditions where proteolysis is minimized (36,37). This possibility must be explored more thoroughly.

Detachment of cells during movement over the substratum mimics the effects of EGTA treatment in that "footprints" of SAM are left behind (15,38). Figure 2B is a primitive model to explain cytoskeletal disorganization based on HA accumulation (24,25). Since HA would only bind to *aggregated* cFN on the surface of the cell and not to pFN on the substratum, this binding would cause disorganization of cFN complexes transmembrane-linked to microfilament elements. Subsequent disarray of microfilament bundles would lead to the formation of a retraction fiber and eventual breaking away from the substratum. An alternative explanation can be offered for some transformed cells that may be defective in *cytosol*-mediated bundle formation by the microfilaments; in this case, HA accumulation in SAM would be unnecessary and coincidental. Construction of these preliminary models has provided us with some testable hypotheses that will be examined below.

---

cell. Although linkage of HS-PG to microfilaments is depicted here as an end-on mechanism, it appears likely that HS-PG may only link to MF laterally (see Section II.C). Also, the membrane-associated form of HS-PG is not depicted in this model, but evidence is accumulating that a membranous form of HS-PG is linked to MF without the participation of processes from the distant outer cell surface into the interior of the cell. Detachment is depicted in B. Hyaluronic acid (HA) and chondroitin sulfate proteoglycans (CS-PG) accumulate at the periphery of the adhesion site, and recent evidence is inconclusive whether they are bound to each other in a complex similar to that identified in cartilage. HA may bind specifically to cell surface fibronectin, a process that competes with HS-PG binding and causes disorder in the microfilaments. With disorder in the cytoskeleton, there is retraction of the elastic plasma membrane leading to pinching off the adhesion site, leaving a "footprint" on the substratum. (Taken from Ref. 24.)

II. ADHESIVE RESPONSES TO SPECIFIC BINDING ACTIVITIES OF FIBRONECTIN

A. Glycosaminoglycan-Binding Proteins: Requirement for Heparan Sulfate Proteoglycan

Fibronectin is a multifunctional protein whose binding activities for various ECM molecules are localized into domains along the polypeptide chain (see Refs. 1-3). Since Balb/c 3T3 cells and its derivatives fail to deposit significant amounts of collagen into SAM (4), we can focus our attention on the significance of two activities in the FN molecule as far as adhesive responses are concerned with these model cells—heparan sulfate-binding activity [since pFN demonstrates little binding activity for the other GAGs, except perhaps for dermatan sulfate (33)] and cell-binding activity (1,39-41), which recognizes some unidentified receptor on the surface of the cell. In this section, we will deal with the significance of HS-binding activity and in Section II.B with cell-binding activity and then arrive at some striking conclusions in Section II.C.

Because the high-affinity C-terminal heparin-binding domain (HBD) of pFN is extremely difficult to purify without any contaminating cell-binding domain (CBD) (1) and because cells respond poorly to the low-affinity N-terminal HBD (39,42), we chose to use another model heparan sulfate-binding protein, platelet factor-4 (PF4), to compare adhesive responses with the intact pFN molecule (43). PF4 has a monomer molecular weight of 7500, with one HS binding site per monomer, and under physiological conditions aggregates into a quatemer (44); it also binds weakly to CS chains. Other GAG-binding proteins were also compared (43)—bovine nasal cartilage chondroitin-keratan sulfate proteoglycan and its "link" glycoprotein (both of which are specific for binding to HA). Surprisingly, Balb/c 3T3 cells failed to attach *at all* to substrata coated with the HA-binding proteins, even though there is plenty of HA on the surface of the cell (43). This may be consistent with a role for cell surface HA in the detachment of cells, rather than attachment.

In contrast to results with the HA-binding proteins, 3T3 cells attached and spread very well on PF4-coated substrata (43). Figure 3 compares the early spreading responses on PF4 versus pFN. The principal difference between the two is the bipolar morphology on pFN and the "pancaked" appearance of cells on PF4 during the initial 2 hr; eventually, the cells on PF4 become bipolar and indistinguishable from pFN-adherent cells (see Ref. 43 for more details). These adhesive responses are shown to be cell surface HS-dependent and not CS-dependent by two different approaches. Adhesive regions on these two substrata were then compared by scanning electron microscopy during EGTA-induced retraction (43). The broad lamellae formed on PF4 (Figure 3E) disintegrate and generate retraction fibers virtually identical to those observed on pFN. The SAM's left by these two populations of cells (Figure 4) are also virtually identical, with filopodiallike fibers at the periphery and punctate material left at more internal regions of the cell. The similarity in responses of these cells to pFN, which has both HS- and cell-binding activities, and to PF4, which has only HS-binding activity, reinforces the arguments that the binding of substratum-bound pFN to cell surface HS-PG plays a central role in the attachment and spreading of fibroblastlike cells.

The nature of the contracting regions of cells on pFN or PF4 was determined then by interference reflection microscopy (20). Tight-focal contacts are very apparent on pFN-adherent cells (Figure 5b) at the ends of microfilamentous stress fibers (Figure 5a). Therefore, pFN contains all the information necessary to generate these specialized con-

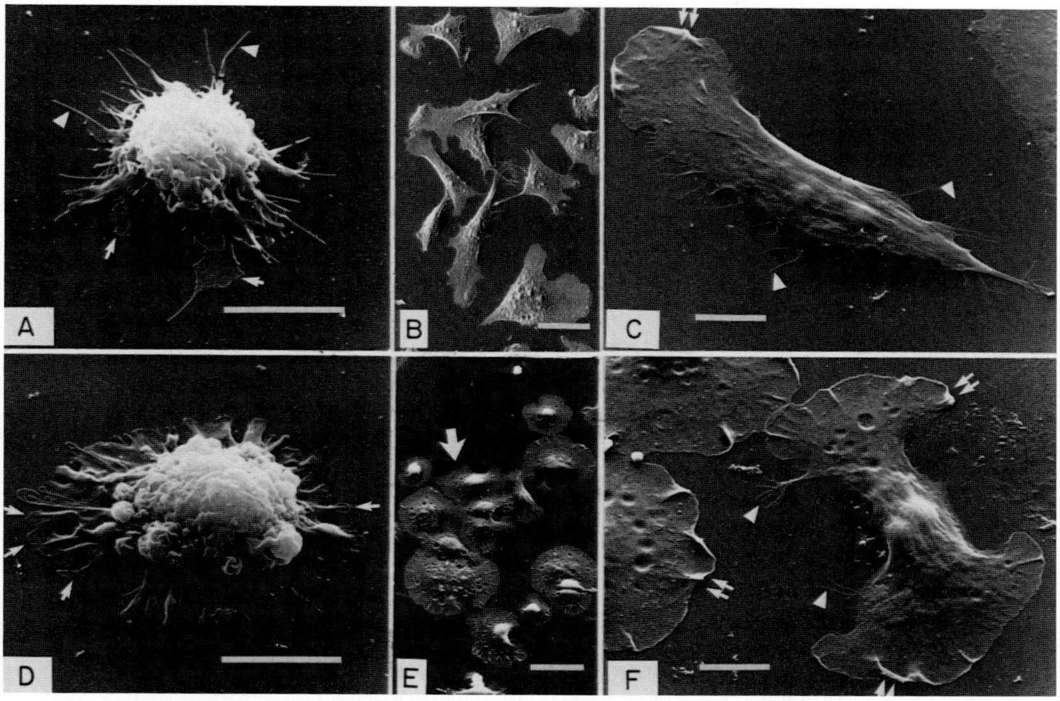

Figure 3 Initial spreading responses of Balb/c 3T3 cells on pFN- or PF4-coated substrata. Thermanox cover slips were coated with pFN (A-C) or PF4 (D-F); then EGTA-detached and well-washed cells were inoculated. After 5 min of adherence (A and D), cells on pFN were extending filopodial contacts (at arrowheads in A), which were spreading into footpad adhesion sites (at arrows in A). Alternatively, cells on PF4 (D) were extending radial pods at their periphery. After 1 hr of adherence, cells on pFN (B) had assumed the typical polygonal morphology while cells on PF4 (E) had formed broad-spreading lamellae. A cell spread for 2 hr on pFN (C) exhibits ruffling leading lamellae (double arrow), a trailing cytoplasmic fiber, and long filopodia (arrowheads). Cells on PF4 at 2 hr (F) also display ruffling lamellae (double arrow) and filopodia (arrowheads). Eventually, cells on PF4 assume a morphology identical to that on pFN (see Ref. 43). Bars A, C, D, and F = 10 μm; B and E = 40 μm. (Taken from Ref. 43.)

tacts. Also, pFN on the substratum must be a component of these contacts (since it is the only protein on the substratum), and recent evidence that FN is not a component of these focal contacts (50,51) may be an artifact of the lack of penetration of antibody reagent into these highly dense matrices (52). In contrast to these cells on pFN, the periphery of the PF4-adherent cells (Figure 5e) looks entirely black as if these regions were entirely tight-focal contact, but this was shown to be an artifact of destructively interfering reflections from the upper cell surface of these very thinly spread cells (20). When a BSA solution with a refractive index higher than that of the cytosol was perfused into cultures (this solution would penetrate under close contacts but not under tight-focal contacts), the periphery became entirely white on PF4 (Figure 5f), indicating that these regions are truly close contacts and not tight-focal contacts. Focal contacts at single arrows in Figure 5c persist on pFN with BSA perfusion. In addition, close contacts are readily apparent on PF4 (double arrow in Figure 5e) at regions of the nucleus where the

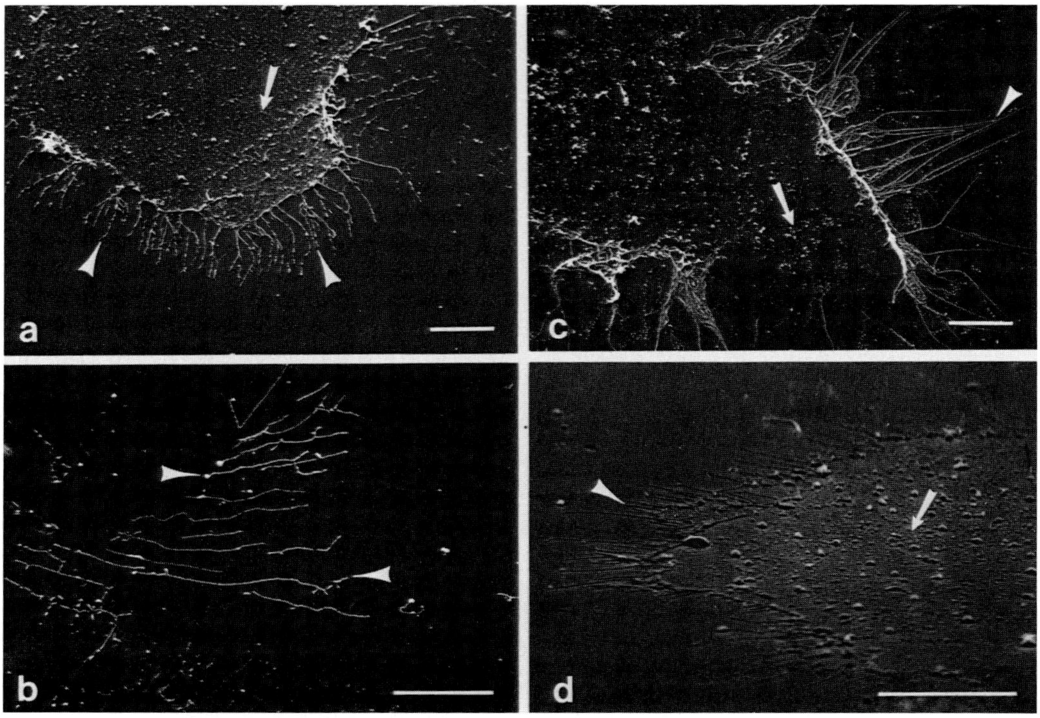

Figure 4 Morphologies of substratum-attached material from cells EGTA-detached from pFN or PF4. Balb/c 3T3 cells, detached from stock cultures with 0.5 mM EGTA in PBS, were inoculated onto coverslips coated with pFN (a, b) or PF4 (c, d). After 2 hr of spreading, the cells were detached from these substrata by EGTA treatment. The SAM-coated coverslips were fixed, dehydrated, critical point-dried, and sputter-coated with gold-palladium as described (20) in preparation for scanning electron microscopy. (a, b, c, d) SAM derived from quadrants of spread cells; the extent and nature of spreading of the cells before EGTA treatment can be deduced from the SAM patterns. Arrowheads indicate linear SAM apparently derived from peripheral lamellae. Arrows indicate smaller flocculent SAM, which was more common under the cell bodies. Bars = 6 μm. (Taken from Ref. 20.)

cell is thicker and reflections from the upper surface are no longer a problem (20). During these experiments, only close contacts were observed on PF4-adherent cells, which indicates that binding of the ECM to cell surface HS-PG is *sufficient by itself* to generate close contacts and that the participation of pFNs unidentified receptor for CBD is somehow required to generate tight-focal contacts (see Section II.C).

It then became important to examine cytoskeletal reorganization during spreading responses on pFN or PF4 (43). Figure 6 illustrates two unusual aspects of these responses on PF4. First, microtubule reorganization occurs throughout the cytosol of pFN-spreading cells (Figure 6A) but only in the central portion of PF4-spreading cells (Figure 6C); the broad lamellae on PF4 are completely devoid of microtubules. Second, NBD-phallocidin stains microfilamentous stress fibers throughout the peripheral cytoplasm of cells on pFN (Figure 6B) but stains short microfilament bundles only in the broad spreading lamellae on PF4 (Figure 6D). These data led to two very important conclusions (43). The

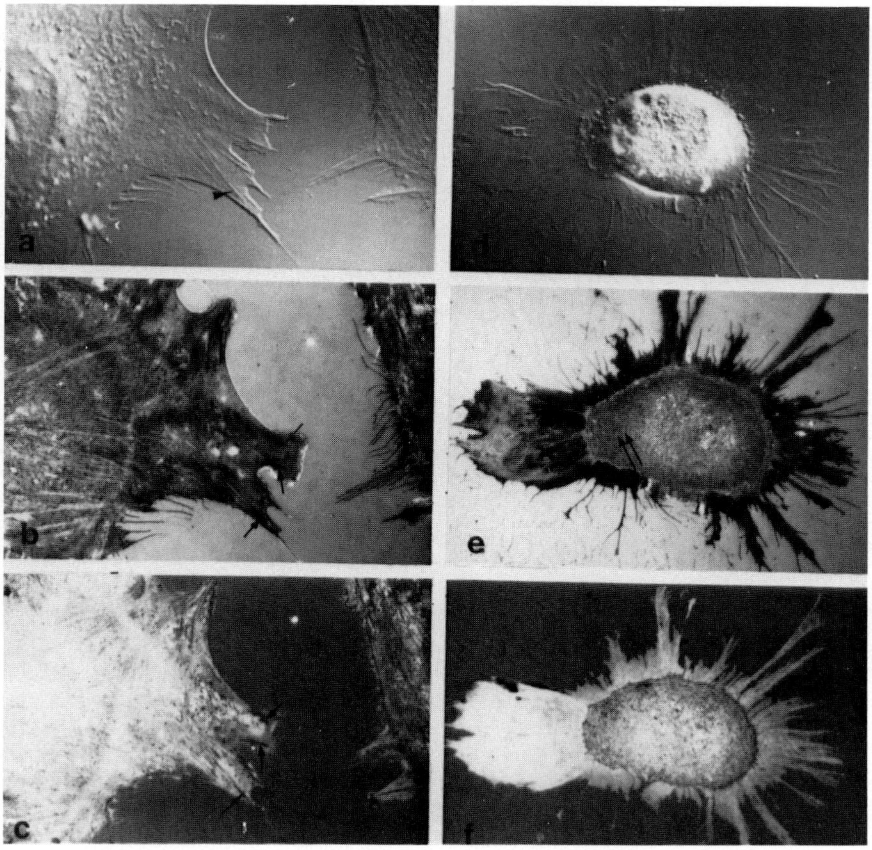

Figure 5 PF4- or pFN-mediated adhesive contacts (interference reflection microscopy). BALB/c 3T3 cells were detached from stock cultures with 0.5 mM EGTA and allowed to spread for 3 hr on (a-c) pFN and (d-f) PF4 in serum-free medium. Coverslips carrying cells were then mounted and observed with both (b, c, e, f) interference reflection and (a, d) differential interference contrast optics, as described (20). Cells on pFN form interference reflection-detectable tight focal contacts (b, arrows) that are surrounded by gray close contacts and by white nonadhesive regions. Tight focal contacts persist as black streaks following perfusion with a solution of MEM X 4 plus BSA having a refractive index of 1.382 (c, arrows), which is greater than the refractive index of cytoplasm. Microfilament bundles (a, arrowhead) extend into tight focal contacts. Cells in PF4 form thin, broadly spread lamellae (d), which appear mostly black (e). Gray areas of close contacts are present under the cell (e, double arrow). That the dark areas (in e) do not contain areas of tight focal contact is indicated by the complete reversal of image contrast following perfusion of high refractive index medium (f). Bar = 20 μm. (Taken from Ref. 20.)

binding of PF4 on the substratum to cell surface HS-PG is *sufficient* to get microfilament bundles reorganized; this demonstrates that cell surface HS-PG is transmembrane-linked to cytoskeletal elements in the cytosol (see Section II.C for further analysis of this point). This evidence also adds considerable weight to the argument that the supramolecular organization of cell surface HS-PG leads to microfilament reorganization in

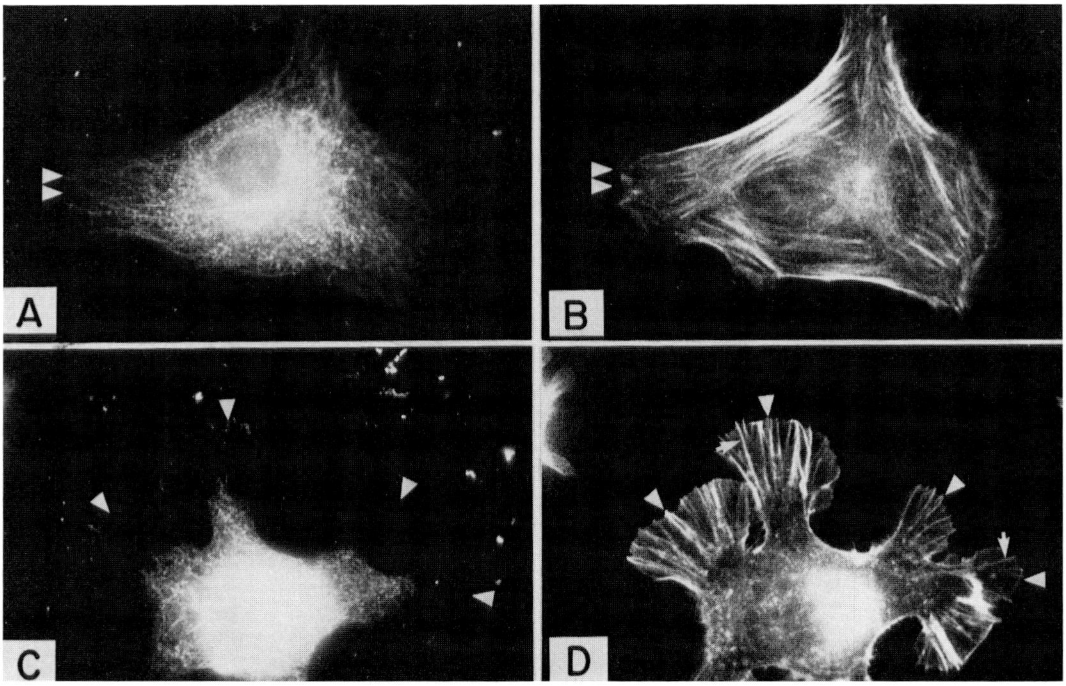

Figure 6 Reorganization of microtubules and microfilaments during spreading on PF4 or pFN. BALB/c 3T3 cells were detached from stock cultures with EGTA and inoculated onto pFN-coated (A and B) or PF4-coated (C and D) glass coverslips in serumless medium as previously described (43). After 2 hr of attachment, the cells were fixed and double-labeled with antitubulin and NBD-phallocidin to visualize both microtubules (A and C) and microfilaments (B and D) in the same cell, respectively. Microtubules of cells on pFN (A) were visualized throughout peripheral lamellae in contrast to the microtubules of cells on PF4 (C), which concentrate more proximal to the nucleus and are not present in well-spread lamellae. The majority of cells on pFN contain well-defined longitudinal stress fibers (B), which are absent on PF4-attached cells (D). Cells on PF4 contain numerous fine filaments of F-actin (D, arrows) throughout lamellae that do not coincide with membraneous folds or ruffles. X800. (Taken from Ref. 43.)

the cytosol as proposed in the model of Figure 2A. Second, these experiments demonstrate that reorganization of microfilaments and microtubules *can occur* in distinctly separate regions of the cytoplasm of the cell and are therefore not spatially interdependent. The molecular details of these important findings must now be resolved.

B. Cell-Binding Domain of Fibronectin

In order to complement the studies on adhesive respones to HS-binding proteins described above, we followed up on the initial report of Pierschbacher et al. (40) that pFN could be cleaved with chymotrypsin in such a way that the cell-binding domain (CBD) could be isolated free of any HS-binding activity. We prepared the 120,000-dalton fragment of pFN, which has CBD activity but no detectable HS-binding activity (46). We could also separate out heparin-binding domains (HBD) of pFN using the fractionation scheme of Pierschbacher et al. (40), but these are contaminated with some cell-binding activity (46) and therefore cannot be used as truly pure HBD. The responses of 3T3 cells were measured on CBD-coated substrata for comparison with those on intact pFN or PF4.

Cells attached as well to CBD as to intact pFN, but their spreading responses were markedly reduced (47). On CBD, cells put out primitive spikelike processes (see Figure 7), which never matured into extending lamellae. Within 24 hr on CBD, the cells became unstable and began to detach; this contrasted with their stability of adherence on pFN or PF4.

Cytoskeletal reorganization on CBD was distinctly different from that on either pFN or PF4 (Figure 7) (47). In the extending "spiky" projections on CBD (Figure $7B_1$), there were high concentrations of NBD-phallocidin-stainable microfilaments, which never organized into bundles characteristic of those on pFN (Figure $7A_1$) or PF4 (Figure $7C_1$). Microtubule reorganization on CBD occurred throughout the cytoplasm (Figure $7B_2$). The responses of cells on CBD as shown in Figure 7 were maximal at 4-6 hr of adherence and slowly disintegrated after 12-16 hr (47). We then measured these responses on substrata coated with varying mixtures of CBD plus PF4 in order to determine whether we could obtain a pFN-like response with combinations of these two binding activities on separate molecules (47). The result was that mixtures gave cytoskeletal reorganization patterns *identical to those on PF4 alone*, even at molar ratios of PF4:CBD as low as 0.1; microfilamentous *stress fibers* could never be demonstrated on mixtures of these two proteins at any time points. These results will be discussed more fully in Section II.C.

Interference reflection microscopy was then used to examine the contracting regions of cells on CBD or HBD (48). Balb/c 3T3 cells on CBD formed *only* close contact regions and were never observed to form tight-focal contacts, consistent with the cytoskeletal studies reported above. Furthermore, in agreement with the IRM studies on PF4, cells on HBD could generate only close contacts (48). We then measured responses to various mixtures of CBD and PF4 as described above; again, only close contacts were observed at any time point. All these results taken together demonstrate that the substratum conditions for forming tight-focal contacts with associated microfilament stress fibers are very special (Section II.C) and indicate that the combination of these two binding activities *in the correct multivalent orientation in the same molecule* is necessary to get stress fiber formation.

C. Transmembrane Linkage of Heparan Sulfate Proteoglycan to the Cell's Microfilament Networks: Requirements for Tight-Focal Contact Formation

The evidence described in Figure 6, originally by Laterra et al. (43), was the first to establish that cell surface heparan sulfate proteoglycans are somehow transmembrane-linked to the microfilament-organizing centers in the cytoplasm of the cell. The responses on PF4-coated substrata are a clear indication of this linkage in that the formation of patching networks of HS-PG in the SAM of cells adhering to PF4 is sufficient to transmit a signal to the cytoplasm that permits bundling of microfilaments over lengthy distances (43). Consistent with this is the preliminary evidence that HS-PG that behaves like an integral membrane protein can bind F-actin (49), an intriguing possibility that will have to be examined more carefully. CBD of pFN could not mediate such an effect, which suggests that the HBDs of pFN play a principal role in cytoskeletal reorganization as originally proposed in the model of Figure 2A (24,25).

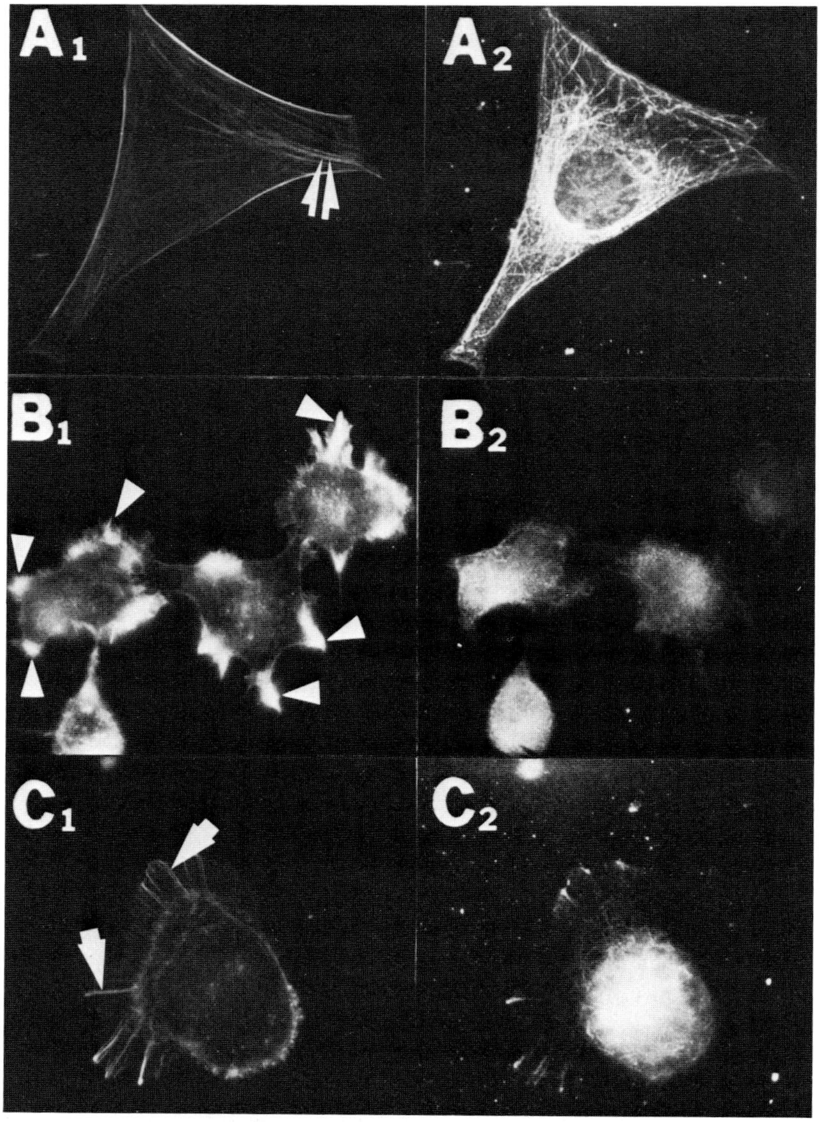

Figure 7 Comparison of cytoskeletal reorganization on pFN, PF4, and the cell-binding domain of pFN. Glass coverslips were coated as described previously (43) with 20 μg/ml of human pFN (A_1 and A_2); CBD of human pFN (B_1 and B_2) prepared as the 120K = mol. wt. fragment after chymotrypsin digestion (40); or PF4 (C_1 and C_2); CBD has no detectable HS-binding activity (46), but promotes adherence and partial spreading of 3T3 cells (47). Coverslips were then postadsorbed with attachment medium (serum-free MEM plus 2 mg/ml BSA) to guarantee coverage of all substratum sites. EGTA-detached and well-washed Balb/c 3T3 cells were permitted to attach and spread in attachment media for 4 hr, when their response was maximal. Cells on CBD became unstable and detached after 16-20 hr, whereas cells on pFN or PF4 remained stably attached and spread in excess of 48 hr (43,47). They were then fixed and double-stained with NBD-phallocidin (series 1) or antitubulin (series 2), as described previously (43). With regard to microfilament reorganization, cells on pFN form well-defined stress fibers (double-arrow

Alternatively, cells could not form stress fibers or tight-focal contacts on the multimeric PF4 molecule. This reveals that pFN's binding to its receptor glycoprotein via CBD must be required to form stress fibers and tight-focal contacts. Mixtures of CBD and PF4 were also incapable of forming these adhesive responses. This indicates that binding activities for HS and the receptor must be present *on the same molecule in a multivalent state* in order to generate these adhesive regions. This should now be testable by synthesizing multifunctional proteins that have these binding activities.

The experiments described here (43,47) (Figure 7) using NBD-phallocidin to stain microfilament bundles suggest interesting possibilities for *two very different mechanisms of linkage* of the HS-PG and the unidentified receptor for FN to the cytoskeleton. That bundles form via a PF4:HS-PG binding reaction suggests that HS-PG may be linked *laterally* to microfilaments and not at their ends. Consistent with this would be the codistribution of HS-PG on the surface with microfilament bundles in the cytoplasm. Then, the intense staining of microfilaments in the "spiky" regions of cells on CBD (Figure 7B_1) indicates that the receptor glycoprotein binding to CBD may promote *end-on* organization of microfilaments, but not lateral linkage. If these two mechanisms were operative, then stress fiber formation may very well require the binding of *both* HS-PG and the unidentified receptor to the same multifunctional protein in order to get this complex response. The use of a variety of adhesion-promoting ligands now offers experimental approaches for testing these hypotheses.

III. STRUCTURAL AND FUNCTIONAL STUDIES OF PROTEOGLYCANS IN SAM

A. Long-Term Culture-Generated Substratum-Attached Material (L-SAM) Versus Newly Formed Substratum-Attached Material from Reattaching Cells (R-SAM)

Since PGs (especially HS-PG) appear to be very important in the adhesion process, this section will focus on the biochemical characterization of these molecules. SAM from cells growing in culture for extended periods of time (72 hr) has been isolated and is referred to as long-term culture-generated SAM (L-SAM) (53). Reattachment of these EGTA-detached and uniformly radiolabeled cells to a fresh substratum for 1 hr results in the formation of newly formed adhesion sites, which can then be isolated as reattaching SAM (R-SAM). Comparison of the molecular composition of the newly formed adhesion sites (R-SAM) to the composition of long-term culture-generated sites (L-SAM) has been used to identify molecules important in the initial attachment of cells to the

---

in A_1) while those on PF4 form long fibers (arrows in C_1) in the well-spread lamellae that fail to develop into stress fibers. On the other hand, cells on CBD generate densely stained concentrations of filaments at their microspikelike projections (arrowheads in B_1), but these fail to develop into lengthy bundles; as cells on CBD become unstable and detach, these dense concentrations of actin-containing filaments disperse. With regard to microtubule reorganization, pFN-attached cells develop these structures throughout their cytoplasm (A_2), whereas CBD- (B_2) and PF4-attached (C_2) cells develop these only in their central cell bodies and are devoid of tubules in peripheral areas of the cytoplasm where microfilaments have developed. X520. (Taken from Ref. 47.)

ECM and to characterize molecules that may play some role in eventual cellular movement and detachment (47).

Studies were initiated to determine whether the GAGs within the adhesion site are bound to core proteins as large multivalent proteoglycans. Using nondegradative isolation techniques, it was determined that the CS and a portion of the HS could be isolated as proteoglycan (54). In a primitive attempt to define the organization of these PG's, they were isolated with high concentrations of guanidine hydrochloride [4 M GdnHCl by analogy with studies of cartilage PG isolation and reaggregation (45)] to dissociate and solubilize them. The extracts were subsequently dialyzed to conditions that should allow for reassociation of the PGs into multimolecular aggregates if they have the *potential* to do so (53). These approaches have been well documented and have been used to characterize proteoglycan interactions in a number of other biological systems (45). It was found that a portion of both HS-PG and CS-PG from the adhesion sites could form a multimolecular aggregates of undefined nature as assayed by their exclusion from Sepharose CL2B columns (53). Using these procedures, it was also found that R-SAM contains a high density HS-PG, as determined by isopycnic CsCl density gradient centrifugation, which is depleted in L-SAM. This was an early indication that there are structural differences in the PGs from R-SAM and L-SAM. Similar results and some notable differences were reported for the KiMSV transformant of Balb/c 3T3 cells (28).

Since these studies indicated that proteoglycans from adhesion sites have the *potential* to aggregate, attempts were made to isolate subclasses of PGs from adhesion sites as "native" aggregates (55). A low concentration of GdnHCl (0.5 M), similar to that which allowed for in vitro PG aggregate formation (56), was initially chosen. Also, the zwitterionic detergent Zwittergent 3-12 was chosen to isolate subclasses of PGs from these sites that may possibly be more intimately associated with the phospholipid bilayer (57). Neither of these reagents was able to isolate large multimolecular complexes from adhesion sites (55). However, these treatments did result in selective extraction of subfractions of molecules and have proven highly beneficial in other studies. For example, 0.5 M GdnHCl solubilized the majority of the hyaluronic acid without its being bound to either HS-PG or CS-PG (55). These same conditions resulted in the solubilization of hyaluronate/CS-PG aggregates from cartilage that have been well characterized (45,58). In contrast to the cartilage system, it is clear that the majority of the CS-PG and hyaluronate in fibroblast adhesion sites do not associate into this type of complex. These studies cannot rule out the possibility that these molecules may interact in some other type of organization, and further studies must be completed to help clarify binding relationships (58).

Using the selective extraction approach described above (55), a change in the organization of CS-PG was observed between L-SAM and R-SAM (59). Only a small portion of the CS-PG in L-SAM was solubilized with 0.5 M GdnHCl, while a much larger amount was extractable from R-SAM under identical conditions. Even though HA was cosolubilized with the CS-PG under these conditions, in neither L-SAM nor R-SAM was the CS-PG associated with HA in a cartilagelike aggregated structure. These results indicate that there is a change with time in the association of the CS-PG with other molecules within the adhesion site.

HS-PG was also analyzed by these selective extraction techniques (55,59). As stated previously, GdnHCl had very little affect on solubilization of the HS moieties from the adhesion sites; however, 1% Zwittergent 3-12 selectively solubilized a subset of HS and glycoprotein from these sites (55,59). The Zwittergent-soluble HS was *not* covalently attached to core proteins as proteoglycans and are therefore single-chain HS molecules. However, these studies have not excluded the possibility that these chains may still have a small peptide covalently linked at their reducing ends. Since Zwittergent extracted only a portion of the total HS from the adhesion sites, the Zwittergent-resistant HS was subsequently solubilized using SDS (59). Characterization of this fraction indicated that it was a large HS-PG with a $K_{av} = 0.55$ on Sepharose CL2B columns after disulfide reduction and under completely dissociative conditions (59). The relative resistance of HS-PG to extraction with GdnHCl under conditions that result in solubilization of both HA and CS-PG confirms the tenacity of binding of this molecule in dense adhesive matrices and explains the lack of penetrability of antibodies to fibronectin in these regions (52). The selectivity of the zwitterionic detergent for solubility of the single-chain HS indicates that the single-chain HS moieties are associated with the adhesion site through a mechanism different that that of the HS-PG (59,60).

B. Generation of HS Chains from HS-PG in SVT2 SAM: Mechanisms and Consequences

Since substratum adhesion sites contain both HS-PG's and single-chain HS molecules, the amount of each of these classes of HS was quantified in L-SAM and R-SAM to determine whether there was a change in their proportion with time that may affect cellular detachment. The results of this analysis (59) indicated that the majority of HS from *R-SAM* was resistant to Zwittergent treatment and was identified as a large proteoglycan, while the majority of the *L-SAM* HS was a single-chain in nature. As described previously, the cell may use the large and multivalent HS-PG, which is complexed into multimolecular aggregates, to form initial stable adhesive bonds with many pFN molecules on the substratum (see Figure 2A). Upon initial attachment, there may be catabolism (see below) of the PG resulting in the accumulation of the single-chain HS. These single-chain HS sequences would be essentially univalent or, at most, bivalent for binding FN and would be much less effective at promoting adhesive bonds; in fact, they may destabilize adhesive bonds by occupying critical FN molecules. (See Table 1.)

The kinetics of turnover of the HS-PG and single-chain HS were therefore examined in some detail to determine whether the single-chain class was the result of catabolism of the large multivalent HS-PG (61). A pulse-chase analysis using $^{35}SO_4^{=}$ was designed to radiolabel the HS moieties and follow their turnover with time (61). These analyses indicated that only a portion of the Zwittergent-resistant HS-PG is catabolized into a Zwittergent-soluble single-chain species prior to its release from the adhesion site. A major portion of the HS-PG is released from the adhesion site without first being catabolized into the single-chain intermediate. It was also observed that the HS-PG and single-chain HS species have different rates of release, with the latter being lost from sites very rapidly. Thus, it is evident that the HS-PG is turned over from the adhesion site with time by at least two very different mechanisms whose significance in altering adhesive processes must be determined.

Two possible types of catabolism come to mind with regard to the formation of the single-chain HS moieties. First, there may be enzymatic cleavage of the HS chains on the PG through endoglycosidase action, as found in a number of other systems (62-64), resulting in the formation of small single-chain HS sequences. Second, there may be proteolytic cleavage of the core protein, which would result in release of intact GAG chains containing small sequences of covalently attached core peptide. Analysis of the Zwittergent-soluble single-chain molecules suggests that endoglycosidase activity may be occur-

Table 1 Heparan Sulfate Extractable with Zwittergent

	Percentage of total heparan sulfate	
	Zwittergent soluble	Zwittergent resistant
L-SAM[a]	70	30
R-SAM[b]	34	66

[a][^{3}H]glucosamine-radiolabeled L-SAM from SVT2 cells was isolated and extracted with Zwittergent 3-12, followed by extraction of the Zwittergent-resistant material with SDS as described by Lark and Culp (59). Both the Zwittergent-soluble and -resistant materials were treated with nitrous acid and chromatographed over a Sepharose CL-6B column in SDS-containing buffer. Heparan sulfate was identified as material that shifted to a further-included position on the Sepharose CL-6B column upon nitrous acid deamination. The percentage of heparan sulfate solubilized with Zwittergent is expressed as the percentage of total nitrous acid-labile material (both solubilized and resistant to Zwittergent extraction) that was extractable with Zwittergent.
[b]$^{35}SO_4^{=}$-radiolabeled R-SAM was isolated from cells treated as described by Lark and Culp (59). The material was extracted with Zwittergent, followed by SDS treatment, and the extracts were treated with nitrous acid to identify heparan sulfate. The percentage of heparan sulfate extractable with Zwittergent was determined identically as described for L-SAM above.

ring, since the chains are smaller in relation to the size of intact HS chains released from the PG core protein using alkaline-borohydride treatment (59). Also, the small chains have a very heterogeneous size distribution, in contrast to a more homogeneous distribution for chains isolated directly from the PG. If the HS-PG is associated with another molecule within the site through binding of its HS chains, endoglycosidase digestion of this sequence could result in eventual release from the adhesion site of a modified proteoglycan (with smaller HS chains). Protease digestion could also result in release of much of the core protein with some of the HS chains, but other HS chains might be maintained within these sites by their binding to other molecules such as FN. If the HS-PG were associated with the plasma membrane or other molecules within the adhesion site through a portion of its core protein, a single proteolytic clip in this region of the molecule could release the HS-containing portion of the PG from the site. A number of experimental approaches can now be used to study the mechanism(s) of catabolism of HS-PG in these sites and consequences of this catabolism upon adhesive events for the cell.

C. Multiple Classes of Heparan Sulfate Proteoglycans and Their Differing Functions: R-SAM → L-SAM

This section will deal with characterization of a number of distinguishable HS-PGs isolated from adhesion sites. The identification of different types of HS-PGs within one biological system is not uncommon (65-68); however, criteria that have been used in other systems to identify different subclasses of PG including differences in size and differential sensitivity to removal with exogenous components, were not useful in this system. As discussed earlier, the HS-PGs from L-SAM or R-SAM appear to be a homogeneous class of molecules as identified by chromatography on Sepharose CL2B columns under stringently dissociative conditions (59). Even though Zwittergent could identify single-chain HS molecules without solubilizing PG, no treatment has yet been found that could distinguish subfractions of the HS-PG. Therefore, a protocol was designed to quantitatively recover the HS-PG from adhesion sites under relatively gentle extraction conditions and to characterize these molecules with regard to their binding affinities for various ligands, including the HS-binding proteins PF4 or pFN and the hydrophobic binding matrix octyl-Sepharose (69). A combination of octyl-glucoside, NaCl, and GdnHCl were found to give maximal solubility of the HS-PG from both R-SAM and L-SAM (69), similar to that recovered using SDS. The advantage to the octyl-glucoside extract was that the detergent and the salt could be removed by dialysis, allowing functional binding assays to be completed on these molecules.

When the HS-PGs from either L-SAM or R-SAM were chromatographed over the PF4 or pFN affinity matrices, a two- to threefold larger amount of the material from R-SAM than from L-SAM bound to either of these ligands (69). Binding patterns to pFN are shown in Figure 9. A similar result was found when extracts from both L-SAM or R-SAM were chromatographed over the hydrophobic binding matrix octyl-Sepharose (Figure 8) (69). Therefore, more of the HS-PG isolated from R-SAM than from L-SAM has the capacity to interact with PF4, pFN, and octyl-Sepharose. It should be noted that the HS-PG isolated from L-SAM and R-SAM bind to PF4 or pFN with different avidities. Also, there were HS-PGs that did *not* bind to these matrices (flow-through fractions in Figures 8 and 9). The binding avidity of these PG's for ECM molecules, such as pFN or PF4, may prove to be quite important in the morphological responses observed upon cell attachment to these proteins (see below).

Since a higher proportion of R-SAM HS-PGs have the ability to bind to all three of these matrices than L-SAM HS-PGs (69), it appears that modifications in the PGs occur in adhesion sites with time. There could be modifications in either or both the carbohydrate chains of the proteoglycan as well as the core protein sequence, which would result in decreased binding. One change that could affect the GAG chains is endoglycosidic digestion of these chains, as discussed previously (62-64). Another possibility is a modification in the GAG chain through a mechanism such as desulfation (70,71) or deacetylation. Presently, the mechanisms by which these PGs bind to PF4 or pFN are unknown since the binding sequences of HS have not been determined. The binding mechanism cannot be based solely on the charge density of the PGs, since the same extracts upon chromatography on DEAE-Sephadex demonstrated quantitative and virtually homogeneous binding or elution (69).

The hydrophobic affinity resin presumably binds proteoglycans containing hydrophobic core protein sequences (65,66,72). Proteolytic cleavage of the hydrophobic core protein sequence away from the remainder of the proteoglycan core protein would result in loss of the ability of HS-PGs to bind to octyl-Sepharose with minimal change overall in the size of the PG. One other possibility must be considered with regard to changes in distribution of the binding and nonbinding classes of HS-PG with time. It is possible that the PF4, pFN, or octyl-Sepharose binding and nonbinding HS-PGs are completely different molecules constructed from different core proteins. Better biochemical characterization of these PGs should resolve these various possibilities.

By rechromatographing fractions of HS-PG that bound to one affinity matrix onto one of the other matrices, a number of classes of HS-PG with overlapping affinities were identified. One of the most striking observations made in these studies was that there is a class of hydrophobic binding HS-PG that can also bind to either PF4 or pFN (69). As discussed at the beginning of this section, the HS-PG in both L-SAM and R-SAM requires detergent treatment for its extraction away from other adhesion site molecules. Since the HS-PG requires treatment with detergent for its extraction and since it binds to a hydrophobic binding resin (Figure 8) (69), it is likely that this particular PG is a mem-

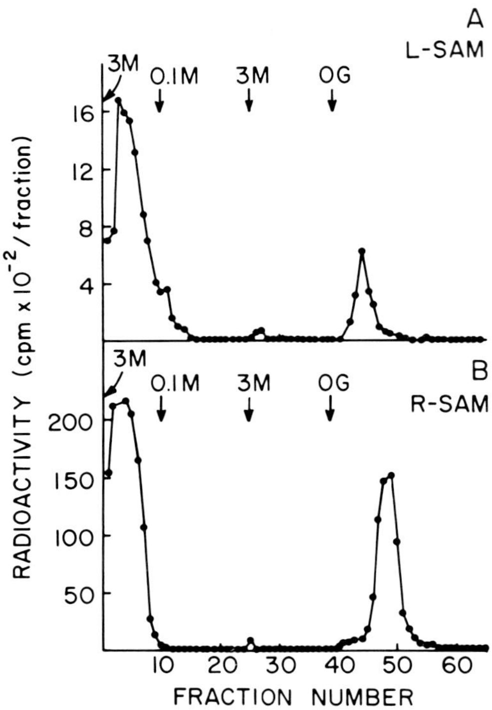

Figure 8 Octyl-Sepharose affinity chromatography of L-SAM or R-SAM extracts. Balb/c SVT2 cells were radiolabeled during 72 hr of growth with $^{35}SO_4^{=}$-containing medium. The cells were then detached with EGTA and the L-SAM quantitatively solubilized with 1% octylglucoside, 0.5 M GdnHCl, and 1 M NaCl in buffer at 37°C for 4 hr. Meanwhile, the radiolabeled cells were washed well and permitted to attach to fresh serum-coated dishes for 1 hr to generate new adhesion sites; these cells were then EGTA-detached and the R-SAM harvested with the octylglucoside mixture. $^{35}SO_4^{=}$-radiolabeled extracts of (A) L-SAM or (B) R-SAM were dialyzed against glass-distilled water and subsequently against 3 M NaCl, 0.5 M GdnHCl, and 0.2MTris (pH 7.3) prior to batch adsorption to octyl-Sepharose, as described in Lark and Culp (69). The resin was then poured into a column and sequentially eluted with 3 M NaCl (3 M); 0.1 M NaCl (0.1); 3 M); and 1% octylglucoside, 3 M NaCl (OG), all eluant buffers containing 0.2 M Tris (pH 7.3). The column was then rinsed with 0.2% SDS in glass-distilled water to elute any material resistant to elution with the buffers described above (none was found). Approximately, 2.5 times more proteoglycan from R-SAM than from L-SAM binds to this hydrophobic matrix, requiring octylglucoside detergent for its elution. (Taken from Ref. 69.)

brane-intercalated molecule responsible for many of the cellular physiological responses described in Section II.A. The ability of this putative membrane-intercalated HS-PG to bind to these ECM molecules allows for a direct link from molecules on the cell's exterior to internal cytoskeletal networks (see Section I.C.). This study also identified HS-PG's that could bind to either the hydrophobic binding resin or to the HS-binding molecules but not to both. Thus, there are putative membrane-intercalated HS-PG's that can bind ECM molecules, putative membrane-intercalated HS-PGs that cannot associate with the binding proteins, and HS-PGs that are not hydrophobic in nature but can bind to ECM molecules.

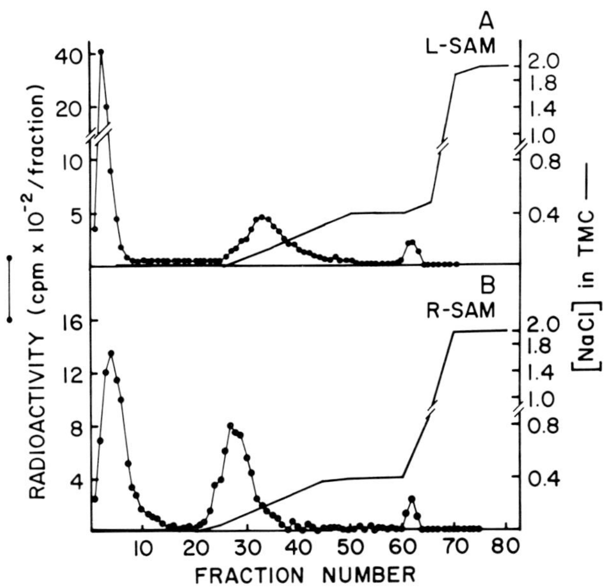

Figure 9 L-SAM and R-SAM extracts affinity-chromographed on a fibronectin-Sepharose column. Balb/c SVT2 cells were $^{35}SO_4^{=}$-radiolabeled and their L-SAM and R-SAM harvested as described in the legend to Figure 8, as originally described by Lark and Culp (69). $^{35}SO_4^{=}$-radiolabeled (A) L-SAM and (B) R-SAM extracts were dialyzed against glass-distilled water and TMC buffer (0.5 M Tris, pH 7.4; 1 mM $CaCl_2$, 1 mM $MgCl_2$) prior to elution through a pFN-Sepharose affinity column. Samples were chromatographed through the column in TMC and eluted with TMC, followed by a linear gradient of 0–0.4 M NaCl in TMC, 0.4 M NaCl in TMC, and finally 2M NaCl in TMC. Salt concentrations of eluted fractions were determined by refractometry. A significantly larger fraction of R-SAM proteoglycan binds to this affinity matrix than L-SAM proteoglycan; also, a small fraction of PG binds with very high affinity (>0.5 M NaCl). (Taken from Ref. 69.)

HS-PGs with overlapping affinities for PF4 and pFN were also identified (69). When the HS-PG from R-SAM that initially bound to pFN-Sepharose was rechromatographed over PF4-Sepharose, over 90% of the pFN-binding HS-PG bound to PF4 (Figure 10). Comparatively, only 50% of the pFN-binding HS-PG from L-SAM also retained the capacity to bind to PF4 (Figure 10). This difference must indicate that (a) significant changes in the heparan sulfate sequences occur as the HS-PG "ages" in the adhesion site and (b) PF4 and pFN have different specificities for HS sequences that bind. The physiological significance of a PG that can bind to pFN or PF4 but not to the other remains to be elucidated.

At the onset of the biochemical studies just discussed, the HS-PG within fibroblast adhesion sites was thought to be a relatively homogeneous molecule. It is now clear that this is not the case and that there are at least several different subclasses of HS-PG within these sites based on a combination of physical chemical and affinity chromatographic studies. Also, there are changes in the structure and/or organization of these subclasses of HS-PG, as well as the CS-PG, with respect to time within these sites. Functional roles for a number of these subclasses can be proposed in the adhesion of fibroblasts to a pFN-containing ECM.

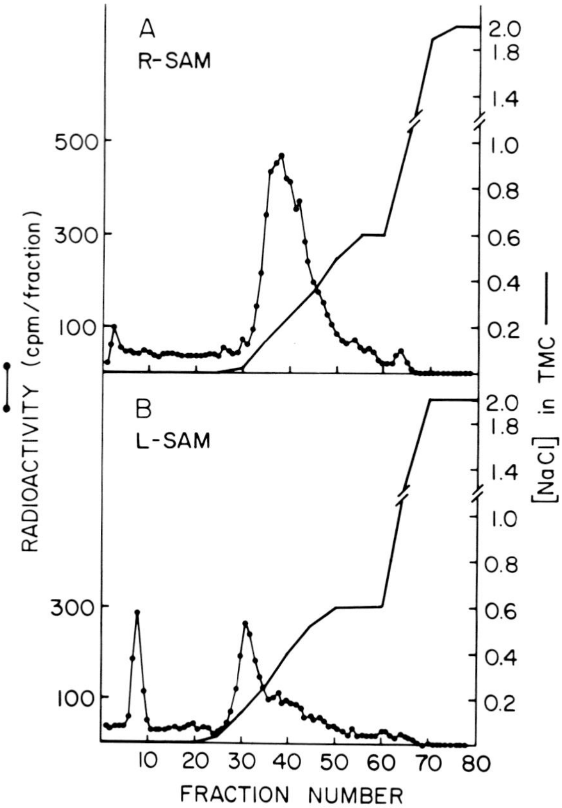

Figure 10 Affinity chromatography of fibronectin-Sepharose-binding fractions on a column of platelet factor-4-Sepharose. The pFN-Sepharose-binding fractions of the $^{35}SO_4^{=}$-radiolabeled (A) R-SAM or (B) L-SAM extract from Figure 9 were eluted from the pFN column, concentrated,and dialyzed against TMC prior to chromatography through a PF4-Sepharose column. Samples were adsorbed to the PF4-Sepharose in TMC and eluted with TMC, followed by a linear gradient of 0–0.6 M NaCl in TMC, 0.6 M NaCl in TMC, and finally 2 M NaCl in TMC. Virtually all of the pFN-binding R-SAM proteoglycan also binds to PF4 (A), whereas only 50–60% of the L-SAM proteoglycan has this capacity (B). This indicates some modification of the heparan sulfate sequences during the "aging" of adhesion sites. (Taken from Ref. 69.)

Early studies indicated that the accumulation of hyaluronic acid and CS-PG result in the detachment of cells from the substratum (see Figure 2B). This model can now be modified since there is not only an accumulation of CS-PG with time but a change in the association of this PG with other molecules within the adhesion site. The accumulation of hyaluronate and CS-PG may affect cellular detachment on different extracellular matrices. For example, in tissue culture where pFN plays a major role in adherence, HA may take a principal role in facilitating detachment, as modeled in Figure 2B, by reacting with aggregated cFN. Under these conditions, CS-PG may play no role at all and its accumulation would be coincidental. Alternatively, on a collagenous substratum in vivo or in vitro, CS-PG may effect detachment by interacting with the collagen directly, thereby inhibiting the binding of other adhesion-promoting molecules (e.g., FN or HS-PG) to the collagen. Therefore, both HA and CS-PG might facilitate detachment of cells, but from different ECMs.

As discussed in Section II, in order to further define a role for HS-PG in the adhesion process, detailed morphological studies were completed on 3T3 cells attaching to the HS-binding proteins pFN or PF4. Two major differences were observed in the morphological events that occur when cells attach to these specific proteins. First, cells attaching to pFN initially form close contact adhesions and eventually tight-focal contacts, while cells attaching to PF4 form only close contacts and never develop tight-focal contacts. Second, cells attaching to PF4 do not form an organized transcytoplasmic stress fiber network similar to that observed with cells on pFN, even though they do form microfilament bundles at their periphery. One of the most interesting observations made in the analyses of the HS-PG from R-SAM of cells grown on pFN was that the majority of this R-SAM HS-PG that binds to pFN also binds to PF4. Thus, the *same* HS-PG may be involved in the attachment of cells to both of these ECMs. Since cells attaching to PF4 or initially attaching to pFN form only close contacts, it is likely that this HS-PG is a "close contact-specific" HS-PG. Characterization of this PG becomes a high priority for future studies.

Biochemical modification of HS-PGs in adhesion sites probably has considerable significance for adhesive changes in these cells. Any or all of these biochemical events may result in the formation of tight-focal contacts, in the dissolution of close contacts, and ultimately in cell detachment. When cells attach to pFN, close contacts at the leading edge of the cell give way to the formation of tight-focal contacts behind the leading edge and at other regions of the periphery. The changes leading to many different subclasses of HS-PG and free HS chains may be causal to these events and may include endoglycosidic or proteolytic digestion of the HS-PG. On the other hand, when cells attach to PF4, the initial association of HS-PG and PF4 occurs similarly to that which occurs between the HS-PG and pFN. However, the adhesion sites of PF4-adherent cells are rather stable structures, and the subsequent metabolism of the HS-PG may occur differently for some unknown reason. Another factor is the higher avidity of PF4 for HS-PG when compared to pFN; this may result in a different series of modification reactions in the adhesion site. Comparison of the HS-PGs in the adhesion sites of cells adhering to pFN or PF4 becomes significant.

IV. SUMMARY AND FUTURE INITIATIVES

A. Requirements for Close and Tight-Focal Adhesive Contact Formation

The studies summarized here provide some insight into the biochemical requirements for forming two types of adhesive contacts with a fibronectin-containing extracellular matrix. Close contacts require only the binding of cell surface HS-PG to the HBD of FN. These contacts are sufficient to generate a rather complete physiological response of the cell. This is demonstrated (43) by the stable and long-term compatibility of Balb/c 3T3 cells on PF4-coated substrata. It will now be important to determine whether cell division can be completed on PF4-coated substrata, verifying that interactions with this artificial substratum are of sufficient quality to get complete cytokinesis. A variety of other cellular physiological responses should also be investigated, including dependence or independence from priming of cells with platelet-derived growth factor (73).

The requirements for forming tight-focal contacts and their associated stress fibers with these cells are much more complicated. Since 3T3 cells could not form these contacts on substrata coated with various mixtures of CBD with either PF4 or HBD, these binding activities must be present *on the same molecule in a multivalent state* in order to achieve the cell's response. As theorized in Section II.C., this may be a reflection of two different transmembrane binding mechanisms for cell surface HS-PG and FNs unidentified receptor with intracellular cytoskeletal elements—the former being a lateral binding while the latter would be an end-on binding to microfilaments. This might explain the inability to obtain tight-focal contacts with mixtures of separate domains; both *end-on* and *lateral* binding must occur to the same pFN molecules on the substratum in order to generate a topographically precise arrangement. A number of experimental approaches can be used to test these hypotheses.

B. Different Roles for Heparan Sulfate Proteoglycans in Adhesion of Fibroblasts Versus Neuronal Cells to Fibronectin-Containing Matrices

Experiments similar to those reported in Section II were performed with rat B104 and human Platt neuroblastoma cells, both of which make neurites on pFN-coated substrata (74), in an effort to discern how neuronal cells would respond similarly to or differently than fibroblasts (46). On PF4, these cells attached well, consistent with their extremely high concentrations of HS-PG in SAM (74); these cells also have been shown to contain much smaller HS-PG's than fibroblasts, which can similarly be partitioned into pFN-, PF4, and octyl-Sepharose-binding and nonbinding fractions (75,76), and described in Section III. Although these neuronal cells spread partially on PF4, they *never* extend neurites on PF4, indicating that cell surface HS-PG is not an obligatory intermediate of growth cone adherence (46). In contrast to these responses, they attach as well to CBD-coated substrata as to pFN and extend neurites on CBD *as efficiently* (46). It was previously shown that the morphologies of cell body and growth cone SAM's were quite different (77), and these cell response studies reveal that the biochemical requirements for growth cone adherence on pFN are different than those for the cell body (46). Growth cone initiation and extension requires the binding of CBD, and not of HBD, to substratum-bound pFN. Comparative biochemical analyses of the unidentified receptors for FN on fibroblasts and neuronal cells become most important.

These conclusions were confirmed when responses of neuroblastoma cells were measured on substrata coated with mixtures of CBD and PF4 (46). The latter HS-binding protein was found to be a potent inhibitor of CBD-mediated neurite outgrowth. This suggests that under some conditions the binding of cell surface HS-PG to an ECM ligand is antagonistic to the formation and/or extension of growth cones.

The dominance or recessiveness for neurite outgrowth was then examined in a topography experiment (46). One region of the substratum was coated with pure CBD and the adjoining region with pure PF4. Neuroblastoma cells were then inoculated to measure their responses at the interface between these two regions; this interface and the cell's location was precisely defined with indirect immunofluorescence and a variety of antibodies. Cells on the CBD region extended neurites very well and on PF4 not at all, as was expected. Cells bridging the interface only extended neurites on the CBD-coated region of the cell's undersurface and never onto PF4. Also, cells adherent at the PF4 regions just adjacent to CBD were incompetent to extend neurites, unless a portion of their undersurface was directly in contact with CBD (46). These experiments reveal the dominance of the chemical nature of the substratum for differentiating growth cones; they also reveal that a cytoplasmic signal cannot be transmitted from a CBD-binding region of the cell to the PF4-binding region that permits neurite extension on the latter.

In conclusion, neuronal cell adherence to pFN-containing matrices is significantly different than that of fibroblasts. The biochemistry of their cell surface HS-PGs are very different. In addition, the binding of their unidentified receptors for FN generate different physiological responses. These studies therefore begin to sort out the biochemical differences between fibroblasts and neuronal cells in their interactions with an FN-containing ECM.

ACKNOWLEDGMENTS

These studies were supported with National Institutes of Health research grants (AM25646, CA27755, and NS17139) to LAC, as well as support for MWL (NIH training grant GM07225).

REFERENCES

1. K. M. Yamada, *Ann. Rev. Biochem., 52*:761 (1983).
2. L. T. Furcht, *Modern Cell Biol., 1*:53 (1983).
3. D. F. Mosher, *Prog. Hemostas. Thromb., 5*:111 (1980).
4. L. A. Culp and H. B. Bensusan, *Nature, 273*:680 (1978).
5. S. A. Aaronson and G. J. Todaro, *Science, 162*:1024 (1968).
6. R. S. Wells, E. W. Campbell, D. W. Swartzendruber, L. M. Holland, and P. M. Kraemer, *J. Natl. Canc. Inst. 69*:415 (1982).
7. J. R. Stephenson and S. A. Aaronson, *J. Exp. Med., 135*:503 (1972).
8. L. A. Culp, *J. Cell Biol., 63*:71 (1974).
9. L. A. Culp and J. Buniel, *J. Cell. Physiol., 88*:89 (1976).
10. L. A. Culp, *Biochemistry, 15*:4094 (1976).
11. R. J. Klebe, *Nature, 250*:248 (1974).
12. E. Pearlstein, *Nature, 262*:497 (1976).
13. M. Höök, K. Rubin, A. Oldberg, B. Obrink, and A. Vaheri, *Biochem. Biophys. Res. Commun., 79*:726 (1977).
14. F. Grinnell and D. G. Hayes, *Exp. Cell Res., 115*:221 (1978).
15. L. A. Culp, *J. Supramol. Str., 5*:239 (1976).
16. J. J. Rosen and L. A. Culp, *Exp. Cell Res., 107*:139 (1977).
17. A. S. G. Curtis, *J. Cell Biol., 20*:199 (1964).
18. C. S. Izzard and L. R. Lochner, *J. Cell Sci., 21*:129 (1976).
19. C. S. Izzard and L. R. Lochner, *J. Cell Sci., 42*:81 (1980).
20. J. Laterra, E. K. Norton, C. S. Izzard, and L. A. Culp, *Exp. Cell Res., 146*:15 (1983).
21. N. S. McNutt, L. A. Culp, and P. H. Black, *J. Cell. Biol., 50*:691 (1971).
22. N. S. McNutt, L. A. Culp, and P. H. Black, *J. Cell Biol., 56*:412 (1973).
23. L. A. Culp, in *Current Topics in Membranes and Transport* (R. Juliano and A. Rothstein, eds.), Vol. II, Academic Press, New York, 1978, p. 327.
24. B. J. Rollins, M. K. Cathcart, and L. A. Culp, in *The Glycoconjugates* (M. I. Horwitz, ed.), vol. III, Academic Press, New York, 1982, p. 289.
25. L. A. Culp, B. Murray, and B. J. Rollins, *J. Supramol. Str., 11*:401 (1979).
26. M. K. Cathcart and L. A. Culp, *Biochim. Biophys. Acta, 556*:331 (1979).
27. B. J. Rollins and L. A. Culp, *Biochemistry, 18*:141 (1979).
28. L. J. Miketo and L. A. Culp, *Arch. Biochem. Biophys., 221*:447 (1983).
29. L. A. Culp, B. J. Rollins, J. Buniel, and S. Hitri, *J. Cell Biol., 79*:788 (1978).
30. N. E. Stathakis and M. W. Mossesson, *J. Clin. Invest., 60*:855 (1977).
31. K. M. Yamada, D. W. Kennedy, K. Kimata, and R. M. Pratt, *J. Biol. Chem., 255*:6055 (1980).

32. E. Ruoslahti and E. Engvall, *Biochim. Biophys. Acta, 631*:350 (1980).
33. J. Laterra, R. Ansbacher, and L. A. Culp, *Proc. Natl. Acad. Sci. USA, 77*:6662 (1980).
34. J. Laterra and L. A. Culp, *J. Biol. Chem., 257*:719 (1982).
35. B. J. Barnhart, S. H. Cox, and P. M. Kraemer, *Exp. Cell Res., 119*:327 (1979).
36. P. J. Gill, J. Adler, C. K. Silbert, and J. E. Silbert, *Biochem. J., 194*:299 (1981).
37. D. D. Harrison and L. A. Culp, *Exp. Cell Res., 146*:29 (1983).
38. W. T. Chen and S. J. Singer, *J. Cell Biol., 95*:205 (1982).
39. K. Sekiguchi and S. Hakomori, *Proc. Natl. Acad. Sci. USA, 77*:2661 (1980).
40. M. D. Pierschbacher, E. G. Hayman, and E. Ruoslahti, *Cell, 26*:259 (1981).
41. M. Pierschbacher, E. G. Hayman, and E. Ruoslahti, *Proc. Natl. Acad. Sci. USA, 80*: 1224 (1983).
42. L. H. E. Hahn and K. M. Yamada, *Cell, 18*:1043 (1979).
43. J. Laterra, J. E. Silbert, and L. A. Culp, *J. Cell Biol., 96*:112 (1983).
44. T. F. Deuel, P. S. Keim, M. Farmer, and R. L. Heinrikson, *Proc. Natl. Acad. Sci. USA, 74*:2256 (1977).
45. V. C. Hascall, *J. Supramol. Str., 7*:101 (1977).
46. S. Tobey, K. J. McClelland, and L. A. Culp, *Exp. Cell. Res., 158*:395 (1985).
47. M. W. Lark, J. Laterra, and L. A. Culp, *Fed. Proc., 44*:394 (1985).
48. C. S. Izzard and L. A. Culp, *J. Cell Biol., 97*:123a (1983).
49. A. Rapraeger and M. Bernfield, in *Extracellular Matrix* (S. Hawkes and J. L. Wang, eds.), Academic Press, New York, 1982, p. 265.
50. C. Birchmeier, T. E. Kreis, H. M. Eppenberger, K. H. Winterhalter, and W. Birchmeier, *Proc. Natl. Acad. Sci. USA, 77*:4108 (1980).
51. W. T. Chen and S. J. Singer, *Proc. Natl. Acad. Sci. USA, 77*:7318 (1980).
52. B. A. Murray and L. A. Culp, *Exp. Cell Res., 131*:237 (1981).
53. J. A. Garner and L. A. Culp, *Biochemistry, 20*:7350 (1981).
54. B. J. Rollins and L. A. Culp, *Biochemistry, 18*:5621 (1979).
55. M. W. Lark and L. A. Culp, *J. Biol. Chem., 23*:14073 (1982).
56. L. L. Faltz, A. H. Reddi, G. K. Hascall, D. Martin, J. C. Pita, and V. C. Hascall, *J. Biol. Chem., 254*:1375 (1979).
57. A. Gonenne, and R. Ernst, *Anal. Biochem., 87*:28 (1978).
58. B. P. Toole, in *Cell Biology of Extracellular Matrix* (E. D. Hay, ed.), Plenum Publishing Corp., New York, 1981, p. 259.
59. M. W. Lark and L. A. Culp, *Biochemistry, 22*:2289 (1983).
60. J. Laterra, M. W. Lark, and L. A. Culp, in *Extracellular Matrix* (S. Hawkes and J. L. Wang, eds.), Academic Press, New York, 1982, p. 197.
61. M. W. Lark and L. A. Culp, *J. Biol. Chem., 259*:212 (1984).
62. H. C. Robinson, A. A. Horner, M. Höök, S. Ogren, and U. Lindahl, *J. Biol. Chem., 253*:6687 (1978).
63. L. Thumberg, G. Backstrom, A. Wasteson, H. C. Robinson, S. Ogren, and U. Lindahl, *J. Biol. Chem., 257*:10278 (1982).
64. J. J. Castellot, L. V. Favreau, M. J. Karnovsky, and R. D. Rosenberg, *J. Biol. Chem., 257*:11256 (1982).
65. L. Kjellen, I. Pettersson, and M. Höök, *Proc. Natl. Acad. Sci. USA, 78*:5371 (1981).
66. A. C. Rapraeger and M. Bernfield, *J. Biol. Chem., 258*:3632 (1983).
67. L. J. Lowe-Krentz and J. M. Keller, *Biochemistry, 22*:4412 (1983).
68. A. Oohira, T. N. Wight, J. McPherson, and P. Bornstein, *J. Cell Biol., 92*:357 (1982).
69. M. W. Lark and L. A. Culp, *J. Biol. Chem., 259*:6773 (1984).
70. T. Diehl, T. Scott-Burden, and W. Gevers, *Biochem. Int., 6*:29 (1983).
71. J. J. Hopwood and J. Elliott, *Biochem. Int., 6*:141 (1983).

72. B. Norling, B. Glimelius, and A. Wasteson, *Biochem. Biophys. Res. Commun., 103*: 1265 (1981).
73. W. J. Pledger, C. A. Hart, and W. R. Wharton, *Adv. Exp. Med. Biol., 138*:287 (1981).
74. L. A. Culp, R. Ansbacher, and C. Domen, *Biochemistry, 19*:5899 (1980).
75. G. Maresh, E. A. G. Chernoff, and L. A. Culp (submitted for publication).
75. G. Maresh, E. A. G. Chernoff, and L. A. Culp, *Arch. Biochem. Biophys., 233*:428 (1984).
76. L. A. Culp and C. Domen, *Arch. Biochem. Biophys., 213*:726 (1982).
77. C. Domen and L. A. Culp, *Exp. Cell Res., 134*:329 (1981).

9

Matrix Interactions of Cell Surface Proteoglycans

Alan Rapraeger,* Joy E. Koda,† and Merton Bernfield
Stanford University Medical Center,
Stanford, California

I. CELL SURFACE HEPARAN SULFATE PROTEOGLYCAN: ROLE IN CELL-MATRIX ANCHORAGE AND CELL BEHAVIOR

A. The Extracellular Matrix Is a Substratum

All adherent cells require a substratum to establish cell shape and to exhibit specific behavior, including gross control of DNA replication and cell division and messenger RNA transcription and translation (1-4). In vivo, the cellular substratum is the extracellular matrix (ECM), a complex of macromolecules that interact with one another to form a highly ordered scaffolding and also interact with cell surface components to provide cell anchorage (5,6). However, particularly in developing organs, the ECM is not static. It undergoes changes in stability and composition directly mediated by interactions between closely apposing tissues, e.g., epithelia and adjacent mesenchyme (7-11). These changes often correlate with changes in cell shape and cell behavior. Ultimately, in differentiated tissues, the specific composition of the substratum, molded by previous tissue interactions, probably contains specific positional information, which not only may exert control over gross cellular functions by providing adhesion, but also may affect, if not direct, the cell's differentiated behavior. Therefore, this cell behavior is dependent on matrix receptors that (a) anchor the cell, (b) recognize specific matrix composition, and (c) translate this information to the cell.

B. Cells Have Multiple Matrix Receptors

Although only a few have been described, cells may have a battery of matrix receptors anchored in the plasma membrane. Reports to date include a receptor specific for

**Present affiliation*: University of Wisconsin, Madison, Wisconsin.
†*Present affiliation*: Hybritech Incorporated, San Diego, California.

laminin (12,13) an adhesive glycoprotein localized to basement membranes, and receptors for interstitial collagens, apparently present on platelets (14), chondrocytes (15), fibroblasts (16,17), hepatocytes (18), and corneal epithelial cells (19). Specific receptors probably exist for other adhesive glycoproteins as well, for example, for fibronectin (20) or chondronectin (21).

The matrix receptors displayed at the cell surface may differ with cell type or during the life of a single cell. Epithelial cells encounter a different matrix than mesenchymal cells and thus may require a special set of receptors to recognize their substratum, and changes in the composition of the ECM during organ development must be reflected in the subset of receptors actively participating in matrix recognition.

C. Cell Surface Proteoglycans Are Potential Matrix Receptors

Cell surface heparan sulfate proteoglycans are excellent candidates to be matrix receptors. Heparan sulfate proteoglycans bind to several ECM components, including fibronectin (22-24), laminin (25,26), interstitial collagen types I, III, and V (24,27-29), and other heparan sulfate proteoglycans (30), and behave as intercalated membrane proteins, thus potentially functioning as integral components of the ECM and the plasma membrane simultaneously (reviewed below). In addition, they may associate with the intracellular cytoskeleton, thus transmitting matrix organization directly into cytoskeletal organization, cell shape, and, ultimately, cell behavior.

Cell surface proteoglycans were originally described by Kraemer (31,32) for CHO cells, and similar reports for hepatic (33), hepatoma (34), glial (35), fibroblastic (36-38), vascular endothelial (39), mammary epithelial (40), and nerve cells (41), to name a few, indicate that the molecules are ubiquitous. Although some cell surface heparan sulfate proteoglycans are peripheral cell surface components, displaced from the surface by heparin (42,43), others are not displaced by heparin, have lipophilic properties, and are believed to be anchored directly into the plasma membrane (40,43-47).

In addition to the ubiquitous cell surface heparan sulfate proteoglycans, other classes of cell surface proteoglycans exist, including chondroitin sulfate (48-50) and hyaluronic acid (51) proteoglycans. Considerable diversity may exist within these classes, possibly reflecting variations in function. For example, heparan sulfate cell surface proteoglycans from the cell types mentioned above range widely in molecular mass, apparently owing to variation in the size and number of the GAG chains and diversity of their core proteins. Also, the heparan sulfate GAG chains themselves can display microheterogeneity, which affects their ability to self-associate (30) as well as their recognition of other biological molecules, including both matrix components and nonmatrix components, such as LDL [see Lindahl and Höök (52); Höök et al., (53)]. The potential diversity increases at the cellular or tissue level if it is considered that many cells contain more than one type of cell surface proteoglycan, that cell surface proteoglycans can display more than one type of GAG chain (54), and that this may change with substratum type or degree of anchorage. For example, blood-borne leukocytes have chondroitin sulfate on their surfaces but apparently replace it with heparan sulfate when they adhere to a substratum (55).

Rather than attempt to encompass this diversity in structure and function, much of which remains unknown, we have focused on a single cell surface proteoglycan from NMuMG mouse mammary epithelial cells, described its properties, and speculated on the function of this proteoglycan, a function that may be extrapolated to other cell surface forms.

II. CELL SURFACE HEPARAN SULFATE PROTEOGLYCAN BEHAVES AS AN INTEGRAL MEMBRANE PROTEIN

Cultured NMuMG mouse mammary epithelial cells incorporate radiosulfate into a variety of heparan sulfate-rich proteoglycans (56,57). The major proteoglycan fraction associated with these monolayers is a lipophilic proteoglycan that behaves as though intercalated into the plasma membrane(40,45,58).

A. Proteoglycan Associates with Lipid Vesicles

When 4 M guanidinium chloride (GdnHCl) extracts of $^{35}SO_4$-labeled cells are chromatographed on Sepharose Cl-4B in 4 M GdnHCl, $^{35}SO_4$-labeled proteoglycan elutes as a single included peak at a K_{av} of 0.33 and as an aggregate in the void volume of the column (Figure 1). The aggregate is destroyed by extraction in organic solvents or by detergents, resulting in elution of all the proteoglycan at K_{av} of 0.33. This suggests that some of the proteoglycan is lipophilic and is extracted in GdnHCl as an aggregate with membrane lipid.

Intercalation into liposomes indicates that the proteoglycan is indeed lipophilic. When delipidated proteoglycan in 4 M GdnHCl is recombined with phospholipid in the presence of a dialyzable detergent, then dialyzed to remove the detergent, the proteoglycan intercalates quantitatively and vectorially into the forming lipid vesicles (Figure 2). This can be assessed either by an increased size, increased sedimentation velocity, or decreased buoyant density of the proteoglycan due to the association with the liposome. For example, as described above, when the lipophilic proteoglycan is extracted with organic solvents, it chromatographs at a K_{av} of 0.33 in 4 M GuHCl in the presence or absence of detergent. However, following insertion of the proteoglycan into liposomes, it elutes once more in the void volume of the column but returns again to a K_{av} of 0.33 in the presence of detergent. Association of the proteoglycan with liposomes is a selective property of only cell-associated proteoglycans; none of the extracellular proteoglycan synthesized by these cells, including proteoglycans from the culture medium or accumulating beneath the cells, will associate with liposomes.

The lipophilic proteoglycan intercalates vectorially into the liposomes. Mild treatment with trypsin (20 μg/ml, 4°C, 5 min) releases all of the GAG-bearing portion of the proteoglycan as a domain that is only slightly smaller in size compared to the parent proteoglycan (Figure 2). However, this released domain will no longer associate with liposomes, suggesting that it has been cleaved from a hydrophobic region of the core protein that anchors the intact proteoglycan in the lipid bilayer. Thus, trypsin cleaves the intercalated proteoglycan into at least two functional domains: a domain that is lipophilic and may anchor the proteoglycan in membranes and an external domain that bears the GAG chains.

B. Lipophilic Proteoglycan Is at the Cell Surface

The majority of the lipophilic proteoglycan is at the cell surface (40). When intact cells are treated with trypsin in a manner identical to the trypsin treatment of liposomes bearing intercalated proteoglycan, 55% of the total cellular proteoglycan is released, indicating that it is at the cell surface (Table 1). Analogous to other cell types, this proteoglycan may be a peripheral cell surface component, e.g., bound to other cell surface molecules via its core protein or GAG chains, or, based on its lipophilic nature, it may be an integral

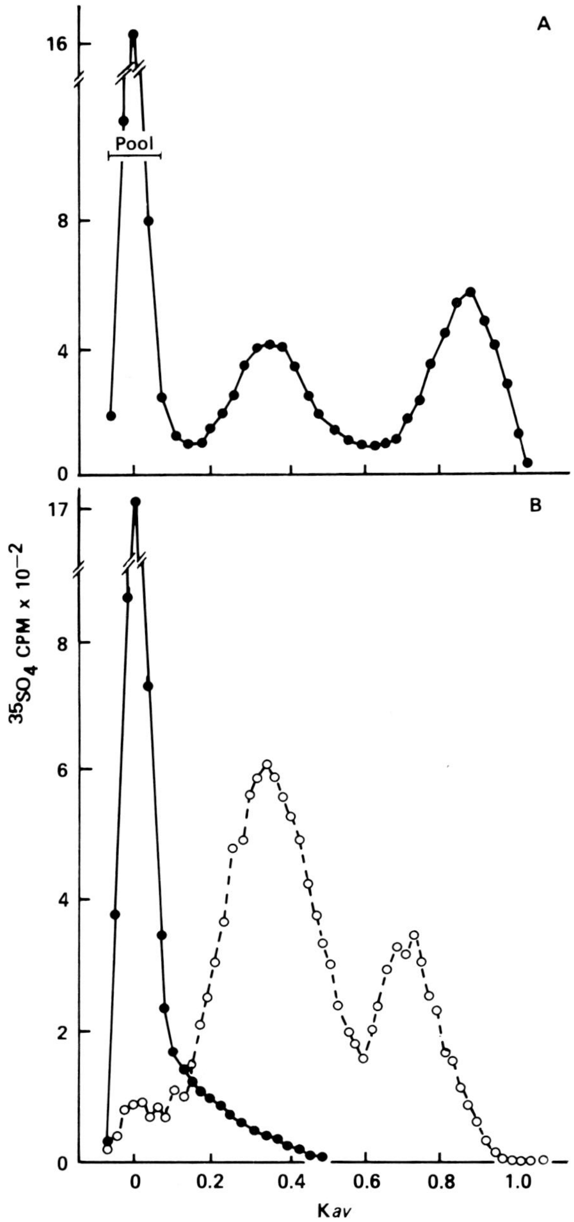

Figure 1 A fraction of the cellular heparan sulfate proteoglycan chromatographs as a detergent-labile aggregate. Monolayers of NMuMG mouse mammary epithelial cells were labeled for 24 hr with radiosulfate, released from the dish, and extracted in acetate-buffered 4 M GdnHCl. (A) The extract was chromatographed in acetate-buffered 4 M GdnHCl on Sepharose Cl-4B. The radiosulfate distributes into three fractions: an aggregate in the void volume (K_{av} of 0), a proteoglycan fraction with a K_{av} of 0.33, and a fraction composed of intracellular heparan sulfate GAG fragments (K_{av} of 0.85) (40,45). (B) The void volume fraction (K_{av} of 0) was pooled as indicated and rechromatographed in acetate-buffered 4 M GdnHCl in the presence (○) or absence (●) of 1% Triton X-100. The detergent treatment resolves the aggregate into a proteoglycan fraction (Kav of 0.33) and an unidentified component that behaves as a sulfated lipid. (From Ref. 45.)

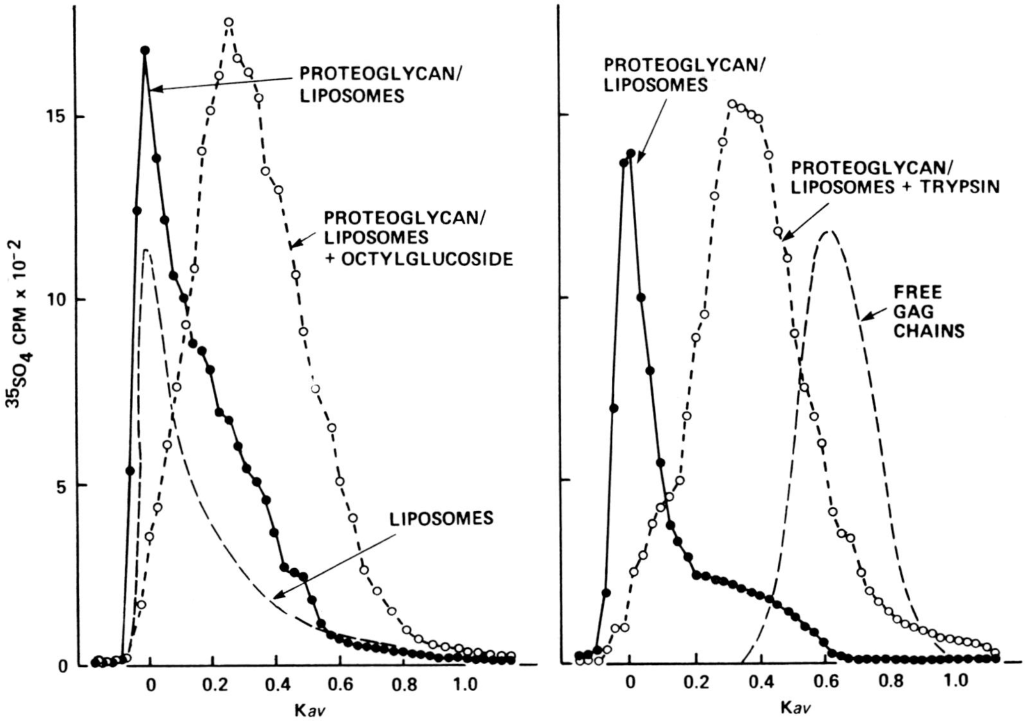

Figure 2 Demonstration of proteoglycan intercalation into liposomes. Proteoglycan was extracted from NMuMG cells labeled with $^{35}SO_4$, delipidated, and combined with octylglucoside and purified phospholipid. When the detergent is removed by dialysis, proteoglycan is associated with the newly formed liposomes, as shown by chromatography on Sepharose Cl-4B in 4 M GdnHCl. On the left, the elution of the proteoglycan liposomes (●—●) is compared with that of liposomes alone (- - -) and with that of the proteoglycan/liposomes chromatographed in octylglucoside, which destroys the liposomes. On the right, the elution of the proteoglycan/liposomes is shown before (●—●) and after (○—○) treatment with trypsin (20 μg/ml, 4°C, 5 min). Profile of free GAG chains is shown for a size comparison. The proteoglycan intercalates with the portion of its core protein bearing the GAG chains exposed to the exterior. Trypsin releases the exposed domain from a putative membrane-intercalated domain. The released domain has not been degraded to free GAG chains; rather, it is a discrete proteoglycan that is just slightly smaller than the parent proteoglycan. (From Ref. 11.)

membrane protein. Peripheral proteoglycans bound to the cell surface via their GAG chains can be displaced by competition with heparin (42,43), a GAG structurally similar to heparan sulfate. However, treatment of the NMuMG cells with heparin or up to 1 M KCl does not displace proteoglycan, suggesting that it is not associated with the cell surface via its GAG chains.

All of the cell surface proteoglycan fraction is lipophilic and therefore may be embedded in the plasma membrane. Estimates based on liposome intercalation indicate that 65% of the total cellular proteoglycan is lipophilic and that 80% of this lipophilic proteoglycan is at the cell surface and susceptible to removal by trypsin (Table 1). Moreover, this fraction accounts for all of the proteoglycan susceptible to trypsin. Trypsin

Table 1 All of the Cell Surface Proteoglycan is Lipophilic

Cellular proteoglycan fraction	% of total
Total cellular proteoglycan	(100)
Lipophilic proteoglycan	
Cell surface	52–55
Intracellular	10–13
Nonlipophilic proteoglycan, intracellular	35

$^{35}SO_4$-labeled NMuMG cells were extracted in 4 M GdnHCl before and after trypsin treatment to cleave cell surface proteoglycan. Lipophilic proteoglycan in the extracts was then quantitated by liposome intercalation. Approximately 80% of the lipophilic proteoglycan is at the cell surface where it comprises all of the trypsin-susceptible cell surface fraction. The remaining proteoglycan is apparently intracellular.
Source: After Rapraeger and Bernfield (40).

releases a proteoglycan domain that is reduced only slightly in size in comparison with the intact cell surface proteoglycan, has lost the ability to associate with liposomes, and is identical in these properties to the domain cleaved by trypsin from liposomes bearing intercalated proteoglycan (Figure 3). We refer to this proteoglycan domain as the "ectodomain" of the cell surface proteoglycan. These data suggest that all of the cell surface proteoglycan from these cells exists as an integral membrane protein.

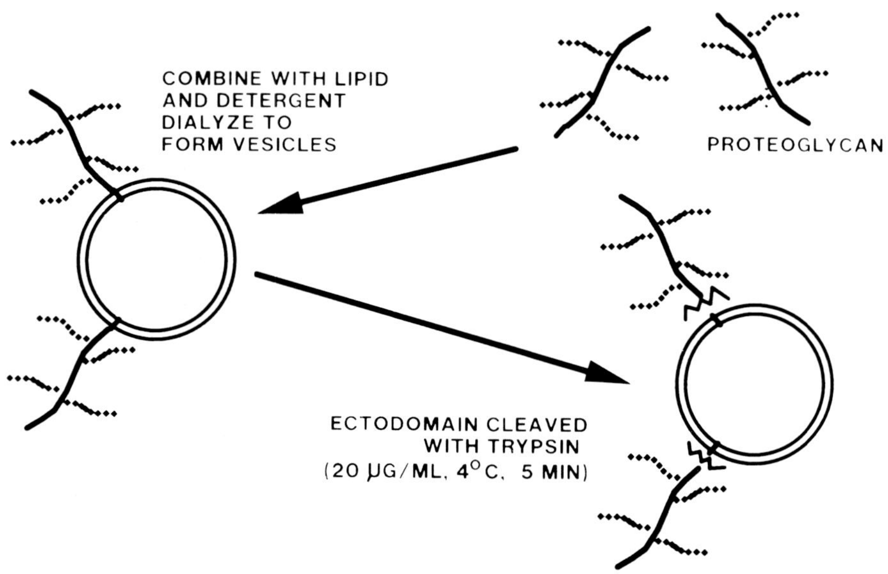

Figure 3 Cell surface proteoglycan intercalates into liposomes and its ectodomain is released by treatment with trypsin. (After Ref. 80.)

C. Immunolocalization of Proteoglycan on the Cell Surface

A monoclonal antibody that recognizes the core protein of the lipophilic proteoglycan binds to the surface of NMuMG cells (58,59). Hybridoma 281-2 was generated by fusing a myeloma cell line with spleen cells isolated from rats previously injected with whole NMuMG epithelial cells. Hybridomas were screened using $^{35}SO_4$-labeled NMuMG cell proteoglycans, leading to the selection of an antibody that reacts with the intact lipophilic proteoglycan as well as its isolated ectodomain. This antibody is localized on cells with a second, fluorescein-conjugated antirat IgG. Fixed cells in monolayers show pericellular staining at their apical surfaces and at their margins where the cells abut (Figure 4A). Mild treatment with trypsin (20 μg/ml, 4°C, 10 min) sufficient to remove the ectodomain of the cell surface proteoglycan abolishes the staining. Enzymatic removal of the GAG chains alone fails to do so, consistent with other evidence that (a) the antigenic determinant is the core protein, and (b) the proteoglycan does not rely on its GAG chains for anchorage to the cell.

Restriction of antibody staining to the apical cell surface of monolayers appears to be due to a lack of accessibility to the basal surface of these cells, probably owing to their formation of lateral junctional complexes. When cells in monolayers are rounded by a brief treatment with 0.5 mM EDTA (37°C, 5 min), the 281-2 antibody stains the entire cell surface. Again, this staining is abolished by mild treatment with trypsin.

D. Cross-Linked Proteoglycan Forms Cell Surface Clusters

The presence of proteoglycan on the apical cell surface allows the study of proteoglycan mobility in the plasma membrane in the absence of a potentially complicating extracellular matrix. This mobility is demonstrated by cross-linking the proteoglycan on unfixed cells with an immune complex generated by the monoclonal antibody 281-2 together with a second, antirat IgG (58). Proteoglycan cross-linked by this complex is assimilated into clusters (Figure 4B). Cluster formation is independent of energy, occurring at 4°C and in the presence of azide, and is not prevented by the cytoskeleton-disrupting drugs cytochalasin D and colchicine. This suggests that the proteoglycan is capable of at least short-range diffusion in the plane of the plasma membrane.

When the cells are warmed to 37°C following cluster formation, the clusters (a) aggregate into larger clumps, (b) are removed from the lateral borders of the cells and accumulate over the nuclear region, and (c) are ultimately lost from the cell surface (Figure 4C). This behavior is similar to classical patching and capping of membrane proteins on motile or suspended cells, such as lymphocytes (60,61). The aggregation of the proteoglycan following cluster formation is energy dependent, being blocked by cold temperature or azide, is inhibited by cytochalasin D (Figure 4D), but is unaffected by colchicine. In addition, the clustered proteoglycans remain immobile even after 4 hr at 37°C as long as cytochalasin D is present. This suggests that the lateral diffusion rate of the proteoglycan is reduced dramatically once assimilated into clusters, or that the clusters are tethered to the cytoskeleton and are immobile if the motive force of the cytoskeleton is inactivated.

The clustering of the proteoglycan at 4°C and the movement that occurs at 37°C requires cross-linking of the proteoglycan. As expected, the monoclonal antibody alone is an ineffective cross-linking ligand owing to its recognition of only a single antigenic determinant on the proteoglycan; following treatment with the monoclonal alone, the proteoglycan remains in a uniform distribution, does not form clusters or aggregates, and

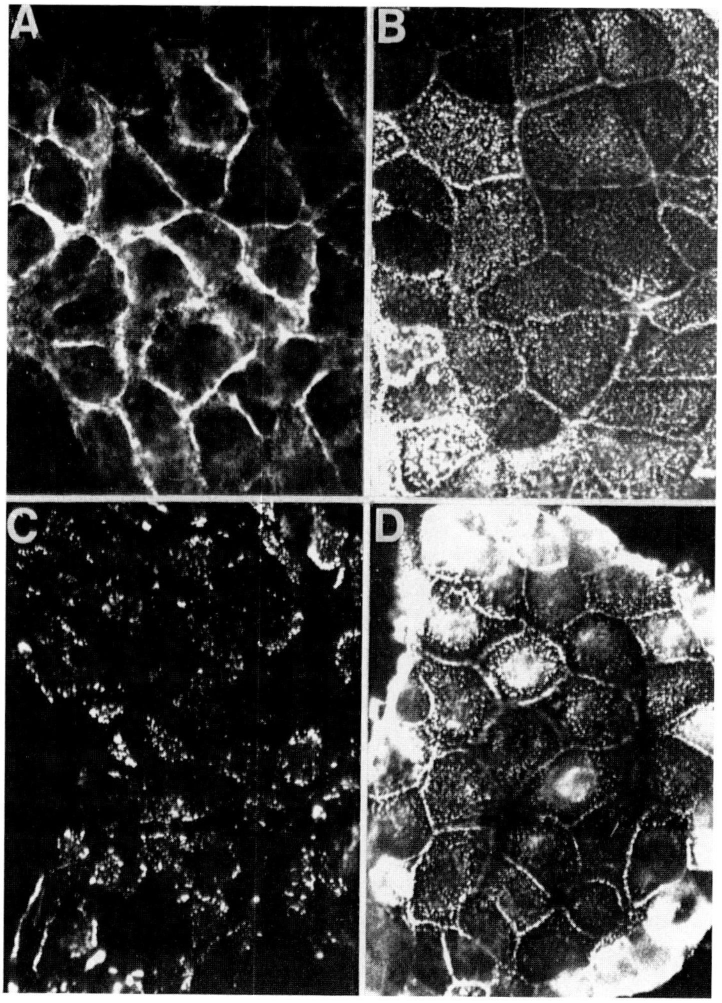

Figure 4 Cross-linked proteoglycan forms cell surface clusters. Proteoglycan is localized on the cell surface by treatment at 4°C with monoclonal antibody 281-2 followed by a second fluorescein-conjugated antibody. (A) Cells are fixed prior to addition of the second antibody; (B) cells are fixed following addition of the second antibody, which effectively cross-links the proteoglycans into cell surface clusters. (C) Cells are warmed to 37°C following generation of the clusters and incubated for 60 min; (D) cells are incubated as in C, but in the presence of cytochalasin D. Cell surface proteoglycan is mobile in the membrane at 4°C, but becomes immobile at this temperature following cluster formation. However, if warmed to 37°C, the clusters form larger aggregates, apparently dependent on the actin-containing cytoskeleton, and then appear to be lost from the surface. (Modified from Ref. 58.)

is not removed from the surface to an extent detectable by this method. Thus, the altered distribution is dependent on cross-linking provided by addition of the second anti-rat polyclonal antibody.

E. Cell Surface Proteoglycan Resists Detergent Extraction

Intercalation of the cell surface proteoglycan into the plasma membrane is suggested by its unique lipophilic nature, its release from the cell surface by proteases that remove its lipophilic region but resistance to release by KCl, heparin, or GAG degrading enzymes, and its mobility when cross-linked by exogenous ligands. However, extraction of the cells with detergents suggests that the proteoglycan is not anchored solely by its association with membrane lipid (58). When NMuMG cells labeled to the steady state with radiosulfate are extracted with 1% Triton X-100 at pH 5.0 to dissolve the membrane lipids, three-quarters of the cellular proteoglycan resists removal and remains bound to the cytoskeleton-rich residue. Intracellular proteoglycan and GAG fragments are released by the Triton treatment, but the cell surface proteoglycan, defined by release from intact cells by trypsin treatment, is not.

The Triton-resistant proteoglycan is released from its anchorage by KCl treatment. Treatment with 0.5 to 1.5 M KCl, together with 1% Triton X-100, removes three-quarters of the cell surface proteoglycan that resists Triton alone. However, treatment with identical KCl concentrations in the presence of saponin, a detergent that only disrupts cholesterol-rich areas of the membrane, thus failing to totally extract the membrane but permeabilizing it to allow KCl entry to the cell interior, does not release the proteoglycan even when followed with Triton. Thus, to be effective, the KCl treatment must follow the Triton extraction, suggesting that KCl must gain access into the membrane itself to release the proteoglycan, perhaps by disrupting an association of the proteoglycan with another integral membrane component.

The resistance to Triton extraction is pH dependent. At pH 7.5, the detergent-resistant cell surface proteoglycan, defined by trypsin treatment, is reduced more than threefold in comparison to extraction at pH 5.0. A similar result is observed when the cell surface proteoglycan is detected with the monoclonal antibody 281-2. Treatment of the cells with the antibody, followed by Triton extraction, fixation, and localization with a fluorescent second antibody, demonstrates that cell surface staining is unaffected by Triton extraction at pH 5.0, but is greatly reduced by extraction at pH 7.5.

The pH dependence of cell surface proteoglycan extraction by Triton is not seen when the cell surface proteoglycan is cross-linked into clusters and aggregates. When clusters are formed by cross-linking with the monoclonal 281-2 and a fluorescent second antibody, the clusters apparently remain unchanged when extracted with Triton at pH 5.0 or 7.5. This suggests that proteoglycan clusters become associated with the cytoskeleton, similar to reports for other membrane proteins (62,63) and consistent with our findings that aggregation of these clusters is dependent on the cytoskeleton. However, whereas cytochalasin D prevents the movement of the clusters into aggregates, treatment with this drug does not cause their release when extracted with Triton. This suggests that their immobilization may be more complex than a simple association with actin. However, the Triton-resistant proteoglycan in clusters is displaced by 0.5 M KCl, similar to nonclustered proteoglycan.

Thus, the cell surface proteoglycan appears to be anchored in the plasma membrane by more than an association with lipid alone. This association is stabilized at pH 5.0 or at higher pH values if the proteoglycan becomes clustered. Consistent with the apparent immobilization of the clusters (Section II.D), cytoskeletal and membrane elements, perhaps including, but not limited to, actin, may have a role in anchorage of the proteoglycan at the cell surface.

III. CELL SURFACE HEPARAN SULFATE PROTEOGLYCAN BINDS THE EXTRACELLULAR MATRIX VIA ITS ECTODOMAIN

A. The Ectodomain Is Released by Mild Treatment with Protease

Trypsin treatment of the lipophilic proteoglycan, whether intercalated into liposomes or at the cell surface, does not reduce the proteoglycan to free GAG chains, but removes a discrete ectodomain bearing all the GAG chains (40) (cf. Figure 3). Chymotrypsin treatment (50 μg/ml, 4°C, 5 min) provides a similar result despite its differing specificity. This suggests that the cell surface proteoglycan contains a protease-labile region between the membrane-anchored and GAG-bearing domains of its core protein (Figure 5). Comparison of the ectodomain with the parent membrane-intercalated proteoglycan on gel permeation chromatography or polyacrylamide gel electrophoresis indicates that it is slightly smaller in size. At least part of the reduction in size is due to the removal of the lipophilic domain, as shown by the failure of the isolated ectodomain to intercalate into liposomes and its increased bouyant density.

B. The Cell Surface Proteoglycan Is a Hybrid

Analysis of $^{35}SO_4$-labeled GAG on the cell surface proteoglycan indicates that it is a unique hybrid molecule, bearing predominantly heparan sulfate, but also chondroitin sulfate (54), at a ratio of one or two chondroitin sulfate chains for every four heparan sulfate chains (Figure 5). Attempts to separate distinct proteoglycans by either sizing, ion exchange or collagen affinity chromatography have been unsuccessful, and more direct experiments indicate that the two GAG species are attached to a common protein: (a) the proteoglycan is reduced in size by the removal of either GAG type, (b) the bouyant density of the chondroitin sulfate moiety is reduced following enzymatic degradation of the dense heparan sulfate chains, (c) the affinity of the proteoglycan for DEAE-cellulose is modified by removal of either GAG type, and (d) the monoclonal antibody 281-2, which recognizes the core protein, isolates proteoglycan bearing both GAG types. Although it is possible that two distinct proteoglycans are initially synthesized, each bearing one or the other GAG species, and are then cross-linked, attempts to abolish disulfide cross-links have no effect on the size of this proteoglycan. Other cross-linking mechanisms have not been ruled out, but the most likely explanation is that the GAGs are initially attached to the same core protein. Because the same glycosidic linkage region links these GAG chains to the core protein (52), this hybrid proteoglycan raises interesting questions as to how the attachment of these chains is controlled. It may be regulated by specific polypeptide sequences in the core protein or by enzymes or precursors specific for the linkage region of one or the other GAG type.

C. The Isolated Ectodomain Binds Interstitial Collagens

The heparan sulfate-rich proteoglycans from the NMuMG cells bind interstitial collagens, demonstrated with the ectodomain of the cell surface proteoglycan (29) and with an extracellular proteoglycan obtained in a soluble form when NMuMG monolayers are suspended by scraping (28). The proteoglycans bind to native collagen fibrils with high

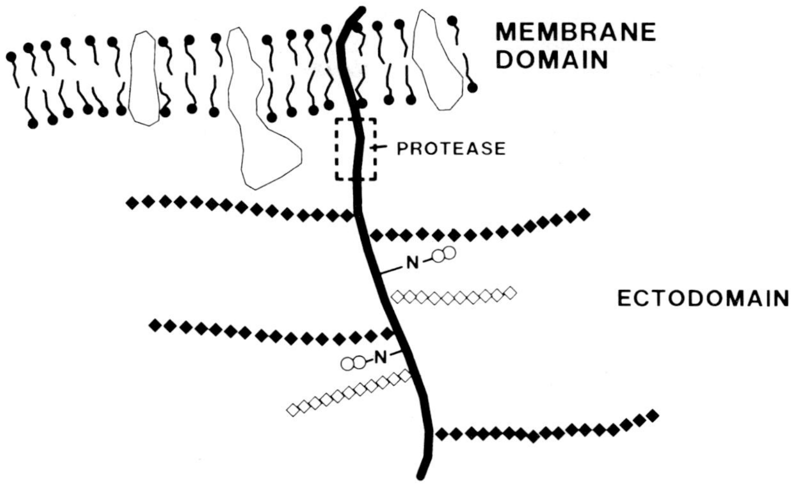

Figure 5 Model of the cell surface heparan sulfate-rich proteoglycan of NMuMG mouse mammary epithelial cells. The proteoglycan is postulated to be anchored in the plasma membrane, possibly by a lipophilic *membrane domain* of its core protein. The *ectodomain*, which bears chondroitin sulfate and heparan sulfate GAG chains, is released from its membrane anchor as a discrete proteoglycan by treatment with mild protease. One or two chondroitin sulfate chains (◇) (M_r ca. 17,000) are attached for every four heparan sulfate chains (◆) (M_r ca. 36,000). The overall size of the cell surface proteoglycan estimated by polyacrylamide gel electrophoresis is ca. 250,000. In addition to the GAG chains, *N*- or *O*-linked oligosaccharides may be attached to the core protein as described for other proteoglycans. (See Ref. 81.)

affinity and specificity. For example, under physiological conditions, proteoglycan will bind to type I collagen fibrils adsorbed to polyvinylchloride wells, but will not recognize heat-denatured collagen (gelatin). Scatchard analysis of proteoglycan binding to various type I collagen fibril concentrations in a centrifugation assay indicates that the dissociation constant for this interaction is circa 10^{-9} M.

Fibrils of native type I collagen or its pepsinized derivative lacking telopeptides bind the proteoglycan equivalently, suggesting that the binding is not to the telopeptide regions. However, binding to pepsinized collagen can be reduced in some cases, e.g., PVC well assays, possibly owing to an inability to maintain a native fibrillar structure.

Whereas the binding parameters have been discerned primarily using type I collagen, the proteoglycan also binds to other interstitial collagen types. For example, under physiological pH and ionic strength conditions, the ectodomain binds to collagen types I, III, and V. However, it fails to bind to types II or IV.

Binding of the proteoglycan to collagen appears to be via the GAG chains. Proteoglycan can be displaced from type I collagen fibril affinity columns by salt, although requiring concentrations in excess of physiological levels, and divalent anions such as sulfate are approximately 10-fold more effective than monovalent anions. However, a select group of sulfated polyanions are the most effective competitors, and these are similar in structure to heparan sulfate. For example, heparin, a highly sulfated polyanion and a close cousin to heparan sulfate, displaces proteoglycan from collagen affinity columns at near-micromolar concentrations, a millionfold more effective than free sulfate alone. Dextran sulfate, a synthetic polymer with a degree of sulfation similar to heparin, is also an effective competitor. However, chondroitin sulfates or dermatan sulfate is totally ineffective, even at 100-fold greater concentrations. These latter GAGs do bind to collagen (52), but apparently at sites distinct from the heparan sulfate proteoglycan binding site.

An explanation for the ineffectiveness of these GAGs to compete for the proteoglycan binding site may lie in the sulfation pattern of the GAGs. Whereas heparin, dextran sulfate, and heparan sulfate potentially have blocks of sulfate residues, sometimes containing three to four sulfates per disaccharide, interrupted by regions of low sulfation, the chondroitin sulfates have a uniform sulfate distribution and rarely more than one per disaccharide. These block regions present in heparan sulfate may recognize highly specific sites on the collagen molecule.

These data suggest that binding of the cell surface proteoglycan is mediated via the heparan sulfate chains. ABCase-removal of the chondroitin sulfate chains does not abolish the binding. Also, the isolated heparan sulfate chains do bind to collagen; however, the salt concentration must be lowered to less than physiological levels to demonstrate this interaction. Therefore, the proteoglycan-collagen interaction (a) requires that both molecules be in an intact, native conformation, suggesting that several heparan sulfate chains bind cooperatively to collagen fibrils, and (b) indicates that the binding region of type I collagen is selective for only heparinlike GAG and an analogous binding region is apparently present on collagen types III and V.

D. Cells Duplicate Proteoglycan Binding to Collagen

Intact, suspended NMuMG cells bearing the cell surface proteoglycan bind to collagen with the same specificity as the isolated ectodomain (29). This is assessed by coating microtiter wells with type I collagen, incubating the wells with cell suspensions, and then sedimenting the unbound cells by centrifugation. In wells where no binding occurs, such as wells coated with BSA or denatured type I collagen, the cells form a "button" at the bottom of the well (Figure 6). However, in wells coated with native collagen fibrils, most of the cells adhere to the entire surface of the well, hence no or a reduced button following centrifugation (Figure 6). Similar to the binding of isolated proteoglycan to collagen affinity columns, the cells fail to bind to collagen-coated wells in the presence of heparin (Figure 6) or dextran sulfate (not shown), but cannot be displaced by chondroitin 4- or 6-sulfate or dermatan sulfate (Figure 6). These data suggest that (a) the cells bind via the cell surface proteoglycan, and (b) binding of the proteoglycan in its native conformation, i.e., intercalated into the plasma membrane, is as effective as the isolated ectodomain alone.

IV. CELL SURFACE PROTEOGLYCAN IS AFFECTED BY CELL-SUBSTRATUM ANCHORAGE

The cell surface proteoglycan of the NMuMG cells turns over at variable rates, apparently depending on the state of the cells' anchorage to the substratum. The mechanism of loss from the surface appears to be twofold: (a) endocytosis followed by intracellular degradation and (b) shedding to the extracellular space.

A. Proteoglycan Is Endocytosed and Degraded

When confluent monolayers of NMuMG cells grown on tissue culture plastic are labeled with $^{35}SO_4$ to the steady state and then chased in the absence of a radioactive precursor,

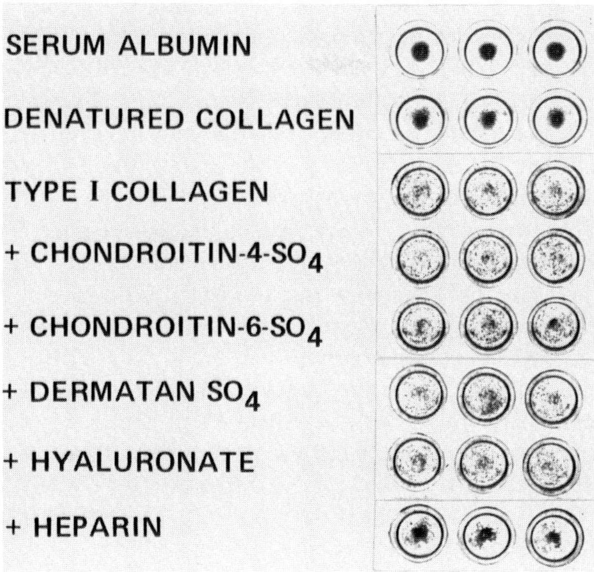

Figure 6 Binding of mammary epithelial cells. NMuMG cells duplicate the collagen-binding properties of the cell surface proteoglycan. Aliquots of suspended cells were incubated with microtiter wells previously coated with bovine serum albumin, denatured type I collagen, or native type I collagen. In addition, the following polyanions were combined with the cells to test for inhibition of binding to wells coated with native type I collagen: 100 μg/ml chondroitin-4-sulfate, 100 μg/ml chondroitin-6-sulfate, 100 μg/ml dermatan sulfate, 100 μg/ml hyaluronic acid, and 1 μg/ml heparin. The microtiter dish was centrifuged, and the cells were fixed and stained. The cells form a "button" at the bottom of the well if binding does not occur, whereas if binding takes place, they coat the surface of the treated well and do form a button. Similarly to isolated cell surface proteoglycan, the cells bind to native but not denatured type I collagen and not to bovine serum albumin. Also, heparin disrupts binding to collagen. (See Ref. 29.)

radiolabeled products accumulate in the medium (56,57). These products include proteoglycan and some peptide-bound GAG chains but primarily inorganic sulfate, indicating that sulfated macromolecules are being degraded. Because the sulfated proteoglycan and GAG found in the medium are neither taken up by the cells nor degraded in the medium, the appearance of inorganic sulfate suggests active intracellular degradation of cell-associated proteoglycans.

Concomitant with the appearance of these products in the medium, the cell surface proteoglycan turns over with a half-life of 6 hr (Rapraeger and Bernfield, in preparation). Pulse label and chase studies indicate that the turnover is accompanied by the appearance of intracellular heparan sulfate GAG fragments (Rapraeger and Bernfield, in preparation), generated by a heparitinase either in endosomes or in lysosomes [see Yanagishita and Hascall, (64)], which are then further degraded in lysosomes, releasing the radiolabel to the medium as inorganic sulfate.

The degradation of the cell surface proteoglycan suggests that it is in a continuous pathway of synthesis and transport to the plasma membrane as a membrane-bound form, with 80% of this form in residence at the cell surface at the steady state, followed by internalization and degradation. This pathway is similar to that described for other integral membrane proteins, primarily cell surface receptors.

Binding of an extracellular ligand may dramatically influence the half-life of the cell surface proteoglycan. For example, cross-linking the proteoglycan by soluble ligands (antibodies) causes cluster formation, aggregation, and ultimate loss from the cell surface (cf. Figure 4), similar to down-regulation described for hormone receptors. In contrast to a soluble ligand, anchorage to an insoluble extracellular matrix may also generate clusters of cell surface proteoglycan, but the insolubility of the matrix would prevent aggregation and loss. When NMuMG cells are grown on hydrated gels of type I collagen, a matrix that avidly binds the proteoglycan, a dense, ruthenium red-staining layer is deposited between the cells and the gel (65), suggesting the accumulation of proteoglycan. Biochemical studies indicate that heparan sulfate proteoglycans do indeed accumulate in response to the collagen and the accumulation is directly due to an decrease in their degradation rate (56), which now approaches a half-life of six to eight days compared to 6 hr on tissue culture plastic. Therefore, the cell surface proteoglycan turns over, apparently via internalization and intracellular degradation and similar to other integral membrane receptors, but the nature of the cell's substratum may have dramatic effects on the rate at which this occurs.

B. Proteoglycan Is Shed from Rounded Cells

When confluent monolayers of NMuMG cells are disrupted by brief incubation with warm EDTA, leading to cell rounding, the cell surface proteoglycan is lost directly to the medium with a half-life of approximately 40 min (unpublished observations). This may be similar to the desquamation of cell surface heparan sulfate by CHO cells preparing to divide (66). The shed proteoglycan is similar in size to the native cell surface proteoglycan, but is not lipophilic, indicating that it is the GAG-bearing ectodomain, presumably released via some cleavage mechanism. Loss is prevented at low temperatures, but is not prevented by azide, EDTA, or any of the protease inhibitors used to date.

Shedding of the cell surface proteoglycan when the cells are rounded may duplicate a process that occurs in vivo. Clearly, anchorage to the extracellular matrix is requisite for generating cytoskeletal organization and maintaining cell shape. But the cell must also divest itself of this anchorage when shape changes are required, e.g., when moving or rounding tc divide, and the cell may accomplish this by cleaving its matrix anchor. This mechanism would involve just a single cell and the temporary loss of a single receptor type rather than destruction of the entire matrix.

V. FUTURE OF THE CELL SURFACE PROTEOGLYCAN

A. Potential Role as a Cell-Basement Membrane Anchor

Examination of the properties of the cell surface heparan sulfate-rich proteoglycan of NMuMG mammary epithelial cells suggests that (a) the proteoglycan is a mobile, integral membrane protein present at the apical and basolateral surface of dividing cells but sequestered basolaterally on quiescent cells, (b) its membrane-anchored domain associates with other integral membrane and/or cytoskeletal components, particularly when the proteoglycan is assimilated into clusters, (c) the proteoglycan binds avidly to interstitial collagens and to other ECM components as well, and (d) its residence at the cell surface is dependent on cell-substratum anchorage. These properties suggest that heparan sulfate proteoglycan at the surface of epithelial cells anchors the epithelial cell sheet to the ECM.

In organs, the ECM of epithelia is organized as a basal lamina, which separates the parenchymal cells from underlying stromal tissue. The basal lamina, which is comprised almost solely of components synthesized by the epithelium, interfaces with the reticular lamina, containing fibronectin and interstitial collagens synthesized by the closely apposed stroma; the basal lamina and reticular lamina together comprise the basement membrane (11) (Figure 7).

The basal lamina is composed of proteoglycan and collagenous and noncollagenous proteins (67,68). The most common proteoglycan, at least in adult tissues, is heparan sulfate (69-71). However, chondroitin sulfate has also been described as a laminar component, as has hyaluronic acid (72,73), although these molecules may be more prevalent in embryonic tissues. The basement membrane proteins include laminin (74), a high-molecular-weight protein composed of two disulfide bonded subunits; type IV collagen (75), postulated to form a two-dimensional latticelike network; entactin (76), a small sulfated glycoprotein; nidogen; and possibly fibronectin, a dimer probably donated by the stroma in which it is more prevalent (20,77). In addition, the stroma contains the fibrillar collagen types I, III, and V (78). It is likely that future work will expand this list of glycoproteins, particularly to include molecules responsible for "fine-tuning" matrix assembly and binding the cells to the matrix.

As described (Section I.C), heparan sulfate GAG (or its close cousin heparin) binds laminin, fibronectin and the interstitial collagen types I, III, and V. Thus, the proteogly-

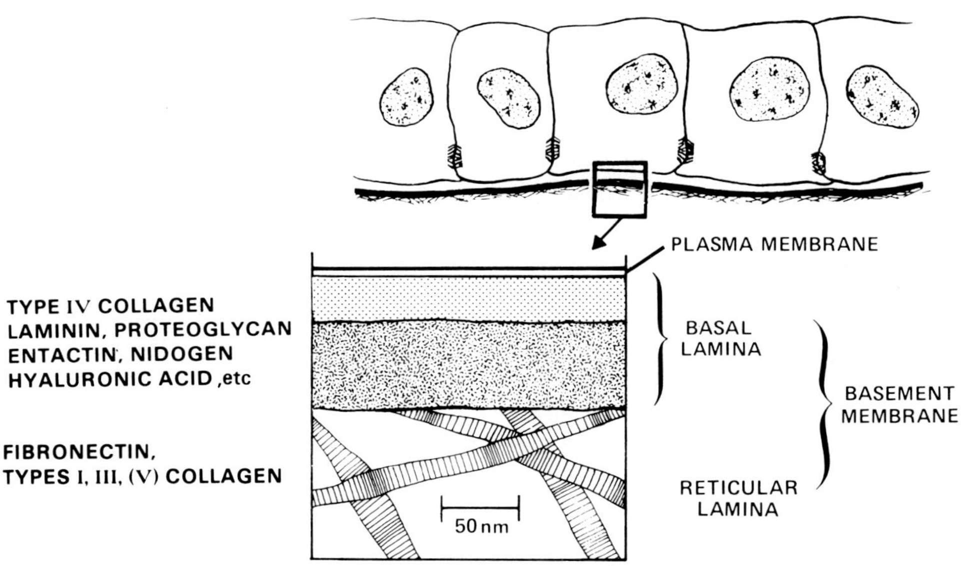

Figure 7 Diagram of the basement membrane present at the basal surface of epithelia. The basement membrane is composed of the basal lamina, comprised of components deposited by the epithelium, and the reticular lamina, deposited by subjacent stromal cells. The basal lamina can be further subdivided into the lamina rara, a less electron-dense region closely apposed to the plasma membrane of the epithelial cell, and the more electron-dense lamina densa, bordering the reticular lamina. Representative components of the basal and reticular laminae are shown on the left. (From Ref. 11.)

can may bind basal lamina components and anchor the basal lamina to the reticular lamina owing to its affinity for interstitial collagens as well as to the cell layer via its integral membrane domain. However, because antibody localization studies in vivo have focused on extracellular heparan sulfate proteoglycan rather than cell surface forms, an anchorage role of cell surface heparan sulfate proteoglycan in the basement membrane remains to be shown conclusively.

B. Proteoglycan May Form Cell Surface Clusters When Cross-Linked by Matrix Components

Cell surface proteoglycan that is mobile in the plasma membrane may be assimilated into clusters when cross-linked by ECM components. These clusters may act as stable adhesion foci between the cell and the ECM, analogous to the heparan sulfate proteoglycan-rich adhesion sites of cells grown in culture (41,47,79). Simultaneously, this cluster would be capable of binding other newly deposited ECM components and would also have the potential to recognize and bind the cytoskeleton on the cytoplasmic side of the membrane. In addition, anchorage of the cluster by the ECM would prevent the subsequent "capping" and loss from the cell surface observed for proteoglycan clusters generated by soluble ligand (antibody) cross-linking. One could speculate that cluster formation might impart additional information to the cell analogous to the biological activity of hormone receptors cross-linked by exogenous ligands, thus eliciting specific cell behavior in the presence of a native ECM.

VI. SUMMARY

At our present level of understanding, heparan sulfate cell surface proteoglycans potentially play an important role in regulating cell behavior. These molecules are ubiquitous on all adherent cells. They are attractive candidates as cell-matrix anchors because they interact with a variety of extracellular matrix components in a multivalent manner and they are also anchored in the plasma membrane. Although the nature of an association with the intracellular cytoskeleton remains to be worked out, initial evidence suggests that such an interaction occurs. Even at this gross level, the potential effects of such interactions on cell shape and behavior are enormous. Regulatory information as well as structural benefits may be derived from this interaction. Finally, because this molecule is at the interface of the cell and the matrix, an interface that is the target of interactions between apposing cell layers (e.g., interactions such as those resulting in alterations of matrix composition and stability during embryogenesis), it is a candidate to transmit and translate these interactions to the epithelial cell.

ACKNOWLEDGMENTS

This work was supported by grants from the National Institutes of Health, GM 07494 and NIRA HD17146 to AR, DE 05273 to JEK, and CA28735 to MB, from the Cystic Fibrosis Foundation to JEK, and from the American Cancer Society, BC-409 to MB.

REFERENCES

1. J. Folkman and A. Moscona, *Nature, 273*:345-349 (1978).
2. A. Ben-Ze'ev, S. R. Farmer, and S. Penman, *Cell, 21*:365-372 (1980).

3. S. C. Wittelsberger, K. Kleene, and S. Penman, *Cell, 24*:859-866 (1981).
4. M. T. Haeuptle, Y. L. M. Suard, E. Bogenmann, H. Reggio, L. Racine, and J. P. Kraehenbuhl, *J. Cell Biol., 96*:1405-1414 (1983).
5. E. D. Hay, in *Cell Biology of the Extracellular Matrix* (E. D. Hay, ed.), Plenum Press, New York, p. 379-409.
6. E. D. Hay, *Am. J. Anat., 165*:1-12 (1982).
7. M. Bernfield, *Dev. Biol., 22*:213-231 (1970).
8. S. D. Banerjee, R. H. Cohn, and M. Bernfield, *J. Cell Biol., 73*:445-463 (1977).
9. M. Bernfield and S. D. Banerjee, *Dev. Biol., 90*:291-305 (1982).
10. R. L. Smith and M. R. Bernfield, *Dev. Biol., 94*:378-390 (1982).
11. M. Bernfield, S. D. Banerjee, J. E. Koda, and A. Rapraeger, in *The Role of the Extracellular Matrix in Development* (R. L. Trelstad, ed.), Alan R. Liss, Inc., New York, 1984, pp. 545-572.
12. N. C. Rao, S. H. Barsky, V. P. Terranova, and L. Liotta, *Biochem. Biophys. Res. Commun., 111*:804-808 (1983).
13. V. P. Terranova, C. N. Rao, T. Kalebic, I. M. Margulies, and L. Liotta, *Proc. Natl. Acad. Sci. USA, 80*:444-448 (1983).
14. T. M. Chiang and A. H. Kang, *J. Biol. Chem., 257*:7581-7586 (1982).
15. J. Mollenhauer and K. von der Mark, *EMBO J., 2*:45-50 (1983).
16. B. Goldberg, *Cell, 16*:265-275 (1979).
17. B. Goldberg and R. E. Burgeson, *J. Cell Biol., 95*:752-756 (1982).
18. K. Rubin, M. Höök, B. Obrink, and R. Timpl, *Cell, 24*:463-470 (1981).
19. S. P. Sugrue and E. D. Hay, *J. Cell Biol., 91*:45-54 (1981).
20. K. M. Yamada, *Annu. Rev. Biochem., 52*:761-799 (1983).
21. A. Hewitt, H. Varner, M. H. Silver, W. Dessau, C. M. Wilkes, and G. R. Martin, *J. Biol. Chem., 257*:2330-2334 (1982).
22. K. M. Yamada, D. W. Kennedy, K. Kimata and R. M. Pratt, *J. Biol. Chem., 255*: 6055-6063 (1980).
23. E. Ruoslahti and E. Engvall, *Biochim. Biophys. Acta, 631*:350-358 (1980).
24. S. Stamatoglou and J. Keller, *Biochim. Biophys. Acta, 719*:90-97 (1982).
25. S. Sakashita, E. Engvall, and E. Ruoslahti, *FEBS Lett., 116*:243-246 (1980).
26. D. T. Woodley, C. N. Rao, J. R.Hassell, L. A. Liotta, G. R. Martin, and H. K. Kleinman, *Biochim. Biophys. Acta, 761*:278-283 (1983).
27. S. Stamatoglou and J. Keller, *J. Cell Biol., 96*:1820-1823 (1983).
28. J. E. Koda and M. Bernfield, *J. Biol. Chem., 259*:11763-11770 (1984).
29. J. E. Koda, A. Rapraeger and M. Bernfield, *J. Biol. Chem., 260*:8157-8162 (1985).
30. L-A. Fransson, I. Carlstedt, L. Coster, and A. Malmstrom, *J. Biol. Chem., 258*:24342-24345 (1983).
31. P. Kraemer, *Biochemistry, 10*:1437-1445 (1971).
32. P. Kraemer, in *Surfaces of Normal and Transformed Cells* (R. O. Hynes, ed.), Wiley & Sons, Ltd., London (1979).
33. A. Oldberg, L. Kjellan, and M. Hook, *J. Biol. Chem., 254*:8505-8510 (1979).
34. S. Mutoh, I. Funakoshi, and I. Yamashina, *J. Biochem., 84*:483-489 (1978).
35. B. Glimelius, B. Norling, B. Westermark, and A. Wasteson, *J. Cell. Physiol., 98*: 527-538 (1979).
36. C. B. Underhill and J. M. Keller, *Biochem. Biophys. Res. Commun., 63*:448-454 (1975).
37. K. G. Vogel and D. W.Peterson, *J. Biol. Chem., 256*:13235-13242 (1981).
38. I. Carlstedt, L. Coster, A. Malmstrom, and L-A. Fransson, *J. Biol. Chem., 258*:11629-11635 (1983).
39. A. Wasteson, B. Glimelius, C. Busch, B. Westermark, C. Heldin, and B. Norling, *Thrombosis Res., 11*:309-321 (1977).

40. A. Rapraeger and M. Bernfield, *J. Biol. Chem.*, *260*:4103-4109 (1985).
41. M. Lark and L. Culp, this volume, Chapter 8.
42. P. Kraemer, *Biochem. Biophys. Res. Commun.*, *78*:1334-1340 (1977).
43. L. Kjellan, A. Oldberg, and M. Höök, *J. Biol. Chem.*, *255*:407-413 (1980).
44. L. Kjellan, I. Petterssen, and M. Höök, *Proc. Natl. Acad. Sci. USA*, *78*:5371-5375 (1981).
45. A. Rapraeger and M. Bernfield, *J. Biol. Chem.*, *258*:3632-3636 (1983).
46. B. Norling, B. Glimelius and A. Wasteson, *Biochem. Biophys. Res. Commun.*, *103*: 1265-1272 (1981).
47. M. Lark and L. Culp, *J. Biol. Chem.*, *259*:6773-6783 (1984).
48. A. Oldberg, E. G. Hayman and E. Ruoslahti, *J. Biol. Chem.*, *256*:10847-10852 (1981).
49. T. F. Bumol, Q. C. Wang, R. A. Reisfeld and N. O. Kaplan, *Proc. Natl. Acad. Sci. USA*, *80*:529-533 (1983).
50. K. Hedman, J. Christner, I. Julkunen, and A. Vaheri, *J. Cell Biol.*, *97*:1288-1293 (1983).
51. Y. Mijuni-Takagaki and B. P. Toole, *J. Biol. Chem.*, *256*:8463-8469 (1981).
52. U. Lindahl and M. Hook, *Ann. Rev. Biochem.*, *47*:385-417 (1978).
53. M. Hook, L. Kjellan, S. Johansson, and J. Robinson, *Annu. Rev. Biochem.*, *53*: 847-869 (1984).
54. A. Rapraeger, M. Jalkanen, E. Endo, J. Koda, and M. Bernfield, *J. Biol. Chem.*, *260*: 11046-11052 (1985).
55. S. Vannucchi, G. Fibbi, R. Cappelletti, M. del Rosso, and. V. Chiarugi, *J. Cell Physiol.*, *111*:149-154 (1982).
56. G. David and M. Bernfield, *Proc. Natl. Acad. Sci. USA*, *76*:786-790 (1979).
57. G. David and M. Bernfield, *J. Cell Biol.*, *91*:281-286 (1981).
58. A. Rapraeger, M. Jalkanen and M. Bernfield, *J. Cell Biol.*, (1986) (submitted for publication.
59. M. Jalkanen, H. Nguyen, A. Rapraeger, N. Kurn, and M. Bernfield, *J. Cell Biol.*, *101*: 976-984 (1985).
60. G. Nicolson, *Biochim. Biophys. Acta*, *457*:57-108 (1976).
61. S. de Petris, in *Dynamic Aspects of Cell Surface Organization*, Cell Surface Reviews, Vol. 3 (G. Poste and G. Nicolson, eds.), North-Holland Pub. Co., Amsterdam, New York, Oxford, 1977, pp. 643-735.
62. J. Flanagan and G. Koch, *Nature*, *273*:278-280 (1978).
63. J. Condeelis, *J. Cell Biol.*, *80*:751-758 (1979).
64. M. Yanagishita and V. C. Hascall, *J. Biol. Chem.*, *259*:10270-10283 (1984).
65. G. David, B. Van der Schueren and M. Bernfield, *JNCI*, *67*:719-728 (1981).
66. P. Kraemer and R. Tobey, *J. Cell Biol.*, *55*:713-717 (1972).
67. G. W. Laurie, C. P. Leblond and G. R. Martin, *J. Cell Biol.*, *95*:340-344 (1982).
68. R. Timpl and G. Martin, in *Immunochemistry of the Extracellular Matrix*, Vol. II (H. Furthmayr, ed.), CRC Press, Boca Raton, FL, 1982, pp. 120-150.
69. J. Gordon and M. Bernfield, *Dev. Biol.*, *74*:118-135 (1980).
70. J. Hassell, P. Robey, H. Barrach, J. Wilczek, S. Rennard, and G. Martin, *Proc. Natl. Acad. Sci. USA*, *77*:4494-4498 (1980).
71. Y. S. Kanwar, V. C. Hascall and M. G. Farquhar, *J. Cell Biol.*, *90*:527-532 (1981).
72. M. Bernfield and S. D. Banerjee, *J. Cell Biol.*, *52*:664-673 (1972).
73. R. H. Cohn, S. D. Banerjee, and M. R. Bernfield, *J. Cell Biol.*, *73*:464-478 (1977).
74. R. Timpl, H. Rohde, P. G. Robey, S. I. Rennard, J-M. Foidart, and G. R. Martin, *J. Biol. Chem.*, *254*:9933-9937 (1979).
75. R. Timpl, H. Weidemann, V. Van Delden, H. Furthmayr and K. Kuhn, *Eur. J. Biochem.*, *120*:203-211 (1981).
76. B. Carlin, R. Jaffe, B. Bender, and A. E. Chung, *J. Biol. Chem.*, *256*:5209-5214 (1981).

77. R. O. Hynes, in *Cell Biology of the Extracellular Matrix* (E. D. Hay, ed.), Plenum Press, New York, 1981, pp. 295-334.
78. K. Kuhn, in *Immunochemistry of the Extracellular Matrix*, Vol. I (H. Furthmayer, ed.), CRC Press, Boca Raton, FL, 1982, pp. 1-29.
79. J. Lattera, J. E. Silbert and L. A. Culp, *J. Cell Biol., 96*:112-123 (1983).
80. A. Rapraeger, J. Koda, and M. Bernfield, in *Proceedings of the 41st Annual Meeting of the Electron Microscopy Society of America* (G. W. Bailey, ed.), San Francisco Press, San Francisco, 1983, pp. 668-671.
81. V. C. Hascall and G. K. Hascall, in *Cell Biology of the Extracellular Matrix* (E. D. Hay, ed.), Plenum Press, New York, 1981, pp. 39-63.

10

Intracellular Degradation as a Modulator of Collagen Production

Donald S. Neblock* and Richard A. Berg
University of Medicine and Dentistry of New Jersey-Rutgers Medical School, Piscataway, New Jersey

I. INTRODUCTION

The extracellular matrix protein, collagen, performs functions necessary for organization of differentiated cells into tissues and organs. Its presence is required to provide structural support and a selective barrier in tissue organization (1) as well as a probable signal for the differentiation and migration of cells during embryogenesis (2). Collagen production must be rigorously controlled to ensure the formation of functional extracellular matrix; many clinically important connective tissue disorders are characterized by the deposition of improper amounts of collagen or collagen that is structurally abnormal (3-5). In addition, the mechanism responsible for control of this normally stable protein must be capable of adjusting to requirements for altered levels of synthesis and degradation, as in periods of rapid growth or during the tissue remodeling that follows injury. It is therefore of interest to understand the molecular mechanisms responsible for control of collagen production.

The possible biosynthetic steps at which collagen production may be controlled are numerous (see Figure 1), including those occurring transcriptionally, translationally, and posttranslationally. It has recently been shown that in addition to the more well-characterized control mechanisms, rapid intracellular degradation of newly synthesized collagen may provide another important way that cells modulate the amount of collagen available for extracellular matrix formation. It is the purpose of this chapter to review recent findings about the cellular mechanisms for intracellular degradation of newly synthesized collagen and the biological importance of this phenomenon.

The intracellular degradation of proteins is a widely observed, but incompletely understood phenomenon. It has been shown to occur in both bacterial and eukaryotic cells

**Present affiliation*: Cistron Technology, Inc., Pine Brook, New Jersey.

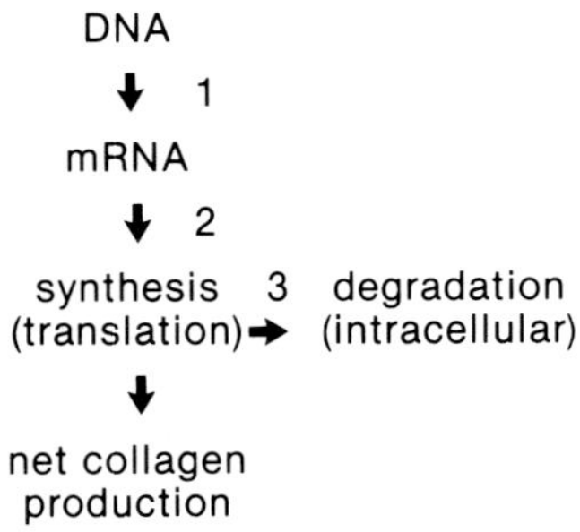

net collagen production = synthesis − degradation

Figure 1 Biosynthetic steps at which collagen production may be regulated. The level of collagen production in any cell may be controlled by the transcription of mRNA (1), posttranscriptional processing of mRNA (2), and adjustment of translation rates (2). In addition, it has been shown that rapid intracellular degradation of newly synthesized collagen is an important posttranslational control mechanism (3).

and may influence a variety of cellular functions by removing defective proteins, adjusting the levels of important enzymes, providing amino acids for protein synthesis or energy production, and modulating the amount of secretory products of specialized cells (for reviews, see Refs. 6-9). Cells responsible for the production of major secretory proteins such as collagen, immunoglobulins, and hormones recently have been the object of considerable study, since many types of secretory proteins have been shown to be subject to significant levels of intracellular degradation en route to secretion (for review, see Ref. 9).

The first of the secretory proteins shown to undergo rapid intracellular degradation was collagen (10,11). Although other secretory proteins have been shown to undergo intracellular degradation, the phenomenon has been most thoroughly characterized in collagen-producing cells and tissues. This is because collagen has a well-defined biosynthetic pathway and contains a relatively large amount of trans-4-hydroxyproline, which is formed posttranslationally by the enzymatic hydroxylation of peptidyl proline (1,12). Hydroxyproline is nearly unique to collagen in most animal tissues, since only low levels of this amino acid are present in a few other proteins (1). After newly synthesized proteins have been labeled with $[^{14}C]$proline, the presence of the trans-4-hydroxy$[^{14}C]$proline marker in degraded collagen fragments allows a sensitive, direct detection of the intracellular degradation products.

Using this basic approach, it has been shown that 10-50% of all newly synthesized collagen may be degraded by cells within a short period of time. Furthermore, the collagen that is degraded has not been secreted and recaptured by the cell, but instead appears to be retained intracellularly through the point at which degradation occurs. The level of degradation observed is characteristic of the cell or tissue being examined, the type of collagen produced, and the culture conditions employed (for reviews see Refs. 9,13,14). What has emerged from the large body of data about the intracellular degradation of newly synthesized collagen is the probably universal occurrence of the phenomenon. Despite the accumulation of evidence supporting the occurrence of intracellular collagen degradation, little is currently understood about the cellular mechanisms responsible for it. Understanding how intracellular degradation contributes to the maintenance of functional extracellular matrix in healthy individuals and what its role might be in various diseases affecting connective tissue remains an important research goal.

II. INTRACELLULAR DEGRADATION OF NEWLY SYNTHESIZED COLLAGEN

While the majority of newly synthesized collagen is processed and secreted along a well-characterized route, 10% or more is degraded in cells grown under normal culture conditions. Intracellular degradation of newly synthesized collagen has now been demonstrated for cells or tissues synthesizing type I (10,11), type II (15), type III, and type IV collagens (16) (see Table 1). The observation of this phenomenon for a variety of collagens provides support for the concept that the degradation is universal in collagen-producing cells. Systems in which intracellular degradation has been observed include cultured fetal human lung (HFL-1) cells (10,17), embryonic chick tendon fibroblasts in monolayer cultures (15) and in suspension cultures (18), cultured human skin fibroblasts and muscle cells (19), and cultured chick chondrocytes (15). Intact tissues such as the lung (11), the bone (20), skin (21), gingiva (22), and tendon (23) grown in organ culture have been shown to degrade collagen intracellularly (see Table 1). The intracellular degradation of newly synthesized collagen has been observed in fibroblasts from patients with osteogenesis imperfecta (24) and in human tumor cells (16). Increased catabolism of newly synthesized collagen has been observed in skins of rats with steptozotocin-induced diabetes (21). These reports indicate the possibility that intracellular degradation may play a role in diseases that directly or indirectly alter collagen function in connective tissues.

III. FUNCTIONAL CLASSIFICATIONS OF INTRACELLULAR DEGRADATION

The intracellular degradation of newly synthesized collagen, like the degradation of cytoplasmic proteins, has been classified by subdivision into two major modes. These have been designated basal and enhanced degradation (9,14). These modes of degradation have defined characteristics that depend on the conditions of culture employed and the levels of degradation determined (7). Basal degradation has been referred to as the release of trichloroacetic acid-soluble amino acids from cellular peptides under nonstressed, fully supplemented culture conditions. In the study of intracellular collagen degradation, basal degradation is recognized to be the release of dialyzable or the ethanol-soluble hydroxyproline from newly synthesized collagen under similarly nonstressed conditions.

Enhanced intracellular degradation is characterized by the release of greater than basal levels of amino acids from cellular peptides when culture conditions have been systematically perturbed by serum removal, addition or removal of hormones, nutritional deprivation, or the addition of specific agents causing the synthesis of conformationally abnormal proteins (see Refs. 6-8 for reviews). Conformation is crucial to the concept of enhanced proteolysis in the study of the intracellular degradation of newly synthesized collagen, since the enhanced mode is defined as the increased release of hydroxyproline caused by conditions that render newly synthesized collagen conformationally abnormal (9).

Table 1 Intracellular Degradation of Newly Synthesized Collagen

Culture system	Principal collagen type	Mode of degradation	Special conditions	Level of degradation (%)	References
Human fetal lung fibroblasts	I, III	Basal	–	20	10, 17
		Enhanced	–Ascorbate	50	59
		Enhanced	+Azetidine-2-COOH	30	10
		Enhanced	Log growth	30	17
		Enhanced	+*cis*-4-Hydroxyproline	30	17
		cAMP enhanced	PGE_1, PGE_2, Isoproterenol	50	73
Rabbit lung slices	I, III	Basal	–	20-40	11
Chicken embryo fibroblasts	I	Basal	Suspension	10	27, 29
		Basal	Cultured	25	15
		Enhanced	+*cis*-4-Hydroxyproline	25-35	27
Embryonic chick tendon	I	Basal	Organ culture	10	23
		Enhanced	Organ culture	25-35	23
Chicken chondrocytes	II	Basal	–	15-20	15,16
Human skin fibroblasts	I, III	Enhanced	Elevated temperature	50	57
		Enhanced	Log growth	50	57
Muscle cells	I, III	Basal	–	10	19
Osteogenesis imperfecta skin fibroblasts	I, III	Basal	–	30	24
		Enhanced?	–	?	91
Human tumor cells	IV	Basal	–	40	16

IV. CELLULAR MECHANISMS FOR BASAL LEVEL INTRACELLULAR DEGRADATION OF NEWLY SYNTHESIZED COLLAGEN

A. Measuring Degradation Levels

Current understanding of basal level intracellular degradation is minimal. As discussed in Section II and outlined in Table 1, 10-20% of all newly synthesized collagen is degraded prior to secretion from the cell. These determinations have been made in cell or tissue cultures incubated in the presence of high-specific-activity [^{3}H]- or [^{14}C]proline to label newly synthesized collagen. The proportion of total radiolabeled hydroxyproline that is dialyzable or soluble in 75% ethanol (25) is used as a measure of the hydroxyproline released by degradation of collagen. Determination of radioactivity in hydroxyproline has been performed by the chromatographic separation of proline and hydroxyproline on conventional (25) or HPLC (26) analytical cation exchange columns.

B. Basal Level Degradation Is Nonlysosomal

Lysosomal proteolysis does not appear to be directly required for basal intracellular degradation, since inhibitors of lysosomal proteolysis such as NH_4Cl, chloroquine, leupeptin, and pepstatin A had no effect on basal levels of degradation in several systems (17,27-29). This was in marked contrast to the ability of lysosomal inhibitors to suppress the enhanced levels of degradation shown to occur when the synthesis of conformationally defective collagen had been induced (17,23,30) (see next section).

In addition to its insensitivity to lysosomal inhibitors, the basal level intracellular degradation mechanism has been shown to be uninhibited by a variety of agents in pharmacological studies. In human fetal lung cells, basal intracellular degradation continued at control levels in the presence of the microfilament disruptor cytochalasin B and the microtubule disrupting agent colchicine (17). The results have been interpreted to indicate the lack of microfilament or microtubule involvement in the basal degradation mechanism. The basal level of degradation in cells has also been shown to be insensitive to the presence of serum or to differences in growth medium pH from 6.8 to 7.8 in cultured human lung fibroblasts (28).

cis-4-Hydroxyproline and various isomers of hydroxyproline can be produced from both free proline and peptidyl proline via nonenzymatic hydroxylations caused by oxygen-derived, free-radical reactions (31,32). For this reason, concern has been raised that dialyzable hydroxyproline present in small peptides or as free amino acid may be produced by free radical reactions operating on proline residues in fibroblasts (32). Although *cis*-4-hydroxyproline is recognized by prolyl tRNA and can be incorporated in place of proline into polypeptide chains (33), it is not synthesized from peptidyl proline in collagen precursors by the enzyme prolyl-4-hydroxylase (34). Prolyl-4-hydroxylase produces the *trans* epimer exclusively. *cis*-4-Hydroxyproline is, however, present in acid hydrolysates of collagen owing to epimerization of trans-4-hydroxyproline during the hydrolysis (35). For this reason, it has been difficult to rule out the possibility that oxygen-derived free radicals were involved in the synthesis of hydroxyproline that was mistakenly identified as resulting from intracellular degradation of newly synthesized collagen. However, a recent study has shown that all detectable *cis*-4-hydroxyproline found in hydrolysates of fibroblast cultures could be attributed to the epimerization of trans-4-hydroxyproline during acid hydrolysis (36). That the *cis* epimer is not produced by fibroblast cultures strongly suggests that the nonenzymatic mechanism for the synthesis of hydroxyproline is not operative in cultured fibroblasts, and that measurements of intracellular degradation of collagen are not compromised by artifacts due to free radical reactions.

C. The Kinetics of Basal Intracellular Degradation

The precise cellular location for the basal degradation mechanism remains to be demonstrated. Also, it has not yet been directly shown that proteinases actually mediate the basal level intracellular degradation of newly synthesized collagen. However, recent data have revealed important clues pertaining to these questions. Kinetic analyses of intracellular collagen degradation have provided evidence in favor of the hypothesis that a metabolic "side pocket" of intracellular collagen exists in fibroblasts that fills with substrate sensitive to basal degradation. While unknown at present, the source of the side-pocket pool may be kinetic or physical, or both.

Intracellular, radiolabeled hydroxyproline derived from the degradation of newly synthesized collagen in human lung fibroblasts (10) and chick tendon fibroblasts (37) could be detected approximately 15 min after addition of label to either culture system. During continuous labeling in chick tendon cells, the kinetics for procollagen secretion indicated that steady-state pools of intracellular procollagen saturated with label in 45 min (37). In contrast, the pathway for intracellular degradation of newly synthesized collagen required more than 240 min to achieve steady state (37). The coexistence of different kinetics for the secretion and degradation processes suggests that there is a branch point in the secretory pathway, and that the degradation occurs in a kinetically distinct compartment. On the basis of these data, the basal degradation pathway appears to require a longer filling time than the major route for secretion of collagen from the cells. Further support for this concept has been obtained from studies that showed that levels of degradation could be increased by changing the length of the labeling time in HFL-1 cells (28). The basal level of degradation was shown to increase from 16% in cells labeled for 8 hr to 23% for a 24-hr labeling period.

D. Role for the Golgi Apparatus in the Basal Degradation Mechanism

Pulse-chase analyses of basal intracellular degradation have also indicated that the collagen fated for degradation may be retained by cells for periods of time longer than required for secretion. After pulse labeling of one to two rounds of procollagen synthesis in chick tendon fibroblasts for 15 min, 15-20% of the total amount of procollagen synthesized was degraded during the chase period (37,38). The degradation was shown to continue for periods of 90-120 min, coincident with the saturation of the intracellular degradation pathway in continuous labeling studies (37). Interrupting the secretion of newly synthesized collagen at the level of the Golgi apparatus using the ionophore monensin (39-41) caused the chick tendon cells to accumulate collagen with a concomitant increase in the amount of collagen degraded (38). The increased degradation shown to occur in the monensin-treated cultures occurred over a longer period of time and did not vary with respect to the rate of degradation, suggesting that the secretory block increased the amount of substrate available for degradation. Since the principal effect of low doses of monensin has been shown to be inhibition of protein secretion at the level of *cis*-Golgi cisternae (40,41), the data point to the involvement of the Golgi apparatus as the site of the basal intracellular degradation mechanism or the site of the branch point leading to it.

The Golgi apparatus has been implicated as the site for limited proteolysis of other secretory proteins (9). The processing of proinsulin to insulin has been shown to occur in Golgi fractions (42,43) as has the conversion of proparathyroid hormone to parathyroid hormone (44). The conversion of proalbumin to albumin has been reported to occur in Golgi fractions and may be mediated by a cathepsin B-like acid proteinase (45). Another report suggested that this conversion occurred only in secretory vesicles (46).

In addition to the limited proteolysis of proteins en route to secretion, it has also been demonstrated that several newly synthesized secretory proteins besides collagen are completely degraded intracellularly before secretion. The milk protein casein has been shown to be degraded intracellularly at levels of at least 55% in organ cultures of mammary explants (47). It was demonstrated that parathyroid hormone may be degraded at levels of up to 70% (48), and as much as 50% of newly synthesized prolactin was shown to be degraded intracellularly (49-51). Intracellular degradation has also been shown to play a role in the regulation of immunoglobulin production (52).

E. Conformationally Defective Collagen Is Degraded in Lysosomes

In contrast to basal degradation, enhanced proteolysis was first recognized as the type of catabolism elicited by stress conditions (such as nutritional deprivation), by the synthesis of conformationally defective proteins, or when protein secretion is inhibited (for reviews, see Refs. 7-9). Conformationally defective proteins and peptides produced in cultured cells include those which have incorporated amino acid analogs or puromycin and error-containing peptides that have arisen from specific mutations (for reviews, see Refs. 8,53).

Collagen containing amino acid analogs was shown to be more susceptible to proteolysis than normal collagen (54), and the degradation of this abnormal collagen was shown to occur intracellularly shortly after the protein was synthesized (18,30). It has been shown that the incorporation of amino acid analogs prevents collagen from assuming a triple-helical conformation (55,56), suggesting that the enhanced degradation of analog-containing collagen is conformation dependent (16-18,57).

This hypothesis has also been supported by studies of intracellular degradation of thermally denatured collagen. Incubation of fibroblasts at 42-43°C caused the degradation of newly synthesized collagen to increase from 12% to 49% in human skin fibroblasts (57). Similar results have been demonstrated for type II collagen in chicken sternal chondrocytes and for type IV collagen in a human tumor cell line, HT-1080 (16). The degradation of type IV collagen was increased from 38% to 62% when incubation temperature was increased from 37°C to 43°C (16).

In addition to the enhanced intracellular degradation brought about by the incorporation of amino acid analogs and thermal denaturation, newly synthesized collagen has also been shown to undergo increased degradation when proline fails to be converted to hydroxyproline. This can be caused by incubating fibroblasts in the presence of the iron chelator α,α'-dipyridyl to inhibit prolyl hydroxylase (12,28,58). Increased intracellular degradation of underhydroxylated collagen was also shown for fibroblasts grown in the absence of ascorbate (59), a cofactor required by prolyl hydroxylase. Early log-phase fibroblasts contain reduced levels of prolyl hydroxylase (60) and were shown to produce collagen that was underhydroxylated and subject to higher levels of degradation than that produced by confluent cells (17,61).

These data emphasize the relationship between the stability of the collagen triple helix and the degree of intracellular degradation. In contrast to fully hydroxylated collagen, underhydroxylated collagen is nonhelical at 37°C and thus may be considered to be defective (62). Originally thought to be accumulated in cells (63), it is secreted at a rate only one-tenth that shown for fully hydroxylated collagen (58). The incorporation of proline analogs such as *cis*-4-hydroxyproline was shown to sterically interfere with the formation of the collagen triple helix (55) and also caused the conformationally defective collagen to be more slowly secreted from the cells (58). Such non-triple-helical collagen has been shown to be subject to enhanced intracellular degradation (17,18,30). The basal degradation level of approximately 10% has been shown to be increased to 40% or more by enhanced degradation under conditions that promote the synthesis of conformationally abnormal collagen.

The function of lysosomal proteinases has been shown to be required for increased levels of degradation of conformationally defective collagen. The enhanced levels of degradation have been shown to be returned to basal or near-basal levels by the lysosomotropic amines chloroquine or NH_4Cl and cathepsin-specific inhibitors such as leupeptin and pepstatin A in HFL-1 (17,28) and chick tendon cells (23).

Direct evidence for the lysosomal destruction of defective collagen has been obtained in studies in which enhanced degradation was stimulated in HFL-1 cells by incorporation of *cis*-4-hydroxyproline into newly synthesized collagen. After inhibition of lysosomal proteolysis with NH_4Cl, the presence of the procollagen marker hydroxyproline was demonstrated in lysosomes obtained by subcellular fractionation in colloidal silica gradients (30).

Extracts of cultured human lung fibroblasts have been shown to possess proteolytic activity against purified [^{14}C] collagenous substrates. These activities were characterized by acidic pH optima, showed a 10-fold preference for denatured substrate, and were partially inhibited by leupeptin or pepstatin A individually (30). Leupeptin plus pepstatin A together inhibited proteolysis by more than 98% (30), indicating that cysteine and aspartic proteinases such as cathepsins B and D, respectively, can degrade nonhelical collagen in vitro. The effects of these agents on intracellular degradation in cultured cells indicate the probable involvement of these classes of proteinases in the enhanced degradation of newly synthesized collagen.

Together, these reports support a model in which the enhanced degradation of nonhelical collagen occurs in lysosomes. Whether lysosomal proteinases mediate the destruction totally, or in conjunction with another pathway (such as the basal degradation mechanism), is unknown. Similarly, it is not known whether the conformation of newly synthesized collagen is the sole recognition signal responsible for the routing of defective collagen to lysosomes. It is possible that an intracellular system similar to the phosphomannosyl receptor for lysosomal hydrolases (64) may participate in the routing of newly synthesized collagen at some point in the secretory pathway.

Precedents exist indicating that lysosomal proteinases are involved in the destruction of at least some classes of defective proteins. Dean (65) has observed that the increased catabolism of analog-containing proteins in murine macrophages was decreased by inhibitors specific for lysosomal proteinases. Similar observations have been made in other systems (66,67).

It has been shown that the lysosomal system participates in the autophagic engulfment of intracellular particles and organelles (68), the degradation of endocytosed particles and molecules (69), and the enhanced intracellular protein degradation seen in nutritional deprivation (7,8,70-72). However, not all types of enhanced degradation occur in lysosomes. Hershko and Ciechanover and others have demonstrated and characterized the ubiquitin-ATP-requiring proteolytic system present in the cytosol of reticulocytes, which is capable of degrading abnormal globin chains and other analog-containing proteins (see Ref. 7 for review). It is thought that this major proteolytic system is present in all cells but participates only in the turnover of cytoplasmic proteins (7). Its relationship to collagen degradation, if any, is unknown.

F. Cyclic AMP Levels in Cells with Elevated Levels of Intracellular Degradation

The intracellular degradation of newly synthesized collagen has been shown to be elevated in cultures treated with hormones that increase cellular cAMP content. Basal degradation levels of 10% were shown to be increased to over 50% in cultured human lung fibroblasts treated with prostaglandins E_1, E_2, or the β-agonist isoproterenol (73). In these studies, a corresponding increase in cAMP values was shown. Cyclic AMP and intracellular degradation were both reduced from the elevated levels caused by the presence of isoproterenol when the β-antagonist propanolol was added to the cells with isoproterenol (74).

In studies of streptozotocin-induced diabetes in rats, it was observed that collagen degradation was increased severalfold relative to that shown for control animals (21). It was not determined whether the soluble hydroxyproline measured in these experiments results from intracellular degradation of newly synthesized collagen or was derived from the turnover of recently deposited matrix collagen. However, the possibility that hormones that affect intermediary metabolism may influence intracellular degradation of newly synthesized collagen should receive further attention. Insulin and amino acids have been shown to cause a reduction in the enhanced proteolysis caused by nutritional deprivation or the presence of glucagon (68,75). Since elevated cyclic AMP levels have been correlated with increased intracellular collagen degradation, it should be considered that hormonally altered intracellular degradation of newly synthesized collagen may contribute to the connective tissue complications associated with long-term diabetes (76).

G. Further Considerations for Cellular Mechanisms

It is likely that cellular proteinases mediate the intracellular degradation of newly synthesized collagen. While lysosomal cathepsins have been shown to degrade nonhelical collagen (17,30) the involvement of proteolysis per se in the basal degradation mechanism has not yet been demonstrated.

Characterization of the products of the basal degradation mechanism may help to elucidate the putative role of cellular proteinases in this phenomenon. It has been shown that basal collagen degradation results in the appearance of mostly free hydroxy[^{14}C]-proline in cultured fibroblasts (10) and cultured muscle cells (19). It has been shown that peptidyl hydroxy[^{14}C]proline as well as the free amino acid was detectable in the basal degradation products from freshly isolated chick tendon fibroblasts (23). The proportion of hydroxy[^{14}C]proline in the peptide-bound fraction was increased in the degradation products obtained after treating the cultures with bestatin (Neblock and Berg, unpublished observations). Since bestatin inhibits several exopeptidases in vitro (77) and in cultured cells (78), this result may indicate a role for initial endoproteolytic attack on newly synthesized substrate followed by the cleavage of low-molecular-weight intermediates by exopeptidases. The appearance of free hydroxy[^{14}C]proline exclusively in other systems (10,19), does not preclude the possibility that such fragments may be generated in those systems as well; rather, the rates of individual proteolytic steps and the length of labeling periods may dictate the amount of intermediates detectable in each system.

It has been suggested (79) that intracellular receptors and coated vesicles may contribute to the processes of sorting and translocation for secretory proteins such as collagen. While this has not yet been demonstrated, it is reasonable to suppose that cellular mechanisms exist that enable the sorting of secretory proteins, membrane-bound glycoproteins, and lysosomal enzymes after synthesis in the RER. The intracellular receptor system for the routing of lysosomal hydrolases has been well characterized (64) and recently was shown to reside in *cis*-Golgi cisternae (80). Evidence continues to accumulate pointing to the existence of additional sorting mechanisms located in the Golgi apparatus or RER that kinetically discriminate the translocation of secretory versus membrane-bound proteins (81-83). In some cases it has been possible to demonstrate kinetically distinct transport of different secretory proteins in the same cell (84,85).

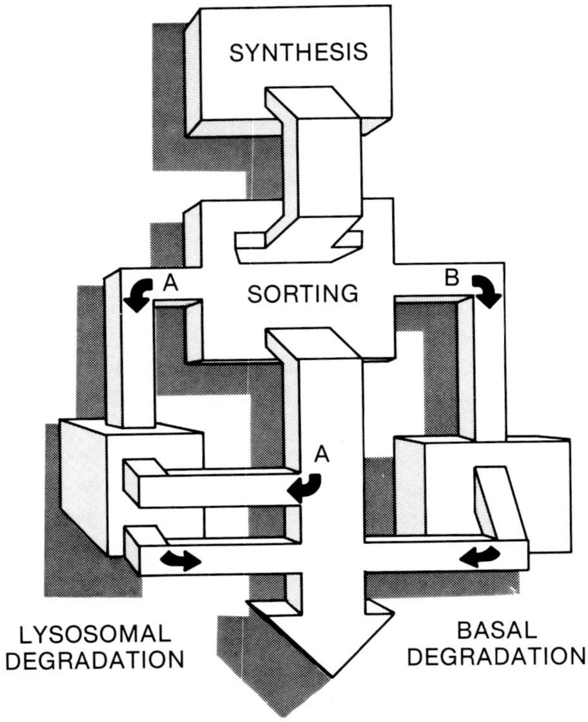

Figure 2 Schematic representation of the proposed cellular mechanisms for sorting of newly synthesized collagen into secretory and degradation pathways. It is postulated that a cellular mechanism exists that is responsible for routing newly synthesized collagen into the lysosomal (A) and basal (B) degradation mechanisms. Current evidence has suggested that the basal sorting mechanism may reside in the Golgi apparatus. It is not known whether routing of defective collagen to the lysosomes occurs in the Golgi apparatus or at other points in the secretory pathway. The products of intracellular degradation may be reutilized by the cells or secreted, as in the case of free hydroxyproline released by the collagen degradation.

These observations suggest that the discovery of an intracellular receptor required for collagen secretion may be inevitable. This prospect appears even more appealing in light of the recently described inhibition of collagen secretion by NH_4Cl and chloroquine (27). The abilities of these agents to interfere with the secretion of several proteins has been described (86) and may be due to their abilities to disrupt receptor recycling (87). If such a receptor system exists, it could provide one step at which collagen destined for intracellular degradation could be selected (see Figure 2), perhaps on the basis of the presence of high- and low-affinity forms of the newly synthesized collagen. Receptor affinity could be based on molecular conformation, posttranslational modifications, or other characteristics of the collagen. Those molecules which bound with lower affinity may be retained longer by the cell and exposed to the mechanism for intracellular degradation. Consistent with this model are the descriptions of a slowly-secreted pool of collagenous peptides in several types of fibroblasts (39,58,88). Similarly, the described kinetics of basal intracellular collagen degradation (37) and the monensin-induced increase in degradation levels (38) are both consistent with the model. The basal degradation-receptor sorting mechanism may also provide the screening step necessary to direct defective collagen molecules to lysosomes, as suggested by Bienkowski (9).

One possible source of the hydroxyproline released by basal intracellular degradation that has received little attention is the *N*-terminal propeptide. In type I collagen, the number of hydroxyproline residues on the *N*-propeptide amounts to about 5% of the total hydroxyproline synthesized in the procollagen molecule (89). The proteolytic modulation of the amount of *N*-propeptide available to regulate collagen mRNA translation (90) has been hypothesized (27,29). If such a mechanism is actually operative in cells, the release of hydroxyproline contained in the *N*-propeptide could account for part of the hydroxyproline used as a measure of the basal level of intracellular degradation of newly synthesized collagen.

V. CONCLUSIONS: MODULATION OF COLLAGEN PRODUCTION BY INTRACELLULAR DEGRADATION IN HEALTH AND DISEASE

The intracellular degradation of newly synthesized collagen seems to be a universal phenomenon in cultured cells and tissues. It can account for 10-50% of the collagenous polypeptides synthesized by cells under a variety of culture conditions. How this phenomenon participates in the control of collagen production at the levels of the cell and tissue in intact organisms should be a primary goal for future research.

Intracellular degradation of newly synthesized collagen has been proposed to serve as a cellular "quality control" mechanism for the detection and destruction of biosynthetic errors (9,17). The collagen biosynthetic pathway contains several steps at which such errors may occur. In type I collagen, for example, both collagen genes contain 50 or more introns (3,4), so that the production of mature messenger RNA requires many splices of the primary transcripts. Translation of incorrectly spliced collagen messages would result in the production of abnormal collagen chains. Random substitution error during translation in a glycine position of the collagenous primary sequence would also result in the loss of triple-helix stability (1). The many posttranslational modifications undergone by collagen also provide points at which errors can cause the loss of conformational stability and the function of the molecule. The consequences of the deposition of improperly formed collagen in the extracellular matrix can be severe or lethal (3-5). Therefore, the concept of a quality control mechanism to detect and destroy biosynthetic errors is reasonable.

Intracellular degradation as a modulator of collagen production probably contributes to a complex regulatory system in cells that produce collagen. Future research in this area will enhance our understanding of the regulation and function of the extracellular matrix in health and disease.

ACKNOWLEDGMENTS

The authors gratefully acknowledge the assistance of Mrs. Moira Schneider and Ms. Pat Clark for their invaluable help in preparing this review. The authors also thank Drs. Ronald Minor and Susan Quinones for their critical reading of the manuscript and helpful suggestions. The original work from the authors' laboratory was supported by NIH research grants AM 31839, AM 16516, and NIH Training grant HL 07467.

REFERENCES

1. P. Bornstein and W. Traub, in *The Proteins*, Vol. 4. (H. Neurath and R. L. Hill, eds.) Academic Press, New York, 1979, pp. 411-632.
2. E. D. Hay, in *Cell Biology of Extracellular Matrix* (E. D. Hay, ed.) Plenum Press, New York, 1981, pp. 379-409.
3. D. J. Prockop and K. I. Kivirikko, *N. Engl. J. Med.*, *311*:376-386 (1984).
4. F. Ramirez, F. O. Sangiorgi, and P. Tsipouras, in *Human Genes and Diseases* (F. Blasi, ed.), John Wiley and Sons, New York, 1986 (in press).
5. R. R. Minor, *Am. J. Pathol.*, *98*:227-280 (1980).
6. G. E. Mortimore, *Nutr. Rev.*, *40*:1-12 (1982).
7. A. Hershko and A. Ciechanover, *Ann. Rev. Biochem.*, *51*:335-364 (1982).
8. A. L. Goldberg and A. C. St. John, *Ann. Rev. Biochem.*, *45*:747-803 (1976).
9. R. S. Bienkowski, *Biochem. J.*, *214*:1-10 (1983).
10. R. S. Bienkowski, B. J. Baum, and R. G. Crystal, *Nature*, *276*:413-416 (1978).
11. R. S. Bienkowski, M. J. Cowan, J. A. McDonald, and R. G. Crystal, *J. Biol. Chem.*, *253*:4356-4363 (1978).
12. D. J. Prockop, R. A. Berg, K. I. Kivirikko, and J. Uitto, in *Biochemistry of Collagen* (G. N. Ramachandran and A. H. Reddi, eds.), Plenum Press, New York, 1976, pp. 163-273.
13. S. I. Rennard, L. E. Stier, and R. G. Crystal, *J. Invest. Dermatol.*, *79*:775-825 (1982).
14. R. A. Berg, in *Biology of the Extracellular Matrix* (R.Mecham, ed.), Academic Press, New York, 1986, pp. 29-52.
15. M. Duchene, S. Wiedebusch, K. Kuhn, and P. K. Muller, *FEBS Lett.*, *135*:119-122 (1981).
16. A. Palotie, *Coll. Rel. Res.*, *3*:105-113 (1983).
17. R. A. Berg, M. L. Schwartz, and R. G. Crystal, *Proc. Natl. Acad. Sci. USA*, *77*:4746-4750 (1980).
18. D. S. Neblock and R. A. Berg, *Conn. Tiss. Res.*, *10*:297-301 (1982).
19. M. Imberman, F. Oppenheim, and C. Franzblau, *Biochim. Biophys. Acta*, *719*:480-487 (1982).
20. M. Sakamoto, S. Sakamoto, D. Brinkley-Parsons, and M. J. Glimcher, *J. Bone Joint Surg.* *61*:1042-1052 (1979).
21. M. Schneir, N. Ramamurthy, and L. Golub, *Diabetes*, *31*:426-431 (1982).
22. M. Schneir, N. Ramamurthy, and L. Golub, *J. Dent. Res.*, *63*:23-27 (1984).
23. D. S. Neblock, Ph.D. thesis, Rutgers University, New Brunswick, NJ (1984).
24. B. Steinmann, G. Martin, B. Baum, and R. G. Crystal, *FEBS Lett.*, *101*:269-272 (1979).
25. R. S. Bienkowski and C. Engels, *Anal. Biochem.*, *116*:414-424 (1981).
26. D. S. Neblock and R. A. Berg (1986) (manuscript in preparation).
27. D. S. Neblock and R. A. Berg, *Biochem. Biophys. Res. Commun.*, *105*:902-908, (1982).
28. R. S. Bienkowski, *J. Cell Physiol.*, *121*:152-158 (1984).
29. D. S. Neblock and R. A. Berg, *Arch. Biochem. Biophys.*, *233*:338-344 (1984).
30. R. A. Berg, M. L. Schwartz, L. Rome, and R. G. Crystal, *Biochemistry*, *23*:2134-2138 (1984).
31. M. Bade and B. S. Gould, *Biochim. Biophys. Acta*, *156*:425-428 (1968).
32. R. L. Trelstad, K. R. Lawley, and L. B. Holmes, *Nature*, *289*:310-312 (1981).
33. J. Rosenbloom and D. J. Prockop, *J. Biol. Chem.*, *246*:1549-1555 (1971).
34. R. Kuttan and A. N. Radhkrishnan, *Adv. Enzymol.*, *37*:273-347 (1973).
35. D. D. Dziewiatkowski, V. C. Hascall, and R. L. Riolo, *Anal. Biochem.*, *49*:550-558 (1972).
36. R. S. Bienkowski, *Arch. Biochem. Biophys.*, *229*:455-458 (1984).

37. R. S. Bienkowski, S. F. Curran, and R. A. Berg, *Biochemistry* (1986) (in press).
38. D. S. Neblock and R. A. Berg, *Ann. NY Acad. Sci. 395*:460 (1984).
39. N. Uchida, P. Smilowitz, P. W. Ledger, and M. L. Tanzer, *J. Biol. Chem., 255*: 8638-8644 (1980).
40. A. M. Tartakoff, *Cell, 32*: 1026-1028 (1983).
41. P. W. Ledger and M. L. Tanzer, *Trends Biochem. Sci., 103*:313-314 (1984).
42. D. F. Steiner, J. L. Clark, C. Nolan, A. H. Rubenstein, E. Margoliash, F. Melani, and R. . E. Oyer in *Thirteenth Nobel Symposium Stockholm*, Interscience, New York, 1970, pp. 123-131.
43. H. S. Tager, D F. Steiner, and C. Patzelt, in *Methods in Cell Biology*, Vol. 23. *Basic Mechanisms of Cellular Secretion* (A. R. Hand and C. Oliver, eds.) Academic Press, New York, 1981, pp. 73-86.
44. J. F. Habener, H. M. Kronenburg, J. T. Potts, and L. Orci, in *Methods in Cell Biology*, Vol. 23. *Basic Mechanisms of Cellular Secretion* (A. R. Hand and C. Oliver, eds.) Academic Press, New York, 1981, pp. 51-71.
45. P. S. Quinn and J. D. Judah, *Biochem. J., 172*:301-309 (1978).
46. C. M. Redman, D. Banerjee, C. Manning, Y. Huang, and K. Green, *J. Cell Biol., 77*: 400-416 (1978).
47. H. H. Razooki-Hassan, D. A. White, and R. J. Mayer, *Biochem. J., 202*:133-138 (1982).
48. J. J. Morrisey and D. Cohn, *J. Cell Biol., 83*:521-528 (1979).
49. R. A. Maurer, *Biochemistry, 19*:3573-3578 (1980).
50. P. S. Dannies and M. S. Rudnick, *J. Biol. Chem., 255*:2776-2781 (1980).
51. P. S. Dannies, *Biochem. Pharmacol., 31*:2845-2849 (1982).
52. B. H. Dulis, T. M. Kloppel, H. M. Grey, and R. T. Kubo, *J. Biol. Chem., 257*:4369-4374 (1982).
53. M. J. Wilson and D. L. Hatfield, *Biochim. Biophys. Acta, 781*:205-215 (1984).
54. J. Uitto and D. J. Prockop, *Biochim. Biophys. Acta, 336*:234-251 (1974).
55. K. Inouye, S. Sakakibara, and D. J. Prockop, *Biochim. Biophys. Acta, 420*:133-141 (1976).
56. K. Inouye, Y. Kobayashi, Y. Kyogoku, Y. Kishida, S. Sakakibara, and D. J. Prockop, *Arch. Biochem. Biophys., 219*:198-202 (1982).
57. B. Steinmann, V. H. Rao, and R. Gitzelmann, *FEBS Lett., 133*:142-144 (1981).
58. W. W-Y. Kao, D. J. Prockop, and R. A. Berg, *J. Biol. Chem., 254*:2234-2243 (1979).
59. R. A. Berg, B. Steinmann, S. I. Rennard, and R. G. Crystal, *Arch. Biochem. Biophys., 226*:681-686 (1983).
60. R. A. Berg, W. W-Y. Kao, and D. J. Prockop, *Biochim. Biophys. Acta, 444*:756-764 (1976).
61. P. Tolstoshev, R. A. Berg, S. I. Rennard, K. H. Bradley, B. C. Trapnell, and R. G. Crystal, *J. Biol. Chem., 256*:3135-3140 (1981).
62. R. A. Berg and D. J. Prockop, *Biochem. Biophys. Res. Commun., 52*:115-120 (1973).
63. S. A. Jimenez, P. Dehm, B. R. Olsen, and D. J. Prockop, *J. Biol. Chem., 248*:720-729 (1973).
64. W. S. Sly and H. D. Fischer, *J. Cell. Biochem., 18*:67-85 (1982).
65. R. T. Dean, *Nature, 257*:414-416 (1975).
66. R. T. Dean and P. A. Riley, *Biochim. Biophys. Acta, 539*:230-237 (1978).
67. M. Hentze, A. Hasilik, and K. von Figura, *Arch. Biochem. Biophys., 230*:375-382 (1984).
68. G. E. Mortimore, A. N. Neely, J. Cox, and R. A. Guinivan, *Biochem. Biophys. Res. Commun., 54*:89-95 (1973).
69. G. Livesy, K. E. Williams, S. E. Knowles, and F. J. Ballard, *Biochem. J., 188*:895-903 (1980).

70. R. T. Dean, *Cellular Degradative Processes*, Halstead Press, New York, 1978, pp. 1-79.
71. G. E. Mortimore and C. M. Schworer, *Nature, 270*:174-176 (1977).
72. F. J. Ballard and J. M. Gunn, *Nutr. Rev., 40*:33-42 (1983).
73. B. J. Baum, J. Moss, S. D. Breul, R. A. Berg, and R. G. Crystal, *J. Biol. Chem., 255*: 2843-2847 (1980).
74. R. A. Berg, J. Moss, B. J. Baum, and R. G. Crystal, *Clin. Invest., 67*:1457-1462 (1981).
75. G. E. Mortimore and C. E. Mondon, *J. Biol. Chem., 245*:2375-2383 (1970).
76. D. W. Foster, in *The Metabolic Basis of Inherited Disease* (J. B. Stanbury, J. B. Wyngaarden, D. S. Fredrickson, J. L. Golstein, and M. S. Brown, eds.), McGraw-Hill, New York, 1983, pp. 99-117.
77. V. Botbol and O. A. Scornik, *J. Biol. Chem., 254*:11254-11257 (1979).
78. V. Botbol and O. A. Scornik, *Proc. Natl. Acad. Sci. USA, 76*:710-713 (1979).
79. J. E. Rothman, *Science, 213*:1212-1219 (1981).
80. W. J. Brown and M. G. Farquhar, *Cell, 36*:295-307 (1984).
81. B. Gumbiner and R. B. Kelly, *Cell, 28*:51-59 (1982).
82. R. Green and D. Shields, *J. Cell Biol., 99*:97-104 (1984).
83. J. L. Guan and J. K. Rose, *Cell, 37*:779-787 (1984).
84. G. J. A. M. Strous and H. F. Lodish, *Cell, 22*:709-717 (1980).
85. H. Lodish, N. Kong, M. Snider, and G. J. A. M. Straus, *Nature, 304*:80-83 (1983).
86. W. Jessup, M. F. Shirazi, and R. T. Dean, *Biochem. Pharmacol., 32*:2703-2710 (1983).
87. R. M. Steinmann, I. S. Mellman, W. A. Muller, and Z. A. Cohn, *J. Cell. Biol., 96*: 1-27 (1983).
88. W. W-Y. Kao, R. A. Berg, and D. J. Prockop, *J. Biol. Chem., 252*:8391-8397 (1977).
89. R. Timpl and R. G. Glanville, *Clin. Orthoped. Rel. Res., 158*:230-248 (1981).
90. D. Horlein, J. McPherson, S. Hangoh, and P. Bornstein, *Proc. Natl. Acad. Sci. USA, 78*:6163-6167 (1981).
91. J. F. Bateman, T. Mascara, D. Chan, and W. G. Cole, *Biochem. J., 215*:103-115 (1984).

III

PATHOLOGY OF THE EXTRACELLULAR MATRIX

11

Disorders Associated with Collagen Biosynthesis

Dermatological diseases as prototypes for collagen abnormalities

Eugene A. Bauer and Daniel J. Santa Cruz
Washington University School of Medicine, St. Louis, Missouri

Jouni Uitto
UCLA School of Medicine and Harbor-UCLA Medical Center, Torrance, California

Arthur Z. Eisen
Washington University School of Medicine, St. Louis, Missouri

I. INTRODUCTION

In human skin, collagen accounts for approximately 70% of the dry weight of the dermis. As in other organs, the biological function of this family of connective tissue macromolecules is to provide the structural framework of the organism. However, in the skin unique functions of providing tensile strength to support the epidermis while allowing substantial mobility must also be fulfilled. Indeed, as will become apparent in the discussions of the diseases that follow, when the integrity of the dermal collagen is compromised, irrespective of the mechanism, these critical functions are lost and with them the essential function of the skin itself—that of serving as an adaptable barrier between the organism and the outside world.

Three major conceptual pathways exist by which diseases of connective tissue might arise: (a) disorders of collagen synthesis, (b) disorders of collagen degradation, and (c) disorders of cellular growth. Research in our laboratory has been directed at the elucidation of certain diseases affecting the connective tissue. Indeed, the pathophysiological mechanisms in three such diseases serve to illustrate the disparate nature of the changes in cell biology that can give rise to these collagen diseases. Thus, the prototype disease for defective collagen biosynthesis is scleroderma, while that for deranged collagen degradation is recessive dystrophic epidermolysis bullosa. Finally, as the prototypic disease for defective regulation of cellular growth in the skin, we shall consider the disorder focal dermal hypoplasia.

II. SCLERODERMA: A DISORDER OF INCREASED COLLAGEN SYNTHESIS

Systemic scleroderma is a clinical condition of unknown etiology characterized by the diffuse accumulation of collagen in the skin, subcutaneous tissues, and a variety of internal organs (1-3). In the skin, the histopathological picture early in the course of the disease is one of a superficial and deep perivascular mononuclear cell infiltrate dominated by plasma cells. This infiltrate is also found at the dermal-subcutaneous junction. In the later stages of the disease (Figure 1) the infiltrate becomes sparse. The collagen bundles coarsen, and the reticular dermis develops a hyalinized appearance. The progressive fibrosis surrounds and replaces the normal adipose cushion around the sweat glands, and the dermal-subcutaneous interface is flattened.

The specific mechanisms for collagen accumulation in the affected tissues in scleroderma have not been clarified. Early studies suggested an increased rate of collagen synthesis in the skin of patients with scleroderma (4) based on the observations that the activity of prolyl hydroxylase, a key enzyme in collagen biosynthesis, was increased in skin biopsy specimens taken from patients with active disease (5,6). In addition, the synthesis of radioactive hydroxyproline was also increased under tissue culture conditions (7,8). The conclusion that collagen biosynthesis might be increased in patients with active scleroderma was further supported by the demonstration of new, labile cross-links in vivo, an observation consistent with excessive synthesis of collagen (9).

However, as implied previously, in a disorder such as scleroderma, the excessive accumulation of collagen in the skin might occur (a) by increased synthesis per cell (i.e., per fibroblast), (b) by decreased degradation of collagen, or (c) by an increased number of matrix cells, each with a normal synthetic rate of collagen. These questions have been addressed using skin fibroblasts in culture. Initially, LeRoy (10,11) suggested that the synthesis of collagen per cell may be increased in patients with scleroderma. Subsequently, Buckingham et al. (12) also observed increased collagen synthesis in scleroderma fibroblasts. Another study, however, indicated that the synthesis of collagen by scleroderma fibroblasts was not different from that of the control cells (13). Furthermore, one investigator has suggested that the activity of collagenase, the enzyme initiating collagen breakdown in vivo, is decreased in scleroderma (14). We have attempted to resolve some of these disparate data on the nature of the collagen defect in scleroderma by using fibroblast cultures to dissect each of the possible general mechanisms in the same cell strains.

For these experiments, skin fibroblast cultures were established from 8 patients with systemic scleroderma (15). As an index of the biosynthetic rate of collagen in vitro, the synthesis of radioactive hydroxyproline was measured relative to DNA or to total cell protein (15). As shown in Table 1, the rate of collagen synthesis in cell lines derived from patients with scleroderma was clearly increased when compared to control cell lines. The mean hydroxyproline synthesis in scleroderma fibroblast lines was about twice that in the corresponding controls. When examined individually, five of the eight scleroderma cell lines demonstrated values significantly higher than those of the controls ($p < 0.05$).

Cultured human skin fibroblasts synthesize and secrete two major genetic types of procollagen, type I and type III (16,17). To estimate the relative synthesis of these procollagens, type I and type III procollagens were isolated by DEAE-cellulose chromatography. The elution patterns were qualitatively similar in samples obtained from both scleroderma and control fibroblasts (15). The distribution of radioactivity between type I and type III procollagen was relatively constant, and the mean ratio of [^{14}C]-hydroxyproline in type I and type III procollagen in scleroderma cell cultures did not differ from the corresponding value observed in control cell lines (Table 2). In addition, the helical stability of the two procollagen types was assessed and found to be equivalent to that seen in control procollagen preparations (15). Thus, in this cell culture system, the observations did

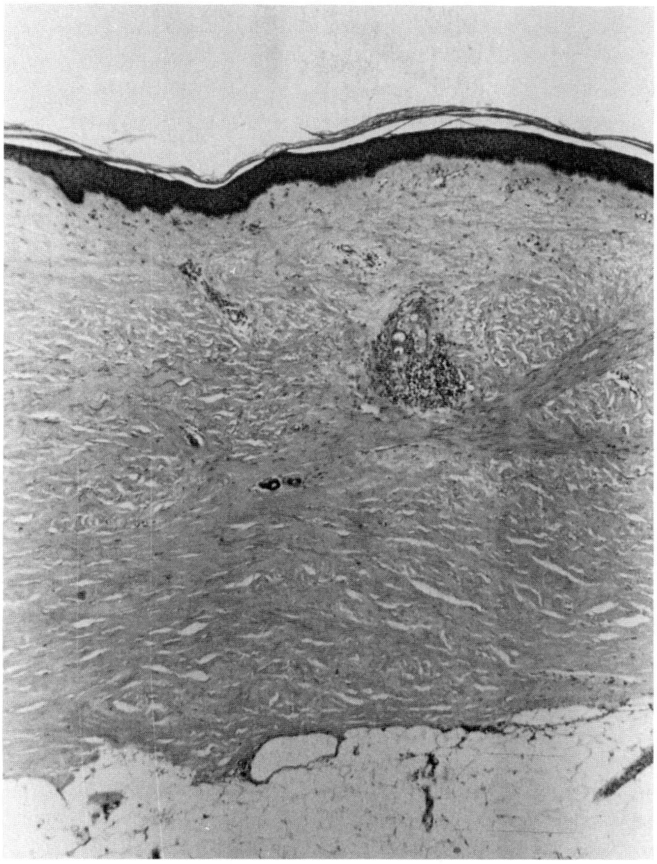

Figure 1 Scleroderma (late stage). The dermis appears hyalinized and a focal mononuclear cell infiltrate is present. The dermal-subcutaneous interface is flattened. (Hematoxylin and eosin; X80).

not corroborate a previous immunofluorescence study, which suggested that the proportions of these collagens may differ depending on the stage of the disease (18).

Recently serum factors have been implicated in the pathogenesis of scleroderma (19-21). In order to determine whether such serum factors were unique to scleroderma, we examined the effects of human serum on the production of type I and type III procollagens of normal and scleroderma skin fibroblasts in culture (22) using sera obtained from normal human subjects as well as from patients with scleroderma. Although the results indicate that normal human serum contains factor(s) that markedly increase the production of types I and III procollagens by the cells, no evidence of a factor that might further stimulate procollagen synthesis was noted in the serum of patients with scleroderma (22).

Since it was possible that the rate of incorporation of radioactive proline into protein might be affected by the growth rates of the cells, this parameter was also compared in normal and scleroderma fibroblasts. The mean population doubling times for control fibroblasts was 3.04 ± 0.11 days compared to 3.00 ± 0.22 days for the scleroderma cells (p = NS). These studies also further indicate that the increased synthesis of procollagen

Table 1 Rate of Procollagen Biosynthesis by Scleroderma Fibroblasts in Culture

	[^{14}C]-Hydroxyproline			
Cell line	dpm/μg DNA/hr	% of controls	dpm/mg protein/hr	% of controls
Controls (n = 11)				
Mean ± SE	127.0 ± 7.2	100	2572.3 ± 202.4	100
Scleroderma[a]				
Case 1	434.2	342	7177.3[b]	279
Case 2	414.1	326	10823.4[b]	420
Case 3	365.6	288	5878.2[b]	229
Case 4	178.0	140	4338.0[b]	169
Case 5	174.6	137	5475.5[b]	213
Case 6	172.3	135	3295.0[b]	128
Case 7	126.3	99	3439.7	134
Case 8	117.6	93	1087.0	42
Mean ± SE	247.8 ± 47.0[c]	195	5189.3 ± 1039.0[c]	202

[a]The individual values represent means of two parallel determinations.
[b]These values are greater than the mean +2 SD of the controls.
[c]Statistically different from the controls ($p < 0.05$).
Source: Adapted from Ref. 15.

(Table 1) cannot be attributed to an increase in an absolute number of matrix cells (15).

Finally, since a decrease in collagenase had been previously proposed as accounting for the increased accumulation of collagen in scleroderma (14), we examined collagenase production in the same cell strains used to quantitate procollagen biosynthesis (15). Immunoreactive collagenase protein was determined using a specific radioimmunoassay (23). As a group, fibroblast cultures from patients with systemic scleroderma displayed a mean value for immunoreactive collagenase of 121 ± 27% of control (mean ± SE) when based on the quantity of medium protein and 113 ± 29% of control when based on the content of cellular protein (15). Individually, only one cell strain displayed a significant decrease in immunoreactive collagenase protein. Because collagenase is found in the cell culture medium in an inactive proenzyme form (24,25), collagenase activity was also de-

Table 2 Relative Synthesis of Type I and Type III Procollagens by Scleroderma Fibroblasts

	Relative amounts of procollagen[a]	
Cell line	Type I	Type III
Controls (n = 9)	83.8 ± 4.0	16.2 ± 0.8
Scleroderma (n = 8)	80.5 ± 6.5[b]	19.5 ± 1.6[b]

[a]The values are percent of [^{14}C]-hydroxyproline in type I and type III procollagen peaks separated by DEAE-cellulose chromatography; mean ± SE.
[b]The values are not statistically different from the control.
Source: Adapted from Ref. 15.

termined in both control and scleroderma fibroblast cultures, after maximal enzyme activation by trypsin proteolysis. Collagenase activity per unit immunoreactive protein was 98 ± 8% of control values, indicating that the catalytic efficiency of the enzyme was apparently normal in the cell cultures derived from scleroderma patients.

Despite these data, which clearly indicate that increased collagen synthesis—not enhanced cell growth or decreased collagen degradation—characterizes fibroblasts derived from patients with scleroderma, the precise mechanisms involved are not fully understood. It has been suggested that immunological mechanisms, mediated through lymphokines, can stimulate collagen synthesis by these cells and, therefore, may play a role in scleroderma (26-30); however, the significance of such lymphocyte-mediated events requires further clarification. For example, Goldberg (31) has shown that normal fibroblasts contain specific receptors for type I collagen and has postulated that the specific binding of collagen to the cell surface might represent a feedback mechanism for the control of collagen synthesis. Indeed, one study has suggested that the addition of amino terminal extension peptides from type I procollagen reduced the excessive synthesis of collagen by scleroderma fibroblasts in vitro (32), thus, lending credibility to the putative feedback mechanism.

III. RECESSIVE DYSTROPHIC EPIDERMOLYSIS BULLOSA: EVIDENCE FOR DEFECTIVE REGULATION OF COLLAGENASE

Recessive dystrophic epidermolysis bullosa is a heritable disorder of the skin and mucous membranes characterized by extensive blistering following minor trauma (33). The early electron microscopic studies of Pearson (34) disclosed marked collagen degeneration and phagocytosis of collagen fibrils by macrophages in the area of blistering, suggesting that mechanisms involved in collagen breakdown might be important in the pathogenesis of blister formation. The essential morphological features of this disorder are shown in Figure 2 and include, in addition to the subepidermal bulla seen by light microscopy, degenerative changes in the collagen of the papillary dermis, a decrease or absence of anchoring fibrils, and the early formation of a sub-basal lamina bulla, which are seen ultrastructurally (33).

Although the etiology of this disorder is unknown, a possible mechanism for blistering has been proposed in the recessive dystrophic epidermolysis bullosa from studies showing that skin obtained from friction blisters produced elevated levels of collagenase in organ culture (35,36). This finding suggested that collagenase could participate in blister formation, and increased activity of the enzyme would be consistent with the electron microscopic findings of collagen degeneration and the phagocytosis of collagen fibrils by dermal macrophages (see above).

It is of interest that in the original organ culture experiments, clinically unaffected skin from recessive dystrophic epidermolysis bullosa patients also manifested increased collagenase activity in vitro (35), suggesting that this might be a characteristic of the disease. However, despite the apparent presence of increased collagenase activity in blistered skin, it was unclear whether the in vitro observations reflected the in vivo pathophysiology of the disease or represented a nonspecific (wound healing) response (35). Indeed, we postulated that if the genetic abnormality were to result in a fundamental biochemical change characterized by an excess of collagenase, this should be reflected in unaffected skin as well (37).

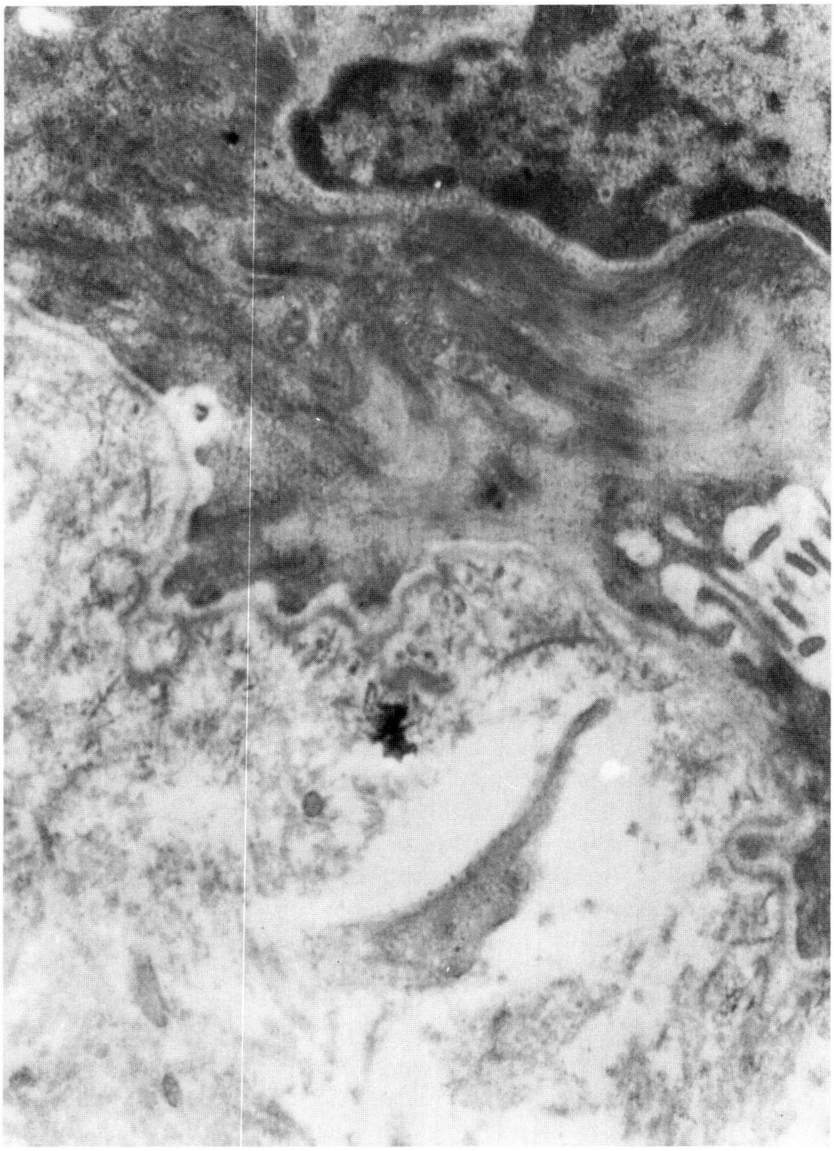

Figure 2 Recessive dystrophic epidermolysis bullosa (dermoepidermal junction). Note the diminished number of anchoring fibrils. The dermis shows a sparsity of collagen fibers, which appear to have been replaced by finely granular material (X 10,300).

In order to approach this question, we sought to measure collagenase in tissue extracts of human skin. However, owing to the presence of a variety of serum and tissue inhibitors of collagenase as well as the secretion of this enzyme in latent form (24,25), the major difficulty in determining the significance of collagenase in vivo has been the inability to detect the active enzyme. Thus, to evaluate the in vivo role of collagenase, we developed a specific radioimmunoassay for quantitating collagenase in biologic materials (23).

When quantitated by immunoassay, tissue extracts of blistered skin contained up to a 10-fold increase in collagenase protein irrespective of whether they were derived from patients with recessive or dominant forms of epidermolysis bullosa (37). Of greater interest, however, was the observation that the clinically unaffected skin from patients with recessive dystrophic epidermolysis bullosa contained a mean fourfold increase in immunoreactive collagenase ($p < 0.001$) and some patients displayed levels that ranged as high as 12-fold greater (37). In contrast, the collagenase content of nonblistered skin from patients with dominantly inherited epidermolysis bullosa was not significantly elevated (37).

Since the synthesis of collagenase appears to represent a major biological function of normal human skin fibroblasts (24,38), we reasoned that if a primary defect in the regulation of synthesis and/or degradation of collagenase were, indeed, basic to the etiology of recessive dystrophic epidermolysis bullosa, it might be expected that this trait would be expressed in fibroblast cultures taken from uninvolved skin. Thus, we examined fibroblast cultures derived from 27 patients with various forms of epidermolysis bullosa for the capacity to synthesize and secrete collagenase (39).

As shown in Figure 3, the mean concentration of immunoreactive collagenase in the fibroblast cultures of 10 patients with recessive dystrophic epidermolysis bullosa was 328% of control ($p < 0.001$). This characteristic was also apparently specific for recessive dystrophic epidermolysis bullosa, since the concentration of immunoreactive collagenase was 95% of control in recessive epidermolysis bullosa letalis, 91% of control in dominant dystrophic epidermolysis bullosa, and 96% of control in dominant epidermolysis bullosa simplex (39). It is important to emphasize that increased collagenase levels were not seen uniformly in fibroblast cultures. Indeed, 2 of the 10 patients with recessive dystrophic epidermolysis bullosa displayed normal in vitro production of immunoreactive collagenase, a finding that suggests that all patients who phenotypically appear to have recessive dystrophic epidermolysis bullosa may not have the same degree of biochemical defect. The biochemical mechanism for the increased amount of collagenase is through increased synthesis (40), a finding that has been substantiated by the observation of increased translatable collagenase messenger RNA in recessive dystrophic epidermolysis bullosa cells (41).

In the cells of two of the patients, and possibly several others as well (42), the increased accumulation of collagenase in culture medium was associated with synthesis of an aberrant form of the enzyme (42). Initially using partially purified procollagenase preparations (42) and more recently electrophoretically homogeneous procollagenase (43), the enzymes derived from the recessive dystrophic epidermolysis bullosa cultures were significantly more thermolabile than control procollagenase preparations (Table 3). The diminished thermal stability was correlated with an altered affinity of the putatively mutant collagenases for Ca^{2+}, a metal cofactor essential for activity and thermal stability (44). The apparent K_m for Ca^{2+} for the control enzymes was about 1 mM compared to about 4 mM for the recessive dystrophic epidermolysis bullosa collagenases (42,43).

It should be pointed out that, although excessive synthesis or activity of an enzyme is an unusual manifestation of any recessively inherited disease, such increases have been observed in a variety of genetic circumstances. Indeed, examples of genetically increased synthesis of abnormal gene products—both intracellular and extracellular—have also been described (45--48). The most detailed study to date has been in the Hektoen variant of glucose-6-phosphate dehydrogenase in which a single amino acid substitution in the enzyme was found to be associated with a fourfold increase in synthesis of the gene pro-

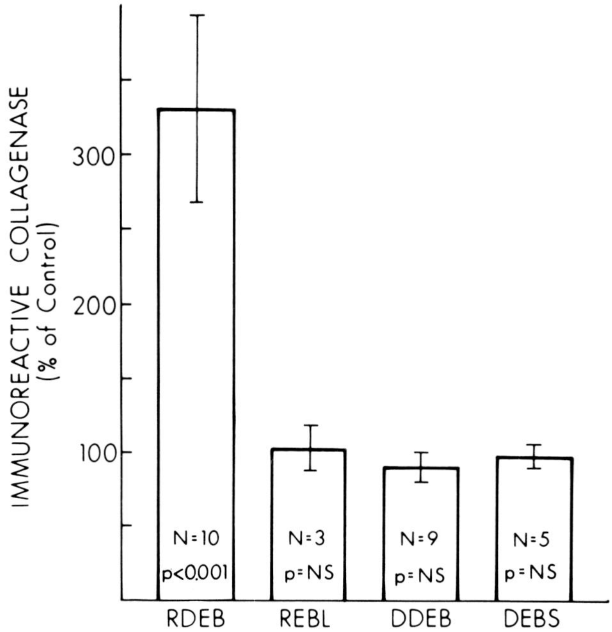

Figure 3 Immunoreactive collagenase concentration in fibroblast cultures from genetically distinct types of epidermolysis bullosa. Collagenase protein was quantitated as ng/mg protein and is expressed as a percent of that found in control cultures. RDEB, recessive dystrophic epidermolysis bullosa (10 patients); REBL, recessive epidermolysis bullosa letalis (three patients); DDEB, dominant dystrophic epidermolysis bullosa (nine patients); DEBS, dominant epidermolysis bullosa simplex (five patients). (Adapted from Ref. 39.)

Table 3 Thermal Stability of Control and Recessive Dystrophic Epidermolysis Bullosa Procollagenase[a]

Procollagenase	Ca^{2+} concentration (mM)	Activity remaining (%)
Control	0	0
RDEB 1	0	0
RDEB 2	0	0
Control	2	77
RDEB 1	4	34
RDEB 2	2	28
Control	10	91
RDEB 1	10	52
RDEB 2	10	48

[a]Portions of purified procollagenase were incubated at 60°C for 5 min after which residual collagenase activity was determined. RDEB, recessive dystrophic epidermolysis bullosa.
Source: Adapted from Ref. 42.

duct (49), representing a so-called quantitative mutant (50). As discussed by Yoshida (49) and Goldberger (51), the data suggest defective autogenous regulation of synthesis, a mechanism that may be relevant in the pathogenesis of recessive dystrophic epidermolysis bullosa.

IV. FOCAL DERMAL HYPOPLASIA: A DEFECT IN CELL GROWTH

Focal dermal hypoplasia is a syndrome characterized by widespread anomalies of cutaneous, osseous, ocular, and dental structures. The syndrome consists of a congenital decrease in the dermal connective tissue in association with other defects involving tissues of ectodermal and mesodermal origin (52,53). It is predominantly seen in females, a fact probably explained by an X-linked dominant inheritance, so that heterozygous females live to demonstrate the congenital abnormalities. In the hemizygous males, however, it has been postulated that the disease is so severe that intrauterine death occurs. Histopathologically (Figure 4) the lesions are characterized by the absence of dermal connective tissue, so that mature fat cells impinge directly on the epidermis, separated only by a thin layer of reticular fibers (54,55). The appearance suggests that there is a decrease in the collagen of the reticular dermis as a result of its being replaced by strands and islands of adipose tissue that follow the dermal vessels.

Although the molecular basis for the absence of connective tissue in focal dermal hypoplasia is unknown, we postulated that the cutaneous abnormalities in this disease might result either from an aberration in the capacity of the cellular elements of the connective tissue to synthesize the major extracellular matrix proteins or from altered growth of these cells. In order to address this problem, we cultured skin fibroblasts from both the affected an unaffected skin of a patient with focal dermal hypoplasia (56).

Interestingly, fibroblasts derived from the lesional skin were strikingly abnormal by phase contrast microscopy and were characterized by a large granular cytoplasm with prominent cytoplasmic vacuoles. The vacuoles were seen only in fibroblast cultures from the affected skin but not in those from the unaffected skin or in control skin fibroblast cultures (56).

In contrast to the observations of biochemical abnormalities made in the previously discussed disorders (see above), two major pieces of evidence indicated that the primary abnormality in focal dermal hypoplasia lies in the proliferative capacity of the fibroblasts. First, the mean population doubling time of cells in cultures initiated from the lesions was about twice as long as that observed in cultures derived from unaffected skin of the same patient or from healthy controls (Table 4). Of additional interest was the finding that the final saturation density at confluence of the cell cultures from the affected skin was one-fifth that of the controls (Table 5). These findings indicated that the cells in the affected skin were characterized by a significantly compromised growth potential. Neither the growth kinetics nor the saturation density changed with passage of the focal dermal hypoplasia cells.

Despite the striking abnormalities in the growth properties of the cells from the affected skin, the rate of procollagen synthesis, when expressed in relation to DNA or cell protein (i.e., on a per cell basis) was not different from that seen in normal control fibroblasts or in fibroblasts derived from unaffected skin (56). Similarly, the synthesis of the two major types of procollagen, type I and type III, occurred in a normal proportion (56). Thus, it appears that fibroblasts in the lesions of focal dermal hypoplasia represent a distinct population of mutant cells. It is possible that the patchy distribution

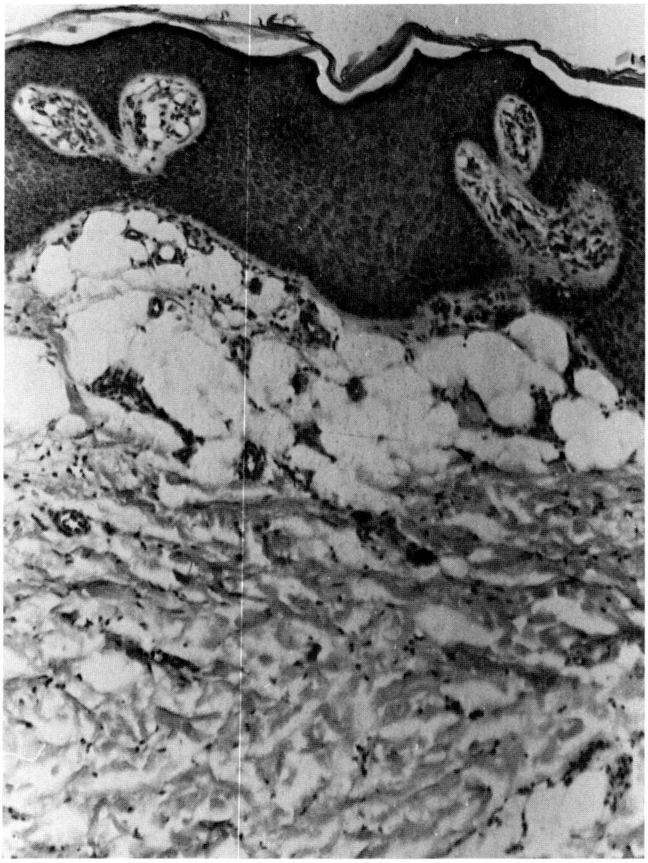

Figure 4 Focal dermal hypoplasia. Note the lobules of mature adipose tissue that impinge on the epidermis and the increase in vascularity. (Hematoxylin and eosin; X 200).

Table 4 Growth Kinetics of Normal and Focal Dermal Hypoplasia Fibroblasts

Culture	Population doubling time[a]
Controls[b]	1.23 ± 0.16
FDH, unaffected skin	1.43 ± 0.21[c]
FDH, affected skin	2.88 ± 0.66[d]

[a]Expressed as mean ± SE days.
[b]Controls represent values from three different normal human skin fibroblast lines. FDH, focal dermal hypoplasia.
[c]p = NS compared to the control cultures.
[d]$p < 0.01$ compared to the control cultures.
Source: Adapted from Ref. 56.

Table 5 Saturation Density of Normal and Focal Dermal Hypoplasia Fibroblasts in Culture

Culture	Saturation density[a] (Cells X $10^3/25\ cm^2$ flask)
Controls[b]	2330 ± 646
FDH, unaffected skin	2335 ± 15[c]
FDH, affected skin	440 ± 66[d]

[a]Expressed as mean ± SE.
[b]Controls represent values from two different normal human skin fibroblast lines. FDH, focal dermal hypoplasia.
[c]p = NS compared to the control cultures.
[d]$p < 0.025$ compared to the control cultures.
Source: Adapted from Ref. 56.

of these cells in the skin of affected females may reflect mosaicism as a consequence of the lyonization process.

V. CONCLUSION

In this discussion, we have used three diseases affecting the connective tissue of the skin to illustrate the marked differences in pathogenetic mechanisms that underlie some types of collagen diseases. In two of these diseases—scleroderma and recessive dystrophic epidermolysis bullosa—the available evidence points to defects in the regulation of synthesis of proteins important to the structural integrity of the skin. In the third disease—focal dermal hypoplasia—the biosynthetic machinery for synthesis of at least the major connective tissue macromolecule, collagen, appears to be intact, while the cellular growth patterns are grossly anomalous.

ACKNOWLEDGMENTS

This work was supported by U.S.P.H.S. grants AM 19537, AM 12129, AM 28450, GM 28833, AM 35927, and TO AM 07284.

REFERENCES

1. R. E. Winkelmann, *Acta Dermato-Venerol., 56*:83-92 (1976).
2. R. Fleischmajer, *Int. J. Dermatol., 16*:310-318 (1977).
3. A. Z. Eisen, J. Uitto and E. A. Bauer, in *Dermatology in General Medicine*, 2nd ed. (T. B. Fitzpatrick, A. Z. Eisen, K. Wolff, I. M. Freedberg, and K. F. Austen, eds.), McGraw-Hill, New York, 1979, pp. 1305-1313.
4. J. Uitto, *Ann. Clin. Res., 3*:250-258 (1972).
5. J. Uitto, J. Halme, M. Hannuksela, P. Peltokallio, and K. I. Kivirikko, *Scand. J. Clin. Lab. Invest., 23*:241-247 (1969).
6. H. R. Keiser, H. D. Stein, and A. Sjoerdsma, *Arch Dermatol., 104*:57-60 (1971).
7. H. R. Keiser and A. Sjoerdsma, *Clin. Chim. Acta, 23*:341-346 (1969).
8. J. Uitto, P. Helin, O. Rasmussen, and I. Lorenzen, *Ann. Clin. Res., 2*:228-234 (1970).

9. C. M. Herbert, K. A. Lindberg, I. V. Jayson, and A. J. Bailey, *Lancet, 1*:187-192 (1974).
10. E. C. LeRoy, *J. Exp. Med., 135*:1351-1362 (1972).
11. E. C. LeRoy, *J. Clin. Invest., 54*:880-889 (1974).
12. R. Buckingham, R. K. Prince, G. P. Rodnan, and F. Taylor, *J. Lab. Clin. Med., 92*: 5-21 (1978).
13. J. S. Perlish, R. I. Bashey, R. E. Stephens, and R. Fleischmajer, *Arthritis Rheum., 19*: 891-901 (1976).
14. A. H. Brady, *J. Clin. Invest., 56*:1175-1180 (1975).
15. J. Uitto, E. A. Bauer, and A. Z. Eisen, *J. Clin. Invest., 64*:921-930 (1979).
16. B. D. Smith, P. H. Byers, and G. R. Martin, *Proc. Natl. Acad. Sci. USA, 69*:3960-3962 (1972).
17. J. R. Lichtenstein, P. H. Byers, B. D. Smith, and G. R. Martin, *Biochemistry, 14*: 1589-1594 (1975).
18. R. Fleischmajer, S. Gay, W. N. Meigel, and J. S. Perlish, *Arthritis Rheum., 21*:418-428 (1978).
19. R. I. Bashey and S. A. Jimenez, *Biochem. Biophys. Res. Commun., 76*:1214-1222 (1977).
20. A. J. Keyser, S. M. Cooper, E. Ruoslahti, M. E. Nimni, and F. P. Quismorio, *Fed. Proc., 38*:1339 (1979).
21. A. J. Keyser, S. M. Cooper, and M. E. Nimni, *Fed. Proc., 39*:874 (1980).
22. E. M. L. Tan, J. Uitto, E. A. Bauer, and A. Z. Eisen, *J. Invest. Dermatol., 76*:462-467 (1981).
23. E. A. Bauer, A. Z. Eisen, and J. J. Jeffrey, *J. Biol. Chem., 247*:6679-6685 (1972).
24. E. A. Bauer, G. P. Stricklin, J. J. Jeffrey, and A. Z. Eisen, *Biochem. Biophys. Res. Commun., 64*:232-240 (1975).
25. G. P. Stricklin, E. A. Bauer, J. J. Jeffrey, and A. Z. Eisen, *Biochemistry, 16*:1607-1615 (1977).
26. J. M. Stuart, E. A. Postlethwaite, and A. H. Kang, *J. Lab. Clin. Med., 88*:601-608 (1976).
27. H. Kondo, B. S. Rabin, and G. P. Rodnan, *J. Clin. Invest., 58*:1388-1394 (1976).
28. A. E. Postlethwaite, R. Snyderman, and A. H. Kang, *J. Exp. Med., 144*:1188-1203 (1976).
29. R. L. Johnson and M. Ziff, *J. Clin. Invest., 48*:240-252 (1976).
30. S. M. Wahl, L. M. Wahl, and J. B. McCarthy, *J. Immunol., 121*:942-946 (1978).
31. B. Goldberg, *Cell, 16*:265-275 (1979).
32. T. Krieg, D. Horlein, M. Wiestner, and P. K. Muller, *Arch. Dermatol. Res., 263*: 171-180 (1978).
33. E. A. Bauer and R. A. Briggaman, in *Dermatology in General Medicine*, 2nd ed. (T. B. Fitzpatrick, A. Z. Eisen, K. Wolff, I. M. Freedberg, and K. F. Austen, eds.), McGraw-Hill, New York, 1979, pp. 334-347.
34. R. W. Pearson, *J. Invest. Dermatol., 39*:551-575 (1962).
35. A. Z. Eisen, *J. Invest. Dermatol., 52*:449-453 (1969).
36. G. S. Lazarus, *J. Invest. Dermatol., 58*:242-248 (1972).
37. E. A. Bauer, T. Gedde-Dahl, and A. Z. Eisen, *J. Invest. Dermatol., 68*:119-124 (1977).
38. K. J. Valle and E. A. Bauer, *J. Biol. Chem., 254*:10115-10122 (1979).
39. E. A. Bauer and A. Z. Eisen, *J. Exp. Med., 148*:1378-1387 (1978).
40. K. J. Valle and E. A. Bauer, *J. Clin. Invest., 66*:176-187 (1980).
41. A. Kronberger, K. J. Valle, A. Z. Eisen, and E. A. Bauer, *J. Invest. Dermatol., 79*: 403-407 (1982).
42. E. A. Bauer, *Proc. Natl. Acad. Sci. USA, 74*:4646-4650 (1977).

43. G. P. Stricklin, H. G. Welgus, and E. A. Bauer, *J. Clin. Invest., 69*:1373-1383 (1982).
44. J. L. Seltzer, H. G. Welgus, J. J. Jeffrey, and A. Z. Eisen, *Arch. Biochem. Biophys., 173*:355-361 (1976).
45. A. Yoshida and A. G. Motulsky, *Am. J. Hum. Genet., 21*:486-498 (1969).
46. F. S. Rosen, C. A. Alper, J. Pensky, M. R. Klemperer, and V. H. Donaldson, *J. Clin. Invest., 50*:2143-2149 (1971).
47. A. B. Laurell, J. Lindegren, I. Malmros, and H. Martensson, *J. Clin. Lab. Invest., 24*: 221-225 (1969).
48. E. Shapira, Y. Ben-Yoseph, F. G. Eyal, and A. Russell, *J. Clin. Invest., 53*:59-63 (1974).
49. A. Yoshida, *J. Mol. Biol., 52*:483-490 (1970).
50. H. E. Sutton and R. P. Wagner, *Annu. Rev. Genet., 9*:187-212 (1975).
51. R. F. Goldberger, *Science, 183*:810-816 (1974).
52. R. W. Goltz, W. C. Peterson, R. J. Gorlin, and H. G. Ravits, *Arch. Dermatol., 86*: 708-711 (1962).
53. R. W. Goltz, R. R. Henderson, J. M. Hitch, and J. E. Ott, *Arch. Dermatol., 101*:1-11 (1970).
54. J. B. Howell, *Arch. Dermatol, 92*:238-248 (1965).
55. W. F. Lever and G. Schaumburg-Lever, in *Histopathology of the Skin*, 5th ed., J. B. Lippincott, Philadelphia, 1975, pp. 67-68.
56. J. Uitto, E. A. Bauer, D. J. Santa Cruz, R. J. Loewinger, and A. Z. Eisen, *J. Invest. Dermatol., 75*:170-175 (1980).

12

Posttranslational Modifications of Collagen and Their Alterations in Heritable Diseases

Kari I. Kivirikko and Helena Kuivaniemi*
Collagen Research Unit, University of Oulu, Oulu, Finland

I. INTRODUCTION

The biosynthesis of collagen resembles that of other secretory proteins but is distinguished by the presence of an unusually large number of cotranslational and posttranslational modifications of the polypeptide chains. Many of these modifications are unique to collagens and a few other proteins with collagenlike amino acid sequences. The basic features of these reactions are now well defined, and most of the enzymes involved have been characterized. Initial information is also available on the regulation of the rates and extents of the modifications, on alterations in some of the modifications in pathological states, and on the possibilities for specific pharmacological control of collagen biosynthesis at the level of some of these events. The purpose of this chapter is to review current information on the posttranslational modifications of collagens, with special emphasis on their alterations in heritable diseases. For more detailed accounts and more complete references, the reader is referred to recent reviews dealing with various aspects of these modifications and of collagen biosynthesis in general (1-9) or with the intracellular modifications (10), hydroxylations (11,12), glycosylations (13), extracellular modifications (14), and lysyl oxidase (15). The reviews describing changes in collagen biosynthesis in pathological states also include information on alterations in some of these modifications in human heritable and acquired diseases (2,4,6,16-24).

II. MAIN FEATURES OF THE MODIFICATIONS

The posttranslational processing of collagens can be regarded as occurring in two stages. Intracellular modifications, together with the synthesis of the polypeptide chains, are

**Present affiliation*: Jefferson Institute of Molecular Medicine and Department of Biochemistry, Thomas Jefferson University, Philadelphia, Pennsylvania.

required to assemble and secrete the procollagen molecules. Extracellular modifications then convert these molecules into collagens and incorporate them into stable, cross-linked fibrils or other supramolecular structures. The main modifications and their biological significance are summarized in Table 1.

A. Intracellular Modifications and Secretion of Procollagen

1. Cleavage of the Signal Sequences

Most secreted proteins in prokaryotes and eukaryotes are synthesized in the form of a precursor containing an initial amino-terminal sequence not found in the secreted product. This extra peptide, which contains about 15-30 amino acid residues, is referred to as the "signal" (or "leader" or "preprotein") sequence. Signal sequences from various secreted proteins have similar structures and usually contain a positively charged amino acid residue near the amino terminus, followed by a core of 10-15 hydrophobic amino acid residues. The signal sequence initiates vectorial transfer of the nascent polypeptide chain across the membrane of the rough endoplasmic reticulum (25-27). The nascent collagen polypeptide chains have similar signal sequences, the size of the signal sequence in the chick preproα1(I), preproα2(I), and preproα1(III) chains being 22-26 amino acid residues (28,29).

The signal sequences are cleaved before chain completion, during or shortly after translocation across the membrane, with a single enzyme probably catalyzing the processing of a number of proteins. The amino acid sequences of different proteins at the cleavage site vary greatly. The "specificity" of the signal peptidases can be summarized as follows: a bond X-Y is hydrolyzed, where Y can be virtually any amino acid and X is an amino acid with a small side chain (26). This cleavage site is a Gly-Gln sequence in the chick preproα1(I) chain, a Ser-Gln or Ser-Glu sequence in the preproα2(I) chain, and probably an Ala-Gln sequence in the preproα1(III) chain (28,29).

Certain mutations in the structural genes of several secretory proteins have been shown to alter their signal sequences so that cleavage is impaired (30). A similar effect is seen when the signal sequences are modified by incorporation of amino acid analogs such as β-hydroxyleucine or β-hydroxynorvaline (9). Most of the resulting altered proteins are either not translocated across the membrane or not released from it, thus demonstrating cleavage to be a requirement for the normal secretory process (9,30). Cleavage probably has a similar function in the case of collagens.

2. Hydroxylation of Proline and Lysine Residues

The hydroxylation of proline and lysine residues is catalyzed by three separate enzymes, prolyl 4-hydroxylase (EC 1.14.11.2, usually termed prolyl hydroxylase), prolyl 3-hydroxylase (EC 1.14.11.7), and lysyl hydroxylase (ED 1.14.11.4). All three hydroxylate proline or lysine in peptide linkage. The corresponding reaction products, 4-hydroxyproline, 3-hydroxyproline, and hydroxylysine, are found within the vertebrate proteins almost exclusively in collagens and a few other proteins with collagenlike amino acid sequences (for detailed studies of hydroxylation, see Refs. 7,10-12).

Prolyl 4-hydroxylase has been purified to homogeneity from several sources, including human tissues, and extensively characterized (12,31). The active enzyme from all these sources is a tetramer ($\alpha_2\beta_2$) with a molecular weight of about 240,000 and consisting of two different types of enzymically inactive monomer with molecular weights of about 64,000 (α subunit) and 60,000 (β subunit). Distinct differences are found be-

Table 1 Posttranslational Events in the Biosynthesis of Collagen and Their Main Functions

Events	Biological significance
Intracellular events	
Cleavage of the signal sequences	Translocation across membrane (?)
4-Hydroxylation of proline residues	Essential for triple-helix formation at 37°C
3-Hydroxylation of proline residues	Unknown
Hydroxylation of lysine residues	Essential for glycosylation of hydroxylysine Essential for stability of cross-links
Glycosylation of hydroxylysine residues	May affect fibril morphology (?)
Glycosylation of propeptides	Unknown
Chain association and disulfide bonding	Needed for rapid triple-helix formation
Triple-helix formation	Needed for normal secretion rate Various properties of the molecules
Extracellular events	
Removal of propeptides	Essential for normal fiber formation
Ordered aggregation	Formation of fibrils or other ordered structures
Cross-link formation	Essential for stable fibrils

tween the α and β subunits in their amino acid and carbohydrate composition and in the peptide fragments produced by several proteases. The presence of two differently glycosylated α subunits has recently been demonstrated in a number of tissues (32). Lysyl hydroxylase has been purified to homogeneity from chick embryos and human placentas (12,31,33). The active enzyme from both sources is a dimer with a molecular weight of about 190,000 by gel filtration and consisting of only one type of monomer with a molecular weight of about 85,000. Prolyl 3-hydroxylase has been purified about 5000-fold from an ammonium sulfate fraction of chick embryo extract (12,31). The molecular weight is about 160,000 by gel filtration, but the subunit structure is unknown. Many observations suggest that the collagen hydroxylases do not have collagen type-specific or tissue-specific isoenzymes (7,9,34,35), but additional studies are required to elucidate this aspect further.

The minimum sequence requirement for interaction with a vertebrate prolyl 4-hydroxylase is an -X-Pro-Gly- triplet, and that for interaction with lysyl hydroxylase an -X-Lys-Gly- triplet (10,12). In some cases, however, both enzymes also hydroxylate certain other triplets. In agreement with these sequence requirements, 4-hydroxyproline and hydroxylysine have been identified in collagens and other vertebrate proteins almost exclusively in the Y positions of -X-Y-Gly- triplets. Prolyl 3-hydroxylase seems to require a -Pro-4Hyp-Gly- triplet, whereas a -Pro-Pro-Gly- sequence is probably not hydroxylated. In agreement with this, 3-hydroxyproline has been identified in collagens only in the sequence -Gly-3Hyp-4Hyp-Gly-. The interaction with prolyl 4-hydroxylase and lysyl hydroxylase is further influenced by the amino acid in the X position of the -X-Y-Gly- triplet that is hydroxylated, and the interaction with all three enzymes is affected by other nearby amino acids, by the peptide chain length, and by the peptide conformation (10,12). The triple-helical conformation prevents all three hydroxylations.

All three hydroxylases require Fe^{2+}, 2-oxoglutarate, O_2, and ascorbate. The 2-oxoglutarate is stoichiometrically decarboxylated and oxidized during hydroxylation, with one atom of the O_2 molecule being incorporated into the succinate while the other is incorporated into the hydroxyl group (Figure 1). The kinetic constants of the three enzymes for their substrates and cosubstrates are very similar, although not identical, and the catalytic center activities of pure prolyl 4-hydroxylase and lysyl hydroxylase are likewise similar (12,31). Immunological studies suggest distinct structural similarities close to or at the active sites of prolyl 4-hydroxylase and lysyl hydroxylase (34). Extensive kinetic studies (12,31) and other data (36,37) are consistent with an ordered binding of Fe^{2+}, 2-oxoglutarate, O_2, and the peptide substrate to the enzymes, and an ordered release of the products, in which Fe^{2+} is not released between most of the catalytic cycles. A detailed stereochemical mechanism involving a pocket model for the active site (Figure 2) has recently been suggested for prolyl 4-hydroxylase (38).

Ascorbate is virtually a specific requirement for the collagen hydroxylases (7,12). It is not consumed stoichiometrically, however, (39,40), and the enzymes can catalyze their reactions for a number of catalytic cycles in the complete absence of ascorbate (37,41,42). On the basis of these findings and the kinetic data (12,31,41), it has been

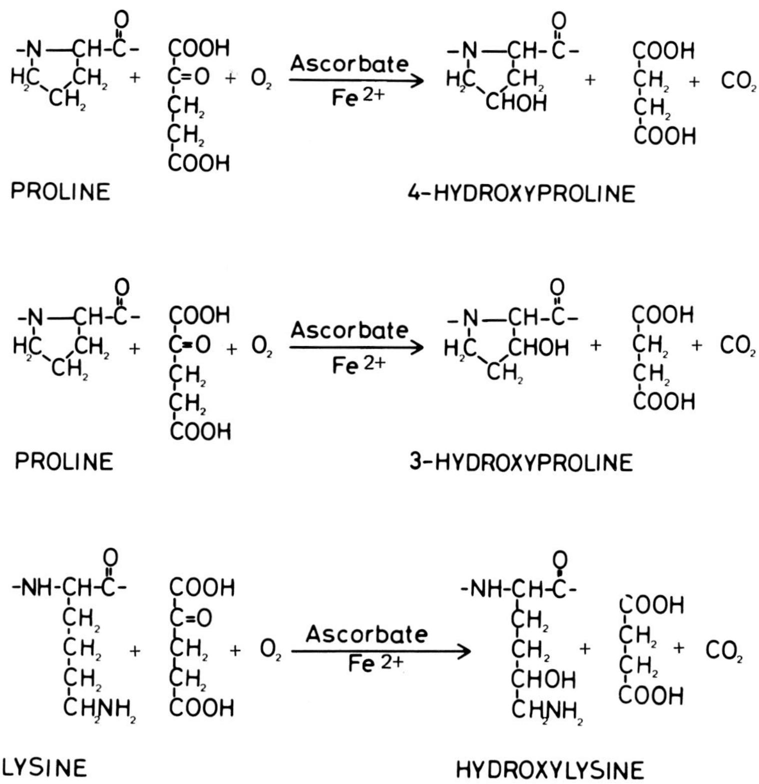

Figure 1 Reactions catalyzed by prolyl 4-hydroxylase, prolyl 3-hydroxylase, and lysyl hydroxylase.

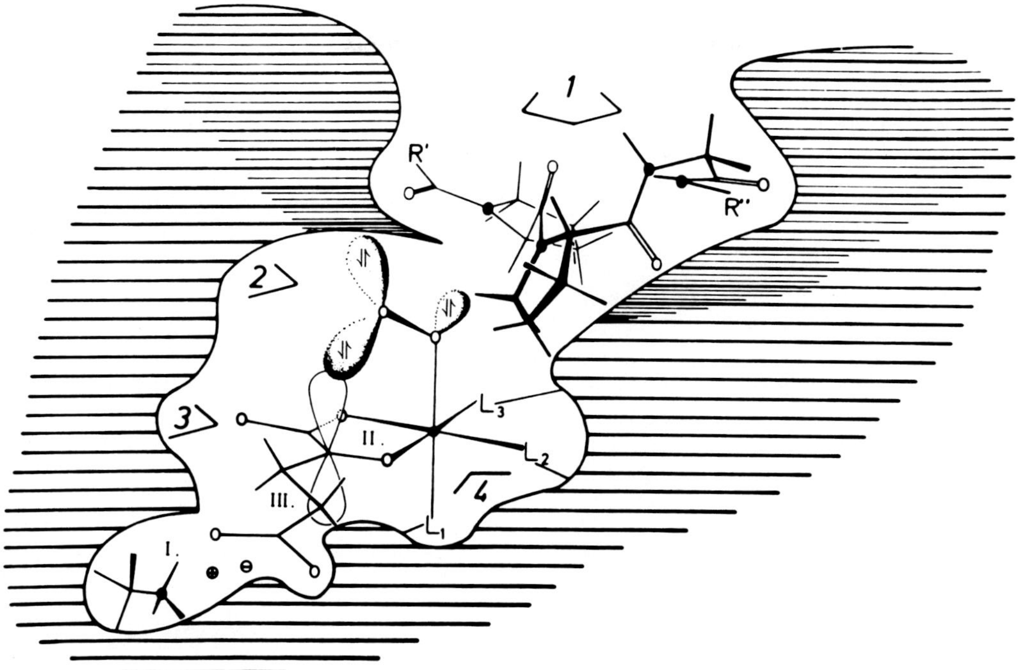

Figure 2 Model for the binding of the peptide substrate, ferrous ion, and cosubstrates at the active site of prolyl 4-hydroxylase at the beginning of the catalytic cycle. The iron (4) is thought to be located in a pocket, and the 2-oxoglutarate (3) to become bound via subsites: (I) a positively charged amino acid residue on the enzyme; (II) two *cis*-positioned equatorial coordination sites of ferrous ion; and (III) a hydrophobic binding site. Molecular oxygen (2) is then thought to be bound end-on in an axial position, yielding a superoxidelike structure. The peptide substrate (1) is sterically oriented to participate stereospecifically in the hydroxylation reaction. (Reproduced from Ref. 148 with permission.)

suggested that ascorbate is required to reduce the enzyme-bound iron and possibly some other groups in the enzyme molecule, which become oxidized in a side reaction during some, but not the majority, of the catalytic cycles. Recent studies have indicated that this side reaction is probably an uncoupled decarboxylation of 2-oxoglutarate. The enzymes are known to catalyze such a reaction at a low rate in the absence of the polypeptide substrate but in the presence of all the cosubstrates (12,31). This uncoupled decarboxylation reaction, unlike the complete hydroxylation reaction, has recently been shown to involve stoichiometric ascorbate consumption (43,44). Peptides such as poly(L-proline) and $(\text{Pro-Ala-Gly})_n$, which do not act as substrates, are known to increase the rate of the uncoupled prolyl 4-hydroxylase reaction (12). Newly synthesized collagenous polypeptide chains contain many other sequences in addition to the amino acid triplets hydroxylated by the enzymes, including -Pro-Ala-Gly-. An interaction between some of these nonhydroxylatable sequences and the enzymes is likely to lead to an uncoupled reaction cycle and thus ascorbate consumption (43). If ascorbate is not present, the uncoupled decarboxylation results in the formation of an oxidized enzyme

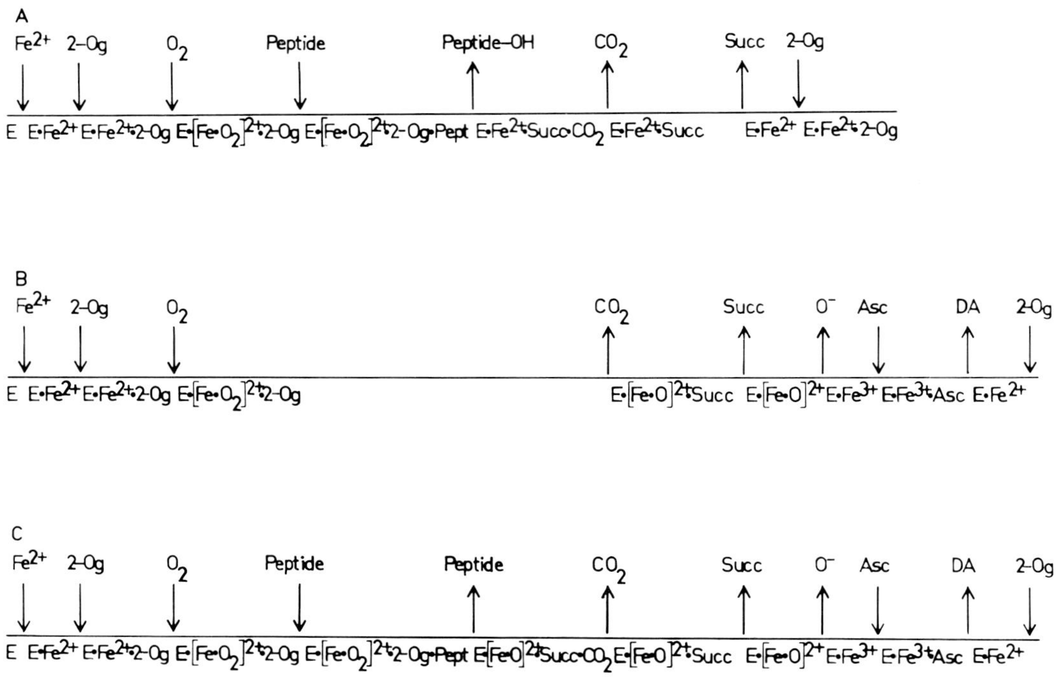

Figure 3 Schematic representation of the mechanism for the prolyl 4-hydroxylase and lysyl hydroxylase reactions. The complete hydroxylation reaction is thought to proceed according to scheme A, in which the order of release of the hydroxylated peptide and CO_2 is uncertain. In the absence of the peptide substrate the enzymes catalyze an uncoupled decarboxylation of 2-oxoglutarate, as shown in scheme B. Certain peptides that do not become hydroxylated are known to increase the rate of the uncoupled decarboxylation (C). In schemes B and C the reactive iron–oxo complex is converted to Fe^{3+}, and ascorbate is needed to reactivate the enzyme.

that is unavailable for new catalytic cycles. It is therefore likely that the main function of ascorbate in the collagen hydroxylase reactions in vivo is to reactivate the enzymes after the uncoupled reaction cycles (Figure 3).

The function of the hydroxyl group of the 4-hydroxyproline residues is to stabilize the collagen triple helix under physiological conditions (10,12). Nonhydroxylated proα chains can fold into the triple helix at low temperatures, but the transition temperature for the unfolding of such molecules is only about 24°C. Thus nonhydroxylated proα chains cannot form triple-helical molecules in vivo, and an almost complete 4-hydroxylation of proline residues in the Y positions of the -X-Y-Gly- triplets is needed to render a molecule stable at 37°C. The role of the 3-hydroxyproline residues is unknown at present. The hydroxylysine residues have two functions: their hydroxyl groups act as the sites of attachment for carbohydrate units, and they are essential for the stability of the intermolecular collagen cross-links (Sections II.A.3 and II.B.3). The significance of hydroxylysine is clearly demonstrated by the profound changes in the mechanical properties of certain tissues that are seen in patients with a genetic deficiency in lysyl hydroxylase activity (Section IV.A).

3. Glycosylation of Hydroxylysine Residues

Formation of the hydroxylysine-linked carbohydrate units is catalyzed by two specific enzymes (13). Hydroxylysyl galactosyltransferase (EC 2.4.1.50) first transfers galactose to some of the hydroxylysine residues, and galactosylhydroxylysyl glucosyltransferase (EC 2.4.1.66) then transfers glucose to some of the galactosylhydroxylysine residues (Figure 4). Part of the carbohydrate is thus found in the form of the monosaccharide galactose and part as the disaccharide glucosylgalactose, these being the only carbohydrate prosthetic groups present in mammalian interstitial collagens.

The highest extents of purification for both enzymes have been reported from chick embryos, the galactosyltransferase having been purified about 1000-fold and the glucosyltransferase having been isolated as a homogeneous protein (13,31). Only preliminary data are available on the molecular properties of the galactosyltransferase, whereas the glucosyltransferase has been found to consist of only one polypeptide chain with a molecular weight of about 70,000. These two transferases, like the three hydroxylases discussed above, probably have no collagen type-specific or tissue-specific isoenzymes (9,45).

Hydroxylysine is formed only as a cotranslational and posttranslational modification, and thus it is apparent that synthesis of the carbohydrate units in vivo takes place on a

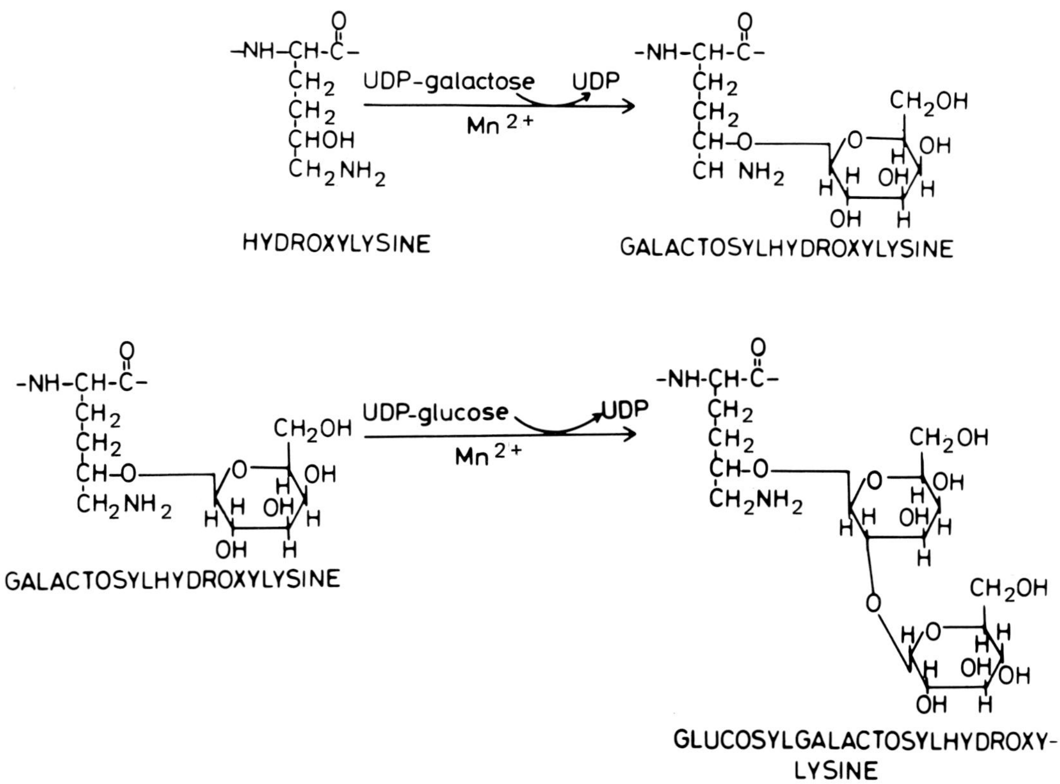

Figure 4 Reactions catalyzed by hydroxylysyl galactosyltransferase and galactosylhydroxylysyl glucosyltransferase.

peptide substrate (13). The free ϵ-amino group of the hydroxylysine residue is an absolute requirement for both transferases. Interaction with both enzymes is further affected by the amino acid sequence of the peptide, the peptide chain length, and the peptide conformation. The reactions are prevented by the triple-helical conformation, which indicates that these two glycosylations, like the three hydroxylations, must occur in the cell before triple-helix formation.

The carbohydrate transferred in these reactions is donated by the corresponding UDP-glycosides (Figure 4). Both enzymes also need a metal cofactor, preferably Mn^{2+} (13). Kinetic studies on the glucosyltransferase reaction suggest an ordered binding of Mn^{2+}, UDP-glucose, and collagen to the enzyme, in that order, and an ordered release of the products, with Mn^{2+} remaining bound to the enzyme between most of the catalytic cycles (13).

The functions of the hydroxylysine-linked carbohydrate units are largely unknown. As these units are the most extrusive groups on the surface of the collagen molecule, they may regulate packing of the molecules into fibrils. In agreement with this suggestion, there is an inverse relationship between carbohydrate content and fibril diameter (13,46). In addition, molecules in native fibrils of type II collagen are spaced farther apart laterally than in those of type I, perhaps because of the higher carbohydrate content of type II collagen (47). Intermolecular collagen cross-links built up from hydroxylysine residues can be glycosylated (13,48), and this may affect their properties (49). The extent of collagen glycosylation possibly also influences the sensitivity of the protein to mammalian collagenases (50), and the carbohydrate units may be involved in the attachment of some cells to type IV collagen (9), but little evidence is available to support these ideas.

4. Glycosylation of Asparagine Residues in the Propeptides

The propeptides of procollagen, unlike the collagen domain of the molecules, contain asparagine-linked carbohydrate units (1-9,13,51). In type I and III procollagens most or all of these are found in the C-terminal propeptides, whereas in type II procollagen such units may also be present in the N-terminal propeptides. The proα1 and proα2 chains of type I procollagen each contain a single high-mannose oligosaccharide unit. The noncollagenous domains of type IV collagen also contain asparagine-linked carbohydrate units (8), but no data are yet available on such units in other procollagen types.

The asparagine-linked carbohydrate units of various glycoproteins are first synthesized on a carrier lipid (52,53). The completed oligosaccharide chain with the structure $Glc_3Man_9GlcNAc_2$ is then transferred en bloc to an asparagine residue in a sequence -Asn-X-Ser/Thr- of the nascent polypeptide chain, and the glucose residues are removed (Figure 5). In the case of complex oligosaccharide units, the high-mannose structure is subsequently processed further. Procollagen glycosylation occurs by a similar mechanism (7,13,51), the asparagine residue being present in the sequence —Asn-Ile-Thr- in the human proα1(I) and proα2(I) chains and the chick proα2(I) and proα1(III) chains and in the sequence -Asn-Val-Thr in the chick proα1(I) chain (9,54-56).

The functions of the asparagine-linked oligosaccharide units of procollagens are unknown. Some data suggest that this glycosylation is needed for a normal rate of procollagen secretion from the cell, but other findings are inconsistent with this suggestion (7,9).

5. Chain Association, Disulfide Bonding and Triple-Helix Formation

The propeptides of procollagens contain both intrachain and interchain disulfide bonds (1-3,57). Intrachain disulfide bonds are found in both the N-terminal and C-terminal pro-

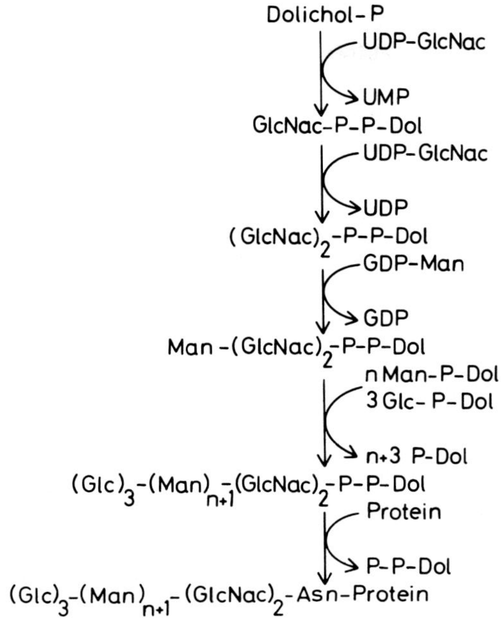

Figure 5 Formation of the asparagine-linked oligosaccharide units in proteins. The glucose residues are subsequently removed, and processing of the carbohydrate unit may also take place within the rough and smooth endoplasmic reticulum and the Golgi complex.

peptides, whereas interchain disulfide bonds are present only between the C-terminal propeptides in type I and II procollagens (Figure 6), and only between the N-terminal propeptides in type IV procollagen (58). In type III procollagen, however, such linkages are found both between the N-terminal propeptides and between the C-terminal propeptides. This procollagen also contains two interchain disulfide bonds at the C-terminal end of the triple-helical region in the collagen domain, and these difficulties may form both intramolecular and intermolecular cross-links (59). The proα2 chains of type V procolla-

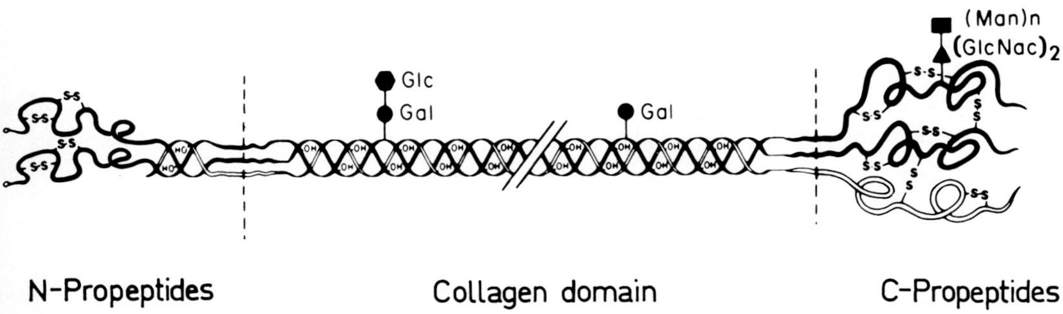

Figure 6 Schematic representation of the structure of the type I procollagen molecule. Glc, glucose; Gal, galactose; Man, mannose; and GlcNac, *N*-acetylglucosamine. (Reproduced from Ref. 2 with permission.)

gens likewise contain interchain disulfide bonds, whereas molecules consisting of three proα1(V) chains do not form such linkages at all (60-63). Some (e.g., Refs. 64,65) but not all (e.g., Ref. 66) of the more recently discovered procollagen types also contain interchain disulfide bonds.

The mechanisms by which disulfide bonds are formed in proteins are poorly understood. In vitro studies on their formation in the reoxidation of reduced proteins indicate that the rate-limiting steps are conformational changes accompanied by rearrangement reactions in which random incorrect disulfide linkages are replaced by native disulfide pairings (67). An enzyme known as protein disulfide isomerase (EC 5.3.4.1), which catalyzes such rearrangements in vitro, is widely distributed in animal tissues, is present in the microsomal fraction, and has a wide specificity for protein disulfide substrates (67,68). This enzyme is present in cells that actively synthesize collagen (67,69). The changes in its activity in the tissues of developing chick embryos (70) and differences between cells synthesizing collagen at markedly different rates (69) agree with those in the rate of collagen synthesis. Thus, protein disulfide isomerase may be involved in the intracellular formation of disulfide bonds in procollagens (69,70).

The function of the intrachain and interchain disulfide bonds is to provide covalent linkages that stabilize protein structures. A further important function of the interchain disulfide bonds between the C-terminal propeptides in procollagens is to facilitate rapid triple-helix formation. Evidence for this has been obtained from model studies in vitro (9) and from the demonstration that reduction of the disulfide bonds in intact cells prevents triple helix formation in a reversible manner (9,10). The bonds between the C-terminal propeptides are formed even when triple helix formation is prevented, whereas those between the N-terminal propeptides present in type III and IV procollagens are not formed in such a situation (58,71). Therefore, the interchain disulfide bonds between the C-terminal propeptides are probably synthesized earlier than those between the N-terminal propeptides, the triple-helical conformation being a requirement for the formation of the latter (9,58,71). The function of the interchain disulfide bonds between the C-terminal propeptides is further supported by the finding (10,13,58,60,61,72) that the time required for triple-helix formation in the biosynthesis of type I and II procollagens is much shorter than in the synthesis of type IV procollagens, which lack these linkages, and type V procollagens, in which only some chains are bridged by disulfide bonds (see above).

The data discussed above and experiments in which puromycin was used to produce shortened proα chains (1,9) suggest that the C-terminal propeptides direct the association between the three proα chains of a procollagen molecule and serve as a point of initiation for triple-helix formation. This suggestion is also supported by recent findings in a patient with osteogenesis imperfecta who has a short deletion mutation in the C-terminal propeptide of the proα2(I) chain and whose proα2(I) chains are not incorporated into triple-helical molecules at all.

Triple-helix formation in vivo cannot occur until almost all the proline residues present in the Y positions of the -X-Y-Gly- triplets have been 4-hydroxylated (Section II.A.2). As discussed above, the time required for this process also depends on whether the nucleation site within the C-terminal propeptides has been stabilized by the formation of the interchain disulfide bonds. Triple-helix formation then proceeds rapidly from a stable nucleation site throughout the chain by a stepwise mechanism limited by the *cis-trans* isomerization of peptide bonds (72).

B. Extracellular Modifications

1. *Cleavage of the Propeptides*

In the extracellular space the procollagen molecules are converted to collagen, and the collagen molecules assemble in the matrix (73). Removal of the propeptides from a procollagen molecule requires two enzymes (EC 3.4.24), a procollagen N-proteinase and a procollagen C-proteinase. The actual number of procollagen proteinases is in fact larger, since the N-proteinase at least has collagen type-specific forms.

A type I procollagen N-proteinase has been purified to apparent homogeneity from chick embryo tendons (74) and to a lesser extent from calf tendons (7,14). A type III procollagen N-proteinase has been purified extensively from human placental tissue (75) and partially purified from cultures of calf tendon fibroblasts (76) and bovine aortic smooth muscle cells (77). The molecular weight of the type I N-proteinase is about 260,000 (74) and that of the type III N-proteinase about 70,000 (75,77). A procollagen C-proteinase has been identified in a number of tissues and cell systems (9) and has been partially purified from chick embryo membranous bones (78). The molecular weight is about 80,000 by gel filtration, but the other molecular properties are at present unknown.

The procollagen proteinases are all endopeptidases, operating at a neutral pH and requiring a divalent cation such as Ca^{2+} for maximal activity (9,74-78). The type I N-proteinase cleaves procollagen types I and II but not III or IV (74), while the type III N-proteinase cleaves procollagen type III but not types I or IV (76,77). Crude and partially purified preparations of the C-proteinase act on procollagen types I, II, III, and V (9,78), but it is not known whether only one enzyme is involved.

The site of cleavage by the type I N-proteinase in the calf proα1(I) chain is a Pro-Gln bond (9,14) and that in the chick proα2(I) chain an Ala-Gln bond (79). The type III N-proteinase likewise cleaves a Pro-Gln sequence in the calf proα1(III) chain (80). In all these polypeptide chains the glutamine subsequently cyclizes, forming a pyroglutamate residue (pyrrolidonecarboxylic acid) (14,80). The C-proteinase cleaves chick and human proα1(I) and proα2(I) chains at an Ala-Asp bond and the chick proα1(III) chain at an Arg-Asp bond (54-56,79). The triple-helical substrate conformation is a requirement for cleavage by the N-proteinases (9,74,77), whereas the C-proteinase has been reported to act on both native and denatured procollagens (9).

There appears to be no obligatory sequence for the removal of the two propeptides from a procollagen molecule (Figure 7). In the synthesis of type I procollagen in many cell and organ culture systems, the N-terminal propeptides are cleaved before the C-terminal propeptides (1,7), whereas in the biosynthesis of type II procollagen, some processing takes place in the opposite order (81). In the case of type III procollagen, the N-terminal propeptides are removed very slowly (7-9), which agrees with the finding that many tissues contain significant amounts of partially processed type III molecules with complete N-terminal propeptides but no C-terminal propeptides (57,82-84). In intact tissues, however, even in the case of type I collagen, embryonic-type thin fibrils, but not mature thick fibrils, contain such partially processed molecules, indicating that even in the synthesis of this collagen type, the N-terminal propeptides may be cleaved after the C-terminal propeptides in some cases (82,84). Type IV collagen is not processed at all in vitro (1-9). Conversion to slightly shorter chains has been demonstrated in vivo in an EHS mouse tumor (85), but it is not known whether any processing occurs in normal tissues. Type V procollagen is converted in at least two stages to final forms which are still larger than the molecules obtained from tissues by pepsin extraction (60-62).

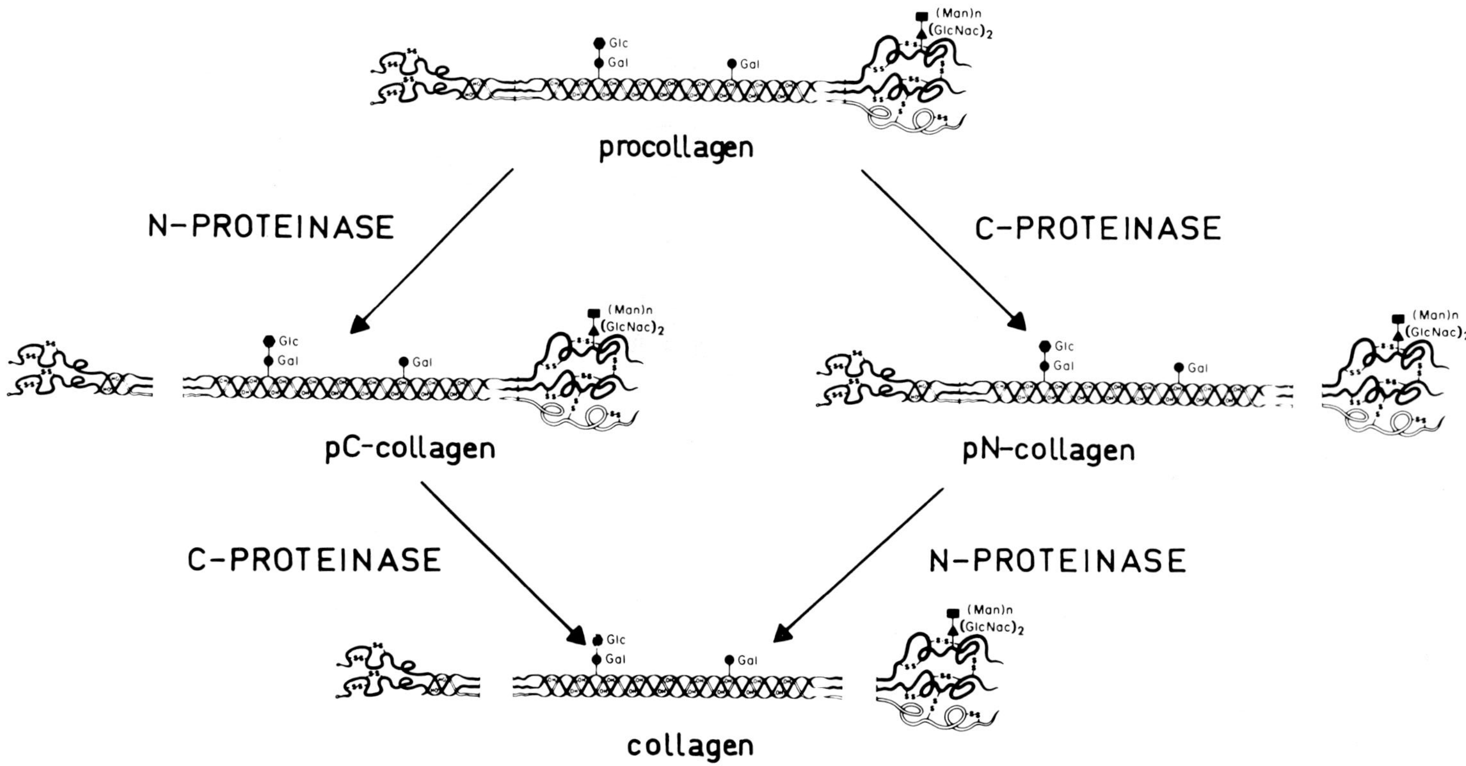

Figure 7 Conversion of procollagen to collagen. There is no obligatory sequence for the removal of the two propeptides from a procollagen molecule, but most of the processing of type I and II procollagens takes place as shown in the left-hand pathway, whereas type III procollagen is primarily processed as shown in the right-hand pathway.

The function of the cleavage of the propeptides is to permit formation of the native type collagen fibers. Removal of the C-terminal propeptides appears to be an absolute requirement for fibril formation (86), whereas partially processed molecules with intact N-terminal propeptides can form thin fibrils (81,84,87). The latter molecules do not form thick fibrils (83), however, and removal of the N-terminal propeptides may thus control the growth of the fibrils (84,88). This has been demonstrated especially in animals with dermatosparaxis and in patients with type VII of the Ehlers-Danlos syndrome, both diseases in which a heritable defect in the removal of the N-terminal propeptides causes impaired fiber formation (Section IVB). After they have been cleaved from the procollagen, the free N-terminal propeptides may also act as feedback regulators of cellular collagen synthesis (7-9,18).

2. *Ordered Aggregation*

The collagen molecules produced by removal of the propeptides have a marked tendency for self-assembly and spontaneous formation of fibrils in the case of types I-III collagens or other ordered structures in the case of the nonfibrillar collagens. This process does not require any enzyme or other factor and will take place easily in collagen solutions in vitro. The resulting fibrils of type I collagen are identical to those found in intact tissues according to criteria based on electron optics and x-ray diffraction, except in their diameter and length. This has made it possible to study collagen self-assembly in vitro, and detailed physicochemical and kinetic data are now available on this process (73,89, 90).

3. *Cross-Link Formation*

The cross-linking of collagen takes place in at least two stages. The initial event is oxidative deamination of the ϵ-amino group in certain lysine and hydroxylysine residues to the corresponding aldehyde in a reaction catalyzed by lysyl oxidase (Figure 8). These aldehydes then participate in the formation of the various cross-links. None of the steps beyond the initial oxidation is known to require any enzyme.

Lysyl oxidase has been purified to homogeneity from several sources (7,14,15), including human tissues (91). Preparations of the enzyme from all these sources are heterogeneous, with multiple peaks of activity in DEAE-cellulose chromatography, the molecular weight of all these enzyme species being about 30,000, or in some studies about 60,000 (15,91,92). The reasons for the multiple forms are unknown, but recent data indicate that there are no definite structural, catalytic, or immunological differences between the four forms of lysyl oxidase present in certain tissues (91,92). Differences between the various enzyme species may exist at the level of some posttranslational modification of the lysyl oxidase protein, but the possibility is not excluded that the multiple enzyme forms may represent products of several very similar but not identical genes.

Lysyl oxidase contains tightly bound copper and possibly pyridoxal and requires molecular oxygen (15). A single enzyme acts on both lysine and hydroxylysine residues, its activity being greater in the latter case. Highly purified enzyme preparations from several sources utilize both collagen and elastin as substrates, there being no difference between the multiple enzyme species in this respect. The highest activity is found with reconstituted collagen fibrils and certain ordered molecular aggregates of elastin, whereas little or no activity is seen with monomeric collagen molecules or denatured collagen α chains (15). No activity is seen with intact procollagen either, which suggests that the

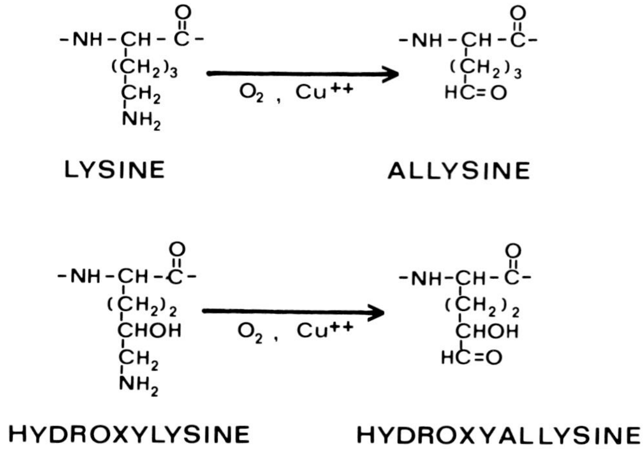

Figure 8 Reactions catalyzed by lysyl oxidase.

reaction with lysyl oxidase in the biosynthesis of fibrillar collagens may not take place until the removal of at least one of the two propeptides (91).

The aldehydes produced in the lysyl oxidase reaction can form two major types of cross-link (1,3,7,8,14,93). Intramolecular cross-links join the polypeptide chains of the same molecule and are formed by aldol condensation of two of the aldehydes. Intermolecular cross-links join the polypeptide chains of adjacent molecules, being products of condensation between one aldehyde and one ϵ-amino group of a second lysine, hydroxylysine or glycosylated hydroxylysine residue (Figure 9). The cross-links formed from a hydroxylysine-derived aldehyde are more stable than those formed from a lysine-derived aldehyde, possibly owing to migration of the double bond to a stable oxo form. The aldol cross-link (Figure 9) can react further with a histidine and hydroxylysine residue from an adjacent collagen molecule to form compounds known as aldolhistidine, hydroxymerodesmosine, and histidinohydroxymerodesmosine, but there is controversy over the extent to which these trifunctional and tetrafunctional cross-links are present in vivo and to what extent they are formed during the procedure used for their isolation (1,3,7,8,14,93).

Further stabilization of collagen cross-links occurs with time. The cross-links in young collagen fibers can be reduced with sodium borohydride, but such reducible cross-links gradually disappear from older collagen. This had led to the hypothesis that the reducible components are only long-lived intermediates that are converted during maturation to more stable nonreducible forms (3,7,8,14,93). The mechanisms for this stabilization have not been established. Reduction of the cross-links does not seem to take place in vivo (3,93), and hence attempts have been made to identify other possible stabilization mechanisms. One suggested recently involves oxidation to yield an isopeptide bond (7,14,93), but this mechanism may not represent any major pathway (93). Another mechanism involves formation of a multivalent cross-link of unknown structure (94). A third mechanism involves synthesis of a compound termed pyridinoline (Figure 10), which is built up from two hydroxylysine-derived aldehydes and one hydroxylysine residue (3,8,95,96). The occurrence of pyridinoline as a major intermolecular cross-link has now been firmly

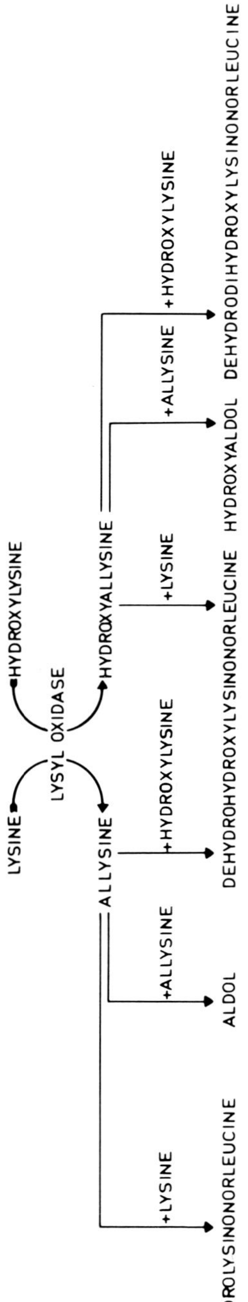

Figure 9 Reducible bifunctional collagen cross-links and their formation.

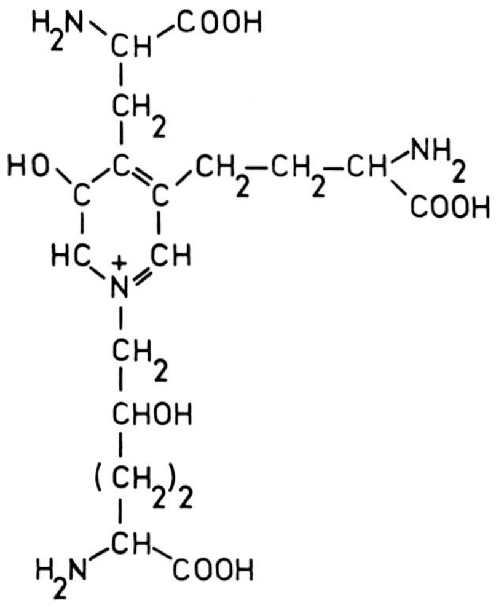

Figure 10 Structure of the pyridinoline cross-link.

established in a number of collagens (3,8,95-97), but it is not known whether this cross-link contributes to increased stability in the mature collagens (96).

The function of the lysine and hydroxylysine-derived cross-links is to give the mechanical strength to fibrils of collagen types I-III and to other structures formed by the nonfibrillar collagens. This function has been demonstrated especially clearly in numerous studies on experimental lathyrism, an acquired molecular disease characterized by a marked fragility of growing connective tissues. Lathyrism can be induced by the administration of β-aminopropionitrile or other substances that inhibit lysyl oxidase activity and hence block the cross-linking of collagens and elastin. Deficient cross-linking is also found in animals with copper deficiency and in mice and patients with heritable abnormalities of copper metabolism leading to an impaired lysyl oxidase activity (Section IV.C).

III. REGULATION OF THE RATES AND EXTENTS OF THE MODIFICATIONS

The rates of the individual modifying reactions and the extents to which the collagen is modified are affected by a large number of different factors and events. In extreme situations, such as when the 4-hydroxylation of proline residues is prevented, the synthesis of collagen fibrils or other ordered structures is inhibited completely. In other instances, the overall rate of collagen formation may be unaffected, but the properties of the fibrils may be influenced profoundly.

Large differences are found between the genetically distinct collagen types in the extents of proline 3-hydroxylation, lysine hydroxylation, hydroxylysine galactosylation, galactosylhydroxylysine glucosylation, and the relative amounts of the various types of cross-link. Smaller, though distinct, differences are also present in the extent of proline 4-hydroxylation (1-10,12,13). Additional variations in the extents of all these reactions and in the proportions of the different kinds of cross-link are found within the same collagen type from different tissues and even the same tissue in many physiological and pathological states, such variations being referred to as changes in the quality of a given collagen type (7,12,13). The 4-hydroxyproline content of the same collagen type varies again only within narrow limits, however, because an almost complete 4-hydroxylation of proline residues in the Y positions of the -X-Y-Gly- triplets is required for formation of a triple-helical molecule at physiological temperature (Section II.A.2).

A. Regulation of Intracellular Modifications

The rates and/or extents of the intracellular modifications are influenced by the amounts of the active enzymes present, the amounts of cofactors and cosubstrates required by the enzymes, the rate of procollagen synthesis, and the time available before the newly synthesized proα chains fold into the triple-helical conformation.

The amount of prolyl 4-hydroxylase activity varies markedly between different cell types and even in the same cell type under different experimental conditions (7,10,12). Since a distinct correlation is found between this enzyme activity and the rate of collagen synthesis in many cells and tissues in numerous physiological and pathological states (10,12), measurements of prolyl 4-hydroxylase activity have been used in a large number of studies for estimating the rate of collagen synthesis. Some situations are known, however, in which the amount of prolyl 4-hydroxylase activity does not correlate with the rate of collagen synthesis (7,12). Some cells have this enzyme activity even though they do not synthesize any collagenous protein, and some have relatively high activity even though only traces of collagenous protein are produced. In addition, cultured fibroblasts have lower prolyl 4-hydroxylase activity during the logarithmic phase of growth than in the stationary phase, even though the logarithmic phase cells actually synthesize slightly more collagen per cell than do confluent fibroblasts (9). The amount of prolyl 4-hydroxylase activity is clearly rate-limiting for the production of procollagen molecules in such logarithmic phase cells. The proline residues in the newly synthesized proα chains are deficient in 4-hydroxyproline, so that the chains fail to form triple-helical molecules and are therefore subject to rapid intracellular degradation (Chapter 10).

The rates of synthesis of the two types of subunit present in the prolyl 4-hydroxylase tetramer are regulated differently. The enzyme protein is found in cells in two forms: active tetramers and an inactive form that corresponds to the β subunit according to a number of criteria (7,12,98,99). Experiments involving radioactive labeling of the enzyme protein in vivo and in cellulo have demonstrated that at least one function of the β-subunit-related protein is to act as a precursor of the β subunit in the assembly of the prolyl 4-hydroxylase tetramer (7,12,98,99). While the α subunit appears to be utilized immediately after its synthesis to form the active enzyme, the β subunit seems to be produced in excess and to enter a precursor pool before being assembled into the tetramer. The synthesis of the α subunit is probably regulated more effectively than that of the β subunit, and this may explain the finding that the ratio of active enzyme tetramer to inactive β-subunit-related protein varies markedly between cell types and even in the same cell type under different conditions (12).

The activities of prolyl 3-hydroxylase, lysyl hydroxylase, and the two hydroxylysyl glycosyltransferases also show wide differences between various cell types and even in the same cell type in many experimental situations (7,12,13). The ratios of the intracellular enzyme activities involved in collagen synthesis are not constant, however, but vary markedly. This seems to explain many of the differences in the extents of the various modifications between the genetically distinct collagen types or in the same collagen type under various conditions (7,100–102). The mechanisms by which the rates of synthesis of the intracellular enzymes are regulated in accordance with the amount, type, and quality of the collagen produced are currently unknown, however.

The intracellular enzymes need several cofactors and cosubstrates, and exclusion of any of these will inhibit the corresponding reactions. This has been demonstrated in the case of the hydroxylases in numerous studies with isolated cells or tissues in which a nitrogen atmosphere or chelation of the ferrous ions has been used to inhibit these modifications completely (10,12). It is not known whether the concentrations of the cofactors and cosubstrates limit the rates of the corresponding reactions in vivo under physiological conditions, but it seems very likely that such a situation exists in certain pathological states (7,9,10,12,13). For example, the K_m of oxygen in all the hydroxylation reactions is of about the same magnitude as the O_2 tensions determined in several types of wounds, suggesting that alterations in O_2 tension may affect the rate of these reactions (10,12). Scurvy provides a classic example of the effect of the concentration of ascorbate. Many findings recorded in this disease are probably due to a reduced rate of collagen synthesis, probably mainly due to deficiency of ascorbate for the hydroxylations.

Changes in the rate of procollagen synthesis affect the ratio of the modifying enzymes to their substrate in the cell. A marked decrease in this rate may in itself increase the extent to which the newly synthesized polypeptide chains are modified. Such a situation has been demonstrated in several types of cultured transformed cells. The rate of procollagen synthesis in these cells is markedly decreased and the extent of modification is distinctly increased, even in cases when the corresponding enzyme activities are unchanged or decreased (100,102).

An additional factor influencing the extents of the intracellular modifications is the time available before the newly synthesized proα chains fold into the triple-helical conformation, as none of the hydroxylases or hydroxylysyl glycosyltransferases can act on triple-helical molecules. Isolated cells can be manipulated so that this time is either shortened or prolonged. Such studies indicate that when helix formation is inhibited, the extent of modification of the newly synthesized procollagen increases, whereas the opposite alterations occur if helix formation is accelerated (7,10,12,13). Major differences are found in the time needed for triple-helix formation between cells synthesizing the various genetically distinct collagens (Section II.A.5), and these help to explain the differences in the extent of modification between the various collagen types. Nevertheless, differences in the timing of triple-helix formation are not the main reason for the greater extent of lysine hydroxylation and hydroxylysine glycosylation in collagen types II and IV when compared with types I and III (101,103). The main reasons seem to be the amounts of the corresponding enzyme activities in the respective cells and the availability of lysine residues in the Y positions of the -X-Y-Gly- triplets in the respective proα chains. The latter factor is particularly important in the case of the high extent of lysine hydroxylation in type IV collagens, since more than 80% of all the lysine residues are hydroxylated.

B. Regulation of Extracellular Modifications

The activity of type I procollagen N-proteinase varies in the medium of cultured human fibroblasts, with maximal values at the stationary phase of growth (7). Little other information is currently available on changes in procollagen proteinase activities.

Lysyl oxidase activity has been assayed in cells and tissues in many physiological and pathological conditions. It is known to vary in the medium of cultured human fibroblasts (15), and production of the enzyme has been reported to be highly responsive to changes in the external environment of cultured rabbit aortic smooth muscle cells (104). Tissue levels of lysyl oxidase activity are low in cases of copper deficiency and are restored by administration of trace amounts of this cation (15), probably owing to induction of new enzyme synthesis rather than activation of preexisting enzyme protein (7,15). Lysyl oxidase activity in implanted sponges correlates with the development of increased wound tensile strength, and in experimental pulmonary and liver fibrosis it precedes or accompanies collagen deposition (15,105-107). Hypophysectomy reduces lysyl oxidase activity, the aldehyde content of newly synthesized collagen, and the degree of cross-link formation (15).

IV. MOLECULAR DEFECTS IN HERITABLE DISEASES

Many heritable connective tissue disorders involve molecular defects in collagens. In most of these disorders the primary defect is probably found in the genes coding for the various proα chains of the different collagen types. Since collagen biosynthesis involves a large number of cotranslational and posttranslational modifications, however, it can also be altered by defects in any of the enzymes involved in these reactions. In fact, much of the information now available about heritable diseases involving collagen concerns changes in the posttranslational modifications, as summarized in Table 2.

A. Deficient Lysyl Hydroxylase Activity in Ehlers-Danlos Syndrome (Type VI)

The Ehlers-Danlos syndrome is characterized by joint hypermobility, skin changes such as hyperextensibility, thinness, and fragility, and other signs of connective tissue involvement. The disorder has been divided into 10 subtypes on clinical, genetic, and other bases, but it is evident that even the subtypes are heterogeneous (6,16-24,108,109).

Type VI of the Ehlers-Danlos syndrome was the first connective tissue disease in man in which a specific defect in the synthesis of procollagen was identified (110,111). The clinical manifestations of this autosomal recessively inherited subtype include all the general characteristics of the syndrome and, as special features, kyphoscoliosis and ocular abnormalities. The latter include fragility of the ocular globe, spontaneous retinal detachment, microcornea, and slightly blue sclerae (110-119).

The biochemical abnormality in most patients with type VI of the syndrome is a deficiency in lysyl hydroxylase activity. Three distinct subtypes have been described, however. In the first subtype there is a marked deficiency in lysyl hydroxylase activity in cultured skin fibroblasts and a substantial decrease in the hydroxylysine content of collagen in the skin and several other tissues (110-112, 115,119). In the second subtype lysyl hydroxylase activity in cultured skin fibroblasts is likewise deficient but, surprisingly, the hydroxylysine content of skin collagen is almost normal (113,119). In the third subtype no evidence of either of these two abnormalities has been found (114-119). Since hydroxylysine is necessary for synthesis of the most stable intermolecular collagen cross-links (Section II.B.3), the low hydroxylysine content probably explains the clinical manifestations in the first subtype, and possibly even in the second. Such an abnormality does not explain the clinical findings in the third subtype, however, and thus the disease must be produced by a different molecular defect in these patients.

Detailed comparisons of the deficiency in lysyl hydroxylase activity between patients with the first subtype of type VI Ehlers-Danlos syndrome have led to the suggestion that

Table 2 Alterations of Collagen Posttranslational Events in Heritable Diseases[a]

Alteration in posttranslational event	Primary molecular defect	Disease
Intracellular events		
Deficient hydroxylation of lysine residues	Lysyl hydroxylase deficiency	EDS VI
Excessive hydroxylation of lysine residues	Mutation in type I procollagen	OI
Excessive glycosylation of hydroxylysine residues	Mutation in type I procollagen	OI
Deficient glucosylation of galactosylhydroxylysine residues	Galactosylhydroxylysyl glucosyltransferase deficiency	Dominant epidermolysis bullosa simplex[b]
Overglycosylation of C-propeptide in type I procollagen	Mutation in the C-propeptide of type I procollagen	OI
Failure of proα2(I) chain to become associated with proα1(I) chains	Mutation in the C-propeptide of the proα2(I) chain	OI
Decreased secretion rate of type I or type III procollagen	Mutation in type I or type III procollagen	OI or EDS IV
Extracellular events		
Impaired cleavage of N-propeptides	A. Type I procollagen N-proteinase deficiency	EDS VIIA
	B. Mutation in type I procollagen	EDS VIIB, OI
Impaired cleavage of C-propeptides	Impaired glycosylation of proteins	EDS-like disorder
Altered morphology of collagen fibers	Many defects	Most types of EDS
Deficient cross-link formation	Alterations in Cu^{2+} metabolism with lysyl oxidase deficiency	EDS IX, Menkes syndrome
Deficient cross-link formation	Cystathionine synthase deficiency	Homocystinuria

[a]For references, see text.
[b]It is not known whether the enzyme deficiency causes the clinical manifestations.
EDS = Ehlers-Danlos syndrome; OI = osteogenesis imperfecta.

the reduced enzyme activity is probably caused by different mutations of the lysyl hydroxylase protein in different families, even within the same subtype (117,118). The mutant enzyme in one family has been reported to have an increased apparent K_m for ascorbate and to deviate from normal in some other properties (117,118,120). No abnormality in the K_m for ascorbate has been found in two other families, however (33,121).

An interesting aspect of type VI Ehlers-Danlos syndrome is that the degree of deficiency in hydroxylysine is not equal in the collagens found in different tissues (110). One explanation might be the presence of tissue-specific or collagen type-specific lysyl hydroxylase isoenzymes, but a number of observations argue against their existence

(33-35). One report suggests that the residual activity of the mutant enzyme is preferentially directed toward type IV collagen rather than type I (122), but work on other patients does not support this suggestion (33). Furthermore, the chains isolated from collagen types I, III, and V synthesized by the affected skin fibroblasts in culture are all deficient in hydroxylysine (119,123,124). The most likely explanation for the varying degree of hydroxylysine deficiency in different tissues is that some tissues have an excess of this enzyme and therefore a partial deficiency may be of no consequence (125). Such an enzyme excess may also be present in all cultured cells, since the extent of lysine hydroxylation in the collagens synthesized by cultured normal fibroblasts is higher than that found in vivo (100) and since the deficiency in hydroxylysine in the collagens synthesized by the affected skin fibroblasts in culture is much milder than that present in the skin of the patients (119,123,124). A large excess of lysyl hydroxylase is known to be present in the macrophages (126), and this may explain the findings that the extent of lysine hydroxylation and the functional activity of the subcomponent Clq of complement are unimpaired in these patients (127).

B. Impaired Cleavage of N-Terminal Propeptides in Ehlers-Danlos Syndrome (Type VII)

Type VII of the Ehlers-Danlos syndrome is characterized by extreme joint laxity producing dislocations of the knees and hips, and soft, nonfragile skin with mild hyperextensibility (128,129). The biochemical abnormality is impaired cleavage of the N-terminal propeptides. Since partially processed procollagen molecules with intact N-terminal propeptides form only thin, irregular fibrils (Section II.B.1), impaired fibril formation probably explains the clinical manifestations. Two subtypes of type VII of the syndrome have been described (Figure 11).

In subtype VIIA, probably inherited as an autosomal recessive disease, the impaired cleavage is due to a deficiency in type I procollagen N-proteinase activity (128). A similar molecular defect has also been identified in dermatosparaxis, an autosomal recessively inherited connective tissue disorder found in cattle, sheep, and cats (Chapter 13). A surprising feature, however, is that the animal disorder is characterized by an extreme skin fragility, whereas the human disease lacks this manifestation.

In subtype VIIB, inherited as an autosomal dominant condition, the activity of type I procollagen N-proteinase is normal, but a structural alteration in the proα2(I) chain prevents its cleavage (129), leading to persistence of partially processed procollagen molecules with an intact N-terminal propeptide in the proα2(I) chain. Impaired cleavage of the N-terminal propeptide of this proα chain has also been reported in two patients with osteogenesis imperfecta, who have a deletion of about 30 amino acids in the proα2(I) chain near the N-terminal end of the α chain domain. These patients likewise have marked looseness of the joints, but they also have blue sclerae and wormian bones, which are not seen in type VIIB of the Ehlers-Danlos syndrome.

C. Altered Copper Metabolism and Deficient Lysyl Oxidase Activity in Ehlers-Danlos Syndrome (Type IX) and in the Menkes Syndrome

Type IX of the Ehlers-Danlos syndrome and the Menkes steely-hair syndrome are X-linked, recessively inherited disorders with alterations in copper metabolism. The secondary products of these alterations are very similar defects in lysyl oxidase activity and hence collagen cross-linking (Table 3).

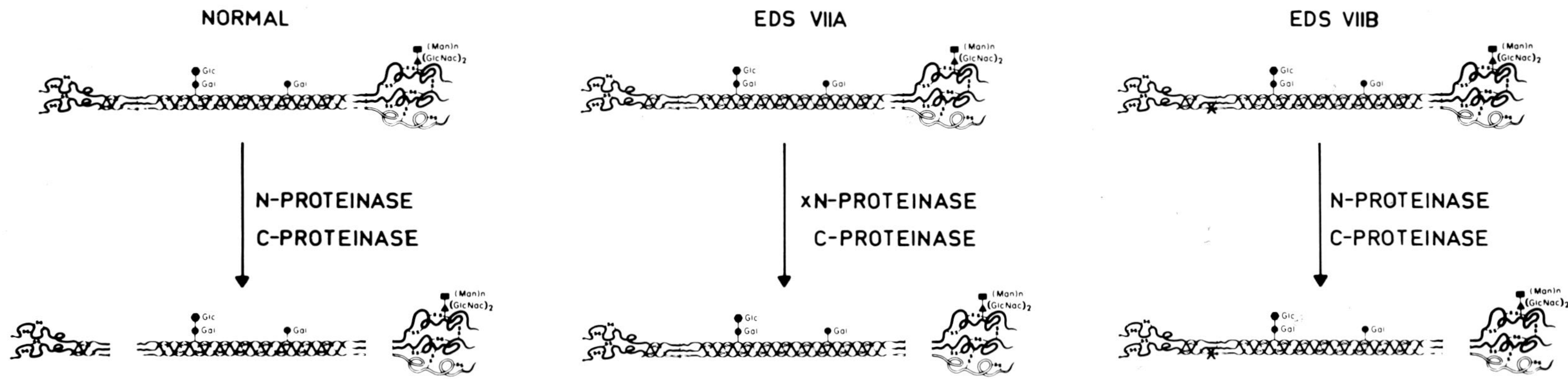

Figure 11 Schematic representation of the impaired cleavage of the N-terminal propeptides found in the two subtypes of type VII Ehlers-Danlos syndrome. X, mutation; EDS, Ehlers-Danlos syndrome.

Table 3 Clinical Findings and Biochemical Alterations in Type IX Ehlers-Danlos Syndrome and in the Menkes Syndrome

Finding or assay	EDS IX	Menkes
X-linked inheritance	Present	Present
Bladder diverticula	Present	Present
Mild skin laxity	Present	Present
Mild skin hyperextensibility	Present	Present
Skeletal abnormalities	Present	Present
Inguinal hernias	Present	Absent
Arterial disease	Absent	Present
Cerebral degeneration	Absent	Present
Pili torti	Absent	Present
Severely reduced life expectancy	Absent	Present
Serum copper	Decreased	Decreased
Serum ceruloplasmin	Decreased	Decreased
Copper in cultured fibroblasts	Increased	Increased
Copper in several other cells	Not studied	Increased
^{64}Cu content after incubation of cells for 24 hr with ^{64}Cu	Increased	Increased
Metallothionein in cultured fibroblasts	Increased	Increased
Lysyl oxidase activity	Decreased	Decreased
Lysyl oxidase protein	Decreased	Decreased
Extractability of collagen	Increased	Increased
Cross-linking of collagen and elastin	Decreased	Not studied

The clinical manifestations of type IX Ehlers-Danlos syndrome include bladder diverticula with spontaneous ruptures, inguinal hernias, slight skin laxity and hyperextensibility, and a number of skeletal abnormalities, a peculiar feature being occipital hornlike exostoses (130-135). Some of these patients were originally classified as having X-linked cutis laxa (130), others an X-linked subtype of the Ehlers-Danlos syndrome (131), and still others a subtype of the Ehlers-Danlos syndrome with abnormal copper metabolism (133). Because of the great similarities of the clinical and biochemical abnormalities, they are now all classified as type IX of the Ehlers-Danlos syndrome (108,132,135). The Menkes steely-hair syndrome also includes bladder diverticula, mildly increased skin laxity and hyperextensibility, and skeletal abnormalities, but unlike type IX Ehlers-Danlos syndrome, it also involves arterial disease and neurological changes so severe that the disorder usually proves lethal by three years of age (135-137).

Copper metabolism shows several very similar alterations in these two disorders. Serum copper and ceruloplasmin concentrations are low (130,133,136), while cultured fibroblasts have markedly elevated concentrations of the cation (133,135-137). The latter abnormality has also been demonstrated in many other cells in the case of the Menkes syndrome (136,137), while in the case of type IX Ehlers-Danlos syndrome, no data are yet available concerning cells other than fibroblasts. In both disorders the excessive copper accumulates in the cation-binding protein metallothionein, which is present in the cultured cells in markedly increased amounts (135-139). The induction of metallothionein may not be the primary alteration, however, since it could be secondary to some unidentified abnormality in cellular copper transport or utilization (135-140).

All the connective tissue manifestations in these two disorders are probably explainable by a secondary reduction in the activity of lysyl oxidase, the copper enzyme that initiates the cross-linking of collagens and elastin (Section II.B.3). This abnormality has been demonstrated in skin specimens (135) and cultured fibroblasts (130,133,135) from patients with type IX Ehlers-Danlos syndrome and in cultured fibroblasts from patients with the Menkes syndrome (135,141,142), although in the latter case one laboratory has failed to show any decrease in this enzyme activity (143). The impaired cross-linking of collagen that results from deficient lysyl oxidase activity has been demonstrated in the collagen produced by cultured skin fibroblasts both indirectly, as increased solubility of the newly synthesized cell layer collagen (130,133,135), and directly (130).

These two human disorders greatly resemble those found in the mottled series of allelic mutant mice. The animal disease shows a wide array of manifestations, some strains of the mutant mice having mainly a neurological disorder without evidence of connective tissue abnormalities and some having mainly signs of the latter, such as reduced tensile strength of the skin, bone abnormalities, and aortic aneurysms (136). The inheritance is X-linked recessive, serum copper and ceruloplasmin concentrations are low, and cultured cells have markedly elevated amounts of copper and metallothionein (136,137). Low lysyl oxidase activity has been demonstrated both in tissue extracts (144,145) and in the medium of cultured fibroblasts (141,145), impaired cross-linking of collagen and elastin having also been established (146).

The mechanisms leading to the deficient lysyl oxidase activity in these disorders are unknown. A high negative correlation has been found in the Menkes syndrome between the cellular copper concentration and the logarithm of lysyl oxidase activity (Figure 12), suggesting a distinct relation between these alterations (135). The deficient enzyme activity may be secondary to a decrease in functional intracellular copper, even though the cellular concentrations of this cation are markedly increased (142). The possibility is not excluded, however, that the genes coding for lysyl oxidase, some other copper-dependent enzymes, and a cellular copper transport protein may be linked by some mechanism so that a single mutation will affect the function of all of them to varying degrees (17,144). Recent data indicate that skin fibroblasts from patients with type IX Ehlers-Danlos syndrome and the Menkes syndrome contain and secrete reduced amounts of the lysyl oxidase protein (147), a finding that would be consistent with the latter alternative. These findings do not exclude the possibility, however, that a copper-deficient enzyme protein may be synthesized and rapidly degraded (147).

The complexity of the disease manifestations in the two human disorders and the various forms of the disease present in the mottled mouse mutants is evidently due to an impaired functioning of certain copper-dependent enzymes (136,137). It is not known, however, why some of the diseases, such as type IX Ehlers-Danlos syndrome have mainly connective tissue abnormalities while others are dominated by neurological findings. Serum copper and ceruloplasmin concentrations are usually distinctly lower in the Menkes syndrome (136) than in type IX Ehlers-Danlos syndrome (130,133), which might suggest that the different clinical manifestations reflect differences in the severity of the alteration in copper metabolism. This suggestion is nevertheless inconsistent with data comparing cellular copper and metallothionein concentrations and cellular copper incorporation between these two disorders (135). Neither does it agree with reports on mild forms of the Menkes syndrome in which the clinical manifestations are distinctly different from those found in type IX Ehlers-Danlos syndrome even though serum copper and ceruloplasmin concentrations are only slightly decreased (136).

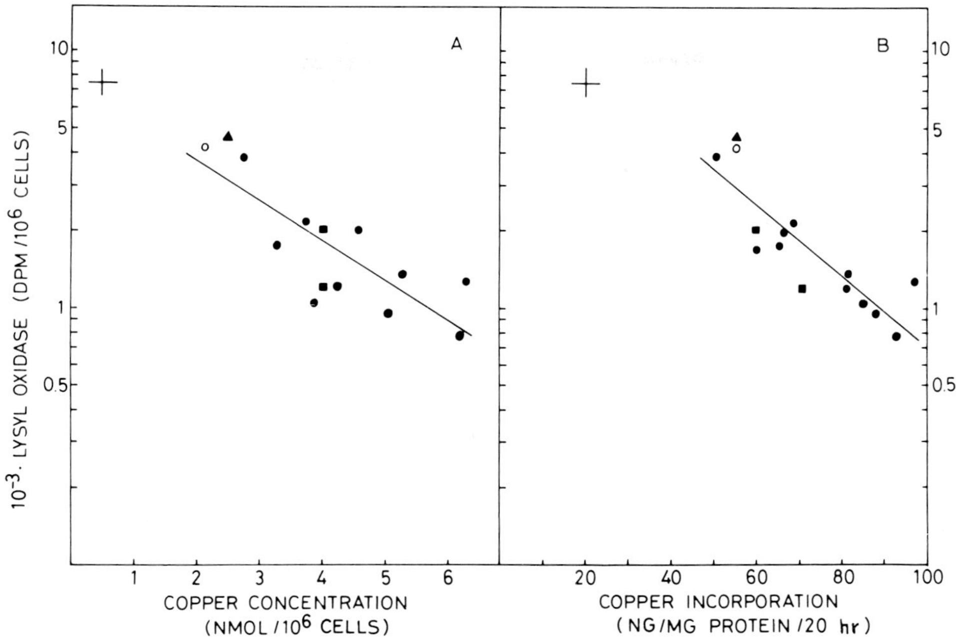

Figure 12 Relationship between lysyl oxidase activity (logarithmic scale) and cellular copper concentration (A) or incorporation (B) in cultures of the Menkes and Ehlers-Danlos syndrome type IX fibroblasts. The regression lines ($p < 0.001$) were calculated for pooled groups of Menkes and Ehlers-Danlos syndrome type IX fibroblasts. (●) Typical Menkes fibroblasts; (○) fetal Menkes fibroblasts; (▲) atypical Menkes fibroblasts; (■) Ehlers-Danlos syndrome type IX fibroblasts. The short horizontal and vertical lines indicate the corresponding control means. (Reproduced from Ref. 135 with permission.)

Various mottled mouse mutants likewise do not show any distinct relation between serum copper concentration and the presence or absence of signs of impairment of copper-dependent enzymes other than lysyl oxidase (144). Therefore type IX Ehlers-Danlos syndrome and the Menkes syndrome do not appear to represent a single biochemical defect with different degrees of severity.

REFERENCES

1. P. Bornstein and W. Traub, in *The Proteins*, Vol. 4 (H. Neurath and R. L. Hill, eds.), Academic Press, New York, 1979, pp. 411-632.
2. D. J. Prockop, K. I. Kivirikko, L. Tuderman, and N. A. Guzman, *N. Engl. J. Med., 301*:13-23, 77-85 (1979).
3. D. R. Eyre, *Science, 207*:1315-1322 (1980).
4. R. R. Minor, *Am. J. Pathol., 98*:226-280 (1980).
5. B. R. Olsen, in *Cell Biology of Extracellular Matrix* (E. D. Hay, ed.), Plenum Press, New York, 1981, pp. 139-177.
6. J. Uitto, L. Ryhänen, and E. M. L. Tan, in *Progress in Diseases of the Skin*, Vol. 1 (R. Fleischmajer, ed.), Grune & Stratton, New York, 1981, pp. 103-141.
7. K. I. Kivirikko and R. Myllylä, in *Collagen in Health and Disease* (J. B. Weiss and M. I. V. Jayson, eds.), Churchill Livingstone, Edinburgh, 1982, pp. 101-120.

8. M. E. Nimni, *Sem. Arthr. Rheum., 13*:1-86 (1983).
9. K. I. Kivirikko and R. Myllylä, in *Extracellular Matrix Biochemistry* (K. A. Piez and A. H. Reddi, eds.), Elsevier, New York, 1984, pp. 83-118.
10. D. J. Prockop, R. A. Berg, K. I. Kivirikko, and J. Uitto, in *Biochemistry of Collagen* (G. N. Ramachandran and A. H. Reddi, eds.), Plenum Press, New York, 1976, pp. 163-273.
11. E. Adams and L. Frank, *Ann. Rev. Biochem., 49*:1005-1061 (1980).
12. K. I. Kivirikko and R. Myllylä, in *The Enzymology of Post-translational Modification of Proteins* (R. B. Freedman and H. C. Hawkins, eds.), Academic Press, London, 1980, pp. 53-104.
13. K. I. Kivirikko and R. Myllylä, *Int. Rev. Connect. Tissue Res., 8*:23-72 (1979).
14. J. G. Heathcote and M. E. Grant, in *The Enzymology of Post-translational Modification of Proteins* (R. B. Freedman and H. C. Hawkins, eds.), Academic Press, London, 1980, pp. 457-506.
15. R. C. Siegel, *Int. Rev. Connect. Tissue Res., 8*:73-118 (1979).
16. P. Bornstein and P. H. Byers, in *Metabolic Control and Disease* (P. K. Bondy and L. E. Rosenberg, eds.), Saunders, Philadelphia, 1980, pp. 1089-1153.
17. P. H. Byers, G. S. Barsh, and K. A. Holbrook, *Collagen Rel. Res., 1*:475-489 (1981).
18. T. Krieg, A. Ihme, L. Weber, E. Kirsch, and P. K. Mueller, *Int. J. Dermatol., 20*:415-425 (1981).
19. D. W. Hollister, P. H. Byers, and K. A. Holbrook, *Adv. Hum. Genet., 12*:1-87 (1982).
20. J. Uitto and E. A. Bauer, in *Collagen in Health and Disease* (J. B. Weiss and M. I. V. Jayson, eds.), Churchill Livingstone, Edinburgh, 1982, pp. 289-312.
21. S. R. Pinnell and S. Murad, in *The Metabolic Basis of Inherited Disease* (J. B. Stanbury, J. B. Wyngaarden, D. S. Fredrickson, J. L. Goldstein, and M. S. Brown, eds.), McGraw-Hill, New York, 1983, pp. 1425-1449.
22. P. H. Byers, in *Clinical Medicine*, Vol. 9 (J. A. Spittel, Jr., ed.), Harper & Row, Philadelphia, 1983, pp. 1-41.
23. S. M. Krane, in *Extracellular Matrix Biochemistry* (K. A. Piez and A. H. Reddi, eds.), Elsevier, New York, 1984, pp. 413-463.
24. D. J. Prockop and K. I. Kivirikko, *N. Engl. J. Med., 311*:376-386 (1984).
25. G. Kreil, *Ann. Rev. Biochem., 50*:317-348 (1981).
26. D. D. Sabatini, G. Kreibich, T. Morimoto, and M. Adesnik, *J. Cell Biol., 92*:1-22 (1982).
27. R. Gilmore and G. Blobel, *Cell, 35*:677-685 (1983).
28. V. Tate, M. Finer, H. Boedtker, and P. Doty, *Cold Spring Harbor Symp. Quant. Biol., 45*:1039-1049 (1983).
29. Y. Yamada, M. Mudryj, and B. de Crombrugghe, *J. Biol. Chem., 258*:14914-14919 (1983).
30. D. Koshland and D. Botstein, *Cell, 30*:893-902 (1982).
31. K. I. Kivirikko and R. Myllylä, *Meth. Enzymol., 82*:245-304 (1982).
32. N. L. Kedersha, J. S. Tkacz, and R. A. Berg, *Biochemistry, 24*:5952-5960 (1985).
33. T. M. Turpeenniemi-Hujanen, U. Puistola, and K. I. Kivirikko, *Collagen Rel. Res., 1*:355-366 (1981).
34. T. M. Turpeenniemi-Hujanen, *Biochem. J., 195*:669-676 (1981).
35. U. Puistola, *Biochem. J., 201*:215-219 (1982).
36. J. J. Nietfeld, L. De Jong, and A. Kemp, *Biochim. Biophys. Acta, 704*:321-325 (1982).
37. L. De Jong, S. P. J. Albracht, and A. Kemp, *Biochim. Biophys. Acta, 704*:326-332 (1982).
38. H. M. Hanauske-Abel and V. Günzler, *J. Theor. Biol., 94*:421-455 (1982).

39. L. Tuderman, R. Myllylä, and K. I. Kivirikko, *Eur. J. Biochem., 80*:341-348 (1977).
40. J. J. Nietfeld and A. Kemp, *Biochim. Biophys. Acta, 657*:159-167 (1981).
41. R. Myllylä, E.-R. Kuutti-Savolainen, and K. I. Kivirikko, *Biochem. Biophys. Res. Commun., 83*:441-448 (1978).
42. U. Puistola, T. M. Turpeenniemi-Hujanen, R. Myllylä, and K. I. Kivirikko, *Biochim. Biophys. Acta, 611*:40-50 (1980).
43. R. Myllylä, K. Majamaa, V. Günzler, H. M. Hanauske-Abel, and K. I. Kivirikko, *J. Biol. Chem., 259*:5403-5405 (1984).
44. L. De Jong and A. Kemp, *Biochim. Biophys. Acta, 787*:105-111 (1984).
45. R. Myllylä, *Biochim. Biophys. Acta, 658*:299-307 (1981).
46. J. J. Harding, M. J. C. Crabbe, and N. A. Panjwani, in *Biochemistry of Normal and Pathological Connective Tissues* (A-M. Robert and L. Robert, eds.), Centre National de la Recherche Scientifique, Paris, 1980, pp. 51-64.
47. M. D. Grynpas, D. R. Eyre, and D. A. Kirschner, *Biochim. Biophys. Acta, 626*:346-355 (1980).
48. S. P. Robins, *Biochem. J., 215*:167-173 (1983).
49. M. Yamauchi, C. Noyes, Y. Kuboki, and G. L. Mechanic, *Proc. Natl. Acad. Sci. USA, 79*:7684-7688 (1982).
50. Z. Werb, in *Collagen in Health and Disease* (J. B. Weiss and M. I. V. Jayson, eds.), Churchill Livingstone, Edinburgh, 1982, pp. 121-134.
51. C. C. Clark, *Meth. Enzymol., 82*:346-360 (1982).
52. S. C. Hubbard and R. J. Ivatt, *Ann. Rev. Biochem., 50*:555-583 (1981).
53. N. Sharon and H. Lis, in *The Proteins*, Vol. 5 (H. Neurath and R. L. Hill, eds.), Academic Press, New York, 1982, pp. 1-144.
54. M. P. Bernard, J. C. Myers, M-L. Chu, F. Ramirez, E. F. Eikenberry, and D. J. Prockop, *Biochemistry, 22*:1139-1145 (1983).
55. M. P. Bernard, M.-L. Chu, J. C. Myers, F. Ramirez, E. F. Eikenberry, and D. J. Prockop, *Biochemistry, 22*:5213-5223 (1983).
56. Y. Yamada, K. Kühn, and B. de Crombrugghe, *Nucl. Acids Res., 11*:2733-2744 (1983).
57. P. Bornstein and H. Sage, *Ann. Rev. Biochem., 49*:957-1003 (1980).
58. L. I. Fessler and J. H. Fessler, *J. Biol. Chem., 257*:9804-9810 (1982).
59. D. T. Cheung, P. DiCesare, P. D. Benya, E. Libaw, and M. E. Nimni, *J. Biol. Chem., 258*:7774-7778 (1983).
60. C. A. Kumamoto and J. H. Fessler, *J. Biol. Chem., 256*:7053-7058 (1981).
61. L. I. Fessler, C. Kumamoto, M. E. Meis, and J. H. Fessler, *J. Biol. Chem., 256*:9640-9645 (1981).
62. L. I. Fessler, W. J. Robinson, and J. H. Fessler, *J. Biol. Chem., 256*:9646-9651 (1981).
63. K. Alitalo, R. Myllylä, H. Sage, P. Pritzl, A. Vaheri, and P. Bornstein, *J. Biol. Chem., 257*:9016-9024 (1982).
64. H. Furthmayr, H. Wiedemann, R. Timpl, E. Odermatt, and J. Engel, *Biochem. J., 211*:303-311 (1983).
65. H. Bentz, N. P. Morris, L. W. Murray, L. Y. Sakai, D. W. Hollister, and R. E. Burgeson, *Proc. Natl. Acad. Sci. USA, 80*:3168-3172 (1983).
66. H. Sage, B. Trüeb, and P. Bornstein, *J. Biol. Chem., 258*:13391-13401 (1983).
67. R. B. Freedman and D. A. Hillson, in *The Enzymology of Post-translational Modification of Proteins* (R. B. Freedman and H. C. Hawkins, eds.), Academic Press, London, 1980, pp. 157-212.
68. B. E. Brockway and R. B. Freedman, *Biochem. J., 219*:51-59 (1984).
69. R. Myllylä, J. Koivu, T. Pihlajaniemi, and K. I. Kivirikko, *Eur. J. Biochem., 134*:7-11 (1983).

70 B. E. Brockway, S. J. Forster, and R. B. Freedman, *Biochem. J., 191*:873-876 (1980).
71. H. P. Bächinger, L. I. Fessler, R. Timpl, and J. H. Fessler, *J. Biol. Chem., 256*: 13193-13199 (1981).
72. P. Bruckner, E. F. Eikenberry, and D. J. Prockop, *Eur. J. Biochem., 118*:607-613 (1981).
73. R. L. Trelstad and F. H. Silver, in *Cell Biology of Extracellular Matrix* (E. D. Hay, ed.), Plenum Press, New York, 1981, pp. 179-215.
74. L. Tuderman and D. J. Prockop, *Eur. J. Biochem., 125*:545-549 (1982).
75. L. Peltonen, R. Halila, and L. Ryhänen, *J. Cell. Biochem., Suppl.* 8B:1586 (1984).
76. B. V. Nusgens, Y. Goebels, H. Shinkai, and C. M. Lapiére, *Biochem. J., 191*:699-706 (1980).
77. R. Halila and L. Peltonen, *Biochemistry, 23*:1251-1256 (1984).
78. F. K. Njieha, T. Morikawa, L. Tuderman, and D. J. Prockop, *Biochemistry, 21*: 757-764 (1982).
79. J. Wozney, D. Hanahan, V. Tate, H. Boedtker, and P. Doty, *Nature, 294*:129-135 (1981).
80. D. Galloway, in *Collagen in Health and Disease* (J. B. Weiss and M. I. V. Jayson, ed.), Churchill Livingstone, Edinburgh, 1982, pp. 528-557.
81. J. Uitto, R. E. Allan, and K. L. Polak, *Eur. J. Biochem., 99*:97-103 (1979).
82. R. Fleischmajer, R. Timpl, L. Tuderman, L. Raisher, M. Wiestner, J. S. Perlish, and P. N. Graves, *Proc. Natl. Acad. Sci. USA, 78*:7360-7364 (1981).
83. K. Hedman, K. Alitalo, S. Lehtinen, R. Timpl, and A. Vaheri, *EMBO J., 1*:47-52 (1982).
84. R. Fleischmajer, B. R. Olsen, R. Timpl, J. S. Perlish, and O. Lovelace, *Proc. Natl. Acad. Sci. USA, 80*:3354-3358 (1983).
85. K. Tryggvason, T. Pihlajaniemi, T. Salo, L. A. Liotta, and K. I. Kivirikko, in *New Trends in Basement Membrane Research* (K. Kühn, H-H. Schöne, and R. Timpl, eds.), Raven Press, New York, 1982, pp. 187-193.
86. M. Miyahara, F. K. Njieha, and D. J. Prockop, *J. Biol. Chem., 257*:8442-8448 (1982).
87. M. Miyahara, P. Bruckner, O. Helle, and D. J. Prockop, *Collagen Rel. Res., 3*:279-293 (1983).
88. D. J. S. Hulmes, *Collagen Rel. Res., 3*:317-321 (1983).
89. R. L. Trelstad, *Cell, 28*:197-198 (1982).
90. K. A. Piez, in *Extracellular Matrix Biochemistry* (K. A. Piez and A. H. Reddi, eds.), Elsevier, New York, 1984, pp. 1-39.
91. H. Kuivaniemi, E.-R. Savolainen, and K. I. Kivirikko, *J. Biol. Chem., 259*:6996-7002 (1984).
92. K. A. Sullivan and H. M. Kagan, *J. Biol. Chem., 257*:13520-13526 (1982).
93. S. P. Robins, in *Collagen in Health and Disease* (J. B. Weiss and M. I. V. Jayson, eds.), Churchill Livingstone, Edinburgh, 1982, pp. 160-178.
94. N. D. Light and A. J. Bailey, *Biochem. J., 185*:373-381 (1980).
95. D. Fujimoto, *Biochem. Biophys. Res. Commun., 93*:948-953 (1980).
96. S. P. Robins and A. Duncan, *Biochem. J., 215*:175-182 (1983).
97. C. Walters and D. R. Eyre, *Calcif. Tissue Int., 35*:401-405 (1983).
98. R. A. Berg, W. W-Y. Kao, and N. L. Kedersha, *Biochem. J., 189*:491-499 (1980).
99. K. Majamaa and J. Oikarinen, *Biochem. J., 204*:737-742 (1982).
100. R. Myllylä, K. Alitalo, A. Vaheri, and K. I. Kivirikko, *Biochem. J., 196*:683-692 (1981).
101. T. Pihlajaniemi, R. Myllylä, K. Alitalo, A. Vaheri, and K. I. Kivirikko, *Biochemistry, 20*:7409-7415 (1981).

102. K. Majamaa, R. Myllylä, K. Alitalo, and A. Vaheri, *Biochem. J., 206*:499-503 (1982).
103. H. Anttinen and A. Hulkko, *Biochim. Biophys. Acta, 632*:417-427 (1980).
104. W. A. Gonnerman, R. Ferrera, and C. Franzblau, *Biochemistry, 20*:3864-3867 (1981).
105. C. O. Chichester, K. C. Palmer, J. A. Hayes, and H. M. Kagan, *Am. Rev. Resp. Dis., 124*:709-713 (1981).
106. D. F. Counts, J. N. Evans, T. A. Dipetrillo, K. M. Sterling, Jr., and J. Kelley, *J. Pharmacol. Exp. Ther., 219*:675-678 (1981).
107. E. A. Carter, M. J. McCarron, E. Alpert, and K. J. Isselbacher, *Gastroenterology, 82*:526-534 (1982).
108. L. N. Cupo, R. E. Pyeritz, J. L. Olson, S. J. McPhee, G. M. Hutchins, and V. A. McKusick, *Am. J. Med., 71*:1051-1058 (1981).
109. S. R. Pinnell, *J. Invest. Dermatol., 79*:90-92 (1982).
110. S. R. Pinnell, S. M. Krane, J. E. Kenzora, and M. J. Glimcher, *N. Engl. J. Med., 286*:1013-1020 (1972).
111. S. M. Krane, S. R. Pinnell, and R. W. Erbe, *Proc. Natl. Acad. Sci. USA, 69*:2899-2903 (1972).
112. M. Sussman, J. R. Lichtenstein, T. P. Nigra, G. R. Martin, and V. A. McKusick, *J. Bone Joint Surg. (Am.), 56A*:1228-1234 (1974).
113. B. Steinmann, R. Gitzelmann, A. Vogel, M. E. Grant, R. Harwood, and C. H. J. Sear, *Helv. Paediatr. Acta, 30*:255-274 (1975).
114. F. G. Judisch, M. Waziri, and J. H. Krachmer, *Arch. Ophthalmol., 94*:1489-1491 (1976).
115. L. J. Elsas, R. L. Miller, and S. R. Pinnell, *J. Pediatr., 92*:378-384 (1978).
116. T. Krieg, U. Feldmann, W. Kessler, and P. K. Müller, *Hum. Genet., 46*:41-49 (1979).
117. S. M. Krane, in *Gene Families of Collagen and Other Proteins* (D. J. Prockop and P. C. Champe, eds.), Elsevier, New York, 1980, pp. 57-69.
118. S. M. Krane, in *American Academy of Orthopaedic Surgeons Symposium on Heritable Disorders of Connective Tissue* (W. H. Akeson, P. Bornstein, and M. J. Glimcher, eds.), C. V. Mosby, St. Louis, 1982, pp. 61-75.
119. A. Ihme, L. Risteli, T. Krieg, J. Risteli, U. Feldmann, K. Kruse, and P. K. Müller, *Eur. J. Clin. Invest., 13*:357-362 (1983).
120. R. S. Quinn and S. M. Krane, *J. Clin. Invest., 57*:83-93 (1976).
121. R. L. Miller, L. J. Elsas, and R. E. Priest, *J. Invest. Dermatol., 72*:241-247 (1979).
122. L. Risteli, J. Risteli, A. Ihme, T. Krieg, and P. K. Müller, *Biochem. Biophys. Res. Commun., 96*:1778-1784 (1980).
123. R. S. Quinn and S. M. Krane, *Biochim. Biophys. Acta, 585*:589-598 (1979).
124. S. Tajima, S. Murad, and S. R. Pinnell, *Collagen Rel. Res., 3*:511-515 (1983).
125. U. Puistola and H. Anttinen, *Biochim. Biophys. Acta, 717*:118-123 (1982).
126. R. Myllylä and H. Seppä, *Biochem. J., 182*:311-316 (1979).
127. H. M. Hanauske-Abel and K-H. Röhm, *FEBS Lett., 110*:73-76 (1980).
128. J. R. Lichtenstein, G. R. Martin, L. D. Kohn, P. H. Byers, and V. A. McKusick, *Science, 182*:298-300 (1973).
129. B. Steinmann, L. Tuderman, L. Peltonen, G. R. Martin, V. A. McKusick, and D. J. Prockop, *J. Biol. Chem., 255*:8887-8893 (1980).
130. P. H. Byers, R. C. Siegel, K. A. Holbrook, A. S. Narayanan, P. Bornstein, and J. G. Hall, *N. Engl. J. Med., 303*:61-65 (1980).
131. J. D. MacFarlane, D. W. Hollister, D. D. Weaver, K. D. Brand, L. Luzzatti, and A. A. Biegel, *Am. J. Hum. Genet., 32*:118A (1980).
132. D. W. Hollister, in *American Academy of Orthopaedic Surgeons Symposium on*

Heritable Disorders of Connective Tissue (W. H. Akeson, P. Bornstein, and M. J. Glimcher, eds.), C. V. Mosby, St. Louis, 1982, pp. 102-113.
133. H. Kuivaniemi, L. Peltonen, A. Palotie, I. Kaitila, and K. I. Kivirikko, *J. Clin. Invest., 69*:730-733 (1982).
134. I. I. Kaitila, L. Peltonen, H. Kuivaniemi, A. Palotie, J. Elo, and K. I. Kivirikko, in *Skeletal Dysplasias* (C. J. Papadotos and C. S. Bartsocas, eds.), Alan R. Liss, New York, 1982, pp. 307-315.
135. L. Peltonen, H. Kuivaniemi, A. Palotie, N. Horn, I. Kaitila, and K. I. Kivirikko, *Biochemistry, 22*:6156-6163 (1983).
136. D. M. Danks, in *The Metabolic Basis of Inherited Disease* (J. B. Stanbury, J. B. Wyngaarden, D. S. Fredrickson, J. L. Goldstein, and M. S. Brown, eds.), McGraw-Hill, New York, 1983, pp. 1251-1268.
137. N. Horn, in *Metabolism of Trace Metals in Man: Developmental Biology and Genetic Implications*, Vol. 2 (O. M. Rennert and W-Y. Chan, eds.), CRS Press, Boca Raton, FL, 1984, pp. 25-52.
138. R. F. Bonewitz and R. R. Howell, *J. Cell. Physiol., 106*:339-348 (1981).
139. G. U. Labadie, K. Hirschhorn, S. Katz, and N. G. Beratis, *Pediatr. Res., 15*:257-261 (1981).
140. C. J. Schmidt, D. H. Hamer, and O. W. McBride, *Science, 224*:1104-1106 (1984).
141. B. Starcher, J. A. Madaras, D. Fisk, F. F. Perry, and C. H. Hill, *J. Nutr., 108*:1229-1233 (1978).
142. P. M. Royce, J. Camakaris, and D. M. Danks, *Biochem. J., 192*:579-586 (1980).
143. K. E. Baerlocher, B. Steinmann, V. H. Rao, R. Gitzelmann, and N. Horn, *J. Inher. Metab. Dis., 2*:(6 Suppl)87-88 (1983).
144. D. W. Rowe, E. B. McGoodwin, G. R. Martin, and D. Grahn, *J. Biol. Chem., 252*:939-942 (1977).
145. B. C. Starcher, J. A. Madaras, and A. S. Tepper, *Biochem. Biophys. Res. Commun., 78*:706-712 (1977).
146. D. W. Rowe, E. B. McGoodwin, G. R. Martin, M. D. Sussman, D. Grahn, B. Faris, and C. Franzblau, *J. Exp. Med., 139*:180-192 (1974).
147. H. Kuivaniemi, L. Peltonen, A. Palotie, and K. I. Kivirikko *Am. J. Hum. Genet., 37*:798-808 (1985).

13

Genetic Diseases of Collagen in Animals

Ronald R. Minor and Joyce A. M. Wootton
New York State College of Veterinary Medicine, Cornell University, Ithaca, New York

Donald F. Patterson
School of Veterinary Medicine, University of Pennsylvania, Philadelphia, Pennsylvania

Jouni Uitto
UCLA School of Medicine and Harbor-UCLA Medical Center, Torrance, California

Donald Bartel
School of Mechanical and Aerospace Engineering, Cornell University, Ithaca, New York

I. INTRODUCTION

The fragile skin disease "dermatosparaxis" in cattle, an analog of the Ehlers-Danlos syndrome (EDS) in humans, was the first "true" collagen disease to be identified (see Refs. 1-3 for review). Dermatosparaxis was discovered simultaneously in cattle in Belgium and Texas (4-7), and it appears that these cases were exploited in an ideal manner. Numerous laboratories were involved in clinical, genetic, morphological, tissue culture, biomechanical, and biochemical studies (4-17). The results of these studies led to exciting and important advances in knowledge of the biology and pathology of collagen. The most exciting result was the discovery that collagen is synthesized as a larger precursor molecule, procollagen, and that the disease is caused by a heritable deficiency of the enzyme procollagen *N*-peptidase (7,8). Since procollagen cannot form normal fibrils, this failure of processing of procollagen blocked fibril formation and accounted for the fragility of the skin of affected cattle (7,8). The discovery of procollagen is one of the important advances in the understanding of collagen and serves as an excellent example of the potential biomedical and biological importance of spontaneous heritable diseases of domestic animals.

Analogs of the Ehlers-Danlos syndrome are now known to occur with moderate regularity in domestic animals (1-3). These analogs include (a) diseases in dogs, cats, and mink that resemble the dominant forms of EDS with major defects in the formation of collagen fibrils and fibers (1-3,18-24); (b) additional cases of dermatosparaxis in sheep (25-27), cats (1,28-30), and dogs (1,31-32) with defects in the processing of procollagen to collagen; (c) the aneurysm-prone mottled mouse, an analog of EDS type V, EDS type IX, or Menkes kinky-hair syndrome with defects in collagen cross-linking (33-36); and (d) sporadic diseases in one horse (36) and three cats with severe skin fragility resulting from increased degradation and lysis of collagen, with phagocytosis of the resulting fragments of fibrils by fibroblasts (36). Thus, diseases like EDS may result from (a) defects in the posttranslational processing of procollagen, (b) structural protein defects that result in defects in collagen fibrillogenesis, (c) regulatory defects that decrease collagen production, (d) defects in collagen cross-linking, or (e) defects that increase collagen degradation.

Other diseases of collagen in animals include the tight skin (Tsk) mutant mouse (37, 38) and osteogenesis imperfecta. Different heritable forms of osteogenesis imperfecta have been identified in the offspring of a bull in Australia (39), an unrelated bull in Texas (40), and a dog in Philadelphia (Dr. P. Jesyk, personal communication). In this disease in cattle a severe deficiency in collagen fibril formation is transmitted as an autosomal dominant trait (39,40). In this chapter we review some of the results of clinical, genetic, biomechanical, morphological, and biochemical studies of collagens in these diseases and show how results of comparative studies can be used to characterize diseases of collagen in both humans and domestic animals.

II. ANIMAL MODELS WITH DISEASES OF COLLAGEN

A. Clinical Studies

To date, most cases of collagen disease in animals have come to the attention of veterinarians with a complaint of lacerations or fractures in a young, immature, or newborn animal with no other signs or history of trauma. Clinical examinations of dogs, cats, and horses with diseases that proved to be analogs of EDS showed that affected tissues were always fragile, and in most cases hyperextensible (Refs. 2, 3 and Figure 1). In most cases the entire skin was fragile and hyperextensible, but in one case in a horse the dorsal back skin was severely affected, while the skin of the extremities was relatively normal (36). In most cases, the severity of the skin fragility and hyperextensibility was sufficient to suggest that the disease was likely to be a heritable defect involving dermal collagen. This suggestion was then confirmed by electron microscopic examination of small skin biopsies and by breeding experiments.

In many cases, owners have donated their animals for research. This has proven to be advantageous for everyone, including the patient. For example, one dog and one cat with autosomal dominant defects in collagen fibrillogenesis are each about 10 years old at this time. They live comfortably, with an ideal environment and diet, as breeding stock, and for studies of age-related changes in analogs of EDS. In other cases, owners permitted their pets to be biopsied for tissue culture and/or electron microscopy, or to be used to breed with unrelated normal animals for genetic studies. Thus, the contributions of enlightened owners and enthusiastic support of practicing veterinarians have been a critical and valuable part of studies of diseases of collagen in domestic animals. Furthermore, studies of the analogs of EDS are facilitated by the fact that the affected animals are generally healthy, pleasant to handle, and have a normal life-span, as long as they are protected from sources of trauma in their environment. In the dogs and cats with dominant traits, skin wounds appeared to heal faster than normal and formed normal scars that disappeared with time, possibly because of a reduction in the intrinsic tension in affected skin. In contrast, the animals with osteogenesis imperfecta are so severely affected that they cannot be maintained in a comfortable, healthy state, so they are studied only as fetal or newborn specimens.

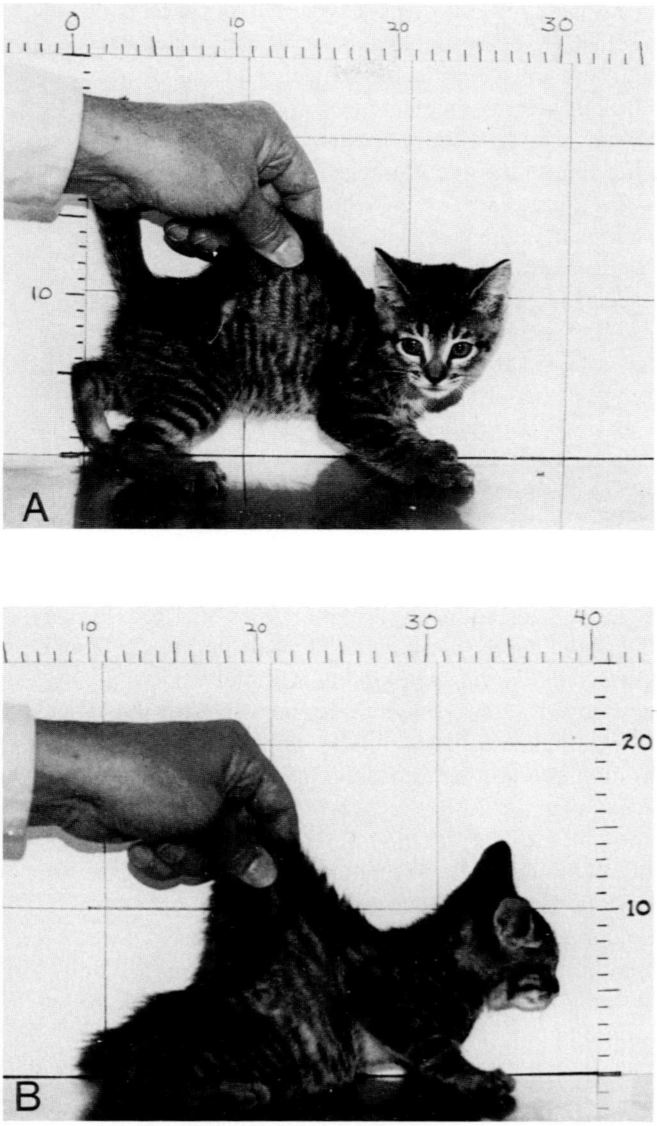

Figure 1 Sibling unaffected (A) and affected (B) kittens from the Philadelphia cat analog of EDS (21). The skin extensibility index varied from 0.1 to 0.15 in unaffected and from 0.18 to 0.25 in affected animals. This index is the ratio of the height of the fully extended fold of thoracolumbar skin shown in each photograph to the length of trunk from the occipital crest to the base of the tail (21).

B. Genetic Studies

1. Autosomal Dominant Traits

Three of the new analogs of EDS in dogs and cats and the one case of bovine osteogenesis imperfecta that we have studied have proven to be transmitted as autosomal dominant traits. In the canine and feline disorders, the propositus in each case appears to have arisen by a de novo mutation in a single germ cell. This is suggested by the facts that (a) the parents were clinically normal, (b) neither siblings or other family members were affected, and (c) the affected individuals produced approximately equal numbers of affected and normal offspring of both sexes when mated to normal individuals (21,24).

In bovine osteogenesis imperfecta, a more unusual situation occurred. Of 64 offspring of a clinically normal bull mated to a series of unrelated cows, 28 (44%) of the calves had the disorder (39). These findings suggest that the mutation occurred in the bull's germ line precursor cells during development giving rise to a population of germ cells that were heterozygous for the mutant allele. Semen from this bull was used to inseminate cows at Cornell University, to produce the paired affected and unaffected fetuses and newborn calves for the biochemical studies described below. Unfortunately, the affected calves are so severely affected at birth that it is not possible to maintain the heterozygotes for breeding to perpetuate this germ line.

Each of the dominant traits that we have studied has had a slightly different morphological defect in collagen fibrillogenesis, and these defects have been present in all affected offspring. We assume that each represents a new and different mutation, and that each involves a protein that in some way regulates lateral growth and parallel packing of collagen fibrils. This pattern of inheritance facilitates the production of age- and sex-matched pairs for experimental studies of both the causes of the defects in collagen packing and effects of the different defects on the physical properties of affected tissues.

To produce homozygous affected animals, we utilized the Philadelphia cat analog of EDS that produced about 50% affected offspring when mated with normal cats (21). The results of these studies suggested that embryos that are homozygous for the affected gene die during early embryonic development. In pregnancies resulting from matings between two heterozygotes, the litter size was reduced, and only 12 of 34 offspring were affected. This is significantly fewer than the 75% expected if heterozygous and homozygous affected animals were both viable ($X^2 = 28.59$, df = 1, $p < 0.001$). Furthermore, when the conception sites produced by these heterozygous × heterozygous matings were examined by uterotomy at 20 and 44 days of a 63-day gestation, approximately one-fourth of the placental sacs were abnormally small and contained no embryo. Since collagen fibrils are thought to play such an important role in embryogenesis, it is not surprising that the homozygous state of a defect in collagen fibrillogenesis would be lethal during early embryonic development. This hypothesis cannot be tested experimentally in cases of EDS in humans, so it is important to learn about the effects of these diseases in other species.

To study whether the autosomal dominant analogs of EDS are also lethal in dogs, we performed two affected × affected matings of offspring of both the Guelph dog and Lynbrook, Long Island dog (24). Unfortunately, none of these four matings produced viable offspring. Conception and implantation were confirmed with ultrasound and/or laparotomy at about 30 days of gestation, but the number of implantation sites was about one-third of that expected, and all of the fetuses were resorbed prior to day 45 of a 63-day gestation. Thus, we do not yet know whether embryonic death resulted from the homozygous state of the mutant allele in any of the embryos, or whether the abnormal collagen fibrils and fibers in the affected uterus contributed to the fetal death.

2. *Autosomal Recessive Traits*

Since autosomal recessive traits require two abnormal alleles to produce clinical disease, prospective genetic studies of these traits are complicated and expensive. However, the fact that half-normal levels of most enzymes are sufficient to prevent clinical disease causes most enzyme deficiencies to be inherited as recessive traits, and this helps to focus experimental studies of these diseases.

Dermatosparaxis in cattle is a good example of an autosomal recessive trait. This disease was disseminated accidentally when a clinically normal bull with the mutation and his clinically normal heterozygous son were both used as sires in an artifical insemination program in Belgium in the 1960s (4,5,13). This accident produced an excellent supply of specimens for research and led to the discovery of procollagen and the enzyme procollagen *N*-peptidase whose deficiency caused this disease (7,8). However, because these clinically normal heterozygous bulls were used in an artificial insemination service, more than 6500 carriers of this allele were identifiable by 1974 (13). This emphasizes the importance of identifying carriers of a deleterious mutant allele for genetic counseling, and the economic importance of these alleles in animal production. At the same time it shows how much progress can result from experimental studies of these biological accidents in domestic animals.

Of the 20 new cases of analogs of EDS that we have been able to study during the last 10 years, only one case, a cat from Rochester, New York, has a history that suggests that its disease is transmitted as an autosomal recessive trait. This female cat is the product of a mother X son mating (Dr. R. Garman, personal communication), and it is apparently infertile. It has been observed to mate with more than one normal cat, but it has never conceived. This is unfortunate, since this cat showed morphological abnormalities in both formation and degradation of collagen fibrils, and the activity of collagenase in explants of its skin was approximately 4 times higher than the activity in comparable explants of control cat skin (Uitto, Bauer, and Minor, unpublished results). Since this cat had extremely thin, fragile and hyperextensible skin as well as bilateral coxofemoral luxations, it would be exciting to learn whether this disease is due to a mutation increasing the activity of collagenase, or a mutation decreasing the activity of an inhibitor of collagenase.

C. Biomechanical Studies

Dumbbell-shaped sections of dorsolateral thoracic skin of dogs and cats were used in uniaxial tension tests to study (a) whether there are directional differences in the physical properties of normal dog or cat skin, (b) whether differences in physical properties are related to characteristic directional differences in the organization of dermal collagen, (c) whether the directional differences in physical properties are affected by the defects in collagen fibrillogenesis, (d) whether the tissue hyperextensibility is relative or absolute, and (e) whether the severity of the tissue fragility varies in the different diseases in dogs and cats. To answer these questions we compared the physical properties of sections oriented perpendicular and parallel to the direction of the hair and studied adjacent sections of skin with light and electron microscopy (24).

1. Effects of Specimen Orientation in Normal Skin

The results of mechanical studies of both dog and cat skin showed that with loading perpendicular to the direction of the hair, there was (a) less elongation per unit increase in load, (b) more load-bearing capacity per milligram of collagen per unit area of skin, and (c) a smaller maximum stretch than with loading parallel to the hair (Figure 2). Conversely, there was more elongation per unit increase in load and a greater maximum stretch in sections cut parallel to the direction of the hair. These directional differences

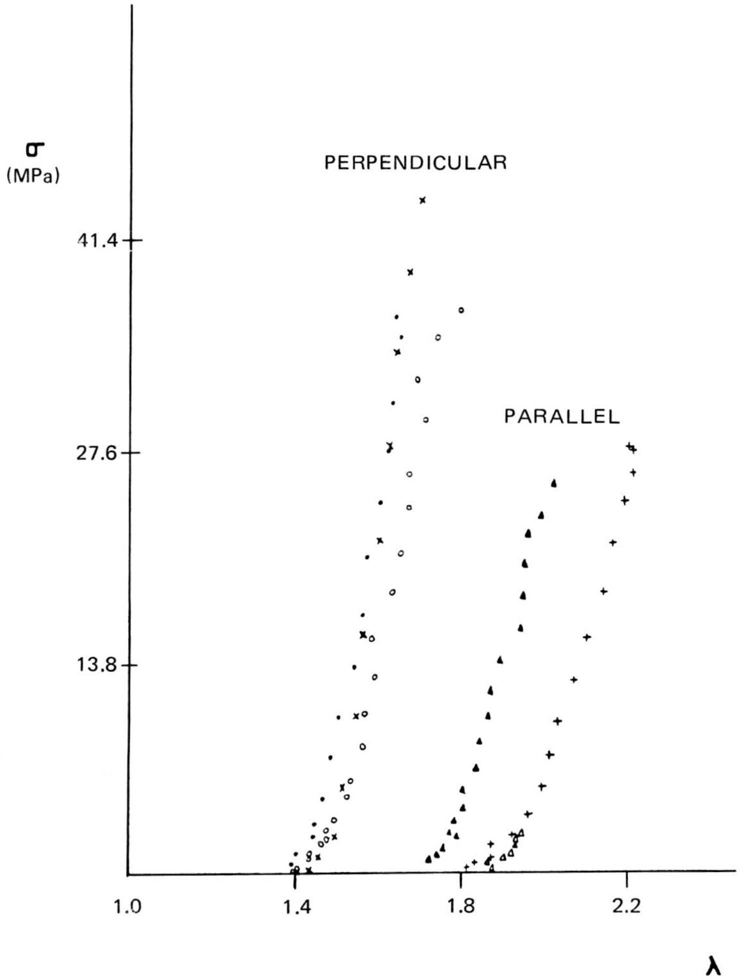

Figure 2 Effects of specimen orientation on the stress-stretch characteristics in uniaxial tension tests of normal dog skin. Dumbbell-shaped biopsies were cut perpendicular to the direction of the hair 2 cm to the left and parallel to the hair 2 cm to the right of the thoracic midline (24). Each biopsy was immediately frozen in liquid nitrogen until testing in an Instron tensile tester (41-43). The specimens were elongated at 2.54 cm/min. Load (Y axis) was recorded on the tensile tester, and elongation (X axis) was recorded by high-speed cinematography of the movement of a matrix of dots on the skin surface. A sonic digitizer tablet and microcomputer were used to analyze the force traces. In this graph load was calculated as Megapascals force per cross-sectional area (MPa), but the force traces were identical when calculated as Newtons force per mg of collagen per mm of skin (see Figure 4). Eight pairs of dogs and four pairs of cats were studied, and the graph shown is representative of the results obtained with both species. The results showed that there are marked directional differences in load versus elongation when the specimens are oriented perpendicular or parallel to the hair. The axis of the speciments cut perpendicular to the hair is parallel to the long axis of the rhomboids formed by the dermal collagen shown in Figure 3.

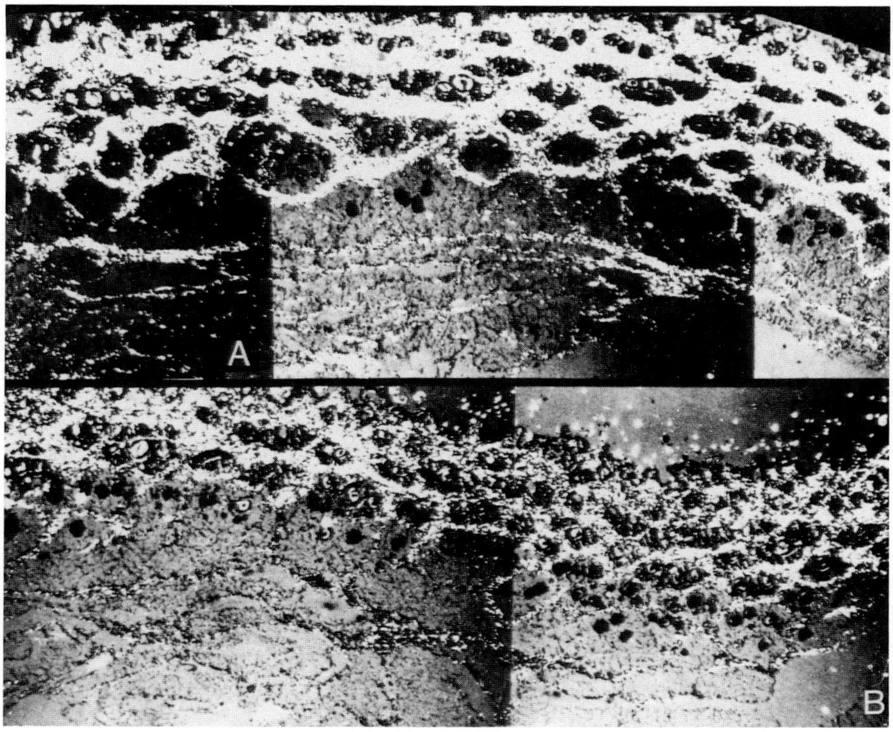

Figure 3 Polarized light micrographs of 5-μm sections of dorsolateral thoracic skin cut perpendicular to the direction of the hair of unaffected (A) and affected (B) sibling dogs from the Guelph dog analog of EDS (24). A rhomboidal pattern of collagen birefringence is clearly evident. The center of each rhomboid is filled with hair follicles, and each hair in these follicles has a circular profile. The intensity of birefringence was greatly reduced in the affected specimen. This must have been due to the disordered packing of fibrils within the abnormal segments of all of the collagen fibers (Figure 6 and Refs. 2,3,24). Extensive chemical analysis showed tht there was absolutely no difference in the amounts of collagen or total protein per unit area of skin of these dogs. The load versus elongation profiles of adjacent sections of skin are shown in Figure 4. Magnification X13.

in the physical properties appeared to be explained by the results of microscopic studies, which showed that there was a basic rhomboidal arrangement of dermal collagen fibers around clusters of hair follicles (Figure 3). This rhomboidal arrangement of dermal collagen had been proposed previously (41,42), and it was clearly shown with polarized light microscopic examination of 5-μm sections of skin cut exactly perpendicular to the hair shafts (Figure 3). The perpendicular orientation of these sections was confirmed by the fact that the cross-sections of the hair shafts were circular.

With loading along the long axis of the rhomboids, perpendicular to the hair, more collagen fibers will be recruited for each unit of elongation. With loading along the short axis of the rhomboid, parallel to the hair, there will be more elongation before a comparable number of fibers are recruited for load bearing. These directional differences in physical properties and organization of collagen fibers in dog and cat skin are comparable to the directional differences associated with Langer's lines in human skin (41). It seems likely that similar directional differences in the organization and function of dermal collagen will be present in the skin of other animals.

2. Abnormalities in Diseased Skin

To characterize the effects of these collagen diseases on the physical properties of skin, we conducted uniaxial tension tests of matched biopsy specimens of dorsolateral thoracic skin of paired sibling dogs and cats with three analogs of EDS. These tests showed that with all three diseases the tensile strength of affected skin ranged from 5 to 10% of normal (Figure 4 and Ref. 24). Even though the rhomboidal arrangement of dermal collagen was clearly evident in microscopic sections, the distinct directional differences in physical properties were absent in the tests of the hyperextensible, fragile skin. It seems likely that this is due to the severity of the fragility of dermal collagen fibers. The maximum load required to tear affected skin appears to be so small that the abnormal collagen fibers tear before they can be reorganized. This is consistent with the suggestion that directional differences in the amount of fiber reorientation during loading are important for the directional differences in the physical properties of normal skin. Furthermore, the results showed that even though there was more elongation at lower loads, there was no increase in the ultimate stretch of affected skin, showing that the clinical hyperextensibility was only relative.

D. Microscopic Studies

1. Defects in Fibril and Fiber Formation

To date, all of the autosomal dominant analogs of EDS have had defects in the regulation of collagen fibril growth, and in the parallel packing of fibrils into fibers (2,3,21,24). Some of these defects were multifocal, with alternating areas of normal and abnormal fibril growth and packing in each collagen fiber (Figures 5-7). In other cases the morphological defects were diffuse, with all collagen fibers being affected (24). The microscopic lesions were similar in all of the affected offspring of each propositus, but they were slightly different in each different disease. However, the range of microscopic lesions was similar in the different cases in dogs, cats, and humans (cf. Figures 5 and 6 and Refs. 2,3,21,24,30-32,44,45).

In cases with multifocal lesions, most electron micrographs of dermal collagen showed that >70% of the cross-sections of collagen fibers were morphologically normal. The fibrils in the normal areas were uniformly cylindrical, had a narrow range of diameters, and were tightly packed in parallel (21,24). In contrast, in the abnormal areas there was fusion and branching of fibrils, resulting in many fibril cross-sections being larger (2-3 times normal diameter) and lobulated (Figures 5 and 6, Refs. 21,24). In addition, the normal dense parallel packing of fibrils was severely disorganized or loose within these abnormal segments of collagen fibers. In trichrome-stained paraffin sections, these abnormal segments of fibers stained red, while all of the normal areas of collagen fibers stained blue (Figure 7 and Ref. 24). However, this red staining of collagen fibers is not pathognomonic for defects in collagen fibrillogenesis (36).

The multifocal distribution of areas of abnormal fibril formation and packing along the length of each collagen fiber appeared to explain the fact that even though >70% of the cross-sections of collagen fibers were morphologically normal, the tensile strength of affected skin was less than 10% of normal (Figure 4). The abnormal segments appeared to act like weak links in a chain, so that the tensile strength of any area of skin was only as strong as the maximum strength of the weakest group of stressed fibers.

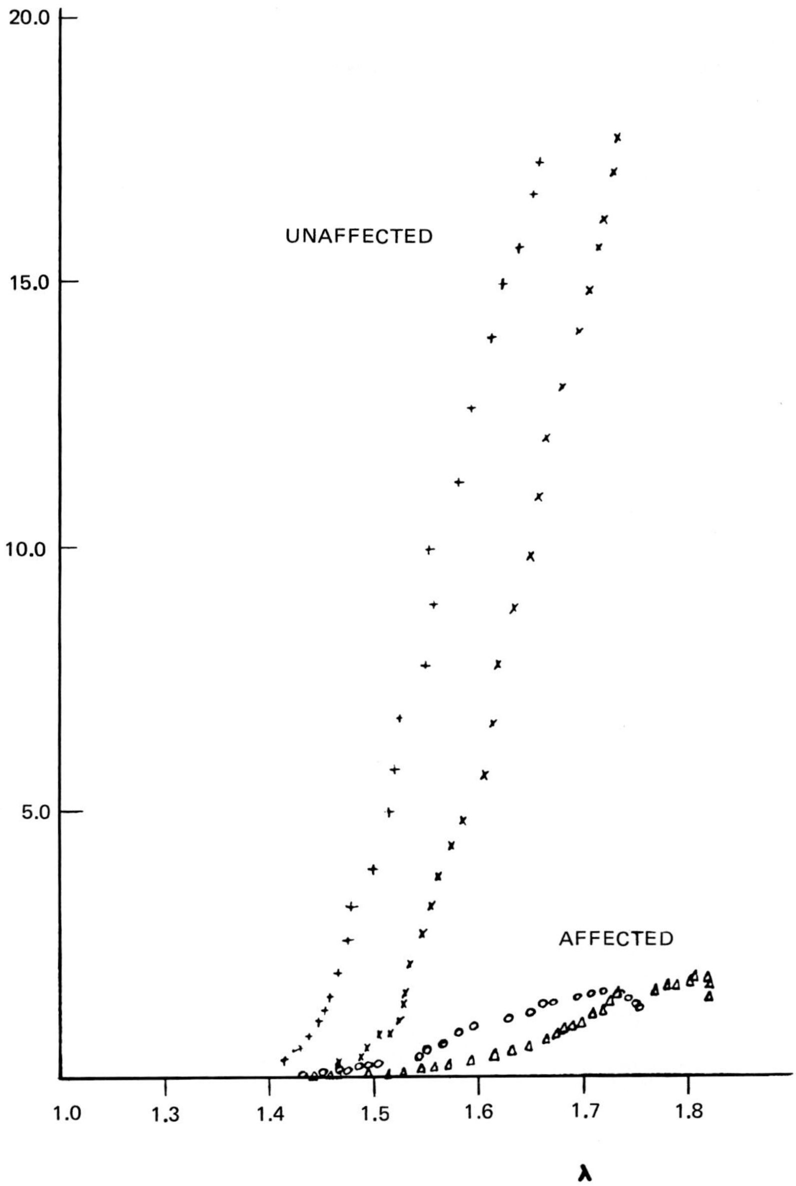

Figure 4 Stress-stretch characteristics of paired specimens of affected and unaffected skin cut perpendicular to the hair of the dogs described in Figure 3. Dumbbell-shaped specimens were tested as described in Figure 2. Elongation (X axis) is the ratio of the elongated length to the original length, and load (Y axis) the Newtons force per mg of collagen per cm of skin. Weight bearing was detected at about 40% elongation in all specimens. In these tests the unaffected specimens tore at 60-70% and affected specimens tore at 75-80% elongation, but in other tests unaffected specimens tore at 175% and affected specimens tore at 160% of the original length. In all tests, affected skin tore at 5-11% of the load required to tear unaffected skin.

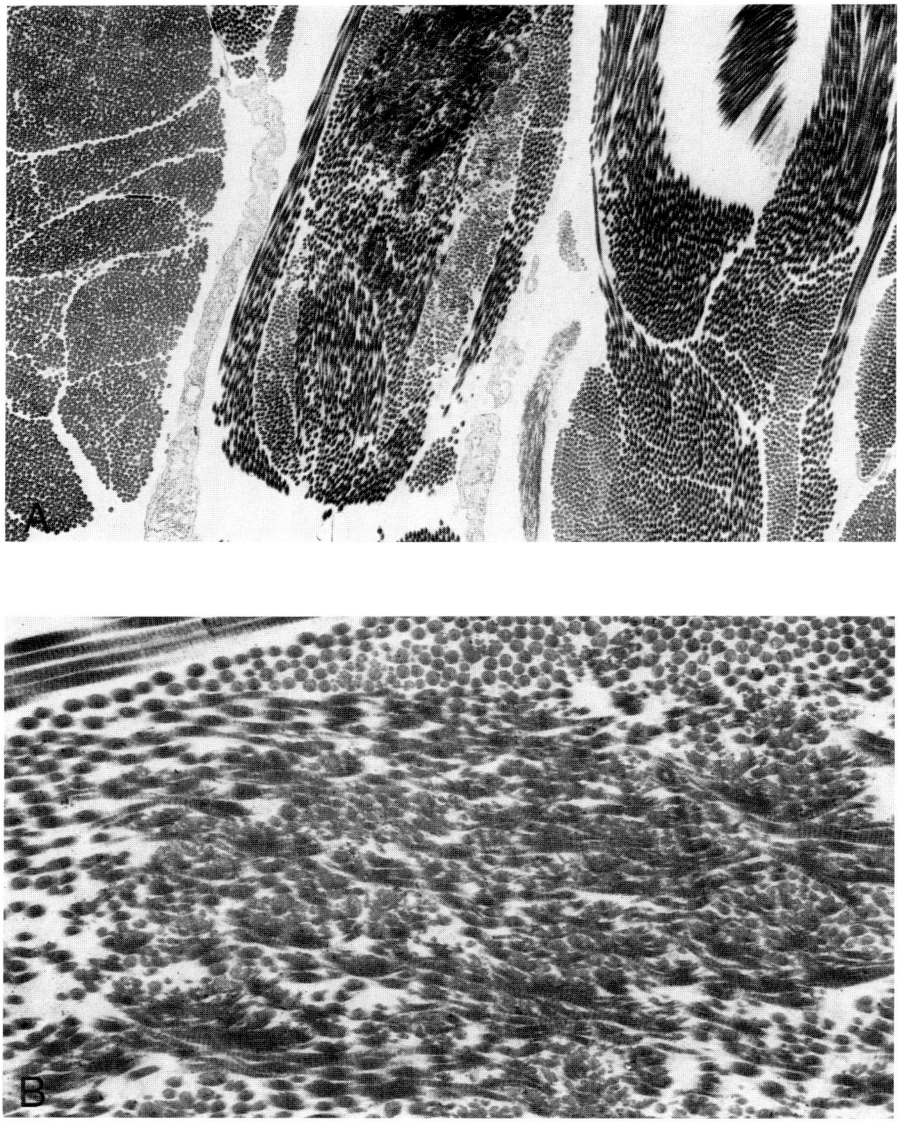

Figure 5 Electron micrographs of skin of a Philadelphia cat analog of EDS (21), showing multifocal defects in packing of collagen fibrils into fibers. In A the sections of the collagen fibers at the left and right are morphologically normal, with a dense parallel packing of uniformly cylindrical fibrils with a narrow range of diameters. In contrast, the fibrils in the section of the centrally located fiber in A are severely disorganized, have a much wider range of diameters, and are often lobulated. This fiber is shown at higher magnification in B. It appears that the increased diameter and lobulation of these abnormal fibrils result from a fusion of short segments of adjoining fibrils. Essentially all the collagen fibers in affected skin contained multiple foci of abnormal fibril packing such as this. In some of the abnormal segments the disorganized packing involved the entire cross-section of the fiber, but in other segments only a small portion of the cross-section was abnormal (cf. Figures 6 and 7). Magnification (A) ×3200; (B) ×13,650.

These weak links also appeared to explain how the tensile strength of affected skin could be <10% or normal, even though the chemical studies described below showed that there was no difference in the extractability or solubility of skin collagen, or in the amount of collagen per unit area of affected and unaffected skin (Table 1). Therefore, these comparative biomechanical, morphological, and biochemical studies suggested for the first time that defects in fibril organization may cause severe fragility of skin in which the amounts, solubility, and molecular structure of dermal collagen are apparently normal.

These studies showed that in two families of dogs with dominantly inherited defects in collagen fibrillogenesis, the fragility of the skin was due to the defects in organization of collagen, and not due to any quantitative decrease in the amount or solubility of dermal collagen. This suggests that in studies of cases with fragile tissues one should first look for evidence of microscopic signs of abnormalities in the formation or degradation of collagen fibrils, and then for chemical evidence of a quantitative decrease either in the cross-linking of collagen molecules within fibrils or in the amount of collagen per unit area of skin. This latter suggestion is prompted by the fact that previous morphological studies showed that the distance between fibrils was too great to be spanned by lysyl-derived cross-links (21), and most diseases of collagen were formerly thought to involve a decrease in the amount of collagen or a defect in cross-linking. These comparative studies are most informative if performed on paired affected and unaffected sibling animals, but the findings contribute directly to our understanding of analogous diseases of humans. These comparisons circumvent many of the limitations of controlled experimental studies of diseases of collagen in humans.

2. *Defects in Fibril Degradation*

To date, we have studied four animals with fragile, hyperextensible skin and signs of increased degradation of banded collagen fibrils. The first case in this category was a cat, from Rochester, New York, with thin, fragile, hyperextensible skin and bilateral coxofemoral luxations. The skin formed cigarette-paper scars during healing. Electron microscopy showed that the dermal collagen fibrils fused and branched and had a severely disorganized packing like those in the other dogs and cats with defects in fibril formation. In addition, there were many signs of collagenolysis, with fragmentation of banded collagen fibrils (Figure 8). Dermal fibroblasts from this cat grew poorly, but no abnormalities were identified in collagen structure or synthesis in the cultured cells. There was, however, a fourfold increase in collagenase activity in explants of biopsies from this skin. Since this cat has proven to be infertile, it may never be possible to determine whether a defect in fibrillogenesis or a defect in the regulation of collagenolysis was the primary defect in this disease. Nevertheless, the fact that the disease appeared to be inherited as an autosomal recessive trait and the observation of both morphological and biochemical signs of increased collagenolysis raise the question of whether the fragility and hyperextensibility of connective tissues in some forms of EDS may result from increased degradation of collagen in the affected tissues. Electron microscopic examination of small biopsies of affected tissues from all patients with these diseases should be used to look for the signs of abnormal collagenolytic activity, as well as for defects in fibril formation.

The second analog of EDS with increased collagenolysis was a horse with fragile, hyperextensible skin, and no signs of a defect in fibrillogenesis (36). There was an extensive accumulation of granular debris and small fragments of collagen fibrils. There was also phagocytosis of these fragments of banded collagen fibrils by fibroblasts, but there

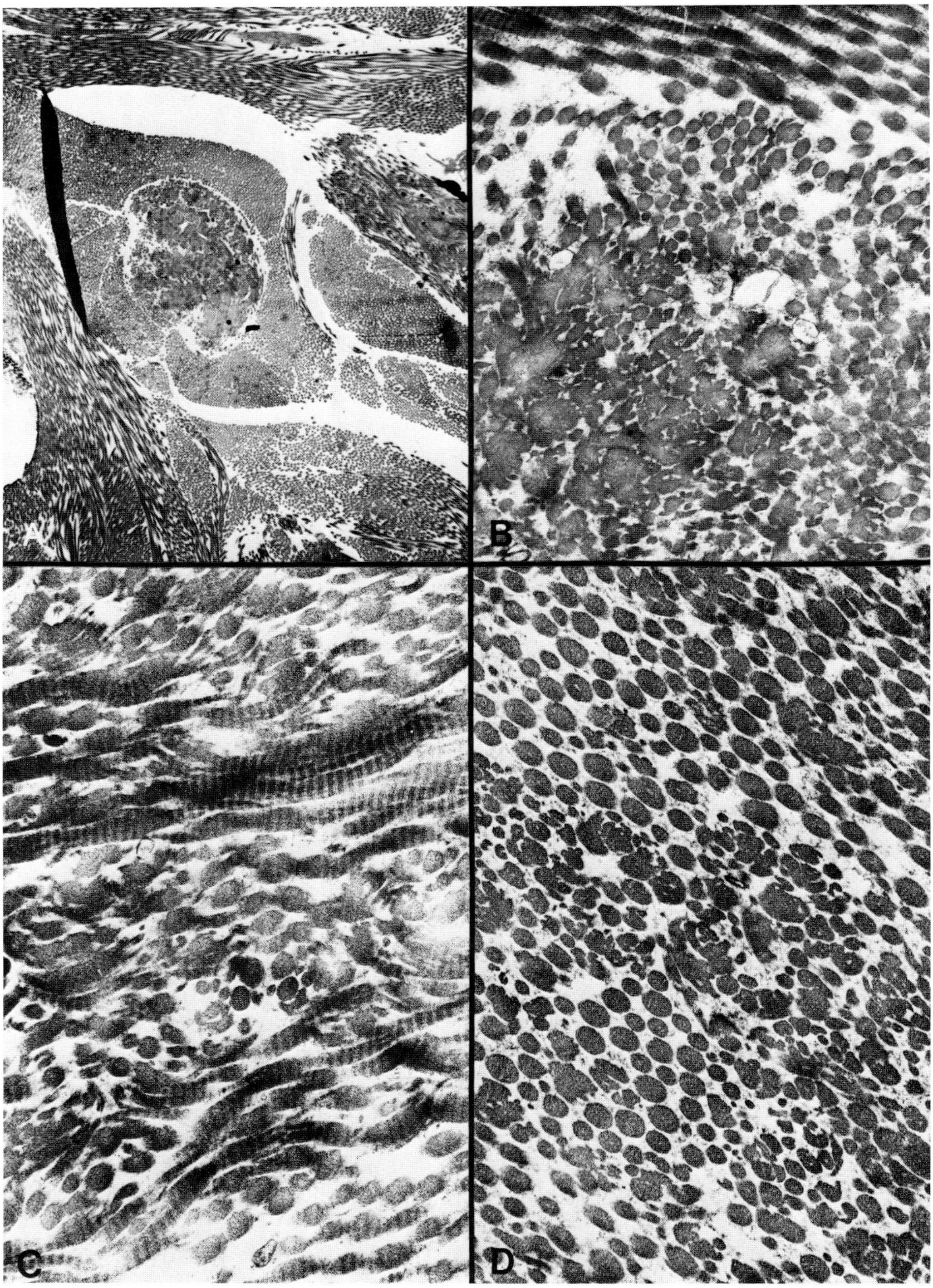

Figure 6

Figure 6 Electron micrographs of focal defects in collagen fibril packing in the skin of the Guelph dog analog (A and B) (24), and a Kingston, New York cat analog of EDS (C and D). These unrelated diseases showed essentially identical multifocal morphological defects in collagen fibrils and fibers. The section of the abnormal fiber in A has a central focus of severe fibril disorganization, but the packing of the fibrils around the periphery of this fiber is morphologically normal. In trichrome-stained sections the focus of abnormal fibril packing stained red, while the peripheral areas of normal fibril packing stained blue (cf. Figure 7). Conversely, the trichrome stain abolished birefringence of polarized light in all the areas of normal fibril packing, but not in the central focus of fibril disorganization. The defects in fibrillogenesis in the Guelph dog and Kingston cat are similar to those in the Philadelphia cat analog of EDS (Figure 5 and Ref. 21). However, as shown in micrographs B-D, there is more evidence of fusion and branching of segments of adjoining fibrils, and many more large aggregates and large lobulated fibrils in the foci of abnormal fibril packing in this disease in the Guelph dog (cf. B-D and Ref. 21). Magnification (A) ×2240; (B) ×24,700; (C) ×47,500; (D) × 47,500.

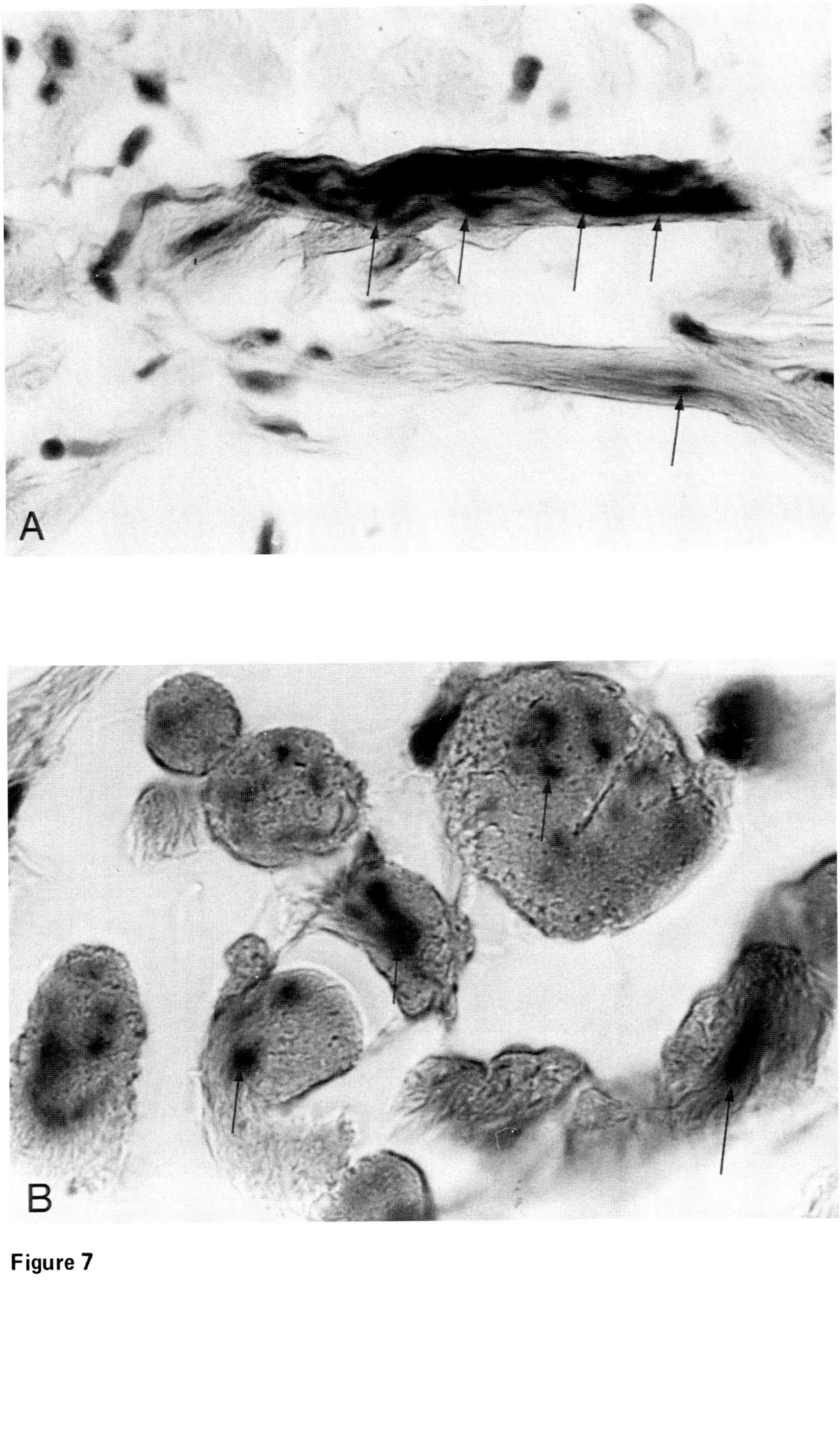

Figure 7

Figure 7 Light micrographs of trichrome-stained, 5-μm sections of affected skin of the Guelph dog analog of EDS. A Kodak #47 B Wrattan filter was used to increase the contrast of abnormal red-stained foci (arrows) with the normal blue-stained areas of collagen fibers. Noncollagenous structures also stained red. Because of the filtration, the red foci appear black, and blue areas are a lighter gray in these micrographs. The red foci within collagen fibers corresponded to the foci of fibril disorganization shown in Figure 6. These red foci occupied the entire cross-section of some fibers, while in others there was no red staining or only a small portion of the cross-section was red. Since these animals were heterozygous for a simple dominant trait, the multifocal distribution of the fibril-packing defect raises the question of whether the cells that produced the collagen in the abnormal foci were expressing only the mutant allele, while those in the areas of normal collagen were expressing only the normal allele. Allelic exclusion such as this has been demonstrated in the transcription of immunoglobulin genes, but not for genes for structural proteins. Magnification (A) ×440; (B) ×1750.

Table 1 Composition of Matched 4-cm^2 Biopsies of Skin and Subcutis of Affected and Unaffected Littermates of the Guelph Dog Analog of EDS

	Wet weight (g)		Water[b] (%)		Lipid[c] (%)		Protein[d] (mg)		Collagen[e] (mg)	
Tissue[a]	U	A	U	A	U	A	U	A	U	A
Skin	2.73	3.32	44.3	46.4	31.3	37.1	524	429	368	296
Subcutis	2.41	5.18	35.3	31.9	50.0	60.8	240	294	56	124
Total mass (mg)	5140	8500	2060	3190	2050	43801	764	723	424	420

[a]A 4-cm² template was used to mark matched areas of skin for dissection from the deep fascia. For chemical analysis the subcutis was dissected from the deep dermis.
[b]Determined as the difference in weight before and after lyophilization.
[c]Determined as the weight of lipid extracted with chloroform and methanol.
[d]Determined by amino acid analysis of delipidated lyophilized tissue.
[e]Determined as hydroxyproline ×9.1.
A, affected; U, unaffected.

were no signs of inflammation in the affected dermis. The morphological studies strongly suggested that the fragility of the connective tissues in this case was due to increased collagenolysis, but the owners of the horse refused permission for any further studies.

The third analog of EDS with increased collagenolysis was truly unusual. This was a cat with fragile skin, but electron microscopy showed that collagenolysis started in the center and spread toward the periphery of cross-sections of large collagen fibers (Figure 9). As the collagen was lysed, it was replaced by a homogeneous osmiophilic material that resembled accumulation of lipids by electron microscopy. As this process continued, the collagen fibers came to resemble doughnuts with a thin rim of collagen fibrils surrounding a large lake of osmiophilic material. Surprisingly, there were no morphological signs of an association of cells or cell processes with the areas of collagenolysis. This process of collagenolysis appeared to result in a severe paucity of collagen in the dermis of this cat, which suggested that the fragility of the skin was primarily due to increased collagenolysis. At this time, however, we have no explanation for the fact that this process appears to begin in the center of collagen fibers, or for the replacement of solubilized fibrils with an accumulation of osmiophilic material. This emphasizes the importance of electron microscopic examination of each case with fragile connective tissues, and at the same time it emphasizes how much we have to learn about these diseases.

Figure 8 Electron micrographs of skin collagen in the Rochester, New York cat with severe defects in both the formation and degradation of collagen fibers. Most fibrils are fragmented and their packing is severely disorganized. Adjoining segments of fibrils are fused, giving rise to larger, lobulated cross-sections. In the areas of degradation the ends of fibrils are frayed and lightly stained, and there is an accumulation of granular debris. Magnification (A) ×11,400; (B) ×38,000.

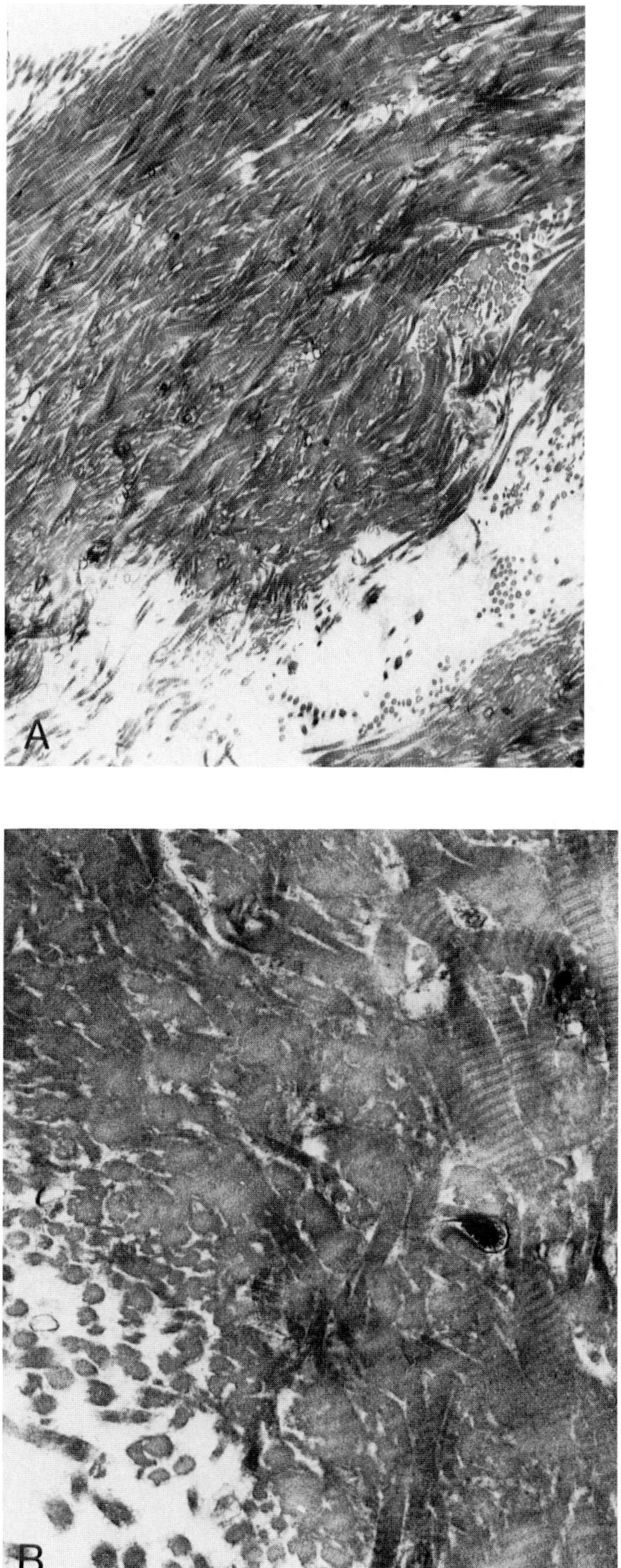

Figure 8

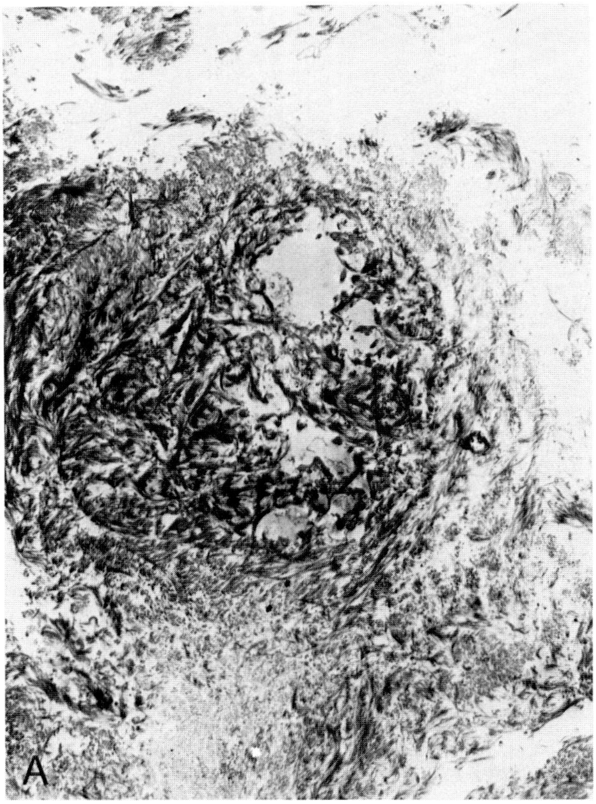

Figure 9 Electron micrographs of collagen fibers in the dermis of a Concord, Massachusetts cat with extremely fragile skin. The collagen fibrils are fragmented, are severely disorganized, and are undergoing lysis in each of these fibers. There was a severe paucity of collagen in the dermis, and all the fibers that remained were undergoing lysis. As shown in A, the lytic process resulted in the appearance of small lakes of osmiophilic material in the center of larger collagen fibers, but at no time were cells or cell processes found in association with this collagenolysis. These lakes gradually coalesced to form a large lake filled with osmiophilic material that occupied the center of each collagen fiber (B). Most of the collagen fibers that remained in the dermis appeared as large lakes of osmiophilic material, surrounded by a thin rim of fragmented collagen fibrils (C). The fragile skin was the only clinical sign, but the cause of these lesions is not known. Magnification (A) ×2100; (B) ×2360; (C) ×2670.

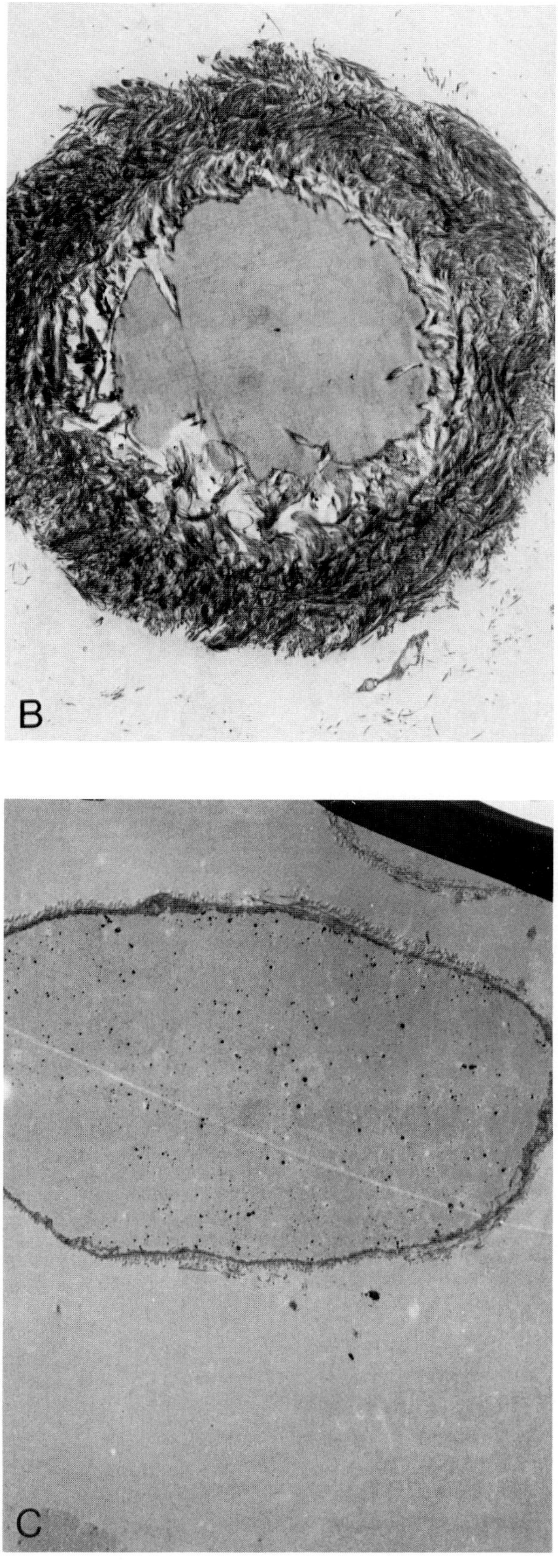
B
C

E. Biochemical Studies

1. *Studies of Tissues with Defects in Fibril Formation*

To learn more about the cause of the tissue fragility and cause of defects in collagen fibrils and fibers we utilized tissues from paired siblings to study the following questions. (a) Are there differences in the amounts of collagen, total protein, total lipid, or total water per unit area of affected and unaffected skin? (b) Are there differences in the extractability or proportions of the different types of procollagen or collagen in affected and unaffected tissues? (c) Is there a difference in the amino acid composition of affected and unaffected tissues, in the purified chains of collagens extracted from these tissues, or in the purified cyanogen bromide peptides derived from these chains? (d) Do differences in the rates and types of collagen synthesis in cultured fibroblasts correspond to differences in the amounts and types of collagen in vivo? (e) Is there a difference in the rates of processing of procollagen to collagen in vivo or in vitro? (f) Is there a difference in the proportions of the chains making up the different types of heteropolymeric collagens in vivo or in vitro? (g) Is there a difference in the electrophoretic mobility of the chains of procollagen or collagen extracted from tissues or synthesized in vitro? (h) Is there a difference in the electrophoretic mobility of any of the peptides derived by cyanogen bromide digestion of any of the purified collagen chains from affected and unaffected tissues? (i) Is there a difference in the resistance of the different types of collagen to digestion with proteases, such as pepsin, trypsin or chymotrypsin? Since collagen is primarily responsible for the tensile strength of tissues, and since collagen fibrils and fibers were affected in each of these diseases, it seemed likely that chemical abnormalities would be identifiable in fibrillar collagens in each of these diseases.

To date, no consistent differences have been identified in any of the collagens in affected and unaffected tissues from the dogs and cats with autosomal dominant traits with defects in collagen fibrillogenesis. Careful quantitative studies of paired areas of skin of affected and unaffected sibling Guelph dogs showed that there was more water and lipid per unit area of affected skin, but there was absolutely no difference in the amount of total protein or collagen per unit area of affected and unaffected skin and subcutis (Table 1). Because of the increased water and lipid, each square centimeter of affected skin was actually heavier than paired specimens of unaffected skin. However, there was no difference in the amino acid composition of paired skin samples, no difference in the skin collagen solubility, resistance to pepsin, electrophoretic mobility, amounts of procollagen, proportions of type I, type III, or type V collagens, or in the cyanogen bromide peptide fragments derived from these collagens from affected and unaffected skin (Figure 10). Similarly, there were no detectable differences in the rates or types of procollagen synthesis, or in the structure of the collagens synthesized by cultured skin fibroblasts from these animals. Thus, there is currently no evidence to suggest that a mutation in a collagen gene is responsible for the defects in collagen fibrillogenesis in these animals.

Recent studies of proteoglycans from the skin of paired littermates of the Guelph dog analog of EDS have shown that the ratio of iduronic acid to glucuronic acid in the proteodermatan sulfate was 6 to 1 in unaffected and 1.9 to 1 in affected skin (Drs. S. Damle and J. Gregory, personal communication). Furthermore, these studies showed that there were no differences in the apparent molecular weight of the proteodermatan sulfate chains, no difference in the ratio of proteoheparan sulfate to proteodermatan sulfate, no difference

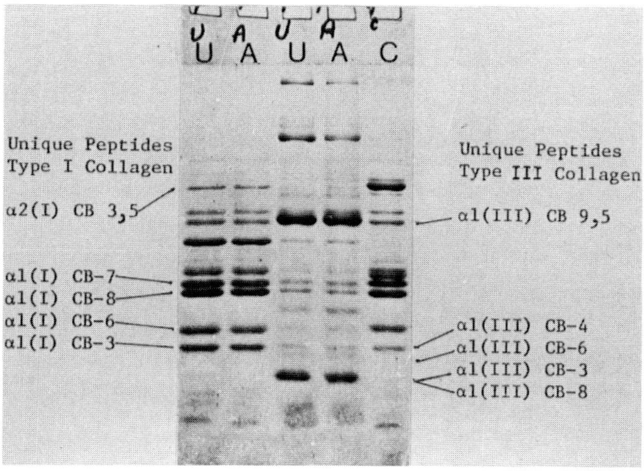

Figure 10 SDS-Polyacrylamide gel electrophoresis of cyanogen bromide digests of purified type I (lanes 1 and 2), and type III (lanes 3 and 4) collagens of unaffected (U) and affected (A) dermis of littermates of the Guelph dog analog of EDS. Lane 5 (C) is a rat tail tendon collagen control. These results showed that there was no detectable difference in the electrophoretic mobility of any of the peptides of these major skin collagens. This suggests that the defects in collagen fibrillogenesis in affected skin may not result from a structural defect in one of the chains of these collagens.

in the molar ratio of sulfate to total hexosamine, and no difference in the proportion of protein (53-55%) in the proteodermatan sulfate in affected and unaffected skin. These values were all similar to those of pig skin proteodermatan sulfate described previously (46).

It will be interesting to learn whether the dramatic differences in epimerization of uronic acid in proteodermatan sulfate are related to the defects in collagen fibrillogenesis in affected Guelph dog skin. Recent studies provide strong evidence that proteodermatan sulfate is intimately involved with collagen fibrils (47,48). There is no known mechanism to explain how a single gene defect in only one allele coding for a core protein can serve to regulate the proportions of iduronic and glucuronic acid in skin proteoglycans, but these animal models offer an excellent opportunity to ask experimental questions about the cause of the differences in epimerization and about the relationship of these differences to the defects in collagen fibrillogenesis. It would be important to learn whether the conformation of the core protein of proteoglycan controls the amount of epimerization of uronic acid residues, and whether the proportions of iduronate and glucuronate serve to regulate collagen fibril growth, prevent the fusion of adjoining fibrils, and affect the tight parallel packing of fibrils within collagen fibers.

Since these collagen-packing defects are constantly present in the heterozygous affected animals, the unlimited supply of paired littermates provides an excellent model system to simultaneously study skin proteoglycans and skin collagens, to search for the mechanisms that are responsible for the loss of control or failure to regulate the narrow range of diameters, uniform cylindrical shape, and tight parallel packing that normally characterize the packing of collagen into fibrils that make up fibers and fiber bundles in all connective tissues.

2. *Studies of Tissues with Defects in Fibril Formation and Degradation*

Since the cat from Rochester, New York showed morphological signs of abnormalities of both collagen fibril formation and degradation, we extracted, purified, and characterized the different types of collagens making up the dermis, as well as those synthesized by cultures of dermal fibroblasts. The initial studies of skin biopsies indicated that the concentration of collagen in the skin of the affected cat was the same as in five healthy control cats (Table 2). The amino acid composition of the dermal collagen was normal, and there was a 6:1 ratio of the type I to type III collagens that were solubilized by limited pepsin proteolysis of both affected and control skin. The solubility of skin collagen in 0.1 N acetic acid was slightly reduced: 8.0% solubilized in affected skin as compared to 12.4% ± 3.9% in control skin (mean ± S.D., $n = 5$). However, most important, the production of collagenase, measured by a radioimmunoassay, was significantly increased in the affected skin explants that were cultured 24 hr (Table 2). Thus, the increased collagenase activity in the affected cat may well contribute to the fragility of skin noted clinically, and to the signs of fibril fragmentation and lysis noted morphologically. To our knowledge this is the first case in an animal with evidence of a heritable tissue fragility associated with increased production of collagenase.

3. *Studies of Tissues from a Case of Bovine Osteogenesis Imperfecta (BOI)*

To learn more about the cause of the fragility of ligaments, tendons, bones, and teeth in the Australian variant of BOI (39), collagens were extracted from skin and tendon of affected and unaffected twin fetuses and of newborn affected and unaffected half-sibling calves produced by Holstein cows at Cornell University. These cows were bred with semen of the Australian bull that produces BOI in 44% of his offspring (39). The extracted skin and tendon collagens were compared to collagens and procollagens produced by cultured fibroblasts from each of these tissues. These studies suggested that the tissue fragility was due to a decrease in the amount of type I collagen, and this was associated with a concomitant decrease in the ratio of type I to type V collagen in affected tissues. Surprisingly, these studies also showed that in SDS-polyacrylamide gels the electrophoretic mobility of all of the collagen chains extracted from tissues or produced by cultured fibroblasts from affected animals was consistently slower than that of the chains from unaffected tissues or produced by unaffected cells in vitro (49). Furthermore, in studies of a case of human OI, all of the chains of type I, type III, and

Table 2 Collagen Content and the In Vitro Production of Collagenase in Skin from the Rochester, New York Cat with Fragile Connective Tissues

	Collagen content[a]		Collagenase production[b]	
Cat	(μg/mg)	(μg/mm^2)	(ng/mg)	(ng/μg hypro)
Affected	137.8 ± 21.3	200.1 ± 45.9	14.5 ± 2.2	0.92 ± 0.16
Controls ($n = 5$)	117.1 ± 17.5	150.9 ± 26.6	3.4 ± 1.0	0.24 ± 0.07

[a]Determined as hydroxyproline ×8.2; values are expressed as μg collagen per mg wet weight of skin or per mm^2 of skin surface area (mean ± SD of three to six determinations).
[b]Values are ng collagenase protein produced in a 24-hr explant culture of 4-mm punch biopsies of skin; values are expressed per mg wet weight of skin or per μg hydroxyproline as an index of collagen (mean ± SD of 10 determinations for the affected cat and 28 determinations for the controls).

type V collagen were overmodified more than those in the case of BOI (49). This increased apparent molecular weight of the collagen chains was present when the posttranslational modifications of collagen were permitted during labeling of cultured fibroblasts, but there was no difference in the apparent molecular weight of any of the collagen chains when the posttranslational modifications were blocked by chelating iron with α,α'-dipyridyl. This suggested that there was an overmodification of all the collagens, but amino acid analysis of purified chains of type I and type V bovine collagens, and of [^{3}H]lysine-labeled newly synthesized collagens in cultures of bovine fibroblasts, showed that there was no difference in the hydroxylation of proline or lysine in either of these collagens in this disease. Thus, the difference in apparent molecular weight of the collagen chains must be due to increased glycosylation of hydroxylysine residues. Preliminary results suggest that the difference may simply be due to the addition of one or two extra glucose residues to galactosyl hydroxylysine in the triple-helical portion of each collagen chain. This, in turn, raises the question of whether the disease is due to a defect in a protein that simply slows triple-helix formation, allowing more time for the glycosylation of hydroxylysine residues in the RER.

The posttranslational overmodification of the chains of at least four different types of collagen suggests that the mutation may not be in a collagen gene, but in a gene for a noncollagen protein that affects the posttranslational processes in the RER. The availability of both paired tissues and cultured fibroblasts will facilitate studies of this suggestion, as well as studies of the relationship of the overmodification to the decrease in type I collagen. At the same time, it is an opportunity to learn how a single gene mutation can result in increased posttranslational modification of more than six different collagen gene products.

III. SUMMARY

In this chapter we have discussed some of the advantages of animal models as research tools to facilitate (a) prospective genetic studies, (b) the production and utilization of paired affected and unaffected littermates for all experimental studies, (c) the production of both heterozygous and homozygous individuals, (d) the production of adequate quantities of tissues for repetitive biomechanical, microscopic, and biochemical studies, on the same tissues that are used for cell culture, (e) the identification of effects of the disease that are not easily demonstrable in humans, (f) the comparison of different diseases at all stages of fetal and postnatal development, as well as during aging, and (g) the utilization of these accidents of nature in domestic animals to study the causes, effects, prevention, and treatment of diseases of collagen in humans.

One direct result of our studies of comparable diseases in different species is the conviction that the characterization and classification of each case of collagen disease should be based on the combined clinical, genetic, microscopic, and biochemical characteristics of the case. It is clear that fragile tissues may result from a variety of causes, and that the distribution of lesions in different fibrous and supporting connective tissues will vary in each different type of disease. It is clear that a similar clinical phenotype may be caused by different mutations, and that different clinical diseases may result from similar mutations. For example, cases of both EDS VII (50) and OI (51) have been shown to result from structural defects in the amino-terminal region of type I collagen chains, while other cases of EDS VII result from a decreased activity of the enzyme procollagen *N*-peptidase (50). This suggests that the current clinical classification of these diseases is inadequate in

Table 3 Criteria for Characterizing and Classifying Diseases of Collagen in Both Humans and Domestic Animals

A. Clinical classification[a]
   1. By the types and patterns of clinical signs and lesions in different tissues and organ systems
   2. By the age of onset
   3. By the pattern of inheritance
      a. Simple autosomal dominant
      b. Simple autosomal recessive
      c. X-linked
      d. Polygenic
      e. Acquired disease
B. Morphological classification[a]
   1. By the presence of morphological signs of defects in the formation of collagen fibrils and fibers
      a. With defects in fibril growth resulting in fibrils that are too thick, too thin, or variable in diameter
      b. With defects in parallel fibril packing resulting in the disorganization of fibrils within fibers
      c. With a quantitative deficiency of fibrils and fibers within affected tissues
   2. By the presence of morphological signs of defects in the degradation of collagen fibrils and fibers
      a. With or without an invasion of leukocytes or activation of macrophages
      b. With or without phagocytosis of fragments of collagen fibrils by fibroblasts
C. Chemical classification[a]
   1. By the presence of a demonstrable mutation(s) in a collagen gene(s)
   2. By the presence of a structural defect(s) in a procollagen chain
   3. By the presence of quantitative abnormalities in the production, secretion, processing, or deposition of collagen
      a. Due to a defect in the transcription, processing or translation of procollagen mRNA
      b. Due to a defect in posttranslational modification, triple-helix formation, intracellular transport, or secretion of procollagen
      c. Due to an increase in intracellular degradation of procollagen
      d. Due to a defect in the processing of procollagen into collagen in the extracellular space
      e. Due to a defect in a noncollagen glycoprotein or proteoglycan that affects the deposition or organization of collagen into fibrils or fibers
   4. By the presence of a defect in the formation or maturation of lysyl-derived intermolecular cross-links within collagen fibrils
   5. By the presence of increased collagenolytic activity
      a. Due to an increase in the production or activity of fibroblast collagenase
      b. Due to a decrease in the activity of a collagenase inhibitor
      c. Due to an increase in the production or activity of a leukocyte or synovial collagenase
      d. Due to a defect in the structure or cross-linking of collagen molecules within fibrils

[a]Characteristics from each method of classification should be grouped to form an appropriate functional name or classification for each disease process.

both humans and domestic animals. Therefore, we suggest that the classification of each case of a collagen disease be based on a combination of the most important of the clinical, genetic, microscopic, and biochemical characteristics from among those shown in Table 3 to develop a functional name or category for each unique disease. A functional classification of each case will undoubtedly lead to more rapid progress in studies of all diseases of collagen.

ACKNOWLEDGMENTS

The original research described in this chapter was supported in part by United States Public Health Service, National Institutes of Health grants AM-20793 and AM-07210 (RRM), GM-32592 (DFP), AM-28450, AM-35297 and GM-28833 (JU), and AM 16416 (P.I.—D. J. Prockop). Studies of bovine osteogenesis imperfecta (BOI) were supported by The Harold Wetterberg Foundation, and General Biomedical Research Support was received from the New York State College of Veterinary Medicine (RRM).

The authors gratefully acknowledge the invaluable contributions of Drs. L. Denholm, C. Hall, and L. Krook in the studies of BOI; D. Gunson in the studies of the horse with increased collagenolysis; D. Lein in the genetic studies of dogs; A. C. Kane, T. G. Porter, A. C. Robinson, and H. Schryver in the biomechanical studies; D. Dinnin and L. Barr in the microscopic studies; B. Hover and S. Wilson in the tissue culture studies; and B. Hover, E. A. Bauer, T. Pihlajaniemi, and M. Sippola-Thiele in the biochemical studies described here. RRM also thanks Dr. Darwin Prockop and the Dept. of Biochemistry, UMDNJ, Rutgers Medical School, for the opportunity to study in their laboratories, to compare the metabolism of collagen in cultured fibroblasts from these animal models with that in cultured fibroblasts from humans with diseases of collagen.

Dogs and cats were kindly donated for these studies by Drs. Robert Garman, Lawrence Freedman, Robert Kirk, Dan Scott, and Wayne Tremper. Dr. Barry Cooper kindly provided the specimens from the Concord, Massachusetts cat.

REFERENCES

1. G. A. Hegreberg, *Prog. Clin. Biol. Res., 94*:229-244 (1982).
2. R. R. Minor, in *Biochemistry and Physiology of the Skin* (L. A. Goldsmith, ed.), Oxford University Press, New York, 1982, pp. 1139-1152.
3. R. R. Minor, *Am. J. Pathol., 98*:225-280 (1980).
4. R. Hanset and M. Ansay, *Ann. Med. Vet., 111*:451-470 (1967).
5. M. Ansay, A. Gillet, and R. Hanset, *Ann. Med. Vet., 112*:449-451 (1968).
6. P. J. O'Hara, W. K. Read, W. M. Romane, and C. H. Bridges, *Lab. Invest., 23*:307-314 (1970).
7. A. Lenaers, M. Ansay, B. V. Nusgens, and C. M. Lapiere, *Eur. J. Biochem., 23*:533-543 (1971).
8. C. M. Lapiere, A. Lenaers, L. D. Kohn, *Proc. Natl. Acad. Sci. USA 68*:3054-3058 (1971).
9. L. J. Simar, and E. H. Betz, *Hoppe Seylers Z. Physiol. Chem., 352*:13 (1971).
10. H. Furthmayer, R. Timpl, M. Stark, C. M. Lapiere, and K. Kuhn, *FEBS Lett., 28*: 247-250 (1900).
11. A. J. Bailey, and C. M. Lapiere, *Eur. J. Biochem., 34*:91-96 (1973).
12. R. Church, M. L. Tanzer, and C. M. Lapiere, *Nature New Biol., 244*:188-190 (1973).
13. R. Hanset and C. M. Lapiere, *J. Hered., 65*:356-358 (1974).

14. J. A. Yaeger, R. L. Church, and M. L. Tanzer, *Vet. Path., 12*:16-31 (1975).
15. G. E. Pierard and C. M. Lapiere, *J. Invest. Dermatol., 66*:2-7 (1976).
16. G. Wick, B. R. Olsen, and R. Timpl, *Lab. Invest., 39*:151-156 (1978).
17. K. Cassidy, E. F. Eikenberry, B. R. Olsen, and B. Brodsky, *Lab. Invest., 43*:542-546 (1980).
18. G. A. Hegreberg, G. A. Padgett, J. R. Gorgam, and J. B. Hansen, *J. Hered., 60*: 249-254 (1969).
19. G. A. Hegreberg, G. A. Padgett, R. L. Ott, and J. B. Hensen, *J. Invest. Dermatol., 54*:377-380 (1970).
20. G. A. Hegreberg, G. A. Padgett, and J. B. Hensen, *Arch. Pathol., 90*:159-166 (1970).
21. D. F. Patterson, and R. R. Minor, *Lab. Invest., 37*:170-179 (1977).
22. D. F. Counts, P. Knighten, and G. A. Hegreberg, *J. Invest. Dermatol., 69*:521-526 (1977).
23. D. F. Counts, *Biochem. Biophys. Res. Commun., 626*:208-217 (1980).
24. R. R. Minor. D. H. Lein, D. F. Patterson, L. Krook, T. G. Porter, and A. C. Kane *J. Am. Vet. Med. Assoc., 182*:142-148 (1983).
25. O. Helle, and N. N. Ness, *Acta Vet. Scand., 13*:443-445 (1972).
26. M. Fjolstad and O. Helle, *J. Pathol., 112*:184-188 (1974).
27. U. Becker, O. Helle, and R. Timpl, *FEBS Lett., 73*:197-200 (1977).
28. L. L. Collier, C. W. Leathers, and D. F. Counts, *Feline Practice, 10*:25-36 (1980).
29. D. F. Counts, P. H. Byers, K. A. Holbrook, and G. A. Hegreberg, *J. Invest Dermatol., 74*:96-99 (1980).
30. K. A. Holbrook, P. H. Byers, D. F. Counts, and G. A. Hegreberg, *J. Invest. Dermatol., 74*:100-104 (1980).
31. D. W. Hollister, P. H. Byers, and K. A. Holbrook, *Adv. Human Genet., 12*:1-87 (1982).
32. K. A. Holbrook and P. H. Byers, *J. Invest. Dermatol., 79*:7s-16s (1982).
33. D. W. Rowe, E. B. McGoodwin, G. R. Martin, M. D. Sussman, D. Grahn, B. Faris, and C. Franzblau, *J. Exp. Med., 139*:180-192 (1974).
34. D. W. Rowe, E. B. McGoodwin, G. R. Martin, and D. Grahn, *J. Biol. Chem., 252*: 939-942 (1977).
35. D. M. Danks, M. B. Campbell, B. J. Stevens, V. Mayne, and E. Cartwright, *Pediatrics, 50*:188-201 (1972).
36. D. E. Gunson, R. A. Halliwell, and R. R. Minor, *Arch. Dermatol., 120*:599-604 (1984).
37. D. N. Menton, R. A. Hess, J. R. Lichtenstein, and A. Z. Eisen, *J. Invest. Dermatol., 70*:4-10 (1978).
38. D. N. Menton and R. A. Hess, *J. Invest. Dermatol., 74*:139-147 (1980).
39. L. J. Denholm and W. G. Cole, *Aust. Vet. J., 60*:9-17 (1983).
40. J. D. Termine, P. Gehron Robey, L. W. Fisher, M. A. Shimokawa, M. A. Drum, K. M. Conn, G. R. Hawkins, J. B. Cruz, and K. G. Thompson, *Proc. Natl. Acad. Sci. USA 81*:2213-2217 (1984).
41. M. D. Ridge and V. Wright, *J. Invest. Dermatol., 46*:341-346 (1966).
42. A. Viidik, *Int. Rev. Connect. Tissue Res., 6*:127-215 (1973).
43. D. R. Veronda and R. A. Westmann, *J. Biomech., 3*:111-124 (1970).
44. A. Vogel, K. A. Holbrook, B. Steinmann, R. Gitzelmann, and P. H. Byers, *Lab. Invest., 40*:201-206 (1979).
45. P. H. Byers, K. A. Holbrook, J. G. Hall, P. Bornstein, and J. W. Chandler, *Hum. Genet., 40*:157-169 (1978).
46. S. P. Damle, L. Coster, and J. D. Gregory, *J. Biol. Chem., 257*:5523-5527 (1982).
47. J. E. Scott and C. R. Orford, *Biochem. J., 197*:213-216 (1981).
48. J. E. Scott, E. W. Hughes, B. Marcyniuk, and J. P. Scott, *Biochem. Soc. Trans., 11*: 770-771 (1983).

49. R. R. Minor, T. Pihlajaniemi, D. J. Prockop, L. J. Denholm, and J. A. M. Wootton, *J. Cell Biol., 99*:405a (1984).
50. B. Steinmann, L. Tuderman, L. Peltonen, G. R. Martin, V. A. McKusick, and D. J. Prockop, *J. Biol. Chem., 255*:8887-8893 (1980).
51. M. Sippola, S. Kaffe, and D. J. Prockop, *J. Biol. Chem., 259*:14094-14100 (1984).

14

The Molecular Pathology of Pulmonary Fibrosis

Joan G. Clark*
Washington University School of Medicine,
St. Louis, Missouri

I. INTRODUCTION

The connective tissue in normal lung confers mechanical strength and elastic recoil properties that allow the lung to undergo repeated cycles of expansion and relaxation. To permit efficient gas exchange the extracellular matrix is necessarily distributed in the alveolar walls such that the barrier between air and blood is minimized (1,2). In diffuse interstitial pulmonary fibrosis, the alveolar structures are thickened, distorted, or obliterated by excessive extracellular matrix (3). These anatomical alterations contribute to the reduced lung volumes, decreased lung compliance, and abnormal gas exchange observed in patients with pulmonary fibrosis.

Pulmonary fibrosis develops as a consequence of a wide variety of insults to the lung (4,5). Some of the known causes are listed in Table 1. In a substantial number of patients, the cause is not identifiable, and the disorder is referred to as "idiopathic pulmonary fibrosis" or "cryptogenic fibrosing alveolitis." Although pulmonary fibrosis may be regarded in some respects as a stereotyped response to lung injury, the inflammatory cell population, distribution of fibrosis, and other histological features are variable and may be characteristic of the etiology (5) (Figure 1). Heterogeneity among patients with pulmonary fibrosis may also be reflected in the activity and rate of progression of disease. Even among patients with pulmonary fibrosis resulting from the same etiology, the disease may remain stable over a number of years or it may progress intermittently or steadily.

Despite clinical and pathological heterogeneity, the hallmark of all forms of pulmonary fibrosis is excessive extracellular matrix (3,5). Characteristically, the interstitium, which occupies the space between the gas-containing alveolus and the pulmonary capil-

**Present affiliation*: The Fred Hutchinson Cancer Research Center and University of Washington, Seattle, Washington.

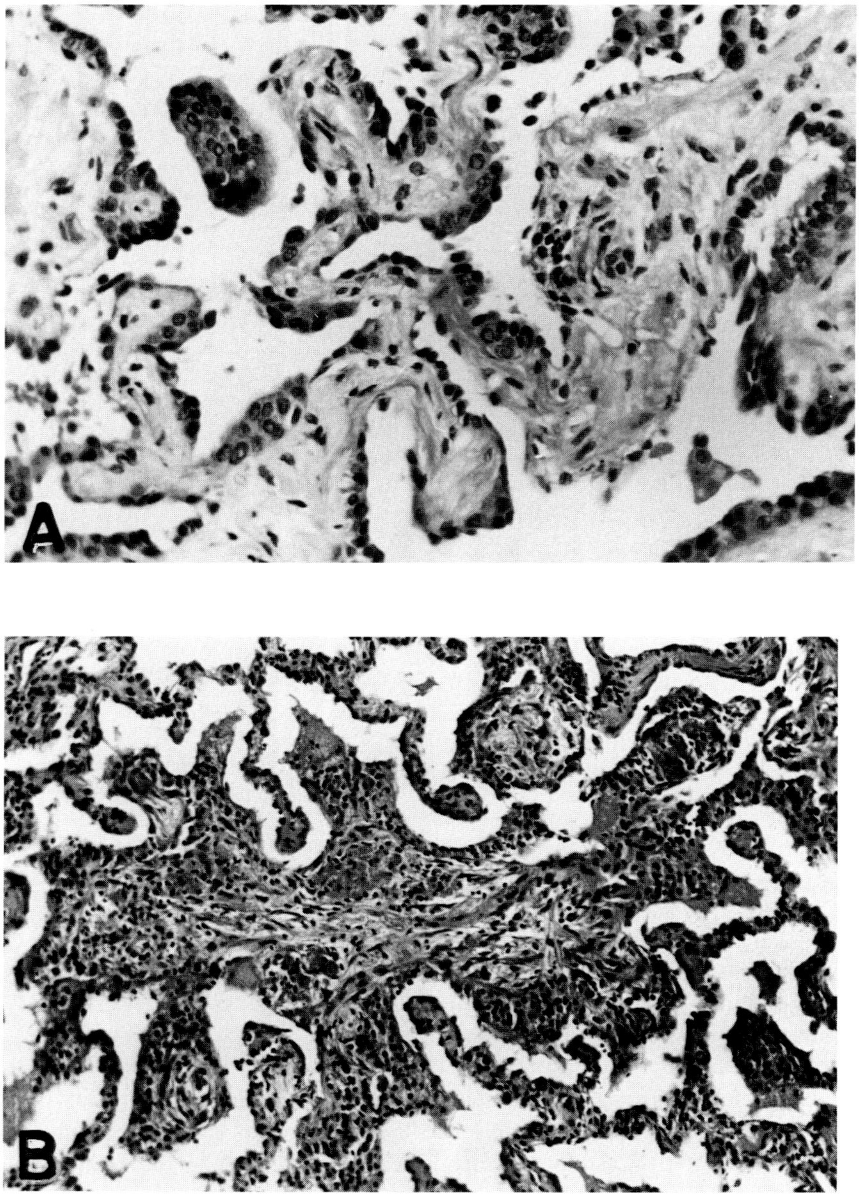

Figure 1 Pathological features in pulmonary fibrosis. (A) Interstitial fibrosis in a typical interstitial pneumonia (idiopathic pulmonary fibrosis). The alveolar septa are widened by dense fibrosis with minimal inflammation. Hyperplastic epithelial cells line the septa. Clusters of macrophages are seen in the air spaces. (B) Intraalveolar fibrosis in idiopathic pulmonary fibrosis. Deposits of connective tissue fill an alveolar duct and extend into the alveolar spaces. The alveolar septa are also thickened by inflammatory cells and extracellular matrix. (C) Silicotic nodules. The nodules are formed by concentric bands of collagen and are demarcated from the surrounding normal parenchyma. Pigment-laden macrophages are seen around the periphery. (D) Asbestosis. There is interstitial fibrosis and mononuclear cell infiltrate. Typical beaded asbestos bodies (asbestos fibers coated by endogenous iron particles) are lying free in the tissue and engulfed by macrophages. (Photomicrographs provided by Dr. Charles Kuhn III, Washington University Medical School.)

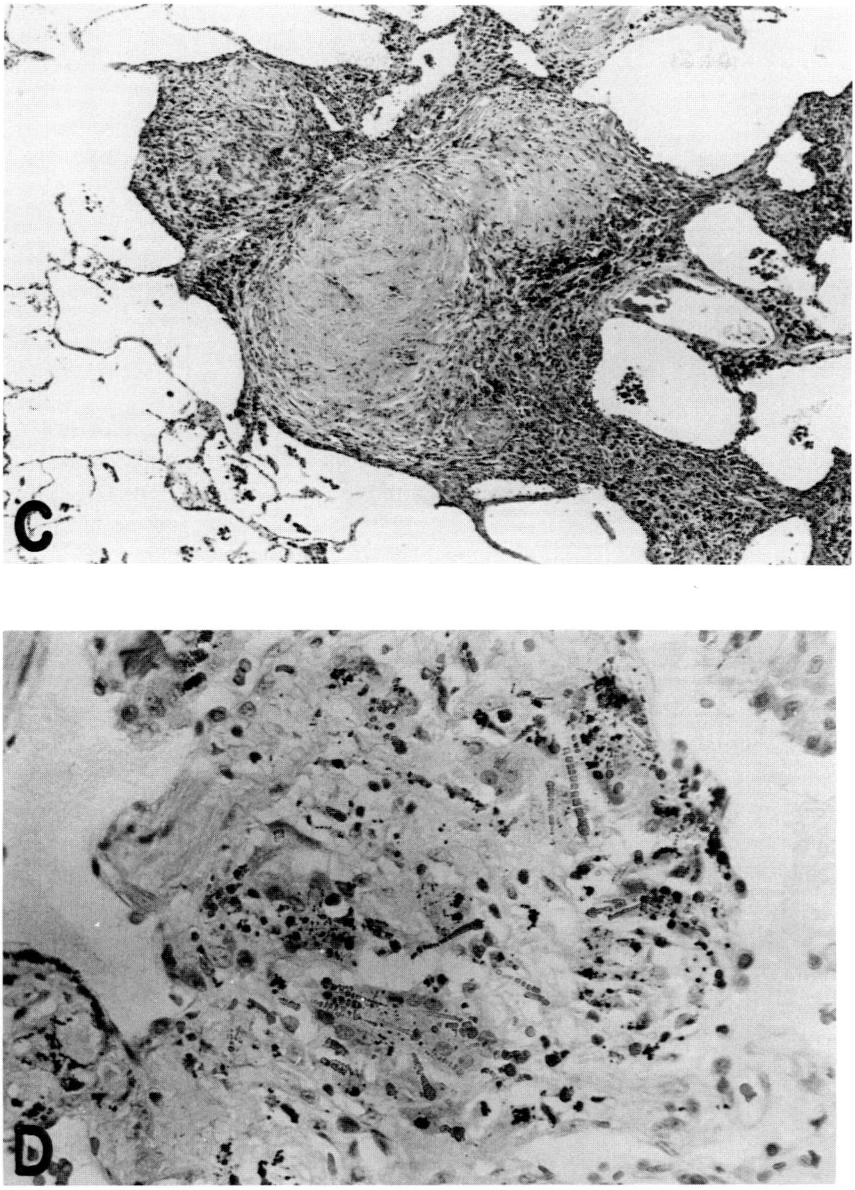

Figure 1 (Continued)

Table 1 Some Causes of Diffuse Interstitial Pulmonary Fibrosis

Categories	Examples
Drugs	Bleomycin, nitrofurantoin
Toxins and chemicals	Paraquat, ozone, oxygen (high concentration)
Irradiation	Therapeutic x-irradiation
Infections	Viral, fungal
Fibrogenic dusts	Silica, asbestos
Granulomatous disorders	Sarcoidosis, hypersensitivity pneumonitis
Immunological disorders	Rheumatoid arthritis, systemic lupus erythematosus, scleroderma

lary blood, is thickened by accumulation of connective tissue. Intraalveolar connective tissue may also be present, and in some instances, it is the predominant finding. In advanced stages of pulmonary fibrosis, the normal architecture of the lung is partially obliterated: the alveoli are replaced by dense, disorganized connective tissue, and the terminal airways are dilated and thickened, giving the lung a "honeycomb" appearance.

The pivotal role of lung connective tissue in the pathogenesis of pulmonary fibrosis has been appreciated for some time. In recent years, the development of experimental models of pulmonary fibrosis, advances in cell biology, and further delineation of the biochemistry and biological activity of macromolecules in the extracellular matrix of lung have enhanced the understanding of the biochemical basis of pulmonary fibrosis. Diseases associated with diffuse pulmonary fibrosis are now recognized as disorders characterized by excessive extracellular matrix in the lung. The accumulation of connective tissue in fibrotic lung reflects changes in the proliferation and metabolic activity of connective tissue-producing lung cells. These complex events are modulated in part by resident lung cells, inflammatory and immune cells, and the extracellular matrix itself. Although many aspects of these events are incompletely understood, the concepts that emerge from recent studies provide important new insights into mechanisms involved in lung extracellular matrix production and deposition in this diverse group of disorders.

II. EXPERIMENTAL APPROACHES TO PULMONARY FIBROSIS

Diverse experimental approaches, utilizing model systems of varying levels of biological complexity, have been employed in investigations aimed at elucidating the biochemical basis of pulmonary fibrosis. The fundamental issues addressed by these studies concern the biochemical changes that occur in extracellular matrix of the lung in pulmonary fibrosis, the metabolic events that result in excess connective tissue, and factors that may be involved in regulation of lung extracellular matrix production and deposition. Some comments regarding the advantages and limitations of certain approaches may serve as a useful introduction to a review of these studies and may assist in integrating data from diverse sources.

A. Human Studies

Clinical and pathological observations in patients have delineated the natural history of pulmonary fibrosis and provided the basis for a variety of subsequent clinical studies as well as investigations in other systems. The importance of clinical studies in human disorders is perhaps self-evident. However, a number of problems are encountered in human studies of pulmonary fibrosis.

First, selection of homogeneous patient groups and/or definition of the population under study may be difficult. Inclusion of patients with different causes of pulmonary fibrosis and in various stages of the disease may make interpretation difficult if these issues are not recognized. Some pathological mechanisms almost certainly differ in pulmonary fibrosis resulting from various etiologies, and the differences possibly occur at a very fundamental level. That is, in some cases pulmonary fibrosis may represent a reparative process in response to lung injury, analogous to dermal wound healing. In other instances, fibrosis may reflect a primary disorder of the regulation of connective tissue production (6). Divergent observations are likely. Another difficulty is the staging of pulmonary fibrosis. As the disease evolves over a period of time, the molecular pathology and pathogenetic mechanisms could be expected to vary. Given the variability in progression rate, parameters such as duration of symptoms or clinical severity may not adequately indicate the pathological stage or activity of the disease (7). Careful consideration of these issues is thus crucial to the interpretation of the findings in clinical studies.

Another problem encountered in human studies is the difficulty in obtaining adequate lung tissue for biochemical evaluation. Samples obtained postmortem generally reflect an advanced stage of pulmonary fibrosis. Findings in such tissue may not be pertinent to early stages of the disorder. Moreover, intervening illness such as heart failure, lung hemorrhage, or infection may confound the interpretation of such studies. Lung biopsy samples may be especially useful in evaluating earlier stages of pulmonary fibrosis (8). However, repeated lung biopsies are infrequently obtained so it is usually not possible to determine sequential changes. In addition, relatively small amounts of tissue are obtained, and sampling errors are likely because the lung is not affected evenly (9). The expression of quantitative results derived from these samples is also problematic as exemplified in studies of collagen in fibrotic lung discussed below.

B. Animal Models

Some of the problems associated with studying pulmonary fibrosis in patients are obviated in studies of experimentally induced pulmonary fibrosis in animals. A number of agents [e.g., bleomycin (10-12), oxygen (13), silica (14,15), and irradiation (16,17)] produce pulmonary fibrosis. Some of the mechanisms involved in the pathogenesis of the fibrosis may differ among species and with the fibrogenic agent, but it is reasonable to expect that some common principles will emerge.

Bleomycin-induced pulmonary fibrosis has been widely studied and may be considered a prototype of experimental lung fibrosis. It is a convenient model because fibrosis is easily and reliably induced with a single intratracheal injection of the drug. Fibrosis develops rapidly and is associated with relatively low mortality, so it is feasible to study fairly large numbers of animals during development of the lesion. Moreover, when small animals such as hamsters or rats are studied, it is possible to study the whole lung. Thus, sampling errors are reduced and quantitative data can be expressed using whole lung as the denominator. Bleomycin-induced pulmonary fibrosis persists for at least three months, but the lesion does not progress after about one month (11). In this respect, it may be appropriate to view bleomycin-induced pulmonary fibrosis as a "normal" reparative response to a fibrogenic lung injury, and some findings may not be pertinent to progressive forms of pulmonary fibrosis (e.g., idiopathic pulmonary fibrosis or silicosis).

Several methodological and technical problems that apply to both human and animal studies are encountered in studies of connective tissue in fibrotic lung. A major obstacle is related to the insolubility of lung connective tissue macromolecules (see Ref. 18). In general, nondenaturing solvents and enzymatic digestion of lung tissue have been unsatisfactory for quantitative studies because of the low yield of intact macromolecules. Quantitative analysis of lung connective tissue usually requires measurement of fragments or amino acids characteristic of collagen, elastin, and proteoglycan (e.g., cyanogen bromide peptides, hydroxyproline, desmosine, and glycosaminoglycans). Although immunohistochemistry has been useful for studying intact in situ connective tissue macromolecules in pulmonary fibrosis, specificity of the antibodies, penetration and migration of probes, and masking of antigenic sites in complex tissue make quantitative results difficult, and interpretations must be cautious (19). Radiolabeled precursors such as proline, valine, and sulfate have also been used to study the metabolism of lung collagen, elastin, and proteoglycan, respectively, in both lung tissue explant cultures and living animals. These are necessarily short-term experiments to avoid problems with reutilization of these precursors. In addition, variability in the precursor pool specific radioactivity must be assessed or controlled (20). This is particularly difficult in complex tissue where the true precursor pool may be impossible to identify or measure.

The "inhomogeneity" of lung imposes limitations on the interpretation of studies of lung connective tissue. Lung comprises several distinct tissues including pleura, airways, blood vessels, and the terminal alveolar-capillary units. The anatomical relationships between these structures preclude dissection to all but a gross extent. Thus, biochemical determinations of lung connective tissue composition and metabolism associated with pulmonary fibrosis represent an average value from a variety of structures that includes a large number of cell types (21). Immunohistochemical studies have been useful in delineating the microscopic pathology, but the cellular basis for changes in the metabolism of connective tissue is difficult to assess in studies of lung tissue.

C. Isolated Lung Cells

Isolated cells in culture represent readily controlled systems in which to study the growth of cells, the synthesis of connective tissue macromolecules and their soluble precursor forms, and the interaction of cells with the extracellular matrix and with other cells. Interpretations of in vitro observations must, of course, be tempered by consideration of the influence of such factors as culture conditions and serial passage of cells on phenotypic expression. Some features of cells that have been isolated from lung are briefly discussed below.

1. Lung Fibroblasts

Lung fibroblasts have been extensively studied as a model, albeit incomplete, of lung connective tissue production. In vitro they synthesize type I and type III procollagen (22-24), the major collagen types in lung (25,26), as well as glycosaminoglycans (27) and fibronectin (28,29). Fibroblasts isolated from other tissues also synthesize type V collagen (30) and elastin (31,32). Although other lung cells certainly contribute to lung connective tissue, fibroblasts are thought to be a major source of these macromolecules in alveolar walls. Both fibroblast proliferation and protein synthesis can be influenced by a variety of agents that may be pertinent to the pathogenesis of pulmonary fibrosis. Some of these studies as well as their limitations are discussed in more detail below.

2. *Endothelial Cells*

The recent development of techniques for maintaining vascular endothelial cells in culture provides an opportunity to study extracellular matrix production by these cells (33). Endothelial cells obtained from large arteries, including pulmonary arteries, synthesize fibronectin (34), laminin (35), glycosaminoglycan (36), elastin (37), and collagen types III, IV, V, and VIII (38,39). Capillary endothelial cells, perhaps more germane to disorders of lung parenchyma, have not been isolated from lung; these cells may possess properties distinct from those derived from large arteries (40).

3. *Epithelial Cells*

Epithelial cells derived from tissue other than lung appear to be an important source of type IV procollagen (41). Type II epithelial cells from lung have recently been cultured and provide an opportunity to evaluate the metabolism of these cells, particularly with respect to the production of basement membrane components (42,43). The prominence of these cells in injured lung (5) suggests that they may play an important role in repair and/or remodeling of lung connective tissue.

4. *Inflammatory and Immune Cells*

Inflammatory and immune cells are prominent cells in pulmonary fibrosis. The biology of these cells, particularly those obtained by bronchoalveolar lavage, has been extensively studied (44). The lavage cell population normally comprises macrophages (more than 90%) and smaller proportions of neutrophils and lymphocytes. The total number of lavage cells and the proportion of cell types are altered in a number of lung diseases, including pulmonary fibrosis (7). These changes in the lavage cell population have been correlated with similar changes in lung parenchyma (45,46), although there may be important differences between lung cells obtained from different sites. For example, prostaglandin production by alveolar macrophages differs from that of mononuclear cells derived from the interstitium (47). Also, the proportion of T-lymphocyte subsets differs in lavage fluid and tissue from bleomycin-treated animals (48). Nonetheless, bronchoalveolar lavage provides relatively pure populations of lung inflammatory cells, and studies of these cells have been useful in delineating their potential role as effectors in pulmonary fibrosis.

D. Other Approaches

Cell-free systems are another model in which lung connective tissue has been studied (e.g., Refs. 49,50). This approach has not been widely applied to studies of pulmonary fibrosis, but information regarding the role of transcription and translation in the regulation of lung connective tissue production may eventually contribute to understanding of the biochemical basis of pulmonary fibrosis.

Much of the fundamental knowledge of the biochemistry of connective tissue macromolecules has been derived from studies of these molecules in isolated forms. Recent in vitro observations regarding the interactions between connective tissue macromolecules (e.g., fibronectin, collagen, and glycosaminoglycans) may have broad implications with respect to the deposition and subsequent accumulation of extracellular matrix in pulmonary fibrosis (51-54). Novel approaches, such as studies utilizing monoclonal antibodies that block intermolecular binding (54), may be useful in elucidating the role of these interactions in complex systems.

III. EXTRACELLULAR MATRIX IN PULMONARY FIBROSIS

A. Collagen

1. Quantity

Collagen is the most abundant connective tissue macromolecule in normal lung, accounting for 20% of lung dry weight (55).It is distributed in the pleura, alveolar septa, bronchial and vessel walls, and basement membranes. In fibrotic lung, collagen is a prominent component in the interstitial and intraalveolar deposits of connective tissue (3), and biochemical analysis have established that total lung collagen content is increased in pulmonary fibrosis. Zapol et al. (56) studied lung obtained at autopsy of patients with severe acute lung injury (adult respiratory distress syndrome), a disorder associated with histologically evident fibrosis. These investigators studied multiple samples of the lung so that sampling errors were reduced, and they expressed their data as the total lung collagen. Collagen was increased within a few days of injury and was twice the normal value after two weeks. When collagen "concentrations" (i.e., collagen per dry weight of lung) were determined, the values were less than or equal to control lung collagen concentrations. This result was due in part to dilution of collagen by nonstructural elements such as blood in lung. In lung biopsy samples from patients with idiopathic pulmonary fibrosis, collagen concentration was similar to control values, but total lung collagen could not be assessed (57). Another study, however, found increased collagen in lung samples from patients with pulmonary fibrosis (58). The discrepancy may be related in part to variability of inflammatory cells and blood in the samples.

Measurement of collagen in lungs of animals with experimentally induced pulmonary fibrosis has consistently disclosed increased lung collagen (11-17). For example, in bleomycin-induced pulmonary fibrosis, collagen content increases steadily to about twice normal values by one month and remains elevated for at least several months thereafter (11,12,59,60).

2. Collagen Synthesis

The metabolic events that result in the increases in collagen content are beginning to be elucidated, primarily in studies of experimental animals. In various forms of experimentally induced pulmonary fibrosis, increased lung collagen synthesis rate has been observed (14,17). In bleomycin-induced pulmonary fibrosis, collagen synthesis rates increase dramatically a few days after administration of the drug (59,60). Between one and two weeks later, synthesis rates are as high as 10 times the control values. In the ensuing weeks, the synthesis rates decline and are restored to control values. Changes in prolyl hydroxylase activity parallel the changes in collagen synthesis rate (59,61). This sequence of events is consistent with the relatively rapid increase in lung collagen content and the subsequent stability of collagen content observed in this model. If one assumes that lung collagen synthesis rates are normally low and that turnover of lung collagen is slow (62), such increases in synthetic activity are required to promptly increase the collagenous matrix. As a corollary, decreased collagen degradation could not by itself effect such a change.

Rates of collagen synthesis have also been measured in lung biopsy samples from patients with idiopathic pulmonary fibrosis (57), utilizing lung explant culture techniques similar to those used in some of the animal studies cited above. The average collagen synthesis rate in these fibrotic lung samples was not different from normal values. Several factors may contribute to this apparent discrepancy with animal studies. First, collagen synthesis rates based on DNA or dry weight may underestimate total lung collagen synthesis rates. This was necessarily the denominator used in the patient study. Second, most patients had been symptomatic for over a year and had advanced disease by physiological and histological criteria. These patients may represent a group in whom the process had evolved to a subacute stage in which collagen synthesis rates are lower than in the initial stages. Such a sequence of events is typical of experimentally induced fibrosis and may occur in pulmonary fibrosis in humans as well. In this regard, it is interesting to note that one patient with mild disease and symptoms for only one-half year had a collagen synthesis rate greater than 3 times the control value. Finally, the method of measuring collagen synthesis rates in lung explant cultures may not be sensitive enough to detect small changes in synthesis rate that could significantly increase lung collagen content over an extended course.

3. Collagen Degradation

The collagen content of normal lung is remarkably constant and reflects a balance of synthesis and degradation. A disturbance of this relationship is the fundamental basis for collagen accumulation in pulmonary fibrosis. While increased collagen synthesis is well established as a mechanism in the pathogenesis of pulmonary fibrosis, the role of collagen degradation is less certain. Collagen degradation could plausibly constitute part of the initial injury that ultimately results in loss of the integrity of lung structure. In addition, the drastic remodeling of lung and loss of normal lung architecture in pulmonary fibrosis would seem to necessitate degradation at some time. In support of this, a preliminary report indicated that lung collagen content was initially decreased in animals exposed to toxic levels of oxygen (63).

Collagen degradation has been described as a potential modulator in lung fibrosis. "Classical" mammalian collagenases are metalloenzymes that function at neutral pH to cleave native helical collagen of types I, II, and III at a specific site three-fourths of the distance from the NH_2-terminal end of the molecule [see Ref. (64)]. In vitro this cleavage occurs at variable rates, depending in part on collagen types (65,66). At physiological temperature, the two products of collagenase action denature and become susceptible to further degradation. Strong circumstantial evidence for a physiological role of collagenase is found in a variety of experiments in which loss of collagen is closely associated with the appearance of collagenase activity (e.g., involution of the pregnant uterus, wound healing, and growth) (64). Collagenase activity as well as neutral protease activity is detectable in bronchoalveolar lavage fluid from patients with idiopathic pulmonary fibrosis, but not in those with sarcoid or in normal controls (67). Collagenase activity has also been detected in lavage fluid from bleomycin-treated animals, and cultured macrophages from these animals release increased amounts of active collagenase (68). Possible sources of collagenase in fibrotic lung include neutrophils and fibroblasts, as well as macrophages.

4. Collagen Types

Collagen is a family of related triple-helical proteins made up of three α chains (41). Different types of collagen are distinguished by their α-chain composition (see Chapter 1 of this volume). Collagen types I, II, III, IV, and V have been isolated from normal lung (25,26,69-73), and with the development of type-specific antibodies, their anatomical distribution has been elucidated (Table 2). The probable cellular origin of the specific

Table 2 Collagen Types in Lung

Type	Location	Probable cellular source
I	Alveolar septa Bronchial and vascular walls	Fibroblasts Smooth muscle cells Endothelial cells
II	Tracheal and bronchial cartilage	Chondrocytes
III	Alveolar septa Vascular walls	Fibroblasts Smooth muscle cells Endothelial cells
IV	Basement membrane	Endothelial cells Epithelial cells
V	Basement membrane Cell membrane Interstitium	Endothelial cells Epithelial cells Smooth muscle cells Fibroblasts

collagen types has been inferred primarily from in vitro demonstration of the production of specific collagen types (see Section II.C and Ref. 31).

Several immunohistochemical studies have examined the distribution of collagen types in fibrotic lung. Madri and Furthmayr (58) described a marked increase in type I collagen in alveolar septa of postmortem lung. Compared to normal, type III collagen was reduced in the septa and seen primarily in a perivascular localization. Type V was increased in the interstitium and located in areas of smooth muscle cell proliferation. There was not apparent change in the amount or localization of type IV collagen.

In a study of advanced fibrosis, Bateman et al. (74) supported these findings. Their study also included samples in which areas of less mature scar and more active fibrosis were present. In these areas, type III collagen predominated. A subsequent study by these investigators involved biopsy specimens from 25 patients with cryptogenic fibrosing alveolitis (idiopathic pulmonary fibrosis). The presence of type III collagen in alveolar walls correlated with active disease (i.e., clinical progression), and 9 of 11 patients who responded to therapy were in this group (75). In contrast, patients with no type III had a stable clinical course and did not respond to treatment. A preponderance of type III collagen in alveolar septa has also been observed in a study that included patients with acute lung injury (76). The investigators suggested that these findings were attributable to the inclusion of samples obtained relatively early in the course of the disease.

The localization of immunolabeled collagen types has also been studied at the ultrastructural level in fibrotic lung (77). Around alveolar septal capillaries the connective tissue appeared loose, giving the pattern of a fibrillary network. In these areas, type III collagen predominated and was closely associated with basement membrane. In other areas showing a low degree of connective tissue organization, type III collagen was seen between type I collagen fibers and associated with small fiber bundles. In areas of dense connective tissue matrix, bundles of type I collagen fibers were observed and type III collagen was scarce. An evolving character of the fibrotic lesion was suggested by these observations; that is, a loose, reticular matrix made up primarily of type III collagen evolves into dense connective tissue composed of fibrillar type I collagen. This view is consistent with observations in granulation tissue (78) and wound healing (79) where, after an initial predominance of type III collagen, type I collagen is predominant in the mature lesion.

Some uncertainties regarding changes in collagen type persist, however, because of the limitations of these studies. First, immunolabeling of collagen is at best a semiquantitative technique, subject to a number of technical problems (see Ref. 19). Moreover, the data have been derived from relatively small numbers of patients, and sampling errors are likely because the lung is usually unevenly involved with fibrosis (9) and different patterns may be present in a single sample (75). Finally, observations regarding "activity" and "maturity" of histological fibrosis are subjective, and may not precisely indicate temporal events.

Quantitative biochemical analyses of lung collagen types circumvent some of these problems and corroborate some of the observations. Seyer et al. (25) isolated cyanogen bromide peptides unique for types I and III collagen and determined that in normal lung type I represents approximately 60% of lung collagen and type III, 30%. They also studied lung from patients who died with pulmonary fibrosis and discovered that the proportion of type I was increased. This finding has been supported by other studies of pulmonary fibrosis in both humans and experimental animals (80-82). An increase in type III collagen was not observed. However, Kirk et al. (83) have recently examined the proportion of types I and III collagen in lung from patients with idiopathic pulmonary fibrosis; the lung samples included both postmortem specimens and biopsy samples. Although there was no difference between the *mean* values in control and fibrotic lung, the percentage of type III collagen decreased with the duration of symptoms and the histological severity of fibrosis. Moreover, there was a positive correlation between percentage of type III and subsequent improvement. These data suggest that increased type III collagen occurs in early fibrosis and may represent a reversible lesion.

A subsequent preliminary report indicated that peptides of type III procollagen were increased in serum of some patients with idiopathic pulmonary fibrosis (84). Patients with high levels of procollagen peptides also had an increased percentage of type III collagen in lung. Moreover, levels of procollagen peptide fell with treatment. Another study of a small group of patients with idiopathic pulmonary fibrosis reported normal values of type III procollagen N-terminal peptides in serum, but elevated levels in lavage fluid (85). Although levels failed to correlate well with most clinical parameters, those patients with the highest levels had evidence of progressive disease. The possible usefulness of analysis of type-specific procollagen peptides for staging of pulmonary fibrosis and evaluation of therapy seems worthy of further study.

The mechanisms involved in the appearance of different collagen types are not known. Shifts could occur as the result of expansion of cell populations dedicated to production of specific collagen types, changing gene expression or translational activity within a cell population, and/or degradation by collagenolytic enzymes that are selective for collagen types. Further delineation of collagen types in pulmonary fibrosis, their biological significance, and regulation would surely improve our understanding of the molecular basis of pulmonary fibrosis.

5. *Posttranslational Changes in Collagen*

Based on the observation of decreased solubility of fibrotic lung collagen, Madri and Furthmayr suggested that cross-linking of collagen may be increased in fibrosis (58). Alternatively, this phenomenon may result from intermolecular interactions with other macromolecules (e.g., glycosaminoglycans) that render collagen less soluble. Other investigators have found increased solubility of fibrotic lung collagen (16,25). This variability in solubility might also be related in part to differences in the rate of collagen synthesis and the proportion of newly synthesized soluble collagen precursors present in the tissue. At present, however, there is little information regarding the posttranslational proteolytic processing of procollagens in normal or fibrotic lung.

Regardless of whether the proportion of cross-linked collagen is increased or decreased, collagen cross-link formation per se appears to be critical in the development of pulmonary fibrosis. Administration of β-aminopropionitrile, an inhibitor of cross-link formation, prevents collagen accumulation and some of the physiological abnormalities associated with bleomycin-induced pulmonary fibrosis (86). Collagen content was also reduced in irradiated animals who received penicillamine, another cross-link inhibitor (87). Also, animals given cadmium chloride developed pulmonary fibrosis, whereas similarly treated animals pretreated with β-aminopropionitrile developed emphysema (88). These data suggest that cross-link formation, perhaps in both collagen and elastin, is a critical element in development of fibrosis in this form of lung injury.

Glycosylation and prolyl hydroxylation of both fibrotic and normal lung collagen were found to be identical in one study (25). A small reduction in hydroxylysine in collagen from fibrotic lung was also observed. The significance of this change is unknown.

6. *Regulation of Collagen Deposition*

A discussion of the regulation of lung collagen content in pulmonary fibrosis may encompass an enormous number of potentially pertinent investigations from diverse areas. In the following sections, I have attempted to identify some general topics that are germane to this subject and, where possible, include evidence that links the observations to the pathogenesis of pulmonary fibrosis.

Fibroblast Regulation. Lung fibroblasts are presumed to be a major source of excess collagen in fibrotic lung. As suggested above, modulation of both fibroblast proliferation and fibroblast collagen production may be involved in the pathogenesis of pulmonary fibrosis. In vitro, fibroblast proliferation and collagen production can be stimulated or inhibited by a variety of substances (23,89-104) (Table 3). In addition, the relative production of collagen types is subject to modulation (23,90,105-109). While these in vitro observations suggest that similar events occur in vivo, cultured fibroblasts are, of course, an incomplete model of lung collagen production. The following considerations may temper the interpretation of in vitro observations. (a) Some substances may have effects on other lung cells that possibly overshadow their direct effects on fibroblasts. For example, steroids that are known to reduce lung collagen synthesis in pulmonary fibrosis (110) might do so by a direct effect on fibroblasts and/or indirectly by their effect on inflammatory cells. (b) The extracellular matrix and other factors in the extracellular milieu may influence the fibroblast's potential to respond to a stimulus (90). In this sense, the fibroblast plays a permissive role in interactions with exogenous agents, and the appropriateness of cell culture conditions is problematic. (c) The phenotype of cultured cells may also be influenced by serial passage (111) or selection of a subset of cells from a heterogeneous population of cells (93); the phenotype of cultured fibroblasts may not be representative of cells in vivo. (d) Finally, the behavior of the fibroblast in vivo is probably the product of multiple influences, some of which have opposing effects. In the extracellular milieu, the predominating influence may be determined by parameters such as local concentration, potency, and duration of effect.

In spite of these reservations, the hypothesis has emerged that lung fibroblast growth and collagen production are subject to stimulatory and inhibitory influences within the

Table 3 Effects of Exogenous Agents on Fibroblast Proliferation and Collagen Production in vitro

Agent	Proliferation	Collagen production
Serum	↑	↑
Steroids		↓
Bleomycin	↓	↑
Proline analogs	↓	↓
Prostaglandin E_2	↓	↓
cAMP	↓	↓
Adrenergic agents	↓	↓
Procollagen peptides		↓
Products derived from		
Macrophages	↑↓	↑↓
Lymphocytes	↑↓	↑↓
Platelets	↑	

alveolus and that these parameters play a central role in the pathogenesis of pulmonary fibrosis. Observations regarding fibroblast biology and responses to exogenous agents in vitro provide important clues to the behavior of these cells in vivo.

Inflammatory Cells. The lung inflammatory cell population is invariably altered in pulmonary fibrosis. The role of these cells in modulating collagen accumulation in pulmonary fibrosis has emerged as a subject of considerable interest. The technique of bronchoalveolar lavage has permitted examination of the inflammatory cells from this lung compartment in both patients with pulmonary fibrosis and animals with experimentally induced pulmonary fibrosis (46,96,112-115). The "alveolitis" as defined by lavage cells has been studied with respect to the number of cells present (or obtainable), the types of cells, and their activity (e.g., mediator release). These data provide the foundation for a number of current theories regarding the regulatory role of inflammatory cells in pulmonary fibrosis. Observations related to the role of inflammatory cells in specific interstitial lung diseases have recently been reviewed (7).

Alveolar macrophages are the predominant cell in normal as well as fibrotic lung. Their phagocytic activity and role in lung defense has long been recognized, and recently their role as effector cells by virtue of their secretory activity has received attention (116). Several products of alveolar macrophages have been identified in patients with idiopathic pulmonary fibrosis and may be pertinent to other forms of pulmonary fibrosis as well. (a) Neutrophil chemotactic factor is spontaneously released by cultured alveolar macrophages from patients with idiopathic pulmonary fibrosis and may thus effect neutrophil accumulation (117). (b) A macrophage-derived growth factor for fibroblasts is also released by cells from these patients and may participate in the expansion of the fibroblast population (95). (c) Alveolar macrophages from these patients release increased amounts of fibronectin which, in addition to its other activities, appears to act as a competence factor in concert with the growth factor for fibroblasts (118,119).

Studies of the role of macrophages in bleomycin-induced pulmonary fibrosis in our laboratory indicate that this cell can also participate in regulation of lung collapse by suppressing fibroblasts. We found that alveolar macrophages, obtained two or three weeks after bleomycin treatment, release increased amounts of a factor that suppresses fibroblast proliferation and collagen production (96). Macrophage-derived fibroblast suppressive factor (MDFSF) increases endogenous fibroblast prostaglandin E_2 production and intracellular cAMP levels, events that subsequently cause suppression of fibroblast growth and collagen production. Increased release of MDFSF occurs when lung collagen synthesis rates are declining (59) and suppressive activity detected in lung explant cultures is increased (119a). Macrophages are also increased in numbers at this time (96). Thus, both expansion of the macrophage population and increased release of suppressive factor may be important in restoring lung collagen synthesis rates to normal after the initial fibrotic response to lung injury.

A dual role for macrophages (i.e., a source of factors that stimulate or inhibit fibroblasts) has also been suggested by studies of experimental silicosis (120-122). As has been previously discussed, the detection of macrophage-derived factors that influence fibroblasts is dependent in part on experimental conditions, especially those of fibroblast culture (96,123). Nonetheless, investigations thus far clearly indicate the potential influence of macrophage secretory production on lung collagen synthesis and suggest that regulation of stimulatory and inhibitory factors derived from macrophages may be an important aspect in modulation of lung collagen synthesis in pulmonary fibrosis.

Other macrophage products of potential importance in pulmonary fibrosis include collagenase (124), prostaglandins such as PGE_2 (47) and leukotriene B_4 (a neutrophil and monocyte chemoattractant) (125), and interleukin 1 (a stimulus for fibroblast collagenase production) (126).

Neutrophils are increased in most patients with idiopathic pulmonary fibrosis (7). They are also prominent in the early stage of bleomycin-induced pulmonary fibrosis (59, 115). The in vitro release of reactive oxidant species (127) and proteases (especially collagenase and elastase) suggests that they may cause damage to lung parenchymal cells and degradation of extracellular matrix. Several lines of evidence, however, indicate that the role of neutrophils may be more complex. First, increase lavage of neutrophils is not invariably associated with progressive fibrosis (128). Second, ablation of neutrophils in bleomycin-induced pulmonary fibrosis results in *increased* lung collagen content and synthesis (129,130). The latter observation suggests that in some instances neutrophils may play a protective role by directly inhibiting collagen synthesis and/or by promoting inhibition of collagen synthesis by other cells [e.g., release of chemoattractants for monocytes which, in turn, suppress fibroblasts (125,131)].

Lymphocytes are especially prominent in bronchoalveolar lavage from patients with sarcoid. In this disorder, the lymphocytes are primarily T cells, expressing surface antigens characteristic of helper T cells (132). They spontaneously secrete interleukin 2 and monocyte chemotactic factors, mediators thought to be important in T-cell proliferation, recruitment of monocytes, and subsequent granuloma formation.

Lymphocytes are also increased in some patients with idiopathic pulmonary fibrosis (114,128). These cells possibly produce immunoglobulins that form immune complexes observed in some patients with idiopathic pulmonary fibrosis (7). Secretory products of lymphocytes may also have direct effects on fibroblast proliferation and collagen synthesis (99-101).

In bleomycin-induced pulmonary fibrosis, steroids, antilymphocyte globulin, and T-cell depletion suppress the fibrotic response (110,133,134). In addition, transient cellular sensitivity to homologous collagen has been demonstrated in bleomycin-treated animals (135). These observations suggest that the immune system is also involved in the development of bleomycin-induced pulmonary fibrosis.

Cell Migration. Recruitment of fibroblasts by directed cell migration in response to chemoattractants has emerged as a subject of considerable interest. Chemotactic factors for fibroblasts may be derived from serum (136), other lung cells (137), and the extracellular matrix itself (138-141). Chemotactic factors for monocytes and neutrophils are also of potential importance, and some are discussed in other sections of this chapter. Although the precise role of chemoattractant substances in pulmonary fibrosis is not known, changes in both the resident and inflammatory cell population of fibrotic lung may clearly be an important determinant of lung collagen production.

Other Regulatory Mechanisms. Regulation of collagen synthesis by lung cell types other than fibroblasts has not been studied in the context of pulmonary fibrosis. A better understanding of their collagen production, especially with respect to collagen types IV, V, and some of the more recently described types, may improve our understanding of connective tissue remodeling in pulmonary fibrosis.

Numerous interactions among macromolecules of the extracellular matrix have been described. The role of these interactions in the context of lung connective tissue deposition and accumulation in pulmonary fibrosis is unclear at the present time. Future studies that elucidate the significance of these interactions in vivo are likely to add important new insights into the pathogenesis of pulmonary fibrosis.

7. Nonstructural Activity of Collagen

Changes in collagen quantity and quality in pulmonary fibrosis undoubtedly play a central role in determining structural and mechanical changes in the fibrotic lung. Collagen may also contribute to the pathogenesis of pulmonary fibrosis by virtue of some of its activities that are not directly related to its mechanical role. Although these functions have not been specifically studied in the context of pulmonary fibrosis, some observations seem potentially important.

Collagen or gelatin may serve in vitro as a substratum for cell attachment and migration, usually mediated by an attachment glycoprotein such as fibronectin. Moreover, a collagen substrate may be required for growth or differentiation in some cells (42,142, 143). The potential role of collagen as a substrate for lung epithelial cell attachment, growth, and perhaps basement membrane repair in lung injury is suggested by studies of type II lung epithelial cells grown on cell-free amnionic membrane (43).

Collagen interacts with blood platelets, and this interaction appears to be a critical step in initiating platelet degranulation and platelet aggregation (144). Platelet granular constituents include a growth factor for fibroblasts (PDGF) (89) and a substance (platelet factor 4) that is chemotactic for neutrophils, monocytes, and fibroblasts (145,146). Injury that exposes collageneous matrix in lung resulting in platelet aggregation and release of platelet products may thus have diverse effects on other lung cells.

Collagen and collagen-derived peptides can also cause directed migration (chemotaxis) of fibroblasts (139). Recruitment of these cells to sites of active collagen synthesis or degradation could contribute to the accumulation and distribution of collagen in pulmonary fibrosis.

B. Elastin

Elastin, representing 5% of lung dry weight, is distributed in alveolar septa, blood vessels, bronchial walls, and pleura (2). The cellular origin of elastin in lung is not known, but potential sources include smoothe muscle cells, endothelial cells, fibroblasts, and myofibroblasts.

In bleomycin-induced pulmonary fibrosis, lung elastin content is increased twofold, approximately paralleling increases in lung collagen (11,12,147). Microscopically, the elastic fibers are thick and tortuous in contrast to the straight and relatively thick fibers of normal lung parenchyma (12). Synthesis rates of elastin have not been measured in pulmonary fibrosis. The concomitant increase in collagen and elastin suggests that the production of collagen and elastin may be subject to similar controlling mechanisms, but this is unproven.

The regulation of elastin synthesis has not been studied in the context of pulmonary fibrosis. However, elastin synthesis by cultured bovine ligamentum nuchae fibroblasts and ligament explant cultures is affected by cell density, serial passage, serum concentration, and drugs (including bleomycin, dexamethasone, and theophylline) (31). Elastin production is also dependent on fetal age in this model, and both synthesis rate and deposition of elastin can be influenced by extracellular matrix (31,148). These interesting observations may eventually prove to be pertinent to an understanding of the phenotypic expression of elastin-producing lung cells and elastin deposition in pulmonary fibrosis.

Lung elastin has been most extensively studied in the context of elastase-induced emphysema (149). These studies have established that elastin is crucial for the mechanical recoil properties of lung. Elastin degradation is a prerequisite for development of experimental emphysema, and resynthesis of elastin follows the drastic elastolysis induced by elastase (150-152). The remodeled lung contains normal quantities of elastin, but the elastic fibers are disorganized. Interference with cross-link formation (e.g., with β-aminopropionitrile) during this time results in more severe emphysematous changes (153). Thus, it appears that connective tissue synthesis is critical to repair of the extracellular matrix, albeit abnormal, following proteolytic injury.

Several models of lung injury indicate that the pathogenesis of pulmonary emphysema and fibrosis may involve some shared "mechanisms." Cadmium chloride administration, for example, results in pulmonary fibrosis, but when cadmium chloride is given concomitantly with β-aminopropionitrile, pulmonary emphysema develops (88). The pulmonary lesion that develops as a consequence of high inspired oxygen concentration also has features of both fibrosis and emphysema (13). These studies suggest that the magnitude and specificity of destruction of the extracellular matrix as well as the "efficiency" of connective tissue resynthesis may be parameters that distinguish lung injury resulting in emphysema from injury causing fibrosis. Thus, issues such as protease-antiprotease balance, generally considered in the context of lung elastin and pulmonary emphysema, may also be pertinent to understanding lung connective tissue remodeling in pulmonary fibrosis.

C. Proteoglycans

Proteoglycans constitute only about 0.5% of the lung's dry weight (154). Isolation and characterization of these complex macromolecules is difficult (155), but the glycosaminoglycan components of lung proteoglycans have been characterized (156). In hamster lung the distribution is approximately 60% heparan sulfate, 23% dermatan sulfate, 14% chondroitin sulfate, and 3% hyaluronic acid. Little or no heparin is present in hamster or human lung.

Specific immunohistological techniques have not been available for localizing proteoglycans in lung. However, ruthenium red, a cationic dye, interacts with anionic glycosaminoglycans and has been used in conjunction with selective enzymatic digestion to examine the distribution of proteoglycans in lung (157,158). Proteoglycans, containing chondroitin or dermatan sulfate, are distributed at regular intervals with collagen in alveolar walls. Heparan sulfate proteoglycans are present in orderly arrays in the lamina rara externa of alveolar basement membrane. In lung capillary basement membrane, heparan sulfate proteoglycans and sulfated proteoglycans overlie the lamina densa. Bleomycin-induced fibrosis is associated with thickening of lung capillary basement membrane but with less than normal amounts of ruthenium red-positive material (159).

Total lung glycosaminoglycan content is increased after bleomycin treatment, and the increase is distributed among all types (156). Glycosaminoglycan synthesis is also increased in experimental pulmonary fibrosis induced by *N*-nitroso-*N*-methylurethane (160). In bleomycin-induced pulmonary fibrosis, sequential analysis showed that glycosaminoglycan synthesis was maximally increased at five days after instillation of the drug, declined markedly at 15 days, and returned to control values at 45 days (161,162). Thus, increases in glycosaminoglycan synthesis may precede collagen synthesis and could perhaps influence collagen synthesis and deposition. These studies also showed that dermatan sulfate synthesis increased to a greater degree than that of other glycosaminoglycans.

The functional significance of proteoglycans in lung is unknown, but several important roles have been suggested including (a) mechanical strength and tissue resistance properties, (b) regulation of the formation and orientation of collagen fibrils, (c) determination of the microenvironment by virtue of their large sphere of hydration, (d) alteration of permeability because of their anionic qualities, and (e) attachment sites for cells during growth and repair. More thorough understanding of proteoglycan biochemistry and biosynthesis, as well as localization and structural characterics of these macromolecules in lung tissue, is likely to clarify their role in normal and remodeled lung.

D. Fibronectin

Fibronectin is a dimeric glycoprotein that is widely distributed in plasma and tissue (see Refs. 2, 53,163 for reviews). Plasma fibronectin may contribute to lung fibronectin, and multiple lung cell types, including fibroblasts (28,29), endothelial cells (34), type II epithelial cells, and alveolar macrophages (164,165), synthesize and secrete fibronectin in vitro. Studies of the intact molecule (isolated from fibroblast cultures and serum) and fragments of fibronectin (released by limited proteolytic cleavage) have resulted in rapid advances in understanding fibronectin structure and function. Fibronectin is codistributed with collagen and is required for normal collagen deposition in the pericellular matrix of cultured fibroblasts (29). In dermal wound healing, fibronectin is present with type III procollagen and appears to play a central role in fibroblast migration and deposition of extracellular matrix (78,166).

By immunofluorescence, fibronectin is found in basement membrane and loose connective tissue stroma of lung (167). At the ultrastructural level, fibronectin is present in alveolar capillary and epithelial basal lamina; in the interstitium, faint staining is associated with collagen fibers and surrounding smooth muscle cells in bronchioles and blood vessels. In addition, fibronectin is present in capillary plasma and bronchoalveolar lavage fluid.

In fibrotic lung, the distribution of fibronectin differs from normal lung in that epithelial and endothelial basal lamina staining is markedly increased. Moreover, collagen fibers are stained with approximately 600 Å periodicity. These alterations and the known biological activities of fibronectin suggest that it may function in fibrotic lung as a cell adhesive molecule in basal lamina, promoting stable attachment of endothelial and epithelial cells, and in guiding the supramolecular organization of the connective tissue molecules such as collagen.

Fibronectin is susceptible to a variety of proteinases. The ubiquity of neutrophils in fibrotic lung and the susceptibility of fibronectin to digestion by neutrophil elastase suggest that fibronectin degradation and subsequent loss of its biological activity could contribute to the alterations observed in lung fibrosis. However, fibronectin fragments retain biological activity in vitro (168). The retention of biological activity by fibronectin fragments may permit expression of the activity at sites apart from proteolytic degradation. Thus, fibronectin degradation may have diverse consequences in lung injury.

IV. CONCLUDING REMARKS

In recent years, substantial progress has been made toward understanding the molecular basis of pulmonary fibrosis. Quantitative and qualitative changes in the macromolecular composition of fibrotic lung have been delineated, and some of the metabolic and biosynthetic events that contribute to biochemical changes in lung extracellular matrix are known. In addition, factors that may regulate connective tissue-producing cells in fibrotic lung are beginning to be appreciated. Many intriguing and important questions remain regarding these tissues as well as other problems that are beginning to be addressed. For example, what factors influence the deposition and accumulation of newly synthesized lung connective tissue macromolecules? Do collagen and proteoglycan types play a special role in the organization of the extracellular matrix in pulmonary fibrosis? How do "nonstructural" macromolecules such as fibronectin participate in the pathogenesis of pulmonary fibrosis? What is the contribution of lung cells such as endothelial and epithelial cells to the extracellular matrix of remodeled lung? Novel and critical approaches will certainly be required to resolve such issues related to the pathogenesis of fibrotic lung diseases, but observations thus far form a sound foundation for understanding the regulation of connective tissue biosynthesis and deposition in fibrotic lung.

ACKNOWLEDGMENTS

I am grateful to Dr. Charles Kuhn III for providing photomicrographs. I also wish to thank Dr. Robert M. Senior for helpful suggestions in the preparation of this chapter.

This work was supported in part by United States Public Health Service Grant HL29594 from the National Institutes of Health.

REFERENCES

1. E. R. Weibel, in *Pulmonary Diseases and Disorders* (A. P. Fishman, ed.), McGraw-Hill, New York, 1980, p. 224.
2. J. G. Clark, C. Kuhn, J. A. McDonald, and R. P. Mecham, *Int. Rev. Connect. Tissue Res., 10*:249 (1983).
3. H. Spencer, in *Pulmonary Diseases and Disorders* (A. P. Fishman, ed.), McGraw-Hill, New York, 1980, p. 648.
4. M. Turner-Warwick, in *Pulmonary Diseases and Disorders* (A. P. Fishman, ed.), McGraw-Hill, New York, 1980, p. 675.
5. A. L. A. Katzenstein and F. B. Askin, *Surgical Pathology of Non-Neoplastic Lung Disease*, W. B. Saunders, Philadelphia, 1982.
6. S. I. Rennard, P. B. Bitterman, and R. G. Crystal, *Chest, 84*:735 (1983).

7. R. G. Crystal, P. B. Bitterman, S. I. Rennard, A. J. Hance, and B. A. Keough, *N. Engl. J. Med., 310*:154, 235 (1984).
8. E. A. Gaensler, M. V. B. Moister, and J. Hamm, *N. Engl. J. Med., 270*:1319 (1964).
9. R. G. Crystal, J. D. Fulmer, W. C. Roberts, M. L. Moss, B. R. Line, and H. Y. Reynolds, *Ann. Intern. Med., 85*:769, (1976).
10. G. L. Snider, J. A. Hayes, and A. L. Korthy, *Am. Rev. Resp. Dis., 117*:1099 (1978).
11. G. L. Snider, B. R. Celli, R. H. Goldstein, J. J. O'Brien, and E. C. Lucey, *Am. Rev. Resp. Dis., 117*:289 (1978).
12. B. C. Starcher, C. Kuhn, and J. E. Overton, *Am. Rev. Resp. Dis., 117*:299 (1978).
13. D. J. Riley, R. A. Berg, N. H. Edelman, and D. J. Prockop, *J. Clin. Invest., 64*:643 (1980).
14. K. M. Reiser, T. W. Hesterberg, W. M. Hascher, and J. A. Last, *Am. J. Pathol., 107*: 176 (1982).
15. J. Halme, J. Uitto, K. Kahanpää, P. Karhunen, and S. Lindy, *J. Lab. Clin. Med., 75*: 535 (1970).
16. J. A. Pickrell, D. V. Harris, F. F. Hahn, J. J. Belasich, and R. K. Jones, *Radiat. Res., 64*:133 (1975).
17. J. F. Collins, W. G. Johanson, Jr., and B. McCullough, *Am. Rev. Resp. Dis., 117*: 1079 (1978).
18. L. W. Cunningham and D. W. Frederickson (eds.), *Methods in Enzymology*, Vol. 82, Part A, Academic Press, New York, 1982.
19. K. von der Mark, *Int. Rev. Connect. Tissue Res., 9*:265 (1981).
20. D. E. Rannels, R. B. Low, T. Youdale, E. Volkin, and W. J. Longmore, *Fed. Proc., 41*:2833 (1982).
21. C. Kuhn, in *The Biochemical Basis of Pulmonary Function* (R. G. Crystal, ed.), Marcel Dekker, New York, 1976.
22. A. J. Hance, K. Bradley, and R. G. Crystal, *J. Clin. Invest., 57*:102 (1976).
23. J. G. Clark, B. C. Starcher, and J. Uitto, *Biochim. Biophys. Acta, 631*:359 (1980).
24. J. G. Clark, C. Kuhn, and J. Uitto, *Am. Rev. Resp. Dis., 122*:971 (1980).
25. J. M. Seyer, E. T. Hutcheson, and A. H. Kang, *J. Clin. Invest., 57*:1498 (1976).
26. B. D. McLees, G. Schleiter, and S. R. Pinnell, *Biochemistry, 16*:185 (1977).
27. C. W. Castor, P. R. Heiss, R. H. Gray, and J. C. Seidman, *Am. Rev. Resp. Dis., 120*: 107 (1979).
28. K. M. Yamada and J. A. Weston, *Proc. Natl. Acad. Sci. USA, 71*:3492 (1974).
29. J. A. McDonald, D. G. Kelley, and T. J. Broekelmann, *J. Cell Biol., 92*:485 (1982).
30. A. S. Narayanan and R. C. Page, *J. Biol. Chem., 258*:11694 (1983).
31. R. P. Mecham, G. Lange, J. Madaras, and B. Starcher, *J. Cell Biol., 90*:332 (1981).
32. J. M. Davidson, M. G. Giro, A. I. Oikarinen, H. Oikarinen, and J. Uitto, *Clin. Res., 32*:137A (1984).
33. U. S. Ryan, E. Clements, D. Habliston, and J. W. Ryan, *Tissue Cell, 10*:535 (1978).
34. F. A. Jaffe and D. F. Mosher, *J. Exp. Med., 147*:1779 (1978).
35. D. Gospodarowicz, G. Greenburg, J. M. Foidart, and N. Savion, *J. Cell. Physiol., 107*: 171 (1981).
36. V. Bunonassissi and M. Root, *Biochim. Biophys. Acta, 385*:1 (1975).
37. R. P. Mecham, J. Madaras, J. A. McDonald, and U. Ryan, *J. Cell. Physiol., 116*:282 (1983).
38. H. Sage, E. Crouch, and P. Bornstein, *Biochemistry, 18*:5433 (1979).
39. H. Sage, P. Pritzl, and P. Bornstein, *Biochemistry, 19*:5747 (1980).
40. H. Sage, P. Pritzl, and P. Bornstein, *Atherosclerosis, 1*:427 (1981).
41. P. Bornstein and H. Sage, *Annu. Rev. Biochem., 49*:957 (1980).
42. E. F. Geppert, M. C. Williams, and R. J. Mason, *Exp. Cell Res., 128*:363 (1980).
43. J. Lwebuga-Mukasa, G. Thulin, J. A. Madri, C. Barrett, and J. Warshaw, *Am. Rev. Resp. Dis., 127*:272A (1983).

44. G. W. Hunninghake, J. E. Gadek, O. Kawanami, V. J. Ferrans, and R. G. Crystal, *Am. J. Pathol., 97*:149 (1979).
45. G. W. Hunninghake, O. Kawanami, V. J. Ferrans, R. C. Young, W. C. Roberts, and R. G. Crystal, *Am. Rev. Resp. Dis., 123*:407 (1981).
46. P. L. Haslam, C. W. G. Turton, B. Heard, A. Lukoszek, J. V. Collins, A. J. Salsbury, and M. Turner-Warwick, *Thorax, 35*:9 (1980).
47. C. A. Rouzer, W. A. Scott, A. L. Hamill, and Z. A. Cohn, *J. Exp. Med., 155*:720 (1982).
48. R. S. Thrall and R. W. Barton, *Am. Rev. Resp. Dis., 129*:279 (1984).
49. P. Tolstoshev, R. A. Berg, S. I. Rennard, K. H. Bradley, B. C. Trapnell, and R. G. Crystal, *J. Biol. Chem., 256*:3135 (1981).
50. P. Tolstoshev, R. Huber, B. C. Trapnell, and R. G. Crystal, *J. Biol. Chem., 256*:9672 (1981).
51. U. Lindahl and M Höök, *Annu. Rev. Biochem., 47*:385 (1978).
52. E. Ruoslahti and E. Engvall, *Biochim. Biophys. Acta, 631*:350 (1980).
53. E. Ruoslahti, E. Engvall, and E. G. Hayman, *Coll. Relat. Res., 1*:95 (1981).
54. J. A. McDonald, D. G. Kelley, and T. J. Broekelmann, *J. Cell Biol., 92*:485 (1982).
55. A. J. Hance and R. G. Crystal, in *The Biochemical Basis of Pulmonary Function* (R. G. Crystal, ed.), Marcel Dekker, New York, 1976.
56. W. M. Zapol, R. L. Trelstad, J. W. Coffey, I. Tsai, and R. A. Salvador, *Am. Rev. Resp. Dis., 119*:547 (1979).
57. J. D. Fulmer, R. S. Bienkowski, M. J. Cowan, S. D. Bruel, K. M. Bradley, V. J. Ferrans, W. C. Roberts, and R. G. Crystal, *Am. Rev. Resp. Dis., 122*:389 (1980).
58. J. A. Madri and H. Furthmayr, *Hum. Pathol., 11*:353 (1981).
59. J. G. Clark, J. E. Overton, B. A. Marino, J. Uitto, and B. C. Starcher, *J. Lab. Clin. Med., 96*:943 (1980).
60. S. H. Phan, R. S. Thrall, and P. A. Ward, *Am. Rev. Resp. Dis., 121*:501 (1980).
61. J. Kelley, R. A. Newman, and J. N. Evans, *J. Lab. Clin. Med., 96*:954 (1980).
62. J. A. Pierce, H. Resnick, and P. H. Henry, *J. Lab. Clin. Med., 69*:485 (1967).
63. D. J. Riley, J. S. Kerr, R. A. Berg, S. Y. Yu, D. J. Glock, N. H. Edelman, and D. J. Prockop, *Am. Rev. Resp. Dis., 125*:231 (1982).
64. J. Gross, in *Biochemistry of Collagen* (G. N. Ramachandran and A. H. Reddi, eds.) Plenum Press, New York, 1976, pp. 275-317.
65. H. Welgus, J. J. Jeffrey, and A. Z. Eisen, *J. Biol. Chem., 256*:9511 (1981).
66. A. L. Horwitz, A. J. Hance, and R. G. Crystal, *Proc. Natl. Acad. Sci. USA, 74*:897 (1977).
67. J. E. Gadek, J. A. Kelman, G. Fells, S. E. Weinberger, A. L. Horwitz, H. Y. Reynolds, J. D. Fulmer, and R. G. Crystal, *N. Engl. J. Med., 301*:737 (1979).
68. Z. Marom, K. S. Weinberg, and B. L. Fanburg, *Am. Rev. Resp. Dis., 121*:859 (1980).
69. K. Bradley, S. McConnell-Bruel, and R. G. Crystal, *J. Clin. Invest., 55*;543 (1975).
70. K. H. Bradley, S. D. McConnell, and R. G. Crystal, *J. Biol. Chem., 249*:2674 (1974).
71. K. H. Bradley, S. McConnell-Bruel, and R. G. Crystal, *Proc. Natl. Acad. Sci. USA, 71*:2828 (1974).
72. N. A. Kefalides and B. Denduchis, *Biochemistry, 8*:4613 (1969).
73. J. A. Madri and H. Furthmayr, *Am. J. Pathol., 94*:323 (1979).
74. E. Bateman, M. Turner-Warwick, and B. C. Adelmann-Grill, *Thorax, 36*:645 (1981).
75. E. D. Bateman, M. Turner-Warwick, P. L. Haslam, and B. C. Adelmann-Grill, *Thorax, 38*:93 (1983).
76. G. Raghu, L. Morel-Maroger, R. Hackman, and G. Stricker, *Fed. Proc., 42*:1025 (1983).
77. C. Takiya, S. Peyrol, J. F. Cordier, and J. A. Grimaud, *Virchows Arch. (Cell. Pathol.), 44*:223 (1983).

78. M. Kurkinen, A. Vaheri, P. J. Roberts, and S. Stenman, *Lab Invest., 43*:47 (1980).
79. J. N. Clore, I. K. Cohen, and R. F. Diegelmann, *Proc. Soc. Exp. Biol. Med., 161*: 337 (1979).
80. K. M. Reiser and J. A. Last, *J. Biol. Chem., 258*:269 (1983).
81. K. M. Reiser and J. A. Last, *Am. Rev. Resp. Dis., 123*:58 (1981).
82. J. A. Last, A. D. Siefkin, and K. M. Reiser, *Thorax, 38*:364 (1983).
83. J. M. E. Kirk, B. E. Heard, I. Kerr, M. Turner-Warwick, and G. J. Laurent, *Clin. Sci., 64*:52 (1983).
84. J. M. E. Kirk, G. J. Laurent, E. Bateman, E. Hahn, P. L. Haslam, and M. Turner-Warwick, *Am. Rev. Resp. Dis., 127*:272A (1983).
85. R. B. Low, K. R. Cutroneo, G. S. Davis, and M. S. Giancola, *Lab. Invest., 48*:755 (1983).
86. D. J. Riley, J. D. Kerr, R. A. Berg, B. D. Ianni, N. H. Edelman, and D. J. Prockop, *Am. Rev. Resp. Dis., 125*:67 (1982).
87. W. F. Ward, A. Shih-Hoellworth, and R. D. Tuttle, *Radiology, 146*:533 (1983).
88. D. E. Niewoehner and J. R. Hoidal, *Science, 217*:359 (1982).
89. R. Ross, C. Nist, B. Kariya, M. J. Rivest, E. Raines, and J. Callis, *J. Cell. Physiol., 97*:497 (1978).
90. B. A. Booth, K. L. Polak, and J. Uitto, *Biochim. Biophys. Acta, 607*:145 (1980).
91. K. M. Sterling, M. J. Harris, J. J. Mitchell, T. A. DiPetrillo, G. L. Delaney, and K. R. Cutroneo, *J. Biol. Chem., 258*:7644 (1983).
92. E. M. L. Tan, L. Ryhänen, and J. Uitto, *J. Invest. Dermatol., 76*:462 (1981).
93. D. S. Ko, R. C. Page, and A. S. Narayanan, *Proc. Natl. Acad. Sci. USA, 74*:3429 (1977).
94. L. E. Saltzman, J. Moss, R. A. Berg, B. Hom, and R. G. Crystal, *Biochem. J., 204*: 25 (1982).
95. P. B. Bitterman, S. Adelberg, and R. G. Crystal, *J. Clin. Invest., 72*:1801 (1983).
96. J. G. Clark, B. A. Marino, and K. M. Kostal, *J. Clin. Invest., 72*:2082 (1983).
97. S. J. Leibovich and R. Ross, *Am. J. Pathol., 84*:501 (1976).
98. J. A. Elias, M. D. Rossman, and R. P. Daniele, *Am. Rev. Resp. Dis., 125*:701 (1982).
99. R. L. Johnson and M. Ziff, *J. Clin. Invest., 58*:240 (1976).
100. S. M. Wahl, L. M. Wahl, and J. B. McCarthy, *J. Immunol., 121*:942 (1978).
101. J. Rosenbloom, W. McArthur, D. Malamud, and S. Jimenez, *Cell. Immunol., 81*: 192 (1983).
102. R. B. Rutherford and R. Ross, *J. Cell Biol., 69*:196 (1976).
103. K. M. Sterling, Jr., T. A. DiPetrillo, J. P. Kotch, and K. R. Cutroneo, *Cancer Res., 42*:3502 (1982).
104. K. M. Sterling, Jr., M. J. Harris, J. J. Mitchell, and K. R. Cutroneo, *J. Biol. Chem., 258*:14438 (1983).
105. L. Thrombley, M. Absher, and J. Kelley, *Am. Rev. Resp. Dis., 123*:694 (1981).
106. B. Steinmann, S. Abe, and G. R. Martin, *Coll. Relat. Res., 2*:185 (1982).
107. A. S. Narayanan and R. C. Page, *FEBS Lett., 80*:221 (1977).
108. S. Abe, B. U. Steinman, L. M. Wald, and G. R. Martin, *Nature, 279*:442 (1979).
109. J. Kelley, L. Thrombley, E. J. Kovacs, G. S. Davis, and M. Absher, *J. Cell. Physiol., 109*:353 (1981).
110. S. H. Phan, R. S. Thrall, and C. Williams, *Am. Rev. Respir. Dis., 124*:428 (1981).
111. J. N. Hildebran, M. Absher, and R. B. Low, *In Vitro, 4*:307 (1983).
112. H. Y. Reynolds, J. D. Fulmer, J. A. Kazierowski, W. C. Roberts, M. M. Frank, and R. G. Crystal, *J. Clin. Invest., 59*:165 (1977).
113. H. Y. Reynolds and H. H. Newball, *J. Lab. Clin. Med., 84*:559 (1974).
114. P. L. Haslam, C. W. G. Turton, and A. Lukoszek, *Thorax, 35*:328 (1980).
115. R. S. Thrall, R. W. Barton, D. A. D'Amato, and S. B. Sulavik, *Am. Rev. Resp. Dis., 126*:488 (1982).

116. C. F. Nathan, H. W. Murray, and Z. A. Cohn, *N. Engl. J. Med., 303*:622 (1980).
117. G. W. Hunninghake, J. E. Gadek, T. J. Lawley, and R. G. Crystal, *J. Clin. Invest., 68*:259 (1981).
118. S. I. Rennard and R. G. Crystal, *J. Clin. Invest., 69*:113 (1981).
119. P. B. Bitterman, S. I. Rennard, S. Adelberg, and R. G. Crystal, *J. Cell Biol., 97*: 1925 (1983).
119a. J. G. Clark, K. M. Kostal, and B. A. Marino, *J. Biol. Chem., 257*:8098 (1982).
120. A. G. Heppleston and J. A. Styles, *Nature, 214*:521 (1967).
121. J. S. Harington, M. Ritchie, P. C. King, and K. Miller, *J. Pathol., 109*:21 (1973).
122. E. Kulonen, M. Aalto, S. Aho, P. Lehtinen, and M. Potila, *Environ. Health Perspec., 51*:119 (1983).
123. S. Kimenez, W. McArthur, and J. Rosenbloom, *J. Exp. Med., 150*:1421 (1979).
124. Z. Werb and S. Gordon, *J. Exp. Med., 142*:346 (1975).
125. T. R. Martin, L. C. Altman, R. K. Albert, and H. R. Henderson, *Am. Rev. Resp. Dis., 128*:106 (1984).
126. A. E. Postlethwaite, L. B. Lachman, C. L. Mainardi, and A. H. Kang, *J. Exp. Med., 157*:801 (1983).
127. S. J. Weiss and A. F. LoBuglio, *Lab. Invest., 47*:5 (1982).
128. R. M. Rudd, P. L. Haslam, and M. Turner-Warwick, *Am. Rev. Resp. Dis., 124*:1 (1981).
129. R. S. Thrall, S. M. Phan, J. R. McCormick, and P. A. Ward, *Am. J. Pathol., 105*: 76 (1981).
130. J. G. Clark and C. Kuhn, *Am. Rev. Resp. Dis., 126*:737 (1982).
131. V. B. Antony, S. A. Sahn, R. N. Harada, and J. E. Repine, *Chest, 5*:95S (1983).
132. G. W. Hunninghake and R. G. Crystal, *N. Engl. J. Med., 305*:429 (1981).
133. R. S. Thrall, J. McCormick, R. Jack, S. H. Phan, and P. A. Ward, *Am. Rev. Resp. Dis., 119*:83 (1979).
134. R. S. Thrall, E. J. Lovett III, R. W. Barton, J. R. McCormick, S. H. Phan, and P. A. Ward, *Am. Rev. Resp. Dis., 121*:99 (1980).
135. D. J. Schrier, S. H. Phan, and P. A. Ward, *J. Immunol., 129*:2156 (1982).
136. A. E. Postlethwaite, R. Snyderman, and A. H. Kang, *J. Clin. Invest., 64*:1379 (1979).
137. A. E. Postlethwaite, R. Snyderman, and A. H. Kang, *J. Exp. Med., 144*:1188 (1976).
138. R. M. Senior, G. L. Griffin, and R. P. Mecham, *J. Clin. Invest., 70*:614 (1982).
139. A. E. Postlethwaite, J. M. Seyer, and A. H. Kang, *Proc. Natl. Acad. Sci. USA, 75*:871 (1977).
140. A. E. Postlethwaite, J. Keski-Oja, G. Balian, and A. H. Kang, *J. Exp. Med., 52*: 494 (1981).
141. H. E. J. Seppä, K. M. Yamada, S. T. Seppä, M. H. Silver, H. K. Kleinman, and E. Schiffmann, *Cell Biol. Int. Rep., 5*:813 (1981).
142. L. A. Liotta, D. Vembu, H. K. Kleinman, G. R. Martin, and C. Boone, *Nature (London), 272*:622 (1978).
143. U. Delvos, C. Gajdusek, H. Sage, L. A. Harker, and S. M. Schwartz, *Lab. Invest., 46*:61 (1982).
144. E. H. Beachey, T. M. Chiang, and A. H. Kang, *Int. Rev. Connect. Tissue Res., 8*: 1 (1981).
145. T. F. Deuel, R. M. Senior, D. Chang, G. L. Griffin, R. L. Heinrikson, and E. J. Kaiser, *Proc. Natl. Acad. Sci. USA, 78*:4584 (1981).
146. R. M. Senior, G. L. Griffen, J. S. Huang, D. A. Walz, and T. F. Deuel, *J. Cell Biol., 96*:382 (1983).
147. B. McCullough, J. F. Collins, W. G. Johanson, Jr., and F. L. Grover, *J. Clin. Invest., 61*:79 (1978).

148. R. P. Mecham, J. G. Madaras, and R. M. Senior, *J. Cell Biol.* (1984) (in press).
149. J. R. Hoidal and D. E. Niewoehner, *Chest, 4*:679 (1983).
150. R. M. Senior, H. Tegner, C. Kuhn, K. Ohlsson, B. C. Starcher, and J. A. Pierce, *Am. Rev. Resp. Dis., 116*:469 (1977).
151. R. A. Goldstein and B. C. Starcher, *J. Clin. Invest., 62*:1286 (1978).
152. C. Kuhn, S. Y. Ku, M. Chraplyvy, R. E. Linder, and R. M. Senior, *Lab. Invest., 34*: 372 (1976).
153. C. Kuhn and B. C. Starcher, *Am. Rev. Resp. Dis., 122*:453 (1980).
154. A. L. Horwitz and R. G. Crystal, *J. Clin. Invest., 46*:1312 (1975).
155. V. C. Hascall and J. H. Kimura, in *Methods in Enzymology*, Vol. 82 (L. W. Cunningham and D. W. Fredrickson, eds.), 1982, p. 769.
156. J. B. Karlinsky, *Am. Rev. Resp. Dis., 125*:85 (1982).
157. C. A. Vaccaro and J. S. Brody, *Am. Rev. Resp. Dis., 120*:901 (1979).
158. C. A. Vaccaro and J. S. Brody, *J. Cell Biol., 91*:427 (1981).
159. C. A. Vaccaro and J. S. Brody, *Fed. Proc., 40*:792 (1981).
160. J. O. Cantor, B. A. Bray, S. F. Ryan, I. Mandl, and G. M. Turino, *Proc. Soc. Exp. Biol. Med., 164*:1 (1980).
161. J. O. Cantor, J. M. Cerreta, M. Osman, S. H. Mott, I. Mandel, and G. M. Turino, *Proc. Soc. Exp. Biol. Med., 174*:172 (1983).
162. J. O. Cantor, M. Osman, J. M. Cerreta, I. Mandl, and G. M. Turino, *Proc. Soc. Exp. Biol. Med., 173*:362 (1983).
163. J. A. McDonald, in *Fibronectin* (D. F. Mosher, ed.), Academic Press, New York, 1984.
164. B. Villiger, D. G. Kelley, W. Engleman, C. Kuhn, and J. A. McDonald, *J. Cell Biol., 90*:711 (1981).
165. S. I. Rennard, G. W. Hunninghake, P. B. Bitterman, and R. G. Crystal, *Proc. Natl. Acad. Sci. USA, 78*:7147 (1981).
166. F. Grinnell, *Am. J. Dermatopathol., 4*:185 (1982).
167. J. A. McDonald, B. Villiger, T. Broekelmann, C. Kuhn, and C. Torikata, *Clin. Res., 39*:499A (1981).
168. J. A. McDonald and D. G. Kelley, *J. Biol. Chem., 255*:8848 (1980).

15

Fibrotic Skin Diseases

Keloids as a model to study the mechanisms of collagen deposition in tissues

R. Patrick Abergel
Harbor-UCLA Medical Center,
Torrance, California

Jouni Uitto
UCLA School of Medicine and Harbor-UCLA Medical Center,
Torrance, California

I. INTRODUCTION

Tissue fibrosis, characterized by excessive deposition of connective tissue components, most notably collagen, is the major pathological feature in various clinical conditions. The collagen deposition can take place in various internal organs, as in pulmonary fibrosis or liver cirrhosis (1,2). Skin is also commonly affected by fibrotic processes, and dermal fibrosis is the clinical pathological hallmark of several acquired and heritable cutaneous disorders (3,4) (Table 1).

In most cases, fibrosis is a reactive process, and several different factors can apparently modulate the pathways leading to tissue fibrosis. Such factors, largely elaborated by the inflammatory tissue reaction, include the local expansion of fibroblast subpopulations, immune modulation of the synthetic functions of fibroblasts, and altered regulation of various metabolic reactions governing the biosynthesis and degradation of the connective tissue components (1,5,6). Thus, the net accumulation of collagen in fibrosis is a result of imbalance between the factors leading to production and deposition or degradation and removal of collagen.

Several model systems have been utilized to examine various components of the pathological reactions leading to tissue fibrosis. These include in vitro studies with isolated tissues or cultured cells, as well as various animal models of fibrosis (1,3). Because of the availability of tissue, fibrosis of the skin provides an excellent opportunity to examine the pathological and biochemical changes taking place in the affected tissues. In particular, keloids, which are characterized by excessive deposition of collagen in dermis and subcutaneous tissue, provide a useful model to study processes leading to fibrosis in human tissues.

In this chapter, we summarize the current knowledge of connective tissue alterations taking place in keloids. Furthermore, we review our recent studies suggesting that the control of collagen synthesis in keloid fibroblasts takes place on the transcriptional level.

Table 1 Classification of Cutaneous Diseases with Dermal Fibrosis

- I. Scleroderma
  - A. Systemic Scleroderma
    1. Progressive systemic sclerosis
    2. CREST syndrome
  - B. Localized scleroderma (morphea)
    1. Circumscribed morphea
    2. Linear morphea
    3. Generalized morphea
    4. Guttate morphea
- II. Scleroderma-like skin changes in association with other clinical conditions
  - A. Connective tissue diseases
    1. Mixed connective tissue disease
    2. Overlap syndromes with lupus erythematosus and dermatomyositis
    3. Eosinophilic fasciitis
    4. Idiopathic acroosteolysis
  - B. Metabolic and immunologic disorders
    1. Chronic graft-versus-host disease
    2. Porphyria cutanea tarda
    3. Phenylketonuria
    4. Carcinoid syndrome
    5. Paraproteinemia
    6. Juvenile-onset diabetes mellitus
  - C. Chemically induced
    1. Bleomycin
    2. Polyvinylchloride
    3. Pentazocine
    4. Silicates
- III. Connective tissue hamartomas of the collagen type
  - A. Inherited
    1. Familial cutaneous collagenoma
    2. Shagreen patches in tuberous sclerosis
  - B. Acquired
    1. Isolated collagenomas
    2. Eruptive collagenomas
- IV. Keloids and hypertrophic scars

Source: Reproduced from Ref. 69.

II. KELOIDS AS A MODEL TO STUDY THE MECHANISMS OF TISSUE FIBROSIS

A. Clinical Features of Keloids

Keloids are dermal nodules with predilection sites on the upper back, shoulders, earlobes, and anterior chest (for review on keloids, see Ref. 7). The onset of the lesions can occur at any age but is most frequent during the second or third decade of life. The development of a keloid is often preceded by trauma at the site of the lesion. Keloids have been reported in most races, but their prevalence is increased in blacks and Chinese. The precise incidence of keloids is unknown; examination of certain African populations has suggested, however, an incidence of approximately 6%. A positive family history of keloids is frequently noted, and in some cases, the pedigree suggests an autosomal dominant inheritance.

The histopathological hallmark of keloids is proliferation of the extracellular matrix of the dermal connective tissue (8). Keloids are often characterized by a marked cellular infiltrate, predominantly fibroblastic, with deposition of dense collagenous meshwork. Transmission electron microscopy demonstrates closely packed collagen fibers often seen adjacent to a fibroblastic cell with a well-developed rough endoplasmic reticulum (Figure 1). Thus, the morphological observations suggest that collagen is the predominant component of keloids, and it can be speculated that the lesions result from a localized loss of control in the regulation of the extracellular matrix production (9).

B. Connective Tissue Composition of Keloids

As indicated above, histopathological and electron microscopic examinations of keloids suggest that collagen is the predominant component of the lesions. Direct biochemical measurements have indeed confirmed this impression. Assay of hydroxyproline as an index of collagen has indicated that this protein comprises approximately 60% of the dry weight of the keloid tissue (9) (Figure 2). This value is, however, not different from the collagen concentration in normal human skin, when expressed per milligram of dry weight of tissue. However, since the thickness of the dermis in keloids is markedly increased in comparison to normal human dermis, there is a large increase in total collagen content. Biochemical assays of elastin and glycosaminoglycans utilizing desmosine and uronic acid as specific markers have shown that the concentration of these two connective tissue components is slightly increased in keloids (Figure 2). These components represent, however, a minor fraction of the bulk of the keloid tissue. Also, the water content of keloids is significantly higher than that in normal skin (9-11).

Several studies have examined the relative concentrations of genetically distinct collagens in keloid tissues. Initial reports suggested that the relative amounts of type III collagen may be increased in keloid scars (12,13), while another study (14) has suggested that type III/I collagen ratio in keloids is the same as in normal dermis. Detailed collagen analyses by us have recently demonstrated that type I collagen represents approximately 95% of the total collagen in keloids; the corresponding value in normal human skin is about 75% (9) (Figure 3). The relative concentration of type III collagen in keloids was 4%, as compared to 21% in the control skin (9). In addition, small amounts of type V collagen could be demonstrated both in keloids and in control skin (9). The discrepancy between different studies may reflect differences in the type of pathological specimens obtained for biochemical analyses. Specifically, increased type III/I collagen ratio may well be found in hypertrophic scars (12,13), while in keloid tissue type I is more predominant (9). Thus, histopathological, ultrastructural, and biochemical evidence indicate that collagen is the major extracellular component of keloids, and type I collagen is the predominant component of these lesions.

C. The Biosynthesis of Collagen in Keloid Fibroblast Cultures

Theoretically, several different mechanisms could lead to excessive collagen deposition in keloids. For example, the deposition of collagen could result from enhanced production of this protein by keloid fibroblasts. Alternatively, reduced degradation of the extracellular collagen fibers could have similar consequences. In addition, local expansion

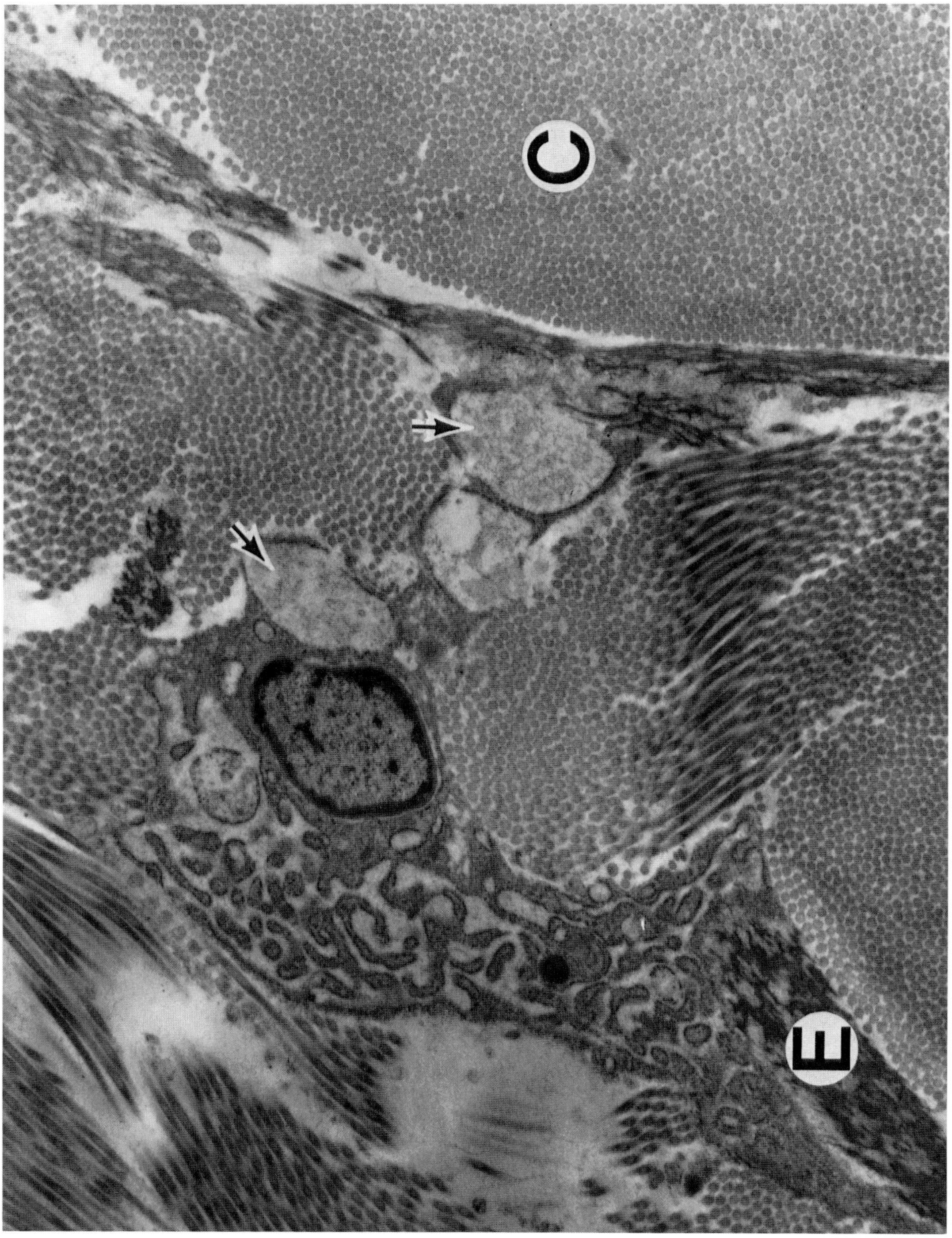

Figure 1

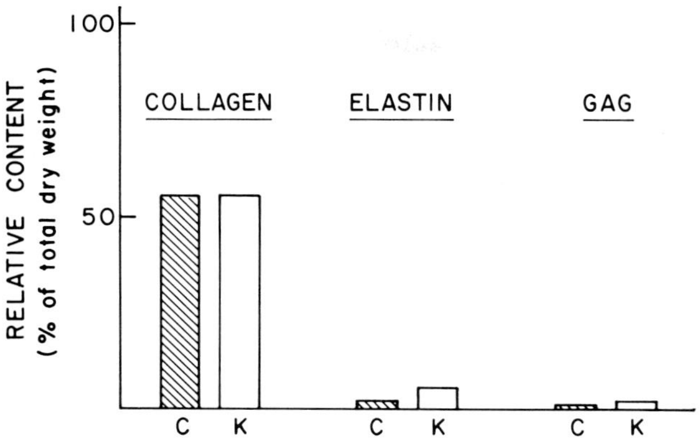

Figure 2 Relative concentrations of collagen, elastin, and glycosaminoglycans (GAG) in keloids (K) and in normal human skin (C). Collagen was determined by assay of hydroxyproline, elastin by determination of desmosine, and glycosaminoglycans by uronic acid analysis. The values are means of eight keloids and three control specimens. (Reproduced with permission from Ref. 9.)

of synthetically active fibroblast populations, even without concomitant increase in collagen production per cell, could lead to collagen accumulation.

Several studies, largely utilizing fibroblast cultures established from keloids, have examined the mechanistic details of collagen accumulation in these lesions. Several studies have indicated that collagen synthesis is increased in some, but not all, keloid fibroblast cultures (9,10,15,16). We have recently demonstrated that five out of nine keloid cell lines studied produced increased amounts of procollagen in culture when compared with age-, sex-, and passage-matched control fibroblast cultures (Figure 4). The largest enhancement of collagen production was up to 3.3-fold over the controls (9). In accordance with these observations, the activity of prolyl hydroxylase, the key enzyme in the intracellular biosynthesis of procollagen, has been shown to be increased in cell lines demonstrating elevated collagen synthetic rates (9,17,18). Thus, on the basis of these observations, keloid fibroblast cultures can be classified as either high or normal producers of collagen.

D. Collagen Degradation

In addition to changes on the biosynthetic level, an altered degradative pathway could result in the accumulation of collagen in tissues. Two different mechanisms of collagen degradation can be considered: (a) extracellular removal of collagen fibers mediated by specific tissue collagenases (19), and (b) intracellular degradation of newly synthesized polypeptides before their transport into the extracellular milieu (20). The possibility that the activity of extracellular collagenase is decreased in keloid tissue appears unlikely, since

---

Figure 1 Transmission electron microscopy of keloids demonstrating closely packed collagen fibers (C) adjacent to a fibroblastic cell. Note the well-developed rough endoplasmic reticulum with dilated cisternae (← →). Irregular electron-dense elastic fibers (E) are also present in the proximity of the fibroblast (original magnification, X11,000). (The figure was kindly provided by Dr. Lois Y. Matsuoka.)

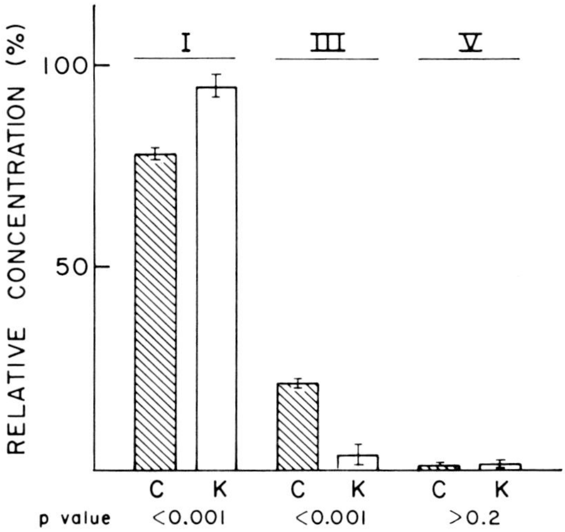

Figure 3 Relative concentrations of collagen types I, III, and V in keloid tissues and in normal human skin. Collagen was solubilized by limited pepsin proteolysis, and the relative concentrations of genetically distinct collagens were determined on the basis of separation of α chains by SDS-polyacrylamide gel electrophoresis. The values are mean ± SD, expressed as percent of the total collagen in the tissue. (Reproduced with permission from Ref. 9.)

the activity of collagenase produced by keloid fibroblast cultures has been shown to be either normal or increased in comparison to control cultures (9,21) (Figure 5). In this context, it is of interest that tissue α-globulins have been suggested to be increased in keloid tissue (22). Since these plasma proteins, and particularly α2-macroglobulin, can inhibit collagenases through irreversible complex formation, an altered α2-macroglobulin/collagenase ratio could lead to reduced collagen degradation in these lesions.

An alternate mechanism for excessive collagen production would entail reduced intracellular degradation of the newly synthesized procollagen polypeptides. Previous studies have demonstrated that a small, but significant, fraction of the newly synthesized proα chains is degraded to small peptides in a pathway probably involving lysosomal enzymes (20). We recently noted that the intracellular degradation of newly synthesized collagen was relatively low in some keloid cell cultures that demonstrated increased collagen production (9) (Figure 6). Thus, reduced degradation of the newly synthesized procollagen might contribute to the elevated collagen production in some keloid fibroblast cultures. Such modulation of procollagen production at the level of intracellular degradation has previously been demonstrated in several experimental situations (20,23,23).

E. Molecular Basis for Increased Procollagen Production in Fibrotic Disease

Several recent observations in the literature have suggested that a major control point in collagen production exists on the transcriptional level (25-27). To examine the possibility that increased procollagen production in keloid fibroblast cultures might reflect an altered transcriptional control, the abundance of procollagen mRNA was determined using human collagen specific cDNA probes (28). Initial analyses indicated that type I

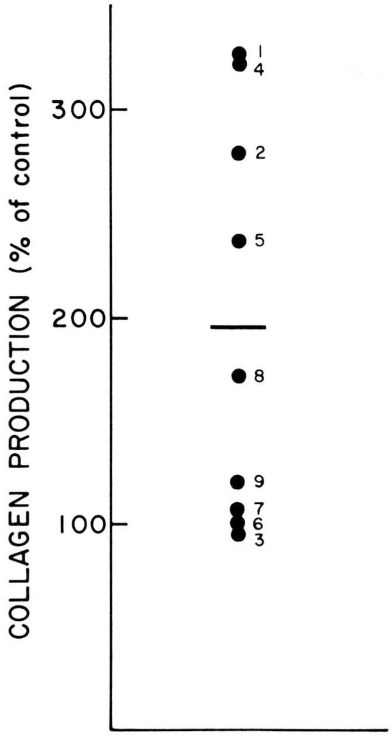

Figure 4 Collagen production in keloid cell cultures in comparison to the age-, sex-, and passage-matched control fibroblast cultures. The cells were incubated with [^{3}H]proline, and the collagen production was assayed by the synthesis of nondialyzable [^{3}H]hydroxyproline. The values are expressed as percent of the corresponding matched controls in each incubation. The individual values are means of three parallel determinations. The numbers by the values refer to the keloid cultures as designated in Ref. 9.

collagen specific mRNA levels in keloid fibroblasts were increased in cultures that were collagen overproducers, and a good correlation between type I procollagen mRNA abundance and type I collagen production was observed (9) (Figure 7). Since the relative concentration of type I collagen in comparison to type III collagen was clearly increased in keloids, it was of interest to examine whether the change might reflect an altered type I/III procollagen mRNA ratio (28). Slot blot hybridizations using human proα1(I), proα2(I), and proα1(III) collagen specific cDNAs indicated that in control fibroblasts, the ratio of type I/III collagen mRNAs was 5.2 ± 2.1, while the corresponding ratio in keloid cells was 22.1 ± 7.8 (Figure 8). In both keloid and control fibroblast cultures, the proα1(I)/proα2(I) collagen mRNA ratio was approximately 2:1 (28). These observations suggest, therefore, that the synthesis of proα chains representing type I and type III procollagens is largely controlled by the corresponding mRNA levels.

At least three plausible mechanisms might explain increased mRNA levels in keloid fibroblasts. First, increased rate of transcription of type I procollagen genes could increase the mRNA levels. Second, reduced degradation and turnover of the corresponding mRNA could have similar consequences. Third, gene amplification in terms of increased type I collagen gene copy number, without a concomitant change in the rate of

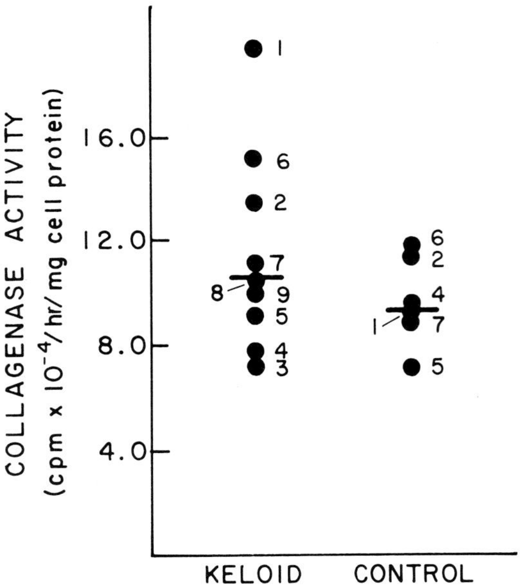

Figure 5 Assay of collagenase activity in keloid and control human skin fibroblast cultures. The cells were incubated in serum-free medium, and collagenase activity was determined by incubation with radioactive collagen substrate, after a brief trypsin activation of the latent enzyme. The values are expressed as cpm [^{3}H] proline-labeled type I collagen substrate degraded $\times$ h^{-1}/mg cell protein. The individual values are means of three parallel determinations, and the mean of each group is indicated by a horizontal line. The numbers by individual values refer to the cultures as designated in Ref. 9.

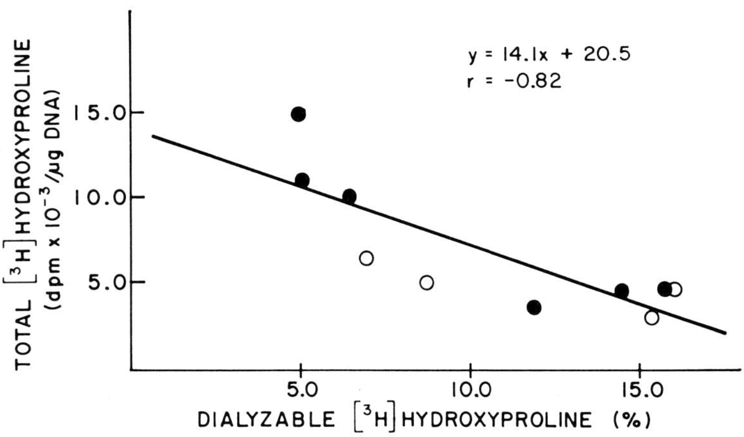

Figure 6 The correlation between the total synthesis of [^{3}H] hydroxyproline and the degradation of newly synthesized procollagen polypeptides in keloid (●) and control (○) fibroblast cultures. The cells were labeled with radioactive proline, and the total radioactive hydroxyproline in dialyzable and nondialyzable form was determined. Linear regression analysis of the data indicate the line of best fit, $Y = (14.1)X + 20.5$, with a correlation coefficient (r) -0.82. (Reproduced with permission from Ref. 9.)

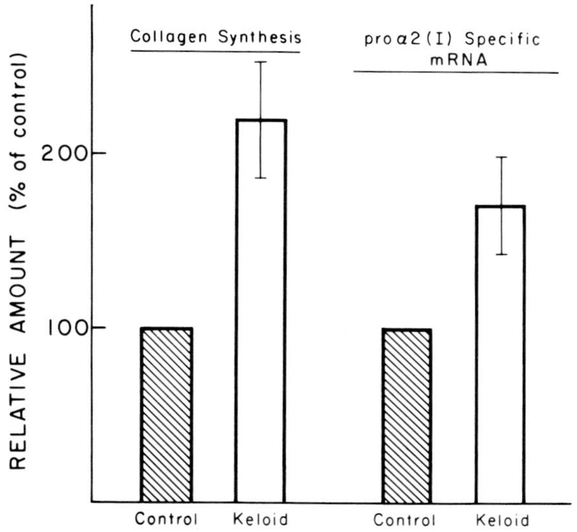

Figure 7 Correlation between collagen production and type I procollagen mRNA abundance in keloid fibroblast cultures. Collagen production was assayed by the synthesis of [^{3}H] hydroxyproline, and procollagen mRNA abundance was assayed by hybridizations with proα2(I) collagen specific ^{32}P-labeled cDNA, as indicated in Ref. 9. The values in keloid cultures are mean ± SEM of four different cell lines, and they are expressed as percent of the corresponding controls.

transcription, could result in the elevated levels of mRNA. We have addressed the last possibility by determining the relative type I procollagen gene copy number in cultured keloid fibroblasts (28). In these studies, the genomic DNA isolated from the fibroblasts was subjected to digestion with EcoRI restriction enzyme, and the resulting fragments were separated by agarose gel electrophoresis. After transfer to nitrocellulose filters by Southern blotting (29), the DNA fragments were hybridized with proα2(I) collagen specific cDNA, together with a human β-globin specific cDNA probe (30) (Figure 9). The β-globin hybridization served as an internal standard in each specimen, with the assumption that the β-globin gene copy number is not altered in keloid fibroblasts (31). The results indicated that the proα2(I) collagen gene copy number was not altered in keloid cells in comparison to the control fibroblasts (28). Since previous studies have established that the proα2(I) collagen gene copy number in human haploid genome is 1 (32,33), our observation excludes gene amplification as an explanation for altered collagen gene expression in keloid cell cultures.

III. GENERAL PATHWAYS LEADING TO TISSUE FIBROSIS

As indicated elsewhere (1-3), the pathways of collagen biosynthesis and degradation involve several distinct steps, many of which are under precise enzymatic control (Table 2). Several of the enzymes participating in the metabolism of collagen are specific for this protein, but they have stringent substrate and cofactor requirements (Table 3). The major steps in the metabolism of collagen, in the sequence in which they are likely to occur in vivo, are presented in Table 2. As indicated, many of the steps leading to colla-

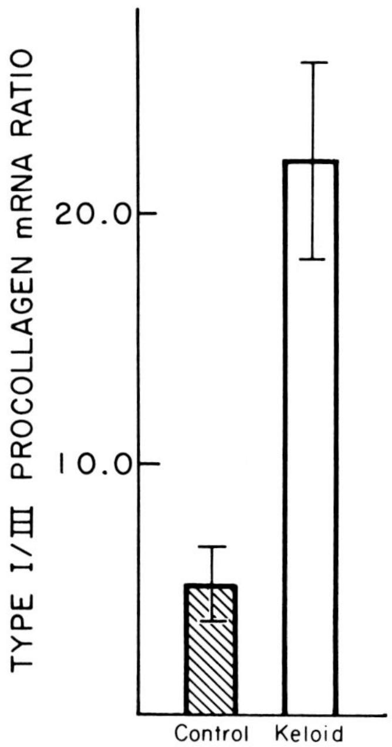

Figure 8 The ratio of type I/III procollagen mRNAs in fibroblast cultures established from normal human skin or from keloids. Poly(A)$^+$-RNA was isolated by oligo-dT cellulose chromatography and dotted on nitrocellulose filters in serial dilutions. Parallel filters were hybridized with human proα1(I) and proα1(III) collagen cDNA probes. The mRNA-[^{32}P] cDNA hybrids were visualized by autoradiography and quantitated by scanning at 700 nm. The values are mean ± SD of six keloid and five control cell cultures. (Reproduced with permission from Ref. 28.)

gen formation are known to have a well-characterized functional significance, so that errors in any of these steps could potentially lead to aberrant collagen formation. Also, many of these steps are subject to modulation by a variety of physiological and pharmacological factors (Table 2). Since the net accumulation of collagen in tissues is a result of careful balance between the production of this protein and removal of the extracellular collagen fibers, it is conceivable that an increased flux through any of the biosynthetic steps could lead to altered amounts of extracellular collagen. This would be particularly true in the case of those steps which under normal circumstances are rate-limiting.

Keloids provide an excellent and easily accessible model to study various mechanisms leading to tissue fibrosis. The studies summarized above have clearly indicated that several distinct mechanisms can lead to collagen accumulation. Such mechanisms include: (a) increased type I collagen production as a result of elevated mRNA level, and (b) reduced intracellular degradation of newly synthesized procollagen polypeptides. Thus, at least these two mechanisms appear to contribute to deposition of collagen in keloids.

The observation that some keloid cell lines demonstrate normal rates of collagen production in culture could have several explanations. For example, previous studies have demonstrated that human skin fibroblasts are a heterogeneous population of cells that can exhibit up to 20-fold differences in their collagen synthesis rates, and such heterogeneity

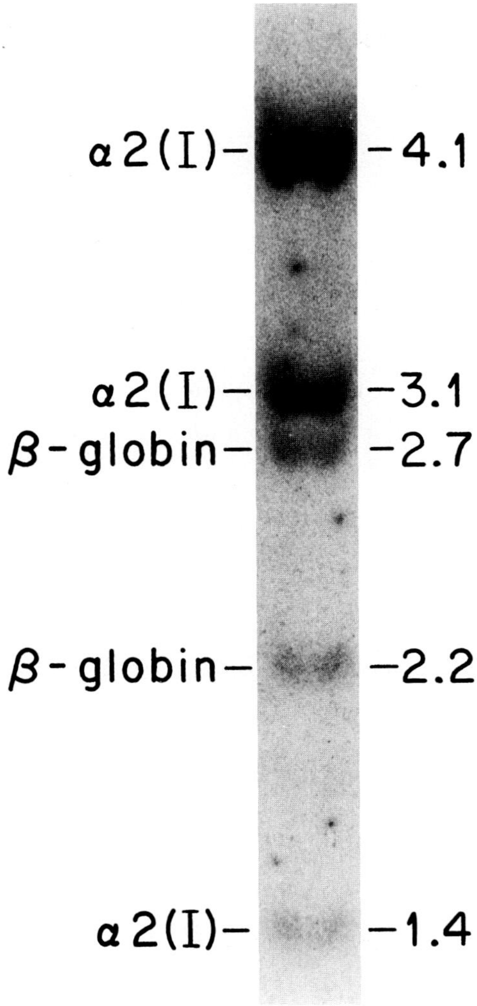

Figure 9 Determination of type I procollagen gene copy number in keloid fibroblast genome. DNA from cultured cells was isolated and subjected to digestion with EcoRI restriction enzyme. The digestion fragments were then separated by agarose gel electrophoresis and transferred to a nitrocellulose filter by Southern blotting. The filters were then hybridized with a mixture of ^{32}P-labeled cDNA probes specific for human proα2(I) collagen and β-globin sequences. The bands hybridizing the proα2 collagen and β-globin probes, as identified in the future with their corresponding sizes (kb), were quantitated by densitometry, and the relative proα2(I) collagen gene copy number in keloid cells in comparison to control fibroblasts was determined as indicated in the text. (Reproduced with permission from Ref. 71.)

Table 2 Major Steps in the Biogenesis and Degradation of Collagen, Their Functional Significance and Modulating Factors

Step	Functional significance	Modulating factors
I. Expression of genes coding for Collagen polypeptides		
a. Gene selection	Determines the collagen isotype to be synthesized	Bromodeoxyuridine, environmental factors
b. Transcription	Formation of mRNA percursor	Glucocorticosteroids, possibly other hormones
c. Processing of mRNA precursor	Formation of functional mRNA	None known
d. Translation	Assembly of polypeptide chains	Feed-back inhibition by extension peptides
e. Control of the rate of transcription, mRNA processing or translation	Determines the amount of polypeptides synthesized	Same as Ib and d
II. Intracellular co- and posttranslational modifications		
a. Removal of the signal sequence	May be necessary for secretion	Amino acid analogues, such as hydroxynorvaline and canavanine
b. Synthesis of 4-hydroxyproline	Stabilization of triple helix	Chelators for Fe^{2+}; dehydroproline-containing polypeptides; poly-(L-proline); some divalent cations
c. Synthesis of 3-hydroxyproline	Unknown	Chelators
d. Synthesis of hydroxylysine	Stabilization of covalent cross-links; attachment site for glycosylation	Chelators; some divalent cations
e. Synthesis of hydroxylysine-O-glycosides	May influence cross-link formation and determine the morphology and stability of the fibers	Co^{2+}; free gal-O-hylys; UDP-gal analogues
f. Glycosylation of the extension peptides	Unknown	Tunicamycin
g. Degradation of nonhelical chains	Removal of defective polypeptides and modulation of collagen production	β-agonists; phosphodiesterase inhibitors
h. Chain association and disulfide bonding	Facilitation of the triple-helix formation	S-S reducing agents
i. Triple-helix formation	Prerequisite for proper secretion	Proline analogs; inhibitors of prolyl 4-hydroxylase

III. Secretion	Transport of procollagen through Golgi apparatus to the extracellular space	Colchicine, cytochalasin B, vinblastine; inhibitors of triple-helix formation; some local anesthetics
IV. Extracellular modifications		
a. Removal of the extension peptides	Necessary for fiber formation	Several amino acids, polyamines, and their analogues; synthetic peptides containing the cleavage sequence; chelators of Ca^{2+}
b. Deamination of certain lysine and hydroxylysine residues	Necessary for cross-link formation	Nitriles; chelators of Cu^{2+}
c. Nonenzymatic glucosylation	May interfere with fiber formation	Glycemic control
V. Fiber formation		
a. Alignment of the molecules	Formation of microfibrils	Inhibitors of procollagen to collagen conversion
b. Formation of cross-links	Stabilization of the fiber structures	D-penicillamine; inhibitors of lysyl oxidase
c. Supramolecular assembly	Architectural organization of collagen in tissues	None known
d. Interactions with other extracellular macromolecules	Determines the physiological properties of tissues	None known
VI. Extracellular degradation		
a. Cleavage by specific collagenases	Rate-limiting step in degradation	Inhibited by chelators of Ca^{2+}, diphenylhydantoin, glucocorticoids, retinoids, α2-macroglobulin, tissue polypeptide inhibitors; activated by sulfhydryl reagents, chaotropic agents, proteolytic activators, monocyte factors
b. Further degradation by peptidases and enzymes metabolizing free amino acids	Removal of degradation products	Chelators of Ca^{2+}; S-S reducing agents

Source: Ref. 40.

Table 3 Characteristics of Enzymes Participating in the Biosynthesis and Degradation of Collagen

Enzyme[a]	Substrate[b]	Product	Cofactors and cosubstrates
1. Signal peptidase	Nascent pre-pro-α chains	pro-α chains	None known
2. Prolyl-4-hydroxylase	Prolyl residue in pro-gly sequence in proα-chains*	4-hydroxyproline	O_2, Fe^{2+}, α-ketoglutarate
3. Prolyl-3-hydroxylase	Prolyl residue in pro-hyp-gly sequence in pro-α chains*	3-hydroxyproline	O_2, Fe^{2+}, α-ketoglutarate, ascorbic acid
4. Lysyl hydroxylase	Lysyl residue in lys-gly, lys-ser or lys-ala-sequence in pro-α chains*	hydroxylysine	O_2, Fe^{2+}, α-ketoglutarate, ascorbic acid
5. Collagen galactosyl transferase	Hydroxylysine in pro-α chains*	gal-O-hydroxylysine	Mn^{2+}, UDP-galactose
6. Collagen glucosyl transferase	Galactosyl-O-hydroxylysine in pro-α chains*	glc-gal-O-hydroxylysine	Mn^{2+}, UDP-glucose
7. Protein disulfide isomerase[c]	Cystine residues in extensions of pro-α chains	S-S-bonds stabilizing the correct protein conformation	Thiols
8. Procollagen N-terminal proteinase	Procollagen or pN-collagen	pC-collagen or collagen[d]	Ca^{2+}
9. Procollagen C-terminal proteinase	Procollagen or pC-collagen	pN-collagen or collagen[d]	Ca^{2+}
10. Lysyl oxidase	Lysyl or hydroxylysyl residue in fibrillar collagen	Aldehyde derivatives of lysine and hydroxylysine	Cu^{2+}, O_2
11. Collagenase	Specific sequences in native collagen	Cleavage products TC^A and TC^B	Ca^{2+}
12. Extracellular metalloproteinases	Denatured collagenous polypeptides	Small peptides and free amino acids	Ca^{2+}
13. Hydroxyproline oxidase	Free hydroxyproline	Δ′pyrroline-3-hydroxy-5-carboxylic acid	O_2
14. Hydroxylysine kinase	Free hydroxylysine	O-phosphohydroxylysine	Mg^{2+}, GTP

[a]The action of these enzymes is relatively specific to collagen; the complete sequences of procollagen synthesis and collagen degradation involve additional, less specific enzymes, such as those of transcription and translation.
[b]The reactions indicated by an * are terminated when the proα chains fold into the triple-helical conformation.
[c]It has not been established whether the formation of interchain disulfides in procollagen involves enzymatic catalysis, as occurs in some other proteins, or whether their synthesis takes place spontaneously.
[d]If intact procollagen is used as a substrate, partially modified products are formed; however, if the partially cleaved proteins serve as substrates, collagen is produced.
Source: Ref. 40.

is conserved over multiple population doublings in vitro (34,35). Thus, the relative proportions of cells in culture characterized either as normal or overproducers of collagen may depend on the selection of cells during initiation of the cultures or may reflect heterogeneity in the disease state at the time the cultures were established. In any event, the data on the overproduction of collagen by cultured cells raise the possibility that keloids represent a process of selection of fibroblast subpopulations programmed to produce increased amounts of collagen.

The increased collagen production characteristic of keloid fibroblast cultures may result from a localized loss of control in regulation of the extracellular matrix production (9). The exact cause of such loss of control is not immediately apparent. Previous studies have indicated that keloid fibroblasts demonstrate normal growth characteristics (16,36), but they may be susceptible to modulation by various environmental factors in a manner different from that in control cells (37-39). The elevated type I procollagen production was accompanied by the increased level of corresponding mRNA, suggesting increased transcriptional activity of type I procollagen genes. Furthermore, we were able to exclude several alternate mechanisms for increased collagen deposition, including reduced collagenase activity and gene amplification in case of keloids. The latter mechanisms may, however, be operative in other fibrotic processes affecting skin and the internal organs.

IV. APPROACHES FOR TREATMENT AND PREVENTION OF TISSUE FIBROSIS

A. Pharmacological Intervention

As indicated above, the biosynthesis of collagen involves several distinct steps that can be modulated by pharmacological means (40,41). Thus, intervention at any of the susceptible steps could lead to inhibition of collagen deposition in fibrotic diseases. Also, enhancement of the activity of collagenase, the enzyme responsible for degradation of the extracellular collagen fibers (42), could similarly decrease the extracellular mass of collagen.

Previously, several pharmacological agents were tested for their effects on collagen metabolism under cell or tissue culture conditions. Some of these agents are already in clinical use or are currently being tested for treatment of fibrotic skin diseases, such as progressive systemic sclerosis and keloids. We summarize here the mechanisms of some of the pharmacological agents known to affect collagen metabolism.

1. Modulation of Transcription or Translation

The early events of collagen gene expression on the transcriptional and translational levels are in general similar to those involved in the production of other eukaryotic proteins. Thus, several agents that interfere with protein synthesis in general could also modulate collagen production, but these agents would not be specific for collagen.

Of particular interest is the use of glucocorticosteroids for treatment of fibrotic connective tissue disorders (43). Several lines of evidence suggest that glucocorticosteroids decrease collagen production, and the inhibition of collagen formation has been shown to be more pronounced than the effects on protein synthesis in general (43). It has also been observed that fluorinated steroids are more potent inhibitors of collagen production than the corresponding nonfluorinated compounds (44,45). This observation would be compatible with clinical findings indicating that dermal atrophy is a more frequent consequence of prolonged use of fluorinated steroids while less frequently encountered in patients treated with nonfluorinated steroids. The exact molecular mechanisms of the inhibition of collagen synthesis by glucocorticosteroids are discussed in Chapter 17.

2. *Inhibition of Co- and Posttranslational Modifications*

The newly synthesized polypeptides of procollagens, proα chains, undergo several co- and posttranslational modification reactions catalyzed by enzymatic means (Tables 2 and 3). Particularly important is the enzymatic synthesis of 4-hydroxyproline, and this reaction has been shown to be critical for normal production of triple-helical collagen at physiological temperature. Since the triple helix is a prerequisite for collagen secretion at the optimal rate, the nonhelical polypeptides that are deficient in 4-hydroxyproline accumulate intracellularly or are released into the extracellular space at the markedly reduced rate. The nonhelical proα chains are also susceptible to degradation by a variety of proteinases. Thus, inhibition of prolyl hydroxylase by pharmacological means could reduce the deposition of collagen in tissues.

Prolyl hydroxylase requires ferrous iron, α-ketoglutarate, molecular oxygen, and ascorbic acid for its activity (46). Several in vitro studies have demonstrated that the formation of hydroxyproline is inhibited by chelators of ferrous iron, such as α,α'-dipyridyl (47). Although this compound or other metal chelators have been proven to be highly effective in limiting the collagen synthesis in isolated tissues or cells in vitro, it is likely that these agents are toxic in whole animals owing to inhibition of other critical enzymes dependent on metal cofactors.

Another approach to inhibit the hydroxylation of prolyl residues employs divalent cations, such as Zn^{2+}, Pb^{2+}, and Co^{2+}. These metals compete with ferrous iron for prolyl hydroxylase, but are unable to serve as cofactors in the hydroxylation reaction (48). Consequently, the procollagen polypeptides are underhydroxylated, are retained intracellularly, and are subject to nonspecific proteolysis. Most promising results in animal studies have been obtained with Zn^{2+}. It has been shown that concentrations of zinc that are in the physiological range can modulate collagen metabolism in pathological fibrosis (49).

Another method for inhibiting prolyl hydroxylase involves incorporation of a proline analog, 3,4-dehydroproline, into the newly synthesized proα chains. The incorporation of this analog into the procollagen polypeptides has several consequences. First, the incorporation of 3,4-dehydroproline into the α chains instead of proline prevents the polypeptides from folding into a stable helical conformation (50,51). Second, the overall hydroxylation of procollagen polypeptides containing the analog is markedly reduced. The inhibition of prolyl hydroxylase by this particular amino acid analog has been explained by tight complex-formation between the enzyme and the analog-containing polypeptides (52). Third, it has been demonstrated that 3,4-dehydroproline increases the error rate of translation of proline codons leading to abnormal protein with markedly reduced imino acid content (53). Thus, as a net result of these three mechanisms, the production of procollagen molecules is markedly reduced in the presence of 3,4-dehydroproline.

3. *Inhibition of Collagen Secretion*

Following the extensive intracellular modifications catalyzed by prolyl and lysyl hydroxylases and by collagen glycosyltransferases, the proα chains fold into a stable triple-helical conformation, which is prerequisite for secretion of procollagen into the extracellular milieu. The procollagen molecules are packaged into Golgi vesicles and transported into the extracellular space by mechanisms involving microtubular functions. Compounds, such as colchicine, that are then able to disrupt the microtubules have been shown to reduce the collagen secretion in isolated cells or organ cultures (54). Colchicine has also been used for treatment of a variety of fibrotic diseases, including progressive systemic sclerosis (55). An additional rationale for the use of this compound for the treatment of fibrotic processes comes from the demonstration that colchicine increases the production of collagenase in cell cultures (56). Thus, as a net result of decreased collagen secretion and increased collagenase activity, the deposition of collagen could be reduced. However, the proof of clinical efficacy of colchicine in treatment of progressive systemic sclerosis or other fibrotic diseases is still limited, and carefully controlled clinical trials are required to demonstrate its efficacy in the future.

4. *Inhibitors of Procollagen to Collagen Conversion*

Following the secretion of procollagen molecules, the precursor forms are converted to functional collagen molecules by the removal of precursor-specific extension peptides (57). The conversion of procollagen to collagen is catalyzed by two specific enzymes, procollagen N- and procollagen C-proteinase. Removal of the extensions is a prerequisite for formation of functional collagen fibers.

Recent studies have demonstrated that the converting enzymes are calcium-dependent metalloproteases functional in the extracellular space. The procollagen C-proteinase, which is responsible for the removal of the C-terminal extension, can be inhibited by a variety of naturally occuring amino acids, polyamines, and structurally related compounds (58). As a consequence, pC collagen, which is unable to form stable, cross-linked collagen fibers, accumulates in tissues. These molecules are then susceptible to degradation by collagenase, and the overall deposition of collagen is reduced.

It has also been demonstrated that procollagen proteases can be inhibited by divalent cations in a competitive manner (59). Specifically, Zn^{2+}, Hg^{2+}, and Cu^{2+} are efficient inhibitors of the procollagen-to-collagen conversion. These observations suggest, therefore, that naturally occurring compounds may modulate the procollagen deposition in tissues, and further development of analogs of these molecules could provide a means of inhibiting collagen production in fibrotic diseases (41).

5. *Inhibition of Collagen Cross-Linking*

The initial step in the formation of covalent cross-links is the deamination of lysyl and hydroxylysyl residues to form corresponding aldehydes (60). This reaction is catalyzed by a copper-dependent enzyme, lysyl oxidase (Table 3). Lysyl oxidase can be inhibited irreversibly by lathyrogenic agents, such as β-aminopropionitrile, and several animal studies have suggested that these compounds would be relatively safe and specific inhibitors of collagen accumulation (40). Encouraging preliminary results have been obtained in limited clinical trials in patients with urethral strictures, tendon adhesions, and pulmonary fibrosis (61). However, the usefulness of nitriles for the treatment of generalized fibrotic conditions, such as progressive systemic sclerosis, is currently unknown.

Another pharmacological agent currently in clinical use for treatment of fibrotic diseases is D-penicillamine. This drug has been employed for treatment of collagen accumulation diseases, as in hepatic fibrosis, progressive systemic sclerosis and morphea (62,63). The potential benefits of D-penicillamine in the treatment of diseases with collagen accumulation theoretically involve two mechanisms: First, D-penicillamine is a copper chelator, and consequently, it can inhibit the activity of lysyl oxidase. Second, D-penicillamine is known to react with lysine- and hydroxylysine-derived aldehydes making them unavailable for collagen cross-linking (64). Thus, D-penicillamine could be a clinically useful drug to interfere with collagen accumulation, but its use is somewhat limited because of both acute and long-term toxicity.

6. Modulation of Collagenase Activity

As indicated above, the net accumulation of collagen is a balance between the synthesis and degradation of this protein. Thus, theoretically an increase in collagenase activity could lead to a reduction of the extracellular collagen in fibrotic diseases. Although several agents have been shown to induce collagenase activity under in vitro conditions, no pharmacological agent has been successful in increasing collagenase activity in vivo (65). Perhaps, further development of analogs to agents such as sulfhydryl reagents, chaotropic agents, and proteolytic enzymes, which have been shown to increase the collagenase activity or enhance the conversion of latent collagenase to active form, could form a basis for future development of an effective pharmacological agent.

B. Laser Treatment of Keloids

Several pharmacological agents have been advocated for the treatment of fibrotic diseases, including keloids, but their clinical efficacy is often compromised by considerable toxicity. In the case of keloids, recent studies have suggested that lasers, and particularly Nd:YAG laser, might provide a new treatment modality (66). This suggestion was originally based on the demonstration that Nd:YAG laser inhibited procollagen production in fibroblast cultures without compromising the cell viability (67) (Figure 10). Also, Nd: YAG laser was shown to reduce collagen accumulation in pig skin in vivo (66) (Figure 11).

Initial clinical studies have indeed suggested the usefulness of Nd:YAG laser for treatment of keloids (67). The results of these studies with a three-year follow up have indicated flattening and softening of the keloid lesions subjected to Nd:YAG laser radiation in a nondestructive manner. It has been suggested that this laser probably acts through a mechanism involving a relatively subtle biological modulation of the fibroblast functions, resulting in selective suppression of the collagen production by these cells (66). The relatively benign nature of the Nd:YAG laser is in contrast to the excisional removal of keloids by CO_2 laser (68). It may well be that effective eradication of keloids could be achieved by a combination of the removal of excess tissue first by conventional surgery by CO_2 laser, followed by inhibition of the collagen production by Nd:YAG laser (66).

V. CONCLUSIONS

In this chapter, we have summarized data indicating that several different mechanisms can lead to excessive accumulation of collagen in diseases. Such mechanisms include:

1. Increased production of procollagen by fibroblasts.
2. Decreased intracellular degradation of the newly synthesized polypeptides.
3. Reduced removal of the extracellular collagen by collagenases.
4. Selection of fibroblast subpopulations active in collagen production.

The available data on keloids suggest that these cutaneous lesions characterized by excessive collagen deposition serve as a useful model to study mechanisms leading to tissue fibrosis. In particular, we have demonstrated that loss of control of the extracellular matrix production by keloid fibroblasts is one of the major factors leading to excessive collagen deposition. Similar mechanisms may well be operative in other fibrotic skin diseases (69).

We have further demonstrated that processes leading to excessive collagen deposition can be modulated by a variety of pharmacological means. Finally, particularly interesting is the demonstration that lasers can control the connective tissue metabolism (66), thus providing a useful modality for arresting collagen deposition in keloids and other fibrotic diseases.

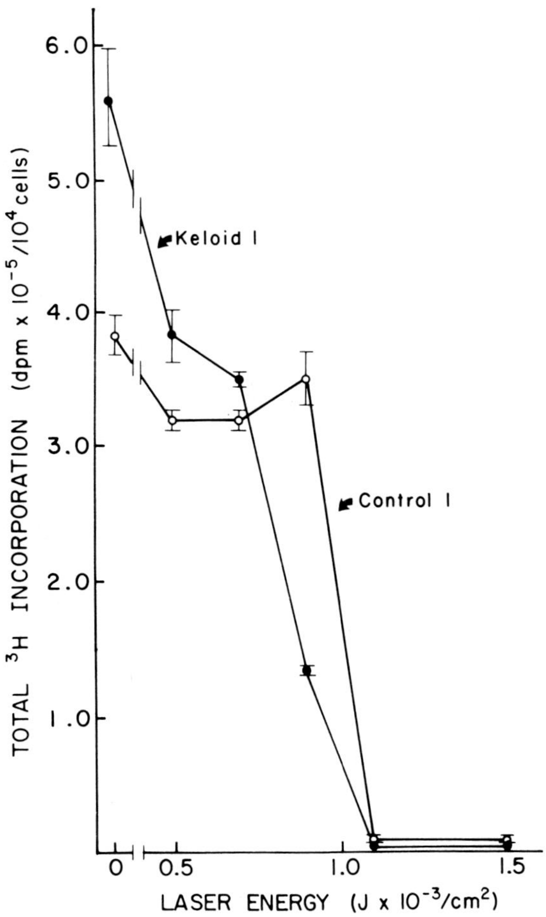

Figure 10 Inhibition of procollagen production in fibroblast cultures, established from normal human skin or from keloids, by Nd:YAG laser. Note that the keloid fibroblast cultures, which initially demonstrate increased procollagen production, are more sensitive to the laser irradiation. The values are mean ± SD. (Reproduced with permission from Ref. 67.)

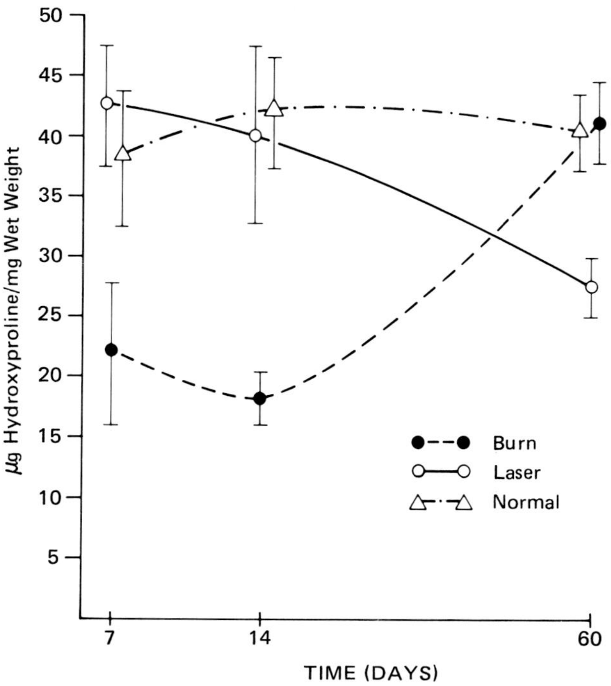

Figure 11 Collagen deposition in pig skin in a wound-healing model induced by Nd:YAG laser irradiation or thermal burn, in comparison to normal skin. Collagen was determined by assay of hydroxyproline at 7, 14, and 60 days after treatment of the skin with the laser, as indicated in the original publication. (Published with permission from Ref. 70.)

ACKNOWLEDGMENTS

The authors thank Dr. Lois Y. Matsuoka for providing the electron micrograph in Figure 1, and Charlene D. Aranda for manuscript preparation.

Supported in part by USPHS, NIH grants GM-28833, AM-35297, and AM-28450. Dr. Abergel was recipient of the Ken Burdick Memorial Fellowship 1984 from Syntex Laboratories awarded by the Dermatology Foundation.

REFERENCES

1. D. J. Prockop, K. I. Kivirikko, L. Tuderman, and N. A. Guzman, *N. Engl. J. Med., 301*:13-23, 77-85 (1979).
2. J. G. Clark, C. Kuhn, J. A. McDonald, and R. P. Mecham, *Int. Rev. Connect. Tissue Res., 10*:249-315 (1983).
3. J. Uitto, L. Ryhänen, and E. M. L. Tan, *Progress in Diseases of the Skin* (R. Fleischmajer, ed.), Grune & Stratton, New York, 1981, pp. 103-141.
4. J. Uitto, D. J. Santa-Cruz, and A. Z. Eisen, *J. Am. Acad. Dermatol., 3*:441-461 (1980).
5. L. Menaker, *Biological Basis of Wound Healing*, Harper & Row, Hagerstown, MD, 1975.
6. Wahl, S. M., *The Surgical Wound* (P. Dinen and G. Hildick-Smith, eds.), Lea & Febiger, Philadelphia, 1981, pp. 63-73.

7. J. C. Murray, S. V. Pollack, and S. R. Pinnell, *J. Am. Acad. Dermatol., 4*:461-470 (1981).
8. W. F. Lever and G. Schaumburg-Lever, *Histopathology of the Skin*, 6th ed., J. B. Lippincott, Philadelphia, 1983.
9. R. P. Abergel, D. Pizzurro, C. A. Meeker, G. Lask, L. Y. Matsuoka, R. R. Minor, M. L. Chu, and J. Uitto, *J. Invest. Dermatol. 84*:384-390 (1985).
10. I. K. Cohen, R. F. Diegelmann, and H. R. Keiser, *The Ultrastructure of Collagen* (J. J. Longacre, ed.), Charles C. Thomas, Springfield, IL, 1973, pp. 199-212.
11. S. Bazin, C. Nicoletis, and A. Delaunay, *Biology of Fibroblast*, (E. Kulonen and J. Pikkarainen, ed.), Academic Press, London, 1973, pp. 571-578.
12. A. J. Bailey, S. Bazin, T. J. Sims, M. LeLous, C. Nicoletis, and A. Delaunay, *Biochim. Biophys. Acta, 405*:412-421 (1975).
13. L. Weber, W. N. Meigel, and W. Spier, *Arch. Dermatol. Res., 261*:63-71 (1978).
14. J. N. Clore, I. K. Cohen, and R. F. Diegelmann, *Biochim. Biophys. Acta, 586*:384-390 (1979).
15. D. P. Craig, J. D. Schofield and D. S. Jackson, *Br. J. Surg., 62*:741-744 (1975).
16. R. F. Diegelmann, I. K. Cohen, and B. J. McCoy, *J. Cell Physiol., 98*:341-346 (1979).
17. I. K. Cohen, H. R. Keiser, and A. Sjoerdsma, *Surg. Forum, 22*:488-489 (1971).
18. P. H. Fleckman, J. J. Jeffrey, and A. Z. Eisen, *J. Invest. Dermatol., 60*:46-52 (1973).
19. E. D. Harris and S. M. Krane, *N. Engl. J. Med., 291*:557-563, 605-609, 652-661 (1974).
20. S. I. Rennard, L. E. Stier, and R. G. Crystal, *J. Invest. Dermatol., 79*(Suppl. 1): 77s-82s (1982).
21. B. J. McCoy and I. K. Cohen, *Connect. Tissue Res., 9*:181-185 (1982).
22. R. F. Diegelmann, C. P. Bryant, and I. K. Cohen, *Plast. Reconstr. Surg., 59*:418-423 (1977).
23. R. S. Bienkowski, M. J. Cowan, J. A. McDonald, and R. G. Crystal, *J. Biol. Chem., 253*:4356-4363 (1978).
24. E. M. L. Tan, L. Ryhänen, and J. Uitto, *J. Invest. Dermatol., 80*:261-267 (1983).
25. D. W. Rowe, R. C. Moen, J. M. Davidson, P. H. Byers, P. Bornstein, and R. D. Palmiter, *Biochemistry, 17*:1581-1590 (1978).
26. R. C. Moen, D. W. Rowe, and R. D. Palmiter, *J. Biol. Chem., 254*:3526-3530 (1979).
27. G. T. Merlino, C. McKeon, B. deCrombrugghe, and I. Pastan, *J. Biol. Chem., 258*: 10041-10048 (1983).
28. J. Uitto, A. F. Perejda, R. P. Abergel, M-L. Chu, and F. Ramirez, *Ann. NY Acad. Sci. 460*:514-516 (1986).
29. E. M. Southern, *J. Mol. Biol., 98*:503-517 (1975).
30. P. S. Thomas, *Proc. Natl. Acad. Sci. USA, 77*:5201-5205 (1980).
31. C. Junien, C. Huerre, and M. O. Rethore, *Am. J. Hum. Genet., 35*:584-591 (1983).
32. R. Dalgleish, B. C. Trapnell, R. G. Crystal, and P. Tolstoshev, *J. Biol. Chem., 257*: 13816-13822 (1982).
33. P. Tsipouras, J. C. Myers, F. Ramirez, and D. J. Prockop, *J. Clin. Invest., 72*:1262-1267 (1983).
34. R. Fleishmajer, J. S. Perlish, T. Krieg, and R. Timpl, *J. Invest. Dermatol., 76*:400-403 (1981).
35. G. R. Botstein, G. K. Sherer, and E. C. LeRoy, *Arthritis Rheum., 25*:189-195 (1982).
36. J. P. Russell and W. S. Witt, *Plast. Reconstr. Surg., 57*:207 (1976).
37. J. D. Russell, S. B. Russell, and K. M. Trupin, *J. Cell Physiol., 93*:389-394 (1977).

38. S. B. Russell, J. D. Russell, and J. S. Trupin, *J. Biol. Chem., 257*:9525-9531 (1982).
39. J. S. Trupin, S. B. Russel, and J. D. Russel, *Collagen Rel. Res., 3*:13-23 (1983).
40. J. Uitto, E. M. L. Tan, and L. Ryhänen, *J. Invest. Dermatol., 79*(Suppl. 1):113s-120s (1982).
41. J. Uitto, L. Ryhänen, E. M. L. Tan, A. I. Oikarinen, and E. Zaragoza, *Fed. Proc., 43*:2815-2820 (1984).
42. E. Harper, *Annu. Rev. Biochem., 49*:1063-1073 (1980).
43. A. Oikarinen and J. Uitto, *Connective Tissue Disease: The Molecular Pathology of the Extracellular Matrix* (J. Uitto and A. J. Perejda, eds.), Marcel Dekker, New York, 1986, Chapter 17, pp. 385-397.
44. I. K. Cohen, R. F. Diegelmann, and M. L. Johnson, *Surgery, 82*:15-20 (1977).
45. J. Uitto and K. K. Mustakallio, *Biochem. Pharmacol., 20*:2495-2504 (1971).
46. K. I. Kivirikko and R. Myllylä, *The Enzymology of Post-Translational Modifications of Proteins* (R. G. Friedman and D. A. Hawkins, eds.), Academic Press, London, 1980, pp. 53-104.
47. D. J. Prockop, R. A. Berg, K. I. Kivirikko, and J. Uitto, *Biochemistry of Collagen* (G. N. Ramachandran and A. H. Reddi, eds.), Plenum, New York, 1976, pp. 163-273.
48. H. Anttinen, L. Ryhänen, and A. Oikarinen, *Biochim. Biophys. Acta, 609*:321-328 (1980).
49. H. Anttinen, L. Ryhänen, U. Puistola, A. Arranto, and A. Oikarinen, *Gastroenterology, 86*:532-539 (1984).
50. J. Rosenbloom and D. J. Prockop, *J. Biol. Chem., 245*:3361-3368 (1970).
51. J. Uitto and D. J. Prockop, *Biochim. Biophys. Acta, 336*:234-251 (1974).
52. J. Rosenbloom and D. J. Prockop, *J. Biol. Chem., 246*:1549-1555 (1971).
53. C. Fiedler-Nagy, P. Bruckner, T. Hayashi, P. P. Fietzek, and D. J. Prockop, *J. Biol. Chem., 257*:4181-4188 (1982).
54. P. Dehm and D. J. Prockop, *Biochim. Biophys. Acta, 264*:375-381 (1972).
55. D. Alarco-Segovia, G. Ibanz, D. Kershenobich, and M. Rojkind, *Lancet, 1*:1054-1055 (1974).
56. E. D. Harris, Z. Werb, E. Cartwright, and J. J. Reynolds, *Scand. J. Rheumatol., Suppl. 8*:15-24, 1975 (abstract).
57. J. H. Fessler and L. I. Fessler, *Annu. Rev. Biochem., 47*:129-162 (1978).
58. L. Ryhänen, E. M. L. Tan, S. Rantala-Ryhänen, and J. Uitto, *Arch. Biochem. Biophys., 215*:230-236 (1982).
59. A. I. Oikarinen, E. J. Zaragoza, L. Ryhänen, and J. Uitto, *Biochem. Pharmacol., 33*:695-697 (1984).
60. D. R. Eyre, *Science, 207*:1315-1322 (1980).
61. A. J. Arem, R. Misiorowski, and M. Chvapil, *J. Surg. Res., 27*:228-232 (1979).
62. G. Asboe-Hansen, *Acta Dermatol.-Venereol. (Stockh.), 59*:465-467 (1979).
63. E. J. Moynahan, *Prosgrad. Med. J., August suppl.*:39-40 (1974).
64. K. Deshmukh and M. E. Nimni, *J. Biol. Chem., 244*:1787-1795 (1969).
65. E. A. Bauer and T. W. Cooper, *Arch. Dermatol., 117*:529-530 (1981).
66. R. P. Abergel, C. A. Meeker, L. S. Lam, R. M. Dwyer, M. A. Lesavoy, and J. Uitto, *J. Am. Acad. Dermatol., 11*:1142-1150 (1984).
67. R. P. Abergel, R. M. Dwyer, C. A. Meeker, G. Lask, A. P. Kelly, and J. Uitto, *Lasers Surg. Med., 4*:291-295 (1984).
68. P. Bailin, *Cutaneous Laser Therapy: Principles and Methods* (K. A. Arndt, J. M. Noe, and S. Rosen, eds.), Wiley, New York, 1983, pp. 187-200.
69. R. P. Abergel, M-L. Chu, E. A. Bauer, and J. Uitto (submitted for publication).
70. D. J. Castro, R. P. Abergel, K. J. Johnston, G. E. Adomian, R. M. Dwyer, J. Uitto, and M. A. Lesavoy, *Ann. Plast. Surg., 11*:131-140 (1983).
71. J. Uitto, A. J. Perejda, R. P. Abergel, M-L. Chu, and F. Ramirer, *Proc. Natl. Acad. Sci. USA, 82*:5935-5939 (1985).

16

Pharmacological and Environmental Modulation of Wound Healing

Oscar M. Alvarez
Bioderm Medical Inc.,
Princeton, New Jersey

I. INTRODUCTION

At the instant of injury, a complex series of events occur that, in the healthy individual, inevitably proceed to repair and eventually to scar formation. Immediately after injury, the wound begins its life as an organ dedicated to cellular regeneration and tissue repair. The major mysteries of repair and regeneration have always been and continue to be: what in the injury stimulates the healing response? And why does the healing stop? Wounds clearly respond to many stimuli. It is well accepted now that the control of repair and regeneration rest in the cells and their reparation behavior.

In the scope of this chapter the events that occur after injury will be briefly described, but greater effort will be made to review the current knowledge about molecular mediators, environmental factors, and pharmacological agents that influence the wound healing process.

II. MORPHOLOGY OF HEALING TISSUE

When tissue is disrupted, vessels are injured, cells are damaged, and platelets and collagen interact. The complement cascade is initiated, and the local microvasculature shows its response to injury. The injured vessels thrombose. Nearby vessels, especially the venules, dilate. Platelets and white cells begin to adhere to the endothelial lining, and leukocytes migrate between the endothelial cells accumulating in the area of injury. Shortly thereafter the edges of the defect are infiltrated with granulocytes and macrophages.

Within a couple of days, depending on the extent of the injury, fibroblasts become visible (1). These fibroblasts increase their population in the local area, and as they do, the amount of total collagen deposited increases.

Angiogenesis is constant throughout repair. Endothelial buds can be seen in new tissue as early as two days in certain wounds, and by the third or fourth day the new circulation can be seen throughout the wound space (2). As the new blood vessels provide nutrition to the regenerating tissues, the epithelium begins its migratory process to resurface the defect. In a partial-thickness wound, if viable epidermis remains in dermal orifices such as hair follicles or sweat ducts, migration of the defect will occur from these epidermal depots as well as from the edges of the defect. In a full-thickness wound, of course, the migration of epidermis is entirely from the edges. In a wound closed by primary intention the epidermis migrates across the incision as early as 24 hr after wounding.

A wound is considered healed when appropriate resurfacing has been achieved, continuity across the wound has been reestablished, and a degree of tissue strength approaching normal has resulted. Resurfacing is achieved by epithelialization and contraction of the wound, while continuity is reestablished by the production of new connective tissue. Tensile strength arises from the deposition of an adequate amount of new and remodeled collagen. Each of these processes is measured differently, and the progress of each may vary with the location, depth, size, and configuration of a wound.

III. THE ACUTE RESPONSE

The vascular response to injury, whether from trauma, physical agents, or bacterial infection, is remarkably uniform. In most injuries, there is a physical interruption of blood vessels and immediate hemorrhage. Small vessels in the area of injury constrict, and at the point of injury, actual vascular occlusion may occur, which impedes further loss of blood. This vasoconstriction usually lasts only 5-10 min and is followed by an active vasodilation (3).

Coincident with vasodilation, and before any cells actually leave the injured vessels, leakage of fluid from venules occurs. This fluid has the same composition as plasma (4). Electron microscopy of wounded vessels indicates that such leakage appears to be facilitated by separation of endothelial cells, which lose direct contact with each other, thus exposing the basement membrane to the luminal contents of the vessel (5).

Lymphatics are more fragile than blood vessels; thus, in any injury, damage to local lymphatics is usually greater than to the vasculature. Leakage of fluid from venules provides fibrinogen and other elements of the blood-clotting system. Fibrin plugs quickly form in damaged lymphatics, effectively stopping any drainage from the injured area.

Local vasodilation, leakage of fluid into the extravascular space, and stoppage of lymphatic drainage produce the classical signs of inflammation—redness, swelling, and heat. Pressure, and perhaps chemical stimulation, produces the fourth sign—pain (6).

IV. LOCAL MEDIATORS OF INFLAMMATION: HISTAMINE, SEROTONIN, KININS, AND PROSTAGLANDINS

The vascular response to injury is probably mediated by several chemical agents. There is substantial evidence that histamine is responsible for the initial, short-lived increase in vascular permeability (7). A number of endogenous compounds will release histamine or cause histamine to be made through the action of histidine decarboxylase. Local increases in capillary permeability are seen following injection of these substances (8). In addition, wound fluid contains appreciable amounts of histamine, and blood histamine levels increase immediately after injury (9).

The major, but not the sole, source of histamine is the mast cell (10). At the time of injury, these cells lose their characteristic granules, which contain a wide variety of bioactive materials including 5-hydroxytryptamine (serotonin), heparin, and histamine (11). Histamine is also found in platelets, but there is wide species variation in platelet histamine content (12).

Histamine action is very short-lived, probably lasting no longer than 30 min (13). Most likely, injury rapidly depletes local sources of histamine so that considerable time is required before sufficient endogenous histamine can be synthesized to bring about further reactions.

It is important to emphasize that the action of these amines, histamine and serotonin, is not on capillaries. As demonstrated by electron microscopy (13), the effect is on vessels 20-30 μm in size on the venous side of capillary loops. Vessels 4-7 μm, true capillaries, are unaffected by these amines (13). These amines modulate vascular permeability by causing a swelling or "rounding" of endothelial cells, leading to a separation of contacts between them. The basement membrane is not visibly affected, but acts as a filter at the points of exposure (14).

Kinins are biologically active peptides that are involved in the inflammatory process and are found in areas of tissue injury. The kinins (bradykinin or kallidin) are released from an α-globulin of plasma, which has also been termed a kininogen, by a plasma enzyme, kallikrein. Kallikrein is activated by Hageman factor (Factor XII), which, in turn, is activated by contact with glass or by other negatively charged insoluble substances, such as collagen. Recently, a kallikrein has been found in granulocytes, and it is suggested that release of this kallikrein may play a role in initiation of the inflammatory response (15).

The kinins appear to act on the microvasculature in a manner quite similar to histamine and serotonin. They are short-lived and are rapidly destroyed by plasma and tissue proteases (16). Thus, their role is probably confined to the early, rather than the later, phase of vascular response during the healing process.

Although the first prostaglandin was discovered in 1930 (17), it was not until nearly 40 years later that the ubiquitous nature of these substances was recognized and the role of certain members of this group of compounds in inflammation delineated (18). Prostagladin E1 and E2 (PGE1 and PGE2) have strong vasodilative, permeability-increasing, and lymphflow-promoting properties. These two prostaglandins, along with PGF1 and PGF2, can induce a wheal-and-flare response in human skin (17). PGE2 also exerts a chemotactic effect on rabbit polymorphonuclear cells (19).

It is now suggested that prostaglandins, particularly PGE1 and PGE2, are terminal mediators of the acute inflammatory response (20). Release of these substances requires the presence of the complement system and follows the phase of kinin formation (21,22). Aspirin and indomethacin are potent inhibitors of prostaglandin synthesis, and this activity probably underlies the antiinflammatory nature of these drugs.

Prostaglandins have been shown to play an important role in producing and prolonging the inflammatory reaction (17). They have also been implicated as mediators of wound healing and scar tissue formation (23,24). Recent studies evaluating the effect of nonsteroidal antiinflammatory agents (cyclooxygenase and lipooxygenase inhibitors) suggest that, in contrast to corticosteroids, such agents do not affect the rate of healing but may have a significant effect on connective tissue remodeling (25).

V. THE REPAIR PHASE: REEPITHELIALIZATION

When a defect is made in the epidermis, the basal cells appear to flatten and move toward the area of deficit. Mitoses appear at the wound edge including the damaged follicle. Migration of epithelium may occur in a number of ways. Most commonly accepted is the "leapfrogging" mechanism. In this process, the migrating epithelial cells appear to move by sliding or rolling over one another (26). During migration, epithelial cells synthesize the components of underlying basement membrane in an organized and orderly fashion. The first such component to be produced is bullous pemphigoid (BP) antigen (27), which can be identified as early as the second day following a partial-thickness wound. Laminin is subsequently synthesized by the basal cell and can be found in the subepidermal location (lamina lucida region) as early as the third day postwounding (27). The last component to be synthesized, type IV collagen, comprises a portion of lamina densa and is not seen until the epidermis has fully covered the defect (4-6 days postwounding). As the wound becomes covered with migrating epithelium, the basal cells begin to assume a more cuboidal shape and start dividing (28). Daughter cells differentiate upward forming once again the characteristic module of the epidermis. Epidermal cells can migrate only over viable tissues since they require blood and a nutritional supply to meed their energy needs. Since the migrating "tongue" of epithelium must burrow between any eschar (crust) and the underlying living tissue, the presence of such a crust impedes resurfacing. Thus, if the wound is kept moist and is protected from the external environment so that desiccation and crust formation are prevented, epithelialization will occur on its surface at a more rapid rate (29,30). Indeed, superficial burns heal more rapidly under an intact blister roof, and a variety of polymeric occlusive or semiocclusive dressings have been shown to enhance epidermal and dermal repair (Table 1). In addition, many topical agents affect wound healing; even those that are thought to be "inert" can significantly influence the healing of superficial wounds (Tables 2-4) (31,32).

Previously, it was thought that atmospheric oxygenation was necessary for epidermal repair, but recent studies have indicated that molecular oxygen (oxygen in the blood) is probably sufficient to provide adequate oxygenation for resurfacing to occur in normally healing tissues (30,33).

Epidermal regeneration and dermal repair most often occur simultaneously and are, to some extent, interdependent. For example, if dermal repair is experimentally altered by using a proline analog that results in the formation of abnormal collagen, the epidermis will not readily migrate over this fraudulent connective tissue (34). On the other hand, evidence for some independence of these processes is seen by the action of hydrocortisone, which inhibits the synthesis of new collagen during repair, yet fails to alter the rate of epidermal resurfacing of superficial wounds (25).

VI. DERMAL REPAIR

The fibroblast synthesizes and secretes many elements of the connective tissue matrix such as collagen, elastin, and proteoglycans. The fibroblast can survive hostile environments, but cannot function optimally in environments that do not include a collagenous or connective tissue matrix. There has been some controversy regarding the origin of fibroblasts in wounds because the type of fibroblast seen in the actively healing wound does not appear to be present in normal tissue (35). This immature fibroblast is more

Table 1 Nonwoven Dressings (Occlusive)

Agent	Relative rate of healing (%)[a]
Saran (polyvinylidene)	+31
Glad (polyethylene)	+24
Op-site (polyurethane)	+18
Tegaderm (polyurethane)	+21
Bioclusive (polyurethane)	+20
Vigilon (hydrogel)	+33
Geliperm (hydrogel)	+34
Spenco (hydrogel)	+31
Stomahesive (hydrocolloid)	+33
Duoderm (hydrocolloid)	+36
Bioplast (hydrocolloid)	+36
Comfeel (hydrocolloid)	+33
Hollihesive (hydrocolloid)	+30
Synthaderm (hydrophylic foam)	+22
Surgihesive (polyurethane foam)	+14
Helicote (collagen/polyurethane)	+25
Telfa (nonwoven-absorbent)	+14

[a]Compared to untreated.

Table 2 Agents that Increase Epidermal Resurfacing

Agent	Relative rate of healing (%)[a]
Silvadene cream	+28
Polysporin cream	+20
Neosporin ointment	+28
Benzoyl peroxide (20%)	+33
SRF (1%)	+29
Shark liver oil	+16
Cod liver oil	+33
Allantoin	+37
Bacitracin zinc	+30
Zinc oxide ointment (Desitin)	+20
Oxybenzone	+29
β-Carotene	+18
Methyl isobutyl xanthine	+31
J & J First aid cream	+20

[a]$p < 0.05$ compared to untreated.

Table 3 Agents That Have Little Effect on Epidermal Migration

Agent	Relative rate of epidermal migration (%)[a]
Low-melting-point petrolatum	+4
Providone iodine solution	+2
70% ETOH	+7
Eucerin	+5
Solubase	+4
Collagen hydrolysate	+2
Collagen matrix	+5
Ibuprofen	+1
Meclofenamate sodium	−5
Hydrocortisone (0.1%)	+3
Benzoyl peroxide (10%)	+7
DMSO (2%, 10%, 50%, 95%, 100%)	+2
Physiological saline	+1
Serum-free tissue culture medium (RPMI 1640, Dulbeccos' MEM)	+2
Distilled water	+1
Ca-Mg-free tissue culture medium	+4
PEG	+3
Ringers' solution	+5
O_2 saturated gel	+7
Prostaglandin E_2	−6
Prostaglandan $F_{2\alpha}$	+2
Thrombostat	−2
Thrombinar	−1

[a]Compared to untreated.

Table 4 Agents That Delay Epidermal Resurfacing

Agent	Relative rate of healing(%)[a]
Triamcinolone acetonide ointment (0.1%)	−34
Furacin	−30
Liquid detergent	−28
L-azetidine-2-carboxylic acid	−78
Panoxyl venicle gel	−11
Benzoyl peroxide (50%)	−39
Ichthammol	−10
Neomycin	− 5
Hebiclens	− 7

[a]$p < 0.05$ compared to untreated.

round and less spindle shaped than the mature fibroblast and does not seem to have the myofibrils associated with movement (35). However, these cells, often called mesenchymal cells or fibrocytes, may become fibroblasts (in most instances), but also may become adipocytes or smooth muscle cells. It should be emphasized that all fibroblasts found at the site of injury originate from wounded tissue, as demonstrated by experiments utilizing radiation inhibition of cells in wounded skin (36). Most repair-associated fibroblasts appear to arise from perivascular cells (36-38). Tissue that is relatively acellular converts to tissue that is very cellular within a couple of days. This response is, in fact, very similar to tumorigenesis, and many investigators have noted the analogy between a healing wound and a vascularized tumor or foreign body (39,40).

The fibroblast has a full complement of metabolic pathways; it has a full range of enzymes for the Krebs cycle, as well as the glycolytic pathways. The varied metabolic requirements of fibroblasts are met by circulating sugars, fats, amino acids, oxygen, ascorbate, and other nutrients. Furthermore, the cell itself is pinocytic and may supply some of its glucose and amino acid needs through hydrolysis of more complex molecules. The mature fibroblast contains an abundance of endoplasmic reticulum, Golgi apparatus, and mitochondria as do other protein-producing cells (35). It is mobile and migrates both in tissue culture and in vivo. The fibroblasts's mobility in vitro is regulated by contact inhibition; however, its movement in vivo is more complex and features a very sophisticated cell-to-matrix interaction.

VII. COLLAGEN

Studies on the composition of certain tissues with respect to genetic collagen type have demonstrated an appreciably higher type III collagen content in embryonic or fetal tissues versus their adult counterparts (41).

In adult human skin, for example, 85% of dermal collagen is type I whereas only 15% is type III. Yet, early in fetal development, both collagens are present in nearly equal amounts (41-43). Subsequently, it has been reported that guinea pig scar collagen contains a high ratio of type III:type I collagen and that collagen formed in foreign body granulomas is predominantly type III. Studies performed on normal human scars revealed that, early in scar formation, type III is the principal collagen synthesized in the wound. During maturation of the scar, type III collagen is gradually replaced by type I collagen (44), and eventually, the normal adult ratio of these collagen types is reestablished. Studies of hypertrophic scars in humans have shown that type III collagen is also deposited early, but its replacement by type I collagen is somehow delayed or even absent (45). These observations suggest that when fibroblasts are first activated in response to a wound, probably by recruitment of more undifferentiated or quiescent mesenchymal cells, the gene for type III collagen is activated while that for type I is suppressed.

New collagen can be found in healing wounds are early as the first day after injury. The peak rate of synthesis in a full-thickness wound seems to occur about the fifth or seventh day after injury (46). However, in partial-thickness skin wounds, the peak rate of collagen biosynthesis is observed somewhat earlier (around the second or third post-injury day) (47). Collagen synthesis may continue at an elevated rate for many months after injury, often up to six months or a year (35).

VIII. FACTORS THAT REGULATE CELL MIGRATION

Cells for the most part respond to chemicals in their environment by altering their metabolism, movement, differentiation, or proliferation. Many cells respond to injury by migration. This movement, which occurs along a chemical gradient, is called chemotaxis.

Upon injury the damaged tissue is inhabited sequentially by specialized cells that repair the defect. The cells involved include blood cells (leukocytes, monocytes, platelets) and cells from neighboring tissues (fibrocytes, endothelial cells, smooth muscle cells). These cells invade the injured tissue in a somewhat regulated sequence and each has a specific function.

Several chemotactic agents have been identified. Many of these agents are produced by the reactions that occur during coagulation (48), inflammation (49), and complement activation (50). Leukoattractants such as C5A, elastic peptides, platelet factor 4, and formylmethionyl peptides are found in wound fluid early during the inflammatory phase (Table 5).

The growth and proliferation of endothelial cells that occurs during the repair process is probably regulated by angiogenesis factors, which may also be potent chemoattractants. Angiogenesis factors have been isolated from normal tissue (51), conditioned medium from adipocytes in culture (52), tumor cells and tumor tissue (53), follicular fluid (54), cultured macrophages (55), and experimentally induced granulation tissue (56).

The matrix component, e.g., collagen, fibronectin, and laminin, may also play a big role in directing the growth and migration of endothelial cells. For instance, endothelial cells grown on type I collagen matrix grow in a monolayer across the matrix substrate (57). When the same endothelial cells are grown in collagen type IV, the cells form a tubelike structure that resembles a capillary bud (58).

Fibroblasts and other connective tissue cells are probably attracted to collagen and/or collagen fragments during the fibroblastic phase of the repair process (59). In addition, different factors derived from platelets may be chemoattractants to connective tissue cells. Platelet-derived growth factor (PDGF) has been shown to be a potent chemotactic agent to fibroblasts and smooth muscle cells (60). PDGF also induces fibroblasts to undergo DNA synthesis and cell division (61).

Table 5 Chemoattractants Present During Wound Repair

Chemoattractants	Target cells	Functions of target cells
C5A, platelet factor 4, elastin peptides, F-Met peptides	Neutrophils, monocytes	Phagocytes
Fibronectin, PDGF	Fibroblasts, smooth muscle cells	Matrix-producing cells
Lymphokines, monokines, complement peptides		
Fibronectin, laminin, monokines	Endothelial cells	Vascular system

Source: Reproduced from Grotendorst et al. (64). Permission has been requested from author and publisher.

Recently, factors that influence epidermal cell migration have also been identified. One such agent is derived from platelets and is yet to be isolated, but is distinctly different from PDGF (62). Stenn (63) has described an epidermal cell spreading factor which has been named epibolin. Epibolin is distinct from other identified agents that influence epidermal cells and is probably synthesized by platelets as well as other blood-derived cells.

IX. FACTORS INVOLVED IN CELL GROWTH DURING WOUND REPAIR

Once connective tissue-producing cells migrate into the wound, they proliferate. The growth of these cells can be influenced by several factors, but polypeptide growth factors are probably the most important promoters of fibroblast and smooth muscle cell proliferation (64) (Table 6). For example, PDGF is produced within minutes after wounding, and the responding cells are fibroblasts.

Other growth factors influencing fibroblasts are monocyte-derived growth factor, endothelial cell-derived growth factor, and alveolar macrophage-derived growth factor (64). The time of appearance of these growth factors varies in accordance with specific wound conditions, but will generally be present in wound fluid between 24 and 96 hr after wounding. These growth factors have been grouped into two functional categories (64,65). The first category consists of competence factors, and the second are called progression factors. The competence factors are involved in the activation of quiescent cells, enabling them to respond to the progression factors (Figure 1).

PDGF is a competence factor present during the very early stages of healing. This factor is released from platelet granules as they aggregate to collagen and to each other. PDGF binds to specific receptors that are present on connective tissue cells.

Macrophages have been reported to release both competence and progression factors. Recently, Glenn and Ross described a macrophage-derived competence factor. This agent does not seem to be synthesized by peripheral blood monocytes without activa-

Table 6 Growth Factors Present During Wound Repair

Factor	Source	Time of appearance	Responsive cells
PDGF[a]	Platelets	Within minutes in smooth muscle	Fibroblasts
MDGF[b]	Monocytes	24-48 hr	Fibroblasts
AMDGF[c]	Alveolar macrophages		
MDECGF[d]	Macrophages	24-48 hr	Endothelial cells
ECDGF[e]	Endothelial cells	5-7 days	Fibroblasts

[a]Platelet-derived growth factor.
[b]Monocyte derived growth factor.
[c]Alveolar macrophage-derived growth factor.
[d]Macrophage-derived endothelial cell growth factor.
[e]Endothelial cell-derived growth factor.
Source: Reproduced from Grotendorst et al. (64). Permission has been requested from author and publisher.

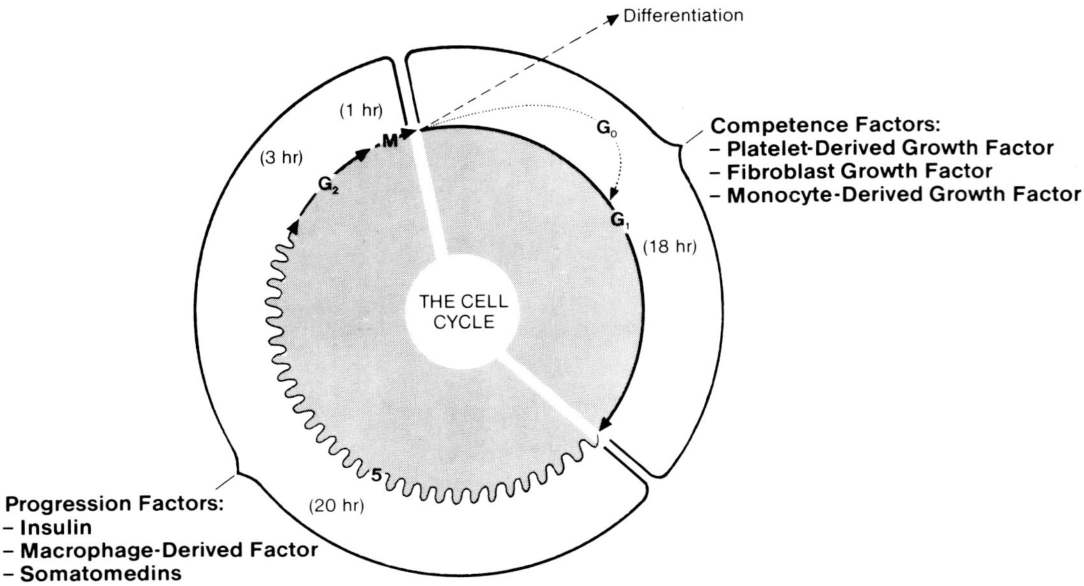

Figure 1 The growth of connective tissue cells is regulated by growth factors that act through specific cell surface receptors. There are two classes of growth factors: "competence" and "progression" factors. Competence factors act during the G0-G1 phase of the cell cycle preparing the cell to enter the S phase. However, the cells cannot enter the S phase unless progression factors are present. Factors from both classes must be present in order for cell replication to occur. (Reproduced from Ref. 64. Permission has been requested.)

tion with substances such as lipopolysaccharides. Therefore, it is assumed that this factor would be present during the inflammatory phase of the healing process. Progression factors derived from alveolar macrophages have also been described by Bitterman et al. (67). Because both types of factors can be produced by macrophages, it is likely that connective tissue cell proliferation can be stimulated by macrophages alone in the absence of any other growth-promoting agent derived from blood.

PDGF has been demonstrated to stimulate the production of collagen in an experimentally induced granuloma in rabbits (64). This stimulation in collagen synthesis is more pronounced during the early days of granulation (day 4 or 5 after implantation) than during the later days (day 20).

X. ENDOTHELIAL CELLS AND ANGIOGENESIS

Neovascularization is seen in most areas that have been injured and undergo an inflammatory reaction. The function of angiogenesis in wounds is to nourish the tissue that has been damaged or the newly synthesized tissue that has replaced eschar. Angiogenesis can occur in three ways during wound healing: (a) by the generation of an entirely new vascular network where a large tissue defect has to be filled, (b) by the coupling of an unused circulation, e.g., when the host bed provides circulation to a skin graft, and (c) by the joining or rejoining of vessels across the span of a healing wound.

Delineation of the mechanisms whereby neovascularization is stimulated during wound healing is being studied in several laboratories. Recently, Knighton and co-workers described the isolation and characterization of an angiogenesis factor from wound fluid (55). It appears that this factor is synthesized by macrophages at the site of injury and may be regulated by oxygen tension.

Endothelial cells do not seem to be affected by PDGF, but have been shown to be stimulated by factors present in brain tissue (68), serum (69), retina (51), and tumors (70). Since angiogenesis requires both the migration and proliferation of endothelial cells, it is likely that factors which act as chemoattractants also promote proliferation.

XI. MATRIX COMPONENTS

Connective tissue matrix is composed of collagens, glycoproteins, and proteoglycans, which interact with each other and with cells native to that tissue. It is well recognized that each component has a particular function that contributes to the overall behavior of the specialized tissue cells. For example, collagens are the structural components that provide the lattice network of the organ. The glycoproteins are primarily attachment factors and seem to be responsible for cell-to-substrate adhesion (71). The proteoglycans appear to have variable functional attributes ranging from hydration to basement membrane stability. Cells do not bind directly to collagen, but interact via glycoproteins called attachment factors (71). These factors are large glycoproteins that contain specific binding sites for collagens, proteoglycans, and cell surface receptors. Fibronectin is the best characterized attachment factor to date. It functions as an attachment protein for fibroblasts and smooth muscle cells to collagen types I and III (72). It also may function in the attachment of macrophages to collagen and to debris during phagocytosis (73).

Fibronectin may be involved in several phases of the wound-healing process. For example, during clotting fibronectin is enzymatically linked to fibrin (72). After cleavage, fibronectin and its fragments have been demonstrated to be chemoattractants for fibroblasts and macrophages (74).

The attachment of endothelial and epithelial cells to basement membrane is mediated by laminin (75). It has been demonstrated that laminin is synthesized by epithelial cells as well as smooth muscle cells (27,77) and is present only in basement membranes.

Attachment factors are believed to influence the phenotypic expression of the cells with which they associate (78). In a wound, for example, the selection of cells that invade the defect, as well as the functional capacity of the cells to which they are attached, may be regulated by adhesive glycoproteins or attachment factors.

XII. WOUND CONTRACTION AND SCAR FORMATION

During contraction the size of a full-thickness wound is diminished by the centripetal movement of the surrounding skin. It is a dermal process whereby the full depth of surviving dermal tissue at the wound edge moves to decrease the size of the defect. The processes of epithelialization and wound contraction are independent; one can occur without the other (79). In contracting wounds, epithelialization may be regarded as a temporary repair mechanism aimed at providing a barrier to infection from without and loss of fluid from within. In a fully contracted wound, much of this migrated epithelium is ultimately lost (80).

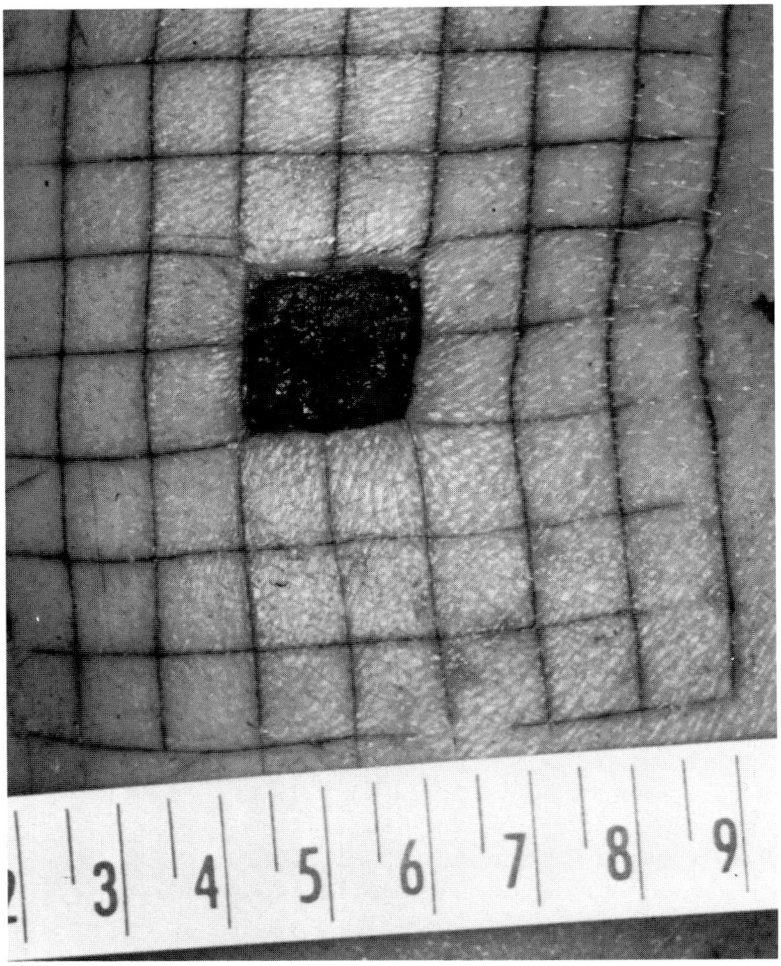

(A)

Figure 2 Photographs of full-thickness excisional wound made on an 8 X 8 cm tattooed grid on domestic pig skin. This sequence depicts healing by both granulation and contraction. (A) Day 5 after wounding. (B) Day 12 after wounding. (C) Day 20 after wounding. (D) Day 45 after wounding. Note the alterations in the shape of each of the grids immediately surrounding the wound.

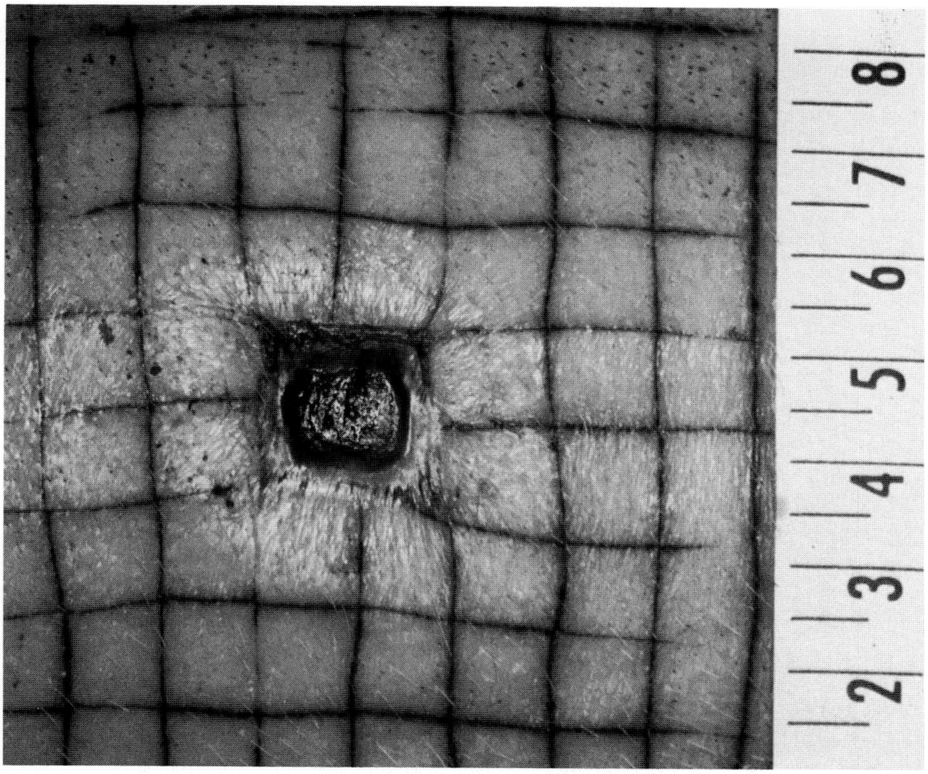

(B)

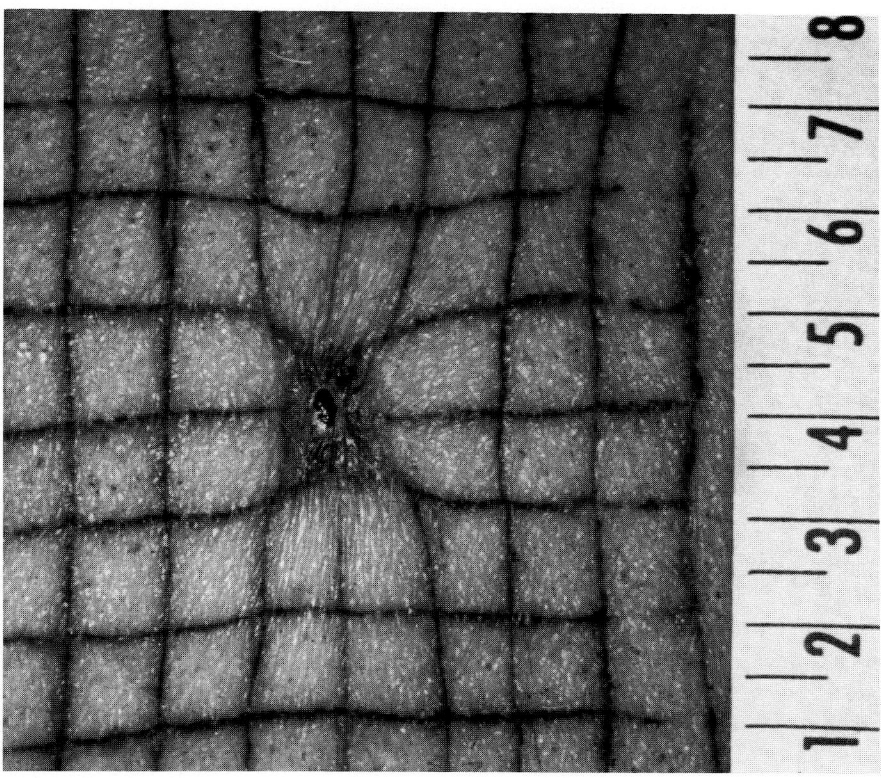

(C)

Figure 2 (Continued)

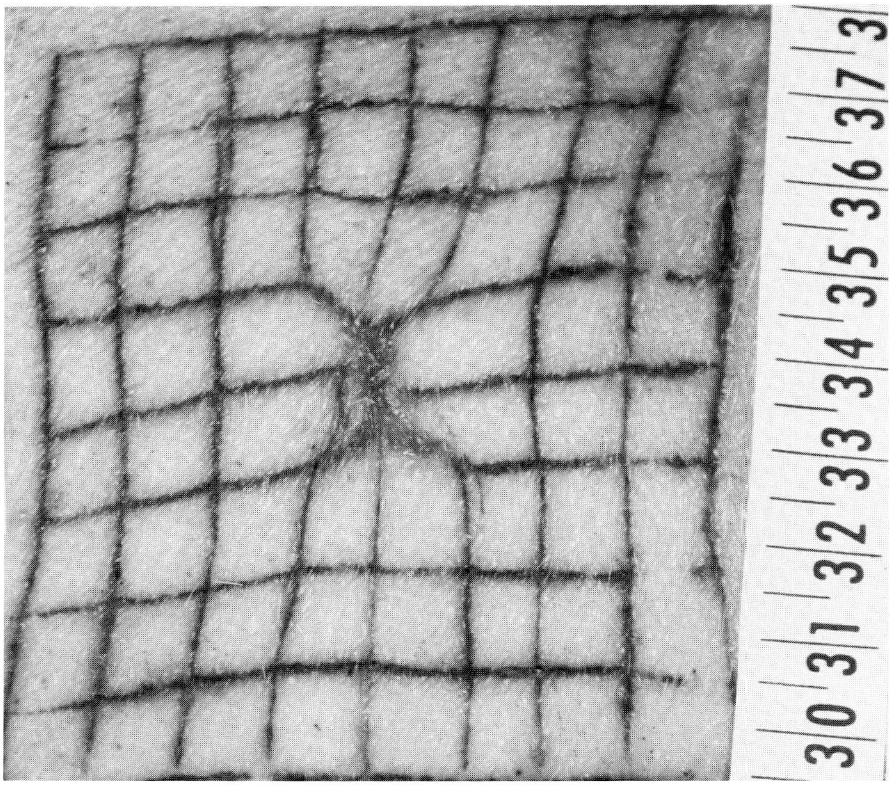

(D)

Figure 2 (Continued)

In lower mammals, contraction is the major process by which surface wounds are healed. In humans, contraction does not often go to completion except in very small wounds; however, it may result in deformity and loss of function, the extent of which depends on the size and location of the original wound. It must be emphasized that contraction involves movement of existing tissue at the wound edge, not the formation of new tissue. It can be appreciated, therefore, that as contraction proceeds, tissue surrounding the wound is thinned, stretched, and under tension. For instance, if a full-thickness square wound is made in the center of a tattooed grid and allowed to heal, one can obtain accurate measurements of changes in wound size and adjacent skin deformation by measuring the alterations in the shape of the grids surrounding the excision (Figure 2). In pigs some degree of wound contraction can be observed during the first five days after wounding. However, between day 5 and day 20 contraction is quite rapid. Wound contraction by grid deformation continues from this point on indefinitely, but is a much slower process (46,81).

The search for agents that may influence the scarring (wound contraction) process actively continues. The most promising to date are collagen lattices or matrices with or without seeded cells. Several investigators, using dermal equivalents comprised of collagen gels and/or fibrous collagen sponges, have demonstrated that the wound contraction process can be significantly inhibited in loose-skin and tight-skin animals (82-84) (Figure

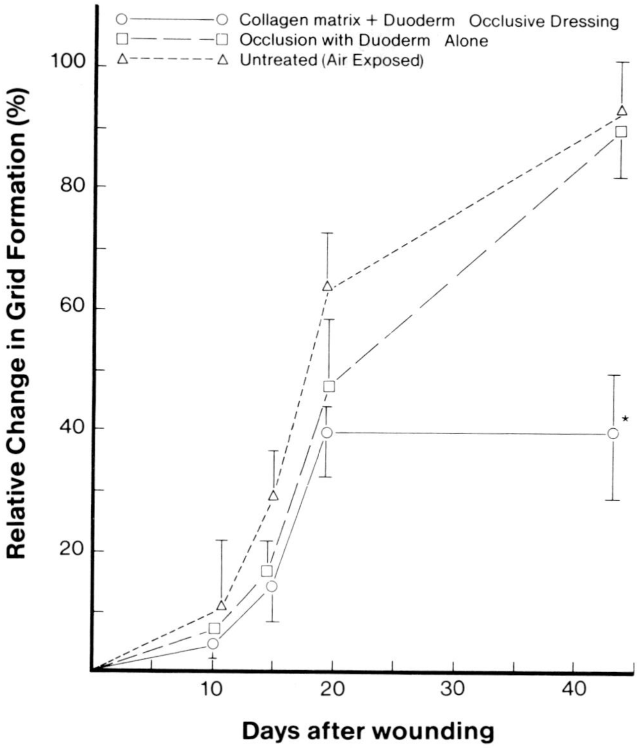

Figure 3 Effect of a type I collagen matrix sponge on total grid deformation as measured by computerized image analysis. $*p < 0.01$ compared to untreated. (Reproduced with permission from Ref. 46.)

3). Grillo et al. (83), in the 1950s, demonstrated that full-thickness wound contraction could be inhibited by connective tissue matrix alone. They found that after lysing of the cells by freezing and thawing, the excised tissue could be transplanted back to the donor and minimize the contraction process. It seems that if a matrix material is provided that favors the ingrowth of endogenous cells while not promoting a foreign-body reaction, wound contraction will be inhibited.

Corticosteroids as well as nonsteroidal antiinflammatory agents have also been shown to prevent the contraction of collagen lattices seeded with fibroblasts in vitro (85).

Pressure, and agents that provide pressure, to healing full-thickness excision wounds have also been implicated with contraction inhibition. However, it is not known whether the pressure provided directly affects the orientation of the collagen bundles, whether the pressure produces decreased vascularization and therefore less collagen production, or whether the pressure inhibits the movement of skin by contractile myofibroblasts.

REFERENCES

1. R. Ross and G. Odland, in *Repair and Regeneration* (J. E.Dunphy and W. Van Winkle, eds.), McGraw-Hill, New York, 1969.
2. P. I. Branemark, *Bibl. Anat.*, *4*:9 (1965).

3. M. J. Karnovsky, *J. Cell. Biol., 35*:213-236 (1967).
4. E. M. Landis and J. R. Pappenheimer, in *Handbook of Physiology*, Sect. 2, Vol. II (W. F. Hamilton and P. Dow, eds.), American Physiology Society, Washington, DC, 1969.
5. G. Majno and G. E. Palade, *J. Biophys. Biochem. Cytol., 11*:571-605 (1961).
6. G. B. Ryan and G. Majno, in *Inflammation* (B. A. Thomas, ed.), The Upjohn Co., Kalamazoo, MI, 1977, pp. 10-12.
7. J. V. Hurley, R. N. Ham, and G. B. Ryan, *J. Pathol., 94*:1-12 (1967).
8. G. B. Ryan, *Beitr. Pathol., 152*:272-291 (1974).
9. S. Bazin, M. Pelletier, and A. Delaunay, *Agents Actions, 3*:317 (1973).
10. B. Uvnas, *Ann. NY Acad. Sci., 116*:880 (1964).
11. W. E. Brocklehurst, *J. Physiol., 131*:416-435 (1960).
12. W. G. Spector and W. J. Tykke, *J. Path. Bact., 92*:103 (1966).
13. G. Majno, G. E. Palade, and G. Schoefl, *J. Biophys. Biochem. Cytol., 11*:607-626 (1961).
14. N. Simonescu, M. Simonescu, and G. E. Palade, *J. Cell. Biol., 64*:586-607 (1975).
15. M. E. Webster and T. V. Pierce, *Ann. NY Acad. Sci., 104*:91 (1963).
16. E. E. Peacock and W. Van Winkle, in *Wound Healing*, 2nd ed., W. B. Saunders, Philadelphia, 1976, pp. 1-21.
17. J. R. Weeks, *Annu. Rev. Pharmacol., 12*:317 (1972).
18. T. E. Pike, *Sci. Amer., 225*(5):84-92 (1971).
19. J. V. Hurley, *Ann. NY Acad. Sci., 116*:918-935 (1964).
20. R. D. R. Camp, *Clin. Exp. Dermatol., 7*:435-444 (1982).
21. A. O. Miles, *Proc. Roy. Soc. Lond. Biol., 173*:341-349 (1969).
22. H. J. Muller-Eberhard, *Annu. Rev. Biochem., 38*:389-414 (1959).
23. N. Blumenkrantz and J. Sondegaard, *Nature New Biol., 239*:246 (1972).
24. C. W. Castor, *J. Lab. Clin. Med., 85*:392-404 (1975).
25. O. M. Alvarez, R. D. Levendorf, R. V. Smerbeck, P. M. Mertz, and W. H. Eaglstein, *Fed. Proc., 43*:2793-2798 (1984).
26. W. S. Krawczyk, *J. Cell. Biol., 49*:247-263 (1971).
27. J. R. Stanley, O. M. Alvarez, E. W. Bere, W. H. Eaglstein, and S. I. Katz, *J. Invest. Dermatol., 77*:240-243 (1981).
28. E. W. Winstanley, *J. Pathol., 114*:155-162 (1974).
29. G. D. Winters, *Nature (London), 193*:293 (1962).
30. O. M. Alvarez, J. M. Hefton, and W. H. Eaglstein, *Infections Surg. 3*:173-181 (1984).
31. W. H. Eaglstein and P. M. Mertz, *J. Invest. Dermatol., 74*:90-91 (1980).
32. R. G. Geronemus, P. M. Mertz, and W. H. Eaglstein, *Arch. Dermatol., 115*:1311-1313 (1979).
33. O. M. Alvarez, P. M. Mertz, and W. H. Eaglstein, *J. Surg. Res., 35*:142-148 (1983).
34. O. M. Alvarez, P. M. Mertz, and W. H. Eaglstein, *Plast. Reconstr. Surg., 69*:284-289 (1982).
35. T. K. Hunt and W. Van Winkle, in *Fundamentals of Wound Management* (T. K. Hunt and J. E. Dunphy, eds.), Appleton-Century-Crofts, New York, 1979, pp. 17-21.
36. W. Van Winkle, Jr., *Surg. Gynecol. Obstet., 124*:369 (1967).
37. H. C. Grillo, *Arch. Surg., 88*:218-224 (1964).
38. H. C. Grillo and M. S. Potsaid, *Am. Surg., 154*:741-750 (1961).
39. A. J. Bailey, et al., *Biochem. Biophys. Res. Commun., 66*:1160-1164 (1975).
40. B. S. Gould, *J. Biochem., 232*:637 (1958).
41. E. H. Epstein, Jr., *J. Biol. Chem., 249*:3225-3231 (1974).
42. E. J. Miller, *Mol. Cell. Biochem., 2*:165–192 (1976).
43. A. J. Bailey, T. J. Sims, M. LeLous, and S. Bazin, *Biochem. Biophys. Res. Commun., 66*:1160-1164 (1975).

44. S. Gay, J. Viljanto, J. Raekallio, and R. Pentinen, *Acta Chir. Scand., 144*:205-211 (1978).
45. H. A. Linares and D. L. Larson, *Plast. Reconstr. Surg., 62*:589-593 (1978).
46. L. S. Leipziger, V. Glushko, B. DiBernardo, F. Shafaic, J. Noble, J. Nichols, and O. M. Alvarez, *J. Am. Acad. Dermatol., 12*:409-419 (1985).
47. O. M. Alvarez, P. M. Mertz, R. V. Smerbeck, and W. H. Eaglstein, *J. Invest. Dermatol., 81*:144-148 (1983).
48. R. Snyderman, J. Phillips, and S. E. Mergenhagen, *Infect. Immunol., 1*:521-525 (1970).
49. A. E. Postlethwaite, R. Snyderman, and A. H. Kang, *J. Exp. Med., 144*:1188-1303 (1976).
50. A. E. Postlethwaite, R. Snyderman, and A. H. Kang, *J. Clin. Invest., 64*:1379-1385 (1979).
51. B. M. Glaser, P. A. D'Amore, H. E. J. Seppa, S. I. Seppa, and E. Schiffmann, *Nature, 288*:483-484 (1980).
52. J. J. Castellot, M. J. Karnovsky, and B. M. Speigelman, *Proc. Natl. Acad. Sci. USA, 79*:5597-5601 (1982).
53. B. R. Zetter, *Nature, 285*:41-43 (1980).
54. J. L. Frederick, T. Shimanuki, and G. S. DiZerega, *Science, 224*:389-390 (1984).
55. D. R. Knighton, T. K. Hunt, H. Schevenstuhl, B. J. Halliday, and Z. Werb, *Science, 221*:1283-1285 (1983).
56. M. J. Banda, D. R. Knighton, T. K. Hunt, and Z. Werb, *Proc. Natl. Acad. Sci. USA, 79*:7773-7777 (1982).
57. T. Maciag, J. Radish, L. Wilkins, M. B. Sternerman, and R. Weinstein, *Cell. Biol., 94*: 511-520 (1982).
58. R. Kramer, P. Davison, and M. Karasek, *J. Invest. Dermatol., 8*:341 (abstract) (1983).
59. A. E. Postlethwaite, J. M. Seyer, and A. H. Kang, *Proc. Natl. Acad. Sci. USA, 75*: 871-875 (1978).
60. G. R. Grotendorst, H. E. J. Seppa, H. K. Kleinman, and G. R. Martin, *Proc. Natl. Acad. Sci. USA, 78*:3669-3672 (1981).
61. H. E. J. Seppa, G. R. Grotendorst, S. E. Seppa, E. Schiffman, and G. R. Martin, *J. Cell. Biol., 92*:584-588 (1982).
62. P. A. Hebda and W. H. Eaglstein, *J. Invest. Dermatol., 82*:410 (abstract) (1984).
63. V. S. Stenn, *Proc. Natl. Acad. Sci. USA, 78*:6907-6911 (1981).
64. G. R. Grotendorst, D. Pencev, G. R. Martin, and J. Sodek, in *Soft and Hard Tissue Repair* (T. K. Hunt, R. B. Heppenstall, E. Pines, and D. Rovee, eds.), Praeger, New York, 1984, pp. 20-40.
65. W. J. Pledger, C. D. Stiles, H. N. Antoniades, and C. D. Scher, *Proc. Natl. Acad. Sci. USA, 74*:4481-4485 (1977).
66. R. C. Glenn and R. Ross, *Cell, 25*:603-615 (1981).
67. P. B. Bitterman, S. I. Rennard, G. W. Hunninghake, and R. G. Crystal, *J. Clin. Invest., 70*:806-822 (1982).
68. D. Gospodarowicz, J. S. Moran, and D. L. Braum, *J. Cell Physiol., 91*:377-386 (1975).
69. D. R. Clemons, W. L. Isley, and M. T. Brown, *Proc. Natl. Acad. Sci. USA, 80*:1641-1645 (1983).
70. J. Folkman, *Ann. NY Acad. Sci., 401*:212-227 (1982).
71. H. K. Kleinman, A. T. Hewitt, et al., in *Current Trends in Prenatal Craniofacial Development* (R. M. Pratt and R. L. Christiensen, eds.), Elseoir-Norok, Holland, NY, 1980, pp. 277-295.
72. E. Pearlstein, L. I. Gold, and A. Garcia-Pordo, *Mol. Cell. Biochem., 29*:103-128 (1980).

73. T. M. Saba, F. A. Blemenstock, P. Weber, and J. E. Kaplan, *NY Acad. Sci., 312*: 43-55 (1978).
74. H. E. J. Seppa, K. M. Yamada, S. I. Seppa, M. H. Silver, H. K. Kleinman, and E. Schiffmann, *Cell Biol. Int. Rep., 5*:813 (1981).
75. R. Timpl, H. Rhode, P. Gehron-Robey, et al., *J. Biol. Chem., 254*:9933-9937 (1979).
76. J. M. Foidart, E. W. Bere, et al., *Lab. Invest., 42*:336-342 (1980).
77. J. M. Foidart, T. J. Berman, et al., *Lab. Invest., 42*:525-532 (1980).
78. L. Saxen, in *Cell and Tissue Interactions* (J. W. Last and M. M. Burger, eds.), Raven Press, New York, 1977, pp. 1-10.
79. O. M. Alvarez and D. G. Biozes, *Clin. Dermatol., 2*:54-67 (1984).
80. E. E. Peacock and W. Van Winkle, eds., *Contraction Wound Repair*, W. B. Saunders, Philadelphia, 1976, pp. 54-80.
81. O. M. Alvarez, J. B. Gaslen, W. H. Eaglstein, H. G. Welgus, and S. P. Stricklin, in *Dermatology in General Medicine* (T. B. Fitzpatrick, et al., eds.), 3rd ed., McGraw-Hill, New York.
82. E. Bell, S. Sher, et al., *J. Invest. Dermatol., 81*:2-10 (1983).
83. H. C. Grillo, G. T. Watts, and J. Gross, *Ann. Surg., 1948*:145-152 (1958).
84. H. P. Ehrlich, in *Soft and Hard Tissue Repair* (T. K. Hunt, R. B. Heppenstall, E. Pines, and D. Rovee, eds.), Praeger, New York, 1984, pp. 533-553.
85. B. Coloumb, L. Dubertret, E. Bell, and R. Touraine, *J. Invest. Dermatol., 82*:341-343 (1984).

17

Molecular Mechanisms of Glucocorticoid Action on Connective Tissue Metabolism

Aarne I. Oikarinen*
Harbor-UCLA Medical Center,
Torrance, California

Jouni Uitto
UCLA School of Medicine and Harbor-UCLA Medical Center,
Torrance, California

I. INTRODUCTION

Glucocorticosteroids are in extensive clinical use for treatment of a variety of both acute and chronic diseases. Although glucocorticosteroids are highly effective for treatment of many of these conditions, their long-term efficacy is often compromised by the development of side effects. Many of the adverse effects involve pathology of the connective tissues. On the basis of these clinical observations, it is clear that glucocorticosteroids can profoundly affect the metabolism of the connective tissue components of the extracellular matrix. In this chapter, we summarize the current knowledge of the mechanisms of action of glucocorticosteroids on connective tissue metabolism based on clinical observations and experimental studies.

II. GENERAL MECHANISMS OF GLUCOCORTICOSTEROID ACTION

Glucocorticosteroids exert a multiplicity of biological effects on cellular functions (for reviews see Refs. 1-4). These effects are mediated by specific intracellular binding proteins in a manner that has been demonstrated for other steroid hormones and some vitamins (5). During the initial stages of hormone action, glucocorticosteroids passively diffuse through the cell membrane into the cytoplasm where the steroid molecules bind to specific receptors (Figure 1). This binding induces conformational changes in the receptor molecule, and the activated steroid-receptor complexes are then translocated to the nucleus where they bind to specific regions of chromatin. The steroid-receptor complexes then influence the transcription of specific sequences of genomic DNA.

The mechanistic details of reactions by which steroid-receptor complexes modulate the transcription of specific genes are under active investigation. It has been shown that in a mouse mammary tumor virus (MMTV), the steroid-receptor complexes selectively

**Present affiliation*: Oulu University Central Hospital, Oulu, Finland.

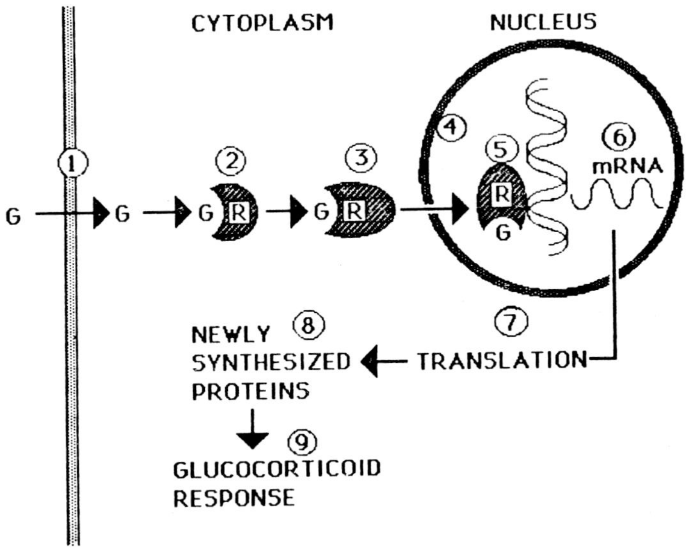

Figure 1 Schematic presentation of the possible mechanisms of glucocorticoid action on cellular metabolism. The glucocorticoid (G) diffuses passively through the cell membrane (1) and binds to an intracellular receptor (R) (2). Following the binding, the steroid-receptor complex undergoes a conformational change (3) and is subsequently translocated to the nucleus of the cell (4); the steroid-receptor complex binds to a specific acceptor site on the chromatin (5), and the interactions between the complex and DNA alter the transcription of the gene (6). The alterations in mRNA abundance affect translation (7) and the synthesis of proteins (8) that elicit the glucocorticoid response (9).

bind to certain nucelotide pairs in the promoter region of the gene (6), thus inducing the initiation of transcription. This model of steroid action on gene activation would then explain the positive effects of steroids in increasing the amount of specific messenger RNA (mRNA) molecules. It is unclear, however, how this model would apply to situations where steroids mediate an inhibitory effect on the synthesis of certain proteins. It has been suggested that the inhibitory effects of glucocorticoids would be mediated by proteins that are specifically induced under the influence of steroid and then subsequently suppress the synthesis of other selected proteins (for reviews see Refs. 1,3,7). An alternate possibility would be that the binding of steroid-receptor complexes to the genome activates some genes, while the expression of other genes would be inhibited. In support of this possibility is the demonstration that glucocorticoids directly reduce the transcription of α-fetoprotein gene, and the glucocorticosteroids may thus have direct inhibitory effects on the transcriptional activity of the cells (8). Finally, steroids may influence the gene expression and protein synthesis at the posttranscriptional level. The potential target sites are the degradation and turnover of mRNA molecules, posttranslational modifications of polypeptides, and degradation and turnover of the proteins.

III. CLINICAL UTILIZATION OF GLUCOCORTICOSTEROIDS AND DEVELOPMENT OF CONNECTIVE TISSUE SIDE EFFECTS

Glucocorticoids are used in clinical situations for treatment of a wide variety of diseases (9,10). Short-term systemic glucocorticoid treatment is often used for acute clinical conditions, such as allergic reactions and anaphylactic shock. Since the treatment for these diseases with steroids is usually of short duration, connective tissue side effects are uncommon. In contrast, several chronic diseases may require long-term glucocorticoid therapy; these include some dermatological conditions, such as pemphigus vulgaris, bullous pemphigoid, systemic lupus erythematosus, and sarcoidosis. Most patients receiving prolonged steroid therapy will develop adverse effects manifesting as a generalized weakness of connective tissue (Table 1). A relatively common side effect is the development of varying degrees of osteoporosis, and postmenopausal women and diabetic patients, as well as elderly and immobilized individuals, are particularly at risk (11). In children, glucocorticoids may cause growth retardation (12). Permanent arrest of growth is unusual, however, since the steroids, in addition to inhibiting bone growth, also delay the epiphyseal closure, thus allowing growth after the steroid therapy is discontinued.

The most common cutaneous side effect of steroid treatment is thinning and atrophy of the skin (13-17). Atrophy of dermis can be seen as a result of topical steroid treatment or following prolonged systemic use of potent synthetic steroids. Clinically, the affected skin is thin, shiny, and translucent, and shows easy bruisability (13,15). Topical or systemic application of glucocorticoids can also lead to formation of striae, cutaneous lesions histologically characterized by loss of the elastic fibers (18). Relatively common and an early side effect of topical application of potent synthetic steroids is the development of telangiectasias (19); these may result from altered properties of the blood vessel walls in tissues.

Glucocorticoids have also been shown to be beneficial for treatment of conditions characterized by excessive accumulation of collagen. In particular, local injections of potent synthetic steroids have been shown to prevent the growth of keloids (21), cutaneous nodules characterized by excessive collagen deposition (see Chapter 15). Glucocorticoids can also alter connective tissue response in diseases characterized by inflammation and immunological alterations. On the basis of these clinical observations, we can conclude

Table 1 Connective Tissue Side Effects of Long-Term Glucocorticosteroid Treatment

Target tissue	Clinical symptoms	Connective tissue pathology
Skeleton and bones	Osteoporosis	Decrease in bone formation, osteoblast function is inhibited, changes in calcium metabolism
Skeleton and bones	Reduced skeletal growth in children	Inhibition or cessation of linear bone growth
Tendon	Fragility	Loss of connective tissue
Skin	Atrophy	Decrease in connective tissue components; decrease in water content
Skin	Striae	Loss of elastin
Skin	Telangiectasia	Expansion of abnormal blood vessels
Subcutis	Atrophy	Loss of connective tissue

that glucocorticosteroids clearly influence the metabolism of connective tissue components in patients subjected to steroid treatment. These clinical observations do not disclose, however, the exact mechanisms of glucocorticoid action on connective tissues.

IV. THE ACTION OF GLUCOCORTICOSTEROIDS ON THE METABOLISM OF CONNECTIVE TISSUES

A. Effects on Collagen Metabolism

On the basis of several lines of clinical evidence summarized above and in Table 1, we can conclude that the metabolism of collagen is clearly influenced by glucocorticosteroids. Based on extensive knowledge of collagen metabolism in normal tissues, we can now visualize several distinct levels at which glucocorticosteroids might interfere with collagen biosynthesis and degradation.

1. Alterations in Collagen Biosynthesis

Animal Studies. Collagen biosynthesis involves several steps that may be influenced by steroids. Several studies have demonstrated that glucocorticoids inhibit collagen synthesis (for review, see Ref. 22); yet relatively little is known about the effects of glucocorticosteroids on collagen metabolism on the gene level, partly owing to the fact that specific methods for molecular studies on the gene level have not been available until recently. Knowledge of the regulation of gene expression by steroids using collagen as a model could be particularly useful in providing insight into the regulation of genes of relatively large size.

The effect of glucocorticoids on the activity and the amount of procollagen mRNA has been studied in various experimental models. In chick embryos, pretreatment with glucocorticoids decreased the activity of type I procollagen mRNA in calvaria (23). Similarly, in rats glucocorticoid treatment decreased polysomal procollagen mRNA activity (24), and cortisol has been shown to decrease the concentration of translatable procollagen mRNA in cultured human skin fibroblasts (25). The amount of type I procollagen specific mRNA has also been studied by Northern blot and dot blot hybridizations. In chick embryo skin, dexamethasone specifically decreased the amount of type I procollagen mRNA (26), and similarly, in a recent study by us, the treatment of 16-day-old chick embryos with 5×10^{-6} M dexamethasone was shown to decrease type I collagen mRNA by about 70% (Figure 2) (27). The decrease was similar in mRNAs coding for both proα1(I) and proα2(I) chains of type I procollagen (27).

There are at least three possible explanations for decreased procollagen mRNA levels: (a) the transcription of the corresponding gene is inhibited; (b) the processing of mRNA precursor to functional mRNA is inhibited; and (c) the degradation of procollagen mRNA is increased. A recent study (28) has suggested that in human skin fibroblast cultures, the degradation of procollagen mRNA may be increased by cortisol. Specifically, it was suggested that glucocorticoids increased the activity of (2′-5′)oligoadenylate synthetase, and the increased concentrations of oligoadenylates would then activate the latent RNAse H, which is capable of degrading mRNAs (29). However, this model does not specifically address the issue of why the effect of steroid would be selective for procollagen mRNA.

Collagen is a unique protein in that its synthesis involves an exceptionally large number of posttranslational modifications, many of which are critical for normal production of the functional collagen fibers (30). Thus, alterations in these modifications could

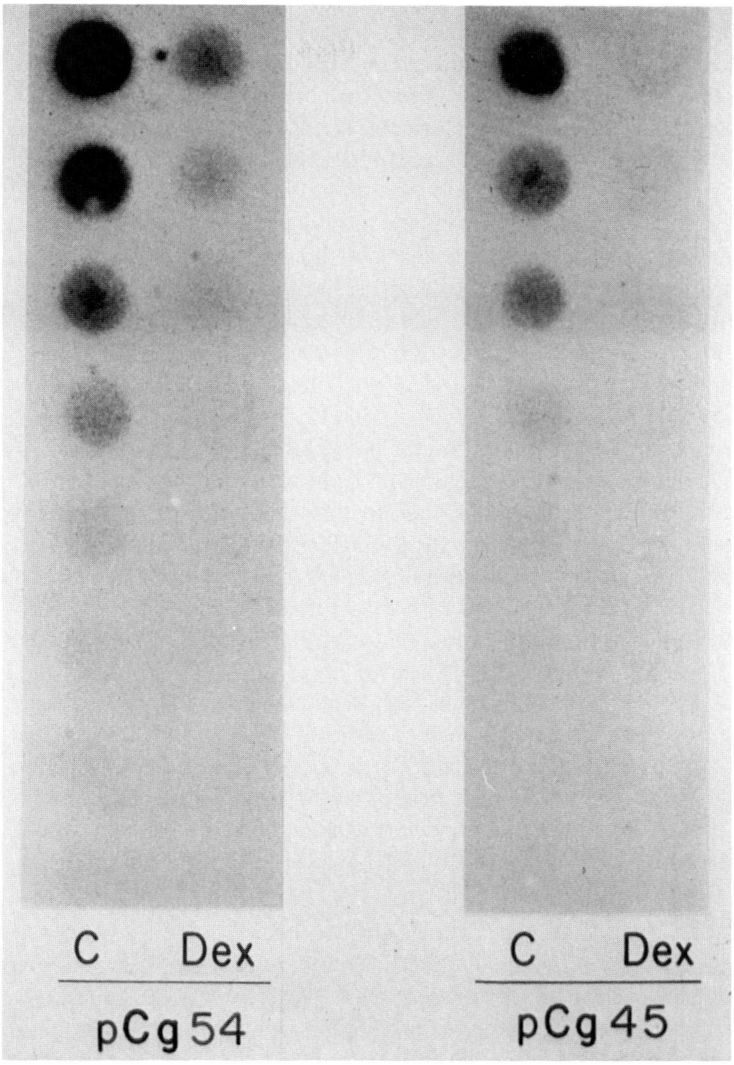

Figure 2 Effect of dexamethasone on collagen mRNA levels in chick embryo tendons: 16-day-old chick embryos were treated with dexamethasone in a final concentration of 5×10^{-6} M. After 24 hr of incubation, tendons were removed, and poly(A)$^+$–RNA was isolated and dotted on nitrocellulose filters in concentrations varying from 100 ng to 0.78 ng, in 1:2 dilutions. The filters were hybridized with ^{32}P-labeled cDNAs corresponding to chick proα1(I) (pCg54) and proα2(I) (pCg45) sequences. (Reproduced from Ref. 27 with permission.)

affect the amount and the quality of collagen being synthesized. The critical steps in the intracellular biosynthesis of procollagen are: (a) proper association of proα chains; (b) interchain disulfide bond formation; (c) hydroxylation of prolyl residues to yield hydroxyprolines in a reaction catalyzed by prolyl hydroxylase (31); and (d) folding into the triple-helical conformation. The inhibition of prolyl hydroxylase leads to synthesis of underhydroxylated collagen, which remains as a nonfunctional protein at physiological temperatures, and which is susceptible to degradation by nonspecific proteases.

In several studies, the activity of prolyl hydroxylase has been shown to be decreased by glucocorticoids (see Ref. 22). For example, glucocorticoids decrease the activity of the prolyl hydroxylase in rat tissues (32,33), chick embryo tendon cells (34,35), human skin (36), and skin fibroblast cultures (37). These studies have also demonstrated that glucocorticoids decrease the amount of immunoreactive prolyl hydroxylase protein, indicating either that the synthesis of prolyl hydroxylase protein is decreased or that the degradation of the enzyme protein is increased. The activities of lysyl hydroxylase and collagen glucosyltransferases, enzymes also participating in the intracellular posttranslational modifications of collagen, are similarly decreased in chick embryo tendon cells (35) and in rat tissues (33). However, the percent hydroxylation of lysyl residues and the glycosylation of hydroxylysyl residues in the newly synthesized procollagens were unaffected (35).

After intracellular biosynthesis and secretion, the procollagen molecules are converted to collagen by extracellular removal of the amino- and carboxy-terminal extensions by two separate enzymes (30). The regulation of these enzymes has not been studied extensively, and it is not known whether glucocorticoids affect their activities. However, the conversion of type I procollagen to corresponding collagen was recently shown to be unaffected by pretreatment of chick embryos with dexamethasone or when dexamethasone was directly added to chick embryo tendon cultures following pulse labeling with [^{14}C] proline (27).

The studies reviewed above have mostly dealt with the metabolism of type I collagen, and relatively little information is available on glucocorticoid effects on the metabolism of other interstitial collagens, such as types II and III. It has been shown, however, that the production of type II procollagen in chick embryo sterna is markedly reduced by dexamethasone (Figure 3) (27). Furthermore, glucocorticoids have been shown to coordinately regulate the synthesis of type I and type III procollagens in rat lung and skin (38).

Fibroblast Culture Studies. As discussed above, glucocorticoids have been shown to decrease collagen synthesis in various animal models. However, studies utilizing cultured human skin fibroblasts have yielded conflicting results (see Refs. 39,40). In some studies, glucocorticoids were found to stimulate collagen production, while in others no effect or an inhibition was noted. The type of effect on collagen synthesis was largely dependent on the cell type used. In general, collagen synthesis was stimulated by glucocorticosteroids in fetal fibroblast cultures, while in adult skin fibroblast cultures an inhibition was noted (see Ref. 39). Also, methodological details, such as the type and dosage of steroid used, the culture conditions, incubation times, and serum concentrations, could explain some of the differences in the results of these studies. Furthermore, the changes in cell number as a result of steroid treatment can also influence the results; this aspect has not been evaluated in most studies (see Ref. 39). Finally, the effects of glucocorticoids on the intracellular specific activities of labeled amino acids could influence the results, since glucocorticoids may alter the transport mechanism of amino acids, leading to changes in their specific activities (41).

The importance of culture conditions in modulating the direction of the glucocorticoid response on cellular metabolism has been demonstrated by previous studies (42). Specifically, it was shown that under deplete culture conditions-glucocorticoids inhibit the proliferation of fibroblasts, while in enriched media, a stimulation can be observed

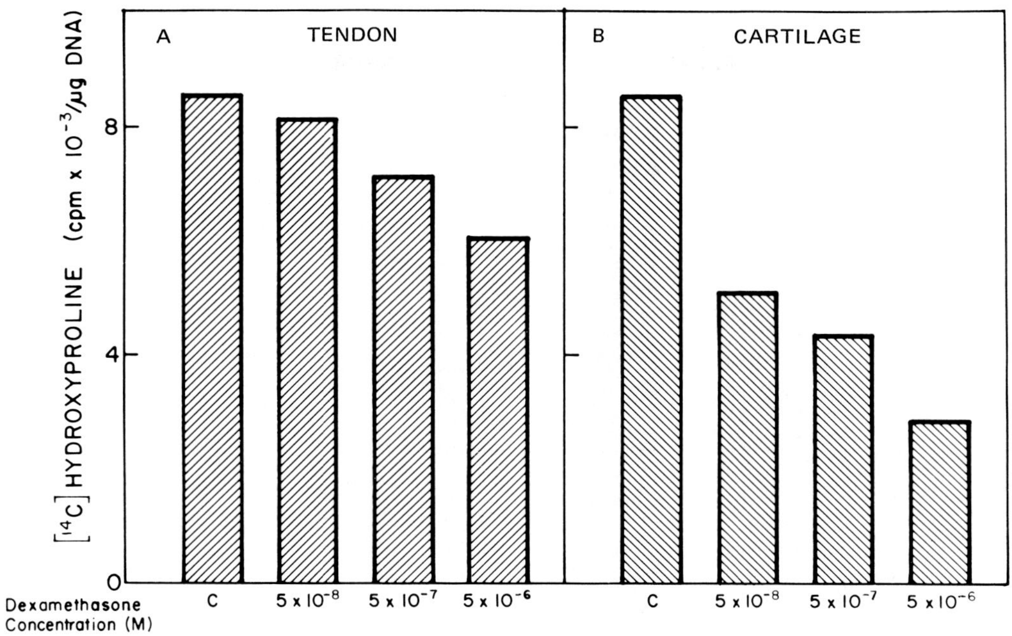

Figure 3 Demonstration that dexamethasone inhibits collagen synthesis in tendons and cartilage of chick embryos. Varying concentrations of dexamethasone were injected onto the chorioallantoic membrane of 16-day-old chick embryos. Tendons and sterna were removed at 24 hr, and collagen synthesis was assayed by incubation with [^{14}C] proline, as indicated in the original study (27). [^{14}C] Hydroxyproline in the nondialyzable fraction was then determined as an index of collagen production. The values are the means of three parallel determinations on two separate samples. (A) Collagen production in tendons. (B) Collagen production in sterna.

(42). It has also been shown that the binding characteristics of glucocorticosteroid receptors can vary during cell cycle, and this parameter could thus affect the glucocorticoid-mediated changes in collagen synthesis (43).

2. *Collagen Degradation*

In general, glucocorticosteroid treatment leads to negative nitrogen balance as a result of catabolic effects of glucocorticoids in many tissues. An example is muscle tissue in which prolonged use of glucocorticoids causes diminution of muscle mass (see Ref. 9). In the case of collagen, it has been shown that injection of cortisone decreases the urinary hydroxyproline excretion in rats, suggesting that glucocorticoids may have predominantly antianabolic, rather than catabolic, effect on collagen metabolism in vivo (44). Some studies have suggested that glucocorticoids may increase the activities of collagenolytic enzymes (45) and increase the degradation of collagen in fibroblast cultures (46). In contrast, several studies have indicated that glucocorticoids are inhibitory to collagenase activity (47-49). Recently, we examined the effects of dexamethasone on collagenase activity in three different human skin fibroblast cell lines, which demonstrated either low, intermediate, or high levels of glucocorticoid receptors per cell (50). Dexamethasone decreased the collagenase activity in all these cell lines, the inhibition varying from 20 to

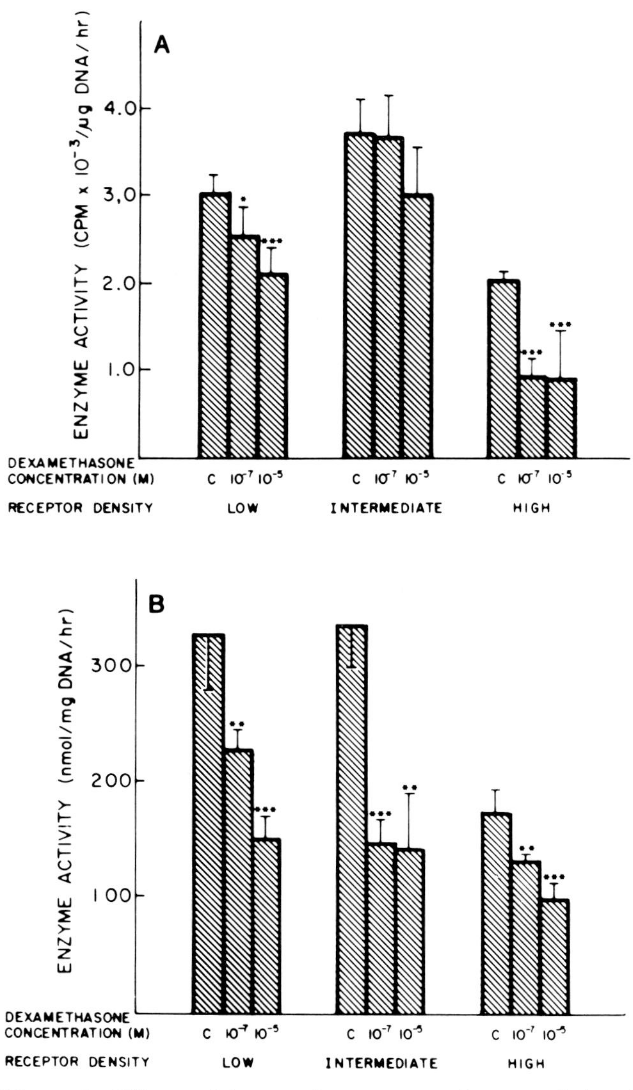

Figure 4 Effects of dexamethasone on collagenase (A) and elastaselike (B) enzyme activity in human skin fibroblast cultures. Fibroblasts containing low, intermediate, or high levels of glucocorticoid receptors were preincubated for 48 hr in the presence of varying concentrations of dexamethasone. The receptor densities (B_{max}) in cells with low, intermediate, and high density were 3.2, 8.8, and 16.8 fmol/μg DNA, respectively. The media were then changed, and the cells were incubated with dexamethasone in serum-free media. After 6 hr of incubation, media were used for assay of collagenase and elastaselike enzyme activities, as described in the original study (27). Collagenase was assayed using type I collagen as the substrate, and the enzyme activity is expressed as degradation of ^{3}H-labeled collagen, cpm/μg DNA/hr. Elastaselike enzyme activity was assayed using a synthetic succinyl-L-(alanyl)$_3$-PNA (SAPNA) as the substrate. The enzyme activity is expressed as hydrolysis of SAPNA, nmol/mg DNA/hr. (*$p < 0.01$ compared to control. **$p < 0.005$ compared to control. ***$p < 0.001$ compared to control).

60% (Figure 4). Since the collagenase activity was determined after a brief trypsin activation, the decrease noted with glucocorticoid appeared to represent a reduction in the amount of total activatable collagenase (50).

B. Effects on Elastin Metabolism

Elastic fibers form a network that is responsible for the elastic properties of organs, such as skin (51-53). The cells synthesizing elastin include smooth muscle cells, endothelial cells, ligamentum nuchae cells, and skin fibroblasts. Several lines of evidence suggest that the synthesis of elastin is under hormonal control. Specifically, glucocorticoids have been shown to stimulate elastin synthesis in embryonic chick aorta (54) and in ligamentum nuchae fibroblast cultures (55). In the latter study, dexamethasone in 10^{-7} M concentration increased the elastin synthesis by about 180% in early passages of cell culture, while the stimulation of elastin synthesis was markedly less when cells were used in older passages (55). The mechanisms of the stimulation of elastin synthesis by glucocorticosteroids are currently unknown.

The extracellular elastic fibers are degraded by elastases, a group of proteolytic enzymes secreted by inflammatory and fibroblastic cells. Glucocorticoids have been shown to decrease the elastase activity released by cultured macrophages, and the inhibition was suggested to be mediated by corticosteroid receptors (56,57). Recently, an elastaselike neutral protease has been demonstrated in human skin fibroblast cultures, and dexamethasone appears to be a potent inhibitor of this enzyme (50) (Figure 4B). Thus, glucocorticosteroids may modulate the amount of elastin in tissues by regulating the activities of enzymes capable of degrading this protein.

C. Effects on Glycosaminoglycans

Although the tissue concentrations of glycosaminoglycans (GAGs) are considerably less than those of collagen, GAGs play an important role in mediating interactions between connective tissue components and in retaining the extracellular water (see Chapter 4). Thus, alterations in the amount of GAGs can cause marked changes in the organization and the mechanical properties of the extracellular connective tissue matrix.

Fibroblasts are primarily responsible for the synthesis of the GAGs in the skin, and these cells thus offer a good model to study the effects of glucocorticosteroids on the synthesis of these macromolecules. In several studies, glucocorticoids have been shown to decrease the synthesis of GAGs. For example, cortisol in 1×10^{-8} M concentration markedly decreased the synthesis of hyaluronic acid in human skin fibroblasts (58). Similarly, cortisol in 10^{-7} M concentration decreased the synthesis of sulfated GAGs (59). Inhibition of hyaluronic acid synthesis by other steroids was obtained in concentrations considerably lower than in the case of cortisol; for example, betamethasone-17-valerate in 10^{-10} M concentration was markedly inhibitory (60). It is of interest that the inhibition of synthesis of hyaluronic acid by different glucocorticoids correlated well with the clinical efficacy of these steroids.

Dermal atrophy is one of the cutaneous side effects of glucocorticoid treatment. Histological and electron microscopic studies have demonstrated marked changes in the structural organization of human skin following the topical application of potent glucocorticosteroids (61). The most significant changes after six weeks of treatment with clobetasol propionate under occlusion were flattening of the dermoepidermal junction and reorganization of the three-dimensional architecture of the dermis (61). The latter finding was explained by the resorption of the ground substance, resulting in decreased interfibrillar spaces between collagen and elastic fibers. As a consequence, the fibrous network of the extracellular matrix collapsed, resulting in reorientation of the collagen and elastic fibers. These findings support the hypothesis that topically applied glucocorticoids first decrease the amount of GAGs in the skin, resulting in atrophic, thin-appearing skin. This change could occur in a relatively short period of time, since the half-life of GAGs in the skin is the range of 2-14 days (62). The relatively rapid turnover of GAGs contrasts the half-lives of elastin and collagen, which are considerably longer, possibly several years (63).

The mechanisms of glucocorticosteroid-mediated decrease in the synthesis of GAGs are not fully understood. The activity of hyaluronate synthetase has been shown to be reduced in the fibroblasts treated with glucocorticoids (60), thus potentially explaining the inhibitory effect of glucocorticoids on hyaluronic acid synthesis.

D. Effects on Basement Membrane Components

Regulation of the metabolism of basement membrane components, such as type IV collagen, laminin, fibronectin, and heparan sulfate proteoglycan, has not been extensively studied. It appears, however, that the metabolism of laminin and type IV collagen are partially regulated by hormones. For example, in rat mammary epithelial cell cultures, hydrocortisone and dexamethasone increased the production of type IV collagen, whereas the degradation of type IV collagen was decreased (64). In human skin, topical application of potent glucocorticosteroids was found to have no apparent effect on the ultrastructural distribution of basement membrane components in dermoepidermal junction (65).

The effects of glucocorticosteroids on fibronectin, which is closely associated with basement membranes, have been studies in virus-transformed fibroblasts. The amount of fibronectin detected by immunofluorescence was markedly increased by glucocorticoids in these cells (66). This effect was rather specific since other steroids (progesterone, testosterone) were ineffective.

E. Glucocorticoid Modulation of Connective Tissue Metabolism Through Inflammatory/Immunological Processes

Glucocorticoids have profound effects on the immune-inflammatory systems of intact animals, consequently affecting the connective tissue metabolism. For example, glucocorticoids decrease the number and migration of inflammatory cells in the inflammatory processes (for review, see Ref. 67). It is also known that inflammatory cells secrete various proteolytic enzymes that can degrade connective tissue components. For example, polymorphonuclear neutrophils (PMN) contain high levels of elastase and collagenase activities, which are capable of degrading insoluble elastin and collagen, respectively. Glucocorticosteroids can inhibit the release of lysosomal enzymes from PMNs, and steroids may thus influence the degradation and turnover of the extracellular matrix.

Monocyte-macrophage functions, such as secretion of monokines, are relatively sensitive to the effects of glucocorticoids (67). Similarly, synthesis and secretion of lymphokines by lymphocytes can be modulated by glucocorticosteroids. Thus, glucocorticoids may indirectly regulate the metabolic responses of fibroblasts by affecting the immune functions.

V. DIRECTIONS FOR FUTURE RESEARCH

Although considerable progress has been made in understanding the molecular mechanisms by which glucocorticosteroids affect the metabolism of extracellular matrix components, several details await clarification. Particularly important are the ongoing studies that would be able to pinpoint the exact interactions between the steroid-receptor complexes and the regulatory DNA sequences of the genes corresponding to the matrix proteins. With the advent of the technology of molecular biology, these questions can now be approached by utilizing purified receptor proteins and isolated promoter sequences corresponding to the genes coding for extracellular matrix proteins, such as collagen (68). Furthermore, the mechanistic details of the gene expression, in terms of transcriptional activity, stability, and turnover of the mRNA precursor molecules, and the processing of the precursor mRNA molecules remain to be elucidated. In addition to the transcriptional level, several unanswered questions concerning the translational activity of the mRNA molecules, as well as the posttranslational modification and turnover of the newly synthesized proteins, are potential targets for future research.

Of considerable clinical importance is the observation that steroids having the highest clinical antiinflammatory efficacy also demonstrate the highest capacity to inhibit the metabolism of the extracellular matrix components. Thus, in general the most efficacious steroids demonstrate the highest risk for connective tissue side effects. One of the future avenues for steroid research would require development of glucocorticosteroids that would selectively demonstrate some of the features usually assigned to the glucocorticosteroids. Specifically, an ultimate antiinflammatory glucocorticosteroid would demonstrate minimal suppression of the connective tissue metabolism, while glucocorticosteroids effective in suppressing collagen production might not demonstrate any antiinflammatory properties (69). The development of such specialized glucocorticosteroids would considerably enhance their utilization for specific clinical conditions in the future.

ACKNOWLEDGMENTS

The original research by Drs. Oikarinen and Uitto summarized in this overview was supported in part by grants from the Medical Research Council of the Academy of Finland, and by grants AM-28450, AM-35297, and GM-28833 from United States Public Health Service, National Institutes of Health. The authors gratefully acknowledge Ms. Charlene D. Aranda for her assistance in preparation of the manuscript.

REFERENCES

1. E. B. Thompson and M. E. Lippman, *Metabolism, 23*:159-202 (1974).
2. K. Leung and A. Munck, *Ann. Rev. Physiol., 37*:245-272 (1975).
3. W. B. Pratt, *J. Invest. Dermatol., 71*:24-55 (1978).
4. A. K. Roy and J. H. Clark, *Gene Regulation by Steroid Hormones. II.* Springer-Verlag, New York.
5. J. D. Baxter and J. W. Funder, *N. Engl. J. Med., 301*:1149 (1979).
6. C. Scheidereit, S. Geisse, H. M. Westphal, and M. Beato, *Nature, 304*:749-752 (1983).
7. K. Kontula, *Ann. Clin. Res., 12*:223-235 (1900).
8. L. Belanger, M. Frain, P. Baril, M-C. Gingras, J. Bartkowiak, and J. M. Sala-Trepat, *Biochemistry, 20*:6665-6672 (1981).

9. S. L. Swartz and R. G. Dluhy, *Drugs, 6*:238-255 (1978).
10. J. A. Miller and D. D. Munro, *Drugs, 19*:119-134 (1980).
11. J. D. Baxter and P. H. Forsham, *Am. J. Med., 53*:573-589 (1972).
12. J. N. Loeb, *N. Engl. J. Med., 295*:547-552 (1976).
13. D. V. Stevanovic, *Br. J. Dermatol., 87*:548-556 (1972).
14. E. Shöpf, *Ann. Clin. Res., 7*:353-367 (1975).
15. J. D. Kirby and D. D. Munro, *Br. J. Dermatol., 94*:111-119 (1976).
16. S. Jablonska, M. Groniowska, and Dabrowski, *Br. J. Dermatol., 100*:193-206 (1979).
17. P. J. Dykes and R. Marks, *Br. J. Dermatol., 101*:599-609 (1979).
18. M. E. Chernosky and J. M. Knox, *Arch. Dermatol., 90*:15-19 (1964).
19. J. G. Smith, R. F. Wehr, and D. K. Chalker, *Arch. Dermatol., 112*:1115-1117 (1978).
20. R. Manthorpe, C. Garbasch, B. Kofod, and I. Lorenzen, *Acta Endocrinol., 86*:437-448 (1977).
21. H. B. Griffiths, *Plastic Reconstruct. Surg., 38*:202-208 (1900).
22. K. R. Cutroneo, R. Rokowski, and D. F. Counts, *Collagen Rel. Res., 1*:557-568 (1981).
23. J. Oikarinen and L. Ryhänen, *Biochem. J., 198*:519-524 (1900).
24. R. J. Rokowski, J. Sheehy, and K. R. Cutroneo, *Arch. Biochem. Biophys., 210*:74-81 (1981).
25. J. Oikarinen, T. Pihlajaniemi, L. Hämäläinen, and K. I. Kivirikko, *Biochim. Biophys. Acta, 741*:297-302 (1983).
26. K. M. Sterling, Jr., M. Harris, T. Dipetrillo, G. Delaney, and K. R. Cutroneo, *J. Biol. Chem., 258*:7644-7647 (1983).
27. A. I. Oikarinen, E. J. Zaragoza, A. Palotie, M-L. Chu, and J. Uitto, *Clin. Res., 32*:73A (1985).
28. L. Hämäläinen, J. Oikarinen, and K. I. Kivirikko, *J. Biol. Chem. 260*:720-725 (1985).
29. J. Oikarinen, *Biochem. Biophys. Res. Commun., 105*:876-881 (1982).
30. D. J. Prockop, K. I. Kivirikko, L. Tuderman, and N. A. Guzman, *N. Engl. J. Med., 301*:12-23 (1979).
31. K. I. Kivirikko and R. Myllylä *Meth. Enzymol., 82*:245-304 (1982).
32. K. R. Cutroneo, F. L. H. Stassen, and G. J. Cardinale, *Mol. Pharmacol., 11*:47-51 (1975).
33. J. Risteli, *Biochem. Pharmacol., 26*:1295-1298 (1977).
34. A. Oikarinen, *Biochem. Pharmacol., 26*:875-879 (1977).
35. A. Oikarinen, *Biochem. J., 169*:533-539 (1977).
36. A. Oikarinen and M. Hannuksela, *Arch. Derm. Res., 267*:79-82 (1980).
37. J. S. Trupin, S. B. Russell, and J. D. Russell, *Collagen Rel. Res., 3*:13-23 (1983).
38. S. Shull and K. R. Cutroneo, *J. Biol. Chem., 258*:3364-3369 (1983).
39. B. A. Booth, E. M. L. Tan, A. Oikarinen, and J. Uitto, *Int. J. Dermatol., 21*:33-337 (1982).
40. M. Poneck, *Int. J. Dermatol., 23*:11-24 (1984).
41. S. B. Russell, J. D. Russell, and J. S. Trupin, *J. Biol. Chem., 257*:9525-9531 (1982).
42. K. L. Johnson, N. C. Lan, and J. D. Baxter, *J. Biol. Chem., 254*:7785-7794 (1979).
43. J. Oikarinen, L. Hämäläinen, and A. Oikarinen, *Biochim. Biophys. Acta, 799*:158-165 (1984).
44. K. Kivirikko and O. Laitinen, *Acta Physiol. Scand., 64*:356-360 (1965).
45. J. C. Houck and V. K. Sharma, in *Inflammation Biochemistry and Drug Interaction*. (A. Bertelli and J. C. Houck, eds.), 1969, pp. 85-93.
46. B. J. McCoy, R. F. Diegelmann, and I. K. Cohen, *Proc. Soc. Exp. Biol. Med., 163*: L216-L222 (1980).
47. T. J. Koob, J. J. Jeffrey, and A. Z. Eisen, *Biochem. Biophys. Res. Commun., 61*: 1083-1088 (1974).

48. J. J. Jeffrey, R. J. Coffey, and A. Z. Eisen, *Biochim. Biophys. Acta, 252*:143 (1971).
49. J. Uitto and A. Z. Eisen, in *Dermatology in General Medicine* (T. B. Fitzpatrick, A. Z. Eisen, K. Wolff, I. M. Freedberg, and K. F. Austen, eds.), McGraw-Hill, New York, 1979, pp. 164-180.
50. A. I. Oikarinen, C. Meeker, H. Oikarinen, E. Tan, and J. Uitto, *J. Invest. Dermatol., 80*:325-326 (1983).
51. L. B. Sandberg, N. T. Soskel, and J. G. Leslie, *N. Engl. J. Med., 304*:566-579 (1981).
52. J. Uitto, L. Ryhänen, P. A. Abraham, and A. Perejda, *J. Invest. Dermatol., 79*:160s-168s (1982).
53. J. Rosenbloom, *Lab. Invest., 51*:605-623 (1984).
54. R. Eichner and J. Rosenbloom, *Arch. Biochem. Biophys., 198*:414-423 (1979).
55. R. P. Mecham, G. Lange, J. Madaras, and B. Starcher, *J. Cell Biol., 90*:332-338 (1981).
56. Z. Werb, *J. Exp. Med., 147*:1695-1712 (1978).
57. Z. Werb, R. Foley, and A. Munck, *J. Immunol., 2*:115-121 (1978).
58. H. Saarni, M. Tammi, and N. S. Doherty, *J. Pharm. Pharmacol., 30*:a200-a201 (1978).
59. J. Saarni, *Biochem. Pharmacol., 26*:1961-1966 (1977).
60. H. Saarni and V. K. Hopsu-Havu, *Br. J. Dermatol., 98*:445-449 (1978).
61. P. Lehmann, P. Zheng, R. M. Lavker, and A. M. Kligman, *J. Invest. Dermatol., 81*: 169-176 (1983).
62. J. E. Silbert, in *Biochemistry and Physiology of the Skin* (L. A. Goldsmith, ed.), Oxford University Press, New York, 1983, Chap. 20, pp. 448-461.
63. K. I. Kivirikko, *Int. Rev. Connect. Tissue Res., 5*:93-163 (1970).
64. D. S. Salomon, L. A. Liotta, and W. R. Kidwell, *Proc. Natl. Acad. Sci. USA, 78*:382-386 (1981).
65. A. Oikarinen, L. Peltonen, J. Hintikka, J. M. Foidart, and U. Kiistala, *Br. J. Dermatol., 108*:171-178 (1983).
66. L. T. Furcht, D. F. Mosher, G. Wendelschafter-Crabb, and J-M. Foidart, *Cancer Res., 39*:2077-2083 (1979).
67. J. E. Parriollo and A. S. Fauci, *Ann. Rev. Pharmacol. Toxicol., 19*:179-201 (1979).
68. D. J. Prockop, in *Gene Families of Collagen and Other Proteins* (D. J. Prockop and P. C. Champe, eds.), Elsevier North-Holland, New York, 1980, pp. 71-87.
69. T. DiPetrillo, H. Lee, and K. R. Cutroneo, *Arch. Dermatol., 120*:878-883 (1984).
70. A. Oikarinen, H. Oikarinen, C. Meeker, E. M. L. Tan, and J. Uitto (1985) (submitted for publication).

18

Pathology of the Elastic Fibers

Jouni Uitto
UCLA School of Medicine and Harbor-UCLA Medical Center,
Torrance, California

Lasse J. Ryhänen
Harbor-UCLA Medical Center,
Torrance, California

I. INTRODUCTION

Elastic fibers are present in various organs, and they apparently play a central role in providing elasticity and resilience to the tissues (for review, see Refs. 1-3). For several years, elucidation of the structure and metabolism of the elastic fibers was hampered by the relative insolubility of these structures. However, with the advent of modern biochemical technology, considerable progress has been made during the past few years in understanding the normal biology and biochemistry of the elastic fibers (for details on normal structure and metabolism of the elastic fibers, see Chapters 2 and 3). At present, we can identify levels of metabolic processes at which elastic fibers can be affected, consequently, resulting in clinical disease. Diseases of the elastic fibers can result either from alterations in the quantity of normal elastic fibers or from abnormalities affecting the molecular structure and organization of these fibers (1-5).

II. NORMAL ELASTIC FIBERS

A. Morphology of Elastic Fibers

Elastic structures, when examined by light microscopy utilizing elastin-specific staining techniques, appear as wavy narrow fibers, which are oriented parallel to the surface of the tissues such as skin and blood vessels (Figure 1). Visualization of the elastic fibers by scanning electron microscopy reveals a network of interlacing fibers of variable thickness (Figure 2A). After the removal of other connective tissue components the elastic fibers appear randomly oriented, suggesting that the interactions with other dermal macromolecules provide the framework for the normal supramolecular organization of these fibers.

Transmission electron microscopy of the elastic fibers demonstrates two distinct components (Figure 2B) (6,7). The major component has an amorphous appearance; this part of the elastic fiber represents elastin, a well-characterized connective tissue protein (2,3,5).

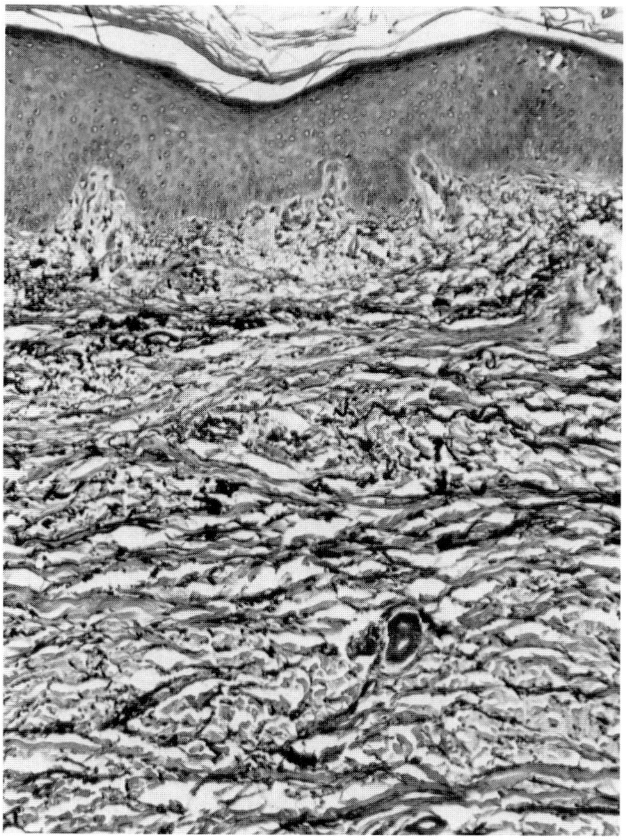

Figure 1 Histology of the elastic fibers in normal human skin. The elastic fibers, which appear as black structures within the dermis, are generally oriented in parallel to the surface of the skin (Verhoeff-van Gieson stain, original magnification X90).

The elastin component is surrounded by distinct fibrillar structures with a uniform diameter of 10–12 nm. These fibrils represent the elastic fiber microfibrillar protein (for details, see Chapter 3). The relative proportions of the two components of the elastic fibers change during development: in the early stages of fetal development, elastic fibers consist mainly of the microfibrillar component, which serves as a scaffold onto which elastin molecules align themselves to form the fiber structure. In a mature, fully developed elastic fiber the elastin represents over 90% of the total content of the fiber (8).

B. Biochemistry of Elastin

1. Structure

The basic molecular unit of elastin is a polypeptide of about 800 amino acids, with an approximate molecular weight of 72,000 daltons. Elastin has a characteristic amino acid composition, rich in glycine, valine, and alanine, while histidine and methionine are absent (Refs. 1-3 and Chapter 2).

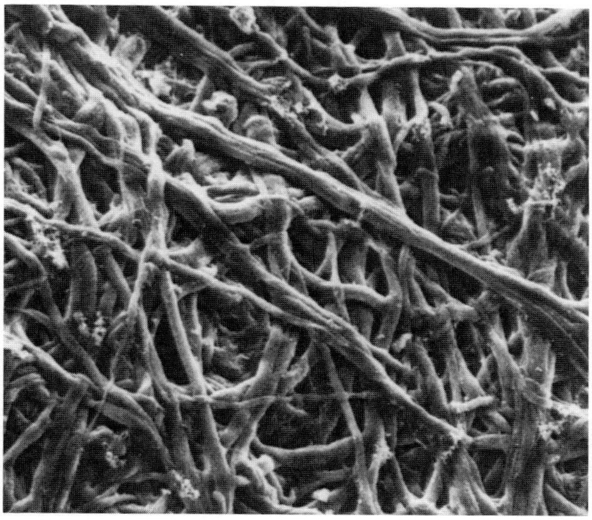

A

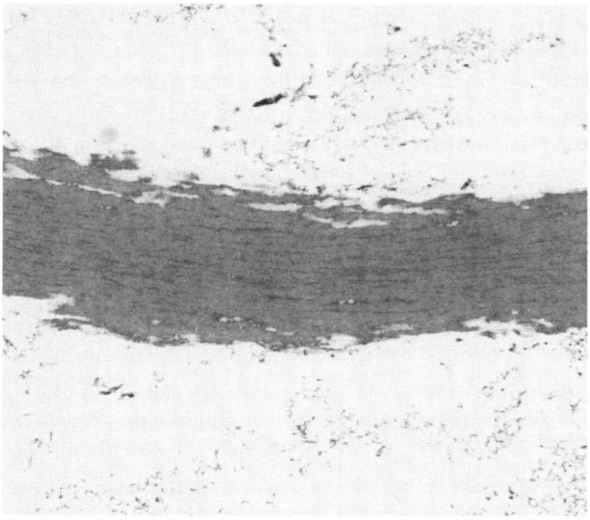

B

Figure 2 Electron microscopic appearance of the elastic fibers in normal human skin. (A) Scanning electron micrograph (original magnification ×670). (B) Transmission electron micrograph (original magnification ×1200).

The insoluble elastin contains two characteristic amino acids, desmosine and isodesmosine, which connect individual elastin polypeptides into a fiber network. The desmosines are not found in any other mammalian protein. The relative concentrations of desmosines are rather invariant in normal elastic fibers, and consequently, desmosines can be assayed as an index of the concentration of properly cross-linked elastic fibers in tissues (see Refs. 1-3,9-11).

The information responsible for the primary amino acid sequence of elastin polypeptides resides in the nucleotide sequence of DNA in the elastin gene (12,13). The messenger RNA molecules coding for elastin are about 3.5 kilobases in size, i.e., considerably larger than the size of 2.4 kilobases that one would predict on the basis of the number of amino acids in an elastin polypeptide. This discrepancy is explained by the fact that elastin messenger RNA contains an unusually long untranslated 3′ end consisting of about 1000 nucleotides (14).

Recent studies have demonstrated that the elastin gene consists of a large number of separate coding sequences (exons) interrupted by long intervening sequences (introns). The exact size and the precise exon-intron organization of the elastin gene are currently being examined. It appears that the entire elastin gene occupies at least 30 kilobases of the genome, and the intron/exon ratio is exceptionally high, being about 15:1 (12-15).

2. *Biosynthesis*

Translation of the elastin messenger RNA takes place in the rough endoplasmic reticulum of smooth muscle cells, as well as of skin, lung, and ligamentum nuchae fibroblasts (16-20). The newly synthesized polypeptides, which initially contain a short signal sequence, are fed into the cisternae of the rough endoplasmic reticulum. Inside the lumina of the rough endoplasmic reticulum, hydroxylation of some prolyl residues takes place in a reaction catalyzed by prolyl 4-hydroxylase. The newly synthesized polypeptides are then transferred from the rough endoplasmic reticulum to the extracellular space by transport mechanisms that involve microtubules of the cell (see Refs. 1-3 and Chapter 2).

3. *Fibrillogenesis*

Following secretion, the elastin polypeptides align into fiber structures, which are stabilized by the formation of desmosines linking individual chains into a fiber network. The formation of the two isomeric forms of desmosine cross-links, desmosine and isodesmosine, requires oxidative deamination of certain lysyl residues to form aldehyde derivatives, allysines. This reaction is catalyzed by lysyl oxidase, an extracellular enzyme that requires copper and molecular oxygen as cofactors. The formation of allysines is the first step in the cross-linking process, followed by the fusion of three allysines and an unmodified lysyl residue into a stable desmosine (1-3,9,21).

Several models for the supramolecular organization of the elastic fibers have been proposed to account for the elastic nature of the fibers. One model depicts elastic fibers as an "oiled-coil," which contains sequences rich in hydrophobic amino acids. During stretching of the elastic fibers, the hydrophobic groups of these amino acids would then be exposed to surrounding aqueous phase, and the energy required for the contraction of the fibers is derived from the return of these groups to a nonpolar environment (22). Alternatively, the elastic fibers have been depicted as rubberlike elastomers without an ordered supramolecular organization (23).

4. *Degradation*

Although the metabolic turnover of elastin is relatively slow in physiological situations, some elastic structures are continuously degraded and replaced by newly synthesized fibers. Also, degradation of elastin can be markedly increased in certain pathological conditions (1-3). The elastic fibers are degraded by elastolytic enzymes, elastases, which are present in several tissues and cells (24,25). Some of the elastases, such as the pancreatic and leukocyte enzymes, are serine proteases that degrade elastin at neutral or slightly alkaline pH. The activity of these enzymes is inhibited by serum factors, such as α1-protease inhibitor and α2-macroglobulin. Furthermore, elastaselike metalloproteases have been demonstrated in macrophages isolated from peritoneal inflammatory exudates, and in human skin fibroblast cultures (25,26). Neutral elastaselike metalloenzymes have been characterized and partially purified from human serum and rheumatoid synovial fluid (27,28).

III. ELASTIC FIBERS IN DISEASES

A. Evidence for Involvement of Elastic Fibers in Diseases

Several acquired and heritable diseases are known to involve abnormalities in the elastic fibers (Tables 1 and 2). The evidence for pathological changes in the elastic fibers has been largely based on histopathological observations, utilizing elastin-specific staining techniques. Furthermore, examination of the elastic fibers by electron microscopy has revealed distinct ultrastructural alterations in diseases. More recently, concomitant with the progress made in understanding the normal biochemistry of the elastic fibers, it has become possible to identify specific molecular defects in elastin diseases (see Refs. 1-3, 29–32). With the advent of the techniques of molecular biology, it is expected that knowledge of changes in elastic fibers in clinical conditions will rapidly accumulate, and the application of the recombinant DNA technology will allow us to link specific gene defects into the clinical phenotype of diseases more closely than has been possible before.

The clinical conditions in which previous morphological and biochemical observations suggest changes in the elastic fibers are listed in Tables 1 and 2. As examples of such diseases, we will discuss in detail selected conditions that serve as prototypes of elastin diseases. In some of these conditions, there is evidence for a specific molecular defect underlying the clinical manifestations.

B. Definition of "Elastin Disease"

As indicated above, the elastic fibers are affected in a variety of both heritable and acquired diseases. The abnormalities in these fibers can result from several different kinds of molecular defects (see below). On the basis of such information, we have suggested that the term "elastin disease" could be applied to pathological conditions in which elastin or enzymes participating in its metabolism are abnormal and the abnormalities can be demonstrated on the molecular level (2). It should be noted that defects in the structure or metabolism of the microfibrillar component can also contribute to the pathology of elastic fibers.

C. Possibilities for Molecular Defects in Elastin Diseases

As discussed above, the elastic fibers consist of two major components, elastin and the microfibillar protein, whose molecular interactions are required for the formation of functional fibers. The synthesis and degradation of both elastin and the microfibrillar protein involve several distinct steps that are under stringent biological control. The events involved in the metabolism of the microfibrillar protein are not well characterized

Table 1 Heritable Diseases with Abnormal Elastic Fibers: Clinical Features, Histopathology, and Associated Biochemical Findings

Disease[a]	Major clinical features	Histopathology	Inheritance[b]	Biochemical findings related to elastic fibers[c]
Pseudoxanthoma elasticum	Yellowish papules coalescing to plaques; cardiovascular problems; angioid streaks	Accumulation of fragmented and calcified elastic fibers in middermis	AR, AD[d]	Deposition of calcium hydroxyapatite crystals; increased desmosine concentration in skin lesions
Elastoderma	Pendulous, irregulary contoured inelastic skin; stretching results in incomplete and slow recoil	Accumulation of pleomorphic elastic fibers in lower dermis and subcutis	?	Increased desmosine content of the skin (Ref. 42)
Endocardial fibroelastosis	Congestive heart failure; cardiac arrhythmias (Ref. 67)	Dense accumulation of elastic fibers in endocardium	AR(?)	
Cutis laxa	Loose, sagging skin	Fragmentation and loss of elastic fibers of the dermis	AD, AR, XR[d]	Increased (metallo)elastase activity in one patient with pulmonary emphysema (Ref. 28)
α1-Antitrypsin deficiency	Pulmonary emphysema	Fragmentation and loss of elastic fibers of the lungs	AR	Imbalance between elastase and its inhibitors

Marfan syndrome	Skeletal and cardiovascular abnormalities; loose-jointedness and hyperextensible skin	Fragmentation of the elastic structures in aorta	AD	Decreased desmosine content in elastin isolated from aorta of some patients (Refs. 68,69)
Menkes syndrome	Tortuosity of the blood vessels; brittle hair; cerebral degeneration; bladder diverticula; mild skin hyperextensibility	Fragmentation and reduplication of internal elastic lamina	XR	Decreased lysyl oxidase activity due to abnormal copper metabolism (Ref. 62)
Ehlers-Danlos syndrome type IX	Inguinal hernias; bladder diverticula; mild skin hyperextensibility, skeletal abnormalities	—	XR	Decreased lysyl oxidase activity due to abnormal copper metabolism (Ref. 60)
Buschke-Ollendorff syndrome	Yellowish dermal papules associated with osteopoikilosis	Increase in thick, branching elastic fibers in the dermis and loss of the microfibrillar component of the elastic fibers in dermis	AD	Increased desmosine content of skin (Ref. 66)

[a]Several of these conditions represent a group of diseases with clinical, genetic, and biochemical heterogeneity.
[b]AD, autosomal dominant; AR, autosomal recessive; XR, X-linked recessive.
[c]These biochemical abnormalities have been demonstrated only in a few patients in each group, and it is not known whether the biochemical defects are the same in each patient with that diagnosis.
[d]Acquired diseases with similar clinical presentations are found (see Table 2).

Table 2 Acquired Diseases with Abnormal Elastic Fibers: Clinical Features, Histopathology, and Associated Biochemical Findings

Disease	Major clinical features	Histopathology	Biochemical findings related to elastic fibers
Actinic elastosis (solar elastosis)	Thickening and furrowing of the skin; loss of elasticity	Accumulation of irregularly thickened elastic fibers in upper dermis	Increased desmosine content of the lesional skin
Isolated elastomas	Dermal papules or nodules	Accumulation of elastic fibers in dermis	–
Elastofibroma dorsi	Deep subcutaneous tumor on subscapular area	Accumulation of globular elastic structures encased in collagenous meshwork	–
Elastosis perforans serpiginosa	Hyperkeratotic papules on the face and neck	Accumulation and transpidermal elimination of elastic fibers	–
Pseudoxanthoma elasticum (see also Table 1)	Yellowish papules in periumbilical distribution; no systemic involvement of the elastic fibers	Accumulation of fragmented and often calcified elastic fibers in middermis	–
Cutis laxa (see also Table 1)	Postinflammatory development of loose and sagging skin	Fragmentation and loss of elastic fibers of the skin; presence of inflammatory cell infiltrates	Release of elastases by inflammatory cells (?)
Anetoderma (Ref. 56)	Localized areas of flaccid, saclike skin	Localized loss and fragmentation of elastic fibers in dermis	–
Pulmonary emphysema (see Chap. 19)	Chronic obstructive pulmonary disease	Loss and fragmentation of elastic fibers in the lung; inflammatory cell infiltrates	Inactivation of α_1 proteinase inhibitor due to oxidation (e.g., by cigarette smoke); increased release of elastase by leukocytes
Atherosclerosis (Ref. 70)	Obstructive arterial disease	Fragmentation and calcification of elastic fibers in the blood vessels	–
Hypothyroid myxedema, pretibial myxedema (Ref. 71)	Diffuse induration of the skin associated with thyroid disease	Decreased amount of elastic fibers with fragmented, irregular appearance; decreased microfibrils	–

(for details see Chapter 3). In contrast, it is clear that the synthesis and degradation of elastin involve several defined steps, many of which are enzymatically catalyzed. Thus, the complexity of biochemistry of the elastic fibers offers several distinct levels at which errors could be introduced to these fibers, and the abnormalities would then manifest as a clinical disease.

Errors could be introduced into the elastic fibers by the following mechanisms:

1. Defects in the structure of elastin or the microfibrillar protein, as detectable in the nucleotide sequences of the genes coding for the corresponding protein.
2. Altered production of structurally normal protein as a result of an error on the transcriptional or translational level of gene expression.
3. Defects in the posttranslational processing of the newly synthesized proteins as a result of aberrant activity of the enzymes participating in the biosynthetic pathway.
4. Aberrations in the degradative pathway, as a result of alterations either in the intracellular degradation of newly synthesized polypeptides or in the extracellular removal of the elastic fibers.

D. Prototypes of Elastin Diseases

1. Diseases of Elastin Accumulation

Several diseases with altered mechanical properties of the skin result from increased elastic fiber content of the dermis, as determined by histopathological and biochemical means. In most cases, the accumulated elastotic material is also ultrastructurally abnormal, and the exact biochemical nature of this material is not known. Nevertheless, the accumulation of elastic material must reflect an imbalance between the rates of production and removal. Increase in the production could result from enhanced expression of the corresponding genes, while impaired removal might reflect reduced elastolytic activity or diminished susceptibility of the elastic fibers to proteolysis.

Several heritable and acquired diseases involve accumulation of the elastotic material in skin and other tissues; these include pseudoxanthoma elasticum (PXE), elastoderma, isolated elastomas, elastofibroma dorsi, actinic elastosis, elastosis perforans serpiginosa, and endocardial fibroelastosis (Tables 1 and 2). As prototypes of the elastin accumulation in diseases, we will discuss PXE, elastoderma, and actinic elastosis in detail.

Pseudoxanthoma Elasticum: PXE is a generalized progressive connective tissue disorder primarily affecting the elastic fibers (33,34). The major clinical finding in PXE is redundant and inelastic skin. The primary cutaneous lesions are small, yellowish papules, which give the affected area a pebbly, "plucked-chicken skin" appearance. The primary lesions coalesce into larger plaques, which with time become thickened and leathery. In addition to cutaneous manifestations, the patients often demonstrate ocular changes and wide-spread vascular abnormalities. Variable severity of these changes leads to presentation of a wide range of clinical pictures.

The ocular changes are characterized by angioid streaks, i.e., grayish or brownish-red, poorly defined streaks radiating across the fundus of the eye. Their development usually starts later than that of the cutaneous lesions, often during the third or fourth decade of life. The ocular changes, often bilateral, frequently lead to impaired vision but rarely result in blindness.

The vascular involvement is manifested as hypertension, weak peripheral pulses, and, occasionally, intermittent claudication. The most devastating complications result from coronary occlusion, cerebral hemorrhage, and, frequently, recurrent bleeding from the gastrointestinal tract.

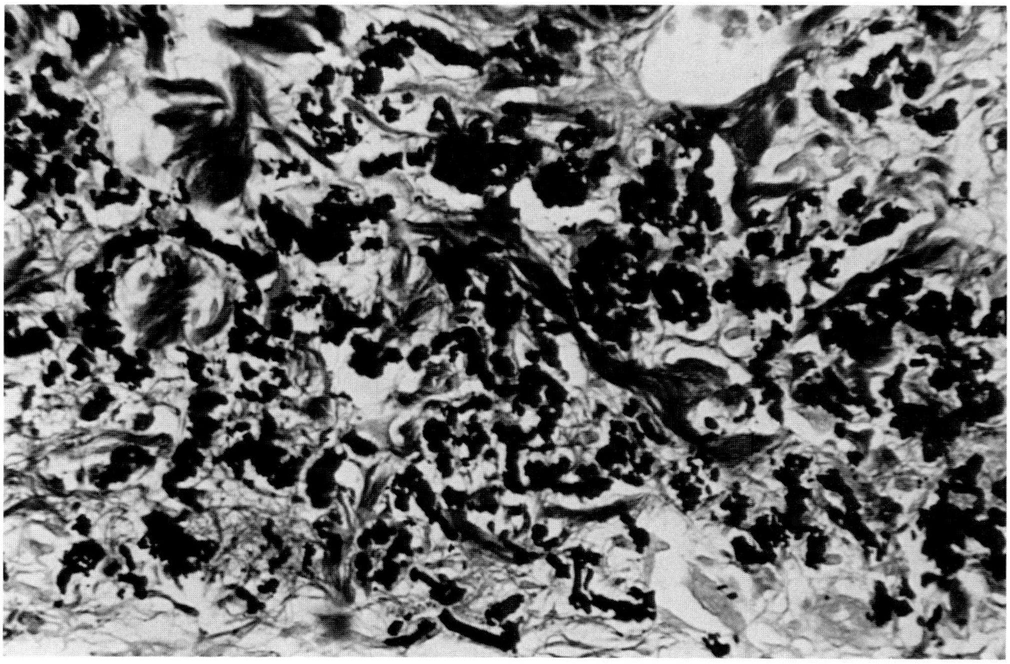

A

Figure 3 Morphology of the elastic fibers in pseudoxanthoma elasticum. (A) Histopathology demonstrates the accumulation of pleomorphic elastotic structures in middermis. (Verhoeff-van Gieson stain, original magnification × 350). (B) Transmission electron micrograph demonstrating an abnormally broad, electron-lucent elastic fiber containing electron-dense deposits of calcium hydroxyapatite (original magnification ×15,000).

In most cases, PXE is inherited as an autosomal recessive condition (33). However, autosomal dominant inheritance has been documented in a few families. The classification proposed by Pope divides the dominantly inherited form of PXE into two categories (35). The type I dominant form is characterized by classic cutaneous changes associated with severe vascular complications. The type II dominant form is characterized by focal cutaneous involvement associated with hyperextensible skin, blue sclerae, arched palate, and loose-jointedness. The type I recessively inherited form of the disease is the classic, most frequently encountered type of PXE with typical cutaneous, vascular, and ocular manifestations. The type II recessively inherited disease is characterized by extensive cutaneous involvement and the absence of vascular and ocular manifestations; this form of PXE is very rare.

In addition to the inherited forms, several cases with cutaneous findings consistent with PXE, but without family history and without vascular or ocular involvement, have been reported (34,36,37). In some of these cases, the development of skin lesions is related to external exposure to chemicals, such as Norwegian saltpeter (36). Recently, an unusual perforating variant of cutaneous PXE has been described (37). In these patients, the lesions have been confined to the abdomen in a periumbilical distribution.

Examination of the skin lesions by light microscopy demonstrates in the middle or lower dermis an accumulation of material that stains as elastin (Figure 3A). In contrast to normal elastic fibers, the elastotic material in PXE appears fragmented and irregularly

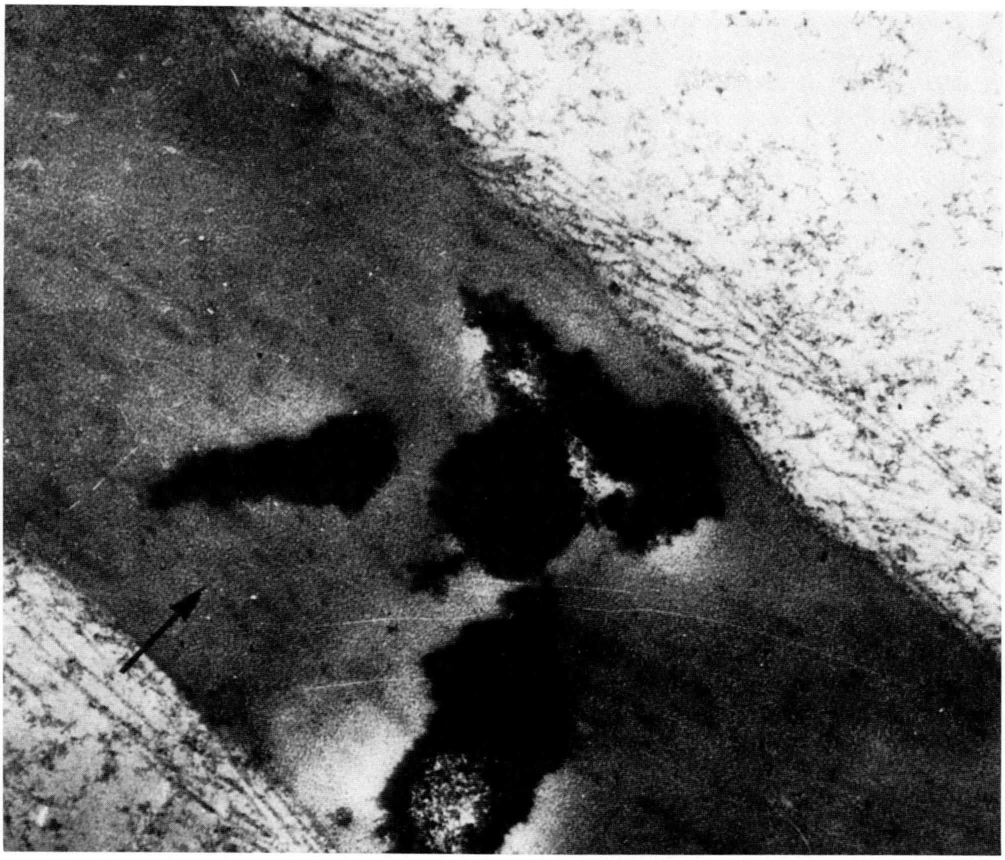

B

Figure 3 (Continued)

clumped. The accumulation of elastotic material has also been quantitated by computerized morphometric analyses and by a radioimmunoassay for desmosine (38). Characteristically, the fragmented elastic fibers contain calcium, which can be demonstrated by calcium-specific stains. Examination of the elastic fibers in the affected area of the skin by electron microscopy has demonstrated that in some areas, the amorphous elastin component has been replaced by bundles of granular material with staining properties different from those of normal elastin (39). In addition, foci containing calcium hydroxyapatite crystals can be detected in the elastic fibers (Figure 3B). These morphological findings thus provide evidence for derangement in the organization of the elastic structures in PXE.

The exact molecular defect in the structure or metabolism of elastin in PXE is currently unknown. It is even unclear whether the calcification of the elastic fibers is a primary or secondary event. Studies employing cultured skin fibroblasts have shown that these cells derived from patients with PXE produce an abnormal polyionic compound, a glycosaminoglycan, which is deposited on the surface of the elastic fibers (40,41). This material could then bind calcium resulting in calcification of the elastic fibers, and subsequently leading to fiber degeneration. Alternatively, an aberration in the primary struc-

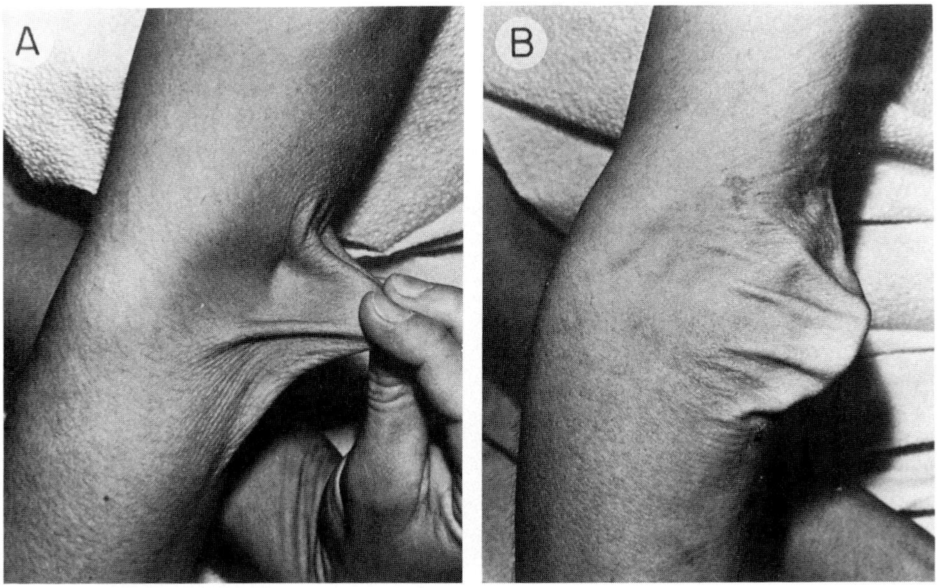

Figure 4 Clinical presentation of elastoderma. Note the laxity of the skin on elbow (A); upon stretching, the skin demonstrates reduced and incomplete recoil (B). (Published with permission from Ref. 42.)

ture of elastin or the enzymes participating in the synthesis and degradation of elastic fibers may be the primary underlying molecular defect in PXE. Finally, the mechanisms of the excessive accumulation of elastotic material in the affected tissues are unclear.

Elastoderma: A novel disease of elastin accumulation within the skin, referred to as elastoderma, has recently been delineated (42). Clinically, the patient was a young female who developed progressive involvement of the skin on the upper trunk and right arm; the skin in these areas became pendulous and irregularly contoured (Figure 4). The affected skin, when grasped between the fingers, felt irregularly firm and lacked the uniform consistency of normal skin. Stretching of the skin was followed by incomplete and delayed recoil (Figure 4B). No evidence of systemic involvement or history of affected family members was available.

Histopathology of the skin demonstrated dense aggregates of elastotic material present throughout the dermis and replacing the subcutaneous adipose tissue (Figure 5A). Unlike normal skin, the elastic fibers were pleomorphic, and only a few normal-appearning elastic fibers were present within the skin. There was no evidence of calcification using the Von Kossa stain. Transmission electron microscopy revealed apposition of abnormal structures in association with, if not in direct extension of, normal elastic fibers (Figure 5B). The abnormal-appearing material was more electron dense than normal elastic fibers. The elastotic structures in the affected areas were unique and contained grapelike globular structures when examined by scanning electron microscopy (Figure 5C). The concentration of desmosine in the lesional skin was significantly higher than that within the unaffected skin of the same patient or in skin from normal, unrelated control subjects.

Thus, by both morphological and biochemical criteria, elastoderma is a disorder of elastin accumulation within the dermis and the subcutaneous tissue. The presence of fibroblastic cells with dilated lumina of rough endoplasmic reticulum, indicating active protein synthesis, suggested that increased elastin production is responsible for elastin accumulation in the lesional skin.

Actinic Elastosis: Actinic elastosis or solar elastosis, which has been variously termed in clinical context as sailor's skin, farmer's skin, and senile elastosis, is characterized by furrowing and thickening of the skin in sun-exposed areas (43). The histopathological changes include basophilic degeneration of collagen, accumulation of glycosaminoglycans, and, most strikingly, deposition of elastotic material (44,45). In early stages of actinic damage, the changes are particularly pronounced in the papillary dermis, and in advanced cases similar pathological alterations are also noted in the reticular dermis (Figure 6).

Transmission electron microscopy of the actinically damaged skin demonstrates an accumulation of elastotic material, which differs in appearance from the elastic fibers seen in normal skin (Figure 7). The main part of the elastotic material consists of a fine granular matrix, and embedded in this matrix there are homogenous, electron-dense inclusions of irregular shape. In skin subjected to relatively mild actinic damage, there is an increase and enlargement of the microfibrillar component of the elastic fibers (Figure 7A), while in severely damaged skin, the microfibrils are absent or they can be recognized only as disorganized electron-dense regions along the periphery of the elastic fibers (Figure 7B) (46-48).

The biochemical alterations accompanying the age-related cutaneous changes are not particularly well delineated. Biochemical analyses of the dermis in the protected area of skin have suggested a small but steady increase in the elastin content between the ages of 20 and 80 years (49). Some studies have suggested that the elastin in solar elastosis is similar to normal dermal elastin in its amino acid composition (49,50). However, detailed analyses of the microfibrillar protein or changes in the elastin cross-link profile during aging have not been carried out.

We have recently performed biochemical analyses of skin obtained from areas that clinically showed severe actinic damage. Histologically, the biopsy specimens demonstrated a massive deposition of elastotic material in superficial and middermis (see Figure 6B). The concentration of desmosine, determined by a radioimmunoassay, was increased fourfold compared to that in unexposed skin from the same individuals. At the same time, the mean concentration of collagen was somewhat decreased. These observations suggest that the elastotic material, which appears to replace collagen in the upper parts of the dermis, contains some cross-linked elastic fibers. However, a fourfold increase in desmosine concentration does not account for the massive deposition of the elastotic material in tissue, indicating that the majority of this material is not composed of normal elastic fibers (32). The fibroblasts present in the actinically damaged skin often demonstrate a pronounced rough endoplasmic reticulum containing amorphous, granular material, suggesting active synthesis of matrix macromolecules (32).

2. *Diseases with Reduced Elastin Content of the Tissues*

Some disease processes involve loss of the elastic fibers with concomitant alterations in the mechanical properties of the tissue (Tables 1 and 2). The best example of such conditions is pulmonary emphysema, which is discussed in detail in Chapter 19. Two cutaneous diseases, cutis laxa and anetoderma, also involve loss and fragmentation of elastic fibers in the dermis.

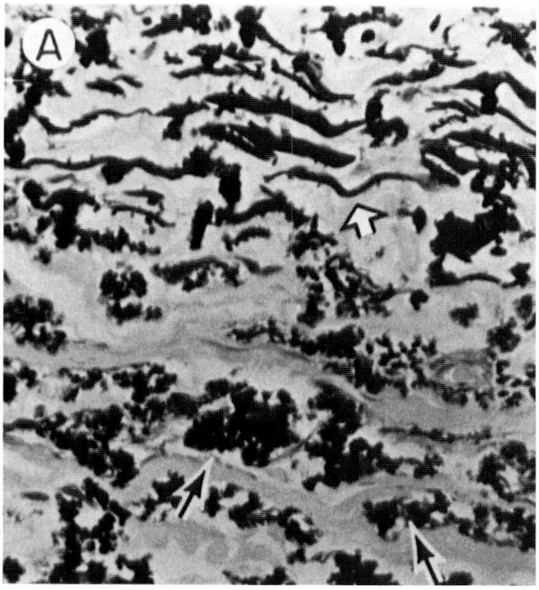

Figure 5 Morphology of the elastotic material that accumulates within the skin in elastoderma. (A) Histopathology reveals few normal-appearing elastic fibers (white arrow), while pleomorphic elastic structures (black arrows) accumulate in the lower dermis and the subcutaneous tissue (Verhoeff-van Gieson stain, original magnification X90). (B) Transmission electron microscopy demonstrates abnormal electron-dense structures (✪) in direct continuity with normal-appearing elastic fibers (✱) (original magnification X 18,000). (C) Scanning electron microscopy of elastic structures isolated by extraction with 0.1 N socium hydroxide at 100°C reveals grapelike structures (✪) in association with normal-appearing elastic fibers (✱) (original magnification X5000). (Published with permission from Ref. 42.)

Cutis Laxa: The characteristic cutaneous feature of cutis laxa is redundant, loose, and pendulous skin, which is inelastic and lacks recoil (33). The skin on the face frequently forms sagging folds, giving the patient a prematurely aged appearance (Figure 8).

Both the clinical presentation and mode of inheritance of cutis laxa demonstrate considerable heterogeneity (51-54). The patients frequently show evidence of a more generalized connective tissue involvement. Most of the cases with the inherited form of the disease suggest an autosomal recessive inheritance, but cases with autosomal dominant or X-linked recessive pattern have been reported (Table 1) (33). In addition to the heritable forms of the disease, typical cutaneous changes can develop as a result of extensive inflammatory reaction in an acquired form of cutis laxa (Table 2) (55).

In several cases of cutis laxa, the amount of elastic fibers has been shown to be reduced, and the remaining fibers present in the skin are fragmented (Figure 9). The loss of elastic fibers could result from several different kinds of defects in the synthesis or degradation of the functional elastic fibers. First, the production of structurally normal elastin by fibroblasts or other dermal cells responsible for elastin synthesis could be markedly reduced. Second, the degradation of elastic fibers could be increased as a result of an imbalance between elastases and their inhibitors. The latter mechanism appears to be

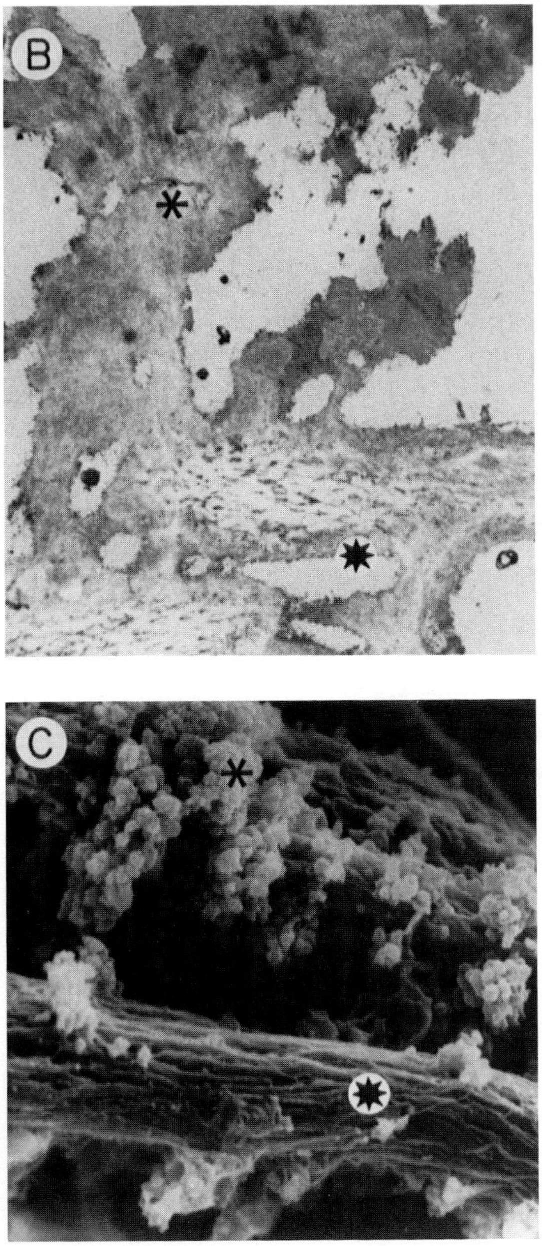

Figure 5 (Continued)

operative in a patient with an autosomal recessive pulmonary emphysema and cutis laxa associated with markedly elevated serum elastaselike metalloproteinase activity (28). Similarly, in cases with postinflammatory elastolysis, the polymorphonuclear leukocytes could be a source of elastases, leading to loss and fragmentation of the elastic fibers. A similar mechanism has been suggested to lead to loss of elastic fibers in anetoderma, a localized cutaneous disease characterized by multiple lesions with soft and sagging skin

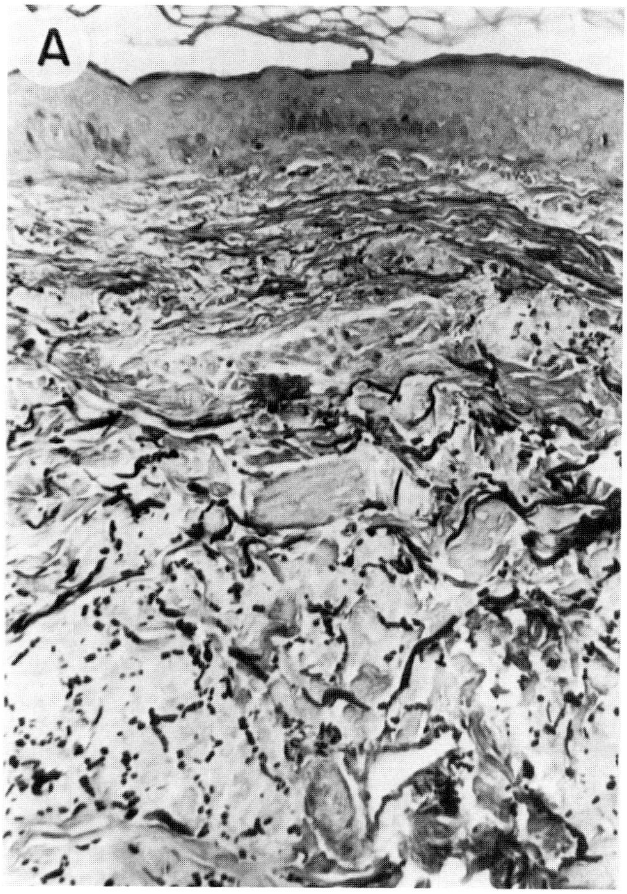

Figure 6 Histopathology of the dermis in actinic elastosis. Note the accumulation of elastotic material in the upper dermis in the early stages of actinic damage (A). Similar, but more severe, pathological alterations can be seen in the upper and middermis in an advanced case (B) (Verhoeff-van Gieson stain, original magnification X90). (Published with permission from Ref. 32.)

(56,57). Finally, the elastin synthesized by cells could be structurally abnormal either as a result of alterations in the primary sequence or owing to altered posttranslational modifications; the abnormal protein would then be more susceptible to proteolysis by elastases.

3. *Defects in the Fibrillogenesis of the Elastic Fibers*

The fibrillogenesis of elastin to form functional fibers requires a delicate balance between the synthesis of elastin polypeptides and the microfibrillar protein component. Furthermore, the stabilization of the fiber structures by desmosines requires several well-controlled reactions, including lysyl oxidase catalyzed conversion of lysines to allysines (see Refs. 1-3). Thus, aberrations in the fibrillogenesis can lead to functionally compromised elastic fibers. Several diseases are known to involve abnormalities in the

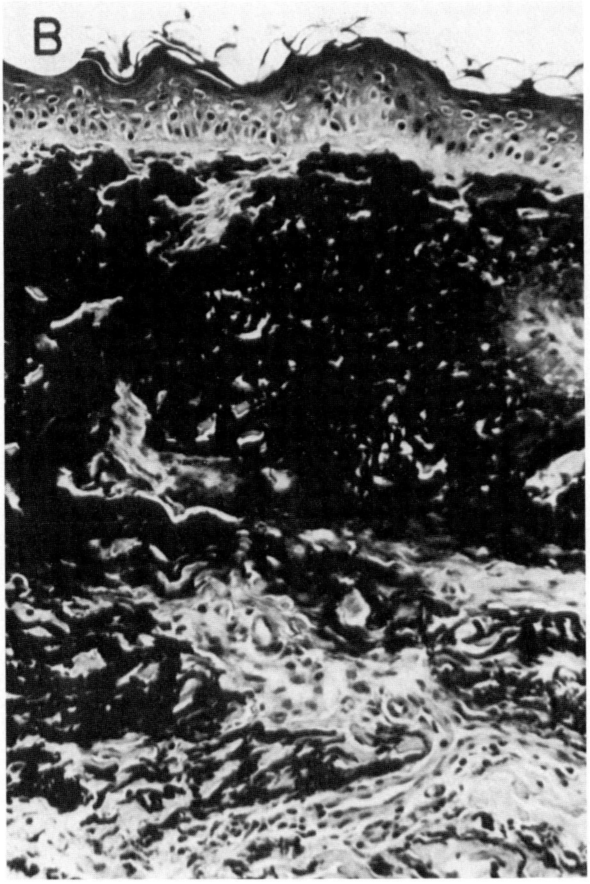

Figure 6 (Continued)

fibrillogenesis of the elastic fibers. Some of them, such as the copper-deficiency syndromes, involve alterations in the activity of lysyl oxidase. Others, such as the Buschke-Ollendorff syndrome, may result from an imbalance between the synthesis of elastin an the microfibrillar protein.

Copper-Deficiency Syndromes: As previously described, lysyl oxidase is critical to the formation of desmosines. Since copper is a necessary cofactor for lysyl oxidase activity, copper deficiency can lead to reduction in the activity of this enzyme and subsequently result in deficient cross-linking (58). Previously, two patients diagnosed as having X-linked recessive cutis laxa have been found to have deficient lysyl oxidase activity (59). More recently, several additional patients with similar clinical observations, including hyperextensible skin, were shown to have reduced lysyl oxidase activity as a result of alterations in the copper metabolism. The latter patients, however, demonstrated clinical signs that are more compatible with the Ehlers-Danlos syndrome than cutis laxa and were diagnosed as having the Ehlers-Danlos syndrome type IX (see Chapter 12). Thus, it is possible that copper deficiency in both reported families would be compatible with the diagnosis of Ehlers-Danlos syndrome (59-61).

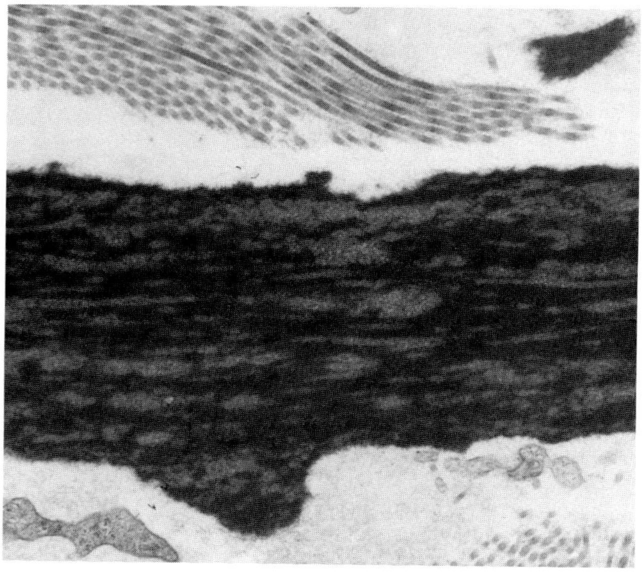

A

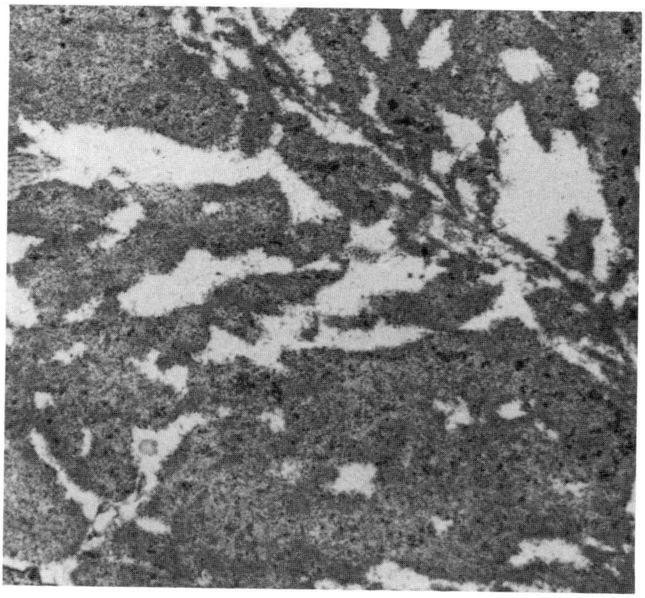

B

Figure 7 Transmission electron microscopy of the elastic structures in actinic elastosis. (A) Increase and enlargement of the microfibrillar component of an elastic fiber (original magnification ×4500). (B) Deposition of fine electron-dense granular material (original magnification ×4500). (Published with permission from Ref. 32.)

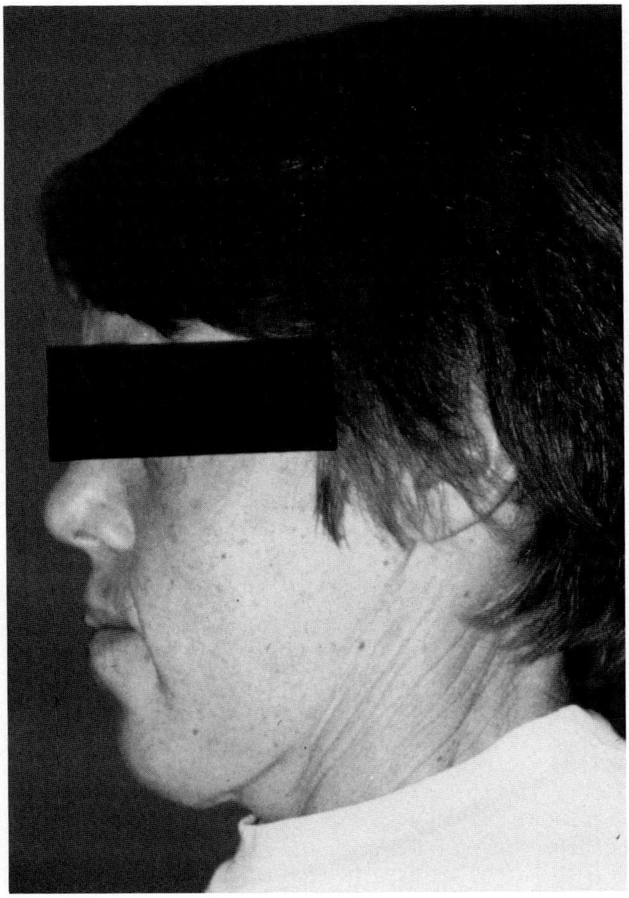

Figure 8 Clinical appearance of a 15-year-old patient with cutis laxa. Note the loose, sagging skin and the premature aging appearance on the face and neck.

Another heritable condition, the Menkes syndrome, also involves abnormalities in the absorption and cellular transport of copper. Patients with the Menkes syndrome clearly demonstrate connective tissue abnormalities including fragmentation and reduplication of the internal elastic lamina of the arterial vessels. Since the activity of lysyl oxidase in fibroblasts derived from patients with Menkes syndrome has been shown to be markedly reduced, the basis of the connective tissue manifestations in this syndrome could well involve defective cross-linking of elastin (61,62). Furthermore, the defect in the copper metabolism seems to be similar in nature in both Menkes' syndrome and Ehlers-Danlos syndrome type IX (61).

Buschke-Ollendorff Syndrome: Clinically, this syndrome consists of typical cutaneous lesions (dermatofibrosis lenticularis disseminata) and bone lesions (osteopoikilosis) (63, 64). The Buschke-Ollendorf syndrome is inherited as an autosomal dominant disease with almost complete penetrance. The expressivity is somewhat variable, and individuals with either skin lesions or bone manifestations have been reported within the families with this syndrome.

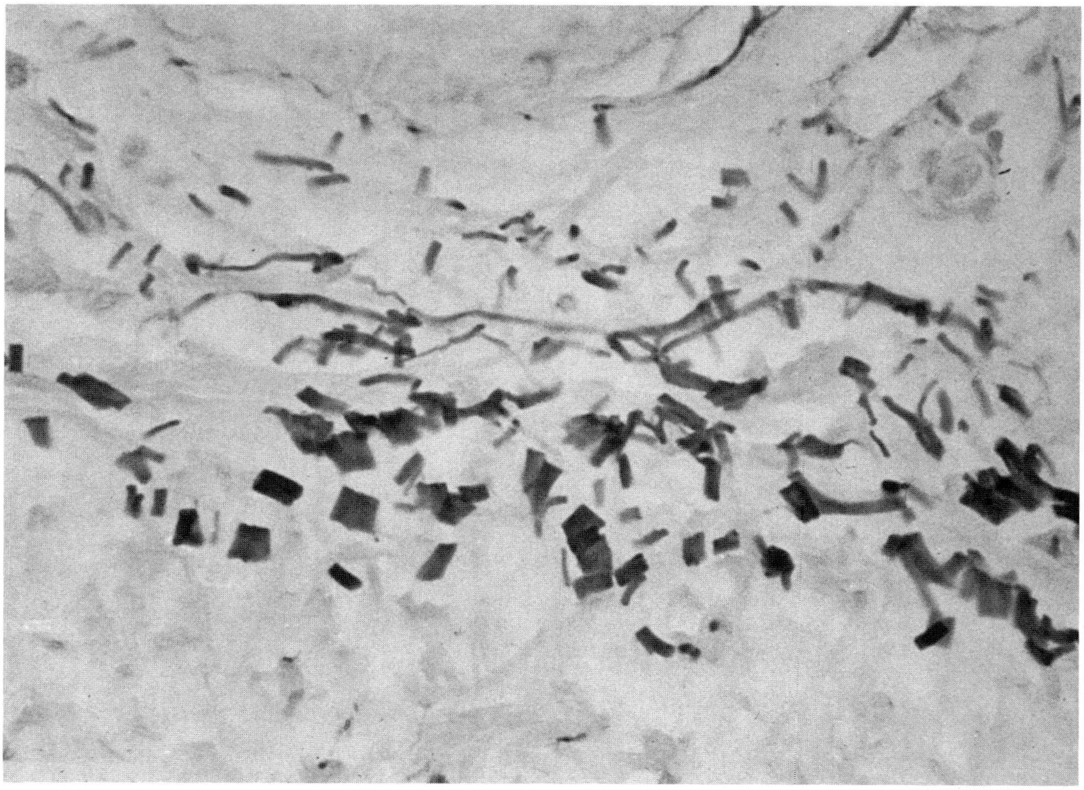

Figure 9 Histopathology of the dermis in a patient with cutis laxa, demonstrating paucity and fragmentation of the elastic fibers (orcein stain, original magnification X120).

The skin lesions are small, yellowish papules, usually distributed symmetrically on the lower trunk or proximal extremities. Histologically, the skin lesions are characterized by an accumulation of broad, branching, and interlacing elastic fibers (Figure 10A) (65,66). Examination of the cutaneous elastic fibers by transmission electron microscopy reveals that they are peculiarly branched and their diameter varies considerably. Of particular interest is the observation that the elastic fibers are largely devoid of the microfibrillar component (Figure 10B). Furthermore, the abnormal elastic fibers are often found in the proximity of a fibroblastic cell with dilated cisternae of the rough endoplasmic reticulum.

The accumulation of elastin in the cutaneous lesions has been verified biochemically, as indicated by a three- to sevenfold increase in desmosine content in comparison to perilesional, unaffected skin (66). The exact molecular basis of the increased elastin content, as defined by morphological and biochemical criteria, is unknown. However, the presence of fibroblasts with dilated cisternae of rough endoplasmic reticulum, indicating active protein synthesis, suggests that production of elastin by these cells is increased. On the other hand, the abnormalities in the fiber structure, as detected by electron microscopy, indicate a faulty fibrillogenesis, which may be a reflection of the paucity of the micro-

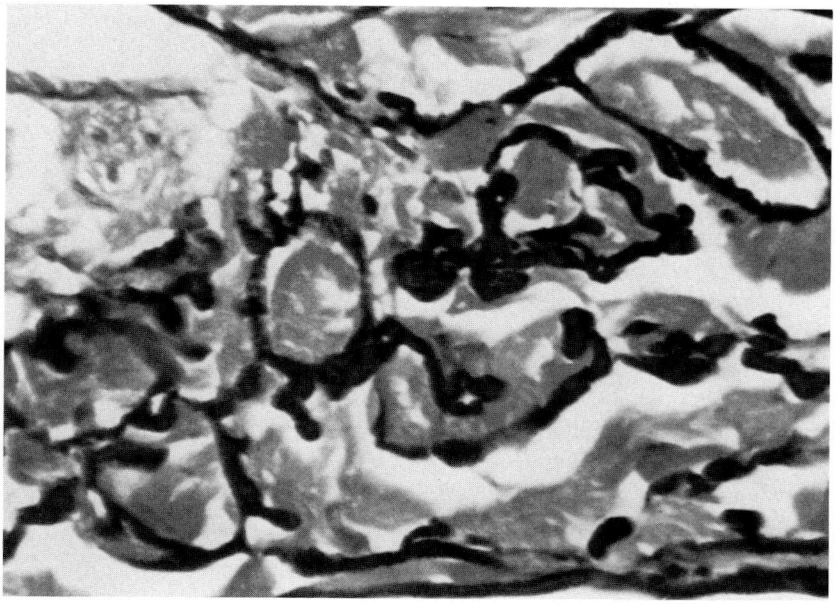

A

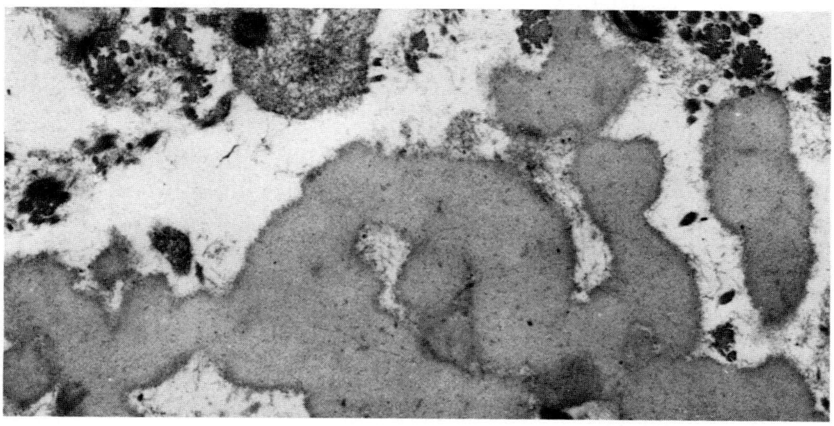

B

Figure 10 Demonstration of abnormal elastic structures in the dermis of a patient with Buschke-Ollendorf syndrome. (A) Histopathology reveals unusually thick, interlacing elastic fibers in the middermis (Verhoeff-van Gieson stain, original magnification ×120). (B) Transmission electron microscopy reveals irregular elastic fibers with variable diameter. Note the paucity of the microfibrillar component (original magnification ×11,000).

fibrillar component. Furthermore, it is conceivable that the overproduction of elastin may well result from the lack of control, perhaps through feedback mechanisms.

IV. CONCLUSIONS

The mechanical properties of the tissues such as skin and blood vessels are largely attributable to the fibrillar components of the extracellular matrix, collagen and elastin. The evidence presented in this overview suggests that an intact elastic fiber network provides resiliency and elasticity to the tissues. In some elastin diseases, for example, actinic elastosis, the histopathological changes consist of an apparent increase in elastotic material, which, however, is abnormal by histological staining characteristics and ultrastructural features. The accumulation of abnormal elastic fibers in these conditions is accompanied by laxity and loss of recoil of the skin. On the other hand, similar functional alterations in the skin can be encountered in diseases with reduced quantity or fragmentation of the elastic fibers, for example, cutis laxa. Thus, we can conclude that the elasticity of the skin is based on the presence of both quantitatively and qualitatively normal elastic fiber network, with proper assembly and physiological interactions with other dermal connective tissue components.

ACKNOWLEDGMENTS

The authors thank Charlene D. Aranda for secretarial assistance. The original research by the authors summarized in this work was supported in part by United States Public Health Service, National Institutes of Health grants AM-28450, AM-35297, GM-28833, and AG-03172.

REFERENCES

1. L. Ryhänen and J. Uitto. In *The Biochemistry and Physiology of the Skin* (L. Goldsmith, ed.), Oxford University Press, New York, 1982, p. 433.
2. J. Uitto, L. Ryhänen, P. Abraham, and A. J. Perejda, *J. Invest. Dermatol., 7*(Suppl. 1):160s (1982).
3. J. Rosenbloom, *Lab. Invest., 51*:605 (1984).
4. J. Uitto, D. J. Santa Cruz, and A. Z. Eisen, *J. Am. Acad. Dermatol., 3*:441 (1980).
5. L. B. Sandberg, N. T. Soskel, J. G. Leslie, *N. Engl. J. Med., 304*:566 (1981).
6. R. Ross and P. Bornstein, *J. Cell. Biol., 40*:366 (1967).
7. D. P. Vardi, *J. Invest. Dermatol., 66*:59 (1976).
8. R. Ross, P. J. Fialkow, and L. K. Altman, *Adv. Exp. Biol. Med., 79*:7 (1977).
9. R. B. Rucker and J. Murray, *Am. J. Clin. Nutr., 31*:1221 (1978).
10. B. C. Starcher, *Anal. Biochem., 79*:11 (1977).
11. B. C. Starcher, and R. A. Goldstein, *J. Lab. Clin. Med., 94*:848 (1979).
12. J. M. Davidson and R. G. Crystal, *J. Invest. Dermatol., 79*(Suppl. 1):133s (1982).
13. L. B. Sandberg and J. M. Davidson, *Peptide Protein Rev., 3*:169 (1984).
14. K. Yoon, M. May, Z. K. Goldstein, K. Indik, L. Oliver, C. Boyd, and J. Rosenbloom, *Biochem. Biophys. Res. Commun., 118*:261 (1984).
15. C. Cicila, K. Yoon, C. Boyd, M. May, and J. Rosenbloom, *Fed. Proc., 43*:1853 (1984).
16. A. S. Narayanan, L. B. Sandberg, R. Ross, and D. L. Layman, *J. Cell. Biol., 68*:411 (1976).

17. M. G. Giro, K. E. Hill, L. B. Sandberg, and J. M. Davidson, *Collagen Rel. Res., 4*: 21 (1984).
18. W. H. Carnes, P. A. Abraham, and V. Buonassisi, *Biochem. Biophys. Res. Commun., 90*:1393 (1979).
19. M. G. Giro, A. I. Oikarinen, H. Oikarinen, G. Sephel, J. Uitto, and J. M. Davidson, *J. Clin. Invest., 75*:672 (1985).
20. R. P. Mecham, G. Lange, J. Madaras, and B. Starcher, *J. Cell. Biol., 90*:323 (1981).
21. H. M. Kagan and K. A. Sullivan, *Methods Enzymol., 82A*:637 (1982).
22. W. R. Gray, L. B. Sandberg, and J. A. Foster, *Nature, 246*:461 (1973).
23. J. M. Gosline, *Adv. Exp. Med. Biol., 79*:621 (1977).
24. J. Bieth, *Front. Matrix Biol., 6*:1 (1978).
25. Z. Werb, M. J. Banda, J. H. McKerrow, and R. A. Sandhaus, *J. Invest. Dermatol., 79*(Suppl. 1):1545 (1982).
26. M. C. Bourdillon, D. Brechemier, M. Blaes, J. C. Derouette, W. Hornebeck, and L. Robert, *Cell. Biol. Int. Rep., 4*:313 (1980).
27. K. Katagiri, K. Ito, M. Miyaji, T. Takuchi, K. Yoshikane, and M. Sasake, *Clin. Chim. Acta, 95*:401 (1979).
28. L. L. Anderson, A. I. Oikarinen, L. Ryhänen, C. E. Anderson, and J. Uitto, *J. Lab. Clin. Med., 105*:537-546 (1985).
29. D. W. Urry, *Perspectives Biol. Med., 21*:265 (1978).
30. J. Uitto, *J. Invest. Dermatol., 72*:1 (1979).
31. J. Uitto, in *Dermatology and General Medicine* (T. B. Fitzpatrick, A. Z. Eisen, K. Wolff, I. M. Freedberg, and K. F. Austen, eds.), McGraw-Hill, New York, 1979, p. 182.
32. J. Uitto, L. Y. Matsuoka, and R. L. Kornberg, in *Problems in Aesthetic Surgery: Biological Causes and Clinical Solutions* (R. Rudolf, ed.), C. V. Mosby, St. Louis, (1986).
33. V. A. McKusick, *Heritable Disorders of Connective Tissue*, C. V. Mosby, St. Louis, 1972.
34. J. Uitto, in *Cecil's Textbook of Medicine,* 17th ed. (J. B. Wyngaarden and L. H. Smith, eds.), W. B. Saunders, Philadelphia, 1985, p. 1152.
35. F. M. Pope, *J. Med. Genet., 11*:152 (1974).
36. A. O. Otkjaer-Nielsen, O. B. Christensen, B. Hentzer, B. Johnson, and T. Kobayasi, *Acta Dermatovener., 58*:323 (1978).
37. K. H. Neldner, A. Martinez-Hernandez, *J. Am. Acad. Dermatol., 1*:523 (1979).
38. J. Uitto, J. L. Paul, K. Brockley, R. H. Pearce, and J. G. Clark, *Lab. Invest., 49*: 499 (1983).
39. L. Danielsen, *Acta Dermatovener., 59*(Suppl. 83):1 (1979).
40. A. Martinez-Hernandez and W. E. Huffer, *Lab. Invest., 31*:181 (1974).
41. A. Martinez-Hernandez, W. E. Huffer, K. Neldner, S. Gordon, and E. B. Reeve, *Arch. Pathol. Lab. Med., 102*:303 (1976).
42. R. L. Kornberg, S. Hendler, A. I. Oikarinen, L. Y. Matsuoka, and J. Uitto, *N. Engl. J. Med., 312*:771 (1985).
43. R. Rudolph and M. Woodward, *Plast. Reconstr. Surg., 67*:295 (1981).
44. W. F. Lever and G. Schaumburg-Lever, *Histopathology of the Skin*, Philadelphia, J. B. Lippincott, 1983, p. 271.
45. A. Oikarinen, J. Karvonen, J. Uitto, and M. Hannuksela, *Photodermatology 2*:15-26 (1985).
46. R. M. Lavker, *J. Invest. Dermatol., 73*:59 (1979).
47. T. Tsuji, *J. Invest. Dermatol., 77*:452 (1981).
48. I. M. Braverman and E. Fonferko, *J. Invest. Dermatol., 78*:434 (1982).

49. R. H. Pearce and B. J. Grimmer, *J. Invest. Dermatol., 58*:347 (1972).
50. J. G. Smith, E. A. Davidson, W. M. Sams, and R. D. Clark, *J. Invest. Dermatol., 39*: 347 (1962).
51. R. W. Goltz, A. Hult, M. Goldfarb, and R. J. Gorlin, *Arch. Dermatol., 92*:373 (1965).
52. K. Hashimoto and R. Kanzaki, *Arch. Dermatol., 111*:861 (1975).
53. R. B. Harris, M. R. Heaphy, and H. O. Perry, *Am. J. Med., 65*:815 (1978).
54. F. R. Brown, K. A. Holbrook, P. H. Byers, D. Stewart, J. Dean, and R. E. Pyeritz, *Johns Hopkins Med. J., 150*:148 (1982).
55. W. B. Reed, *Arch. Dermatol., 103*:661 (1971).
56. A. I. Oikarinen, R. Palatsi, G. E. Adomian, H. Oikarinen, J. G. Clark, and J. Uitto, *J. Am. Acad. Dermatol., 11*:64 (1984).
57. P. Y. Venencie and R. K. Winkelmann, *Arch. Dermatol., 120*:1040 (1984).
58. W. H. Carnes, *Int. Rev. Connective Tissue Res., 4*:197 (1968).
59. P. H. Byers, R. C. Siegal, K. A. Holbrook, A. S. Narayanan, P. Bornstein, and J. G. Hall, *N. Engl. J. Med., 303*:61 (1980).
60. H. Kuivaniemi, L. Peltonen, A. Palotie, I. Kaitila, and K. I. Kivirikko, *J. Clin. Invest., 69*:730 (1982).
61. L. Peltonen, H. Kuivaniemi, A. Palotie, N. Horn, I. Kaitila, and K. I. Kivirikko, *Biochemistry, 22*:6156 (1983).
62. P. M. Royce, J. Camakaris, and D. M. Danks, *Biochem. J., 192*:579 (1980).
63. W. F. Schorr, J. M. Optiz, and C. N. Reyes, *Arch. Dermatol., 106*:208 (1972).
64. L. Danielsen, K. Mitgaard, and H. E. Christensen, *Arch. Dermatol., 100*:465 (1969).
65. J. G. L. Morrison, E. J. Wilson Jones, and D. M. McDonald, *Br. J. Dermatol., 97*: 417 (1977).
66. J. Uitto, D. J. Santa Cruz, B. C. Starcher, M. P. Whyte, and W. A. Murphy, *J. Invest. Dermatol., 76*:284 (1981).
67. V. S. Behar, in *Cecil's Textbook of Medicine,* (J. B. Wyngaarden and L. H. Smith, eds.), W. B. Saunders, Philadelphia, 1985, p. 332.
68. P. A. Abraham, A. J. Perejda, W. H. Carnes, and J. Uitto, *J. Clin. Invest., 70*:1245, (1982).
69. T. Halme, T. Vihersaari, T. Savunen, J. Niinikoski, N. Inberg, and R. Penttinen, *Biochim. Biophys. Acta., 171*:105 (1982).
70. R. Ross, *Arteriosclerosis, 1*:293 (1981).
71. L. Y. Matsuoka, J. Wortsman, J. Uitto, J. Hashimoto, C. E. Kupchella, A. M. Eng, and J. E. Dietrich, *Arch. Intern. Med., 145*:117 (1985).

19

Pulmonary Emphysema
From animal models to human diseases

Norman T. Soskel
Veterans Administration Medical Center and University of Tennessee,
Memphis, Tennessee

Lawrence B. Sandberg*
University of Utah Medical Center,
Salt Lake City, Utah

I. INTRODUCTION

Emphysema is the only disease among the four major forms of chronic obstructive lung disease (the others being chronic bronchitis, asthma, and bronchiolitis) that is not associated with normal alveolar walls (1). Because a specific etiology is not known, and because the disease is defined on the basis of anatomical findings, a definite diagnosis cannot be made during life. In spite of the difficulty this poses on epidemiological studies, emphysema has been found to be second only to heart disease in causing major disability in the United States (2). Cigarette smoking is cited as the major potential cause (2), but other factors most surely coexist.

In the past three decades there has been a tremendous increase in studies related to the etiology of emphysema. This chapter will bring together that information in order to provide an overview of the current concepts relating to possible causes and pathogenesis of this disease. Normal structure and function of the lung will first be described, followed by the manner in which the alteration in structure affects pulmonary function in emphsyema. Next is a discussion of potential mechanisms involved in producing emphysema in humans based on in vitro as well as in vivo experiments, many of which are derived from animal studies. The conclusions are based on present understanding and on prevailing opinions. Some of these may be disproven in time; however, every effort has been made to be reasonably complete, accurate, and current.

**Present affiliation*: Loma Linda Veterans Administration Medical Center, Loma Linda, California.

II. STRUCTURE/FUNCTION RELATIONSHIPS IN THE LUNG

A. Structure of the Normal Lung (Figures 1 and 2)

The normal lung is a complex organ originating as an outpouching of the foregut and composed of endodermal epithelium and splanchnic mesenchyme (3,4). After ingrowth of mesoderm and weeks of maturation, the delicate capillary membrane of the alveolar wall is formed. The capillary component is so narrow that in some places only one red blood cell can pass through at a time. The proximity and interdigitation of the capillary to the alveolar wall provide a minimal barrier to the diffusion of gases between the alveolar space and the red cell's interior (Figure 1). It is this meshing of two embryonic germ layers that makes the lung a truly remarkable organ capable of performing its major function of distributing oxygen from the atmosphere to the blood and returning carbon dioxide from the blood to the atmosphere. This merging of structure and function is demonstrated in the lung, perhaps more dramatically than in any other part of the human body, and underscores the essentiality of their interplay and their interdependence.

The normal lung is composed mainly of connective tissue macromolecules, lipids, over 40 cell types, and a high percentage of air (85-90% by volume). The cell types and lipids are reviewed in another monograph (5). Grossly, the lung is divided into three right lobes and two left lobes. Within each lobe the airways branch in a fashion that is usually dichotomous. Branching continues for 23 generations, resulting in over 300 million alveoli (6,7), but this development is not complete until age eight.

Larger airways (bronchi) are lined with ciliated columnar epithelium and mucus-secreting cells. Their walls contain cartilaginous plates composed mainly of collagen and proteoglycans. Smaller airways (bronchioles) do not have cartilage, and distal to the terminal bronchioles, the respiratory bronchioles, which contain alveoli budding off their walls, further lack cilia and mucus-secreting cells. Distal to the third order of respiratory bronchioles, alveolar ducts lead to alveolar sacs from which bud individual alveoli (Figure 2).

Accompanying the bronchi are pulmonary arteries that arise from the right ventricle of the heart. Pulmonary veins are located in the connective tissue septae between individual lobules. Bronchial arteries (which arise from the aorta and supply blood to the lung parenchyma) and lymphatic vessels are also found in the connective tissue septae. The septae divide a lobe into smaller and smaller units until they surround individual lung lobules. Each lobule (or acinus) is defined as the unit of parenchymal tissue supplied by a terminal bronchiole (Figure 2) and bounded by connective tissue septae.

Alveoli comprise the gas-exchanging region of the lung. The alveolar wall alternates between thick regions, composed of connective tissue and cells, and thin regions, containing an attenuated alveolar capillary, composed only of an endothelial cell and the contents of the capillary (red cell and plasma) (Figure 1). It is across the thin regions of the alveolar wall that gas exchange occurs (i.e., influx of oxygen and outflux of carbon dioxide). In disease states, the thickness, composition, and integrity of the alveolar wall may be altered and may affect distribution and diffusion of gases.

The alveolus is lined with an epithelium comprised of type I and type II pneumocytes. The former is considered mainly structural and contains few organelles. The latter contains many organelles, is metabolically active, and produces the surface-active material (surfactant) that helps keep alveoli patent by lowering surface tension. The importance of the epithelium as a barrier to prevent penetration of agents that may cause alveolar wall

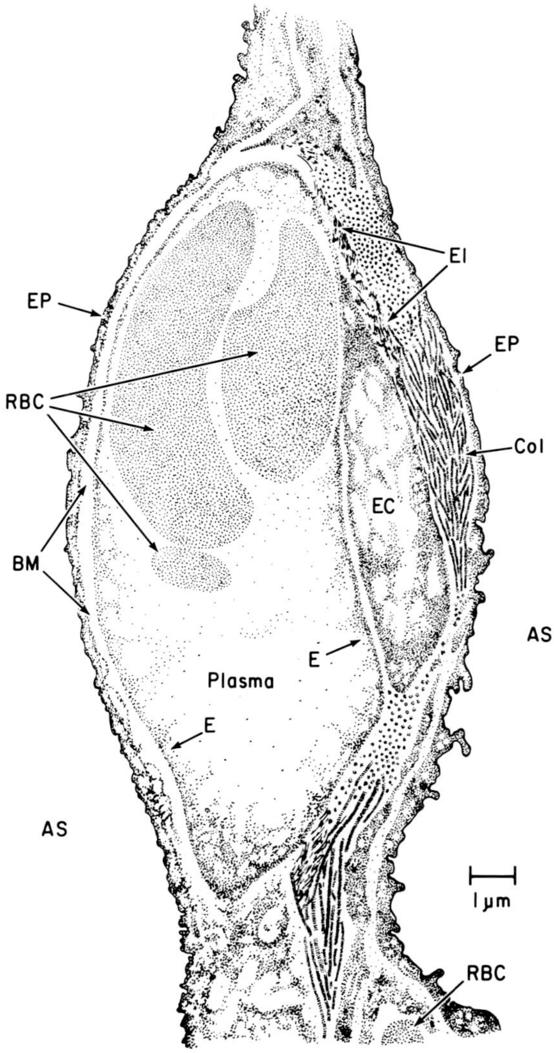

Figure 1 Structure of the alveolar-capillary membrane. An alveolar capillary is shown bordering on alveolar walls. The walls, which are of varying thickness, contain matrix molecules only in the thicker regions. Gas exchange is thought to occur across membranes in the thinner regions. Col = collagen, El = elastin, EP = alveolar epithelial cell membranes, EC = capillary endothelial cell nucleus, E = endothelial cell membranes, RBC = erythrocyte, AS = alveolar space, BM = basement membrane.

damage (e.g., emphysema) is currently under intense investigation.

Alveolar walls contain holes called "pores of Kohn" (8). These are usually circular or oval and only a few microns in diameter. They provide channels for collateral ventilation of small lung units (9,10). These pores are distinct from the alveolar wall fenestrations observed in emphysematous lungs (vide infra).

The vast majority of the connective tissue of the lung is made up of elastin and collagen. Elastin comprises about 1% of the dry weight of the lung (11,12). Elastic fibers are thought to be responsible for the forces involved in the ability of the lung tissue to stretch when

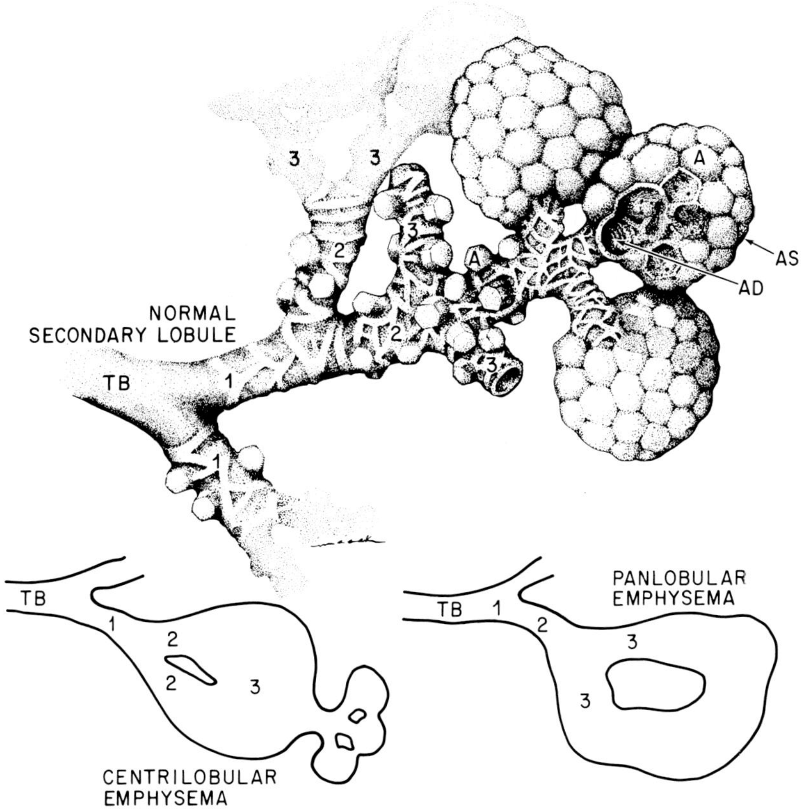

Figure 2 The pulmonary acinus. The normal acinus or secondary lobule originates at the terminal bronchiole (TB), which branches into respiratory bronchioles, containing alveoli budding off their walls. Respiratory bronchioles branch for about three divisions or orders before merging with an alveolar duct (AD). The alveolar duct leads to an alveolar sac (AS) and to individual alveoli (A), which arise circumferentially. The alveoli are roughly hexagonal and mimic a honeycomb in arrangement. (Lower left) Acinus in centrilobular emphysema. In centrilobular emphysema the second- and third-order respiratory bronchioles are primarily affected. In this region structures appear to merge. The more distal structures are preserved, initially; however, later in the course of the disease there too may be damage. (Lower right) Acinus in panlobular emphysema. In panlobular emphysema the distal regions of the acinus are involved along with those of the more proximal regions. As a result, distinctions among various components become less discernible.

surface forces are eliminated (13). Indeed, disturbances in elastic tissue result in mechanical alterations in excised lungs manifested by loss in elastic recoil (increased compliance) (Figure 3) (14–17).

Collagen, which represents about 10% of the dry weight of the lung, exists in multiple forms, each representing true genetic subtypes. Type I collagen has high tensile strength and lacks the ability to stretch, making it ideal for limiting lung expansion at high volumes.

Elastin and collagen are arranged as a spiral network of fibers that encircles alveoli, alveolar ducts, bronchioles, and bronchi. It is believed that the fibers represent a connective tissue continuum (18) that interconnects airspaces and airways with each other and with the connective tissue septae and vascular sheaths. Collagen and elastin codistri-

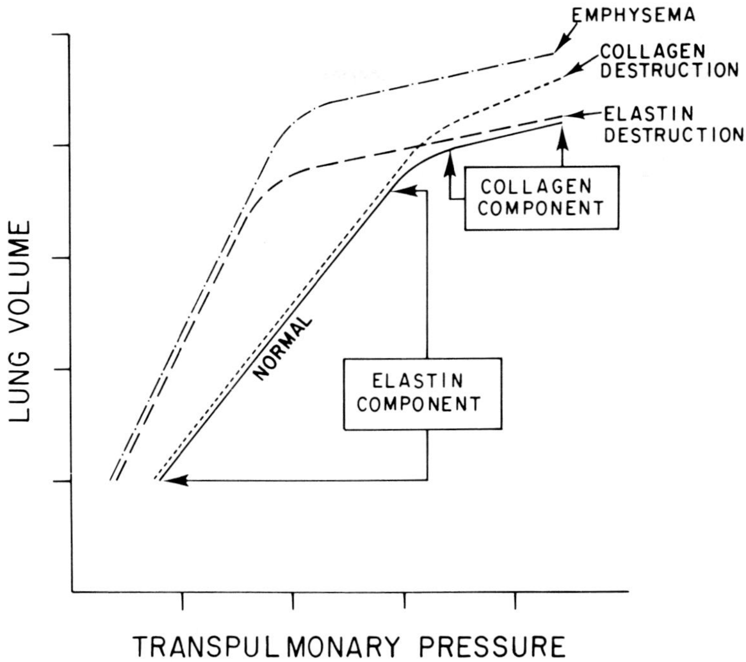

Figure 3 Connective tissue involvement in lung mechanics. The normal saline pressure-volume curve is denoted by the solid line. The relative contributions of collagen and elastin are shown. If elastin is selectively destroyed, the lower one-third of the curve is shifted toward the left; i.e., compliance is increased at lower lung volumes (dashed line). Conversely, if collagen is selectively destroyed, the upper, flattened part of the curve is shifted upward, demonstrating increased compliance at high lung volumes (dotted line). This may increase maximal lung volume (total lung capacity) as well (14–16). In emphysema, the lower part of the curve is shifted leftward, and the upper part is shifted upward (dot-dashed line) with respect to normal; however, the slope of the upper part of the curve is unchanged.

bute in the lung so that alterations in the structure of either macromolecule are likely to produce alterations in the mechanical function. Fibronectin, laminin, and proteoglycans are also structural components of alveolar walls, but make up less than 1% of the connective tissue. Their relationship to function of the lung is not well established.

B. Structure of the Emphysematous Lung

1. Definition of Emphysema

The common features of the various forms of pulmonary emphysema are *abnormal dilatation* and *tissue destruction*. The currently accepted definition of emphysema is "a condition of the lung characterized by abnormal, permanent enlargement of air spaces distal to the terminal bronchiole, accompanied by the destruction of their walls, and without obvious fibrosis" (19a). This definition represents the best definition available to date, but it has limitations (vide infra).

2. Description of Emphysema

Laennec in 1835 provided one of the earliest descriptions of emphysema (20). His subdivision, vesicular emphysema, corresponds to the disease we usually refer to as emphysema today. He described the disease as simply dilatation of alveoli, but remarked that severe cases demonstrated their rupture.

Rainey in 1848 focused on destruction of the pulmonary capillary membrane in emphysema and described remnant fibers that may have been frayed connective tissue fibers (21).

Most recent descriptions have merely amplified the details of these earlier reports in conjunction with technical improvements in microscopy and histological staining. Currently, two main forms of emphysema can be distinguished, centrilobular and panlobular (22). Centrilobular emphysema involves primary destruction at the level of the second and third branches of the respiratory bronchioles (Figure 2) with relative sparing of more distal structures. However, the disease may progress to involve all lung structures in severe cases. Centrilobular emphysema (the more common variety) is the form usually associated with cigarette smoking, but in severe cases the lungs may be indistinguishable from those manifesting as panlobular emphysema (22). Panlobular emphysema, a more diffuse destruction of all structures in the lobule (Figure 2), is the form that occurs in patients with deficiency of α-1-protease inhibitor and in experimental animal models of emphysema induced by intratracheal enzyme instillation.

3. *Quantitation of Emphysema*

Early methods of quantitating emphysema employed grading its severity and distribution using thin whole lung sections mounted on paper (23,24). Although this method was useful, precise quantitation of degree of disease was not possible. It has been replaced by more exact point-counting techniques (25). In addition to increasing precision, these newer methods allow automation with computer assistance (26).

Point-counting methods (morphometric techniques) can reliably distinguish destructive enlargement of airspaces from overdistention (19). For performance of these techniques, lungs must be adequately fixed in an inflated state (27,28) and lung volumes determined, usually by water displacement (29). Currently, a mean alveolar dimension, the mean linear intercept (Lm), is used to determine the degree of distention. Coupled with the measurement of the alveolar internal surface area (ISA), the loss of surface area (destruction) of the alveolar-capillary membrane can be determined (30). Lm and ISA have been used routinely to determine the degree of emphysema. These parameters have been especially useful in describing lesions produced in animal models of emphysema (30,31). Morphometric techniques have been extended to quantitating elastic fiber dimensions (32), although this has not gained widespread use. They have also been used to predict physiological parameters such as gas diffusion across the lung (33,34).

C. Structural Determinants of Function

Normally, excised lungs resist expansion owing to their "recoil properties," which can be accounted for mainly by surface forces (surface tension) and tissue forces (recoil pressure) (35-37). When surface forces are eliminated by filling lungs with saline, inflation is achieved by generating pressure that equals the intrinsic tissue force tending to collapse the lung. Dividing the volume by the pressure (tissue force) gives the compliance of the lung, which describes the ability of the lung to stretch. Pressure-volume curves graphically portray this relationship (Figure 3). The compliance of the normal lung is greater at lower lung volumes than at higher lung volumes. In diseases such as emphysema in which the tissue is destroyed, the lung is more compliant (less stiff) than normal at lower lung volumes, and the maximal achievable volume is increased above normal. That is, for a given pressure, a larger-than-normal lung volume is obtained in emphysema, and a larger maximal lung volume may be demonstrated.

In attempting to define the structural counterparts of function, Setnikar destroyed elastin in excised lungs (15). In so doing, he demonstrated increased compliance in the lower one-third of the pressure-volume curves (Figure 3). In contrast, when he destroyed the collagenous components, only the flattened upper portion of the curve was shifted. Because the formerly treated lungs behaved mechanically similar to those removed from patients dying of emphysema, he concluded that elastin must be destroyed in emphysema. Mead amplified the theory (16) to suggest that as the lung expands, the elastin fibers begin to stretch, accounting for the lower portion of the curve. This is reasonable because elastin fibers can stretch 130–140% of their resting length (38). In contrast, the coiled collagen fibers begin to uncoil, and they limit maximal expansion since these fibers have high tensile strength (39). Thus the upper portion of the pressure-volume curve is flat. Although these observations and speculations may appear simplistic, the results provide a theoretical construct on which to build.

Further evidence to support this theory includes the demonstration that length-tension curves of lung strips are shifted leftward if elastin is destroyed and upward if collagen is destroyed (40). Intratracheal elastase instillation in vivo results in the development of emphysema in hamsters (vide infra), but collagenase instillation does not (41,42). The lower one-third of the pressure-volume curve is shifted leftward with elastase, and the upper flattened part is shifted upward with collagenase. These results implicate elastin and collagen as the major connective tissue determinants of structural and functional impairment in emphysema.

Although histological staining demonstrates diminution in the amount of elastin in lungs of emphysematous patients compared to controls (18), biochemical quantitation of whole lung elastin and collagen (43) reveals normal amounts of these parenchymal matrix macromolecules per lung. Furthermore, amino acid analysis of insoluble lung elastin is qualitatively similar in normal and in emphysematous lungs (44). As a result of these and the animal studies described below, most investigators currently believe that elastin is destroyed in emphysema but is resynthesized and laid down in an architecturally deranged fashion (45).

III. PATHOGENESIS

A. Historical Perspective

Most of the early theories of the pathogenesis of emphysema are reviewed admirably by Strawbridge (46). It is curious that the words "cigarette" and "smoking" or references to them do not appear in this excellent review. Furthermore, the initial report (19) from which our current definition (19a) is derived only mentions in passing that the profile of a patient with emphysema includes smoking cigarettes. In spite of this, a few reports mention the proposed importance of cigarette smoking, and the prevailing opinion among clinicians attending patients with emphysema was that if cigarette smoking could be eliminated, so could emphysema (47). In 1964, the Surgeon General's report stated that cigarette smoking is associated with emphysema, although causality had not yet been established (48). Now there is much more epidemiological data available to link cigarette smoking to emphysema (49-54).

A fundamental question relating to the pathogenesis of emphysema has been whether the primary destructive event occurs first at the level of the bronchioles, where inflammation and obstruction are observed histologically, or at the level of the alveoli, where rupture of walls is seen. The earlier investigators felt that bronchiolar obstruction with subsequent production distally of high pressures in relation to increased effort associated with breathing produced alveolar dilatation and eventual rupture. Investigators expressed the view that alveolar destruction could be explained if bronchiolar obstruction occurred first, but that the bronchiolar lesion would be difficult to explain on the basis of a primary disruptive alveolar event (55,56).

We now know that obstructive phenomena represent at least two distinct and separate clinical diseases that often occur in the same individual (19,56,57). Chronic bronchitis represents a primary bronchial disease defined on a clinical basis and is discussed in several excellent recent reviews (58,59). Emphysema, the focus of this chapter, represents a primary abnormality of the alveolar wall. The obstructive component of this disease that is detected by functional tests currently is believed to result mainly from loss of recoil properties of the lung tissue and only minimally from increased compressibility of the airways (60). However, mucus plugging with structural obstruction is seen frequently at autopsy in patients with emphysema (52). In contrast to chronic bronchitis, emphysema is believed *not* to result initially from bronchial wall changes leading to structural obstruction, but instead from primary alveolar wall damage. Emphysema is defined by pathological criteria that can be determined with certainty only at autopsy or upon viewing a surgically excised lung. In considering the diagnosis of emphysema in patients, we must remember the caveat of Bates and Christie (quoted by Fraser and Paré) that "clinical emphysema... is a clinical diagnosis expressing the *probability* that signs and symptoms may be ascribed to morphological emphysema" (61) (emphasis added).

B. Protease/Antiprotease Theory

The modern era in understanding the pathogenesis of emphysema was ushered in by two major discoveries, one in humans and one in an animal model. In 1963, Laurell and Eriksson described the absence of an alpha-1 globulin in serum suggesting an association with degenerative lung disease (62). Subsequently, Eriksson described the association of the absence of α-1-protease inhibitor (α1-Pi) with a hereditary form of emphysema (63,64). The former designation, α-1-antitrypsin, was discarded when the inhibitor's more global activity was determined. In 1964, Gross et al. described a model of emphysema induced in rats by intratracheal instillation of the proteolytic enzyme papain (65,66). These two discoveries have resulted in a burst of scientific effort that continues today.

1. Induction and Description of the Lesion in Animals

The initial phases of investigation in animals accomplished several major goals. First, the primary objective of inducing the disease in animals was accomplished using proteolytic enzymes (67–69). Not surprisingly, the degree of emphysema produced appeared to correlate with the elastolytic activity of the enzyme preparation used (70,71). Other enzymes such as collagenase, which have no elastolytic activity, were unable to produce emphysema, although other changes were noted (41,42). The intratracheal route appeared more suitable for induction, presumably owing to the rapid inhibition of parenterally administered exogenous enzyme by endogenous protease inhibitors (72,73). Of some surprise to those demonstrating a normal amount of elastin in emphysematous lungs was the accumulation of evidence in favor of the hypothesis that *destruction of elastin is a prerequisite to the development of emphysema.*

The initial studies raised a number of questions concerning the pathogenesis of emphysema and pointed out the need to critically evaluate the methods employed. These early studies focused on quantitating the total amount of insoluble elastin and collagen or total accumulation of these proteins. While presenting needed information, they overlooked the fact that total accumulation of a particular protein at a given point in time is a static measurement representing a balance between synthesis and degradation. Furthermore, because immature soluble precursors are more susceptible to degradation, the amount of a matrix macromolecule that is determined by many of these static measurements may underestimate the total amount either present or being made.

A second question concerns the way in which these parameters are expressed or normalized. Various investigators normalize quantitative parameters to protein, DNA (or cell number), or lung volume, or simply determine values per lung. Because there is no generally accepted standard, it is often difficult to compare studies. Furthermore, inherent errors may occur when studying diseases. For example, the number of cells may be increased in more proliferative diseases or decreased in very degradative disease; the lung volume is likely to decrease in severe fibrotic diseases and to increase in bullous emphysema; and the lung weight or volume may not always reflect body size as it does in normals. Which parameter is more important depends on the specific questions being addressed. What is most important is a clear understanding of the implications of the findings with respect to the normalizing variable. Often it may be advisable to normalize to several parameters and discuss the significance of each.

Two other important questions arose from these studies in animals. First, because exogenous elastases can cause emphysema in animals, where would comparable enzymes arise endogenously? As a corollary to this question, in experimental emphysema, why does the lesion progress long after disappearance of exogenous elastase (74)? Second, because synthesis is apparently an important part of the equation describing the instantaneous state of connective tissue, what role is played in the pathogenesis of emphysema by the repair processes of the body? These questions are addressed below.

2. *Evidence of Elastin Damage*

Currently, the protease-antiprotease theory of tissue damage is generally believed to most accurately explain the pathogenesis of emphysema (17,45,75–79). Many aspects of this theory were derived from studies utilizing animal models of emphysema and in vitro experiments demonstrating mechanisms of lung elastin damage and protection (31). Early models utilized plant enzymes (65,66,68). Subsequently, cells and cell extracts (72,80), porcine pancreatic elastase (elastase 2) (67,69), and, more recently, leukocyte elastase (81,82) were used. It was demonstrated that within 4 hr of intratracheal administration of enzyme, the lungs develop a severe hemorrhagic and inflammatory response (67). This reaction subsides within the first 24–48 hr. During this time lung elastin content diminishes, but by 4 hr the majority of the instilled elastase is absent from lung tissue. The early emphysematous lesions are visible by about two weeks after the instillation of enzyme. Desmosine appears in the urine of elastase-treated hamsters in amounts that correspond quantitatively to a diminution of lung elastin during the early phases of lung injury (83). However, shortly after the appearance of the initial lesions, elastin and collagen contents return to control levels (84). Elastin synthesis is maintained at greater than control levels (85), suggesting that degradation of elastin must also be increased. Despite the apparent disappearance of the majority of elastolytic activity and the return of elastin content to control levels, the lesion progresses for at least one year (74). These results, along with older observations of frayed elastic fibers (11,12,18), substantiate the hypothesis that, in emphysema, elastin is destroyed and resynthesized, but that it is laid down in an abnormal fashion resulting in an architecturally deranged lung (45,84).

Table 1 Enzymes and Inhibitors Potentially Affecting Lung Strucutres

Enzyme	Effects on insoluble elastin	Effects on native collagen	pH optimum	Type	Natural inhibitors (113 a)	Activators
Pancreatic						
Elastase 1 (86)	Minimal	None	8.6	Serine	α-1-Pi, α-2-M	Trypsin
Elastasez (86)	Much	None	8.6	Serine	α-1-Pi, α-2-M, BSI-E	Trypsin
PMN (neutrophil)						
Elastase (88,91,220)	< Elastase 2	None	8.4	Serine	α-1-Pi, α-2-M,[a] BSI-TE, BSI-E[b]	PF-4 (95) —
Cathepsin G (88,221)	< Elastase 2	—	8.4	Serine, Chymotrypsin-like	α-1-Pi, α-achy,[a] α-2-M, BSI-TE	—
Collagenase (96–98)	None	3/4–1/4 split	8.4 (?)	Metallo-endopeptidase	β-1-AC, α-2-M, PF-4	Activator protease of PMNs

Macrophage						
Elastase (102,103)	< Elastase 2	—	7.6-7.8	Metallo-endopeptidase	α-2-M	Divalent calcium
Collagenase (110)	None	3/4–1/4 split	7.4	Metallo-endopeptidase	α-2-M, β-1-AC, PF-4	—
Monocyte						
Elastase (106,107)	< Elastase 2	—	8.4	Serine	α-1-Pi	—
Pseudemonas						
Elastase (112 a)	Active	—	7.0	Metallo-protease	α-2-M	—
Other cathepsins						
B (222)	Active	None	6.0	Cysteine	α-1-CPI (weak), α-2-M	Thiols
H (222)	Active	None	6.8	Cysteine, Amino- and endopeptidase	α-1-CPI, α-2-M	Thiols
L (222)	Active	None	5.5	Cysteine	α-1-CPI, α-2-M (probably)	Thiols

[a]Preferred.
[b]Also inhibited by long-chain fatty acids, sulfated polysaccharides, and elastatinal.

3. Sources of Enzymes (See Table 1)

Although emphysema *can* be produced by administering exogenous enzymes that manifest elastolytic activity, a fundamental question concerns what naturally occurring endogenous reservoirs might be potentially instrumental in producing the disease in the absence of external sources. Endogenous sources of elastase include the pancreas, neutrophils, and macrophages. Bacterial enzymes provide an exogenous source.

Pancreatic Elastases. The human pancreas contains at least two elastaselike enzymes. Elastase 1 (protease E, mw 30 kd) has very little activity against the insoluble elastic fiber owing to lack of adsorption commonly ascribed to its anionic character (86). Pancreatic elastase 2 (mw 25 kd) vigorously digests insoluble elastic fibers probably owing to its cationic nature, which facilitates adsorption to the elastin matrix. Zymogens of these two elastases have been described (87). Both enzymes are stable at neutral or slightly alkaline pH, having similar amino acid composition, are inhibited by serum α-1-Pi, and demonstrate equivalent protelytic activity when tested against protein substrates other than elastin (86). Although most of the experimental work on enzyme-induced emphysema has involved porcine pancreatic elastase 2, pancreatic elastases are not considered major contenders for producing emphysema in humans because of limited access to the lung while they are active.

Neutrophil Proteases. Neutrophil granules contain the major enzymes elastase and cathepsin G (cat G) in the form of endoproteases. Both have several isoenzyme forms (88). Both are basic proteins with isoelectric points at pH 11 or greater and appear to have similar N-terminal amino acid sequences. Optimum activity appears to lie between pH 7.5 and pH 10 (89). These enzymes, however, have less activity than pancreatic elastase 2 in solubilizing lung elastin. α-1-Pi controls the activity of PMN elastase, whereas α-1-antichymotrypsin controls that of cathepsin G. However, under certain conditions α-1-Pi can inactivate cat G. Both PMN elastase and cat G can degreade immunoglobins, collagen, cartilage matrix proteins, fibrinogen, and elastin (88). The rate of hydrolysis of lung tissue by cat G is 12% of that caused by human leukocyte elastase (90).

Leukocyte elastase binds rapidly to alveolar macrophages, and whereas pancreatic and bacterial proteases require the presence of α-2-M for binding to macrophages, leukocyte elastase does not (91). Macrophages appear to clear and at least partially inactivate neutrophil elastase in the lung. However, a significant fraction of the ingested elastase remains in an enzymatically active form, and, thus, macrophages may serve as an elastase reservoir (92). In vitro cultures have revealed that elastase released into conditioned medium by alveolar macrophages is considerably greater in total activity than the initial elastase activity ingested, suggesting that the enzymatic activity of endocytosed neutrophil elastase is masked by an intracellular inhibitor (93). The fate of neutrophil elastase ingested by alveolar macrophages appears to be modulated by the local tissue environment. Noxious agents enhance the release of bound neutrophil elastase, suggesting that alveolar macrophages may participate in the pathogenesis of pulmonary emphysema by serving as vectors for these enzymes (94).

Platelet Factor IV (PF-4) greatly enhances human leukocyte elastase (HLE) activity against synthetic substrates, but not against nonelastin proteins (95). One explanation is that PF-4 and the fragments of PF-4 generated by HLE may cause a conformtional change in elastin, opening up previously unavailable sites for attack by HLE.

Although the role of leukocyte collagenase in the pathogenesis of lung injury is not known, the granule fraction of leukocytes contains a specific collagenase. This enzyme cleaves collagen molecules into two fragments (96), but alone is incapable of solubilizing collagen fibrils (97). An activator protease, also present in the neutrophil, appears to be necessary for activation of the zymogen of this collagenase (97,98) and for maximal collagen fibril degradation. The significance of the neutrophil collagenase and its activator in the development of lung destruction is not known at this time.

The importance of the leukocyte as a mediator of tissue injury cannot be disputed; however, recent studies cast doubt on its supreme role in the pathogenesis of emphysema. Elastase-induced emphysema and the usual nonhereditary forms of human emphysema are thought by many investigators to be perpetuated by elastase release from neutrophils either directly or by internalization and slow release from alveolar macrophages as discussed above. However, leukocyte depletion in animals does not prevent the lesion and may, in fact, exacerbate the injury induced by elastase or papain (99,100). Furthermore, when the elastase load delivered to the lung is diminished by administering antineutrophil antibodies to hamsters treated with cadmium plus BAPN (101) (see also discussion of other models below), the same degree of emphysema develops as without antineutrophil antibodies despite the markedly lowered elastase activity recovered by bronchoalveolar lavage. This study suggests that leukocytes may not be essential in mediating the progression of connective tissue damage in the lung. This does not imply, however, that leukocytes could not be involved in producing lung injury; it merely suggests that emphysemalike lung injury *can* be produced in the absence of leukocytes.

Macrophage/Monocyte Proteases. Human alveolar macrophages synthesize a calcium-dependent elastase quite different from the PMN enzyme (102). It has a pH optimum of 7.6 to 7.8 and is inhibited by human serum. Its properties are identical to those of the enzyme found in mouse peritoneal and alveolar macrophages, but the pH optimum differs from those of leukocyte elastase (pH 8.4) and pancreatic elastase (pH 8.6) (102). Macrophage elastase is not a serine protease but rather a metalloendopeptidase. It is inhibited by α-2-M but not by α-1-Pi (102,103), and the mouse peritoneal macrophage enzyme can degrade α-1-Pi (104). Three chromatographically distinct forms of the latter enzyme have been isolated (103). The major peak of activity from sucrose-velocity-gradient analysis has a molecular size of 22,000 daltons. An endogenous inhibitor (which may be α-2-M) masks more than 80% of the secreted activity (105).

Human blood monocytes also secrete an elastase that has catalytic and immunological properties similar to those of neutrophil elastase, is a serine protease, and may exist in at least two distinct forms (106,107). It is primarily an intracellular enzyme, whereas true macrophage elastase is found principally in culture medium. Lungs of animals exposed to cigarette smoke contain macrophages that may be derived from migration of bone marrow monocytes rather than from in situ division of resident macrophages (108). It is of specific interest that monocyte lysates contain an elastase identical to neutrophil elastase (107,109), but that monocyte-derived macrophages contain a metalloprotease identical to that found in mouse peritoneal macrophages and capable of degrading α-1-Pi (104). This suggests that as the monocytes differentiate into macrophages, they also undergo a differentiation in expression of their intrinsic elastase.

Macrophages also contain a collagenase that is distinct from the leukocyte elastase in its properties (110). This enzyme is apparently secreted in increased amounts from activated macrophages.

Macrophages also secrete a plasminogen activator (111), which may play a role in degradation of matrix macromolecules even in the presence of protease inhibitors (vide infra).

Plasminogen/Plasminogen Activator/Plasmin. The potential role in lung destruction of the circulating serine protease plasmin has been largely neglected. This two-chain, trypsin-like protease hydrolyzes proteins at arginyl and lysyl peptide bonds, basic amino acids, and amides, and it is inhibited by diisopropyl-phosphofluoridate (DFP) as well as L-1-chloro-3-tosylamido-7-amino-2-heptanone (TLCK), both of which react with the serine and histidine residues, respectively, found at the active site on the B chain (112). Plasmin exists predominantly as the zymogen plasminogen, which requires plasminogen activator for cleavage of a single arginine-valine bond to form the active protease plasmin (113). Plasminogen activator (PA) is ubiquitous, being found in most tissues and plasma, although the tissue activator is somewhat different from the plasma activator. PA also is a serine protease inhibited by DFP and epsilon, amino caproic acid (EACA) but not by TLCK. Plasminogen is the only substrate against which PA has any activity.

4. Natural Inhibitors of Elastase Activity in Body Fluids

In order to prevent unregulated proteolysis by endogenous enzymes, natural inhibitors exist in various cellular and extracellular compartments.

Plasma Inhibitors. The subject of plasma elastase inhibitors has been admirably reviewed in a recent monograph on plasma enzyme inhibitors (113 a). The following section summarizes this topic. The natural plasma inhibitors of elastase appear to be α-1-Pi, α-2-M, α-1-antichymotrypsin (α-1-achy), and α-cysteine protease inhibitor (α-CPi). α-1-Pi is a single-chain glycoprotein having a molecular weight of 53,000 daltson. It inactivates all mammalian serine proteses (endopeptidases) including pancreatic and neutrophil elastase and, to a lesser degree, neutrophil cathepsin G. The interaction of serine proteases with α-1-Pi occurs through limited proteolysis on a "bait region" of the inhibitor. Many investigators believe that its primary role is to control elastolytic-type activity secreted by neutrophils.

Inactivation of α-1-Pi may play an important role in modulating tissue injury. Lowering of this important protective "screen" is believed to produce an "acquired" deficiency in α-1-Pi, which may lead to lung destruction and emphysema. The integrity of a methionyl residue at the active site appears to be important for maintaining activity of α-1-Pi and may provide a regulatory mechanism for proteolysis. Recently, a preliminary report suggested that a tyrosine residue in the molecule may have a similar function (113 b). Once inactivated, besides being incapable of protecting against protease damage, α-1-Pi itself is more susceptible to proteolytic cleavage by human leukocyte and porcine pancreatic elastases (113 c).

Alpha-1-Pi can be inactivated by oxidation of active site methionine residues to methionine sulfoxide (113 c, 114). Of note is the observation that oxidative inactivation of α-1-Pi is not complete, but the rate of its association with PMN elastase is diminished to 1/2000 that of unoxidized inhibitor (115).

Cysteine-active site proteases such as cat B1 may also play a role in inactivation of α-1-Pi, although this is not well established (116).

A recently developed electrophoretic method may be useful for determining whether α-1-Pi inactivation is due to proteolysis or oxidation (117). Accordingly, oxidation should increase the negative charge and increase electrophoretic mobility of the inhibitor. In contrast, proteolytic enzymes have two effects: the formation of an enzyme-inhibitor complex in a gradual fashion along with limited proteolysis and a decreased mobility, compared to native α-1-Pi. Exposure of α-1-Pi to an aqueous solution of cigarette smoke results in an increased mobility suggesting oxidation, whereas exposure to proteolytic enzymes (regardless of exposure to cigarette smoke) diminishes mobility, as predicted for proteolytic cleavage. This method may prove very useful for studies involving various enzyme-inhibitor interactions; however, in vivo studies do not correlate with these in vitro results. When applied to bronchial secretions from smokers, the results were confusing but suggested that cigarette smoke may exert different effects in vivo, that there may be a temporal relationship between smoking cigarettes and inactivation, and that macrophage recruitment with subsequent enzymatic cleavage of α-1-Pi may be important (104).

The critical role of the primary structure of the antiproteases was recently demonstrated when a fatal bleeding disorder was observed in a child (118). The patient's α-1-Pi contained a mutation (active site methionine to arginine). Because of widespread sequence homology, the mutated antiprotease resembled antithrombin III in its structure and activity (118) and resulted in the hematological abnormality.

α-2-M is a 75,000 dalton molecular weight protein (113a) with the unique ability to interact with all classes of proteases: serine, cysteine, aspartic, and metallo- (113a,119). This function is tied to its rapid clearance by the reticuloendothelial cells, suggesting that its main role is to remove enzymes and that inhibition of enzymes is incidental (113a,120). Furthermore, specific alpha-2-M receptors on macrophages are involved in the endocytotic clearance process (121). Another unique feature of α-2-M is the retention of proteolytic activity of bound proteases against small substrates (122). Because of its reactivity with so many proteases with different functions and sources, the physiological role of α-2-M is poorly understood (113 a).

The respective roles of α-1-Pi and α-2-M in counteracting proteolysis of matrix proteins differ from each other considerably. α-1-Pi is more prevalent in serum and binds elastases more tightly than does α-2-M. Furthermore, α-1-Pi irreversibly inactivates the enzymes preferentially (123), and, owing to its large size, the enzyme-inhibitor complexes formed are cleared through the reticuloendothelial system rapidly (120). However, enzymes bound to α-2-M retain activity against small-molecular-weight peptides. Furthermore, α-1-Pi complexes may dissociate and transfer enzymes to α-2-M for clearance. HLE binds α-1-Pi preferentially over α-2-M. Complexes of enzymes with α-1-Pi and α-2-M demonstrate 1:1 stoichiometry, as do inhibitor-collagenase complexes and α-1-achy-PMN protease complexes.

The role of α-2-M-enzyme complex formation has recently been addressed (123–126). Both elastolytic activity and complexes have been found in bronchoalveolar lavage fluid up to 15 days after intratracheal elastase instillation in hamsters (124). It is presumed that the activity was associated with the bound enzyme. Therefore tissue destruction may not be dependent on prior observations that α-2-M complexes are incapable of degrading large-molecular-weight substrates such as trypsin (123) and tropoelastin (125). Because macrophages can secrete α-2-M (126), whether or not this inhibitor reaches tissue spaces from the blood may be less important than previously believed. It is unclear at this time whether α-2-M contributes to protective or destructive events or both in the pathogenesis of alveolar wall damage.

α-1-Achy forms stable complexes with human chymotrypsin, human neutrophil cat G, and mast cell "chymases" (127) and specifically inactivates cat G (90). The interaction with cat G may be important since this protease may be involved in elastolysis. α-1-Achy has no effect on neutrophil elastase, but, in fact, neutrophil elastase may be involved in turnover of α-1-achy. Because α-1-achy levels rise dramatically during the acute phases of inflammation, its protective role may be significant at this time (128).

α-CPi inactivates cathepsins H and L, but reacts slowly with cathepsin B (113 a). Recently this inhibitor has been shown to inhibit cat G preferentially over the inhibitory effect of α-1-Pi (129). It is believed that these proteases, which normally are only intracellular, are released into tissue spaces when a cell is lysed and can thus cause significant degradation of proteins, including elastin. It is also believed that the circulating α-CPi may be involved in regulation of elastolysis by diffusing from the blood into tissue paces in a similar fashion as α-1-Pi (113 a).

Inhibitors in the Tracheobronchial Tree. The larger airways require protection from proteases because proteolytic, specifically elastolytic, activity is markedly increased in bronchial secretions during inflammatory states (129). Characterization of inhibitors in sputum is not as definitive as that in plasma. Recent studies on sputum have demonstrated the presence of the serum-derived inhibitor α-1-Pi, a small amount of α-1-achy, and very little α-2-M owing to its large molecular weight (130). In addition, at least three bronchial inhibitors, one of which is an active fragment derived from a serum inhibitor, have been described (vide infra).

The protease inhibitory activity initially detected in sputum (131) was subsequently found associated with a protein having a molecular weight of 11 kd and having inhibitory activity toward leukocyte proteases (132). This low-molecular-weight protein is distinct from α-1-Pi and is acid stable. The inhibitor, which has been called BSI-TE, antileukoprotease, or bronchial mucus inhibitor, represents 70% of the protease inhibitory capacity of bronchial secretions (132). Similar inhibitory activity has been found in seminal fluid and cervical mucus but is believed to occur in all mucous secretions (130). The inhibitor is active against leukocyte elastase, cat G, trypsin, and chymotrypsin and contrasts with the activity of the inhibitor recently isolated from bronchial mucus (BSI-E), which inhibits porcine pancreatic elastase (PPE) and leukocyte elastase but not trypsin, chymotrypsin, or leukocyte cat G (133). The latter inhibitor is active only under acidic conditions.

In addition to these inhibitors, HI-30, a trypsin-chymotrypsin inhibitor that is cleaved from serum inter-α-trypsin inhibitor, is found in bronchial mucus (134). Because it has no affinity for porcine pancreatic elastase or human leukocyte cat G and very weak affinity for HLE, its biological function is presently undetermined (133).

5. *Anticollagenases*

In addition to inhibitors of elastase, anticollagenases may also be important in tissue remodeling during injury and repair. Anticollagenases have been found in plasma and in a number of tissues, most notably skin, synovial and amniotic fluids, polymorphonuclear leukocytes, and chondrocytes (135–138). Three inhibitors: the circulating beta-1-anticollagenase, tissue inhibitor of metalloproteins, and collagenase inhibitor from human polymorphonuclear leukocytes may have access to the lung, but no direct evidence has established a relationship to modulation of lung injury.

An anticollagenase in the β-1 fraction of serum, named β-1-anticollagenase (β-1-AC), specifically inhibits the human collagenases to different degrees (gastric mucosal > skin > rheumatoid > synovial > granulocyte) but does not inhibit other enzymes (139). This inhibitor has a molecular weight of 30,500, and its activity is dependent on the integrity of its two thiol groups. It reacts with human leukocyte collagenase in a 1:1 stoichiometric fashion by disulfide/thiol interchange. The complex, often referred to as the latent enzyme, can be activated by disulfide-containing compounds as well as by the oxidized glutathione-generating system myeloperoxidase/hydrogen peroxide glutathione (140).

Tissue inhibitor of metalloproteinases (TIMP) is found in many tissues and, in addition to collagenases, inhibits other metalloproteinases that degrade gelatin and proteoglycans (141,142).

The specific collagenase found in human polymorphonuclear leukocytes has properties similar to β-1-AC and TIMP (137). Its amino acid composition is similar but not identical to that of the former.

All these inhibitors are belived to be important in regulation of various collagenases under physiological conditions, and TIMP may similarly affect additional metalloproteinase-mediated processes (143). Specific studies related to lung injury and emphysema, however, have not yet been published.

It is of additional interest with regard to collagenase control, that α-2-M and platelet factor 4 can inhibit collagenase (144), as can cationic proteins from cartilage and aorta (145,146). The initial demonstration of the ability of α-1-Pi to inhibit collagenase (121, 147) has not been confirmed (137,148–150).

C. Role of Glycosaminoglycans and Proteoglycans

Very few studies concerning emphysema have included glycosaminoglycans (GAGs) and proteoglycans (PGs), which have been thought to play a role in conjunction with collagen and elastin in determining the mechanical properties of the lungs (151). It is unclear whether hyaluronidase is increased, decreased, or unchanged in human emphysematous lung tissue (152–154), but hyaluronic acid appears diminished (155). Sex hormones appear to correlate with GAG content, but the relevance of this observation is speculative at this time.

In the elastase animal model of emphysema, an imbalance was demonstrated in accumulation among collagen, elastin, and heparan sulfate in lung explants taken over a one-year period (156), providing evidence of interrelationships in synthesis and accumulation among several matrix macromolecules. The relative paucity of heparan sulfate suggests that defective repair of basement membranes might account for structural derangements and functional abnormalities in gas exchange that are commonly observed in emphysema.

In the same model, there is increased glucosamine incorporation into a fraction of lung tissue GAGs (157), heparan sulfate, and hyaluronic acid (158). The significance of these observations is unclear but may signify elastase-induced stimulation of GAG synthesis despite apparent relative dimunution in quantity. Studies involving GAGs and PGs are difficult to perform and even more difficult to design with relevance to disease, partially accounting for the paucity of published data concerning the role of these macromolecular matrix components in the development of emphysema.

Although proteoglycanases have been described in various tissues and body fluids (159, 160), no direct studies have demonstrated lung activity or have addressed potential effects on lung tissue.

D. Oxidant/Antioxidant Systems

1. Sources of Oxidants

Recent studies have demonstrated the potential involvement of oxidative mechanisms in the pathogenesis of emphysema (2,79).

In vitro oxidation of α-1-Pi can occur as a result of contact with chloramine T in animals (161) or as a result of contact with endogenous reactive oxygen species (162). These reactive species may be produced endogenously by PMNs, monocytes, and macrophages during the respiratory burst that occurs during phagocytosis and microbial killing (163, 164). Phagocyte-related oxidation can also inactivate bronchial mucus inhibitor (BSI-TE) (165).

Tobacco smoke contains reactive oxygen species (166,167), peroxides (167,168), and other oxidants (166,167). In vitro, tobacco smoke components, both the crude tar (169) and aqueous solutions (170), are capable of inactivating α-1-Pi. The rate of inactivation appears to be dependent on the oxidizing capability of cigarettes, which varies among different commercially available brands (168). Bronchial mucus protease inhibitor (BSI-TE) is also inactivated by aqueous solutions of cigarette smoke (165).

Tissue injury may occur when endogenous scavengers of reactive oxygen species and antioxidants (enzymatic or chemical) are diminished. In such a case, even normal cellular production of reactive oxygen may lead to inactivation of α-1-Pi or may cause direct tissue damage by cleaving peptide bonds in structural proteins such as elastin (unproven) or collagen (171,172).

2. Sources of Antioxidants (Figure 4)

The reactive oxygen species normally produced by metabolizing cells might destroy body tissues in the absence of protective antioxidant systems. Several such systems have been described (173). Superoxide dismutase (SOD) catalyzes dismutation of superoxide anion to hydrogen peroxide and oxygen. Catalases convert hydrogen peroxide into water and oxygen. Peroxidases, e.g., glutathione peroxidase, convert hydrogen peroxide into water. Furthermore, removal of superoxide and hydrogen peroxide prevents the formation of hydroxyl radical. Alternatively, the hydrogen peroxide produced by the superoxide dismutation reaction may be converted to hydroxyl radical and water via the controversial Haber-Weiss reaction requiring ferric iron (174), potentially leading to more severe hydroxyl radical-mediated tissue injury (175).

Superoxide dismutase exists in two forms in mammalian cells. The cytosolic copper-zinc form (CuZn SOD) is the major species; the mitochondrial manganese (Mn SOD) form exists in lesser amounts (176). The relative protective role of each form is currently unknown. Experiments designed to test the efficacy of superoxide dismutase in protecting against oxygen-induced lung injury have recently been reported. Some protection may be afforded by aerosolization of SOD enclosed in liposomes in rat lungs and in cultured endothelial cells (177,178). Theoretical questions concerning an intracellular enzyme's ability to catalyze dismutation of extracellular free radicals (produced by inflammatory cells, cigarette smoke, etc.) are somewhat satisfied by the demonstration that SOD inside red-cell membranes can dismute superoxide generated outside the membrane or vice versa (179,180). It is possible, although not yet proven, that emphysemalike lung injury may be modulated using these systems to protect α-1-Pi from being oxidized.

The selenium-dependent glutathione peroxidase may also protect against lung injury, but its role is controversial. Some studies demonstrate protection (181). Other studies demonstrate diminution in its activity upon exposure to high concentrations of oxygen (182). At the current time, it is not possible to determine the efficacy of glutathione peroxidase in protecting against oxidant-induced injury and emphysema.

Ceruloplasmin, the multifunctional (183) acute-phase-reaction protein (184) in serum carrying over 95% of the circulating copper in mammals (183), is a superoxide scavenger (185). Ceruloplasmin also inhibits myeloperoxidase-dependent inactivation of α-1-Pi and the resultant leukocyte elastase-mediated elastolysis in vitro (186). The superoxide scavenging capability of ceruloplasmin is dramatically less than that of CuZn-SOD (183).

1. NATURAL DISMUTATION OF SUPEROXIDE ANION

$$O_2^{\cdot -} + O_2^{\cdot -} + 2H^+ \xrightarrow[\text{Spontaneous}]{} O_2 + H_2O_2 \qquad Km = 1 \times 10^4 M^{-1}S^{-1}$$

2. SUPEROXIDE DISMUTASE CATALYZED DISMUTATION OF SUPEROXIDE ANION

$$SOD-Cu^{2+} + O_2^{\cdot -} \longrightarrow SOD-Cu^{1+} + O_2$$

$$SOD-Cu^{1+} + O_2^{\cdot -} + 2H^+ \longrightarrow SOD-Cu^{2+} + H_2O_2$$

$$O_2^{\cdot -} + O_2^{\cdot -} + 2H^- \xrightarrow{SOD-Cu^{2+}} O_2 + H_2O_2 \qquad Km = 2 \times 10^9 M^{-1}S^{-1}$$

3. REMOVAL OF PEROXIDE BY CATALASES AND PEROXIDASES

$$H_2O_2 \xrightarrow[\text{Catalases}]{+H_2O_2} 2H_2O + O_2$$

$$H_2O_2 \xrightarrow[\text{Peroxidases}]{+RH_2} 2H_2O + R$$

4. FORMATION OF HYDROXYL RADICAL AND HYDROXIDE

$$O_2^{\cdot -} + Fe^{3+}-\text{Chelate} \longrightarrow O_2 + Fe^{2+}-\text{Chelate}$$

$$Fe^{2+}-\text{Chelate} + H_2O_2 \longrightarrow Fe^{3+}-\text{Chelate} + OH^- + OH\cdot$$

$$O_2^{\cdot -} + H_2O_2 \longrightarrow O_2 + OH^- + OH\cdot \text{ (Haber–Weiss Reaction)}$$

Figure 4 Enzymatic free-radical scavengers. (1) Superoxide produced by normal cellular oxidative metabolism is dismuted at a reasonable rate by endogenous nonenzymatic measures (e.g., citrate, metals, etc.) (173). (2) Superoxide dismutase neutralizes superoxide by a two-step reaction in which oxygen and hydrogen peroxide are formed. The rate of this reaction is about 10,000 times faster than the nonenzymatic rate (176). (3) Peroxide is removed by catalases and peroxidases (173). (4) Hydrogen peroxide and superoxide, if not removed, can react to form oxygen, hydroxyl anion, and the hydroxyl radical by the controversial Haber-Weiss reaction (174,176). The hydroxyl radical is very reactive and difficult to control. As a result, the hydroxyl radical is probably more damaging to tissues than other, less reactive species (175).

However, because the circulating protein reaches extracellular tissue spaces, ceruloplasmin's antioxidant protective role may complement that of the intracellularly located superoxide dismutase(s) with respect to tissue injury. Plasma ceruloplasmin is diminished in copper-deficient hamsters that develop emphysema, but its role in the pathogenesis of emphysema in this model is not known.

A new system has been described that may play an additional protective role (187). Methionine sulfoxide peptide reductase can reactivate α-1-Pi that has been oxidatively inactivated by chloramine T or by myeloperoxidase from canine PMNs (188). This enzyme has been found in human lung homogenates, human PMN lysates, and rabbit alveolar type II cells (189). Initially, it appeared to provide another mechanism of protection against oxidative inactivation of α-1-Pi; however, this enzyme does not consistently reactivate α-1-Pi that has been inactivated by an aqueous solution of cigarette smoke and peroxide (188). It is speculated that cigarette smoke might modify nonmethionine groups required for α-1-Pi activity or that cigarette smoke might oxidize a larger number of methionines per molecule of α-1-Pi than the other oxidizing systems previously employed. The findings that chemical oxidation produces 2 moles of methionine sulfoxide/mole of inhibitor (114), but that alpha-1-Pi recovered from smokers contains 4 moles/mole of inhibitor (190), appear to support the latter possibility. Currently, the significance of this enzyme in protecting against lung injury is not known.

E. Cell-Matrix Contact

Two lines of investigation suggest that cell contact with matrix macromolecules may be important in promoting matrix degradation. PMNs in contact with fibronectin are capable of causing elastase-mediated degradation of that matrix macromolecule even in the presence of the serum protease inhibitors α-1-Pi and α-2-M (191). A possible explanation is that serum inhibitors are partially excluded from the microenvironment created by close cell-matrix contact. Furthermore, prevention of connective tissue degradation by therapeutic maneuvers to increase inhibitor activity (such as replacement therapy with α-1-Pi) may not be adequate if the inhibitor is unable to reach the microenvironment where matrix degradation is occurring.

Along similar lines, the degradative effects of human alveolar macrophage-derived PA on fibrin and elastin matrices have been studied (192). PA has recently been found in lung lavage fluid and in human alveolar macrophages (193). Furthermore, human alveolar macrophages secrete PA when stimulated (192). Macrophages degrade fibrin in the presence of serum that contains protease inhibitors only if the cells are in contact with the fibrin matrix. This may be important to lung remodeling during repair.

Similarly, macrophages degrade elastin in the same milieu as mentioned above only if they are in contact with the elastin matrix. If matrices contain fibrin and elastin, cells cannot degrade elastin unless plasminogen is added to the medium, indicating the importance of plasminogen activator in macrophage-mediated matrix degradation in this system. Although macrophages can secrete PA (111), the mechanism of elastin degradation is unclear because plasmin alone does not enhance elastin solubilization. Therefore, it is possible that plasmin activates a latent macrophage elastase activity not yet described (192,194). Furthermore, macrophage-mediated elastin degradation may involve multiple proteases, including a metalloprotease (e.g., macrophage elastase), a thiol protease, and perhaps a neutrophil elastase. The relative role of macrophage-mediated elastolysis in producing lung injury is unclear; however, macrophages cultured in serum containing protease inhibitors are capable of degrading 5 times more elastin than are neutrophils (194).

The ability of tumor cells to degrade alveolar basement membranes in hamster lungs appears to be dependent on the presence of plasminogen and, hence, on plasminogen activator activity (195). This is thought to partially explain the ability of tumor cells to invade tissue. Perhaps macrophage invasion into the interstitium may be facilitated in a similar fashion, bringing them into contact with matrix macromolecules. On the other hand, macrophages secrete a protein resembling α-1-Pi, perhaps identical to it (196), which would tend to neutralize any elastase that may be secreted by or activated within macrophages. The overall significance in the destruction of lung tissue in diseases such as emphysema is unknown.

F. Role of the Repair Process

Recent information suggests that repair processes and their inhibition may be equal in importance of degradative processes in the pathogenesis of emphysema and lung injury in general.

As we have mentioned, accumulation of a protein represents a balance between synthesis and degradation. We have seen that degradation occurs in elastase-induced emphysema; yet, synthesis is increased, and total accumulation approaches control values. Nevertheless, the lung is architecturally and functionally deranged.

Kuhn and Starcher (196 a) fed adult hamsters β-aminopropionitrile (BAPN), a known lathyrogen that inhibits the cross-linking process in collagen and elastin, presumably by inhibition of lysyl oxidase (197). Emphysema was then induced in these animals using intratracheal instillation of elastase. The hamsters were studied six to eight weeks later using the morphometric parameters mean linear intercept (Lm) and internal surface area (ISA). Animals given only elastase develop emphysema as expected. Animals given either saline or BAPN alone develop none. However, animals fed BAPN *and* given elastase develop a more severe degree of emphysema than those given only elastase. Furthermore, the desmosine/lysine ratio is reduced along with other elastin aldehydes in the BAPN + elastase group of animals, and, more important by the restoration of desmosines and aldehydes to normal (as occurs in the group only given elastase) is impaired. This is an important study that underscores the role of the repair process in attempting to protect the lung from structural derangements caused by proteolysis at least in the early stages of enzyme-induced disease.

Recent evidence demonstrates the ability of cigarette smoke to inhibit repair mechanisms. Laurent et al. (198) demonstrated that various solutions of cigarette smoke inhibit lysyl oxidase activity and lysyl oxidase-mediated desmosine formation during posttranslational modification of tropoelastin. That the effect is substrate directed is demonstrated by the fact that lysyl oxidase-mediated oxidation of diaminopentane is not affected, whereas conversion of lysines in tropoelastin to desmosine is inhibited. It was suggested that this effect could interfere with elastin repair and contribute to the development of emphysema in smokers. Osman et al. confirmed these results by demonstrating that in elastase-induced emphysema in hamsters, cigarette smoke exposure causes decreased incorporation of [^{14}C] lysine (administered intraperitoneally) into desmosine and isodesmosine in lungs, compared to elastase-treated animals not exposed to cigarette smoke (199).

Further studies regarding the interplay between cigarette smoke and elastase in the development of emphysema were performed by Niewoehner and Hoidal (200). Porcine pancreatic elastase was given to hamsters in a dose (2 units) known to cause minimal emphysema. Several groups of hamsters were exposed to cigarette smoke in closed chambers for three weeks prior to elastase treatment, four weeks after elastase treatment, or continuously. Controls received continuous smoke exposure plus intratracheal saline, elastase plus sham exposure, or normal saline plus sham exposure. Controls and animals exposed to cigarette smoke after elastase administration developed mild or no emphysema. Animals exposed to cigarette smoke prior to receiving elastase developed severe emphysema. The results implicate four possible mechanisms: (a) α-1-Pi is inactivated by cigarette smoke or by inflammatory cells recruited to the lung in response to cigarette smoke; (b) protease levels are increased owing to recruited cells; (c) cigarette smoke inhibits cross-linking and hence the repair process; or (d) other smoke-induced changes occur that enhance susceptibility to elastase damage.

The fact that starvation potentiates elastase-induced emphysema in rats (201) and that refeeding partially restores connective tissue accumulation (202) suggests that the repair process may be altered during starvation by invoking one of the selective repair mechanisms discussed in this chapter.

A genetic deficiency of lysyl oxidase in mice (202a,b) may in fact represent failure to adequately resynthesize normally cross-linked collagen and elastin and result in the observed lesions.

The cadmium-treated hamster model represents a special case. Cadmium aerosolization has been reported to produce centrilobular emphysema in rats (203), but other investigators have had difficulty reproducing emphysema in this model (101,204). Furthermore, pulmonary fibrosis is rather severe in some animals receiving cadmium (151,205). More recently, however, cadmium aerosolization has been reported to change the character of elastase activity from that initially inhibited by chloromethylketones (CMK) to that inhibited by EDTA but not by CMK, suggesting an initial role for leukocyte elastase and a subsequent, perhaps chronic, role for alveolar macrophage elastase, a metalloenzyme (206). Data suggesting that cadmium inhibits lysyl oxidase-mediated cross-link formation in collagen (207) appear to contradict the observations, mentioned above, that fibrosis occurs in rats and hamsters exposed to cadmium (101,151). However, this observation is in agreement with the recent demonstration that cadmium-induced fibrosis in hamsters no longer results when BAPN is added to the animals' diet (205). Instead, emphysema is produced, implying again that the repair process is important in modulating the form of lung injury that develops.

IV. HUMAN STUDIES

Most of the investigative advances in understanding the pathogenesis of emphysema have come from studies of animal models, especially the elastase-induced hamster model. Because direct extrapolation of mechanisms from animal models to human pathogenesis is not always warranted, human studies are, of necessity, extremely important. However, the study of a disease which is chronic, often taking most of the individual's life-span to develop, and which is currently defined by anatomical criteria discernible only at autopsy or after surgical resection makes the design of meaningful human studies a challenge. This challenge is further compounded by the fact that no causative agent has been definitively identified in humans, with the possible exception of the hereditary diminution or absence of α-1-Pi in a relatively less common form of emphysema. Although cigarette smoke is strongly implicated, there is no direct evidence demonstrating that cigarette smoke causes human emphysema.

Until fairly recently, the biochemical studies of emphysema have been limited to analyses performed on ex situ lung tissue. However, with the introduction of bronchoalveolar lavage (BAL), other possibilities arose. Reservations exist as to the accuracy of BAL in sampling interstitial events, because the technique involves bathing alveolar surfaces that are located on the opposite side of the alveolar capillary membrane. BAL may only reflect interstitial events in disease states in which the membrane is violated but not in normal patients with structurally intact alveolar walls. It is impossible to control the dispersion of the fluid, a factor of importance in emphysema, in which the lesions are irregularly distributed. The optimum normalizing variable for BAL components is not universally agreed upon (208). Recent studies suggest normalizing soluble constituents to albumin (208a) because protein concentrations in BAL appear to be constant (209). Although BAL is currently a valuable research technique (210), it may become clinically useful in the future.

BAL studies have been used to test the protease/antiprotease and oxidant/antioxidant theories in humans. These studies have provided conflicting results concerning whether smokers contain higher elastolytic or less functional α-1-Pi in the lower respiratory tract compared to nonsmokers (211–217). The origins of BAL constituents are in question, because α-1-Pi in BAL from smokers may have come from serum and elastolytic activity may have come from macrophage-sequestered neutrophil elastase (215). Timing of studies with respect to cigarette smoking may be a factor contributing to the variability of measured enzyme and inhibitor levels (216).

BAL fluid taken from smokers reveals oxidation of α-1-Pi's active site methionine to methionine sulfoxide (4 moles/mole of inhibitor) (190). This can be partially reversed by methionine sulfoxide peptide reductase (189). In contrast to cigarette smoke inactivation of α-1-Pi, chloramine T- and phagocyte-mediated oxidation are completely restored by the reductase (188). This latter discrepancy might be explained by the fact that the chloramine T- and leukocyte-mediated inactivation each produce only 2 moles of methionine sulfoxide per mole of inhibitor (79,113c) and, therefore, require only half as much reduction as the effects of cigarette smoke.

In addition, alveolar macrophages recovered in BAL fluid from asymptomatic cigarette smokers release elevated amounts of superoxide anion despite the presence of elevated levels of intracellular superoxide dismutase, most of which is MnSOD (218).

V. PERSPECTIVES, SUMMARY, AND CONCLUSIONS

The gains made in understanding emphysema have been substantial, but questions of similar or greater magnitude still remain. Emphysema has been defined; however, there are limitations to the current definition, especially with respect to experimentally produced lesions in animals. Refinement of the definition may await determination of a precise etiology. Other diseases of unknown etiology are defined by more detailed descriptive paragraphs (219). Perhaps this approach would be useful with respect to emphysema at this time and might allow inclusion of a description of the lesion induced in animals.

Functional correlates of structural alterations produced in emphysema are being revealed, although interest in molecular mechanisms has prevailed more recently. Elastin appears to contribute to the mechanical properties of lung tissue differently from collagen. The ability to refine these relational observations rests on the ability to make careful and precise measurements physiologically, histologically, and biochemically. It is tempting to disregard the importance of making these correlations in the face of the rapid increases in our knowledge of molecular processes.

Understanding of pathogenetic mechanisms of emphysema has greatly increased in the last 25 years. However, direct evidence that these mechanisms actually cause emphysema in humans is lacking. This lack of direct proof stems partially from the chronic incipient nature of most forms of emphysema and the lack of consistent development of the disease in individuals exposed to proposed etiological agents such as cigarette smoke. Furthermore, owing to the variability of protective mechanisms and differential susceptibility to lung injury of various animal species, mechanisms involved in animal models of emphysema cannot uniformly be expected to apply to the human counterpart. Even further removed from clinical application are the in vitro studies of injury or protection in explant or cell culture. What a tissue or cell *can* do in vitro, removed from its native "milieu," may not reflect its normal activity in vivo.

Elastin damage has been demonstrated in end-stage emphysema in humans and in enzyme-induced lesions in animals, but not in actively developing lesions in humans. The importance of the repair mechanism has been demonstrated in animals. A number of endogenous and exogenous proteases have been demonstrated that are capable of degrading lung connective tissue in humans, as have a number of endogenous protective systems such as antiproteases. A number of endogenous and exogenous oxidative species have also been enumerated that are capable of inactivating endogenous protective mechanisms, capable of stimulating endogenous degradative mechanisms, and perhaps capable of directly damaging matrix molecules. Direct evidence of their role in lung damage in humans is lacking. This area of investigation has revealed a more direct association between cigarette smoking and the development of emphysema. If oxidation proves to be relevant, methods of delivering free-radical scavenging enzymes or antioxidants to potential areas of inflammation may be developed as an extension of animal studies.

Studies to date have concentrated on static measurements of connective tissue macromolecule (most often collagen and elastin) accumulation in lung tissue in various circumstances. A few studies have addressed synthesis, and only more recently has degradation been studied. Control of degradation and repair may be the most important aspects of this disease with respect to regulating tissue damage by therapeutic intervention. There is an obvious need to extend connective tissue studies to include synthesis and degradation and to include other nonelastin and noncollagen alveolar wall matrix molecules, such as glycosaminoglycans, proteoglycans, fibronectin, and other basement membrane components that are destroyed in emphysema. Future studies must also be directed toward determining which cell(s) in the lung make(s) elastin so that subsequent efforts can elucidate modalities by which elastin synthesis can be controlled. Molecular biological techniques may aid in localization as well as determining the level of control (DNA, mRNA, tRNA, etc.). Similar studies of the elastase gene may allow exertion of control over elastase production and further regulation of elastin degradation.

The role of the epithelial barrier (air-tissue interface) of the tracheobronchial tree has been studied to a much lesser degree than the blood-tissue interface. This occurred because the initial insights into the pathogenesis of emphysema in humans arose from the observations of lowered blood levels of α-1-Pi. Although subsequent induction of emphysema in animals was achieved via intratracheal enzyme instillation, pathogenetic theories have usually focused on aspects that might link blood antiprotease/protease interactions with the interstitial damage observed in that disease. Few emphysema studies have addressed the role of the epithelial barrier's interaction with toxins and subsequent encroachment on the interstitium.

The future rewards in pulmonary investigation and pulmonary medicine will be reaped in part by those addressing molecular mechanisms and applying their findings to help us understand structural changes affecting normal and disordered lung function. As these relationships are understood, the patient will benefit as knowledge extends toward developing the means to prevent disease or halt its progression.

ACKNOWLEDGMENTS

We thank Dr. A. D. Renzetti for his valuable comments concerning early perspectives of cigarette smoke's relationship to emphysema and Dr. J. R. Hoidal for his critical review of the manuscript and more current perspectives on the pathogenesis of emphysema.

This work was supported by a grant from the American Lung Association and a Contract from E. I. Dupont de Nemours. N.T.S. is the recipient of a Pulmonary Academic Award from the National Heart, Lung and Blood Institute of the National Institutes of Health.

REFERENCES

1. L. Reid, *Pulmonary Diseases and Disorders* (A. P. Fishman, ed.), McGraw-Hill, New York, 1980, pp. 503–525.
2. J. R. Hoidal and D. E. Niewoehner, *Chest, 83*:679–685 (1983).
3. R. G. Fraser and J. A. P. Pare, *Diagnosis of Diseases of the Chest,* 2nd ed., Vol. I, W. B. Saunders, Philadelphia, 1977, p. 116.
4. H. Spencer, *Pathology of the Lung,* 3rd ed., Vol. I, Pergamon Press, Philadelphia, 1977, p. 7.
5. R. G. Crystal, *The Biochemical Basis of Pulmonary Function,* Marcel Dekker, New York, 1976.
6. E. R. Weibel, *Morphometry of the Human Lung,* Academic Press, New York, 1963, pp. 63, 134.
7. M. S. Dunnill, *Thorax, 17*:329–333 (1962).
8. C. G. Loosli, *Arch. Pathol., 24*:743–776 (1937).
9. P. T. Macklem, *Physiol. Rev., 51*:368–436 (1971).
10. H. A. Menkes and Traystman R. J., *Am. Rev. Resp. Dis., 116*:287-309 (1977).
11. H. Eppinger, *Vrtljschr. Prakt. Heilk., 4*:1–80 (1876).
12. F. Orsos, *Beitr. Pathol., 41*:95–121 (1907).
13. N. P. Tenderloo, *Erbegn. Inn. Med. U. Kinderh, 6*:1–28 (1910).
14. A. C. Burton, *Physiol. Rev. 34*:619–642 (1954).
15. I. Setnikar, *Arch. Fisiol., 55*:349–374 (1955).
16. J. Mead, *Physiol. Rev., 41*:281–330 (1961).
17. L. B. Sandberg, N. T. Soskel, and J. G. Leslie, *N. Engl. J. Med., 304*:566–579 (1981).
18. R. R. Wright, *Am. J. Pathol., 39*:355–367 (1961).
19. G. R. Meneely, A. D. Renzetti, Jr., J. D. Steele, J. P. Wyatt, and H. W. Harris, *Am. Rev. Resp. Dis., 85*:762–768 (1962).

19a. G. Snider, J. Kleinerman, W. M. Thurlbeck, et al., *Am. Rev. Respir. Dis., 132*:182-185 (1985).

20. R. T. H. Laennec, *A Treatise on the Diseases of the Chest and on Mediate Auscultation,* 4th ed. (J. Forbes, trans.), Wood, New York, 1835.
21. G. Rainey, *Med.-Chir. Trans. (London), 31*:297–304 (1848).
22. R. G. Fraser and J. A. P. Parè, *Diagnosis of Diseases of the Chest,* 2nd ed., Vol. III, W. B. Saunders, Philadelphia, 1979, p. 1370.
23. Ciba Guest Symposium, *Thorax, 14*:286–299 (1959).
24. J. Gough and J. E. Wentworth, *J. Roy. Micr. Soc., 69*:231–235 (1949).
25. E. R. Weibel and D. M. Gomez, *J. Appl. Physiol., 17*:343–348 (1962).
26. R. DeB. Johnston, M. D. Levine and W. M. Thurlbeck, *IEEE Trans. Biomed. Eng., BME-19*:422–428 (1972).
27. E. R. Weibel, *Morphometry of the Human Lung,* Academic Press, New York, 1963, pp. 41–45.
28. G. Hayatdavoudi, J. D. Crapo, F. J. Miller, and J. J. O'Neil, *J. Appl. Physiol., 48*:389-393 (1980).
29. W. Scherle, *Mikroskopie, 26*:57–60 (1970).
30. G. L. Snider and A. L. Korthy, *Am. Rev. Resp. Dis., 117*:685–693 (1978).
31. J. B. Karlinsky and G. L. Snider, *Am. Rev. Resp. Dis., 117*:1109–1133 (1978).
32. D. E. Niewoehner and J. Kleinerman, *Am. Rev. Resp. Dis., 115*:15–21 (1977).
33. E. R. Weibel, *Stereological Methods,* Vol. I. *Practical Methods for Biological Morphometry,* Academic Press, New York, 1979.
34. E. W. Weibel, *Am. Rev. Resp. Dis., 128*:752–760 (1983).
35. K. von Neergaard, *Z. Ges. Exp. Med., 66*:373–394 (1929).
36. J. A. Clements, R. F. Hustead, R. P. Johnson, and I. Gribetz, *J. Appl. Physiol., 16*:444–450 (1961).
37. J. A. Clements, *Sci. Am., 207*:120–129 (1962).

38. R. W. Carton, J. Dainauskas, and J. W. Clark, *J. Appl. Physiol., 17*:547–551 (1962).
39. D. D. Stromberg and C. A. Wiederhielm, *J. Appl. Physiol., 26*:857–862 (1969).
40. R. M. Senior, D. R. Bielefeld, and M. K. Abensohn, *Am. Rev. Resp. Dis., 111*:184–188 (1975).
41. W. G. Johanson and A. K. Pierce, *J. Clin. Invest., 51*:288–293 (1972).
42. G. L. Snider, C. B. Sherter, K. W. Koo, J. B. Karlinsky, J. A. Hayes, and C. Franzblau, *J. Appl. Physiol., 42*:206–215 (1977).
43. J. A. Pierce, J. B. Hocott, and R. V. Ebert, *Ann. Intern. Med., 55*:210–222 (1961).
44. M. Fitzpatrick and V. D. Hospelhorn, *J. Lab. Clin. Med., 60*:799–810 (1962).
45. C. Kuhn and R. M. Senior, *Lung, 155*:185-197 (1978).
46. H. T. G. Strawbridge, *Am. J. Pathol., 37*:161–174 (1960).
47. A. D. Renzetti, Jr., Personal communication.
48. U.S. Department of Health, Education and Welfare Public Health Service Center for Disease Control Publication No. 1103, 1964.
49. O. Auerbach, A. P. Stout, E. C. Hammond, and L. Garfinkel, *N. Engl. J. Med., 269*: 1045–1054 (1963).
50. O. Auerbach, E. C. Hammond, L. Garfinkel, and C. Benanie, *N. Engl. J. Med., 286*: 853–857 (1972).
51. O. Auerbach, L. Garfinkel, and E. C. Hammond, *Chest, 65*:29–35 (1974).
52. W. M. Thurlbeck, *Chronic Airflow Obstruction,* W. B. Saunders, Philadelphia, 1976.
53. M. G. Cosio, K. A. Hale, and D. E. Niewoehner, *Am. Rev. Resp. Dis., 122*:265–271 (1980).
54. W. M. Thurlbeck, J. A. Henderson, R. G. Fraser, and D. V. Bates, *Medicine, 49*:81–145 (1970).
55. J. B. Amberson and D. M. Spain, *Trans. Assoc. Am. Physicians, 60*:92–101 (1947).
56. D. M. Spain, H. Siegel, and V. A. Bradess, *JAMA, 224*:332–325 (1973).
57. *WHO Tech. Rep. Ser., 213*:15 (1961).
58. G. L. Snider, *Med. Clin. North Am., 65*:647–665 (1981).
59. W. M. Thurlbeck and M. R. ABell, *The Lung, Structure, Function and Disease,* Williams & Wilkins, Baltimore, 1978.
60. A. P. Fishman, *Pulmonary Diseases and Disorders,* McGraw-Hill, New York, 1980, p. 459.
61. R. G. Fraser and J. A. P. Paré, *Diagnosis of Diseases of the Chest*, 2nd ed., Vol. III, W. B. Saunders, Philadelphia, 1979, p. 1369.
62. C-B. Laurell and S. Eriksson, *Scand. J. Clin. Lab. Invest., 15*:132–140 (1963).
63. S. Eriksson, *Acta Med. Scand., 175*:197–205 (1964).
64. S. Eriksson, *Acta Med. Scand., 177* (Suppl. 432):1–85 (1965).
65. P. Gross, M. A. Babyak, E. Tolker, and M. Kaschak, *J. Occup. Med., 6*:481–484 (1964).
66. P. Gross, E. A. Pfitzer, E. Tolker, M. A. Babyak, and M. Kaschak, *Arch. Environ. Health, 11*:50–58 (1965).
67. P. D. Kaplan, C. Kuhn, and J. A. Pierce, *J. Lab. Clin. Med., 82*:349–356 (1973).
68. W. G. Johanson, Jr., A. K. Pierce, and R. C. Reynolds, *J. Lab. Clin. Med., 78*:599–607 (1971).
69. J. A. Hayes, A. Korthy, and G. L. Snider, *J. Pathol., 117*:1–14 (1975).
70. G. E. Blackwood, R. Hosannah, E. Perman, S. Keller, and I. Mandl, *Proc. Soc. Exp. Biol. Med., 144*:450–454 (1973).
71. G. L. Snider, J. A. Hayes, C. Franzblau, H. M. Kagan, P. S. Stone, and A. L. Korthy, *Am. Rev. Resp. Dis., 110*:254–262 (1974).
72. G. Weinbaum, V. Marco, T. Ikeda, B. Mass, D. R. Meranze, and P. Kimbel, *Am. Rev. Resp. Dis., 109*:351–357 (1974).
73. M. R. Schuyler, D. J. Rynbrandt, and J. Kleinerman, *Am. Rev. Resp. Dis., 117*:97–102 (1978).

74. G. L. Snider and C. B. Sherter, *J. Appl. Phyisol., 43*:721–729 (1977).
75. J. E. Gadek, G. W. Hunninghake, G. A. Fells, R. L. Zimmerman, B. A. Keogh, and R. G. Crystal, *Bull. Eur. Physiopathol. Resp., 16*(Suppl.):27–40 (1980).
76. G. L. Snider, *Med. Clin. North Am., 65*:647–655 (1981).
77. G. L. Snider, *Am. Rev. Resp. Dis., 124*:321–324 (1981).
78. A. Janoff, *Chest, 83*(Suppl.):54s–58s (1983).
79. A. Janoff, *J. Appl. Physiol., 55*:285–293 (1983).
80. V. Marco, B. Mass, D. R. Meranze, G. Weinbaum, and P. Kimbel, *Am. Rev. Resp. Dis., 104*:595–598 (1971).
81. R. M. Senior, T. Tegner, C. Kuhn, K. Ohlsson, B. C. Starcher, and J. A. Pierce, *Am. Rev. Resp. Dis., 116*:469–475 (1977).
82. A. Janoff, B. Sloan, G. Weinbaum, V. Damiano, R. A. Sandhaus, J. Elias, and P. Kimbel, *Am. Rev. Resp. Dis., 115*:461–478 (1977).
83. R. A. Goldstein and B. C. Starcher, *J. Clin. Invest., 61*:1286–1290 (1978).
84. C. Kuhn, S-Y. Yu, M. Chraplyvy, H. E. Linder, and R. M. Senior, *Lab. Invest., 34*: 372–380 (1976).
85. S. Y. Yu., N. R. Keller, and A. Yoshida, *Proc. Soc. Exp. Biol. Med., 157*:369–373 (1978).
86. P. A. Mallory and J. Travis, *Biochemistry, 14*:722–730 (1975).
87. F. Clemente, A. DeCaro, and C. Figarella, *Eur. J. Biochem., 31*:186–193 (1972).
88. J. Travis, P. J. Giles, L. Porcelli, C. F. Reilly, R. Baugh, and J. Powers, *Ciba Foundation Symp., 75*:51–68 (1980).
89. P. H. Pryca-Jones, J. Saklatvala, and G. C. Wood, *Clin. Sci. Mol. Med., 47*:403–414 (1974).
90. C. F. Reilly and J. Travis, *Biochim. Biophys. Acta, 621*:147–157 (1980).
91. E. J. Campbell, R. R. White, R. M. Senior, R. J. Rodriguez, and C. Kuhn, *J. Clin. Invest., 64*:824–833 (1979).
92. S. E. McGowan, P. J. Stone, J. D. Calore, G. L. Snider, and C. Franzblau, *Am. Rev. Resp. Dis., 127*:449–455 (1983).
93. E. J. Campbell and M. S. Wald, *J. Lab. Clin. Med., 101*:527–536 (1983).
94. E. J. Campbell and M. S. Wald, *Am. Rev. Resp. Dis., 127*:631–635 (1983).
95. S. A. Lonky and H. Wohl, *J. Clin. Invest., 67*:817–826 (1981).
96. G. S. Lazarus, R. S. Brown, J. R. Daniels, and H. M. Fullmer, *Science, 29*:1483–1485 (1968).
97. G. Lazarus, J. R. Daniels, J. Lian, and M. C. Burleigh, *Am. J. Pathol., 68*:565–578 (1972).
98. E. Harper, *Annu. Rev. Biochem., 49*:1063–1078 (1980).
99. P. A. Martorana, N. N. Share, and J. W. Richard, *Am. Rev. Resp. Dis., 116*:57–63 (1977).
100. C. Kuhn III, J. Slodkowska, T. Smith, and B. Starcher, *Bull. Eur. Physiopathol. Resp., 16*(Suppl.):127–137 (1980).
101. J. R. Hoidal and D. E. Niewoehner, *Chest, 83*(Suppl.):58s–59s (1983).
102. C. Kuhn III, R. White, R. Reppun, and H. S. Lin, Technical Information Center, Energy Research and Development Administration, 1977.
103. M. J. Banda and Z. Werb, *Biochem. J., 193*:589–605 (1981).
104. M. J. Banda, E. J. Clark, and Z. Werb, *J. Exp. Med., 152*:1563–1570 (1980).
105. R. White, G. S. Habicht, H. P. Godfrey, A. Janoff, E. Barton, and C. Fox, *J. Lab. Clin. Med., 97*:718–729 (1981).
106. R. M. Senior, E. J. Campbell, J. A. Landis, F. R. Cox, C. Kuhn, and H. S. Koren, *J. Clin. Invest., 69*:384–393 (1982).
107. R. A. Sandhaus, K. M. McCarthy, R. A. Musson, and P. M. Henson, *Chest, 83* (Suppl.):60s–62s (1983).

108. D. H. Matulionis, *Arch. Environ. Health, 34*:298–301 (1979).
109. Hinman L. M. Stevens C. A. Matthay R. A., and Gee J. B. L., *Am. Rev. Resp. Dis., 121*:263–271 (1980).
110. Z. Werb and S. Gordon, *J. Exp. Med., 142*:346–360 (1975).
111. J. C. Unkless, S. Gordon, and E. Reich, *J. Exp. Med., 139*:834–850 (1974).
112. K. C. Robbins and L. Summaria, *Meth. Enzymol., 45*:257–273 (1976).
112a. K. Morihara, H. Tsuzuki, and K. Oda, *Infect. Immun., 24*:188–193 (1979).
113. G. H. Barlow, *Meth. Enzymol., 45*:239–244 (1976).
113a. J. Travis and G. S. Salvesen, *Ann. Rev. Biochem., 52*:655–709 (1983).
113b. A. S. Feste and J. C. Gan, *Fed. Proc., 39*:1689, (1980).
113c. D. Johnson and J. Travis, *J. Biol. Chem., 254*:4022–4026 (1979).
114. D. Johnson and J. Travis, *J. Biol. Chem., 253*:7142–7144 (1978).
115. K. Beatty, J. Bieth, and J. Travis, *J. Biol. Chem., 255*:3931–3934 (1980).
116. D. Johnson and J. Travis, *Biochem. J., 163*:639–641 (1977).
117. R. A. Stockley and S. C. Afford, *Clin. Sci., 64*:223–230 (1983).
118. M. C. Owen, S. O. Brennan, J. H. Lewis, and R. W. Carrell, *N. Engl. J. Med., 309*: 694–698 (1983).
119. A. J. Barrett and P. M. Starkey, *Biochem. J., 133*:709–724 (1973).
120. K. Ohlsson, *Ann. NY Acad. Sci., 256*:409–419 (1975).
121. M. T. Debanne, R. Bell, and J. Dolovich, *Biochim. Biophys. Acta, 411*:295–304 (1975).
122. B. J. Haverback, B. Dyce, H. F. Bundy, S. K. Wirtschafter, and H. A. Edmonson, *J. Clin. Invest., 41*:972–980 (1962).
123. E-L. Gustavsson, K. Ohlsson, and A-S. Olsson, *Hoppe-Seyler's Z. Physiol. Chem., 361*:169–176 (1980).
124. P. J. Stone, J. D. Calore, G. L. Snider, and C. Franzblau, *J. Clin. Invest., 69*:920–931 (1982).
125. F. Kueppers, W. R. Abrams, G. Weinbaum, and J. Rosenbloom, *Arch. Biochem. Biophys., 211*:143–150 (1981).
126. R. White, A. Janoff, and H. P. Godfrey, *Lung, 158*:9–14 (1980).
127. J. Travis, J. Bowen, and R. Baugh, *Biochemistry, 17*:5651–5656 (1978).
128. Travis J., Garner D., and Bowen J., *Biochemistry, 17*:5647–5651 (1978).
129. J. Lieberman, B. M. Trimmer, and N. B. Kurnick, *Lab. Invest., 14*:249–257 (1965).
130. H. Schiessler, K. Hochstrasser, and K. Ohlsson, *Neutral Proteases of Human Polymorphonuclear Leukocytes* (K. Havemann and A. Janoff, eds.), Urban & Schwarzenberg, Baltimore-Munich, 1978, pp. 195–207.
131. K. Hochstrasser, H. Haendle, R. Reichert, E. Werele, and S. Schwarz, *Hoppe-Seyler's Z. Physiol. Chem., 353*:954–958 (1971).
132. K. Hochstrasser, R. Reichert, S. Schwarz, and E. Werle, *Hoppe-Seyler's Z. Physiol. Chem., 353*:221–226 (1972).
133. K. Hochstrasser, G. J. Albrecht, O. L. Schonberger, B. Rasche, and K. Lempart, *Hopper-Seyler's Z. Physiol. Chem., 362*:1369–1375 (1981).
134. K. Hochstrasser O. L. Schonberger I. Rossmanith, and E. Wachter, *Hoppe-Seyler's Z. Physiol. Chem., 362*:1357–1362 (1981).
135. H. G. Welgus and G. P. Stricklin, *J. Biol. Chem., 258*:12259–12264 (1983).
136. T. E. Cawston, E. Mercer, M. DeSilva, and B. L. Hazleman, *Arthritis Rheum., 27*: 285–290 (1984).
137. H. W. Macartney and J. Tschesche, *Eur. J. Biochem., 130*:79–83 (1983).
138. T. I. Morales, K. E. Kuettner, D. S. Howell, and J. F. Woessner, *Biochim. Biophys. Acta, 760*:221–229 (1983).
139. D. E. Woolley, D. R. Roberts, and J. M. Evanson, *Nature, 261*:325–327 (1976).
140. H. W. Marcartney and H. Tschesche, *Eur. J. Biochem., 130*:85–92 (1983).

141. A. Sellers, G. Murphy, M. C. Meikle, et al., *Biochem. Biophys. Res. Commun., 87*: 581–587 (1979).
142. T. E. Cawston, W. A. Galloway, E. Mercer, G. Murphy, and J. J. Reynolds, *Biochem. J., 195*:159–165 (1981).
143. G. Murphy, T. E. Cawston, and J. J. Reynolds, *Biochem. J.*, 195:167–170 (1980).
144. J. Hiti-Harper, H. Wohl, and E. Harper, *Science, 199*:991–992 (1978).
145. K. E. Kuettner, J. Hiti, P. Eisenstein, and E. Harper, *Biochem. Biophys. Res. Commun., 72*:40–46 (1976).
146. K. E. Kuettner, L. Soble, R. L. Croxen, B. Merczynska, J. Hiti, and E. Harper, *Science, 196*:653–654 (1977).
147. K. Ohlsson and I. Olsson, *Eur. J. Biochem., 36*:473–481 (1973).
148. Y. Nagi, *Mol. Cell. Biochem., 1*:137–145 (1973).
149. D. E. Woolley, R. W. Glanville, M. J. Crossley, and J. M. Evanson, *Eur. J. Biochem., 54*:611–622 (1975).
150. M. B. Berman, J. C. Barber, R. C. Talamo, and C. E. Langley, *Invest. Ophthalmol., 12*:759–770 (1973).
151. G. L. Snider and J. B. Karlinsky, *Pathobiol. Annu., 7*:115–142 (1977).
152. L. J. Pecora, W. R. Manne, G. L. Baum, D. P. Feldman, and S. Recavarren, *Am. Rev. Resp. Dis., 95*:623–629 (1967).
153. C. D. Laros, C. M. A. Kuyper, and H. M. J. Janssen, *Respiration, 29*:458–467 (1972).
154. C. M. A. Kuyper and C. D. Laros, *Bull. Eur. Physiopathol. Resp., 16*(Suppl.):107–113 (1980).
155. K. Konno, A. Hideo, M. Motomiya, H. Nagai, M. Ito, H. Sato, and K. Satoh, *Am. Rev. Resp. Dis., 126*:979–801 (1982).
156. J. Karlinsky, J. Fredette, G. Davidovits, A. Catanese, R. Snider, B. Faris, G. L. Snider, and C. Franzblau, *J. Lab. Clin. Med., 102*:151–162 (1983).
157. M. Moczar, C. Lafuma, F. Lange, J. Bignon, L. Robert, and E. Moczar, *Bull. Eur. Physiopathol. Resp., 16*(Suppl.):99–104 (1980).
158. C. Lafuma, F. Lange, M. Moczar, J. Bignon, and L. Robert, *Bull. Eur. Physiopathol. Resp., 16*(Suppl.):91–96 (1980).
159. S. G. Gordon, L. L. Hinkle, and E. Shaw, *J. Lab. Clin. Med., 102*:400–410 (1983).
160. W. A. Galloway, G. Murphy, J. D. Sandy, et al., *Biochem. J., 209*:741–752 (1983).
161. W. R. Abrams, A. Eliraz, P. Kimbel, and G. Weinbaum, *Exp. Lung Res., 1*:211–223 (1980).
162. H. Carp and A. Janoff, *J. Clin. Invest., 66*:987–995 (1980).
163. N. R. Matheson, A. Janoff, and J. Travis, *Mol. Cell. Biochem., 45*:65–71 (1982).
164. B. M. Babior, *N. Engl. J. Med., 298*:659–668, 721–725 (1978).
165. H. Carp and A. Janoff, *Exp. Lung Res., 1*:225–237 (1980).
166. R. L. Stedman, *Chem. Rev., 68*:153–207 (1968).
167. W. A. Pryor, *Molecular Basis of Environmental Toxicity* (R. S. Bhatnagar, ed.), Ann Arbor Science, Ann Arbor, MI, 1980, pp. 3–36.
168. A. B. Cohen and H. L. James, *Am. Rev. Resp. Dis., 126*:25–30 (1982).
169. A. Janoff and H. Carp, *Am. Rev. Resp. Dis., 116*:65–72 (1977).
170. H. Carp and A. Janoff, *Am. Rev. Resp. Dis., 118*:617–621 (1978).
171. J. P. Borel, P. Braquet, J. C. Menboise, and A. Randoux, Third International Conference on Superoxide and Superoxide Dismutase, Ellenville, NJ, 1982.
172. S. Curran, M. Amoruso, B. Goldstein, D. Riley, N. Edelman, and R. Berg, 26th Aspen Lung Conference, Aspen, CO, June 9, 1983.
173. Fridovich, *Science, 201*:875–880 (1978).
174. F. Haber and J. Weiss, *Proc. R. Soc. Lond. A. Math. Phys. Sci., 147*:332–351 (1934).

175. B. H. J. Bielski and H. W. Richter, *J. Am. Chem. Soc., 99*:3019–3023 (1977).
176. H. M. Hassan, *Ciba Symp., 79*:125–142 (1980).
177. R. Padmonabhan, S. R. Gudapaty, I. E. Liener, B. A. Schwartz, and J. R. Hoidal, *Clin. Res., 30*:750A (1982).
178. B. A. Freeman, J. F. Turrens, J. D. Crapo, and S. L. Young, Third International Conference on Superoxide and Superoxide Dismutase, Ellenville, NJ October 3–9, 1982.
179. R. E. Lynch and I. Fridovich, *J. Biol. Chem., 253*:1838–1845 (1978).
180. R. E. Lynch and I. Fridovich, *J. Biol. Chem., 253*:4697–4699 (1978).
181. A. F. Junod, C. Ody, F. Dargent, Y. Jean-Mairet, and B. Housset, *Bull. Eur. Physiopathol. Resp., 18*:143p (1982).
182. R. A. Lawrence and R. F. Burk, *Biochem. Biophys. Res. Commun., 71*:952–958 (1976).
183. E. Frieden, *Ciba Symp., 79*:93–124 (1980).
184. H. S. Hsieh and E. Frieden, *Biochem. Biophys. Res. Commun., 67*:1326–1331 (1975).
185. I. M. Goldstein, H. B. Kaplan, H. S. Edelson, and G. Weissmann, *J. Biol. Chem., 254*:4040–4045 (1979).
186. J. C. Taylor and L. Oey, *Am. Rev. Resp. Dis., 126*:476–482 (1982).
187. N. Brot, L. Weissbach, J. Werth, and H. Weissbach, *Proc. Natl. Acad. Sci. USA, 78*:2155–2158 (1981).
188. W. R. Abrams, G. Weinbaum, L. Weissbach, H. Wiessbach, and N. Brot, *Proc. Natl. Acad. Sci. USA, 78*:7483–7486 (1981).
189. H. Carp, A. Janoff, W. R. Abrams, R. R. Drew, G. Weinbaum, H. Wiessbach, and N. Brot, *Am. Rev. Resp. Dis., 127*:301–305 (1983).
190. H. Carp, F. Miller, J. R. Hoidal, and A. Janoff, *Proc. Natl. Acad. Sci. USA, 79*:2041–2045 (1982).
191. E. J. Campbell, R. M. Senior, and J. A. McDonald, *J. Clin. Invest., 70*:845–852 (1982).
192. H. A. Chapman, O. L. Stone, and Z. Vavrin, *J. Clin. Invest., 73*:806–815 (1984).
193. R. M. Bowen, R. D. Estensen, and J. R. Hoidal, *Am. Rev. Resp. Dis., 127*:280a (1983).
194. H. A. Chapman and O. L. Stone, *Biochem. J., 222*:721–728 (1984).
195. S. Sheela and J. C. Barnett, *Carcinogenesis, 3*:363–369 (1982).
196. R. White, D. Lee, G. S. Habicht, and A. Janoff, *Am. Rev. Resp. Dis., 123*:447–449 (1981).
196a. C. Kuhn III and B. C. Starcher, *Am. Rev. Resp. Dis., 122*:453–460 (1980).
197. M. V. Barrow, C. F. Simpson, and E. J. Miller, *Q. Rev. Biol., 49*:101–128 (1974).
198. P. Laurent, A. Janoff, and H. M. Kagan, *Am. Rev. Resp. Dis., 127*:189–192 (1983).
199. M. Osman, J. Cantor, S. Roffman, G. M. Turino, and I. Mandl, *Am. Rev. Resp. Dis., 125*:213 (1982).
200. D. E. Niewoehner and J. R. Hoidal, *Chest, 83*(Suppl.):62s–63s (1983).
201. H. Sahebjami and C. L. Vassallo, *J. Appl. Physiol., 48*:284–288 (1980).
202. H. Sahebjami and J. Macgee, *Am. Rev. Resp. Dis., 128*:644–647 (1983).
202a. D. E. Fisk and C. Kuhn, *Am. Rev. Resp. Dis., 113*:787–797 (1976).
202b. D. W. Rowe, E. B. McGoodwin, G. R. Martin, and D. Grahn, *J. Biol. Chem., 252*:939–942 (1977).
203. G. L. Snider, J. A. Hayes, A. L. Korthy, and G. P. Lewis, *Am. Rev. Resp. Dis., 108*:40–48 (1973).
204. D. Smith, personal communication.
205. D. E. Niewoehner and J. R. Hoidal, *Science, 217*:359–361 (1982).

206. R. V. Padmanabhan, S. R. Gudapaty, I. E. Liener, and J. R. Hoidal, *Environ. Res., 29*:90–96 (1982).
207. G. K. Panigrahy and B. K. Patnaik, *Exp. Gerontol., 10*:305–307 (1975).
208. G. S. Davis, M. S. Gianocola, M. C. Costanza, and R. B. Low, *Am. Rev. Resp. Dis., 126*:611–616 (1982).
208a. N. T. Soskel, S. Watanabe, E. Hammond, L. B. Sandberg, A. D. Renzetti, and J. D. Crapo, *Am. Rev. Resp. Dis., 126*:316–325 (1982).
209. W. Merrill, E. O'Hearn, J. Rankin, G. Naegel, R. A. Matthay, and H. Y. Reynolds, *Am. Rev. Resp. Dis., 126*:617–620 (1982).
210. J. D. Fulmer, *Am. Rev. Resp. Dis., 126*:961–963 (1982).
211. J. W. Gadek, G. W. Hunninghake, G. A. Fells, R. L. Zimmerman, B. A. Keogh, and R. G. Crystal, *Bull. Eur. Physiopathol., Resp., 16*(Suppl.):27–40 (1980).
212. G. W. Hunninghake, J. E. Gadek, and R. G. Crystal, *Chest, 77*(Suppl.):273 (1980).
213. D. W. Golde, *J. Reticuloendothel. Soc., 22*:223–230 (1977).
214. J. E. Cadek, G. A. Fells, and R. G. Crystal, *Science, 206*:1315–1316 (1979).
215. A. Janoff, L. Raju, and R. Dearing, *Am. Rev. Resp. Dis., 127*:540–544 (1983).
216. R. T. Abboud, T. Fera, and A. Richter, *Am. Rev. Resp. Dis., 127*:294 (1983).
217. P. J. Stone, J. D. Calore, S. E. McGowan, J. Bernardo, G. L. Snider, and C. Franzblau, *Science, 221*:1187–1189 (1983).
218. J. R. Hoidal, R. B. Fox, P. A. LeMarbe, R. Perri, and J. E. Repine, *Am. Rev. Resp. Dis., 123*:85–89 (1981).
219. D. G. James, J. Turiaf, Y. Hosoda, W. J. Williams, H. L. Israel, A. C. Douglas, and L. E. Siltzbach, *Ann. NY Acad. Sci., 278*:742 (1976).
220. A. J. Barrett, *Meth. Enzymol., 80*:581–588 (1981).
221. A. J. Barrett, *Meth. Enzymol., 80*:561–565 (1981).
222. A. J. Barrett, H. Kirschke, and Cathepsin B, *Meth. Enzymol., 80*:535–561 (1981).

20

Fibronectin and Disease Processes

Joanne E. Murphy-Ullrich and Deane F. Mosher
University of Wisconsin, Madison, Wisconsin

I. INTRODUCTION

Fibronectin is a glycoprotein that is present in various body fluids and widely distributed in tissues (reviewed in Refs. 1-9). Fibronectin is evolutionarily conserved and has been identified in mammals, birds, and several invertebrate species, including the sea urchin embryo, and fresh and saltwater sponges (10). There are extensive immunological cross-reactivities among fibronectins of different species (11) and between the plasma and tissue forms of the molecule (12), although there are antigenic (13) and structural (14,15) differences between the plasma and cellular forms of the molecule.

Because fibronectin is a major and almost ubiquitous component of the body, there has been intense interest in how it may be related to various pathological states. The resulting literature has been reviewed several times recently (16-18). In this chapter, we try to be interpretative rather than encyclopedic and to focus on several issues that have arisen from the studies of fibronectin in diseases.

II. MOLECULAR INTERACTIONS AND STRUCTURE

One of the striking characteristics of fibronectin is its specific interactions with a number of macromolecules. These interactions are mediated by different parts or domains of the fibronectin subunit. Fibronectin is known to specifically bind to such matrix or cell surface molecules as collagen, heparan sulfate, hyaluronic acid, and gangliosides (19-23). It also binds to actin (24). It interacts with certain bacteria, DNA, complement components Clq and C3, heparin, fibrin(ogen), aggregated serum amyloid P component, and thrombospondin (25-33). It is a substrate for plasma transglutaminase (Factor XIIIa) (34). Finally, when coated onto tissue culture plastic, it has cell-adhesive activity (35).

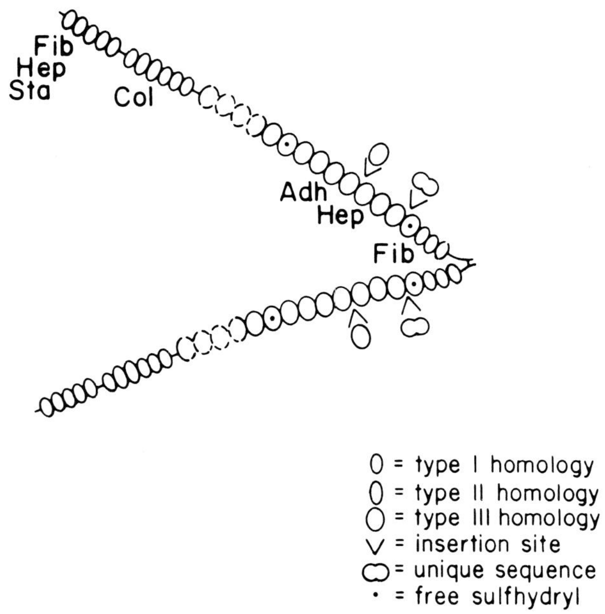

Figure 1 Diagram of plasma fibronectin dimer. The two subunits are depicted as two arms. Amino-termini are to the left; carboxy-termini are to the right. The two arms are joined through disulfide bonds at their extreme carboxy-termini. All except the region shown as dashed lines have been analyzed by protein sequencing (36,57) or cDNA sequencing (37). The 45-residue type I homology units are shown as small beads, the 50-residue type II homology units are shown as slightly larger beads. Because of their distinctive pattern of disulfide loops, the type I homology units have been called "fingers" (36). A protease-sensitive insertion in the type III homology region is shown as a peanut. The location of a second insertion, which has type III homology and has been identified in mRNA from fibroblasts, is indicated by the bead at the insertion site. Free sulfhydryls are indicated by dots. The following binding activities are localized: for fibrin (Fib), for heparin (Hep), for *Staphylococcus aureus* cell surface (Sta), for collagen and gelatin (Col), and for cell adhesion (Adh).

As depicted in Figure 1, fibronectin is the paradigm of a modular protein (36-38). It is a dimer of identical or nearly identical subunits, each approximately 2000 residues long. It is coded by a series of exons, each able to code for approximately 45 amino acids (39). Three types of homologous sequences are present: I, II, and III. There are 12 repeats of type I sequences and two repeats of type II sequences (38). The type I units are approximately 45 residues long, and the type II units are approximately 60 residues long. Both these homology units are characterized by a pair of disulfide bridges in a 1-3, 2-4 arrangement. The second and third cysteines are close together in the sequence, and thus each unit has a major and minor loop (36,38). The major loops of the type I units look like fingers on a hand when drawn out. Accordingly, the type I units have been called "fingers." Both type II repeats and nine of the type I repeats are at the carboxy-terminal end (36,38). In between are at least 10 and perhaps up to 15 repeats of type III sequences, each approximately 90 residues long. The type III homology units do not contain disulfide bridges, but two of the units do contain free sulfhydryl groups. The homologies among type III sequences are less strict than the homologies among type I sequences.

A variety of electron microscopic images have been obtained of fibronectin deposited on various surfaces (40–45). Fibronectin is a dimer held together at the extreme carboxy-termini of the two subunits by disulfide bonds (36). Under certain conditions, it deposits on electron microscopic grids as V-shaped structures with arms 60–70 nm in length, which come together at an angle of approximately 70° (40). Under other conditions, deposited molecules appear as tangled webs (42–45). The differences are due to conditions under which protein is deposited on the surface or to the surface itself. The hydrodynamic behavior of fibronectin in solution also varies depending on its solvent. (44,46–49). Depolarization studies of probes introduced at several sites indicate that all parts of protein exhibit considerable flexibility (47,50), and the persistence length of the extended form is 20–30 nm (47). Thus, fibronectin can be modeled as a "string of beads," which, when fully extended on an electron microscopic grid or in solution, is 120–140 nm long and 2–3 nm in diameter (40,41,47). Each one of the beads would correspond to one of the homology units (Figure 1). Under physiological conditions of pH and ionic strength, the string folds up into more compact structures. Models have been proposed in which the compact structures have an order that is favored by ionic interactions among different parts of the protein (47,51,52). It has also been suggested that the compact form of fibronectin represents a variety of tangled structures that arise randomly upon collapse of the extended form (48).

Subtle differences have been found between the two subunits of plasma fibronectin (14,53–55) and between plasma fibronectin and fibronectin secreted by cultured cells (13,14,56). Only one hybridization band was found by Southern blotting when cDNA probes were reacted with genomic DNA digested with several different restriction enzymes (37,57). Several different mRNAs, however, have been identified (37,57). These mRNAs contain inserts that account for the observed heterogeneity in the protein products (Figure 1). In addition, some cells contain an additional nucleotide insert that encodes exactly one type III homology repeat (37). In other words, the usual mRNA for fibronectin does not contain every possible exon present in the gene for fibronectin. Thus, the differences between the subunits of fibronectin or among fibronectin, synthesized by different cells arise from transcriptional or posttranscriptional events.

III. SOURCES

Plasma fibronectin is present in concentrations of approximately 300 μg/ml in the blood with slightly higher levels in males (9). Serum levels are lower because 20–50% of the fibronectin is incorporated into the clot through interactions with fibrin (30). Fibronectin is also present in amniotic fluid, in cerebrospinal fluid (1–3 μg/ml), and in synovial fluid at concentrations often above the concentrations in plasma (5,58). Amniotic fluid fibronectin differs from the plasma form in that the former has 7–9.5% carbohydrate compared to the latter with about 5% (15).

Many different cell types are known to synthesize fibronectin in vitro: epithelial cells, endothelial cells, fibroblasts, chondrocytes, myoblasts, macrophages, amniotic cells, embryonic tissues, teratocarcinoma cells, Schwann cells, and astroglial cells (5,9). Several types of evidence suggest that most, if not all, circulating fibronectin is produced by hepatocytes. Hepatocytes in culture secrete fibronectin (59). Hepatocyte fibronectin co-migrates in electrophoretic systems with plasma fibronectin; fibronectin secreted by cultured fibroblasts does not comigrate (56). The isolated perfused rat liver produces enough fibronectin to account for fibronectin in plasma (60). Finally, patients with decompensated cirrhosis (61) or fulminant hepatic failure (62) have low concentrations of plasma fibronectin. Endothelial cells have been shown to contribute to fibronectin synthesis in vivo, especially when local vascular injury occurs (63). Other cell types known to synthesize fibronectin in vitro presumably also synthesize fibronectin when appropriately stimulated in vivo. Finally, activated platelets secrete fibronectin from their α granules and bind it to their cell surfaces (64).

IV. DESCRIPTION OF FIBRONECTIN IN THE EXTRACELLULAR MATRIX

In cultured cells, fibronectin is located on cell surfaces, but is primarily present as fibrils in the extracellular matrix formed by cells in vitro (65). The fibronectin-containing matrix that has been best studied is the extensive filamentous network of human skin fibroblasts in culture (66,67). Immunocytochemical techniques have localized fibronectin and types I and II procollagens to 5–10 nm filaments. Some of these filaments decorate with reaction product at regular 70–100 nm intervals. Although it has been suggested that the periodicity may be due to fibronectin molecules bound to collagen microfibrils at regular intervals (67), the filaments are trypsin-sensitive and not collagenase-sensitive. In addition, all the filaments react with antifibronectin whereas only some react with antiprocollagen (67). Treatment of cells with ascorbate causes formation of thicker, collagenase-sensitive filaments, which are more clearly collagenous (66,68). Thus, it seems likely that fibronectin is a major structural component of the early, thinner filaments that are seen around cultured cells. Similar thin filaments have been found in granulation tissue of wounded skin (69).

Immunochemical and biochemical analyses of the detergent-insoluble extracellular matrix of cultured cells indicate that the matrix, in addition to fibronectin, contains collagen, heparan sulfate proteoglycans, hyaluronic acid, and other glycoproteins (70–76). Transformed cells often lack a prominent fibronectin matrix (77). Although fibronectin is secreted in vitro as a soluble dimer, insoluble fibronectin is present in matrices as large, disulfide-bonded aggregates (78–80). Evidence suggests that fibronectin is present in the cell layer of confluent fibroblasts in two distinct pools (78,81), deoxycholate-soluble and deoxycholate-insoluble. Initial binding is to the deoxycholate-soluble pool. Such binding is reversible and presumably to specific cell surface receptors (81). After time, fibronectin becomes irreversibly bound to the matrix in the form of deoxycholate-insoluble, disulfide-bonded, high-molecular-weight aggregates.

V. LOCALIZATION OF FIBRONECTIN IN TISSUES

In vivo, fibronectin is present in most connective tissue matrices of the body. Stenman and Vaheri originally described the localization of fibronectin in human tissues using the indirect immunofluorescence technique (82). They found fibronectin in most basement membranes, in basal laminae surrounding smooth muscle cells and the sarcolemma of striated muscle fibers, in the stroma of lymphatic tissue, and in the loose connective tissues of nerves, the dermis, and blood vessels. Fibronectin was associated with the intimal subendothelial basement membranes of both capillaries and in larger vessels but was not present in the internal elastic lamina itself. Fibronectin surrounded the medial smooth muscle cells of muscular arteries.

There are several controversies concerning the localization of fibronectin. For example, numerous studies have been undertaken to determine the localization of fibronectin in the kidney (83-91). It is agreed that fibronectin is present in large amounts in the glomerular mesangium, especially at the mesangial-endothelial interface, and that it is also present in smaller amounts in the tubular basement membranes, in the basement membranes of peritubular capillaries, and along Bowman's capsule. However, there is disagreement as to whether fibronectin is present along the glomerular basement membrane. In studies at the ultrastructural level, some investigators were unable to demonstrate fibronectin in the glomerular basement membrane (88), while others found fibronectin to be present in the lamina rara interna and externa (83,87,91). Others have demonstrated staining of the lamina densa, but only after treatment with sodium borohydride (92). Staining of the lamina rara has been interpreted to be the result of nonspecific trapping of plasma fibronectin between the endothelium and glomerular basement membrane (90). When rat kidneys were perfused free of plasma proteins before fixation and immunostaining, however, staining for fibronectin was seen in the lamina rara, whereas staining for albumin was negative (91). Furthermore, using the immunoferritin technique at the electron microscopic level to localize fibronectin on isolated glomerular basement membrane, clusters of ferritin were detected on the lamina rara at a distance >20 nm from the lamina densa and about 100 nm from each other (91). The disagreements over the localization of fibronectin in glomerular basement membrane are probably due to differences in methods of tissue fixation, which would influence preservation of antigenicity and the accessibility of antigens to immunological probes; sizes of immunological probes, which would influence their accessibility to antigens; the species from which the fibronectin was purified, in which the antiserum was prepared, and from which the kidneys to be examined were prepared; and exact methods of preparation of fibronectin prior to immunization. The importance of these variables should be kept in mind when evaluating studies in which the distribution of fibronectin in diseased tissues differs from normal.

Another example of controversy concerns the localization of fibronectin in cartilage. Fibronectin was initially thought to be a component of immature, but not adult, cartilage. However, recent studies, in which tissues were treated with hyaluronidase prior to immunostaining, report localization of fibronectin in adult cartilage (93-95). Binding of fibronectin to hyaluronic acid appears to mask the antigenic accessibility of fibronectin. This may account for the negative results obtained by others who did not use enzymatic digestion (96-98). Even with the digestion, there is a discrepancy as to the localization of fibronectin. In normal human cartilage, Clemmensen et al. (93) localized fibronectin throughout the cartilage in a diffuse pattern in tissues that had been digested with hyaluronidase. Staining was more intense surrounding the chondrocytes. In undigested tissues, only the acellular cartilage surface stained. Shiozawa et al. (95) also found that prior treatment of normal human cartilage with hyaluronidase resulted in increased staining of the articular surface as compared to untreated tissues. However, they did not observe staining near chondrocytes or in the deeper regions of the cartilage. In bovine nasal cartilage, fibronectin was detectable in the interterritorial matrix of both untreated and hyaluronidase-digested sections but not in the region surrounding the chondrocytes (94). In normal mouse joints, staining for fibronectin was seen in chondrocytes, but not in the matrix in untreated tissue sections (99). The reason for these discrepancies is not clear, although there may be species heterogeneity with respect to the cartilage matrix.

In normal skin, fibronectin is absent from the epidermis, but present at the dermal-epidermal junction and in both reticular and papillary dermis (100–102). The distribution of fibronectin in the deeper reticular dermis is more prominently fibrillar in association with thicker collagen fibrils as compared to the finer fibrillar and mostly intracellular fibroblast staining for fibronectin in the papillary dermis (100,101,103). The connective tissue sheaths of hair follicles, nerve bundles, and subdermal muscle bundles also stain for fibronectin (100–102).

VI. ALTERED LEVELS OF FIBRONECTIN IN BODY FLUIDS

A. Plasma

The concentration of immunologically detected fibronectin in plasma varies according to sex and age. Plasma levels in younger women are approximately 20% less than levels in younger men (104,105). Labat-Robert et al. reported levels of 1100 ± 260 μg/ml in people 70–90 years old as compared to 274 ± 130 μg/ml in people 0–50 years old (106). Others have also reported fibronectin to be increased with age, especially in women (107), but not to the extent reported by Labat-Robert et al. The reason for the increase in plasma fibronectin levels with age is not known, although it has been suggested that there is increased fibronectin synthesis to compensate for possible defective fibronectin synthesized by senescent cells (106).

Concentrations of fibronectin in the plasma are known to vary in various disease states. Patients with liver metastases have decreased plasma fibronectin levels (107), whereas patients with primary biliary cirrhosis, nephrotic syndrome, or obstructive jaundice have increased fibronectin concentrations (107). Significantly elevated levels of plasma fibronectin were reported in patients with colon or breast carcinomas regardless of whether metastases were present or the patient had had chemotherapy (108). Levels of plasma fibronectin in patients with acute leukemia were normal prior to chemotherapy, but induction of chemotherapy resulted in decreased levels of fibronectin followed by a subsequent increase in fibronectin concenration to at least normal levels by day 5 (108).

Fibronectin levels are known to be decreased in patients with disseminated intravascular coagulation, probably owing to consumption of fibronectin by incorporation into fibrin clots (105). Fibronectin levels have been shown to be decreased also after sepsis, trauma, burns, and cardiac bypass surgery (109–112), especially in sicker patients. Decreased plasma fibronectin levels are seen in patients after transfusion of stored packed-red blood cells (113). Removal of microaggregate debris by use of filters or washing red cells alleviates this decrease in plasma fibronectin concentration (113).

B. Synovial Fluid

Fibronectin is present in synovial fluids from normal persons at concentrations of 172 ± 69 μg/ml (114). Substantially elevated levels are found in synovial fluids from patients with rheumatoid arthritis, while moderately elevated levels are detected in fluids from patients with osteoarthritis (58,114,115). Although the exact values differ for each investigator, the levels are greater than those expected if plasma fibronectin had simply diffused into the synovial cavity. Consequently, it is generally held that synoviocytes, themselves, synthesize much of the fibronectin present in the synovial fluid (116). The recent report that synovial fluid fibronectin migrates more slowly than plasma fibronectin in an agarose gel system and generates different molecular weight fragments when treated with thermolysin (114) is further evidence that synovial fluid fibronectin is synthesized locally. Others have reported the presence of fibronectin degradation products in synovial fluids (117).

Fibronectin is known to be a component of cryoglobulins (118,119). It interacts with Clq-containing complexes from patients with systemic lupus erythematosus (120). Fibronectin is also a component of synovial fluid cryoproteins (115,121).

C. Bronchoalveolar Fluid

Fibronectin is present in bronchoalveolar lavage fluid (122), with significantly elevated levels detected in smokers as compared to nonsmokers, possibly owing to local synthesis.

VII. DISTRIBUTION OF FIBRONECTIN IN DISEASED TISSUES

A. Kidney

Several investigators have examined the distribution of fibronectin in the renal glomerulus in various renal diseases at the light microscopic level (84–86,123,124). As described above, fibronectin is present primarily in the mesangium of the normal kidney. Increased amounts of fibronectin were detected in diseases involving mesangial expansion (acute proliferative glomerulonephritis, membranoproliferative glomerulonephritis, diabetic glomerulopathy, and transplant glomerulopathy), although both Linder et al. (86) and Scheinman et al. (84) reported decreased amounts of immunoreactive fibronectin in the mesangia of patients with membranoproliferative glomerulonephritis. Increased staining for fibronectin was observed in the glomerular capillary walls in patients with acute proliferative glomerulonephritis, membranous glomerulonephritis, membranoproliferative glomerulonephritis, and diabetic glomerulopathy (85,123,124). Fibronectin was also found to be prominent in glomerular crescents in association with fibrin (85,123,124). It was speculated that the fibrin-fibronectin complex may serve as a substratum for subsequent fibrotic scar formation (124). The source of increased glomerular fibronectin in diseased kidneys is not known. It has been suggested that the increase is due to increased adsorption of fibronectin from the circulation, which may be binding to deposited fibrin or exposed collagen (86,123). Alternatively, the increase may be due to increased fibronectin synthesis by glomerular cells in response to injury (85,86,123). Similarly, Clark et al. have reported a local increase in endothelial cell synthesis of fibronectin at sites of injury (63).

B. Synovium and Cartilage

Increased amounts of immunologically detectable fibronectin have been reported to be present in synovial tissues from patients with rheumatoid arthritis (58,95,98,116,125). In the normal human synovial membranes, fibronectin is localized in a fibrillar network surrounding synoviocytes. In the subsynovial connective tissue, fibronectin is present only in the vascular connective tissue (116). Vartio et al. (58) detected faint staining for fibronectin in the subsynovial connective tissue. Various investigators have observed increased amounts of fibronectin surrounding synovial cells, and in the subsynovial connective tissue in specimens from both humans with rheumatoid arthritis (58,95,116) and rabbits with experimentally induced rheumatoid arthritis (126). Scott et al. also observed intracellular staining for fibronectin in the normal synovium distant from the rheumatoid pannus (98) and in normal rabbit synovial cells (127). Fibronectin was also observed to be prominently distributed throughout the rheumatoid pannus (98,125), primarily as a coarse extracellular network surrounding the invading inflammatory cells (98). At the ultrastructural level, fibronectin formed a layer bordering both fibroblastlike and macrophagelike cells in the pannus, and also appeared to be codistributed with collagen fibrils (125). Only weak, diffuse staining of fibronectin was observed at the cartilage-pannus junction (98,125).

The sequential change in distribution of fibronectin during the induction of experimental arthritis in rabbits has been studied by both Holund et al. (126) and Scott et al. (127). Both investigators report an initial increase in fibronectin surrounding the synoviocytes, with a diffuse distribution throughout the pannus. Scott et al. report a decrease in extracellular distribution of fibronectin, with positive intracellular staining for fibronectin within synoviocytes and the fibroblasts and endothelial cells of the pannus at 8 and 15 days postimmunization, whereas Holund et al. report a moderate increase in the amount of fibronectin in the matrix surrounding the synovial cells, but no certain intracellular staining between 11 and 28 days. In the chronic stage of the disease (8–13 weeks), fibronectin was present in subsynovial and perisynovial matrices. Fibronectin was present in a fibrillar arrangement in association with fine collagen but was no longer detectable near coarse (mature) collagen fibers. Both studies also documented fibronectin staining of the endothelial cells, vascular connective tissue, and fibroblasts of the pannus, similar to the distribution of fibronectin in experimental granulation tissue (128,129). Scott et al. also observed that fibronectin was codistributed in areas of the pannus with fibrin deposits. No correlation, however, was found between the codistribution of fibronectin and either IgG or the C3 component of complement (127).

Wurster and Lust extracted a high-molecular-weight glycoprotein from osteoarthritic canine articular cartilage that had the same apparent molecular weight as fibronectin and reacted with monoclonal and polyclonal antibodies to fibronectin (130). They detected increased amounts (200–900 ng/mg wet weight) of fibronectin in osteoarthritic cartilage compared to normal cartilage (20–50 ng/mg wet weight) (130). The increased amounts of fibronectin found in degenerating cartilage and in areas of cartilage containing decreased amounts of proteoglycans correlate with the finding that fibronectin is seen in undifferentiated cartilage matrices using immunolocalization techniques that do not detect fibronectin in mature chondrocyte matrices in vitro (96,131,132). Addition of fibronectin to the medium of chondrocytes inhibits their differentiation (133). Whether fibronectin is present in increased amounts in degenerating cartilage as a consequence of injury and subsequent repair or as a phenotypic characteristic of undifferentiated cartilage remains to be elucidated.

C. Skin Diseases

Similar to the increased amounts of fibronectin present in diseased kidneys and joints, increased amounts and/or abnormal distributions of fibronectin are described for a variety of diseases afflicting the skin. Some of these changes are probably related to cellular proliferative activity. Increased amounts of fibronectin were found to be present in fetal rat dermis, in association with type III collagen fine fibrils, as compared to adult rat skin (134). In psoriatic skin, the stratum corneum stains for fibronectin, and there is an increase in staining in the dermal-epidermal junction and papillary dermis, characterized by a more fibrillar pattern (101). Increases in dermal fibronectin are also described in evolving but not stabilized lesions of scleroderma (135,136) and in dermatitis herpetiformis (137). Skin affected by lichen planus and discoid lupus shows an increase in fibronectin in the dermal blood vessels and at the dermal-epidermal junction (138,139). Increased amounts of fibronectin were detected, along with serum amyloid component P, in the thickened blood vessels of erythropoietic protoporphyria skin (140). When the distribution of fibronectin in skin of insulin-dependent diabetic males under 50 years of age was examined, fibronectin was present in the vascular and dermal-epidermal basement membranes and the papillary dermis in markedly elevated amounts (141). Examination of the distribution of fibronectin in basal cell epitheliomas revealed linear staining for fibronectin at the margin of tumor lobules and fine filaments in the central portion of the tumor nodule (142). There was no staining for either fibronectin or von Willebrand factor within the tumor tissue; the authors interpreted this to mean that the fibronectin located in the tumor is not derived from plasma or the microvasculature of the tumor (von Willebrand factor would be expected to be present in endothelial cells) and suggested that the source of fibronectin is the epithelioma cells, themselves (140). Recent evidence shows that under conditions inhibiting terminal differentiation, human foreskin keratinocytes synthesize and incorporate fibronectin into a pericellular matrix in vitro (143).

VIII. SOURCES OF FIBRONECTIN IN DISEASED TISSUES

In disease processes that involve cell dedifferentiation and proliferation, fibronectin may be derived in situ from the proliferating cells themselves. This will be discussed in more detail in section IX. The sources of fibronectin deposited in other diseases are not clear. However, it has been suggested that fibronectin is derived by extravasation from the leaky microvasculature in diseases in which fibronectin codistributes with fibrin (e.g., glomerular crescents, rheumatoid pannus, dermatitis herpetiformis) (144). Endothelial cells increase fibronectin synthesis in response to local injury (63). Such local synthesis of fibronectin by endothelial cells may explain the increase in staining for fibronectin in various tissues in which vascular injury is a component of the disease process (e.g., diabetes, erythropoietic protoporphyria, membranous glomerulonephritis).

IX. FIBRONECTIN AND WOUND HEALING

Wound healing involves several different processes. These include: (a) hemostasis at the site of injury, (b) clearance of cellular and bacterial debris from the wound area, (c) formation of a provisional scaffolding or matrix, (d) recruitment of vascular and parenchymal cells, and (e) stimulation of cellular proliferation. Evidence suggests that fibronectin plays a role in each of these processes.

Fibronectin interacts with several components involved in the hemostatic process, including fibrin(ogen), platelets, and another platelet α granule glycoprotein, thrombospondin. Serum concentrations of fibronectin are 20–50% less than plasma concentrations due to factor XIIIa-catalyzed covalent cross-linking of fibronectin to fibrin at 22-37°C (145). The cross-linked fibronectin mediates cell adhesion to fibrin (146). Platelets activated by thrombin have fibronectin on their surface (147). Nonactivated platelets do not express surface fibronectin. It is believed that fibronectin is released from the platelet α granules upon activation (64). Specific receptors for fibronectin are expressed on the surfaces of platelets following activation (148). Fibronectin enhances spreading of platelets on collagen surfaces (149). It has been postulated that fibronectin is involved in platelet adhesion to the subendothelium in injured blood vessels through its interactions with both activated platelet membranes and collagen (150). Inasmuch as fibronectin from plasma, on platelet surfaces, and in vessel walls could all be involved, however, the situation in a wound area must be complex and dynamic. In addition, the effects of fibronectin on platelet adhesion are probably only additive to the effects of the major adhesive protein for platelets, von Willebrand factor (151,152). The effects of fibronectin may also be mediated by another platelet protein, thrombospondin. The two proteins interact when platelets spread on a fibronectin-coated substratum (32).

Fibronectin is thought to aid in the clearance of cellular and also bacterial debris from wound areas through its function as a nonspecific opsonin for macrophages, and possibly neutrophils (7,153–155). Circulating macrophages and peritoneal and alveolar macrophages synthesize and secrete fibronectin (156,157); however, they do not form a fibronectin matrix or have fibronectin on their cell surfaces, except in regions of adherence to gelatin-coated latex particles (122). Monocytes (158) and polymorphonuclear leukocytes (159) have cell surface receptors of fibronectin. Fibronectin is thought to play a role in clearance of bacterial, cellular, fibrinous, and collagenous debris by the mononuclear phagocytic system (160). Fibronectin potentially can specifically interact with the aforementioned components and, through its interaction with the fibronectin receptors on phagocytic cells, clear debris through the mononuclear phagocytic system. In addition, fibronectin interacts with two components of the complement system, Clq and C3, and may also be a factor in immune-mediated opsonization (19,27,28). Fibronectin with heparin as a cofactor has been shown to promote phagocytosis of gelatin-coated particles by murine macrophages in an in vitro assay (161). However, a recent report states that macrophage phagocytosis of bacteria is independent of fibronectin (162).

A 180-kd fragment of fibronectin, which enhances phagocytosis of zymosan particles and erythrocytes by human monocytes at 100-fold higher activity than the intact fibronectin dimer, has been described (163,164). Using rat peritoneal macrophages, Rourke et al. found that 170–210-kd gelatin-binding fragments of fibronectin retained the ability to bind and ingest gelatin-coated formalin-fixed or tanned sheep erythrocytes. However, they found that the intact dimer was more active than the fragments (165). Fibronectin will not induce phagocytosis of C3b-coated erythrocytes by polymorphonuclear leukocytes, unless the leukocytes are activated by prior stimulation with chemotactic factors such as C5a or the synthetic peptide fmet-leu-phe (159). Patients with trauma and/or sepsis clear gelatinized particles more slowly than normal (160). Such conditions are associated with low levels of plasma fibronectin, suggesting that fibronectin is "consumed" in the clearance of debris presented to the mononuclear phagocytic system. Thus, "blockade" of the mononuclear phagocytic system may be due not to saturation of phagocytic cells with particulates, but to depletion of fibronectin, thus resulting in a depressed presentation of debris to the phagocytes (7,160). Because of the therapeutic implications of this hypothesis, it will be important to learn exactly what roles fibronectin does play in phagocytosis.

Several studies have examined the temporal distribution of fibronectin in experimental wounds in vivo. Similar results were obtained in studies that induced granulation tissue formation in mice by subcutaneous implantation of cellulose sponges (129) and from experiments using full-thickness skin wounds in guinea pigs (128). In the guinea pig model, Grinnell et al. describe the codistribution of fibronectin and fibrin 5 hr after wounding. In other studies, this group also showed that Factor XIIIa-mediated cross-linking of fibronectin to fibrin provided a better substrate for fibroblast attachment and spreading than did a fibrin substrate alone (146). Similarly, patients with Factor XIII deficiency have been shown to have defective wound healing (166). In addition to providing a provisional matrix for attachment of invading fibroblasts, fibronectin also is a chemoattractant for both endothelial cells (167) and fibroblasts (168,169). Thus, fibronectin can direct the migration of fibroblasts, endothelial cells, and perhaps epithelial cells over a wound substratum. Variant endothelial cells stimulated by angiogenesis factors deposit fibronectin "tracks" on which normal endothelial cells migrate (170). Fibronectin has been shown to support reepithelialization of wound sites in adult newt hindlimbs (171), rabbit cornea (172,173), and rodent epidermal wounds (174). Administration of fibronectin-containing eyedrops accelerated the rate of epithelial healing in rabbit corneal wounds (175). Fibronectin produced by human alveolar macrophages is chemotactic for human lung fibroblasts and may facilitate fibrosis in lung disease (176).

Fibronectin in granulation tissue is associated primarily with the fine, reticular fibers composed of type III collagen (8,128,129). This is consistent with the finding that of all the collagen types in their native state, fibronectin has the strongest interaction with type III collagen (177). In later stages of wound repair, when thicker bundles of type I collagen fibers appear, fibronectin is no longer present (129). This correlates with the situation in developing skin. Considerable amounts of fibronectin are present in fetal skin in association with type III collagen fibrils in the dermis. Lesser amounts of fibronectin are present in adult dermis in association with type I collagen fibers (102,128,129). Similarly, once reepithelialization of corneal wounds occurs, fibronectin and fibrin are no longer detected at the wound site (172).

Clark et al. have employed a model utilizing species-specific antibodies and rat-to-immune-suppressed-mouse skin grafts to determine the sources of fibronectin that accumulates at sites of injury. In some experiments, they injected ^{125}I-labeled-fibronectin into guinea pigs after induction of delayed type hypersensitivity reactions (144). They found maximal extravascular accumulation of labeled fibronectin just subsequent to the occurrence of maximal vascular permeability, suggesting plasma as the source of initial fibronectin accumulation. In epithelial wounds made in rat skin grafted onto thymectomized mice, the fibronectin deposited in the wound during the first four days after injury was derived from the host mouse (174). This finding is also consistent with the conclusion that the fibronectin that is initially deposited at the site of injury is derived from plasma. It has been shown both in vivo (178) and in vitro (81) that soluble fibronectin can be incorporated into insoluble extracellular matrices. Subsequent production of fibronectin at sites of injury appears to occur locally. Clark et al. showed that proliferating endothelial cells adjacent to the injured site synthesize fibronectin (63,144). It has also been reported that plasma fibronectin levels are significantly increased after venostasis that induces endothelial cell damage (179). Seven days after wounding, epithelial migration occurs over a substratum derived from either exclusively donor (rat) fibronectin or a combination of donor and host (rat and mouse) fibronectins (174).

In situations characterized by excessive wound repair, such as hypertrophic scars and keloid formation, increased amounts of fibronectin are localized in the nodular regions of these lesions (180). Dermal fibroblasts from these lesions accumulated more extracellular matrix in vitro as compared to normal dermal fibroblasts (180). However, since these cells were cultured in the presence of 10% fetal calf serum, it is not clear how the increased staining for fibronectin in hypertrophic scars and keloids is related to the increased accumulation of fibronectin-containing matrix in vitro. Furthermore, it is not known whether the increased accumulations of fibronectin result from an increased ability to synthesize fibronectin or an increased ability to incorporate exogenous fibronectin into matrices or some combination of the two.

X. FIBRONECTIN AND THE IMMUNE RESPONSE

In addition to its putative opsonic functions, fibronectin has been found to have least four distinct activities that are related to the immune response. Remold et al. identified a fibronectinlike macrophage surface component from guinea pig peritoneal exudate that augmented the response of macrophages to the lymphokine migration inhibition factor (181). Although absorption of peritoneal exudate with antibodies to plasma fibronectin or preincubation of macrophages with plasma fibronectin abrogated the response to migration inhibition factor, soluble plasma fibronectin was unable to substitute for migration inhibition factor enhancing activity in this system (181). Whether the molecule described in these studies is fibronectin or an antigenically similar component that may compete for similar binding sites is not clear (181). Moreover, fibronectin may need to undergo a conformational rearrangement (e.g., binding to macrophage cell surface) in order to be able to interact with migration inhibition factor. Such an explanation may confer some specificity to this interaction: if soluble fibronectin should substitute for the active surface component, migration inhibition factor would become complexed to circulating plasma fibronectin instead of selectively recognizing the macrophage surface. Godfrey et al. have identified a guinea pig T-cell-dependent lymphokine, macrophage agglutination factor, with physical properties similar to those of fibronectin. Macrophage agglutination factor is recognized by polyclonal antobodies to human fibronectin and exhibits reversible binding to gelatin (182,183). In a recent report, the macrophage agglutination factor activity of fibronectin was localized to a 52,000-dalton gelatin-binding region on guinea pig fibronectin (184). It was speculated that lymphokine fibronectin may mediate monocyte/macrophage response in delayed-type hypersensitivity reactions (184). Fibronectin has been shown to stimulate T-lymphocyte blast transformation in the presence of a critical number of macrophages (185). The effect of fibronectin on T lymphocytes appears to be mediated through its interaction with accessory macrophages, because fibronectin was found not to bind directly to T lymphoctyes (185). Fibronectin has been shown to be capable of regulating lymphocyte responsiveness in the mixed lymphocyte assay (186). Finally, plasma fibronectin also can stimulate the production of monocyte/macrophage-derived growth factors (187).

Ataxia-telangiectasia is an autosomal recessive disease characterized by abnormal DNA synthesis and by deficiencies in cellular immunity and a selective decrease in IgA levels. It was shown that fibroblasts from patients with ataxia-telangiectasia demonstrate overproduction of secreted proteins (188). Threefold higher levels of fibronectin were synthesized by ataxia-telangiectasia fibroblast as compared to normal cells. Furthermore, fibronectin isolated from growing ataxia-telangiectasia cultures was approximately 10,000 daltons smaller than fibronectin from normal fibroblasts or fibronectin isolated from nongrowing ataxia-telangiectasia cells. Whether the altered fibronectin is a product of the genetic defects of the disease or whether the abnormalities in DNA synthesis are due to yet unidentified regulation of DNA synthesis by defective extracellular matrix proteins remains to be clarified.

XI. FIBRONECTIN AS A TARGET FOR IMMUNOLOGICAL INJURY

Because of its widespread distribution in tissues, fibronectin is a potential antigen for autoimmune-mediated tissue injury. Zanetti and Takami demonstrated binding of injected rabbit antirat fibronectin antibodies to the glomerular mesangium of rats, with subsequent deposition of rat IgG in the mesangium (189). Twenty-four hours after injection of the rabbit antifibronectin antibodies, electron-dense deposits were observed in the mesangium and in the subendothelial region of the glomerular capillary in the rat kidneys. The localization of these deposits was specific, since no deposits were observed in rats injected with normal rabbit IgG (189). Murphy-Ullrich et al. demonstrated the production of autoantibodies to rabbit fibronectin in rabbits immunized with denatured human plasma fibronectin monomer (190). All rabbits immunized with fibronectin developed mesangial lesions. One of the rabbits also developed atypical intramembranous dense deposits. This animal produced antibody to the 27,000-dalton amino-terminal fragment of rabbit fibronectin (190). In a subsequent study, mice immunized with mouse fibronectins in native dimeric or various denatured forms developed moderate titers of circulating autoantibodies to mouse fibronectin and significantly elevated levels of circulating immune complexes (191). Mice immunized with either native dimeric or denatured multimeric and monomeric fibronectins developed focal glomerular lesions. These lesions were characterizied by mesangial dense deposits and sclerosis and lamellation of the glomerular basement membrane. Although these lesions were focal in nature, glomerular basement membrane lesions were observed more frequently in mice immunized with reduced and alkylated fibronectin monomer than in mice immunized with either native dimeric or denatured multimeric fibronectins. As demonstrated by both competitive inhibition assays and immunoblotting, mice immunized with monomer produced antibodies that preferentially reacted with the monomeric form of fibronectin. These studies demonstrate that fibronectin is a potential autoantigen that may be involved in immune-mediated tissue injury, especially in the renal glomerulus. The results also suggest that fibronectins have different conformations, which can be recognized immunologically. It is interesting to speculate that fibronectin in different connective tissue structures may have distinct conformational arrangements as a result of the heterogeneity of the macromolecular composition and structure of various tissue matrices.

XII. CONCLUDING REMARKS

Fibronectin is a multifunctional protein able to interact with a variety of other physiologically significant substances. Because it is a major soluble protein of body fluids and a major insoluble protein of connective tissue matrices, it seems likely that fibronectin has a number of physiological functions that will turn out to be of pathophysiological significance. On the basis of its presence, absence, or altered structure or concentration, fibronectin has been implicated in a number of diseases. At present, however, most of the evidence that implicates fibronectin in pathological processes is circumstantial, and it is not known whether abnormal quantities and/or distributions of fibronectin are a nonspecific response of cells to injury or whether such abnormalities contribute to initiation or propagation of the pathological state. The continuing studies by a number of investigators of fibronectin in human disorders, in vivo animal models, and in vitro systems should lead to a more exact understanding or mechanisms involving fibronectin and disease processes.

ACKNOWLEDGMENTS

This work has been supported by NIH grants F32HL-06819, HL-21644, and HL-29586. We thank Ms. Nancy Turman for her excellent assistance in preparing this manuscript.

REFERENCES

1. K. M. Yamada and K. Olden, *Nature, 275*:179–184 (1978).
2. E. Pearlstein, L. I. Gold, and A. Garcia-Pardo, *Mol. Cell. Biochem., 29*:103–128 (1980).
3. M. W. Mosesson and D. L. Amrani, *Blood, 56*:145–158 (1980).
4. D. F. Mosher, *Prog. Hemost. Thromb., 5*:111–151 (1980).
5. E. Ruoslahti, E. Engvall, and E. G. Hayman, *Coll. Relat. Res., 1*:95–128 (1981).
6. D. F. Mosher, and L. T. Furcht, *J. Invest. Dermatol., 77*:175–180.
7. D. F. Mosher, R. A. Proctor, and J. E. Grossman, *Adv. Inflammation Res., 2*:187–207 (1981).
8. F. Grinnell, *Am. J. Dermatopathol., 4*:185–188 (1982).
9. R. O. Hynes, and K. M. Yamada, *J. Cell Biol., 95*:369–377 (1982).
10. S. K. Akiyama, K. M. Yamada, and M. Hayashi, *J. Supramol, Struct. Cell Biochem., 16*:345–348 (1981).
11. P. Kuusela, E. Ruoslahti, E. Engvall, and A. Vaheri, *Immunochemistry, 13*:639–642 (1976).
12. K. M. Yamada, and D. W. Kennedy, *J. Cell. Biol., 80*:492–498 (1979).
13. B. T. Atherton, and R. O. Hynes, *Cell, 25*:133–141 (1981).
14. M. Hayashi, and K. M. Yamada, *J. Biol. Chem., 258*:3332-3340 (1983).
15. G. Balian, E. Crouch, E. M. Click, W. G. Carter, and P. Bornstein, *J. Supramol. Struct., 12*:505–516 (1979).
16. S. K. Akiyama, and K. M. Yamada, *Connective Tissue Diseases,* Williams & Wilkens, Baltimore, 1983, pp. 55–96.
17. A. J. D'Ardenne, and JO'D. McGee, *J. Pathol., 142*:235–251 (1984).
18. G. Zerlauth, and G. Wolf, *Am. J. Med., 77*:685–689 (1984).
19. L. H. E. Hahn, and K. M. Yamada, *Cell, 18*: 1043–1051 (1979).
20. M. B. Furie, A. B. Frey, and D. B. Rifkin, *J. Biol. Chem., 255*:4391–4394 (1980).
21. E. Ruoslahti, E. G. Hayman, P. Kuusela, J. E. Shively, and E. Engvall, *J. Biol. Chem., 254*:6054–6059 (1979).
22. S. S. Yamada, K. M. Yamada, and M. C. Willingham, *J. Histochem. Cytochem., 28*: 953–960 (1980).
23. H. K. Kleinman, A. T. Hewitt, J. C. Murray, L. A. Liotta, S. I. Rennard, J. P. Pennypacker, E. F. McGoodwin, G. R. Martin, and P. H. Fishman, *J. Supramol. Struct., 11*:69–78 (1979).
24. J. Keski-Oja, A. Sen, and G. J. Todaro, *J. Cell Biol., 85*:527–533 (1980).
25. R. A. Proctor, D. F. Mosher, and P. J. Olbrantz, *J. Biol. Chem., 257*:14788–14794 (1982).
26. L. Zardi, A. Siri, B. Carnemolla, L. Santi, W. D. Gardner, and S. O. Hoch, *Cell, 18*: 649–657 (1979).
27. D. H. Bing, S. Almeda, H. Isliker, J. Lahav, and R. O. Hynes, *Proc. Natl. Acad. Sci. USA, 79*:4198–4201(1982).
28. A. Hautanen, and J. Keski-Oja, *Scand. J. Immunol, 17*:225–230 (1983).
29. M. Hayashi, D. H. Schlesinger, D. W. Kennedy, and K. M. Yamada, *J. Biol. Chem., 255*: 10017–10020 (1980).
30. N. E. Stathakis, and M. W. Mosesson, *J. Clin. Invest., 60*:855–865 (1977).
31. R. B. Colvin, P. I. Gardner, R. O. Roblin, E. L. Verderber, J. M. Lanigan, and M. W. Mosesson, *Lab. Invest., 41*:464–473 (1979).
32. J. Lahav, M. A. Schwartz, and R. O. Hynes, *Cell, 31*:253–262 (1982).
33. F. C. DeBeer, M. L. Baltz, S. Holford, A. Feinstein, and M. B. Pepys, *J. Exp. Med., 154*:1134–1139 (1981).
34. D. F. Mosher, P. E. Schad, and J. M. Vann, *J. Biol. Chem., 255*:1181–1188 (1980).

35. M. D. Pierschbacher, E. G. Hayman, and E. Ruoslahti, *Cell, 26*:259–267 (1981).
36. T. E. Petersen, H. C. Thogersen, K. Skorstengaard, K. Vibe-Pedersen, P. Sahl, L. Sottrup-Jensen, and S. Magnusson, *Proc. Natl. Acad. Sci. USA, 80*:137–141 (1983).
37. A. R. Kornblihtt, K. Vibe-Pedersen, and F. E. Baralle, *Nucl. Acids Res., 12*:5853–5868 (1984).
38. K. Skorstengaard, H. C. Thogersen, and T. E. Petersen, *Eur. J. Biochem., 140*:235–243 (1984).
39. H. Hirano, Y. Yamada, M. Sullivan, B. DeCrombrugghe, I. Pastan, and K. M. Yamada, *Proc. Natl. Acad. Sci. USA, 80*:46–50 (1983).
40. J. Engel, E. Odermatt, A. Engel, J. A. Madri, H. Furthmayr, H. Rohde, and R. Timpl, *J. Mol. Biol., 150*:97–120 (1981).
41. H. P. Erickson, N. Carrell, and J. McDonagh, *J. Cell Biol. 91*:673–678 (1981).
42. V. E. Koteliansky, E. L. Arsenyeva, G. T. Bogacheva, M. A. Chernousov, M. A. Glukhova, A. R. Ibraghimov, M. L. Metsis, M. N. Petrosyan, and O. V. Rokhlin, *FEBS Lett, 142*:199–202 (1982).
43. T. M. Price, M. L. Rudee, M. Pierschbacher, and E. Ruoslahti, *Eur. J. Biochem., 129*:359–363 (1982).
44. H. P. Erickson, and N. A. Carrell, *J. Biol. Chem., 258*:14539–14544 (1983).
45. N. M. Tooney, M. W. Mosesson, D. L. Amrani, J. F. Hainfeld, and J. S. Wall, *J. Cell Biol., 97*:1686–1692 (1983).
46. S. S. Alexander, G. Colonna, and H. Edelhoch, *J. Biol. Chem., 254*:1501–1505 (1979).
47. E. C. Williams, P. A. Janmey, J. D. Ferry, and D. F. Mosher, *J. Biol. Chem., 257*:14973–14978 (1982).
48. M. Rocco, M. Carson, R. Hantgan, J. McDonagh, and J. Hermans, *J. Biol. Chem., 258*:14545–14549 (1983).
49. Z. Markovič, A. Lustig, J. Engel, H. Richter, and H. Hörmann, *Hoppe Seyler's Z. Physiol. Chem., 364*:1795–1804 (1983).
50. C. S. Lai, and N. M. Tooney, *Arch. Biochem. Biophys., 228*:465–473 (1984).
51. H. Hormann, *Klin. Wochenschr., 60*:1265–1277 (1981).
52. E. Odermatt, J. Engle, H. Richter, and H. Hormann, *J. Mol. Biol., 159*:109–123 (1982).
53. D. F. Mosher, and E. R. Blout, *J. Biol. Chem., 248*:6896–6903 (1973).
54. H. Richter, M. Seidl, and H. Hörmann, *Hoppe Seyler's Z. Physiol. Chem., 362*:399–408 (1981).
55. K. Sekiguchi, and S. Hakomori, *Biochemistry, 22*:1415–1422 (1983).
56. J. W. Tamkun, and R. O. Hynes, *J. Biol. Chem., 258*:4641–4647 (1983).
57. J. E. Schwarzbauer, J. W. Tamkun, I. R. Lemischka, and R. O. Hynes, *Cell, 35*:421–431 (1983).
58. T. Vartio, A. Vaheri, R. Von Essen, H. Isomaki, and S. Stenman, *Eur. J. Clin. Invest., 11*:207–212 (1981).
59. B. Voss, S. Allan, J. Rauterberg, K. Ullrich, V. Gieselmann, and K. VonFigura, *Biochem. Biophys. Res. Commun., 90*:1348–1354 (1979).
60. M. R. Owens, and C. D. Cimino, *Blood, 59*:1305–1309 (1982).
61. M. Matsuda, T. Yamanaka, and A. Matsuda, *Clin. Chim. Acta, 118*:191–199 (1982).
62. J. Gonzalez-Calvin, M. F. Scully, Y. Sanger, J. Fok, V. V. Kakkar, R. D. Hughes, A. E. Gimson, and R. Williams, *Br. Med. J., 285*:1231–1232 (1982).
63. R. A. Clark, J. H. Quinn, H. J. Winn, J. M. Lanigan, P. Dellepella, and R. B. Colvin, *J. Exp. Med., 156*:646–651 (1982).
64. M. H. Ginsberg, E. F. Plow, and J. Forsyth, *J. Supramol. Struct. Cell. Biochem., 17*:91–98 (1981).
65. K. Hedman, A. Vaheri, and J. Wartiovaara, *J. Cell Biol., 76*:748–760 (1978).

66. L. T. Furcht, G. Wendelschafer-Crabb, D. F. Mosher, and J. M. Foidart, *J. Supramol. Struct., 13*:15–33 (1980).
67. P. S. Irish, and D. L. Hasty, *J. Histochem, Cytochem., 31*:69–77 (1983).
68. L. T. Furcht, G. Wendelschafer-Crabb, D. F. Mosher, and J. M. Foidart, *Prog. Clin. Biol. Res., 41*:829–843 (1980).
69. I. I. Singer, D. W. Kawka, D. M. Kazazis, and R. A. F. Clark, *J. Cell Biol., 98*:2091–2106 (1984).
70. L. B. Chen, K. Burridge, A. Murray, M. L. Walsh, C. D. Copple, A. Bushnell, J. K. McDougall, and P. H. Gallimore, *Ann. NY Acad. Sci. 312*:366–381 (1978).
71. K. Hedman, M. Kurkinen, K. Alitalo, A. Vaheri, S. Johansson, and M. Höök, *J. Cell Biol., 81*:83–91 (1979).
72. M. E. Perkins, T. H. Ji., and R. O. Hynes, *Cell, 16*:941–952 (1979).
73. W. G. Carter, and S. I. Hakomori, *J. Biol. Chem., 256*:6953–6960 (1981).
74. E. G. Hayman, A. Oldberg, G. R. Martin, and E. Ruoslahti, *J. Cell Biol., 94*:28–35 (1982).
75. K. Hedman, S. Johansson, T. Vartio, L. Kjellen, A. Vaheri, and M. Höök, *Cell, 28*: 663–671 (1982).
76. J. A. McDonald, D. G. Kelley, and T. J. Broekelmann, *J. Cell Biol., 92*:485–492 (1982).
77. A. Vaheri, and D. F. Mosher, *Biochim. Biophys. Acta, 516*:1–25 (1978).
78. M. G. Choi, and R. O. Hynes, *J. Biol. Chem., 254*:12050–12055 (1979).
79. R. O. Hynes, and A. Destree, *Proc. Natl. Acad, Sci. USA, 74*:2855–2859 (1977).
80. D. F. Mosher, and R. B. Johnson, *J. Biol. Chem., 258*:6595 (1983).
81. P. J. McKeown-Longo, and D. F. Mosher, *J. Cell Biol., 97*:466–472 (1983).
82. S. Stenman, and A. Vaheri, *J. Exp. Med., 147*:1054–1064 (1978).
83. Y. S. Kanwar, and M. G. Farquhar, *Proc. Natl. Acad. Sci. USA, 76*:1303–1307 (1979).
84. J. I. Scheinman, A. J. Fish, A. M. Matas, and A. F. Michael, *Am. J. Pathol., 90*:71–84 (1978).
85. M. A. Weiss, B. S. Ooi, Y. M. Ooi, E. Engvall, and E. Ruoslahti, *Lab. Invest., 41*: 340–347 (1979).
86. E. Lindner, A. Miettinen, and T. Tornroth, *Lab. Invest., 42*:70–75 (1980).
87. T. D. Oberley, D. F. Mosher, and M. D. Mills, *Am. J. Pathol., 96*:651–672 (1979).
88. J. A. Madri, F. J. Roll, J. Furthmayr, and J. M. Foidart, *J. Cell Biol., 86*:682–687 (1980).
89. J. M. Boselli, E. J. Macarak, C. C. Clark, A. G. Brownell, and A. Martinez-Hernandez, *Coll. Relat. Res., 1*:391–404 (1981).
90. A. Martinez-Hernandez, C. A. Marsh, C. C. Clark, E. J. Macarak, and A. G. Brownell, *Coll. Res., 5*:405–418 (1981).
91. P. F. Courtoy, R. Timpl, and M. G. Farquhar, *J. Histochem. Cytochem., 30*:874–886 (1982).
92. G. W. Laurie, C. P. Leblond, S. Inoque, G. R. Martin, and A. Chung, *Am. J. Anat., 169*:463–481 (1984).
93. I. Clemmensen, B. Holund, N. Johansen, and R. B. Andersen, *Histochemistry, 76*: 51–56 (1982).
94. H. B. Evans, S. Ayad, M. Z. Abedin, S. Hopkins, K. Morgan, K. W. Walton, J. B. Weiss, and P. J. L. Holt, *Ann. Rheum. Dis., 42*:575–581 (1983).
95. K. Shiozawa, S. Shiozawa, S. Shimizu, and T. Fujita, *Arthritis Rheum., 27*:615–622 (1984).
96. W. Dessau, J. Sasse, R. Timpl. F. Jilek, and K. Von Der Mark, *J. Cell Biol., 79*: 342–355 (1978).
97. R. E. Weiss, and A. H. Reddi, *J. Cell Biol., 88*:630–636 (1981).

98. D. L. Scott, J. P. Delamere, and K. W. Walton, *Br. J. Exp. Pathol., 62*:362–368 (1981).
99. G. Linck, S. Stocker, J. A. Grimaud, and A. Porte, *Histochemistry, 77*:323–328 (1983).
100. O. Fyrand, *Br. J. Dermatol., 101*:263–270 (1979).
101. O. Fyrand, *Arch. Dermatol. Res., 226*:33 (1979).
102. J. R. Couchman, W. T. Gibson, D. Thom, A. C. Weaver, D. A. Rees, and W. E. Parish, *Arch. Dermatol. Res., 266*:295–310 (1979).
103. R. A. F. Clark, *J. Invest. Dermatol., 81*:473–479 (1983).
104. O. Fyrand, and N. D. Solum, *Thromb. Res., 8*:659–672 (1976).
105. D. F. Mosher, and E. M. Williams, *J. Lab. Clin. Med. 91*:729–735 (1978).
106. J. Labat-Robert, J. P. Potazman, J. C. Derouette, and L. Robert, *Cell Biol. Int. Rep., 5*:969–973 (1981).
107. N. Stathakis, A. Fountas, and E. Tsianos, *J. Clin. Pathol., 34*:504–508 (1981).
108. J. J. Choate, and D. F. Mosher, *Cancer, 51*:1142–1147 (1983).
109. E. L. Snyder, P. G. Barash, D. F. Mosher, and S. D. Walter, *J. Lab. Clin. Med., 102*:881–889 (1983).
110. T. M. Saba, F. A. Blumenstock, P. Weber, and J. E. Kaplan, *Ann. NY Acad. Sci., 312*:43–55 (1978).
111. W. A. Scovill, T. M. Saba, J. E. Kaplan, H. Bernard, and S. Powers, *J. Trauma, 16*:898–904 (1976).
112. W. A. Scovill, T. M. Saba, F. A. Blumenstock, H. Bernard, and S. R. Powers, *Ann Surg. 188*:521 (1978).
113. E. L. Snyder, D. F. Mosher, A. Hezzey, and G. Golenwsky, *J. Lab. Clin. Med., 98*:336–341 (1981).
114. B. Carnemolla, M. Cutolo, P. Castellani, E. Balza, S. Raffanti, and L. Zardi, *Arthritis Rheum., 27*:913–921 (1984).
115. M. Lu-Steffes, A. J. Iammartino, F. R. Schmid, C. W. Castor, L. Davis, R. Entwistle, and B. Anderson, *Ann. Clin. Lab. Sci., 12*:178–185 (1982).
116. I. Clemmensen, B. Holund, and R. B. Andersen, *Arthritis Rheum., 26*:479–485 (1983).
117. I. Clemmensen, and R. B. Andersen, *Arthritis Rheum., 25*:25–31 (1982).
118. B. Anderson, M. Rucker, R. Entwistle, F. R. Schmid, and G. W. Wood, *Ann. Rheum. Dis., 40*:50–54 (1981).
119. G. Wood, M. Rucker, J. W. Davis, R. Entwistle, and B. Anderson, *Clin. Exp. Immunol. 40*:358–364 (1980).
120. I. Kono, T. Sakurai, T. Kabashima, K. Yamane, and H. Kashiwagi, *Clin. Exp. Immunol., 52*:305–310 (1983).
121. S. Carsons, B. B. Lavietes, and H. S. Diamond, *J. Lab. Clin. Med., 102*:721–731 (1983).
122. B. Villiger, T. Broekelmann, D. Kelley, G. J. Heymach, and J. A. McDonald, *Am. Rev. Respir. Dis., 124*:652–654 (1981).
123. A. J. Dixon, J. Burns, M. S. Punnill, and J. O'D. McGee, *J. Clin. Pathol., 33*:1021–1028 (1980).
124. E. E. Pettersson, and R. B. Colvin, *Clin. Immunol. Immunopathol., 11*:425–436 (1978).
125. S. Shiozawa, and M. Ziff, *Ann. Rheum. Dis., 42*:254–263 (1983).
126. B. Holund, I. Clemmensen, and M. Wanning, *Histochemistry, 80*:39–44 (1984).
127. D. L. Scott, T. J. Almond, K. W. Walton, and I. M. Hunneyball, *J. Pathol., 141*:56 (1983).
128. F. Grinnell, R. E. Billingham, and L. Burgess, *J. Invest. Dermatol., 76*:181–189 (1981).

129. M. Kurkinen, T. Vartio, and A. Vaheri, *Biochim. Biophys. Acta, 624*:490–498 (1980).
130. N. B. Wurster, and G. Lust, *Biochem. Biophys. Res. Commun., 109*:1094–1101 (1982).
131. R. E. Weiss, and A. H. Reddi, *Proc. Natl. Acad. Sci. USA, 77*:2074–2078 (1980).
132. W. Dessau, B. M. Vertel, H. Von Der Mark, and K. Von Der Mark, *J. Cell Biol., 90*: 78–83 (1981).
133. J. P. Pennypacker, J. R. Hassell, K. M. Yamada, and R. M. Pratt, *Exp. Cell Res., 121*:411–415 (1979).
134. W. T. Gibson, J. R. Couchman, and A. C. Weaver, *J. Invest. Dermatol., 81*:480–485 (1983).
135. P. Birembaut, P. Cahuzao, H. Delhomme, Y. Caron, J. Labat-Robert, L. Robert, and B. Kalis, *Ann. Dermatol. Venereol., 109*:933–937 (1982).
136. S. M. Cooper, A. J. Keyser, A. D. Beaulieu, E. Ruoslahti, M. E. Nimni, and FP. Quismorio, Jr., *Arthritis Rheum., 22*:983–987 (1979).
137. S. Reitamo, T. Reunala, Y. T. Konttinen, O. Saksela, and O. P. Salo, *Br. J. Dermatol., 105*:167–177 (1981).
138. O. Fyrand, *Acta. Dermatol. Venereol., 59*:487 (1979).
139. O. Fyrand, *Br. J. Dermatol., 102*:167–171 (1980).
140. S. M. Breathnach, B. Bhogal, F. C. DeBeer, S. M. Melrose, M. M. Black, and M. B. Pepys, *Br. J. Dermatol., 108*:267–275 (1983).
141. M. Leutenegger, P. Birembaut, J. P. Poynard, J. P. Eschard, Y. Ricard, Y. Caron, L. Robert, M. Szendroi, and J. Labat-Robert, *Pathol. Biol., 31*:45–48 (1983).
142. R. E. Grinwood, J. C. Huff, J. W. Harbell, and R. A. Clark, *J. Invest. Dermatol., 82*:145–149 (1984).
143. M. Kubo, D. A. Norris, S. E. Howell, S. R. Ryan, and R. A. F. Clark, *J. Invest.Dermatol., 82*: 580–586 (1984).
144. R. A. Clark, H. F. Dvorak, and R. B. Colvin, *J. Immunol., 126*:787–793 (1981).
145. D. F. Mosher, *J. Biol. Chem., 250*:6614–6621 (1975).
146. F. Grinnell, M. Feld, and D. Minter, *Cell, 19*:517–525 (1980).
147. M. H. Ginsberg, R. G. Painter, J. Forsyth, C. Birdwell, and E. F. Plow, *Proc. Natl. Acad. Sci. USA, 77*:1049–1053 (1980).
148. E. F. Plow, and M. H. Ginsberg, *J. Biol. Bhem., 256*:9477–9482 (1981).
149. F. Grinnell, M. Feld, and W. Snell, *Cell. Biol. Int. Rep., 3*:585–592 (1979).
150. H. B. Bensusan, T. L. Koh, K. G. Henry, B. A. Murray, and L. A. Culp, *Proc. Natl. Acad. Sci. USA, 75*:5864–5868 (1978).
151. D. F. Mosher, *Interaction of the Blood with Natural or Artificial Surfaces,* Marcel Dekker, New York, 1981, pp. 85–101.
152. J. Lahav, and R. O. Hynes, *J. Supramol. Struct. Cell Biochem., 17*:299–311 (1981).
153. F. A. Blumenstock, P. B. Weber, T. M. Saba, and R. Laffin, *Am. J. Physiol., 232*: 80-87 (1977).
154. J. Molnar, F. B. Gelder, M. Z. Lai, G. E. Siefring, R. B. Credo, and L. Lorand, *Biochemistry, 18*:3909–3916 (1979).
155. O. D. Zinkevich, R. I. Litvinov, and M. S. Kuravskaya, *Bull. Exp. Biol. Med., 94*: 86–88 (1982).
156. S. Johansson, K. Rubin, M. Höök, T. Ahlgren, and R. Seljelid, *FEBS Lett., 105*: 313–316 (1979).
157. K. Alitalo, T. Hovi, and A. Vaheri, *J. Exp. Med., 151*:602–613 (1980).
158. M. P. Bevilacqua, D. Amrani, M. W. Mosesson, and C. Bianco, *J. Exp. Med., 153*:42–60 (1981).
159. C. G. Pommier, J. O'Shea, T. Chused, K. Yancey, M. M. Frank, T. Takahashi, and E. J. Brown, *J. Exp. Med., 159*:137–151 (1984).

160. T. M. Saba, and E. Jaffe, *Am. J. Med., 68*:577–594 (1980).
161. L. Van de Water, S. Schroeder, E. B. Crenshaw, and R. O. Hynes, *J. Cell Biol., 90*: 32–39 (1981).
162. L. Van de Water, A. T. Destree, and R. O. Hynes, *Science, 220*:201–204 (1982).
163. J. K. Czop, J. L. Kadish, and K. F. Austen, *Proc. Natl. Acad, Sci. USA, 78*:3649–3653 (1981).
164. J. K. Czop, S. E. McGowan, and D. M. Center, *Am. Rev. Resp. Dis., 125*:607–609 (1982).
165. F. J. Rourke, F. A. Blumenstock, and J. E. Kaplan, *J. Immunol., 132*:1931–1936 (1984).
166. F. Duckert, *Ann. NY Acad. Sci., 202*:190–199 (1972).
167. J. C. Bowersox, and N. Sorgente, *Cancer Res., 42*:2547–2551 (1982).
168. A. E. Postlethwaite, J. Keski-Oja, G. Balian, and A. H. Kang, *J. Exp. Med., 153*: 494–499 (1981).
169. V. Gauss-Muller, H. K. Kleinman, G. R. Martin, and E. Schiffmann, *J. Lab. Clin. Med., 96*:1071–1080 (1980).
170. B. R. McAuslan, G. N. Hannan, W. Reilly, and F. H. C. Stewart, *J. Cell. Physiol., 104*:177–186 (1980).
171. D. J. Donaldson, and J. T. Mahan, *J. Cell Sci., 62*:117–127 (1983).
172. L. S. Fujikawa, C. S. Foster, T. J. Harrist, J. M. Lanigan, and R. B. Colvin, *Lab. Invest., 45*:120–129 (1981).
173. T. Nishida, S. Nakagawa, T. Awata, Y. Ohashi, K. Watanabe, and R. Manabe, *J. Cell Biol., 97*:1653–1657 (1983).
174. R. A. F. Clark, H. J. Winn, H. F. Dvorak, and R. B. Colvin, *J. Invest. Dermatol., 80* (Suppl.6):26s–30s (1983).
175. T. Nishida, S. Nakagawa, C. Nishibayashi, H. Tanaka, and R. Manabe, *Arch. Ophthalmol., 102*:455–456 (1984).
176. S. I. Rennard, G. W. Hunninghake, P. B. Bitterman, and R. G. Crystal, *Proc. Natl. Acad. Sci. USA 78*:7147–7151 (1981).
177. E. Engvall, E. Ruoslahti, and E. J. Miller, *J. Exp. Med., 147*:1584–1595 (1978).
178. E. Oh, M. Pierschbacher, and E. Ruoslahti, *Proc. Natl. Acad. Sci. USA 78*:3218–3221 (1981).
179. H. P. Schwarz, A. Luger, H. Graf, H. K. Stummvoll, and P. Aiginger, *Thromb. Haemostas,. 48*:345 (1982).
180. C. W. Kischer, and M. J. C. Hendrix, *Cell Tissue Res., 231*:29–37 (1983).
181. H. G. Remold, J. E. Shaw, and J. R. David, *Cell. Immunol., 58*:175–187 (1981).
182. H. P. Godfrey, and A. Purohit, *Immunology, 46*:507–514 (1982).
183. H. P. Godfrey, and A. Purohit, *Immunology, 46*:515–526 (1982).
184. H. P. Godfrey, C. V. Angadi, P. A. Wolstencroft, and C. Bianco, *J. Immunol., 133*: 1417–1423 (1984).
185. D. B. Lause, D. H. Beezhold, and J. E. Doran, *J. Immunol., 132*:1294–1299 (1984).
186. D. B. Lause, J. E. Doran, and J. A. Houston, *Transplantation, 34*:147–150 (1982).
187. B. M. Martin, M. A. Gimbrone, F. R. Majeau, E. R. Unanue, and R. S. Cotran, *Am. J. Pathol., 111*:367 (1983).
188. J. P. Murnane, and R. B. Painter, *Biochemistry, 22*:1217–1222 (1983).
189. M. Zanetti, and T. Takami, *Clin. Immunol. Immunopathol., 31*:353–363 (1984).
190. J. E. Murphy-Ullrich, T. D. Oberley, and D. F. Mosher, *Am. J. Pathol., 117*:1–11 (1984).
191. J. E. Murphy-Ullrich, T. D. Oberley, and D. F. Mosher (submitted for publication).

21

Vascular and Cutaneous Complications of Diabetes Mellitus

The role of nonenzymatic glucosylation of collagens

Andrea J. Perejda
UCLA School of Medicine and Harbor-UCLA Medical Center,
Torrance, California

I. INTRODUCTION

Diabetes mellitus is a metabolic disorder characterized by hyperglycemia secondary to relative or absolute insulin deficiency (1,2). The disease includes two broadly defined phenotypic expressions: type I, or juvenile-onset diabetes, and type II, or maturity-onset, diabetes. Although both types of diabetes exhibit variable age of onset and may be the result of polygenic inheritance, the juvenile-onset form can generally be considered ketosis-prone, while the maturity-onset type is frequently ketosis-resistant.

With the advent of treatment of diabetes mellitus by insulin therapy, oral hypoglycemic agents, or dietary control, the life expectancy of the diabetic patient has dramatically increased. As a result, however, the development of complications affecting virtually all organ systems in the body to which glucose is freely permeable has become a source of major concern.

II. COMPLICATIONS OF DIABETES MELLITUS

Diabetic complications variably affect the cardiovascular, nervous, renal, and ocular systems (3). While the pathogenesis of individual diabetic complications is still the subject of extensive investigation, it is clear that altered metabolism as a result of insulin deficiency provides a common basis for development of all complications (1-3). The brief review that follows describes some of the clinical and biochemical features of various diabetic complications, which suggest that the metabolic consequences of hyperglycemia may explain the development of a wide variety of seemingly unrelated aberrations.

A. Cardiovascular Complications

Generalized cardiovascular dysfunction consequent to diabetes mellitus consists of a wide variety of interrelated aberrations. First, hematological alterations such as increased erythrocyte aggregation (4), decreased erythrocyte deformability (5), increased amounts of hemoglobin A_{1c} having high oxygen affinity (6,7), and decreased apparent levels of 2,3-diphosphoglycerate (8) are commonly found. These abnormalities may play a significant role in the development of diabetic complications related to hypoxic tissue injury (7). Determination of the relative proportion of glycosylated hemoglobin in the blood of diabetic patients has proven to be a useful index of the degree of long-term control of hyperglycemia in these patients (7,9). Other hematological aberrations include defective polymorphonuclear leukocyte function (10), increased platelet adhesiveness (11,12), and plasma protein abnormalities (13,14).

Second, intrinsic disorders of the cardiovascular system per se, including increased incidence of atheroma formation, cerebral vascular accidents, myocardial infarction, and glomerulosclerosis (15–17), comprise the basis for widespread morbidity and mortality among diabetic patients. Large-vessel disease (macroangiopathy) is associated with lipoprotein abnormalities (18), exacerbated by hyperglycemia and/or insulin deficiency. Endothelial cell injury and increased vascular permeability, followed by platelet adhesion, result in proliferation of smooth muscle cells and consequent scar and atheroma formation (19,20). Small-vessel disease (microangiopathy) is most clinically apparent as renal disease in diabetic patients. Glomerular basement membrane thickening is progressive and associated with proteinuria, renal enlargement, and increased glomerular filtration surface area (21). Functional alterations such as elevated glomerular volume and increased filtration surface area can be reversed by insulin treatment (22,23). However, no decrease in glomerular basement membrane thickness can be detected with insulin therapy, probably owing to the stability of glomerular basement membrane in vivo (24).

B. Ocular Complications

Diabetic complications also include cataract formation (25) and/or retinopathy (26) due to ischemia, microaneurysm formation, and increased capillary permeability (27,28). Neovascularization and retinal capillary basement membrane thickening are also important features of diabetic retinopathy (26). It has been demonstrated in diabetic patients that with early tight control of hyperglycemia, increased capillary permeability and subsequent retinopathy can be reduced (29).

C. Neuropathic Complications

Peripheral neuropathy is manifested as paresthesia and anesthesia. Segmental demyelination (30), endoneurial edema (31), and decreased nerve conduction velocity (32) have all been noted in patients with peripheral neuropathy. Recent evidence suggests progressive demyelination correlates to the degree of hyperglycemia in rats (33,34). In addition, significant basement membrane thickening and evidence of endothelial cell swelling are common features of neuronal vascular injury (35). Improved nerve function has been demonstrated in both humans and experimental animals following normalization of hyperglycemia (36,37).

D. Cutaneous Complications

Cutaneous manifestations of diabetes mellitus may reflect underlying vascular disease (i.e., erysipelaslike erythema, gangrene), neuropathy (i.e., ulcerations of the skin and foot disease), or lowered resistance to infection (38). In addition to these dermatological disorders, however, are several disorders that appear to be the result of altered connective tissue metabolism. One such disorder is scleredema diabeticorum, in which nonpitting induration of the skin occurs on the upper back and neck of maturity-onset diabetics (39–42). Diabetic scleredema may or may not occur following infection (40,41), and the dermis is characterized by increased collagen and altered glycosaminoglycan contents (40, 41,43). Patients with scleredema diabeticorum demonstrate a significantly increased incidence of angina pectoris in comparison to diabetic patients without scleredema (44).

A second syndrome affecting the skin and joints of juvenile-onset diabetics is digital sclerosis, which may be associated with joint contractures (45–49). The sclerodermalike skin changes over the digital, metacarpal, and carpal surfaces of the hand and forearm appear to progress from thickening to tightness to waxiness of the skin with increasing duration of diabetes (50). Patients with more advanced sclerodermalike skin changes also suffer from limited joint mobility, the occurrence and severity of which correlate significantly with the incidence of retinopathy and nephropathy (46,51). This syndrome may also be associated with a higher incidence of restrictive pulmonary disease (50). Histopathological examination of the affected skin in these patients reveals collagen accumulation in the lower dermis and subcutaneous tissue in a manner similar to that seen in scleroderma (49,52). One report indicates that a reduction in skin thickness in these patients may be attained by close control of hyperglycemia utilizing pump administration of insulin (53).

From this brief overview of the major diabetic complications, it is apparent that the development of many complications may be directly related to the extent of control of the patients' hyperglycemia. In fact, some recent reports suggest that amelioration of some complications, such as increased vascular permeability in the eye and glomerulus, peripheral nerve dysfunction, and skin thickening, can be achieved by insulin therapy with reduction of the degree of hyperglycemia (22,23,29,34,36,37,53,54). These reports are not undisputed, however,(55–57). This etiology would also explain the increased severity of diabetic complications usually observed with increased duration of the disease. It is also clear that pathological and functional alterations associated with abnormal collagen metabolism are associated with diabetic complications where vascular injury, increased vascular permeability, and associated capillary basement membrane thickening occur, as well as in cutaneous diseases specific to diabetes mellitus. The primary focus of the remainder of this chapter will be to examine the aberrations of collagen metabolism in diabetes and to examine the relationship of nonenzymatic glucosylation of proteins, including collagen, to the development of diabetic complications.

III. NONENZYMATIC GLUCOSYLATION OF PROTEINS IN DIABETES MELLITUS

Allen, Schroeder, and Balog in 1958 (58) observed that a small, but significant, proportion of hemoglobin exists as negatively charged hemoglobins, HbA_{1a}, HbA_{1b} and HbA_{1c}. These minor hemoglobin components differ from HbA only in the covalent linkage of a hexose on the N-terminal amino group of the β chain. For example, HbA_{1a_1} is an adduct of HbA with β-*N*-fructose-1, 6-diphosphate, while HbA_{1a_2} is an adduct of β-*N*-glucose-6-phosphate. Interest in HbA_{1c}, which is modified by a β-*N*-glucose adduct, has recently increased with the demonstration that HbA_{1c} levels are significantly increased in patients with diabetes mellitus, and that HbA_{1c} levels can be correlated with the degree of control of hyperglycemia (7,9).

Over the past few years, it has been established that many other proteins can be nonenzymatically glucosylated, both in vitro and in the diabetic milieu. The addition of glucose occurs via condensation of the aldehydic carbon of glucose with a free amino group, which is either the α-amino group of the N-terminal amino acid (valine, in the case of Hb), or an ϵ-amino group of an internal peptidyl lysyl or hydroxylysyl residue (see Figure 1). The initial Schiff base formation is rapid and reversible; irreversible conversion of this aldimine intermediate to a stable ketoamine linkage via an Amadori rearrangement occurs more slowly (59). It has been demonstrated in vitro that nonenzymatic glucosylation is linear with respect to glucose concentration, pH, time, and temperature of the reaction (59–62). The amount of glucose bound to protein by ketoamine linkage can be determined by using the colorimetric assay of Flückiger and Winterhalter (60). Hexose bound to protein is first converted to 5-hydroxymethylfurfural by heating under acidic conditions; following reaction with 2-thiobarbituric acid, the 5-hydroxymethylfurfural content is measured spectrophotometrically. Limitations of this technique are protein concentration and presence of free glucose (63).

The quantity of hexose bound to protein can also be determined by chromatographic separation of glucosylated lysyl or hydroxylysyl residues from each other and from unmodified lysyl and hydroxylysyl residues after reduction with sodium borohydride or sodium cyanoborohydride (64,65). Separation can also be achieved using boronic acid-affinity chromatographic techniques. Alternatively, glucosylation can be monitored by incorporation of [^{14}C]D-glucose into purified protein in vitro, after appropriate purification of the radioactive glucose (59).

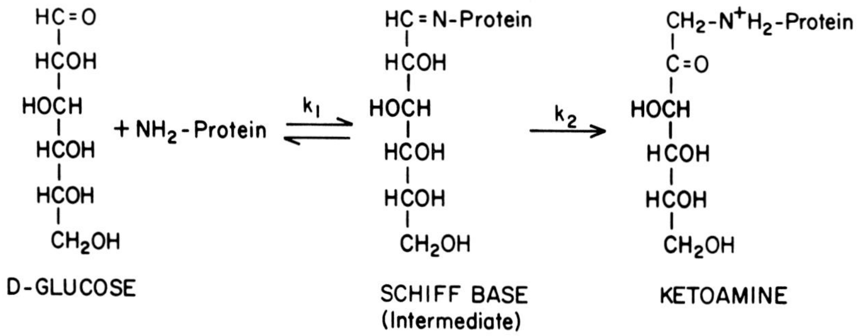

Figure 1 Schematic diagram depicting the mechanisms of nonenzymatic glucosylation of proteins. Initial condensation of D-glucose with an amino group results in the formation of a Schiff base intermediate, which is then rearranged to form a stable ketoamine. (Reprinted with permission from *Collagen and Related Research,* 2:81, 1982.)

The nonenzymatic glucosylation of many proteins, both in vitro and in vivo, has been demonstrated by similar techniques. In addition to HbA_{1c}, extracellular proteins such as serum albumin (66,67), α2-macroglobulin (68), collagens (69,70), immunoglobulins (68), and fibrinogen (71,72) can be nonenzymatically glucosylated. Also, lipoproteins such as LDL and HDL are nonenzymatically glucosylated under hyperglycemic conditions (73–75). Cellular proteins, such as erythrocyte membrane proteins (76,77), peripheral nerve proteins (78), tubulin (79), and lens crystallin (25,80), can be modified by nonenzymatic glucosylation as well.

Although it is clear that posttranslational, nonenzymatic glucosylation of many proteins does occur, functional alterations have been described for only a handful. For example, nonenzymatic glucosylation of hemoglobin significantly alters its oxygen affinity (6,7), possibly contributing to hypoxic injury in the diabetic patient. Receptor-mediated uptake of glucosylated LDL by human skin fibroblasts is significantly decreased as compared to unmodified LDL (75,81–83), perhaps providing an explanation for elevated plasma LDL levels noted in patients with diabetes mellitus. Nonenzymatic glucosylation of HDL, on the other hand, results in an increased rate of non-macrophage-dependent clearance when injected into guinea pigs (74). The degradation of nonenzymatically glucosylated fibrin by plasmin is also apparently impaired (72). Finally, inoculation of rats with soluble, glucosylated rat skin collagen elicits antibody production against the collagen; however, inoculation of unmodified collagen does not (84). Aside from these examples of specific functional aberrations as a direct result of nonenzymatic glucosylation of a specific protein, no other clear-cut demonstrations of altered function due to glucosylation exist as of this writing. These few examples are suggestive, however, that nonenzymatic glucosylation of many proteins may have a significant effect on their metabolism and activities.

IV. COLLAGEN METABOLISM IN DIABETES MELLITUS

A. Vascular Collagen Metabolism

One of the major extracellular matrix alterations associated with complications of diabetes mellitus is the thickening of basement membranes around retinal, skeletal muscle, neuronal, and glomerular capillaries (see Figure 2) (85–88). Basement membrane thickening progresses with the duration of diabetes and is greater in capillaries exposed to greater venous hydrostatic pressure (89). The pathogenetic mechanisms responsible for increased basement membrane thickness in skeletal muscle capillaries are speculative; basement membrane synthesis or degradation may be altered as a result of cellular metabolic aberrations associated with (a) hyperglycemia or insulin deficiency, (b) endothelial cell or pericyte injury and necrosis, and/or (c) increased vascular hydrostatic pressure. Skeletal muscle capillary basement membrane thickening in diabetes is associated with increased vascular permeability (28,90,91), as well as decreased vascular compliance (92). The correlation between the extent of basement membrane thickening in diabetes and the clinical presentation of proteinuria and decreased glomerular filtration rate is well known (21).

Basement membranes are complex extracellular matrices comprised of type IV collagen and associated noncollagenous proteins including proteoglycans, laminin, fibronectin, and entactin, which are secreted by epithelial cells, endothelial cells, and/or pericytes (93). Thickening of the basement membrane in diabetes is primarily the result of an increased accumulation of type IV collagen (94,95). Although early studies suggested that diabetic

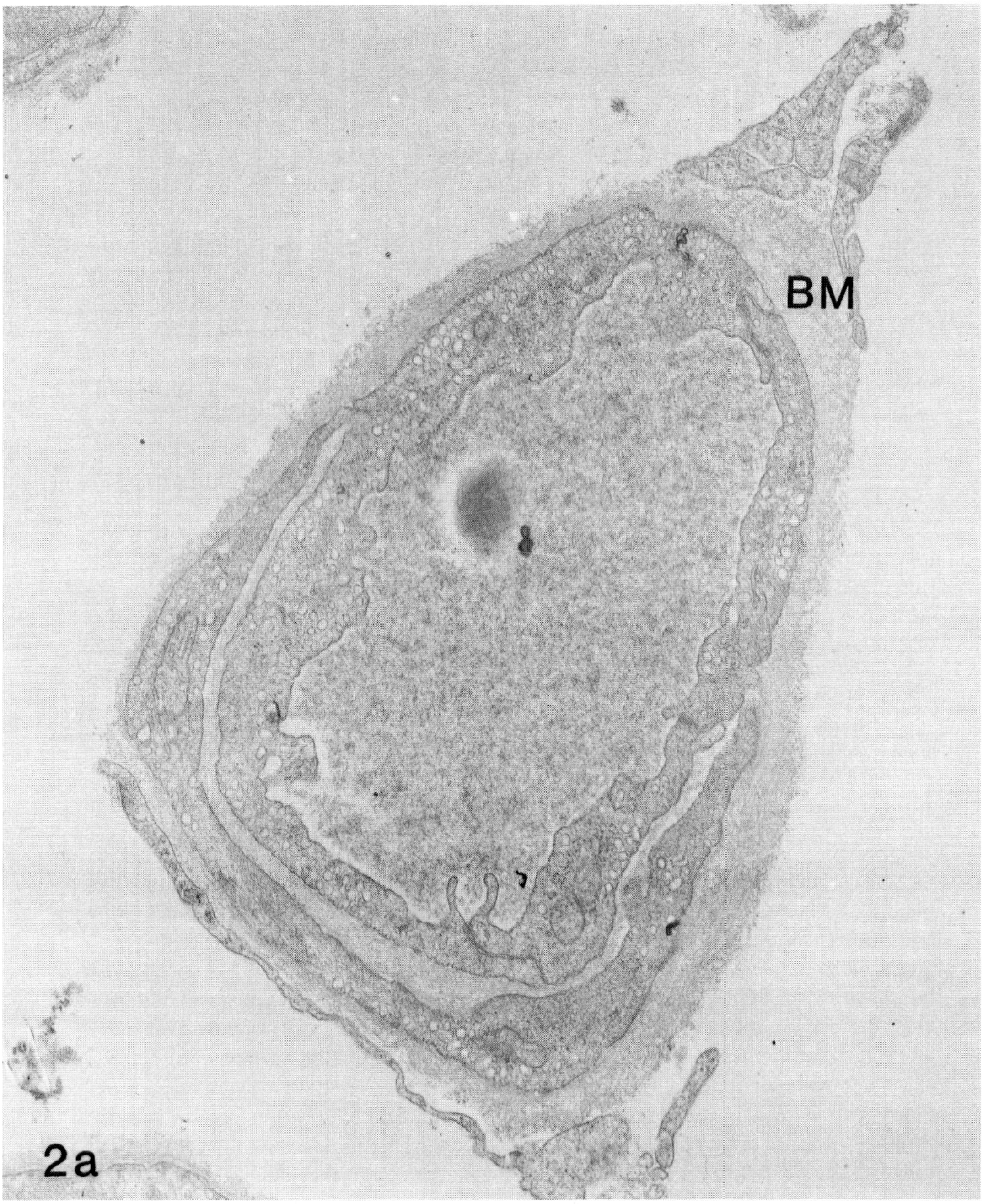

Figure 2 Transmission electron micrographs of skeletal muscle capillaries from gastrocnemius muscles of a healthy human subject (a) and a diabetic subject (b). The capillary basement membrane (BM) surrounding the vessel from the diabetic subject is significantly thicker than that from the nondiabetic subject. (a) X 21,220. (b) X 22,400.

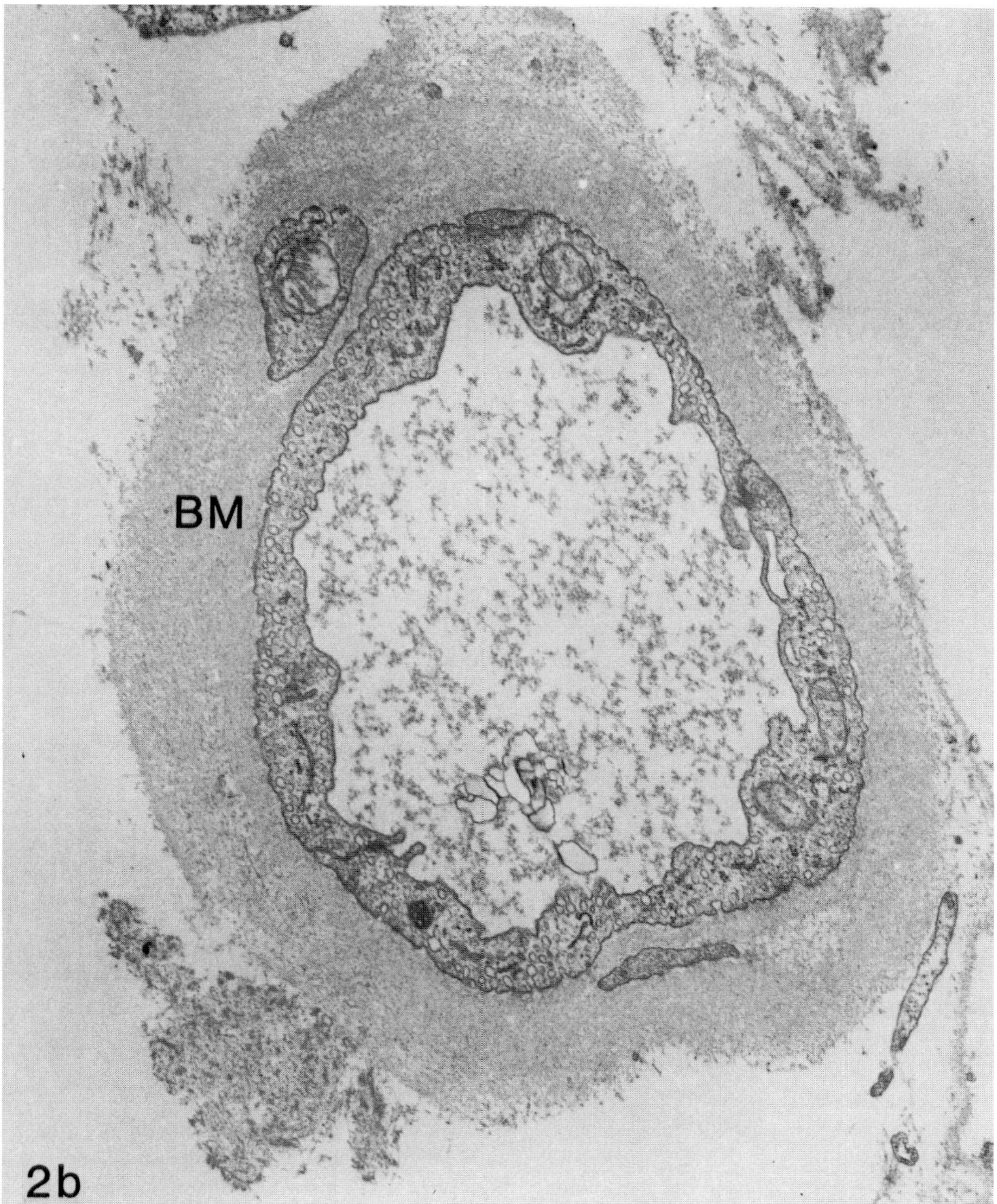

Figure 2 (Continued)

glomerular basement membrane had a decreased lysine content and an increase in the hydroxylysine:lysine ratio (96), these studies could not be duplicated (94,95). The disparity of those results is probably due to the heterogeneous nature of basement membrane preparations. A more recent study (70), in which glomerular basement membrane collagen was isolated from diabetic and nondiabetic human glomeruli by limited pepsin solubilization, definitively demonstrated that there is no difference in the amino acid composition of the collagen between the two sources, nor is there any difference in the relative proportions of 3-hydroxyproline, 4-hydroxyproline, or hydroxylysine. No differences in the hydroxylysine-*O*-glycoside contents were noted; the ratio of the disaccharide glucosylgalactosyl-*O*-hydroxylysine to the monosaccharide galactosyl-*O*-hydroxylysine was unchanged, and in both instances the percentage of hydroxylysine residues glycosylated was the same. In contrast, the quantity of ketoamine-linked glucose bound to glomerular basement membrane collagen, as assayed by the thiobarbituric acid assay, was 2.0–2.5-fold higher in diabetic collagen than in normal collagen. In experimental diabetes induced in rats, increased quantities of urea-insoluble glomerular basement membrane collagen (97) have been demonstrated. Time-, temperature-, and glucose concentration-dependent glucosylation of glomerular basement membrane in vitro has also been demonstrated (98).

Additional studies have demonstrated that increased synthesis of glomerular basement membrane collagen occurs in several different experimental diabetic conditions. First, increased incorporation of [^{14}C]proline or [^{14}C]lysine into [^{14}C]hydroxyproline or [^{14}C]hydroxylysine was measured in glomeruli from streptozotocin diabetic rats incubated with label in vitro when compared to glomeruli from control rats (99–101). In vivo labeling experiments in streptozotocin diabetic rats have suggested that increased basement membrane synthesis occurs in diabetic rats (99,100), and one study concluded that decreased degradation of glomerular basement membrane collagen may also contribute to collagen accumulation in diabetic glomerular basement membrane (102).

Abnormalities have also been described for collagens isolated from other sites in the cardiovascular system. Characterization of collagen types obtained from normal arteries and from atherosclerotic plaques of nondiabetic patients has suggested an increase in the type I to type III collagen ratio in atherosclerotic lesions (103–105). This alteration may be significant with regard to the prognosis of diabetic vascular complications, owing to the accelerated rate of atherosclerosis in diabetic patients.

In separate studies, the amount of ketoamine-linked hexose in proteins isolated from streptozotocin-diabetic rat aortae was approximately doubled, as compared to controls (106). It should be noted, however, that collagen preparations in that study were obtained by autoclaving aortae, it is therefore unclear as to what extent collagens were specifically glucosylated.

Increased total myocardial collagen content has been reported in two separate investigations of diabetic dog and monkey hearts (107,108). Not only was the total myocardial collagen content larger, but the solubility of collagen in 0.5 M acetic acid was also significantly less in diabetic monkey heart than in controls (107). Conflicting results were obtained, however, in a third study of hearts from streptozotocin diabetic rats (109).

Studies demonstrating increased synthesis of collagens in diabetes led to the suggestion that hyperglycemia might stimulate increased basement membrane collagen biosynthesis. When lens capsules isolated from normal rats were incubated with [^{3}H]proline, in the presence of high glucose (10 or 20 mM), increased incorporation of radioactivity into basement membrane-associated macromolecules was noted (110). In separate experiments, the effects of glucose, insulin, and streptozotocin diabetes on type IV collagen metabolism in murine basement membrane-forming EHS tumor tissue were measured (111). No differences were noted in amino acid compositions or intracellular enzyme activities of tumors grown in diabetic or normal mice (111). However, high glucose concentrations resulted in increased total protein (including collagen) synthesis; type IV collagen biosynthesis was increased to the same extent as total protein synthesis. Insulin exerted no specific effect on type IV collagen biosynthesis in this system. The pro-α1(IV) and pro-α2(IV) collagen chains synthesized under all circumstances were qualitatively identical.

Studies of intracellular enzyme activities specific to collagen biosynthesis in kidneys of diabetic mice, however, seem to reflect the overall stimulation of collagen synthesis in this tissue (111). Prolyl 4-hydroxylase and lysyl hydroxylase, as well as hydroxylysyl galactosyltransferase and glucosyltransferase, activities were slightly increased per unit protein during the first three weeks following induction of diabetes by streptozotocin injection (111). In alloxan diabetic rats, significant increases in collagen glucosyltransferase in kidney were observed (112). Interestingly, Khalifa and Cohen (113,114) have reported that increased lysyl hydroxylase activities can be restored to normal levels by insulin therapy.

Glomerular basement membrane accumulation may also be related to the reported glucose-mediated inactivation of a collagen-specific α-glucosidase (115), an enzyme presumed partially responsible for degradation of basement membrane collagen (116). β-Glycosidase activities are also reported to be low in diabetic rats (117,118). An altered basement membrane synthesis or degradation rate may also be reflected by the recent finding that the terminal cross-linking domain of type IV collagen, the 7S region, has been identified in increased quantities in serum from diabetic rats when compared to normals (119).

Thus, the regulation of glomerular basement membrane collagen synthesis and degradation is multifactorial. Increased biosynthesis and secretion of collagens in diabetes may occur owing to hyperglycemia and resultant alterations in glucose metabolism. It is important to remember that diabetic complications associated with hyperglycemia occur in tissues that are freely permeable to glucose. Evidence that decreased degradation of glomerular basement membrane collagen contributes to glomerular basement membrane thickening is indirect at this time. The suggestion has been made that increased nonenzymatic glucosylation of glomerular basement membrane collagen may play a role in the pathogenesis of glomerular basement membrane thickening. Indeed, it has been suggested that increased enzymatic or nonenzymatic glucosylation of collagens may interfere with their degradation by specific mammalian collagenases (120–122).

B. Nonvascular Collagen Metabolism

Collagen accumulation is the hallmark of the diabetic cutaneous sclerotic syndromes scleredema and digital sclerosis (39,46,49). In both these disorders the thickness of the dermis is noticeably enlarged, and collagen accumulation extends into the subcutaneous tissue. It is clear that in scleredema diabeticorum, the cutaneous collagen content is increased per mm^2 skin surface area; however, no change in the amount of collagen per mg dry weight of skin is observed (43). Mean diameters of collagen fibers in sclerederma, as measured by electron microscopic morphometric techniques, are identical to controls (43). While quantitation of the amount of hexose bound nonenzymatically to collagen has not been determined for skin from patients with scleredema, recent studies have demonstrated it is increased up to 13-fold in three patients with digital sclerosis (49). In those same patients, HbA_{1c} values range from 12 to 16% of the total hemoglobin, normal values being about 3.5%.

In studies involving cultured human skin fibroblasts obtained from diabetic patients (123), cells derived from diabetics incorporated 2.0–2.5 times more [^{14}C]proline and [^{14}C]glycine into total protein (including collagen and fibronectin) when grown in the presence of Dulbecco's medium, with no additional glucose supplementation. The amount of collagenase-sensitive protein synthesized by diabetic fibroblasts was also increased approximately fourfold compared to normal cell lines.

Quantitation of collagen production by human skin fibroblasts cultured from diabetic patients with sclerodermalike skin changes has shown that the total collagen synthesis rate is reduced in these cells (52). However, if collagen production is normalized to the DNA synthetic rate, as measured by thymidine incorporation, diabetic fibroblasts are no different from controls. Fibroblast collagenase activity secreted into the medium was identical for diabetic and control cells in this study.

In vitro studies comparing the nonenzymatic glucosylation of type I and II collagens have recently shown that both collagens can be nonenzymatically glucosylated (62). Both the $\alpha1$(I) and $\alpha2$(I) chains of type I collagen are subject to nonenzymatic glucosylation in the native, triple-helical conformation and in the denatured state. After 144 hr incubation in the presence of 500 mg/dL glucose, 18.9% of lysyl residues and 36.5% of hydroxylysyl residues in type I collagen were glucosylated. In contrast, under identical conditions, 47.9% of lysyl and 68.1% of hydroxylysyl residues in type II collagen were modified. It is interesting to note that in general, hydroxylysine was more readily glucosylated than lysine, and that type II collagen was a better substrate for nonenzymatic glucosylation than type I collagen. Perhaps the greater hydroxylysine-*O*-glycoside content of type II collagen confers properties on the molecule that make it a better substrate for nonenzymatic glucosylation (62).

Further studies indicate that type I collagen solubility is frequently decreased in diabetes. The proportion of neutral salt-soluble collagen was reduced as compared to controls in human skin biopsy specimens examined in two separate studies (124,125), as well as in dura mater (126). The solubility of collagen in 0.5 M acetic acid is also significantly decreased in normal diabetic skin (125), as well as in sclerotic skin obtained from two of three patients with sclerodermalike skin changes and joint contractures (49). Investigations by Chang et al. (127) and Golub et al. (128) also demonstrated that in experimental diabetes, rat skin and tendon collagens are less soluble than in controls.

The generalized decrease in collagen solubility found in diabetes has been attributed by some to reflect an increase in collagen cross-linking (125,127). This hypothesis has been indirectly corroborated by several findings. First, the quantity of pepsin-soluble collagen measured in some studies of diabetic human skin is decreased (125). Second, lysyl oxidase activity is markedly increased in lungs from diabetic animals (129). Lysyl oxidase is the enzyme responsible for catalyzing the oxidative deamination of lysine and hydroxylysine to allysine or hydroxyallysine, which can either condense to form aldol condensation products (intramolecular cross-links) or condense with ϵ-NH_2 groups of other lysyl or hydroxylysyl residues to form intermolecular cross-links. Third, the susceptibility of insoluble collagen to digestion by bacterial collagenase is decreased in diaphragmatic tendon collagen obtained from diabetic patients at autopsy (130,131). Studies of the susceptibility of soluble collagen from diabetic patients to digestion by specific vertebrate (i.e., fibroblast or polymorphonuclear leukocyte) collagenases have not been reported. However, we recently reported that soluble type I collagen isolated from chick tendon, which was nonenzymatically glucosylated in vitro, was digested as readily as was control type I collagen by human leukocyte collagenase (62). The studies of Chang et al. (127) provide convincing evidence that decreased solubility of collagen in diabetes is due to altered or increased lysine-derived cross-link formation. In this investigation, administration of β-aminopropionitrile or D-penicillamine (lathyrogens that prevent lysine-derived cross-link formation) to streptozotocin diabetic rats abolished the observed difference in collagen solubility between diabetic and control animals.

Finally, investigations of mechanical properties of rat tail tendons or tail skin (132,133) have shown that the maximum tension of contraction is increased after at least four weeks of diabetes, as compared to controls. According to Yue et al. (134), the increased thermal rupture time observed for tail tendon collagen correlated with its degree of nonenzymatic glucosylation. Increased nonenzymatic glucosylation of tendon collagen also correlated positively with increased maximum contractile force in a second investigation (132); however, these changes did not correlate with any change in the density of reducible cross-links. Thus, the changes in mechanical properties found in diabetic collagen are the same changes as those found during aging of normal collagen and suggest increased stability of the molecule, usually associated with an increase in the number of reducible cross-links (132,135). However, the observed increase in stability of collagen fibers from diabetics appears to have a basis other than that found in normal aging, related in some unknown way to the increased amount of glucose bound in ketoamine linkage to lysyl and hydroxylysyl residues.

In this context, studies of lens crystallin are of interest, since they provide another potential mechanism by which to explain collagen fiber stability observed in diabetes (136,137). The discovery that human cataractous lenses contain increased numbers of disulfide bonds, coupled with the finding that disulfide-bonded, high-molecular-weight aggregates are found in diabetic lenses, led to the hypothesis that formation of these aggregates is facilitated by an increased susceptibility of lens protein to sulfhydryl oxidation, resulting from increased nonenzymatic glucosylation (25,138). Putative cross-linked products formed are the result of the Maillard or the nonenzymatic browning reaction. Evidence for such products has been identified in diabetic lens crystallin (136), in insoluble collagen obtained from diabetic human dura mater (126), and in diabetic DNA (139), by measurement of fluorescence-excitation spectra between 370 and 440 nm. The role of disulfide-bonded, browning reaction products in the development of altered properties of collagen in diabetes is still unclear, and the putative intermolecular cross-link has not yet been isolated and identified. The paucity of sulfur-contining amino acids in collagens opens to question the suggestion that collagens have reduced solubility due to disulfide bonding among collagen molecules.

Functional alterations of collagens as a result of nonenzymatic glucosylation of peptidyl lysyl and hydroxylysyl residues thus may include decreased turnover and increased tensile strength (102,129,132,134,140). Additional bases for the development of collagen-related complications in diabetes may also reside in the immunological properties of glucosylated collagens. Recent evidence suggests that acid-soluble collagen obtained from rat skin that is nonenzymatically glucosylated in vitro and subsequently reduced is antigenic when inoculated into rats (48). The antibody detected was apparently directed against glucitol-lysyl residues (84). In another study, data was obtained which suggested that nonenzymatic glucosylation of soluble calf skin collagen, which is then immobilized on agarose beads, is capable of binding albumin and IgG. In situ formation of immune complexes was also accomplished using this technique, suggesting that immune complex formation and the tissue damage associated with it may be, at least in part, related to nonenzymatic glucosylation of long-lived structural proteins, including collagen (141).

V. SUMMARY

In conclusion, it is clear that nonenzymatic glucosylation of collagens can and does occur. For several reasons, collagens are excellent candidates for nonenzymatic glucosylation, contributing to the development of diabetic complications. Their amino acid compositions are rich in lysine and hydroxylysine, the sum of these amino acids being as high as 50-60 residues per 1000 in type IV collagen (142). The enrichment in these amino acids containing ϵ-amino groups results in the existence of many potential sites for glucose attachment. Second, collagen turnover is relatively slow in tissues; the average half-life of the bulk of body collagen is estimated to be 2.5 years (143,144), and that of type IV collagen is in excess of 100 days (24). The extreme stability of collagens in vivo, coupled with their extracellular locations, permits long-term exposure of existing molecules to ambient interstitial glucose levels.

The relationship between the nonenzymatic glucosylation of collagen and its accumulation in tissues in diabetes mellitus is enigmatic at the present time. Increased collagen synthesis may occur in response to a hyperanabolic effect of hyperglycemia. Alternatively, the effects of tissue hypoxia or cellular injury may result in abnormal collagen metabolism. These stimuli, coupled with the possible decreased susceptibility of collagen to degradation, may contribute to the wide variety of collagen-related complications noted in diabetes. Thus, overglucosylation of collagen appears to result in multiple metabolic and functional consequences in diabetes mellitus.

ACKNOWLEDGMENTS

This work was supported in part by the United States Public Health Service, National Institutes of Health grants AM-32637 and AM-28450 and the American Diabetes Association, Southern California Affiliate.

The author thanks Dr. Joseph R. Williamson for kindly providing the micrographs depicting skeletal muscle capillaries.

REFERENCES

1. J. Steinke and J. S. Soeldner, in *Harrison's Principles of Internal Medicine* (G. W. Thorn, R. D. Adams, E. Braunwald, K. J. Isselbacher, and R. G. Petersdorf, eds.), McGraw-Hill, New York, 1977, p. 563.
2. G. M. Grodsky, C. E. Anderson, D. L. Coleman, J. E. Craighead, G. C. Gerritsen, C. T. Hansen, L. Herberg, C. R. Howard, Jr., A. Lernmark, F. M. Matschinsky, E. Rayfield, W. J. Riley, and A. A. Rossini, *Diabetes, 31*(Suppl. 1):45 (1982).
3. M. Brownlee and A. Cerami, *Ann. Rev. Biochem., 50*:385 (1981).
4. H. L. Little and A. H. Sacks, *Trans. Am. Acad. Ophthalmol. Otolaryngol., 83*:522 (1977).
5. H. Schmid-Schönbein and E. Volger, *Diabetes, 25*(Suppl.2):897 (1976).
6. J. Ditzel and E. Standl, *Acta Med. Scand., 278*(Suppl.):49 (1975).
7. H. F. Bunn, K. H. Gabbay, and P. M. Gallop, *Science, 200*:21 (1978).
8. E. Standl and H. J. Kolb, *Diabetologia, 9*:461 (1973).
9. R. J. Koenig, C. M. Peterson, C. Kilo, A. Cerami, and J. R. Williamson, *Diabetes, 25*:230 (1976).

10. J. D. Bagdade, M. Stewart, and E. Walters, *Diabetes, 27*:677 (1978).
11. J. A. Colwell, P. V. Halushka, K. E. Sarji, and J. Sagel, *Med. Clin. North Am., 62*: 753 (1978).
12. D. Bensoussan, S. Levy Toledano, P. Passa, J. Caen, and J. Canivet, *Diabetologia, 11*: 307 (1975).
13. L. Monnier, G. Follea, and J. Mirouze, *Horm. Metab. Res., 10*:470 (1978).
14. L. O. Almer M. Pandolfi, and I. M. Nilsson, *Diabetes, 24*:529 (1975).
15. K. M. West, *Epidemiology of Diabetes and Its Vascular Lesions,* Elsevier North-Holland, New York, 1978.
16. J. Jarret, *Clin. Endocrinol. Metab., 6*:389 (1977).
17. P. Kimmelstiel and C. Wilson, *Am. J. Pathol., 12*:83 (1936).
18. B. V. Howard, P. J. Savage, L. J. Bennion, and P. H. Bennett, *Atherosclerosis, 30*: 153 (1978).
19. R. Ross and J. A. Glomset, *Science, 180*:1332 (1973).
20. R. Ross and L. Harker, *Science, 193*:1094 (1976).
21. C. E. Mogensen, R. Osterby, and H. J. G. Gunderson, *Diabetologia, 17*:71 (1979).
22. R. Rasch, *Diabetologia, 16*:319 (1979).
23. R. Rasch, *Diabetologia, 17*:243 (1979).
24. R. G. Price and R. G. Spiro, *J. Biol. Chem., 252*:8597 (1977).
25. V. J. Stevens, C. A. Rouzer, V. M. Monnier, and A. Cerami, *Proc. Natl. Acad. Sci. USA, 75*:2918 (1978).
26. H. L. Little, R. L. Jack, A. Patz, and P. H. Forsham, eds., *Diabetic Retinopathy,* Theime-Stratton, Inc., New York, 1983.
27. J. G. Cunha-Vaz, J. R. Fonseca, J. R. deAbreau, and J. J. Lima, *Arch. Ophthalmol., 96*:809 (1978).
28. J. G. Cunha-Vaz, J. R. Faria deAbreau, A. J. Campos, and G. M. Figo, *Br. J. Ophthalmol., 59*:649 (1975).
29. S. Waltman, T. Krupin, S. Hanish, C. Oestrich, and B. Becker, *Arch. Ophthalmol., 96*:878 (1978).
30. R. S. Clements, *Diabetes, 28*:604 (1978).
31. J. Jakobsen, *Diabetologia, 14*:113 (1978).
32. R. J. Graf, J. B. Halter, E. Halar, and D. Porte, *Ann. Intern. Med., 90*:298 (1979).
33. J. S. Chopra, B. B. Sawhney, and R. N. Chakravorty, *J. Neurol. Sci., 32*:53 (1977).
34. S. Yagihashi, K. Kudo, and M. Nishihira, *Tohoku J. Exp. Med., 127*:35 (1979).
35. W. R. Timperley, E. Williams, J. D. Ward, and F. E. Preston, *Diabetologia, 13*:437 (1977).
36. R. R. Holman, T. L. Dornan, V. Mayon-White, J. Howard-Williams, C. Orde-Peckar, L. Jenkins, J. Steemson, R. Rolfe, B. Smith, D. Barbour, K. McPherson, P. Poon, C. Rizza, J. I. Mann, A. H. Knight, A. J. Bron, and R. C. Turner, *Lancet, 1*:204 (1983).
37. D. A. Greene, P. V. DeJesus, and A. I. Winegrad, *J. Clin. Invest., 55*:1326 (1975).
38. A. C. Huntley, *J. Am. Acad. Dermatol., 7*:427 (1982).
39. R. Fleischmajer and J. V. Lara, *Arch. Dermatol., 92*:643 (1965).
40. R. Fleischmajer, G. Faludi, and S. Krol, *Arch. Dermatol., 101*:21 (1970).
41. B. A. Cohn, C. E. Wheeler, and R. A. Briggaman, *Arch. Dermatol., 101*:27 (1970).
42. G. W. Cole, S. J. Handler, and K. Burnett, *J. Clin. Ultrasound, 9*:501 (1981).
43. A. Perejda, G. Cole, L. Anderson, and J. Uitto, *Clin. Res., 31*:149A (1983).
44. G. W. Cole, J. Headley, and R. Skowsky, *Diab. Care 6*:189 (1983).
45. A. Grgic, A. L. Rosenbloom, F. T. Weber, B. Giordano, J. I. Malone, and J. J. Shuster, *J. Pediatr., 88*:584 (1976).
46. A. L. Rosenbloom, J. H. Silverstein, D. C. Lezotte, K. Richardson, and M. McCallum, *N. Engl. J. Med., 305*:191 (1981).

47. A. L. Rosenbloom, J. H. Silverstein, D. C. Lezotte, W. J. Riley, and N. K. Maclaren, *J. Pediatr., 101*:874 (1982).
48. J. R. Seibold, *Arthr. Rheum., 25*:1357 (1982).
49. B. A. Buckingham, J. Uitto, C. Sandborg, T. Keens, T. Roe, G. Costin, F. Kaufman, B. Bernstein, B. Landing, and A. Castellano, *Diab. Care 7,*:163 (1984).
50. B. A. Buckingham, A. J. Perejda, C. Sandborg, J. Errett, A. Kershnar, and J. Uitto, in *Pediatric and Adolescent Endocrinology*: The 6th International Beilinson Symposium on Future Trends in Juvenile Diabetes (in press).
51. P. B. Costello, P. K. Tambar, and F. A. Green, *J. Rheumatol., 11*:62 (1984).
52. J. R. Seibold, J. Uitto, B. B. Dorwart, and D. J. Prockop, *Arthr. Rheum., 27*:586 (1984).
53. L. S. Lieberman, A. L. Rosenbloom, W. J. Riley, and J. H. Silverstein, *N. Engl. J. Med., 303*:940 (1980).
54. P. Raskin, A. O. Pietri, R. Unger, and W. J. Shannon, Jr., *N. Engl. J. Med., 309*:1546 (1983).
55. A. E. Puklin, W. V. Tamborlane, P. Felig, M. Genel, and R. S. Sherwin, *Am. Acad. Ophthalmol., 89*:735 (1982).
56. W. V. Tamborlane, J. E. Puklin, M. Bergman, C. Verdonk, M. C. Rudolf, P. Felig, M. Genel, and R. Sherwin, *Diab. Care, 5*:58 (1982).
57. Kroc Collaborative Study Group, *Diabetes, 32*(Suppl.1): 10A (1983).
58. D. W. Allen, W. A. Schroeder, and J. Balog, *J. Am. Chem. Soc., 80*:1628 (1958).
59. P. J. Higgins and H. F. Bunn, *J. Biol. Chem., 256*: 5204 (1981).
60. R. Flückiger and K. H. Winterhalter, *FEBS Lett., 71*:356 (1976).
61. K. M. Spicer, R. C. Allen, D. Hallett, and M. G. Buse, *J. Clin. Invest., 64*:40 (1979).
62. A. J. Perejda, E. J. Zaragoza, E. Eriksen, and J. Uitto, *Coll. Rel. Res., 4*:427 (1984).
63. R. Dolhöfer and O. H. Wieland, *Clin. Chim. Acta, 112*:197 (1981).
64. S. H. Chiou, L. T. Chylack, W. H. Tung, and H. F. Bunn, *J. Biol. Chem., 256*:5176 (1981).
65. S. P. Robins and A. J. Bailey, *FEBS Lett., 38*:334 (1974).
66. C. E. Guthrow, M. A. Morris, J. F. Day, S. R. Thorpe, and J. W. Baynes, *Proc. Natl. Acad. Sci. (USA), 76*:4258 (1979).
67. R. Dolhöfer and O. H. Wieland, *FEBS Lett., 103*:282 (1979).
68. K. F. McFarland, E. W. Catalano, J. F. Day, S. R. Thorpe, and J. W. Baynes, *Diabetes, 28*:1011 (1979).
69. M. L. Tanzer, R. Fairweather, and P. M. Gallop, *Arch. Biochem. Biophys., 151*:137 (1972).
70. J. Uitto, A. J. Perejda, G. A. Grant, E. A. Rowold, C. Kilo, and J. R. Williamson, *Conn. Tiss. Res., 10*:287 (1982).
71. B. A. McVerry, S. Thorpe, F. Joe, P. Gaffney, and E. R. Huehns, *Haemostasis, 10*: 261 (1981).
72. M. Brownlee, H. Vlassara, and A. Cerami, *Diabetes, 32*:680 (1983).
73. B. Gonen, J. Baenziger, G. Schonfeld, D. Jacobsen, and P. Farrar, *Diabetes, 30*:875 (1981).
74. J. L. Witzum, M. Fisher, T. Pietro, U. P. Steinbrecher, and R. L. Elam, *Diabetes, 31*:1029 (1982).
75. J. L. Witzum, E. M. Mahoney, M. J. Branks, M. Fisher, R. Elam, and D. Steinberg, *Diabetes, 31*:283 (1982).
76. A. J. Bailey, S. P. Robins, and M. J. A. Tanner, *Biochim. Biophys. Acta, 434*:51 (1976).
77. J. A. Miller, E. Gravallese, and H. F. Bunn, *J. Clin. Invest., 65*:896 (1980).
78. B. W. Vogt, E. D. Schleicher, and O. H. Wieland, *Diabetes, 31*:1123 (1982).
79. S. K. Williams, N. L. Howarth, J J. Devenny, and M. W. Bitensky, *Proc. Natl. Acad. Sci. (USA), 79*:6546 (1982).

80. N. H. Ansari, Y. L. Awasthi, and S. K. Srivastava, *Exp. Eye Res., 31*:9 (1980).
81. J. Sasaki and G. L. Cottam, *Biochim. Biophys. Acta, 713*:199 (1982).
82. J. Sasaki, T. Okamura, and G. L. Cottam, *Eur. J. Biochem., 131*:535 (1983).
83. U. A. Kesäniemi, J. L. Witzum, and U. P. Steinbrecher, *J. Clin. Invest., 71*:950 (1983).
84. A. R. Bassiouny, H. Rosenberg, and T. L. McDonald, *Diabetes, 32*:1182 (1983).
85. N. A. Kefalides, *Coll. Rel. Res., 1*:295 (1981).
86. R. Osterby and H. J. G. Gunderson, *Diabetologia, 18*:493 (1980).
87. J. R. Williamson, N. J. Vogler, and C. Kilo, *Acta Diabet. Lat., 8*(Suppl.1):117 (1971).
88. J. R. Williamson, N. J. Vogler, and C. Kilo, in *Proceedings of the 4th International Congress of Endocrinology* (R. O. Scow, ed.), Excerpta Medica, Amsterdam, 1973, p. 1122.
89. J. R. Williamson, N. J. Vogler, and C. Kilo, *Am. J. Pathol., 63*:359 (1971).
90. J. R. Williamson and C. Kilo, *Diabetes, 25*(Suppl.2):925 (1976).
91. H. H. Parving and N. Rossing, *Scand. J. Clin. Lab. Invest., 32*:239 (1973).
92. M. E. Murphy and P. C. Johnson, *Microvasc. Res., 9*:242 (1975).
93. D. T. Woodley, this volume, Chapter 6.
94. N. G. Westberg and A. F. Michael, *Acta Med. Scand., 194*:39 (1973).
95. N. A. Kefalides, *J. Clin. Invest., 53*: 403 (1974).
96. P. J. Beisswenger and R. G. Spiro, *Diabetes 22*:180 (1973).
97. A. LePape, J-D. Guitton, and J-P. Muh, *Biochem. Biophys. Res. Commun., 100*: 1214 (1981).
98. M. P. Cohen, E. Urdanivia, M. Surma, and C. J. Ciborowski, *Diabetes, 30*:367 (1981).
99. M. E. Grant, R. Harwood, and I. F. Williams, *J. Physiol., 257*:56P (1976).
100. M. P. Cohen, A. Dasmahapatra, and V-Y. Wu, *Nephron, 27*:146 (1981).
101. M. Brownlee and R. G. Spiro, *Diabetes, 28*:121 (1979).
102. M. P. Cohen, M. L. Surma, and V-Y. Wu, *Am. J. Physiol., 242*:F385 (1982).
103. K. G. McCullagh and G. Balian, *Nature, 258*:73 (1975).
104. K' G. McCullagh, V. C. Duance, and K. A. Bishop, *J. Pathol., 130*:45 (1980).
105. R. K. Rhodes, in *Collagen in Health and Disease* (J. B. Weiss and M. I. V. Jayson, eds.), Churchill Livingstone, Edinburgh, 1982, p. 376.
106. H. Rosenberg, J. B. Modrak, J. M. Hassing, W. A. Al-Turk, and S. J. Stohs, *Biochem. Biophys. Res. Commun., 91*:498 (1979).
107. B. Haider, C-K. Yeh, G. Thomas, H. A. Oldewurtel, M. M. Lyons, and T. J. Regan, *Trans. Assoc. Am. Physicians, 91*:197 (1978).
108. T. J. Regan, C. F. Wu, C-K. Yeh, H. A. Oldewurtel, and B. Haider, *Circ. Res., 49*: 1268 (1981).
109. J. Modrak, *Diabetes, 29*:547 (1980).
110. M. P. Cohen, C. J. Ciborowski, and M. L. Surma, *Diabetes 31,*:1084 (1982).
111. T. Pihlajaniemi, R. Myllylä, K. I. Kivirikko, and K. Tryggvason, *J. Biol. Chem., 257*:14914 (1982).
112. R. G. Spiro,and M. J. Spiro, *Diabetes, 20*:641 (1971).
113. A. Khalifa and M. P. Cohen, *Biochim. Biophys. Acta, 386*:332 (1975).
114. M. P. Cohen and A. Khalifa, *Biochim. Biophys. Acta, 496*:88 (1977).
115. M. Sternberg and J. Andre, *FEBS Lett., 139*:53 (1982).
116. M. Sternberg and R. G. Spiro, *J. Biol. Chem., 54*:10329 (1979).
117. A. Y. Chang, *Biochim. Biophys. Acta, 522*:491 (1978).
118. H. Fushimi and S. Tarvi, *J. Biochem., 79*:265 (1976).
119. Ch. Hasslacher, R. Reichenbacher, F. Gechter, and R. Timpl, *Diabetologia, 26*:150 (1984).

120. V-J. Uitto, D. Schwartz, and A. Veis, *Eur. J. Biochem., 105*:409 (1980).
121. C. A. Vater, E. D. Harris, and R. C. Siegel, *Biochem. J., 181*:639 (1979).
122. G. Lubec and A. Pollak, *Renal Physiol., 3*:4 (1980).
123. B. D. Smith and C. K. Silbert, *Biochem. Biophys. Res. Commun., 100*:275 (1981).
124. M. J. O. Francis, J. P. Ellis, and T. D. R. Hockaday, *Proc. R. Soc. Med., 67*:35 (1974).
125. S. L. Schnider and R. R. Kohn, *J. Clin. Invest., 67*:1630 (1981).
126. S. L. Schnider and R. R. Kohn, *Exp. Gerontol., 17*:185 (1982).
127. K. Chang, J. Uitto, E. A. Rowold, G. A. Grant, C. Kilo, and J. R. Williamson, *Diabetes, 29*:778 (1980).
128. L. M. Golub, R. A. Greenwald, E. J. Zebrowski, and N. S. Ramamurthy, *Biochim. Biophys. Acta, 534*:73 (1978).
129. A. M. Madia, S. J. Rozovski, and H. M. Kagan, *Biochim. Biophys. Acta, 585*:481 (1979).
130. C. R. Hamlin, R. R. Kohn, and J. H. Luschin, *Diabetes,24*:902 (1975).
131. S. L. Schnider and R. R. Kohn, *J. Clin. Invest., 66*:1179 (1980).
132. T. T. Andreassen, K. Seyer-Hansen, and A. J. Bailey, *Biochim. Biophys. Acta, 677*: 313 (1981).
133. S. F. Yosha, H. R. Elden, A. Rabinovitch, D. H. Mint, and R. J. Boucek, *Diabetes, 32*:739 (1983).
134. D. K. Yue, S. McLennan, L. Delbridge, D. J. Handelsman, T. Reeve, and J. R. Turtle, *Diabetologia, 24*:282 (1983).
135. A. Lepape, J-P. Muh, and A. J. Bailey, *Biochem. J., 197*:405 (1981).
136. V. M. Monnier, V. J. Stevens, and A. Cerami, *J. Exp. Med., 150*: 1098 (1979).
137. V. M. Monnier, R. R. Kohn, and A. Cerami, *Proc. Natl. Acad. Sci. (USA), 81*:583 (1984).
138. A. Cerami, V. J. Stevens, and V. M. Monnier, *Metabolism 28*(Suppl.1):431 (1979).
139. R. Bucala, P. Model, and A. Cerami, *Proc. Natl. Acad. Sci. (USA), 81*:105 (1984).
140. A. J. Perejda and J. Uitto, *Coll. Rel. Res., 2*:81 (1982).
141. M. Brownlee, S. Pongor, and A. Cerami, *J. Exp. Med., 158*:1739 (1983).
142. N. A. Kefalides, in *Biology and Chemistry of Basement Membranes* (N. A. Kefalides, ed.), Academic Press, New York, 1978, p. 215
143. R. Pelkonen and K. I. Kivirikko, *N. Engl. J. Med., 283*:451 (1970).
144. K. I. Kivirikko, *Int. Rev. Connective Tiss. Res., 5*:93 (1973).

22

Diseases of Glycosaminoglycan and Proteoglycan Metabolism

Arvan L. Fluharty
MRRC–UCLA Research Group at Lanterman Developmental Center,
University of California, Los Angeles,
Los Angeles, California

I. INTRODUCTION

Acidic polysaccharides with their high hydrodynamic volumes constitute one of the critical elements of most connective tissues. They are thought to control hydration, compressibility, and diffusion of soluble nutrients, and to influence the organization of fibrous elements. In most cases, the acidity of the polymer is imparted by uronic acid residues and/or sulfate substituents. The major animal polysaccharides are linear repeats of a uronic acid-amino sugar disaccharide. These acidic polysaccharides are the glycosaminoglycans (GAGs), formerly known as mucopolysaccharides or acid mucopolysaccharides. Most GAG do not occur as such in organized connective tissue; rather, the oligosaccharide chains are attached to a core protein that may have multiple GAG attachment sites. The number, size, and arrangement of substituents will vary from tissue to tissue within certain limits. These proteoglycans (PG) are further organized into aggregates by noncovalent associations of varying degrees of strength and specificity. The subtleties of PG composition and organization are only beginning to be worked out, but it is clear that these supermacromolecules play an important role in the organization and structural properties of connective tissue. Moreover, aberrations in their metabolism are associated with a diverse range of pathological processes. Details on the structure and diversity of the PG and their constituent GAG are considered elsewhere in this volume (Chapter 4). In addition, a number of excellent chapters and books on these macromolecules have appeared in the past few years (1-6).

The most detailed picture of metabolic processes involved with this class of macromolecules has come from studies on a group of human genetic disorders usually referred to as the mucopolysaccharidoses. Over the past 15-20 years, much has been learned about specific enzyme deficiencies involved in the mucopolysaccharidoses, and, from this, much has been inferred about the catabolism of GAG. Consideration of the pathological consequences of faulty GAG catabolism has permitted some insight into the roles these macromolecules must play in normal connective tissue structure and regulation of cellular interactions. Much of this chapter will be devoted to the mucopolysaccharidoses and closely related lysosomal enzyme defects.

II. REVIEW OF GLYCOSAMINOGLYCAN/ PROTEOGLYCAN STRUCTURES

While the structures of GAGs and proteoglycans are discussed in Chapter 4, a brief review of certain generalizations will probably be helpful. The simplest substructure characteristic of the GAG is a disaccharide-repeating unit. The disaccharide subunits and their specific pattern of sulfate substitution define six or seven GAG subtypes generally recognized in animal tissues, as summarized in Figure 1. These can be grouped into four biosynthetic subfamilies, the two chondroitin sulfates (C4S and C6S) and dermatan sulfate (DS) being related, as are heparin (Hep) and heparan sulfate (HS). Alternatively, two degradative subgroupings can be designated based on susceptibility to endolytic cleavage by tissue hyaluronidases (endo-*β*-*N*-acetylhexosaminidases). It is the hyaluronidase resistant subtypes (DS, KS, and HS/Hep) that accumulate in most mucopolysaccharide storage diseases and are therefore associated with characteristic patterns of connective tissue abnormalities.

The GAG themselves are linear chains of disaccharide units of variable length (typically 20-60 disaccharide repeats). Mixed chains between biosynthetic subtypes (i.e., KS and HS) have not been reported, but mixed chains within a biosynthetic subtype may be fairly common (i.e., C4S-DS copolymers). GAG chains are attached to protein cores through special linkage sequences. A given core protein, and there appear to be a number of types, has multiple GAG attachment points, and GAG of more than one biosynthetic chain type may be attached to a single protein core. The most studied PG, that of cartilage, may have more than 100 CS and KS chains, with the ratio of the two chain types varying with age (2). In contrast, a DS containing PG from ovarian granulosa cells has only two to six GAG chains (7). In addition, numerous *N*-linked and *O*-linked oligosaccharides similar to those found in glycoproteins are also present on protein cores. Extensive phosphorylation of both protein backbone and sugars in the GAG linkage region has recently been reported (8). The HA and Hep normally present in tissues are not attached to protein, although it is generally believed that they are synthesized as PG and subsequently processed to protein-free forms.

Higher-order structures form outside the cell by way of nonconvalent interactions. The best documented example is the association of the major cartilage PG with HA (6). This interaction occurs through a specific HA-binding domain of the core protein. The PG-HA interaction is further stabilized by an additional small glycoprotein, the so-called link protein. Other PG types interact with cell surface sites, bind to structural proteins such as fibronectin, or self-associate into larger macromolecular complexes. Ultimately, PG containing aggregates become enveloped in, or direct the organization of, the fibrous elements of connective tissue. The hierarchy of structures involving GAG and PG are summarized in Figure 2.

III. GENERAL METABOLIC PATTERN FOR GAG AND PG

Recent advances suggest a general model for PG biosynthesis and degradation. This "pathway" is only a rough approximation and certainly does not encompass all the biological

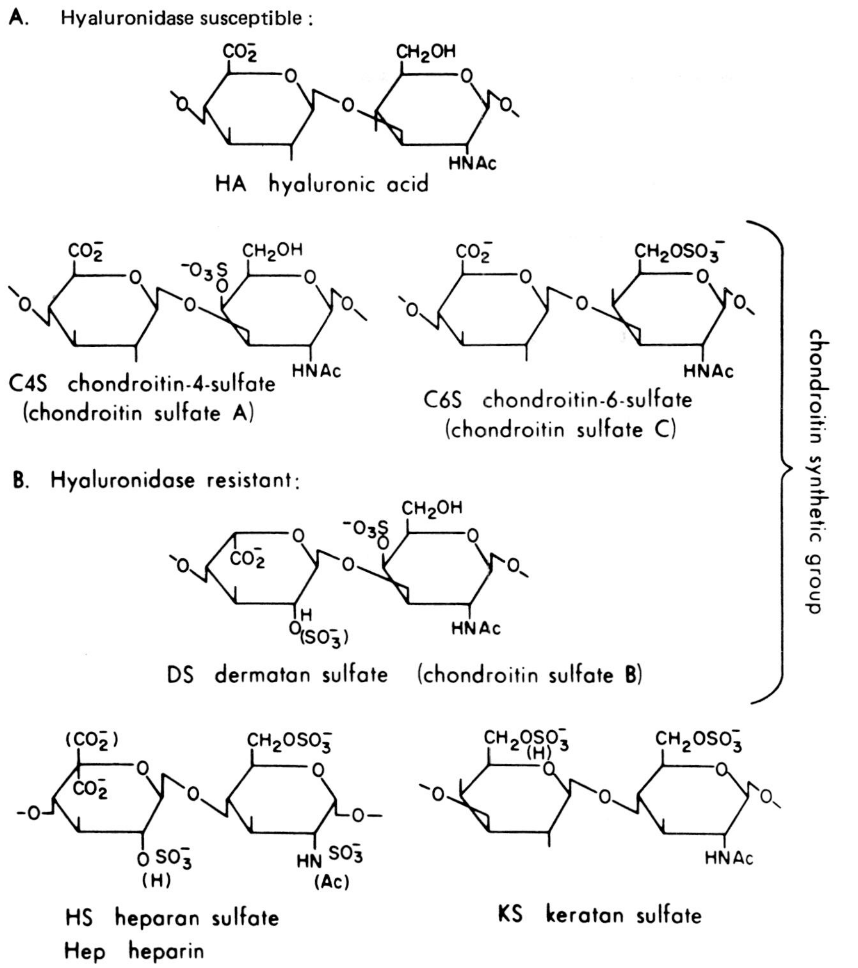

Figure 1 Glycosaminoglycan dimeric repeat units. Parenthetical structures indicate common alternate substituents.

roles that this class of molecules has acquired. The pathway is largely based on structures of cartilaginous connective tissue and the assembly of glycosylated excretory proteins (9). Enough information is available from other tissues to suggest that the pathway is generally valid, although much remains to be learned about the sequence of processing reactions and specific subcellular sites.

All GAG are probably assembled while attached to protein. Ultimate control of synthesis is manifest through the rates of core protein production and the activity of various assembly enzymes. Prevention of core protein synthesis will eventually result in cessation of GAG production. Over the short run, however, oligosaccharide assembly is independent of protein synthesis, and available stocks of core protein and unfinished polysaccharide chains permit GAG assembly to proceed in the presence of protein synthesis inhibitors. There is no evidence at present for the preassembly of blocks of oligosaccharide attached to a nucleoside or lipoidal intermediate, as is the case for bacterial cell wall and asparagine-linked glycoprotein assembly.

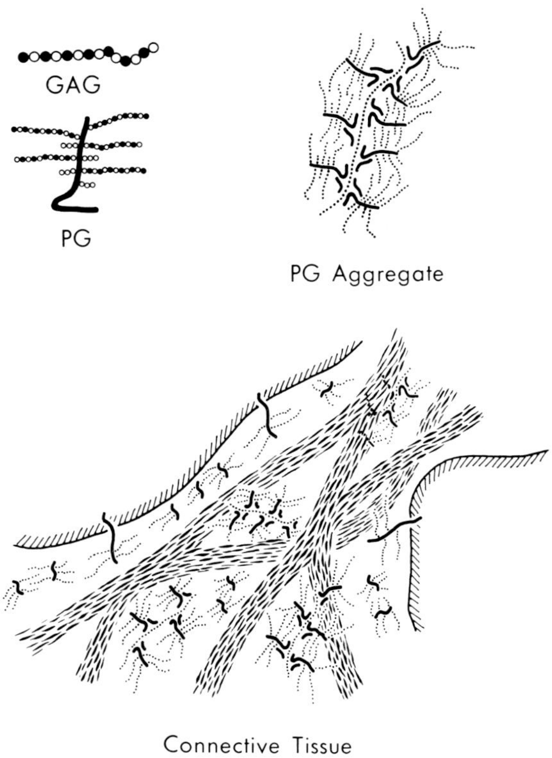

Figure 2 Hierarchy of glycosaminoglycan structures.

Core proteins may have from a few to more than 100 GAG attachment sites, depending on the specific type of PG. Linkage is typically between the hydroxyl group of a serine (or threonine) and D-xylose (KS is an exception). This critical coupling is promoted by a xylosyl transferase located within the cisternal space of the rough endoplasmic reticulum or the Golgi complex. (Glycoprotein and mucin-type oligosaccharides are also attached as *N*-asparagine and *O*-serine/threonine derivatives within the same subcellular regions.) A characteristic digalactoside sequence is then added to the xylose. Next, the repeating GAG disaccharide is assembled by an alternating action of transferase enzymes. Sulfation and inversion of D-glucouronic acid to L-iduronic acid occurs subsequent to chain assembly. Although chain modifications and oligosaccharide assembly may be occurring on different parts of the same chain in some cases, there is evidence for a highly-ordered sequence of events in Hep synthesis (10). Factors controlling length of GAG chains are unclear at present. After assembly, there may be some subsequent trimming of saccharide chains and/or protein core modifications. This may continue after excretion of the completed macromolecule into the extracellular space. Considerable heterogeneity can, therefore, exist in the final population of macromolecules incorporated into the connective tissue matrix.

PG core proteins often have distinct functional regions or domains, only one of which is rich in GAG attachment sites. Other domains may be involved in interaction with cell walls or with macromolecules of the extracellular milieu. The presence of such special-

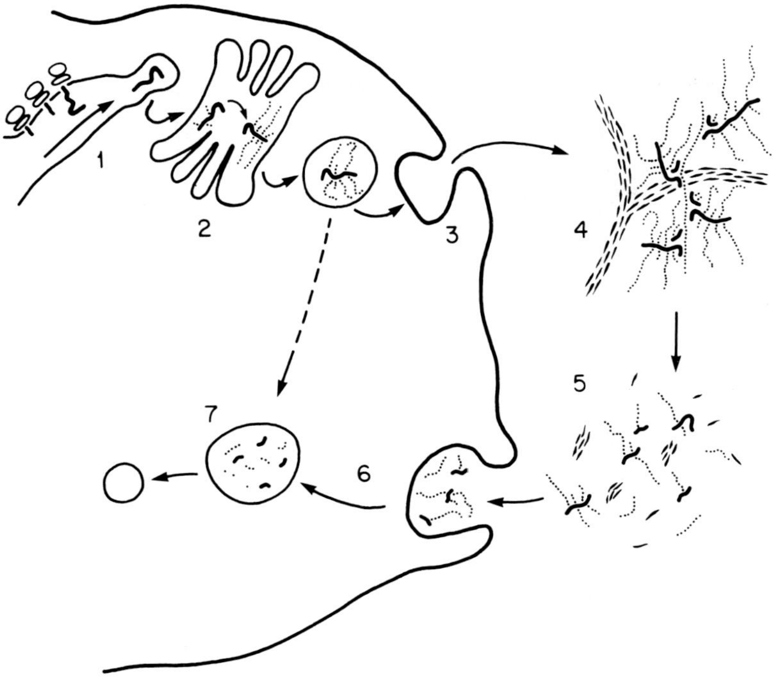

Figure 3 Summary of GAG/PG metabolism. Core proteins are synthesized in the RER and secreted into the cisternal space (1). The protein is transported to the Golgi, where glycoprotein oligosaccharides are added and GAG chains assembled (2). Completed PG are trimmed and excreted (3). Aggregates form and become integrated into connective tissue (4). Eventually, PG are nicked and fragmented by extracellular enzymes (5) and pieces taken up into cells by endocytosis (6). In secondary lysosomes, the polymers are degraded to monomolecular substituents.

ized regions allows the newly secreted PG to organize into specific large macromolecular aggregates. Additional core protein domains interact with link protein, fibronectin, or other organizing principles of the extracellular milieu.

The highly hydrated PG aggregates are held in a compressed conformation by a meshwork of fibrous proteins such as collagen and elastin. The fine structure of fibrous proteins and PG aggregates are highly interdependent, each interacting with the other in a variety of specific and nonspecific ways. The highly charged PG complexes are capable of binding enormous quantities of water and would expand to immense molecular dimensions if uncontained. By ordering water structure, they restrict solvent flow and control the diffusion of nutrients and salts. The end result is a spectrum of connective tissue types with diverse structural and metabolic properties. The relationship between structure and function of GAG, PG, and complex aggregates has been considered by Comper and Laurent (1).

The breakdown of matrix structures appears to be initiated by glycosidases and proteases acting pretty much at random throughout the meshwork (11). Actually, very little is known about the details of the primary extracellular events, particularly during normal matrix turnover. Eventually, the accumulating breaks result in disaggregation of the matrix into large PG fragments, which can then be taken up by phagocytic cells. Once within

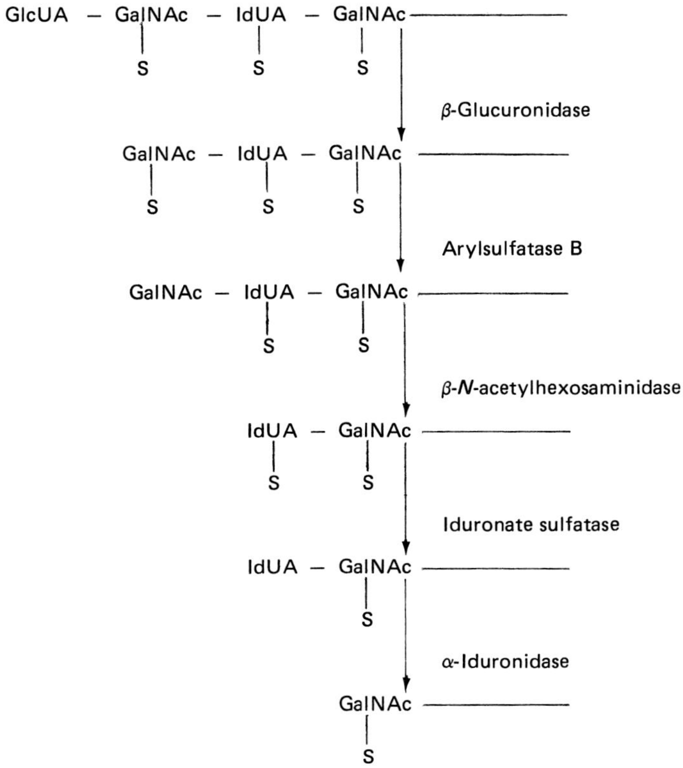

Figure 4 Sequential DS breakdown.

the cell, they enter lysosomes for further digestion. The initial stages of the lysosomal degradation probably involve random endolytic attack on both polysaccharide and core protein remnants—a continuation of the extracellular events. At some stage of disaggregation, the disassembly of the GAG fragments by a stepwise exolytic pathway becomes dominant. Sulfate and sugar residues are removed sequentially from the nonreducing end of the oligosaccharide fragments. The overall sequence of events in GAG and PG turnover is summarized in Figure 3.

Endolytic enzymes are not present for all GAG chains. Extensive fragments of DS, KS, and HS must be disassembled by the sequential exolytic pathways. In these cases, the failure of any single step prevents further breakdown. Much of our information concerning these final exolytic sequences has been inferred from studies on the human storage diseases. Figures 4 through 6 depict our present ideas on the minimal sequence of enzymatic steps needed for the exolytic hydrolysis of DS, KS, and HS polymers.

IV. DISEASES OF PROTEOGLYCAN METABOLISM

The mucopolysaccharidoses are a group of human genetic conditions associated with the storage and urinary excretion of abnormal quantities of GAG. Clinically, the hallmarks of the mucopolysaccharidoses have been coarse features, skeletal malformation, and anomalies in connective tissue-rich structural elements. Severe mental impairment is encountered

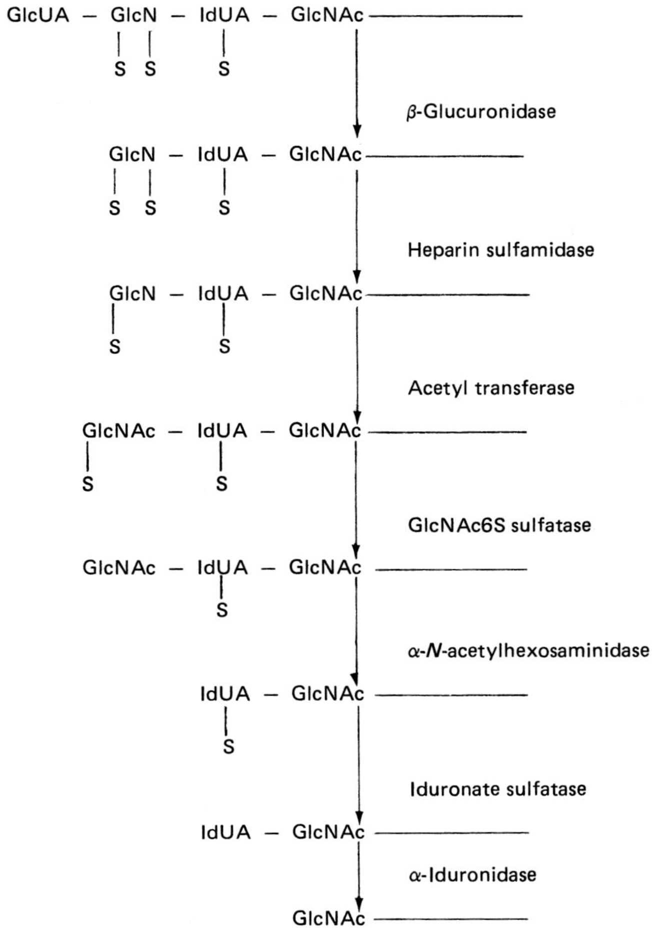

Figure 5 Sequential Hep/HS breakdown.

in some, but not all, of these conditions. A number of names have been associated with this group of disorders over the years, with gargoylism, lipochondrodystrophy, dysostosis multiplex, and Hurler (or Hurler/Hunter) disease being the most noteworthy. A number of recent reviews treat the mucopolysaccharidoses from several different prospectives (11-17).

In recent years a rapid increase has occurred in our understanding and differentiation of these diseases at the biochemical level. Of germinal importance was the demonstration that GAG accumulated in the tissues and were excreted in the urine of such patients. Moreover, the types of GAG being stored or excreted could be used to differentiate among a number of clinical subtypes. This led McKusick et al. to propose a widely adopted classification system for the mucopolysaccharidoses (18), which greatly facilitated the subsequent identification of a series of specific enzyme deficiencies. While knowledge of primary enzymatic defects has resulted in some modifications of the basic McKusick classification (12), it remains the most effective way for combining clinical, genetic, and biochemical information. While the organization scheme in Table 1 emphasizes specific enzyme anomalies, the basic McKusick arrangement has been maintained.

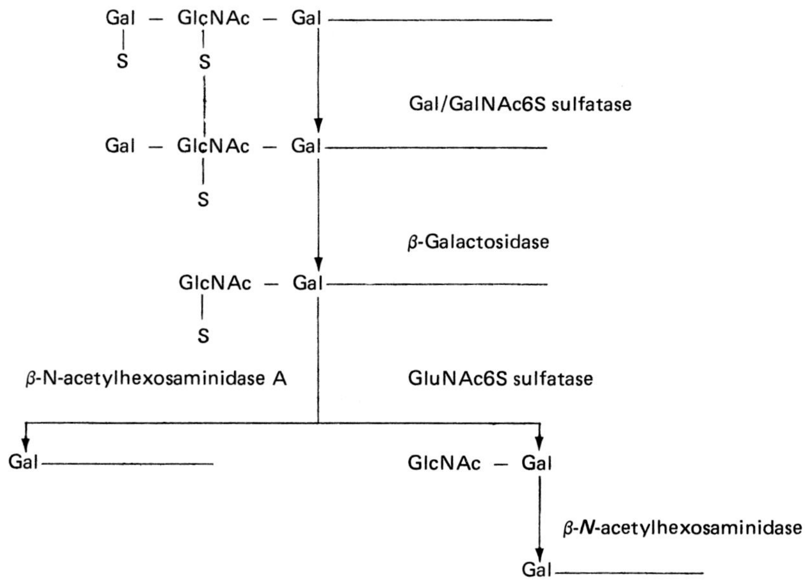

Figure 6 Sequential KS breakdown.

In addition to the true mucopolysaccharidoses—those conditions in which a primary enzyme defect occurs in a GAG degradative pathway and GAG storage and excretion are readily demonstrated—there are a number of clinically similar conditions where urinary GAG excretion is normal (19-23). Many of these have proven to be anomalies in glycoprotein oligosaccharide metabolism. The clinical similarities between these oligosaccharidoses (or mucolipidoses) and the mucopolysaccharidoses suggest caution in ascribing particular clinical features to the storage of any specific material. The importance of the glycoprotein portions of PG tends to be overlooked in most discussions of connective tissue structure and function. They may be as important as the GAG substituents in controlling matrix organization.

There appear to be few good estimates of the frequency of occurrence of the mucopolysaccharidoses and oligosaccharidoses. The most frequently quoted number is 1:100,000 for MPS I, which was estimated by Lowry and Renwick for the predominantly white population of British Columbia (24). Present enzymatic methods of diagnosis would certainly increase the yield of cases, particularly if all allelic and variant forms were included. However, experience suggests that enzymatic evaluation is unlikely to more than double clinically obvious cases. The frequency of MPS II has received renewed interest since Schaap and Bach suggested this condition might be enriched among Jews in Israel (25). About 1:150,000 was reported for British Columbia (24), 1:132,000 for the United Kingdom (26), and 1:67,500 for Israeli Jews (25). Chakravarti and Bale consider these differences to be within expectations for genetic drift and not to reflect selection, as had been suggested (27). Overall, the Sanfilippo syndrome may be more frequent than other types, with the A form somewhat more common than MPS III-B. MPS III-C and MPS III-D are rarer, although this may simply reflect their more recent recognition as distinct defects. In the Netherlands, van de Kamp et al. found an overall frequency of MPS

III of 1:24,000, with the A form accounting for about 50% of the cases, the B form about 30%, and the C form about 20% (28). This may represent a specific geographic pattern and not accurately reflect other populations (29).

From all of this, one can suggest that the overall frequency of the mucopolysaccharidoses/oligosaccharidoses group of diseases may be as high as 1:20,000. It is unlikely that the combined occurrence exceeds 1:10,000 since diseases of this frequency (e.g., PKU) are clearly more prevalent. Such considerations imply that the overall carrier frequency for the mucopolysaccharidoses is somewhere between 1:50 and 1:100, with specific enzyme genes being mutant in 1:150 to 1:500 individuals. No striking geographic, racial, or ethnic concentrations have been noted.

A. The Mucopolysaccharidoses

1. α-L-Iduronidase Deficiency

L-Iduronic acid is a rare sugar that occurs almost exclusively in GAG oligosaccharides. It is not incorporated, as such, into the growing oligosaccharide chain, but derives from D-glucuronic acid residues by an inversion in the linkage of the carboxylate carbon in the intact polysaccharide (30). The presence of L-iduronic acid differentiates DS from CS, and it occurs in varying abundance in Hep and HS chains. In DS, at least, it modifies the susceptibility of the chain to endolytic cleavage by hyaluronidase-type enzymes.

α-L-Iduronidase is the lysosomal enzyme responsible for removing iduronic acid end groups, and in its absence neither DS nor HS fragments can be fully catabolized. This is the situation in Hurler's syndrome (MPS I or MPS II-H), where α-L-iduronidase is completely deficient (31,32). Both DS and HS accumulate in tissues and are excreted in the urine. The eventual consequences are abnormal skeletal growth, dwarfing, cardiac anomalies, herniation, skin changes, hirsutism, corneal clouding, and mental retardation. At the cellular level, abundant intracellular inclusions can be found in most tissues (33). These stain metachromatically and are electron lucent or finely granular in electron microscopic preparations. A single limiting membrane surrounds the storage granules, and they are acid phosphatase-positive, implying that the structures are GAG-engorged lysosomes. In some tissues, particularly liver and brain, other types of storage granules are also frequent. These "zebra bodies" have characteristic lamellar structures. They are thought to result from secondary storage of membrane glycolipids that accumulate because of an inhibitory action of GAG on some lysosomal hydrolases (34-36). This secondary enzyme inhibition may be a general phenomenon in the mucopolysaccharidoses and increases the difficulty in ascribing a given clinical symptom to the accumulation of a given GAG.

Less severe forms of α-L-iduronidase deficiency occur in which some residual enzyme activity remains. One such condition, Scheie syndrome (MPS V or MPS I-S), was originally classified separately from Hurler syndrome on the basis of clearly different clinical manifestations. Scheie syndrome is now regarded as a late-onset form of MPS I resulting from an allelic mutation in the α-L-iduronidase structural gene (31). While certain mucopolysaccharidosis characteristics are present, such as coarse features, there is no mental retardation and limited skeletal deformation. Intermediate forms have been inferred to be Hurler/Scheie compounds (37). However, the occurrence of a series of allelic mutations resulting in enzymes of varying degrees of stability and/or catalytic efficiency is coming to be regarded as the norm for most enzyme deficiency conditions (38). Thus, the MPS I-H, MPS I-H/S, and MPS I-S constellation can be considered as a parallel to the infantile, juvenile, and adult subforms of other lysosomal enzyme diseases such as metachromatic

Table 1 Enzymatic Subtypes of the Mucopolysaccharidoses

Deficient enzyme	Syndrome eponym	McKusick subtype	GAG stored	Characteristics of principal interest
α-Iduronidase	Hurler	MPS I-H	DS, HS	Skeletal defects; coarse features; mental retardation; heart defects; hepatosplenomegaly; corneal opacity; death in the first decade
	Scheie	MPS I-S (MPS V)	DS, HS	Milder allelic form; normal intelligence; survival into adulthood
				Intermediate forms also reported
Iduronate sulfatase	Hunter	MPS II	DS, HS	X-linked; similar to Hurler but clear corneas and somewhat milder course; death in the second decade
				Milder allelic form with little or no intellectual impairment; survival into adulthood
Heparin *N*-sulfatase	Sanfilippo A	MPS III-A	HS	Progressive mental retardation; clear corneas; mild skeletal and connective tissue involvement; death in the second decade
α-*N*-acetylglucosaminidase	Sanfilippo B	MPS III-B	HS	Same as for Sanfilippo A
Lysosomal *N*-acetylase	Sanfilippo C	MPS III-C	HS	Same as for Sanfilippo A

N-acetylglucosamine-6 sulfatase	Sanfilippo D	MPS III-D	HS	Same as for Sanfilippo A
Galactose/*N*-acetyl-galactosamine-6-sulfatase	Morquio A	MPS IV-A	KS (may be absent)	Severe skeletal deformities; corneal clouding; normal intelligence; survival into adulthood
β-Galactosidase	Generalized gangliosidosis	(OS 4a-c)	KS	Skeletal abnormalities; coarse features; severe neurological defect; glycolipid storage; evident at birth or early infancy; early death
	GM_1 gangliosidosis			Late-onset forms with mild or absent skeletal manifestations
	Morquio B	MPS IV-B (OS 4d)	KS	Moderate skeletal defects; corneal clouding; normal intelligence; near-normal GM_1 ganglioside hydrolysis; survival into adulthood
Arylsulfatase B, *N*-acetylgalactosamine-4-sulfatase	Maroteaux-Lamy	MPS VI	DS	Skeletal deformities; coarse features; corneal clouding; normal intelligence; death in first decade
				Mild allelic form with survival into adulthood
β-Glucuronidase	Sly-Neufeld	MPS VIII	DS, HS	Skeletal deformities; mental retardation; few cases with clinical variability

leukodystrophy (39). The number of individual mutations affecting α-L-iduronidase is uncertain, but this question may soon be approachable by gene-mapping and DNA-sequencing techniques.

The recognition of α-L-iduronidase as the enzyme defect in MPS I exemplifies much of the recent history of the mucopolysaccharidoses. The fact that both DS and HS accumulate directed attention to features common to these two GAG types but absent from others. A β-galactosidase deficiency demonstrated in liver from mucopolysaccharidoses patients of different diagnostic types was therefore considered an unlikely primary defect since it could not explain the unique excretion patterns (40). A pivotal advance was a demonstration by Danes and Bearn of metachromatically staining inclusions in fibroblast cultures from mucopolysaccharidoses patients (41). Neufeld's group extended this observation by utilizing [^{35}S] sulfate as a marker for GAG in cultured cells (42). Accumulation of radiolabeled GAG in mucopolysaccharidoses fibroblasts was shown to be due to faulty degradation rather than enhanced synthesis, implicating a lysosomal enzyme defect. When cells from different diagnostic classifications were grown together, neither cell type showed its typical ^{35}S-GAG accumulation pattern (43). The factor(s) responsible for this metabolic correction could be transferred through the culture media and had the properties expected of proteins (44). It was also noted that cells from certain diagnostic categories would not correct each other, while cells within other categories could be differentiated into cross-correcting subgroups. Thus, the allelism of the Hurler and Scheie syndromes and the enzymatic heterogeneity of the Sanfilippo (MPS III) grouping (see below) was anticipated.

Another critical element in establishing α-L-iduronidase as the enzyme deficient in MPS I was the development of substrates to assay the enzyme. Neufeld's group had carried out a considerable purification of the Hurler corrective factor. Although they suspected that it was a lysosomal enzyme, it did not correspond to any of the then-assayable activities. Dorfman's group suggested an α-L-iduronidase defect and was able to support this idea with DS and HS oligosaccharide fragments rich in iduronic acid end groups (32). However, it was only after Friedman and Weissmann synthesized a phenyl-α-iduronide substrate (45) that the enzymatic activity of the corrective factor could be confirmed and the MPS I enzymatic defect readily assayed. More recently, methylumbelliferyl and oligosaccharide substrates have also become available for the assay of this enzyme (46). Still, the enzymatic diagnosis of this and several other mucopolysaccharidoses is only available in specialized laboratories.

It is also valuable to note that subsequent investigations on the cellular uptake of α-L-iduronidase and other lysosomal enzymes have provided substantial contributions to our understanding of cell surface recognition markers, lysosomal enzyme maturation, and the packaging of enzymes into their appropriate subcellular organelles (47).

2. *Iduronate Sulfatase Deficiency*

Hunter syndrome (MPS II) is unique among the mucopolysaccharidoses in having an X-linked mode of inheritance. While similar clinical features and identical GAG excretion patterns of Hurler and Hunter syndromes suggested closely related genetic diseases, the different inheritance patterns implied distinct primary enzymatic defects. A second distinctive structural feature shared by DS and HS is sulfation at the 2 position of iduronic acid. The majority of the iduronic acid residues in Hep and HS are sulfated, while only about 10% of those in DS carry this substituent. Nonetheless, the exolytic degradation of neither polymer type can proceed to completion without the action of a sulfatase to remove this sulfate residue. A defect in such an iduronate sulfatase has proven to be responsible for MPS II (48,49).

The *N*-sulfates of Hep are labile in mild acid, rendering the polymer susceptible to cleavage by nitrous acid (50). This reaction was exploited for the preparation of a disaccharide with a sulfated iduronic acid terminus (51). The introduction of a ^3H label into this material provided a convenient substrate for assaying iduronate sulfatase activity in small tissue samples and has allowed the enzyme deficiency in MPS II to be firmly established (52). However, this substrate for the assay of iduronate sulfatase is not commercially available, and the enzymatic diagnosis of this disorder is available in only a small number of laboratories.

Clinically, Hunter syndrome is much like Hurler syndrome. Subtle differences such as the lack of corneal clouding have been emphasized in distinguishing the two mucopolysaccharidosis subtypes. However, genetic criteria were the only certain means of differentiation before cross-correction and specific enzyme diagnoses became available. The close similarity of phenotypes does suggest that it is the common storage products and not the specific enzyme defect per se which produces the characteristic patterns of connective tissue defects, skeletal growth abnormalities, and intellectual insufficiency.

There is clinical heterogeneity within the MPS II-iduronate sulfatase deficiency classification. Mild and severe variants of the Hunter syndrome have been described, with the particular subtype usually "breeding true" within a given family (53). However, reports of significant clinical variation within an occasional family do suggest that other aspects of the specific genetic or environmental milieu can modify the effect of a low or absent iduronate sulfatase activity (54).

The occurrence of iduronate sulfatase deficiency in female mucopolysaccharidosis patients has been reported, a phenomenon that has not been adequately explained (55). A new mutation in the paternal X chromosome and a carrier mother could result in a homozygous female. Alternatively, an extremely unequal degree of X-chromosome inactivation in a heterozygous female could be invoked. In either circumstance, typical Hunter syndrome would be expected in male relatives. Conversely, an autosomal recessive form of the defect might be considered for isolated cases. Perhaps a defect in a secondary locus such as is known or suspected in I-cell and multiple sulfatase deficiency diseases will provide an explanation.

3. *Heparin Sulfamidase Deficiency*

HS storage and excretion is the unique pathochemical feature in MPS III or the Sanfilippo syndrome. GAG in the HS biosynthetic group have two distinctive structural features. One is an α rather than a β linkage between the amino sugar and the adjacent uronic acid; the other is the presence of a sulfate rather than an acetate substituent on nitrogen in a portion of the amino sugar residues. Enzyme defects are associated with both features. One of these involves an enzyme specific for the hydrolysis of the amino sugar sulfamide. This so-called heparin sulfamidase or heparin *N*-sulfatase is absent in a subgroup of patients with MPS III (29).

The substrate most commonly employed for the GAG sulfamidase assay has been a commercially available [^{35}S] Hep. All of the *N*-sulfate residues in the molecule have been replaced with radiolabeled sulfate resulting in both terminal and interior hexosamine residues being labeled. As a consequence, only 2-5% of the radioactivity in the molecule is directly susceptible to removal by the sulfamidase (56). This serves to substantially restrict the sensitivity of the assay. Moreover, the half-life of ^{35}S is only three months, necessitating frequent replacement of substrate. These properties limit availability of the assay, and relatively little work has been carried out on the heparin sulfamidase enzyme itself (57). Recently, oligosaccharide substrates for this enzyme have been introduced, which should alleviate some of these problems (58).

Clinically, MPS III-A is not distinguishable from other types of the Sanfilippo syndrome, although there may be a somewhat earlier onset (28). All share an inability to degrade HS, resulting in tissue accumulation and urinary excretion. Presumably, Hep catabolism is also defective, but its fragments are not quantitatively significant enough to be recognized in the presence of high levels of HS. Type III MPS differs from types I and II by the relative mildness of the skeletal abnormalities and severity of the intellectual impairment. However, certain connective tissue abnormalities are evident early, and by the time of death, typically in the second decade, coarse features, hirsutism, and skeletal deformations are usually present. Deposits of partially degraded HS can be found in most tissues. The nature of the stored material suggests that some endolytic cleavage of HS/Hep type GAG does occur, and both endohexosaminidases and endohexuronidases have been implicated (59,60). There is also evidence for storage of membrane lipids, particularly in brain and liver, which is secondary to HS inhibition of other lysosomal enzymes. There are no specific elements of MPS III pathology that might be attributed to faulty Hep catabolism.

4. α-N-*Acetylglucosaminidase Deficiency*

A second form of Sanfilippo syndrome results from a failure to cleave the α-*N*-acetylglucosamine linkage found in HS. When it was established that the hexosamines were α-linked in these polymers, the α-*N*-acetylglucosaminidase was considered as a likely enzyme defect in the Sanfilippo syndrome. A colorimetric substrate was available, and this prediction was soon shown correct by several laboratories, which reported the enzyme deficiency in patients of the second complementation group defined by Neufeld's laboratory (61,62).

α-*N*-acetylglucosaminidase is assayed by many laboratories, and MPS III-B can readily be diagnosed. Type A disease was often inferred if the clinical picture and urinary excretion suggested MPS III and α-*N*-acetylglucosaminidase levels were normal. Only recently has it been realized that additional defects also produce the Sanfilippo syndrome and that the presence or absence of α-*N*-acetylglucosaminidase is not sufficient to establish the diagnosis.

5. *Lysosomal* N-*Acetyltransferase Deficiency*

The discovery that some patients with an apparent Sanfilippo syndrome lacked neither the heparin sulfamidase nor the α-*N*-acetylglucosaminidase led to the elucidation of a somewhat unexpected step in the catabolism of HS (63). After the sulfamide group is hydrolyzed, the resulting glucosamine end group cannot be removed until the free amine has been acetylated. To accomplish this, what would normally be considered a biosynthetic reaction occurs in the lysosome. Acetate and coenzyme A are converted to acetyl coenzyme A, which then serves as substrate for a lysosomal acetate CoA: oligosaccharide glucosamine *N*-acetyl transferase (64). A deficiency of this enzyme prevents HS catabolism and results in a clinical condition indistinguishable from MPS III-A and MPS III-B. This defect, which has been designated MPS III-C, accounted for about 20% of 73 cases of the Sanfilippo syndrome reviewed by van de Kamp (28).

6. N-*Acetylglucosamine-6-Sulfatase Deficiency*

Another enzymatically distinct form of the Sanfilippo syndrome has been recognized more recently (65,66). In this case, the deficient enzyme hydrolyzes a sulfate residue at the 6 position of a terminal *N*-acetylglucosamine. The assay substrate for this enzyme was obtained by partial acid hydrolysis of Hep followed by *N*-acetylation of the now unsubstituted sugar amines. Only a few cases of this MPS III-D have been reported, and little is known about the missing enzyme. This sulfatase does not appear to be necessary for removal of the sulfate from the 6 position of the β-linked *N*-acetylglucosamine in KS. It is also clearly different from the enzyme desulfating the 6 positions of galactose in KS and *N*-acetylgalactosamine in C6S.

The Sanfilippo syndrome provides an interesting example of phenocopies. Four separate enzyme deficiencies prevent the catabolism of the same macromolecular species. The differences in end-group structure with different enzyme defects do not appear to influence the physiological consequence of the HS accumulation. Perhaps careful clinical differentiation will reveal reliable differences between the four different genetic diseases, but evidence presently available suggests that biological variation within enzyme deficiency subgroups will mask any subtle differences between subgroups.

7. *Galactose/*N-*Acetylgalactosamine-6-Sulfatase Deficiency*

KS has a number of unique structural features, one of which is a 6-*O*-sulfated galactose rather than a uronic acid alternating with the amino sugar. Desulfation of this galactose requires a sulfatase active toward both galactose and *N*-acetylgalactosamine-6-sulfates (67-69). The latter occurs in C6S, but C6S can be endolytically cleaved and does not accumulate to any great extent in the enzyme's absence. However, the sulfatase is usually assayed with a substrate obtained by partial degradation of C6S and is, therefore, most often referred to as *N*-acetylgalactosamine-6-sulfatase. The deficiency disease is MPS IV-A or Morquio-A syndrome, and it is the failure to cleave 6-sulfated galactose which is critical.

Among the mucopolysaccharidoses, Morquio syndrome is associated with the most severe skeletal defects, kyphoscoliosis and dwarfing being the dominant characteristics. Corneal clouding is often present but intelligence is usually normal. Galactose/*N*-acetylgalactosamine-6-sulfatase deficient patients (A type) can have phenotypes ranging from severe to relatively minor, suggesting multiple mutant enzymes with varying degrees of functional deficit. Excess KS is not always found in the urine, particularly in milder forms or at older ages (70).

8. *β-Galactosidase Deficiency*

Lysosomal β-galactosidase deficiencies are involved in a complex constellation of clinical syndromes (71). The predominant acid β-galactosidase (type I) occurs in tissue extracts in monomeric and aggregate forms. This enzyme appears to be necessary for KS, glycoprotein oligosaccharide, and GM_1 ganglioside hydrolysis. In its most severe form, the deficiency of type I β-galactosidase results in severe brain damage, which is evident in early infancy and causes death by about two years of age. While KS and oligosaccharides appear in the urine and coarse features are present, the dominant pathology results from glycolipid storage in the brain. Thus, the defect is usually classified among the lipidoses and referred to as GM_1 or generalized gangliosidosis. Late-onset forms of GM_1 gangliosidosis are also known.

A defect in the same type I β-galactosidoses is also observed in some patients with a relatively mild form of Morquio syndrome that has been designated type B (MPS IV-B) (72,73). In this instance, it is a KS hydrolytic defect that seems to dominate the clinical picture. It has been postulated that the β-galactosidase mutation in MPS IV-B more severely affects the KS hydrolytic capacity than the gangliosidase activity (74). Substantial GM_1 ganglioside hydrolytic activity can be demonstrated in the presence of GM_1 activator (75). This emphasizes the importance of investigating a full range of synthetic and physiological substrates when trying to correlate clinical and laboratory studies on enzyme deficiency diseases. Differences in enzyme processing in GM_1 gangliosidosis and MPS IV-B have recently been compared by Hoogeveen et al. (76).

To further complicate the situation, type I β-galactosidase and sialidase are both deficient in one type of oligosaccharidoses (to be considered below). β-Galactosidase can also be substantially depressed in tissues from patients with MPS types I, II, and III (40). This indirect enzyme deficiency may play a role in the lipid storage in liver and brain seen with these diseases. There is also a second, genetically distinct, lysosomal β-galactosidase, cerebroside β-galactosidase, which is deficient in Krabbe leukodystrophy (77). The activity of the latter enzyme is relatively low and does not contribute much to the overall activity measured with synthetic β-galactosidase substrates. The ganglioside β-galactosidase does not appear to have a role in normal GAG metabolism, but a possible modulating role in deficiencies of the type I enzyme should not be overlooked.

9. Arylsulfatase B Deficiency

Both C4S and DS have 4-sulfated *N*-acetylgalactosamine residues. Desulfation of this structure appears to be the major physiological role of arylsulfatase B, an enzyme that has been extensively investigated for its activity with synthetic phenolic substrates (78). Desulfation of the terminal *N*-acetylgalactosamine residue is necessary before β-*N*-acetylhexosaminidase can act and, therefore, is a necessary step in exolytic degradation of these GAG. Since endolytic routes are available for C4S, only DS accumulates and is excreted in the absence of arylsulfatase B. The enzyme is also known as *N*-acetylhexosamine-4-sulfate sulfatase. Selective assay protocols for this enzyme in human tissues were well developed before the deficiency syndrome was recognized, making this one of the more easily diagnosed mucopolysaccharidoses subtypes (79). Nucleotide and oligosaccharide "physiological" substrates for arylsulfatase B have been investigated (80), but none so far has provided any advantage over the readily available colorimetric and fluorogenic substrates for investigating MPS VI, the Maroteaux-Lamy syndrome.

Clinically, the Maroteaux-Lamy syndrome resembles Hurler syndrome in its structural manifestations, but lacks the intellectual deficits (12). Only DS is found in the urine, leading to the generalization that DS accumulation has primarily connective tissue and skeletal repercussions. Both severe and mild variants (MPS IV-A and MPS IV-B) have been reported, but cannot at present be differentiated on enzymatic criteria. MPS VI is somewhat unusual in that cells cultured from Maroteaux-Lamy patients have a substantial (5-20%) residual arylsulfatase B activity (80). The enzyme responsible for the residual activity has so far proven indistinguishable from the arylsulfatase B of normal cells in both catalytic and structural properties. This may simply mean that a complete arylsulfatase B deficiency is incompatible with embryonic or fetal development. However, alternate explanations must also be entertained, such as multiple arylsulfatase B genes, selective stabilization, or nonprimacy of the observed enzyme deficiency. Much more experimental work will be needed before the basic biochemical pathology of MPS VI can be considered to be adequately explained.

10. β-Glucuronidase Deficiency

Glucuronic acid occurs in all GAG subtypes except KS. β-Glucuronidase, therefore, must be involved in exolytic breakdown of almost all GAG. Endolytic pathways are also available for most β-glucuronide-containing structures resulting in a somewhat less well-defined excretion pattern in the β-glucuronidase deficiency disorders. In the original patient of Sly et al. (81) and several subsequent cases, C4S and C6S were the urinary excretion products. In one recent case, mucopolysacchariduria was not present at nine months but by 22 months was easily demonstrated (82). In some instances, DS and/or HS appear to dominate the excretion pattern (83). The β-glucuronidase deficiency is designated MPS VII and has been referred to as the Sly or Sly/Neufeld syndrome.

The number of cases of β-glucuronidase deficiency so far reported is small, and a "typical" clinical picture is difficult to formulate. Bony abnormalities, retarded growth, enlarged liver and spleen, and mucopolysaccharidosislike facies seem to be the most generally recognized features (83). Mental retardation is also a part of the syndrome, but has varied considerably in severity. Older-onset variants have been reported. The major importance of the β-glucuronidase defect has been in establishing the role of carrier-mediated uptake in lysosomal enzyme localization (84).

*11. β-*N*-Acetylhexosaminidase—The Missing Enzyme Deficiency*

One enzyme that must be involved in the breakdown of GAG has not yet been associated with a GAG storage disease—that is, β-*N*-acetylhexosaminidase. Two lysosomal β-*N*-acetylhexosaminidases are known (types A and B), and their deficiency states, Tay-Sachs and Sandhoff diseases, are among the most extensively investigated (71). The absence of β-*N*-acetylhexosaminidase A is associated with Tay-Sachs disease and later-onset variants of GM_2 gangliosidosis. The combined absence of both A and B enzyme forms is responsible for Sandhoff disease, a somewhat more generalized glycolipid storage disease. There are no clinical indications of GAG storage in either of these conditions, although oligosaccharide excretion does occur in Sandhoff disease (21). Accumulation of sulfated GAG by fibroblasts in culture has been reported with Sandhoff disease, but not Tay-Sachs cells, suggesting a role for the B enzyme in GAG catabolism (85). In contrast, Thompson et al. showed that extracts from neither Tay-Sachs or Sandhoff fibroblasts could remove the end group from a CS heptasaccharide, implicating *N*-acetylhexosaminidase A in GAG breakdown (86). Studies utilizing oligosaccharide substrates with purified enzymes show that both A and B enzymes can act on GAG-derived material, with the A form showing relatively higher activities with CS-based substrates and the B form favoring KS-derived materials (80,87). Perhaps additional enzymes hydrolyzing the β-*N*-acetylhexosaminide linkages in GAG will be discovered and thereby explain the lack of any deficiency disease involving this rather critical structural element.

12. N*-Acetylglucosamine-6-Sulfatase (KS-Type) Deficiency*

Some confusion still exists concerning the enzyme responsible for the removal of sulfate from 6-sulfated *N*-acetylglucosamine residues in KS and its involvement in a distinct mucopolysaccharidosis. The enzyme deficient in MPS III-D appears to be specific for the α-linked amino sugar since activity against β-linked *N*-acetylglucosamine-6-sulfate substrate was not deficient in the fibroblasts from MPS III-D patients (65). This implied the presence of another enzyme desulfating the β-linked *N*-acetylglucosamine of KS. Di Ferrante's group reported a patient excreting both KS and HS and suggested the deficiency of an enzyme acting on the 6-sulfated *N*-acetylglucosamines found in both GAG. However, further examination of fibroblasts from this patient revealed a heparin sulfami-

Table 2 Enzymatic Subtypes of the Oligosaccharidoses with Mucopolysaccharidosislike Symptoms

Deficient enzyme	Designation	Leroy subtype	ML subtype	Characteristics of principal interest
Glycoprotein neuraminidase	Spranger syndrome Sialidosis type II Dysmorphic sialidosis	OS-1a	ML I	Skeletal deformities; coarse features; mental retardation; blindness (cherry-red spots); onset at birth to one year; death in first decade
				Mild allelic form; survival into adulthood
GluNAc phosphotransferase	I-cell disease	OS-2a	ML II	Severe skeletal defects; coarse features; corneal opacities; mental retardation; no urinary GAG; multiple lysosomal enzyme defects in cultured fibroblasts; onset in first year; death in first decade
	Pseudo-Hurler	OS-2b	ML III	As above but later onset and milder symptoms; survival into adulthood
				Genetically complex: OS-2a and OS-2b can be allelic, but 2 or 3 complementation groups in OS-2
β-Galactosidase and glycoprotein neuraminidase	Galactosialidosis Goldberg-Wenger syndrome	OS-3a	ML I	Mild skeletal abnormalities; some facial coarsening, retardation, and visual deficits; onset in first or second decade; survival into adulthood

β-Galactosidase	GM_1 gangliosidosis	OS-4a-c	See Table 1
α-Mannosidase	Mannosidosis	OS-6	Skeletal deformities; coarse features; corneal opacities; mental retardation; no urinary GAG; death in first decade
			Milder allelic form with survival into adulthood
α-Fucosidase	Fucosidosis	OS-7	Skeletal abnormalities; facial coarsening; mental retardation; no urinary GAG; onset about one year; death in first or second decade
			Possibly a milder allelic form with survival into adulthood
Multiple sulfatases	Multiple sulfatase deficiency disease (MSDD) Mucosulfatidosis Austin-variant MLD	OS-8	Progressive neurological disease—leukodystrophy; mild skeletal and facial dysmorphosis; ichthyosis; elevated urinary GAG; unknown primary defect
Aspartylglucosamine amidohydrolase	Aspartyl glycosaminuria	OS-10	Coarse facies; sagging skin; lens opacities; mental deterioration; sun sensitivity; no urinary GAG excretion; survival into adulthood

dase deficiency and no *N*-acetylglucosamine-6-sulfatase defect (88). Subsequently, Matalon et al. reported KS and HS excretion in a child with normal development to 18 months, followed by rapid deterioration accompanied by several symptoms typical of the mucopolysaccharidoses (89). Cells cultured from this patient could not desulfate an oligosaccharide substrate derived from KS with a β-*N*-acetylglucosamine-6-sulfate terminus. Unfortunately, activity against corresponding HS-derived material was not evaluated.

Kresse et al. attempted to purify the KS-type enzyme and found it to be indistinguishable from β-*N*-acetylhexosaminidase A, the enzyme missing in Tay-Sachs disease (90). They suggested that β-*N*-acetylhexosaminidase A participates in the catabolism of KS by removing sulfated sugar residues, thus bypassing a 6-sulfatase requirement. This would then be an exception to the general requirement for desulfation before deglycosylation, which appears to be obeyed by all other GAG polymers. At present, the existence of a separate KS- and HS-excreting mucopolysaccharidosis remains unresolved.

B. The Oligosaccharidoses

In addition to the "true" mucopolysaccharidoses—single-enzyme defects associated with enhanced GAG excretion—there are a number of diseases with mucopolysaccharidosislike clinical presentations. Most of these are due to deficits in enzymes involved in the processing or catabolism of glycoprotein oligosaccharides. Several of these conditions were included in the mucolipidoses classification introduced by Spranger and Wiedemann (91). Recently, Leroy proposed a somewhat broader classification of the oligosaccharidoses (20). Since this is a more descriptive and encompassing scheme, it has been adopted for the present discussion (see Table 2). Both mucolipidosis (ML) and oligosaccharidosis (OS) designations are provided in the text. These conditions have also recently been reviewed by Warner and O'Brien (21).

The reasons for considering the oligosaccharidoses in the present context are clinical similarities with the mucopolysaccharidoses suggesting similar alterations in connective tissue metabolism. Since proteoglycans have mucin- and glycoprotein-type oligosaccharide constituents in addition to GAG, oligosaccharides could provide the common element underlying similarities in phenotypic expression.

A good example is provided by I-cell disease (OS-2a, ML-II), which is extremely "Hurler-like" in its clinical presentation (22). There is no GAG excretion, but Coppa has provided a thoughtful discussion of possible GAG storage (92). The favored postulate is that the phenotype reflects a localized GAG accumulation in connective tissue cells rather than the generalized storage encountered in the true mucopolysaccharidoses. However, I-cell disease could also mean that connective tissue aberrations in the mucopolysaccharidoses are not direct consequences of GAG accumulation but reflect a secondary glycoprotein metabolic defect similar to the secondary glycolipid storage known to occur in the mucopolysaccharidoses (93).

1. Multienzyme Deficiencies

The most interesting oligosaccharidoses from the standpoint of molecular mechanism are the multiple enzyme deficiencies. Best characterized are the UDP-*N*-acetylglucosamine: glycoprotein *N*-acetylglucosamine phosphotransferase defects in I-cell disease and pseudo-Hurler polydystrophy (OS-2a and OS-2b, ML-II and ML-III) (22). This nucleotide sugar phosphate transfer is the first step in the phosphorylation of the glycoprotein mannose residues that mark enzymes destined for lysosomal localization (84). In the absence of the phosphotransferase, lysosomal hydrolases are processed as excretory glycoproteins and are excreted from the cell. This results in inadequate enzyme levels within cellular lysosomes despite high levels in serum and tissues. In effect, these conditions are defects of enzyme localization. Various enzymes of GAG catabolism are affected, which might explain the "Hurler-like" phenotype although no GAG excretion occurs. Actually, the molecular mechanisms of this group of conditions may prove to be more complex since two or three distinct complementation groups have been recognized (94).

Another multienzyme defect where the molecular mechanism is beginning to be understood is OS-3a—the combined glycoprotein sialidase and β-galactosidase deficiency. β-Galactosidase fails to form a high-molecular-weight aggregate, the predominant form of the enzyme in normal tissue extracts. The unaggregated enzyme is more rapidly degraded resulting in a functional enzyme deficit in the cells. This failure in enzyme aggregation is due to the absence of a 32,000-molecular-weight "protective" protein that is thought to function as an organizer and stabilizer (95). The glycoprotein sialidase associates with the β-galactosidase aggregates and may thereby be stabilized by the same factor, although this has not yet been demonstrated. The phenotypic expression is similar to the primary oligosaccharide sialidase deficiency and the milder form of GM_1 gangliosidosis, suggesting that the hydrolase defects are important in the clinical syndrome. However, the enzyme deficiencies are secondary to what can best be described as a structural protein defect. Very little is known about the structural organization of enzymes within the lysosomes, and additional examples of this general phenomenon will probably be found.

Multiple sulfatase deficiency (OS-8) actually fulfills the definition of a mucopolysaccharidosis, there being GAG storage and excretion (39). Most of the sulfatases involved in GAG catabolism are deficient in many tissues, and substantial alteration in GAG breakdown is understandable. Although it is now reasonably certain that the disorder does not represent a defect in a common sulfatase subunit and that the structural genes for the various sulfatases are intact (96), no real clues to the nature of the primary defect and the mechanism of disease are yet available.

2. *Oligosaccharide Glycohydrolase Deficiencies*

Leroy also included mannosidosis (OS-6) and fucosidosis (OS-7) among the oligosaccharidoses (20). Both disorders were originally considered to be mucopolysaccharidoses, although urinary GAG could not be demonstrated. Coarse facies, mental retardation, corneal opacities, and dysostosis multiplex still suggests a close pathological relationship to the mucopolysaccharidosis group (19,23). Ultrastructural studies on various tissues were taken to indicate the simultaneous accumulation of polysaccharide and lipoidal materials in lysosomelike structures, leading to the mucolipidosis designation. Since the recognition of the specific enzyme defects, urinary excretion of distinctive patterns of oligosaccharides has been demonstrated (21). Over the same period of time, an increasingly sophisticated knowledge of glycoprotein biochemistry has developed, and the urinary oligosaccharides have been recognized as remnants of these structures. Mannosidosis and fucosidosis are now accepted as primary defects of glycoprotein oligosaccharide catabolism. However, the close clinical similarity to the mucopolysaccharidoses, particularly the similar spectrum of connective tissue defects, has not been explained. Some common biological mechanism must be operating at the phenotypic level.

Aspartylglycosaminuria (OS-10) might also be considered within the same context. Features such as coarsening of the facies, hernias, sagging skin folds, lens opacities, and a mild dysostosis multiplex imply considerable connective tissue involvement (23). There appears to be considerable phenotypic overlap with the milder forms of the mucopolysaccharidoses.

C. Possible Diseases of GAG-Proteoglycan Biosynthesis

The mucopolysaccharidoses are the best understood disorders of GAG-PG metabolism at the biochemical level, but they represent only one facet of the metabolic system. Abnormalities in PG synthesis and biological regulation will certainly be equally important. Lowe's syndrome is one possible example (97). In this condition, there are defects in a number of organ systems, including visual (cataracts and glaucoma), intellectual (mental retardation), and renal (tubular defects). The condition is sometimes referred to as the oculocerebrorenal syndrome. Yokoi et al. reported an abnormally *low* excretion of HS by Lowe's syndrome patients, although the total GAG excretion was reasonably normal (98). This was subsequently associated with a lowered synthesis of intracellular HS by Lowe's syndrome fibroblasts (99). Other investigators reported the excretion of an undersulfated chondroitin sulfate in Lowe's syndrome (100). Cultured fibroblasts produced substantially undersulfated GAG. It was suggested that this was due to depressed levels of PAPS, the biological sulfating agent (101). The lowered PAPS appeared to result from a substantial increase in the activity of a nucleotide pyrophosphatase having the ability to hydrolyze PAPS (102). Although fundamental differences between the two research groups are presently difficult to reconcile by a single mechanism, evidence seems strong for some defect in PG biosynthesis in Lowe's syndrome.

Another disorder where faulty proteoglycan metabolism has been suggested is the spondyloepiphyseal dysplasia reported by Maurao and co-workers (103). They demonstrated a low PAPS-CS sulfotransferase activity in serum and an undersulfated CS in urine. The authors suggest a relationship between this disorder and that of brachymorphic mice where there is a poorly sulfated cartilage GAG, and the fundamental defect is believed to be in the synthesis of PAPS (104). Weber-Christian disease, a disorder involving recurrent fever and inflammation of connective tissues, has also been associated with an increase in undersulfated CS (and, to a lesser degree, HS) in urine (105). In progressive systemic sclerosis, a moderately increased GAG excretion enriched in HS and a shift in the C4S/C6S ratio was indicative of perturbed connective tissue metabolism (106). The development of sensitive indices for subtle alterations in PG metabolism needs further work, and many additional examples of pathobiology of PG biosynthesis can be expected. The details of PG synthesis and connective tissue organization are only now beginning to be understood, and an increasing recognition of specific enzyme defects associated with these processes should follow.

V. PREVENTION AND TREATMENT

Medical intervention in the mucopolysaccharidoses and oligosaccharidoses had little impact on the insidious progression or frequency of occurrence until very recently. Even now, more can be done toward prevention through carrier identification, genetic counseling, and prenatal diagnosis than can be done to slow disease progression or alleviate major symptoms. Efforts at enzyme replacement have not proven particularly successful. Certain cell transplantation approaches appear promising at present, while gene replacement remains for the future.

A. Carrier Detection

The early work on fibroblast metachromasia by Danes and Bearn demonstrated that cells from mucopolysaccharidoses heterozygotes show evidence for GAG storage in culture

(41). However, the use of this method for specific carrier detection was effectively undermined when carriers for a variety of conditions were found to give positive reactions (107). A reinvestigation of this basic approach with a more sophisticated cell-loading techniques and intracellular GAG analyses might prove more fruitful.

Gene-dose effects have been observed with most lysosomal enzymes (108). However, there is often a substantial overlap between controls and obligate heterozygotes when large samples have been investigated. Within a given family, groupings of high and intermediate enzyme levels can often be recognized, and reasonable predictions as to who are carriers and noncarriers are often possible. Reliable diagnosis of heterozygotes for any of these diseases in a general population is much more uncertain. While the Tay-Sachs experience establishes the power of carrier screening by enzymatic methods (109), its effectiveness was favored by special features such as homogeneous mutation (vida infra) and a target population at reasonably high risk. Such conditions cannot be duplicated for any of the diseases considered here, and general population screening has not been considered feasible.

A possible exception is suggested by the work of Vance and co-workers with MPS III-B, the α-*N*-acetylglucosaminidase deficiency (110). Only minimal overlap of serum enzyme activities was encountered between 12 obligate heterozygotes and 360 controls. Some 175 individuals from six MPS III-B families distributed in a manner consistent with their probability of being carriers. One control was clearly within the carrier range and may have been a heterozygote, one or more carriers being unexpected in a sample of this size.

Hunter syndrome (MPS II) poses a special problem in carrier detection. It is the only X-linked mucopolysaccharidosis, and carrier females transmit the defect to their sons at high frequency. Close female relatives of affected individuals are at high risk for being carriers and having affected children. Because of the clonal inactivation of one or the other X chromosome in female somatic cells, each cell in a carrier will be either enzyme deficient or normal. This was shown to be true for MPS II carriers with cloned fibroblasts (111), providing one of the earliest proofs for the Lyon hypothesis. However, fibroblast cloning is too difficult for routine carrier screening and is employed only sparingly. Selective survival of enzyme-deficient cells on freezing cultures for storage has been suggested (112), but has not proven reliable enough for diagnostic application. Cells from individual hair root bulbs represent a nearly clonal population, and plucked hair roots have been investigated for carrier diagnosis of several X-linked disorders, including Hunter syndrome. The iduronate sulfatase assay was sensitive enough to detect the enzyme in single hair root samples, and reasonable distributions of deficient samples were detected in obligate carriers (113,114). Tonnesen et al. have suggested the use of an inhibitor of lysosomal enzyme uptake to suppress cross-correction of the deficient cells in a mixed culture (115). This portion of the total cell population could then be detected by GAG accumulation. Archer et al. recently presented an extensive study utilizing serum and hair root iduronate sulfatase assessments (116). They found that the hair root technique gave better differentiation than did serum enzyme levels, but using information from both assays improved results.

Better carrier recognition can be anticipated in the future, as gene-mapping techniques become available for the lysosomal enzymes. One must remember, however, that the range of normal variation will probably be large, particularly at the gene level, and substantial ambiguities could remain.

B. Prenatal Diagnosis

The possibility of detecting mucopolysaccharidoses and oligosaccharidoses prenatally has given families at high risk a chance for normal children while avoiding additional affected children. Three approaches to prenatal diagnosis in the mucopolysaccharidoses have proven effective. One is to look for accumulated materials (GAG) in the amniotic fluid; the second is to look for accumulated materials in cultured amniotic fluid cells ($^{35}SO_4$ loading); and the third is to analyze enzyme activities in cultured amniotic fluid cells. A number of alternatives have also been tried, such as assay of enzymes in the amniotic fluid or analysis of uncultured cells, but they have not proven effective. In one instance, fetal leukocytes were employed for enzymatic diagnosis after amniotic cell culture failed on two separate attempts (117).

Direct determination of GAG in amniotic fluid has several potential advantages, the greatest of which is speed. Analyses can be completed within a few days. Amniotic fluid composition is, however, variable and changes considerably with the stage of fetal development (118). Amniotic fluid GAG patterns suggesting fetal disease have been found in several instances, but failures have also been reported. Greater problems seem to have been associated with the earlier amniocenteses (119), which are necessary for cell culture. Whiteman and Henderson presented a simple two-dimensional electrophoretic method for looking at GAG patterns in amniotic fluids (120), which was used by Mossmand and Patrick in 72 pregnancies at risk for mucopolysaccharidoses (121). The procedure correctly diagnosed a MPS I fetus when amniotic cell cultures had given erroneous results, apparently caused by maternal cell contamination. When used in conjunction with cell culture information, direct amniotic fluid GAG measurement adds a valuable dimension to prenatal diagnosis in this group of disorders.

The $^{35}SO_4$ GAG accumulation by cultured amniotic fluid cells was employed for prenatal diagnosis of the mucopolysaccharidoses before specific enzyme defects were known (122) and continues to be used in some laboratories. It has the advantage of being relatively nonspecific in those instances when the exact diagnosis of the index case is unclear. $^{35}SO_4$ incorporation is useful in interpreting marginal enzyme values, since it is a measure of the functional enzyme capacity within the cell. Kleijer et al. found it particularly valuable in a series of 40 pregnancies at risk for MPS I-H when enzyme levels in the low heterozygote range were encountered (123).

Specific enzyme analysis on cultured amniotic fluid cells has now been reported for nearly all enzymes deficient in the mucopolysaccharidoses and the oligosaccharidoses groups (12). Most have actually been utilized to diagnose affected fetuses (e.g., Ref. 124), although the total number of reported cases is small for most diagnostic subgroups. Extensive series of cases, such as those recently reported by Mossman and Patrick (121) and Kleijer et al. (123), are rare, making it difficult to assess the reliability of the methodology.

One problem often encountered is that enzyme levels in amniotic fluid cultures are lower than in skin fibroblasts, particularly when early-passage amniotic cell cultures are being assessed (125). Differing culture morphology and growth characteristics can also be encountered. It is, therefore, difficult to decide on an appropriate point of reference for comparison of cultures. Control cells with a matched history in culture are extremely important, although they are often difficult to obtain. The use of internal enzyme markers and enzyme ratios may provide more rational reference points, but much more information on the "natural history" of lysosomal enzymes in early passage amniotic fluid cell cultures is needed.

Available evidence suggests that enzyme-based prenatal diagnosis of this group of diseases has been exceptionally reliable, but it remains within the purview of a small number of research oriented laboratories and cannot at present be considered a generally available service.

VI. GLYCOSAMINOGLYCANS AND PROTEOGLYCANS IN OTHER DISEASES

The end result of most disease is tissue damage. This is usually followed by some attempt at restructuring and repair. Pathogens and body defense mechanisms are strongly influenced by properties of the cell surface and extracellular matrix. Reproductive processes, organization of tissue, hormonal controls, and aging, all involve substantial connective tissue changes. It is, therefore, not surprising that GAG and their higher-order structures have been investigated in a vast spectrum of disease contexts. Two recent books have considered this subject in great detail (4,5), and only a few broad generalizations will be considered in the present context. It should be remembered that our present emphasis on the mucopolysaccharidoses only reflects the current state of knowledge: GAG metabolism has a much broader biomedical significance.

One major focus of GAG-PG research has been cartilage and bony skeleton (6,126). Particular attention has been directed toward joint cartilage and its destruction in arthritis. The GAG composition of cartilage has been found to reflect the kind and degree of stress imposed. Variations in GAG ratios occur over the surface of epiphyseal cartilage, for example, and changes in stress patterns are altered. The inflammatory response involves extensive PG breakdown, and remodeling GAG may play a significant role in stimulating inflammatory and repair processes.

GAG have also received considerable attention in cardiovascular diseases, particularly atherosclerosis (127). The primary events in the formation of vascular lesions may be changes in connective tissue structures. GAG, GAG derivatives, and hyaluronidase have been evaluated as possible therapeutic agents. Although substantial therapeutic value has been claimed, this approach has not received general acceptance. Heart valves exhibit a unique combination of toughness and flexibility and are relatively rich in HA and DS. GAG turnover is relatively high, and considerable variation occurs between valves. Damage to vessels, valves, or myocardium results in changes in GAG content and ratios.

Hormonal responses often involve changes in tissue structure and thereby alterations in connective tissue (128). In the ovulatory cycle, a complex sequence of structural changes occurs in response to hormonal variation. Ovulation, fertilization, implantation, and birth itself, all involve substantial alterations in GAG- and PG-containing structures. Because of its capacity to bind water, HA is important in the development of a variety of secondary sexual characteristics, such as the comb of the cock or the sexual skin of some primates. Somatotropin tends to stimulate sulfated GAG production, while ACTH and glucocorticoids tend to depress GAG synthesis, although considerable variation occurs between tissues. The myxedema and exophthalmos of thyroid deficiencies reflect enhanced extracellular GAG, particularly HA (129). Insulin stimulates uptake of GAG precursors in diabetic animals (130), and diabetes mellitus is characterized by decreased synthesis and turnover of GAG.

Each organ system, tissue, and tissue subdivision has its own pattern of PG and GAG types, and each responds with characteristic changes in disease. Lung function is highly dependent on structural integrity (131,132). Increased amounts of GAG, particularly

CS and DS, occur in fibrotic lung disease, while smoking lowered HA and CS levels in guinea pig lungs. Information on emphysema appears contradictory, with surprisingly little change in GAG patterns being apparent. In the eye, several distinct anatomical and functional structures have unique PG and GAG compositions. Changes accompany a variety of visual system defects. The skin constitutes one of the largest and most complex of the connective tissue-rich organs (133,134). Patterns of PG and GAG composition and turnover vary greatly with the skin of a single individual. Hair distribution, glandular excretions, solar irradiation, abrasion, pressure, heat, chemicals, and pathogen exposure, all influence detailed chemical composition. Almost all connective tissue undergoes age-related changes, usually with loss of flexibility or functional adaptability (135,136). Recovery of GAG becomes more difficult with age, and there is generally an increase in the proportion of KS, a shift in the C4S/C6S ratio, an increase in unsulfated disaccharide residues, and a decreased GAG turnover. One suspects that alterations in GAG-PG metabolism underlie a great deal of structural pathology throughout the body, but it has so far been difficult to dissect primary events from secondary manifestations.

Since PG are thought to play a role in cell-cell interaction and tissue organization, they would be expected to be important in cancer (132,138). Neoplasms tend to show a "less differentiated" metabolic pattern, which is frequently marked by HA in their basic matrix, but many differentiated tumors continue to have GAG patterns characteristic of their parental tissue.

VII. SUMMARY

The GAG and their more complex parent substances the PG constitute fundamental building blocks of the extracellular environment. The character of a particular connective tissue is dependent on this class of macromolecules, and GAG have been implicated in a wide variety of pathological processes. However, our knowledge of the biochemistry and functional physiology of the PG remains relatively limited. General patterns for both biosynthesis and catabolism can now be formulated, but tissue-specific details are not clear. Our most detailed metabolic information concerns the catabolism of the GAG. This information has mostly come from studying human lysosomal enzyme deficiency diseases. Essentially complete descriptions of the catabolic pathways can now be deduced for both GAG and glycoprotein oligosaccharides. PG biosynthetic defects are beginning to emerge, and it is hoped that they will provide as effective a window on synthesis as the mucopolysaccharidoses have on catabolism. A better understanding of the relationship between clinical symptoms and the materials that accumulate should provide insight into the physiological roles of the complex carbohydrates. This, in turn, will guide us in our search for effective means of treating and preventing diseases of the connective tissues.

ACKNOWLEDGMENTS

I would like to thank Dr. Hayato Kihara for his collaboration and support, Dr. Leonard Rome for reviewing the illustrations, and Mrs. Cindy Southard for help in preparation of the manuscript. Supported in part by NIH grants HD-04512 and NS-11665.

REFERENCES

1. W. D. Comper and T. C. Laurent, *Physiol. Rev., 58*:255–315 (1978).
2. V. C. Hascall, in *Proteoglycans: Structure and Function in Biology of Carbohydrates,* Vol. 1 (V. Ginsburg and P. Robbins, eds.), Wiley, New York, 1981, pp. 1–49.

3. L. Rodin, in *The Biochemistry of Glycoproteins and Proteoglycans* (W. J. Lennarz, ed.), Plenum Press, New York, 1980, pp. 267-371.
4. R. Varma and R. S. Varma, in *Mucopolysaccharides–Glycosaminoglycans–of Body Fluids in Health and Disease,* Walter de Gruyter, Berlin, 1983.
5. R. S. Varma, and R. Varma, eds., in *Glycosaminoglycans and Proteoglycans in Physiological and Pathological Processes of Body Systems,* Karger, Basel, 1982, p. 520.
6. L. C. Rosenberg, V. C. Mow, J. A. Buckwalter, and A. R. Poole. in *American Acacemy of Orthopaedic Surgeons Symposium on Heritable Disorders of Connective Tissue* (W. H. Akeson, P. Bornstein, and M. J. Glimcher, eds.), C. V. Mosby, St. Louis, 1982, pp. 285-305.
7. M. Yanagishita and V. C. Hascall, *J. Biol. Chem., 258*:12847-12856 (1983).
8. T. R. Oegema, E. L. Kraft, G. W. Jourdian, and T. R. Van Valen, *J. Biol. Chem., 259*: 1720-1726 (1984).
9. W. J. Lennarz, ed., in *The Biochemistry of Glycoproteins and Proteoglycans,* Plenum Press, New York, 1980, p. 381.
10. I. Jacobsson, U. Lindahl, J. W. Jensen, L. Rodin, H. Prehar, and D. S. Feingold, *J. Biol. Chem., 259*:1056-1063 (1984).
11. A. L. Fluharty, *J. Invest. Dermatol., 79*:38s-44s (1982).
12. V. A. McKusick and E. F. Neufeld, in *The Metabolic Basis of Inherited Disease,* 5th ed., (J. B. Stanbury, J. B. Wyngaarden, D. S. Fredrickson, J. L. Goldstein, and M. S. Brown, eds.), McGraw-Hill, New York, 1983, pp. 751-777.
13. H. Kresse, M. Cantz, K. von Figura, J. Glossl, and E. Paschke, *Klin. Wochenschr., 59*:867-876 (1981).
14. G. J. M. Hooghwinkel, in *Glycosaminoglycans and Proteoglycans in Physiological and Pathological Processes of Body Systems,* Karger, Basel, 1982, pp. 330-353.
15. J. N. Thompson, *Ala. J. Med. Sci., 19*:168-173 (1982).
16. A. L. Horwitz, *Am. J. Ment. Defic., 84*:113-123 (1979).
17. L. J. Shapiro, *Birth Defects: Original Article Series, 17*:191-203 (1981).
18. V. A. McKusick, D. Kaplan, D. Wise, W. B. Hanley, S. B. Suddarth, M. E. Sevick, and A. E. Maumanee, *Medicine, 44*:445-483 (1965).
19. A. C. Sewell, *Eur. J. Pediatr., 134*:183-194 (1980).
20. J. G. Leroy, *Birth Defects: Original Article Series, 18*:3-12 (1982).
21. T. G. Warner and J. S. O'Brien, *Ann. Rev. Genet., 17*:395-441 (1983).
22. E. F. Neufeld and V. A. McKusick, in *The Metabolic Basis of Inherited Disease,* 5th ed. (J. B. Stanbury, J. B. Wyngaarden, D. S. Fredrickson, J. L. Goldstein, and M. S. Brown, eds.), McGraw-Hill, New York, 1983, pp. 778-787.
23. A. L. Beaudet, in *The Metabolic Basis of Inherited Disease,* 5th ed. (J. B. Stanbury, J. B. Wyngaarden, D. S. Fredrickson, J. L. Goldstein, and M. S. Brown, eds.), McGraw-Hill, New York, 1983, pp. 788-802.
24. R. B. Lowry and D. H. G. Renwick, *N. Engl. J. Med., 284*:221 (1971).
25. T. Schaap and G. Bach. *Hum. Genet., 56*:221-223 (1980).
26. I. D. Young and P. S. Harper, *Hum. Genet., 60*:391-392 (1982).
27. A. Chakravarti and S. J. Bale, *Am. J. Hum. Genet., 35*:1252-1257 (1983).
28. J. J. P. van de Kamp, M. F. Niermeijer, K. von Figura, and M. A. H. Giesberts, *Clin. Genet., 20*:152-160 (1981).
29. P. Whiteman and E. Young, *Clin. Chim. Acta, 76*:139-147 (1977).
30. A. Malmström, *J. Biol. Chem., 259*:161-165 (1984).
31. G. Bach, R. Friedman, B. Wiessman, and E. F. Neufeld, *Proc. Natl. Acad. Sci. USA, 69*:2048-2051 (1972).
32. R. Matalon and A. Dorfman, *Biochem. Biophys. Res. Commun., 42*:340-345 (1972)
33. J. Crow, D. A. Gibbs, W. Cozens, E. Spellacy, and R. W. E. Watts, *J. Clin. Pathol., 36*:415-430 (1983).
34. J. L. Avila and J. Convit, *Biochem. J., 152*:57-64 (1975).

35. J. A. Kint, G. Dacremont, D. Carton, E. Orye, and C. Hooft, *Science, 181*:352–354 (1973).
36. J. Baumkotter and M. Cantz, *Biochim. Biophys. Acta, 761*:163–170 (1983).
37. P. R. Winters, M. J. Harrod, S. A. Molenich-Heetred, J. Kirkpatrick, and R. N. Rosenberg. *Neurology, 26*:1003–1007 (1976).
38. N. Kaibara, I. Katsuki, T. Hotokebuchi, K. Takagishi, and T. Kure, *Clin. Orthop. Rel. Res., 175*:233–236 (1983).
39. E. H. Kolodny and H. W. Moser, in *The Metabolic Basis of Inherited Disease,* 5th ed. (J. B. Stanbury, J. B. Wyngaarden, D. S. Fredrickson, J. L. Goldstein, and M. S. Brown, eds.), McGraw-Hill, New York, 1983, pp. 881-905.
40. P. A. Ockerman, B. Hultberg, and O. Erickson, *Clin. Chim. Acta, 25*:97–102 (1969).
41. B. S. Danes and A. G. Bearn, *Science, 149*:987–989 (1965).
42. J. C. Fratantoni, C. W. Hall, and E. F. Neufeld, *Proc. Natl. Acad. Sci. USA, 60*: 669–706 (1968).
43. J. C. Fratantoni, C. W. Hall, and E. F. Neufeld, *Science, 162*:570–572 (1968).
44. E. F. Neufeld and M. J. Cantz, *Ann. NY Acad. Sci., 179*:580–587 (1971).
45. R. Friedman and B. Weissmann, *Carbohydr. Res., 24*:123–131 (1972).
46. J. Butterworth and D. M. Broadhead, *J. Inher. Metab. Dis., 2*:71–74 (1979).
47. E. F. Neufeld, in *Lysosomes and Lysosomal Storage Diseases* (J. W. Callahan and J. A. Lowden, eds.), Raven Press, New York, 1981, pp. 115–129.
48. G. Bach, F. Eisenberg, M. Cantz, and E. F. Neufeld, *Proc. Natl. Acad. Sci USA, 70*: 2134–2138 (1973).
49. I. Sjoberg, L. A. Fransson, R. Matalon, and A. Dorfman, *Biochem. Biophys. Res. Commun., 54*:1125 (1973).
50. J. A. Cifonelli, *Carbohydr. Res., 8*:233 (1968).
51. T. W. Lim, I. G. Leder, G. Bach, and E. F. Neufeld, *Carbohydr. Res., 37*:103–109 (1974).
52. I. Liebaers and E. F. Neufeld, *Pediatr. Res., 10*:733–736 (1976).
53. I. D. Young, P. S. Harper, R. G. Newcombe, and I. M. Archer, *J. Med. Genet., 19*: 408–411 (1982).
54. S. Yatziv, R. P. Erickson, and C. J. Epstein, *Clin. Genet., 11*:319–326 (1977).
55. E. F. Neufeld, I. Liebaers, C. J. Epstein, S. Yatziv, A. Milunsky, and B. R. Migeon, *Am. J. Hum. Genet., 29*:455–461 (1977).
56. C. W. Hall, I. Liebaers, P. Di Natale, and E. F. Neufeld, *Meth. Enzymol., 50*:439–456 (1978).
57. R. Schmidt, K. von Figura, E. Paschke, and H. Kresse, *Clin. Chim. Acta, 80*:7–15 (1977).
58. J. J. Hopwood and H. Elliott, *Clin. Chim. Acta, 123*:241–250 (1982).
59. U. Klein, H. Kresse, and K. von Figura, *Biochem. Biophys. Res. Commun., 69*:158–166 (1976).
60. A. Oldberg, C-H Heldin, A. Wasteson, C. Busch, and M. Hook, *Biochemistry, 19*: 5755–5762 (1980).
61. J. S. O'Brien, *Proc. Natl. Acad. Sci. USA, 69*:1720–1722 (1972).
62. K. von Figura and H. Kresse, *Biochem. Biophys. Res. Commun., 48*:262–269 (1972).
63. U. Klein, H. Kresse, and K. von Figura, *Proc. Natl. Acad. Sci. USA, 75*:5185-5189 (1978).
64. J. J. Hopwood, C. Freeman, P. R. Clements, R. Stein, and A. L. Miller, *Biochem. Int., 6*:823–830 (1983).
65. H. Kresse, E. Paschke, K. von Figura, W. Gilberg, and W. Fuchs, *Proc. Natl. Acad. Sci. USA, 77*:6822–6826 (1980).

66. R. Gatti, C. Borrone, P. Durand, S. De Virgilis, G. Sanna, A. Cao, K. von Figura, H. Kresse, and E. Paschke, *Eur. J. Pediatr., 138*:168–171 (1982).
67. R. Matalon, B. Arbogast, P. Justice, I. Brandt, and A. Dorfman, *Biochem. Biophys. Res. Commun., 61*:759–765 (1974).
68. J. Singh, N. Di Ferrante, P. Niebes, and D. Travella. *J. Clin. Invest., 57*:1036–1040 (1976).
69. T. Yutaka, S. Okada, T. Kato, K. Inui, and H. Yabuuchi. *Clin. Chim. Acta, 122*:169–180 (1982).
70. A. Fujimoto and A. L. Horwitz, *Am. J. Med. Genet., 15*:265–273 (1983).
71. J. S. O'Brien, in *The Metabolic Basis of Inherited Disease,* 5th ed. (J. B. Stanbury, J. B. Wyngaarden, D. S. Fredrickson, J. L. Goldstein, and M. S. Brown, eds.), McGraw-Hill, New York, 1983, pp. 945–969.
72. J. S. O'Brien, E. Gugler, A. Giedion, U. Wiessmann, N. Herschkowitz, C. Meier, and J. Leroy, *Clin. Genet., 9*:495–504 (1976).
73. A. I. Arbisser, K. A. Donnelly, C. I. Scott, Jr., N. Di Ferrante, J. Singh, R. E. Stevenson, A. S. Aylesworth, and R. R. Howell. *Am. J. Med. Genet., 1*:195–205 (1977).
74. G. T. J. van der Horst, W. J. Kleijer, A. T. Hoogeveen, J. G. M. Huijmans, W. Blom, and O. P. van Diggelen, *Am. J. Med. Genet., 16*:261–275 (1983).
75. E. Paschke and H. Kresse, *Biochem. Biophys. Res. Commun., 109*:568–575 (1982).
76. A. T. Hoogeveen, H. Graham-Kawashima, A. d'Azzo, and H. Galjaard, *J. Biol. Chem., 259*:1974–1977 (1984).
77. K. Suzuki and Y. Suzuki, in *The Metabolic Basis of Inherited Disease,* 5th ed. (J. B. Stanbury, J. B. Wyngaarden, D. S. Fredrickson, J. L. Goldstein, and M. S. Brown, eds.), McGraw-Hill, New York, 1983, pp. 857–880.
78. D. F. Stumpf, J. H. Austin, A. C. Crocker, and M. LaFrance, *Am. J. Dis. Child., 26*: 747–755 (1973).
79. H. Baum, K. S. Dodgson, and B. Spencer, *Clin. Chim. Acta, 4*:453–455 (1959).
80. A. L. Fluharty, in *Lysosomes and Lysosomal Storage Diseases* (J. W. Callahan and J. A. Lowden, eds.), Raven Press, New York, 1981, pp. 249–261.
81. W. S. Sly, B. A. Quinton, W. H. McAlister, and D. Rimoin, *J. Pediatr., 82*:249–257 (1973).
82. H. E. Hoyme, K. L. Jones, M. C. Higginbottom, and J. S. O'Brien, *J. Med. Genet., 18*:237–239 (1981).
83. A. C. Sewell, J. Gehler, G. Mittermaier, and E. Meyer, *Clin. Genet., 21*:366–373 (1982).
84. W. S. Sly, M. Natowiez, A. Gonzalez-Noriega, J. H. Grubb, and H. D. Fischer, in *Lysosomes and Lysosomal Storage Diseases* (J. W. Callahan and J. A. Lowden, eds.), Raven Press, New York, 1981, pp. 131–146.
85. M. Cantz and H. Kresse, *Eur. J. Biochem., 47*:581–590 (1974).
86. J. N. Thompson, A. C. Stoolmiller, R. Matalon, and A. Dorfman, *Science, 181*: 866–867 (1973).
87. E. Werries, I. Neue, and E. Buddecke, *Hoppe-Seyler's Z. Physiol. Chem., 356*:953–960 (1975).
88. N. Di Ferrante, *Science, 210*:448 (1980).
89. R. Matalon, R. Wappner, M. Deanching, I. K. Brandt, and A. Horwitz, *Ann. Clin. Lab. Sci., 12*:234–238 (1982).
90. H. Kresse, W. Fuchs, J. Glossl, D. Holtfrerich, and W. Gilberg, *J. Biol. Chem., 256*: 12926–12932 (1981).
91. J. W. Spranger and H-R. Wiedemann, *Humangenetik, 9*:113–139 (1970).
92. G. V. Coppa, in *Glycosaminoglycans and Proteoglycans in Physiological and Pathological Processes of Body Systems,* Karger, Basel, 1982, pp. 309–315.

93. A. R. Rushton and G. Dawson, *Clin. Chim. Acta, 80*:133–139 (1977).
94. O. T. Mueller, N. K. Honey, L. E. Little, A. L. Miller, and T. B. Shows, *J. Clin. Invest., 72*:1016–1023 (1983).
95. A. T. Hoogeveen, F. W. Verheijen, and H. Galjaard, *J. Biol. Chem., 258*:12143–12146 (1983).
96. H. Kihara, *Am. J. Hum. Genet., 34*:171–181 (1982).
97. C. U. Lowe, M. Terrey, and E. A. MacLachlan, *Am. J. Dis. Child., 83*:164–184 (1952).
98. T. Yokoi, T. Uozaki, M. Kasei, T. Sato, and N. Taniguchi, *Clin. Chim. Acta, 116*: 153–160 (1981).
99. T. Yokoi and N. Taniguchi, *J. Lab. Clin. Med., 100*:461–468 (1982).
100. M. Akasaki, S. Fukui, T. Sakano, T. Tanaka, T. Usui, and I. Yamashina, *Clin. Chim. Acta, 89*:119–125 (1978).
101. S. Fukui, H. Yoshida, T. Tanaka, T. Sakano, T. Usui, and I. Yamashina, *J. Biol. Chem., 256*:10313-10318 (1981).
102. H. Yoshida, S. Fukui, I. Yamashina, T. Tanaka, T. Sakano, T. Usui, T. Shimotsuji, H. Yabuuchi, M. Owada, and T. Kitagawa, *Biochem. Biophys. Res. Commun., 107*: 1144–1150 (1982).
103. P. A. S. Maurao, S. Kato, and P. V. Donnelly, *Biochem. Biophys. Res. Commun., 98*:388–396 (1981).
104. J. P. Pennypacker, K. Kimata, and K. S. Brown, *Dev. Biol., 81*:280–287 (1981).
105. K. Murata, Y. Yukiyama, and Y. Horiuchi, *Clin. Chim. Acta, 49*:129–139 (1973).
106. K. Murata and M. Takeda, *Clin. Chim. Acta, 108*:49–59 (1980).
107. K. Taysi, M. L. Kistenmacher, H. H. Punnett, and W. J. Mellman, *N. Engl. J. Med., 281*:1108–1111 (1969).
108. L. J. Shapiro, in *Lysosomes and Lysosomal Storage Diseases* (J. W. Callahan and J. A. Lowden, eds.), Raven Press, New York, 1981, pp. 343–355.
109. M. M. Kaback, in *Lysosomes and Lysosomal Storage Diseases* (J. W. Callahan and J. A. Lowden, eds.), Raven Press, New York, 1981, pp. 331–342.
110. J. M. Vance, P. M. Conneally, R. S. Wappner, P. L. Yu, I. K. Brandt, and M. A. Pericak-Vance, *Clin. Genet., 20*:135–140 (1981).
111. B. R. Migeon, J. A. Sprenkle, I. Liebaers, J. F. Scott, and E. F. Neufeld, *Am. J. Hum. Genet., 29*:448–454 (1977).
112. C. W. Booth and H. L. Nadler, *Pediatrics, 53*:396–399 (1974).
113. N. Nwokoro and E. F. Neufeld, *Am. J. Hum. Genet., 31*:42–49 (1979).
114. T. Yutaka, A. L. Fluharty, R. L. Stevens, and H. Kihara, *Am. J. Hum. Genet., 30*: 575–582 (1978).
115. T. Tonnesen, C. Lykkelund, and F. Guttler, *Hum. Genet., 60*:167–171 (1982).
116. I. M. Archer, I. D. Young, D. W. Rees, A. Oladimeji, F. S. Wusteman, and P. S. Harper, *Am. J. Med. Genet., 16*:61–69 (1983).
117. J. Mossman, A. D. Patrick, A. H. Fensom, L. R. Tansley, P. F. Benson, V. M. Der Kaloustian, and G. Dudin, *Prenatal Diagnosis, 1*:121–124 (1981).
118. L. Dahl, J. J. Hopwood, U. B. G. Laurent, K. Lilja, and A. Tengblad, *Biochem. Med., 30*:280–283 (1983).
119. S. K. Mitra and K. Blau, *Clin. Chim. Acta, 89*:127–134 (1978).
120. P. Whiteman and H. Henderson, *Clin. Chim. Acta, 79*:99–105 (1977).
121. J. Mossman and A. D. Patrick, *Prenatal Diagnosis, 2*:169–176 (1982).
122. J. C. Fratantoni, E. F. Neufeld, B. S. Uhlendorf, and C. B. Jacobson, *N. Engl. J. Med., 280*:686–688 (1969).
123. W. J. Kleijer, E. J. Thompson, and M. F. Niermeijer, *Prenatal Diagnosis, 3*:179–186 (1983).
124. D. L. Van Dyke, A. L. Fluharty, I. A. Schafer, L. J. Shapiro, H. Kihara, and L. Weiss, *Am. J. Med. Genet., 8*:235–242 (1981).

125. A. Fujimoto and A. L. Fluharty, *Clin. Chim. Acta, 90*:157–161 (1978).
126. H. Muir and T. E. Hardingham, in *Biochemistry of Carbohydrates* (W. J. Whelan, ed.), MTP International Review of Science, Biochemistry, Series One, Vol. 5, University Park Press, Baltimore, 1975, pp. 153–222.
127. L. M. Morrison and R. Varma, in *Glycosaminoglycans and Proteoglycans in Physiological and Pathological Processes of Body Systems,* Karger, Basel, 1982, pp. 214–230.
128. I. N. Likar, R. W. Robinson, and L. J. Likar, in *Glycosaminoglycans and Proteoglycans in Physiological and Pathological Processes of Body Systems,* Karger, Basel, 1982, pp. 412–439.
129. H-H. Parving, G. Helin, C. Garbarsch, A. A. Johansen, B. A. Jensen, P. Helin, P. Lund, and J. Lyngsoe, *Clin. Endocrinol., 16*:207–210 (1982).
130. L. Kjellen, D. Bielefeld, and M. Hook, *Diabetes, 32*:337–342 (1983).
131. M. Motomiya, H. Arai, and K. Konno, in *Glycosaminoglycans and Proteoglycans in Physiological and Pathological Processes of Body Systems,* Karger, Basel, 1982, pp. 165–179.
132. L. A. Thet, A. C. Howell, and G. Han, *Biochem. Biophys. Res. Commun., 117*: 71–77 (1983).
133. R. S. Varma and R. Varma, in *Glycosaminoglycans and Proteoglycans in Physiological and Pathological Processes of Body Systems,* Karger, Basel, 1982, pp. 151–164.
134. K. A. Latif, I. Amla, and P. B. R. Rao, *Clin. Chim. Acta, 122*:317–325 (1982).
135. L. Robert and M. Moczar, in *Glycosaminoglycans and Proteoglycans in Physiological and Pathological Processes of Body Systems,* Karger, Basel, 1982, pp. 440–460.
136. S. Mohan and E. Radha, *Exp. Gerontol., 16*:385–392 (1981).
137. M. Nakajima, T. Irimura, D. Di Ferrante, N. Di Ferrante, and G. L. Nicolson, *Science, 220*:611–613 (1983).
138. L. O. Sampaio, C. P. Dietrich, and O. G. Filho, *Biochim. Biophys. Acta, 498*:123–131 (1977).

23

The Role of Connective Tissue Degrading Enzymes in Human Pathology

Carlo L. Mainardi*
University of Medicine and Dentistry of New Jersey-Rutgers Medical School,
Piscataway, New Jersey

I. INTRODUCTION

The remodeling of the connective tissue matrix is an essential process in human physiology and normally depends on a precise balance between the breakdown of existing connective tissue and the deposition of newly synthesized matrix elements. In most tissues, the same cell is responsible for the regulation of the two arms of the process (cartilage, tendon, etc.), and this cell is generally a differentiated mesenchymal cell (chondrocyte, fibroblast, etc.). In certain tissues (e.g., bone) different cells are responsible for each of these components of remodeling. In other specialized tissues (basement membrane), the remodeling process is orchestrated by epithelial or endothelial cells. Thus, the resident connective tissue cells possess the enzymatic potential for degrading the individual components of the extracellular matrix. In addition, certain cells that are frequently recruited to sites of acute or chronic inflammation (neutrophils and macrophages) also have this capacity. Therefore, the enzymatic breakdown of the matrix must be controlled by strict regulatory mechanisms.

In many pathological conditions, qualitative or quantitative alterations of the matrix result from the disease process and develop because of an imbalance between the synthetic and degradative components of tissue remodeling. In some conditions, the imbalance involves only the resident connective tissue cell (e.g., epidermolysis bullosa). In other conditions, the cell responsible for the matrix breakdown is one that is not normally involved in the degradation of the particular tissue. Examples of the latter circumstance include rheumatoid arthritis, in which inflammatory exudate cells and native synovial lining cells participate in the breakdown of the articular cartilage.

**Present affiliation*: University of Tennessee Center for the Health Sciences and Veterans Administration Medical Center, Memphis, Tennessee.

With the above considerations in mind, it is logical to hypothesize that abnormal connective tissue degradation occurs to some degree in most, if not all, pathological conditions associated with matrix abnormalities. A comprehensive treatment of all such disorders is obviously beyond the scope of this chapter. In the first part of this chapter, the mechanisms of connective tissue breakdown are reviewed with a focus on collagen degradation. Following this basic review, human pathological conditions that have been associated with connective tissue degrading enzymes are briefly discussed.

II. ENZYMATIC MECHANISMS IN EXTRACELLULAR MATRIX DEGRADATION

As outlined in this book, the extracellular matrix is composed of several different protein and nonprotein elements. Furthermore, the matrix composition varies greatly from tissue to tissue. In discussing the mechanisms involved in matrix degradation, one must consider the individual components of the connective tissue, which include the following:

1. Proteoglycans (PGs).
2. Glycoproteins (which include fibronectin in the interstitium and laminin in basement membrane).
3. Collagen (of which there are at least ten known types).
4. Elastin.

The role of each of these constituents has been partially defined, but the result of the selective depletion of any of these from the matrix is less well known. There is some support for the concept that degradation of the structural components (collagen and elastin) is more likely to result in irreversible alterations in tissue architecture than the depletion of the other components.

Several general principles of matrix degradation apply to all connective tissue elements. First, the pH of the extracellular milieu is near neutral; therefore, the enzymes must be active in this pH range. In some cases, partially degraded matrix elements are taken up by the cell and further degraded. PG side chains and collagen fibrils can be broken down in this manner. The intracellular environment can support the action of enzymes with an acidic pH optimum, and these enzymes probably contribute to matrix degradation in this context only. Second, many matrix degrading proteinases, especially metalloproteinases, are released from the cells in a latent form. Thus, a mechanism must exist to convert these latent enzymes to their active counterparts. Third, inhibitors of these enzymes exist in the tissues as well as the bloodstream and represent important counterregulatory factors in matrix breakdown. An imbalance must exist between active enzyme and accessible inhibitor(s) in the degradative process.

With these generalizations in mind, a discussion of the breakdown of the individual components of the matrix is in order.

A. Proteoglycan Degradation

The structure and breakdown of PGs is reviewed in detail in Chapters 4 and 22. The protein components of PGs are sensitive to the action of most proteolytic enzymes, whereas the enzymes that degrade the side chains are active at acid pH (1). Thus, the breakdown of PG in acute and chronic inflammation is probably initiated by extracellular proteolysis and is followed by internalization and intracellular enzymatic degradation of the side chains.

It has recently been shown that hyaluronic acid could be degraded by products of oxidative reactions, specifically, by oxygen radicals (2,3). This has relevance for inflammatory states in that neutrophils and macrophages are capable of generating such radicals. This nonenzymatic degradation of matrix components represents an important new concept in the study of connective tissue alterations.

B. Glycoprotein Degradation

Like the protein core of PGs, the glycoproteins fibronectin and laminin are relatively sensitive to the action of proteinases, and, therefore, the presence of an active specific proteinase is probably not a rate-limiting factor in the depletion of these proteins from the matrix. Degradation of one or both of these proteins can be accomplished by enzymes such as trypsin and chymotrypsin as well as neutrophil elastase (4-7). It is of interest that both these proteins contain large proteinase-resistant domains that retain some of the biological functions of parent molecule (4-7). These include functions such as fibroblast chemotaxis (8,9), and the solubilization of such biologically active fragments during proteolysis of fibronectin may contribute to the pathological process.

C. Collagen Degradation

Investigations into the role of connective tissue degradation in human pathology have focused primarily on the degradation of collagen and elastin. This is partly due to the fact that these structural proteins are resistant to general proteolysis. This resistance is related to the triple-helical configuration of native collagen and to the cross-linked state of native insoluble elastin. In fact, standard procedure for the isolation of these proteins has taken advantage of this phenomenon.

As described in detail below, the interstitial collagens (types I, II, and III) have similar mechanisms of degradation, and until recently, a unified model for collagen breakdown was accepted (10). However, the demonstration that noninterstitial and basement membrane collagens were degraded in manners distinct from the interstitial collagens has led to the concept of type-specific collagenolysis (11-14). Thus, the topic of collagen degradation will be discussed in this light.

1. Interstitial Collagen Degradation

General Considerations: This term includes types I, II, and III collagens. Types I and III collagen are the major collagenous components of stromal tissues, whereas type II is exclusively found in hyaline cartilage and vitreous. Degradation of the interstitial collagens is dependent on the presence of a specific metalloproteinase that makes a single clip in all three chains of the triple helix at a site three-fourths of the distance from the amino-terminus. This results in two helical fragments that denature at physiological temperatures and are subsequently degraded to oligopeptides by other proteinases (15). The cleavage site is in precisely the same location in all four collagen chains involved (residue 772-773), and the enzyme results in the hydrolysis of a glycyl-isoleucyl bond. The enzyme responsible for this cleavage has been referred to as mammalian or vertebrate collagenase but will be referred to in this chapter as interstitial collagenase (IC).

It is of interest that the collagenase cleavage site represents only one of three glycyl-isoleucyl bonds in the $\alpha 1$ (I) chain but is the only one hydrolyzed by the enzyme. This suggests that tertiary structure of the substrate is important in the action of the enzyme and that the helix may be "loose" in the cleavage site region. Experimental evidence sup-

Table 1 Sources of Interstitial Collagenase

A. Tissue extracts
Tadpole tail
Rat uterus (involution phase)
Isolated granulocytes
Isolated eosinophils
Gingiva
Rabbit VX-2 ascites cell tumor
B. Tissue or organ culture (various species)
Bone
Uterus
Cartilage
Rheumatoid synovium
Skin
C. Cell culture
Fibroblasts
Skin
Synovial
Gingival
Uterine
Neutrophils
Macrophages and monocytes
Peritoneal
Alveolar
Kupffer cells
Peripheral blood monocytes
Tumor cells
Endothelial cells
Chondrocytes
Hepatocytes

ports this concept (15). Furthermore, there is evidence that IC cleaves this bond more rapidly if helical configuration is preserved than if the chains are denatured (16,17).

Type III collagen is somewhat unique among interstitial collagens in that it can be specifically cleaved by enzymes other than IC. Neutrophil elastase, trypsin, and thermolysin have been reported to cleave type III collagen, and the cleavage sites are all in the same region of the molecule as the IC cleavage site (18-21). Neutrophil elastase probably further contributes to the degradation of the interstitial collagens by cleaving in the telopeptide region of the molecule and solubilizing entire collagen molecules (22). The role of this phenomenon is minimal compared to that of IC (23).

Characteristics of IC: IC has been identified from a variety of sources, which are listed in Table 1. The best characterized of these are human skin collagenase and rabbit synovial collagenase. Despite the variety of cell types, the enzymes are remarkably similar. Neutrophil collagenase is an exception in that it has some unique characteristics. Thus, the basic characteristics of IC can be discussed in a general sense and are based on studies of skin fibroblast and synovial cell enzymes.

The enzyme is a metalloproteinase that is calcium dependent and inhibited by chelating agents (10). It is synthesized and released by cultured cells without intracellular stor-

Table 2 Properties of Interstitial Collagenases

Metalloproteinase
Inhibited by chelating agents, α-2 macroglobulin, β-1 anticollagenase, and a variety of tissue inhibitors
Released from cultured mesenchymal cells without intracellular storage
Stored intracellularly in neutrophils
Released in a latent form and can be activated by certain proteinases, organomercurials, and chaotropic agents
The molecular weight of latent fibroblast IC is 59,000 (neutrophil IC is larger)
Activation is associated with a decrease in molecular weight of approximately 10,000 daltons
Make a single cleavage through the three respective chains of the interstitial collagens
Native collagen is a better substrate than gelatin
Inactive against types IV and V collagens

age in lysosomes (except for neutrophil collagenase). The synthesis and release of IC can be manipulated in vitro by a variety of compounds. Agents that can stimulate its synthesis include cytochalasin B (24), urate crystals (25), and the phorbol ester tetradecanoyl phorbol acetate (26,27). In the case of the latter two agents, the stimulation occurs at the level of transcription as they result in an increase in mRNA synthesis (28). The regulating substance most germane to the study of human disease is interleukin 1. This is a protein with a mol. wt. of 16,000 which is a product of macrophage/monocytes (29, 30). It causes a marked stimulation of collagenase release and may be a major factor in pathological conditions characterized by collagen breakdown. In addition, a factor from neutrophils capable of stimulating IC release from synovial cells has been reported (31). Release of IC by macrophages can be stimulated by bacterial lipopolysaccharide and by lymphokines, and this may be regulated through endogenous prostaglandin production (32-34).

Corticosteroids inhibit the release of IC from synovial cells as well as macrophages (35-37). In the case of the latter, the effect is apparent in vivo as well as in vitro (37). Retinoids inhibit IC synthesis by synovial or skin fibroblasts (38,39). This phenomenon has been used as rationale for the therapeutic use of this class of agents. Inhibition of prostaglandin synthesis by indomethacin, an inhibitor of cyclooxygenase, results in an abrogation of the interleukin I induction of IC release by synovial cells (34,40).

The enzyme is released in a latent form requiring "activation" for the manifestation of collagenolytic activity. Agents that convert latent IC to active IC are of three general categories: proteinases (trypsin, plasmin); organomercurial compounds (aminophenylmercuric acetate, etc.); and chaotropic agents (NaI, NaSCN) (41-45). Activation with the first two classes of compounds is associated with a decrease in relative molecular mass of approximately 10 kd, although activation is observed before the molecular weight change (46,47). In the case of synovial collagenase, a proactivator protein has been isolated. In this system, the activating agent (trypsin, organic mercurial compound) acts on the proactivator, which in turn activates IC, and original agent is ineffective in activating IC in the absence of proactivator. Thus, rabbit synovial IC is activated by a cascade phenomenon (48).

When analyzed by gel electrophoresis, the latent enzyme appears as a doublet with a mol. wt. of 57-59 kd. The more slowly migrating species is glycosylated (49,50).

The rate of cleavage of the collagen types by IC (human skin IC) is not uniform. Type III collagen is degraded most rapidly, followed by type I collagen. Type II is degraded very slowly by this enzyme (12). In addition, kinetic studies have indicated an exceptionally high energy of activation (51). This observation supports earlier studies that demonstrated a marked increase in the rate of collagenolysis with modest increases in reaction temperatures (52).

Natural inhibitors of IC include the serum proteins α-2 macroglobulin and β-1 anticollagenase (53,54). In addition, tissue inhibitors have been identified from cartilage, aorta, tendon, fibroblast culture, organ culture of bone and uterus, amniotic fluid, and homogenates of human neutrophils (55-63). These inhibitors probably represent an important regulatory factor in collagen breakdown. It is of interest that this latter group represents a homogeneous family of metalloproteinase inhibitors referred to as TIMP (tissue inhibitor of metalloproteinases). In addition, ovostatin, a large-molecular-weight metalloproteinase inhibitor, has been isolated from egg white (64).

Neutrophil IC: The neutrophil is often present at sites of inflammation and, therefore, is a likely participant in collagen degradation in inflammatory states. This cell contains an interstitial collagenase, which is distinct from the enzymes described above. First, the mature neutrophil no longer synthesizes proteins. Thus, IC is probably synthesized during maturation and is packaged in the specific granule compartment. The enzyme is released from the cell under a variety of stimuli, including calcium ionophore, concanavalin A, phagocytosis, and phorbol myristate acetate (65), and, like the other IC, the enzyme is latent. The enzyme is activated by the same agents as described above with the exception that trypsin is a poor activator. Furthermore, the purified enzyme is larger with forms identified with molecular weights of between 57 and 90 kilodaltons (66,67).

A monoclonal antibody raised against human neutrophil IC failed to react with human skin or human synovial IC, whereas a polyclonal antibody did cross-react with human skin IC (68). These experiments suggest that some, but not all, of the antigenic determinants are preserved. In other studies, a polyclonal antibody raised against human synovial IC did not react with human neutrophil IC (69). Finally, it has been shown that this enzyme degraded type I collagen more rapidly than type III, which is in contrast to the skin IC (70).

Thus, it has become clear that neutrophil IC is different from skin or synovial IC. Discriminating monoclonal antibodies should help define the relative role of these enzymes in pathological conditions.

2. Basement Membrane Collagen Degradation

The study of collagen degradation expanded when several investigators independently discovered that both type IV (basement membrane) collagen and type V (pericellular) collagen were resistant to the action of the interstitial collagenase (11,12,14). This was followed by the identification of a type IV-specific collagenase from murine tumor cells by Liotta and his colleagues (13). Thus, the concept of type-specific collagenolysis has gained acceptance among investigators.

Although type IV collagen comprises a small part of the total matrix, its role as a major structural component of basement membranes suggests a critical role for it in the cellular arrangement within a given tissue. Furthermore, the barrier function of basement membrane may be dependent on the integrity of its collagenous component. Thus, the enzymatic degradation of type IV collagen may result in profound alterations in tissue architecture and function.

The structure of type IV collagen is described in detail in Chapter 6, but it is important to stress that this collagen is quite different from the interstitial collagens. The constituent chains are longer with a molecular weight of 170-185 kd (71). Also, the gly-X-Y sequence is interrupted in certain locations. Thus, it is not surprising that its breakdown is dependent on different mechanisms (72).

Type IV collagen degrading enzymes have been identified from several sources. The first report of such an enzyme was by Liotta et al. and has been described as a metalloproteinase (13). Further purification and characterization has revealed that the enzyme exists in culture medium in a partially latent form and can be activated by trypsin or mercuric chloride (73). The purified enzyme has a molecular weight of approximately 68,000. The enzyme makes a single cleavage in the substrate near the amino terminus of the molecule (74).

Type IV collagen can be degraded by other proteinases. Mast cell proteinase, a chymotrypsinlike serine proteinase, has been shown to cleave pepsin-extracted type IV collagen (75). Human neutrophil elastase, a well-characterized serine proteinase, degrades native, insoluble type IV collagen as well as intact, acid-extracted collagen (76). Type IV collagen-degrading activity has been reported to be present in monocytes and macrophages (77). Metallogelatinases from human neutrophils and rabbit bone have been shown to cleave pepsin-extracted type IV collagen (78,79). Endothelial cells have been shown to degrade type IV collagen under conditions that favor cell migration (80).

The study of type IV collagen degradation suffers from the lack of a universally accepted assay system, and different substrates have been used in the above-outlined studies. Some investigators have used biosynthetically labeled type IV collagen from a benign murine tumor (EHS sarcoma) (13). Basement membrane collagen can be prepared from organ culture of this tumor without the use of pepsin, which is a distinct advantage. Others have used insoluble bovine lens capsule (76), which has the advantage of preserving natural molecular packing, but this approach suffers from a lack of sensitivity. Pepsin-extracted collagen has also been used and is less than the ideal substrate as the molecule has been altered by this proteinase.

Future studies should delineate the role of these enzymes in pathological conditions and characterize the pathophysiological implications of type IV collagen degradation.

3. *Type V Collagen Degradation and the Role of Gelatin-Specific Metalloproteinases*

Type V collagen, another minor component of the matrix, has been localized to the pericellular area and is felt to be a component of the exocytoskeleton (81). Like type IV collagen, there is at least one noncollagenous domain that is cleaved by pepsin (82-84), and therefore, the use of this acid proteinase results in an altered molecule. Although the precise function of this collagen is not known, it is possible that its enzymatic removal from the matrix may result in severe tissue dysfunction.

Type V collagen degrading enzymes have been isolated from various sources, including rabbit macrophages (14), human tumor cells (85), migrating endothelial cells (80), human neutrophils (79), rabbit bone cultures (78), and human sputum from patients with chronic bronchitis (86). In addition, the coagulation serine proteinase thrombin has been reported to degrade type V collagen (87).

Like the interstitial collagenases, similarities have been found among many of these type V collagenases. They are metalloproteinases released from the cells in a latent form, are separable from the interstitial collagenase by ion-exchange chromatography, and tend to be greater in relative molecular mass. Furthermore, they are inactive against native interstitial collagens but rapidly degrade denatured collagen chains and are identical to previously described gelatinases. The fact that denatured collagen chains can be degraded by many nonspecific proteinases has led some investigators to question the role these metallogelatinases play in matrix degradation. At present there is evidence to suggest three distinct roles for these enzymes in collagen breakdown.

1. After the interstitial collagenases cleave the triple helix, the reaction products denature and may be completed digested by a gelatinase.
2. They selectively degrade native type V collagen and, therefore, can be classified as collagenases.
3. They potentiate the action of interstitial collagenase (67). The mechanism of this potentiation is not clear but is significant when reconstituted type I collagen fibrils are used as substrate. When soluble collagen is incubated with a mixture of the IC and gelatinase from human neutrophils at a temperature that preserves the native configuration of the IC-generated fragments, the native three-fourths fragment is cleaved by the gelatinase, resulting in a shortening of approximately 10-20%. This phenomenon may, at least in part, explain the potentiating effect.

D. Elastin Degradation

Like collagen, elastin, in its native matrix form, is relatively resistant to proteolysis. Thus, solubilization from the extracellular matrix of cross-linked elastin requires the action of a "specific" elastase (for a review of elastolysis, see Ref. 88). The best characterized vertebrate elastases are derived from pancreas, neutrophils, and macrophages. Pancreatic elastase is a digestive enzyme and is not associated with pathological processes except for pancreatitis. It is a serine proteinase that hydrolyzes peptide bonds in which a hydrophobic residue (ala, val, leu, etc.) is the amino constituent (88). Granulocyte elastase is a serine proteinase present in the azurophil granule of the human polymorphonuclear neutrophil (89). This enzyme is a potent general proteinase that cleaves other matrix components including types III and IV collagen as well as fibronectin (7,20,21,76). This enzyme has been directly incriminated in the development of disease states in the human including pulmonary emphysema (90). Macrophage elastase is distinct from these other elastases in that it is a metalloproteinase (91). Like the neutrophil enzyme, macrophage elastase is not a specific elastase and could contribute to pathological processes by degrading other proteins. One substrate of this enzyme is α-1 proteinase inhibitor, which is cleaved and inactivated by mouse macrophage elastase (92). This is of interest as α-1 proteinase inhibitor is a major counterregulatory protein controlling the action of neutrophil elastase.

Elastaselike enzymes have been described from other cellular sources including fibroblasts, smooth muscle cells, and platelets (93-95). The designation "elastaselike" is based on the hydrolysis of synthetic tri- and tetrapeptide substrates as well as soluble and insoluble elastin, although activity against insoluble elastin is variable. These enzymes have been partially characterized, and their roles in physiological and pathological processes remain to be determined.

III. MATRIX DEGRADING PROTEINASES IN PATHOLOGICAL PROCESSES

A. General Considerations

It is logical to assume that matrix degradation contributes to any pathological process that results in an alteration in the architecture of a given tissue or organ. This includes virtually all the chronic inflammatory, fibrotic, or neoplastic diseases. It is beyond the scope of this chapter to provide a comprehensive list of these disorders. Instead, certain representative conditions will be reviewed briefly with a focus on collagen degradation, and an attempt will be made to minimize overlap with other chapters.

In these disorders there is some question as to the cellular source of the matrix degrading proteinases. Specifically, in inflammatory and neoplastic diseases, matrix degradation may occur by two major mechanisms: (a) the invading inflammatory cells (neutrophils, macrophages), tumor cells, or vascular cells (endothelial cells in neovascularization) release proteinases into the extracellular space; or (b) the invading cell(s) influence the resident cells (fibroblasts, epithelial cells, smooth muscle, etc.) to release these enzymes.

Indeed, there is evidence to support both these mechanisms in disease processes. Investigators have demonstrated cell-cell interactions that influence matrix breakdown between macrophage/monocytes and fibroblasts (29,30,96), epithelial cells and fibroblasts (97), tumor cells and fibroblasts (98,99), and tumor cells and macrophages (100). In the case of the macrophage/monocyte, this is mediated by interleukin 1, a soluble factor that stimulates collagenase synthesis by target cells (29,30). Similarly, interstitial collagenase secretion by fibroblasts can be stimulated by the presence of epithelial and tumor cells (97-100). In general, the fibroblast has a greater capacity for synthesis and release of IC than the macrophage, and it is likely that the native fibroblast makes the major contribution to collagenolysis in inflammatory conditions. In vitro studies support this concept (96).

Another mechanism by which matrix breakdown can be enhanced is the relative deficiency of a counterregulatory mechanism. An example of this is the well-known model of α-1 proteinase inhibitor deficiency and emphysema. This subject is covered in Chapter 19. Genetic abnormalities in matrix degrading proteinases can lead to pathological processes as in recessive dystrophic epidermolysis bullosa.

Decreased activity of these enzymes can also lead to abnormalities. Decreased interstitial collagenase has been found in a connective tissue nevus of the skin (101), and a decrease of active enzyme has been associated with diphenylhydantoin-induced gingival hyperplasia (102). The latter observation suggests that the abnormality exists at the level of enzyme activation. It is logical to hypothesize that conditions associated with an excess of connective tissue (i.e., liver cirrhosis, lung fibrosis, progressive systemic sclerosis) might be associated with a relative deficiency of matrix degrading proteolysis. This has not been the case. Inflammatory fibrotic disorders are generally associated with an increase in collagenolytic activity (vide infra), whereas IC has been found in normal quantities in scleroderma fibroblasts (103).

Although it is not the subject of this chapter, matrix degradation plays a role in certain nonpathological conditions. In embryogenesis and tissue remodeling, precisely controlled enzymatic mechanisms are probably essential. In wound healing, enzymatic debridement occurs during an early inflammatory phase, and during the later reparative phase, enzymatic mechanisms must be involved in remodeling. There is evidence to indicate that matrix degradation plays a role at various stages in the reproductive process, including ovulation (104) and uterine involution (105). It is possible that abnormalities in connective tissue degradation may be associated with disorders of fertility and reproduction, although none have been defined.

B. Arthritis

Chronic arthritis due to a variety of causes frequently results in a lesion characterized by destruction of articular cartilage, subchondral bone, and other juxtaarticular connective tissue structures. Thus, investigators interested in the pathogenesis of these diseases have had more than a casual interest in the study of connective tissue breakdown. A simplistic approach to the study of the pathogenesis of arthritis is to categorize the lesions into two groups: (a) those in which inflammation and synovial proliferation are prevalent (e.g., rheumatoid arthritis), and (b) those in which inflammation is minimal or absent (e.g., osteoarthritis).

The model for inflammatory arthropathy is rheumatoid arthritis, in which the joint is the setting for most of the known facets of acute and chronic inflammation. In the early lesion, the synovial lining becomes infiltrated with monocytes and lymphocytes, and locally produced immune complexes activate complement and attract granulocytes into the synovial fluid. As the disease progresses, cell-mediated immune events prevail, and the synovial lining becomes hyperplastic and thickened. This synovial proliferation becomes polarized and invades the articular cartilage and subchondral bone, destroying the existing matrix in its path. In addition to the articular destruction associated with pannus invasion, there is thinning of the articular cartilage at sites distant from the invading tissue. Owing to the variety of cell types present in the rheumatoid lesion, there is some controversy and confusion as to the relative role of the individual cells in the destructive process. Preservation of the collagenous framework of cartilage is essential to regeneration of the tissue after injury. Therefore, the focus of investigation has been on collagen degradation.

The initial observations were made by Evanson et al. (106), who found collagenolytic activity in culture medium from explants of rheumatoid synovium, and by Lazarus et al. (107), who identified a collagenase in human neutrophils. These early reports established the fact that the potential for collagenolysis existed in the rheumatoid lesion. In a later study of collagenase activity in rheumatoid synovial fluid, there was a suggestion that two distinct enzymes were present (108). Subsequently, Dayer and his colleagues characterized a population of adherent synovial cells that could be isolated from rheumatoid synovium and maintained in culture (109). These cultures secreted impressive quantities of collagenase and PGE_2 in early passage, with a subsequent decrease in these markers with passage. Further studies identified a mononuclear cell factor that could stimulate the release of these two substances from these cultures and this factor was shown to be identical to interleukin 1 (29). Immunolocalization studies of cultured rheumatoid synovial cells revealed that the stellate or dendritic cells frequently present in large numbers were particularly rich in immunoreactive collagenase (110). The importance of synovial cell collagenase in cartilage destruction was strongly suggested by the study of Woolley et al. (111) in which immunoreactive collagenase was localized at the invading front of the cartilage/pannus junction of the rheumatoid lesion. The antibody used was not cross-reactive with human neutrophil collagenase.

Rheumatoid synovial collagenase is released from the cell in a latent form. It has been shown that plasmin can activate the latent enzyme, and rheumatoid synovial cells secrete a plasminogen activator. Thus, in the presence of plasminogen, active enzyme can be generated (35). This phenomenon probably occurs through the activation of a proactivator by plasmin (48). Studies of synovial fluid have indicated that plasma kallikrein may be important in the activation of latent IC (112).

The establishment of a role for synovial cell collagenase in the rheumatoid lesion does not preclude a role for other proteinases of different origin. Neutrophils are found in large quantities in rheumatoid synovial fluid, and proteinases from these cells reach the extracellular space. These enzymes probably contribute to the joint destruction by degrading both collagen and noncollagenous components. Neutrophil elastase has been found in extracts of rheumatoid articular cartilage, suggesting that it diffuses into the cartilage matrix where it could degrade connective tissue elements (113). This enzyme has been identified at the cartilage-pannus junction in the rheumatoid lesion (114).

Recent investigations have focused on the role of the chondrocyte in the breakdown of the cartilage matrix. The fact that the chondrocyte is the only cell present in normal cartilage indicates that it must possess the necessary enzymes to break down its matrix during normal tissue remodeling. In fact, cultured cartilage slices as well as isolated chondrocytes secrete collagenase and a proteoglycan-degrading neutral proteinase (115-117), and the secretion of these enzymes can be stimulated by a macrophage factor (117). The observation that cartilage matrix was degraded when vital cartilage was cocultured with pig synovial tissue, whereas devitalized cartilage was unaffected, led to the conclusion that chondrocytes could be stimulated to accelerate the resorption of cartilage (118). This phenomenon has been attributed to an acidic protein with M_r 21,000, called "catabolin," which has been found in synovium or mononuclear leukocytes from either human or porcine sources (119-122). The relationship of this protein to interleukin 1 remains to be elucidated.

Thus, the pathological process involved in the rheumatoid joint is an extremely complex phenomenon with a myriad of cell-cell interactions. The net result is a destructive lesion with little reparative response. The cells directly involved in matrix breakdown include the synovial cell, the neutrophil, and the chondrocyte, and it is likely that all play a major role in the end-stage lesion.

In osteoarthritis, little or no inflammation accompanies the process, and, therefore, some investigators have concluded that cartilage erosion is a result of mechanical forces. Recent evidence has implicated chondrocyte proteinases in the formation of the areas of denuded cartilage (123,124).

Another joint disease that has been associated with collagen degradation is "Milwaukee shoulder." This is characterized by decreased mobility, rotator cuff abnormalities, and degeneration of the glenohumoral joint (125). The synovial fluid is characteristically noninflammatory with microspheroids containing hydroxyapatite crystals and active collagenase and neutral proteinase. The source of these enzymes, as well as their role in the pathogenesis, requires clarification.

C. Matrix Degradation and Periodontal Disease

Chronic periodontal inflammation is a common condition in our society and is a leading cause of tooth loss. The disease is initiated by dental plaque which induces an inflammatory reaction in the gingiva. If the bacterial plaque persists, the inflammatory reaction becomes chronic with proliferation of epithelial cells and inflammatory cell infiltration. This lesion is associated with resorption of the connective tissue structures that maintain the tooth in the jaw, including the alveolar bone and periodontal membrane (126).

Like the rheumatoid lesion, collagenase and PG-degrading gelatinase have been associated with this process (127), but the cellular response is heterogeneous and the specific role of each cell is not clear. Explant cultures of gingiva release these enzymes, as do isolated cultured gingival fibroblasts (127-130). The interstitial collagenase has been partially characterized and is similar, if not identical, to the IC from synovium and skin (127-131). Collagenolysis is greater in inflamed gingiva than in healthy gingiva (132). There is evidence to support a role for the neutrophil, the epithelial cell, and the fibroblast in the process (127,133). It is likely that, as in rheumatoid arthritis, all these cells contribute and cell-cell interactions similar to those described above are operative in chronic periodontal disease.

D. Corneal Disease

Corneal ulceration can be due to several causes, including alkali burns, bacterial and viral infections, vitamin A deficiency, and systemic connective tissue disease. In view of the dense connective tissue matrix of this structure, it is logical to assume that ulceration must be associated with collagen breakdown (134). Experimental evidence exists to support this in that organ cultures of ulcerating corneal tissue secrete an interstitial collagenase into the culture medium (135). Early studies suggested that epithelial cells may be the source of this collagenase because epithelial explants from ulcerating corneas were capable of lysing collagen gels (134). Subsequent studies suggested a role for PMNs owing to the observation that prevention of PMN influx by a glued-on contact lens prevented the development of corneal ulceration in a model of vitamin A deficiency (136). Stromal fibroblasts also produce a collagenase (97). Gordon et al., in an immunochemical study of human corneas, identified immunoreactive interstitial collagenase exclusively in the stromal region of ulcerating corneas, but no enzyme could be detected in nonulcerating corneas. Of further interest was the fact that enzyme could be detected in nonulcerating corneas after organ culture, suggesting that the process of organ culture alone is sufficient to induce IC release (137). Thus, it appears that the stromal fibroblast is responsible for secretion of the collagenase in corneal injury. The work of Johnson-Wint (97) has suggested that the epithelial cell may play a major role in orchestrating the collagenolysis by secreting soluble factor(s) that stimulate the release of IC by fibroblasts. This provides an attractive model in which the injured epithelial cell stimulates the adjacent stromal fibroblast to degrade the connective tissue in ulcer formation.

Keratoconus is an unusual condition characterized by central protrusion of the cornea with thinning of the stroma. Cultured corneas from affected individuals secrete significantly more IC and gelatinase than control corneas, suggesting a major role for these enzymes in the pathogenesis of the disease (138).

E. Skin Disease

The best characterized relationship between matrix degradation and skin disease is the rare blistering disorder recessive dystrophic epidermolysis bullosa, which is discussed in detail in Chapter 11. This has led others to investigate the role of matrix-degrading proteinases in other blistering diseases. Oikarinen and colleagues examined blister fluid from a variety of bullous diseases for elastase and IC activity (139). IC activity was found in association with bullous pemphigoid, dermatitis herpetiformis, and chronic bullous disease of childhood. Elastase activity was found in fluid from patients with the above disorders as well as with pemphigus vulgaris. In dermatitis herpetiformis the elastase was found to be a serine proteinase, suggesting that it was derived from neutrophils, whereas the elastolytic activity from bullous pemphigoid was calcium dependent, suggesting another cellular source. Other investigators have presented in vitro evidence that epidermal cell proteinase(s) are responsible for cellular dyshesion and that pemphigus antibody induces this phenomenon (140).

Decreased collagenolytic activity and immunoreactive interstitial collagenase was found in association with regional fibroblasts from a patient with connective tissue nevi of the collagen type (101).

F. Tumor Invasion and Metastasis

The local growth of tumors involves replacement of normal tissue with tumor tissue. This process involves remodeling of the matrix and, therefore, connective tissue degradation. Ultrastructural studies have provided evidence of collagen breakdown in the stromal tissues surrouding invading tumors, and the collagenolytic potential of whole tumors and isolated tumor cells has been documented (141).

Local invasion of tumors and the associated stromal remodeling is dependent on interstitial collagen degradation and the action of interstitial collagenase. There is evidence to support two mechanisms of interstitial collagen breakdown in tumor invasion. First, the tumor cell can participate in the process directly by secretion of an IC. Alternatively, the tumor cell interacts with the host stromal cells stimulating the local release of IC. The former is supported by the observation that cultured mammary carcinoma cells can secrete IC and plasminogen activator (PA) and that plasmin which is generated from plasminogen by PA could activate IC (142). The second hypothesis is supported by an immunolocalization study of basal cell carcinomas in which the enzyme was found exclusively in the stromal tissue adjacent to the tumor (143). Biswas (98), as well as Dabbous and his colleagues (99), using the rabbit VX-2 carcinoma model, have provided evidence that isolated, cultured epithelial tumor cells secreted an IC but also secreted a substance(s) that could enhance IC secretion by isolated tumor fibroblasts. Henry et al. have shown that tumor cells could be stimulated to degrade type I collagen when cocultured with macrophages (100). These studies underscore the complexity of the potential cell-cell interactions in the invasion of the stroma by tumors, and it is likely that all these possible mechanisms are operative in the process.

Metastasis of tumors is dependent on the invasion of tumor cells into vascular structures. The extracellular matrix of blood vessels is different from that of stromal tissues, and, therefore, the invasion of these structures is felt to be dependent on the degradation of basement membrane constituents and elastin. Liotta and his colleagues demonstrated that certain tumor cells secrete an enzyme that specifically degrades type IV or basement membrane collagen (13). This enzyme has been purified to homogeneity and partially characterized (73). Further studies have correlated metastatic potential with type IV collagenase secretion (144). Latent type IV collagenase could be activated by plasmin, indicating a cascade phenomenon similar to that described for IC (145). Type IV collagenase was secreted only by malignant cells, but PA was secreted by malignant and certain nonmalignant cells.

Other investigators have devised an in vitro model for tumor cell penetration of the bloodstream using monolayers of vascular endothelial cells. Metastatic tumor cells adhere to these monolayers, induce a disruption in the endothelial cell junctions, and migrate to the substratum of the monolayers (146). This invasive process has been associated with the enzymatic degradation of fibronectin and sulfated proteoglycans (147).

G. Hepatic Fibrosis

Chronic liver injury results in a fibrotic response and ultimately a disruption of normal hepatic architecture. This has been associated with an increase in hepatic collagen content, which is mostly represented by type I collagen. Investigations into the role of collagen degradation in experimental and human liver fibrosis have generally demonstrated an increased presence of IC, although the advanced lesion in humans has been associated with a decrease in collagenolytic activity (148,149). The cellular source of the enzyme is not certain as IC has been found in cultures of hepatocytes and Kupffer cells, although the native hepatic fibroblasts may contribute to the process (150,151). The increase in IC activity suggests that the development of hepatic fibrosis is related to an ongoing remodeling of the connective tissue framework of this organ.

H. Lung Disease

The role of connective tissue proteinases in lung injury as it relates to emphysema has been covered in depth in Chapter 19. Interstitial collagenolytic activity in the lower respiratory tree has been associated with idiopathic pulmonary fibrosis (152), the pathogenesis of which is covered in Chapter 14. Recently, the role of connective tissue degradation in other pulmonary diseases has been investigated. Intratracheal instillation of cadmium to laboratory animals generally results in development of interstitial fibrosis. However, if tissue repair is altered by rendering the animals lathyritic, emphysema results (153). This suggests that these two diverse pathological lesions can result from the same initial insult. Adult respiratory distress syndrome is a life-threatening situation in which the barrier function of alveolar basement membrane becomes acutely impaired. Elastase activity has been found in the pulmonary lavage fluid of patients with this condition (154), but the role of proteolysis in the pathogenesis of this condition remains to be established.

I. Other Conditions

Resorption of bone is a complex situation in which demineralization plays a role that may be more important than proteolysis of the connective tissue elements. IC is released by bone explants in culture and has been felt to be important in the process of resorption (for review, see Ref. 155). Recent studies have questioned the role of IC in the process owing to a lack of correlation between IC release and in vitro bone resorption (156). This issue remains controversial.

Peptic ulcer disease has traditionally been associated with gastric acid production. A neutral collagenase has been identified from gastric mucosa (157), and recent evidence suggests a role for this enzyme in the development of ulcers (158).

Ultrastructural evidence of collagen degradation in atherosclerosis and varicose veins has suggested a role for connective tissue degrading enzymes in these common disorders (159,160).

REFERENCES

1. A. Baici, in *Proteinases and Tumor Invasion* (P. Strauli, A. J. Barrett, and A. Baici, eds.), Raven Press, New York, 1980, p. 17.
2. J. M. McCord, *Science, 185*:529 (1974).

3. R. A. Greenwald and W. W. Moy, *Arthritis Rheum., 23*:455 (1980).
4. E. Ruoslahti and E. G. Hayman, *FEBS Let., 97*:221 (1979).
5. G. Balian, E. M. Click, E. Crouch, J. M. Davidson, and P. Bornstein, *J. Biol. Chem., 254*:1429 (1979).
6. L-H. E. Hahn and K. M. Yamada, *Cell, 18*:1043 (1979).
7. J. A. McDonald and D. G. Kelley, *J. Biol. Chem., 255*:8848 (1980).
8. A. E. Postlethwaite, G. Keski-Oja, G. Balian, and A. H. Kang, *J. Exp. Med., 153*: 494 (1981).
9. V. Gauss-Muller, H. K. Kleinman, G. R. Martin, and E. Schiffman, *J. Lab. Clin. Med., 96*:1076 (1980).
10. E. D. Harris, Jr. and E. C. Cartwright, in *Proteinases in Mammalian Cells and Tissues* (A. J. Barrett, ed.), Elsevier, New York, 1977.
11. D. E. Woolley, R. W. Glanville, D. R. Roberts, and J. M. Evanson, *Biochem. J., 169*: 265 (1978).
12. H. G. Welgus, J. J. Jeffrey, and A. Z. Eisen, *J. Biol. Chem., 255*:9511 (1981).
13. L. A. Liotta, S. Abe, P. G. Robey, and G. R. Martin, *Proc. Natl. Acad. Sci. USA, 76*:2268 (1979).
14. C. L. Mainardi, J. M. Seyer, and A. H. Kang, *Biochem. Biophys. Res. Commun. 97*: 1108 (1980).
15. J. Gross, J. H. Highberger, B. Johnson-Wint, and C. Biswas in *Collagenase in Normal and Pathological Connective Tissue* (D. E. Wooley and J. M. Evanson, eds.) Wiley, New York, 1980, p. 11.
16. P. A. McCroskery, S. Wood, and E. D. Harris, Jr., *Science 182*:70 (1973).
17. H. G. Welgus, J. J. Jeffrey, G. P. Stricklin, and A. Z. Eisen, *J. Biol. Chem., 257*: 11534 (1982).
18. H-M. Wang, J. Chang, D. W. Pettigrew, and J. Sodek, *Biochim. Biophys. Acta, 533*: 270 (1978).
19. E. J. Miller, J. E. Finch, Jr., Chung, E., W. T. Butler, and P. B. Robertson, *Arch. Biochem. Biophys. 173*:631 (1976).
20. C. L. Mainardi, D. L. Hasty, J. M. Seyer, and A. H. Kang, *J. Biol. Chem., 255*: 12006 (1980).
21. J. E. Gadek, G. A. Fells, D. G. Wright, and R. G. Crystal, *Biochem. Biophys. Res. Commun., 85*:1815 (1980).
22. P. M. Starkey, A. J. Barrett, and M. C. Burleigh, *Biochim. Biophys. Acta., 483*: 386 (1977).
23. E. D. Harris, Jr., C. A. Vater, C. L. Mainardi, and R. C. Siegel, in *Biology of Articular Cartilage in Health and Disease* (H. Gastpar, ed.), Verlag, Stuttgart, 1980, p. 175.
24. E. D. Harris, Jr., J. J. Reynolds, and Z. Werb, *Nature, 257*:243 (1975).
25. R. M. McMillan, C. A. Vater, P. Hasselbacher, and E. D. Harris, Jr., *J. Pharm. Pharmacol., 33*:382 (1981).
26. C. E. Brinckerhoff, R. M. McMillan, J. V. Fahey, and E. D. Harris, Jr., *Arthritis Rheum., 22*:1109 (1979).
27. D. Moscatelli, E. Jaffe, and D. B. Rifkin, *Cell, 20*:343 (1980).
28. C. E. Brinckerhoff, R. H. Gross, H. Nagase, L. Sheldon, R. C. Jackson, and E. D. Harris, Jr., *Biochemistry, 27*:2079 (1982).
29. S. B. Mizel, J-M. Dayer, S. M. Krane, and S. E. Mergenhagen, *Proc. Natl. Acad. Sci. USA, 78*:2474 (1980).
30. A. E. Postlethwaite, L. B. Lachman, C. L. Mainardi, and A. H. Kang, *J. Exp. Med., 157*:801 (1983).
31. R. Hashida, K. Terato, K. Miyamoto, T. Morimoto, H. Hori, and Y. Nagai, *Biomed. Res., 3*:506 (1982).

32. L. M. Wahl, S. M. Wahl, S. E. Mergenhagen, and G. R. Martin, *Science, 187*:261 (1974).
33. L. M. Wahl, S. M. Wahl, S. E. Mergenhagen, and G. R. Martin, *Proc. Natl. Acad. Sci. USA, 71*:3593 (1975).
34. L. M. Wahl, C. E. Olsen, A. L. Sandberg, and S. E. Mergenhagen, *Proc. Natl. Acad. Sci. USA, 74*:4955 (1977).
35. Z. Werb, C. L. Mainardi, C. A. Vater, and E. D. Harris, Jr., *N. Engl. J. Med., 296*: 1017 (1977).
36. Z. Werb, *J. Exp. Med., 147*:1695 (1978).
37. S. W. Crawford, A. H. Kang, and C. L. Mainardi, *Am. Rev. Respir. Dis., 127*:46 (1983).
38. C. E. Brinckerhoff, R. M. McMillan, J-M. Dayer, and E. D. Harris, Jr., *N. Engl. J. Med., 303*:432 (1980).
39. E. A. Bauer, J. L. Seltzer, A. Z. Eisen, *J. Invest. Dermatol. 81*:162 (1982).
40. J-M. Dayer, S. R. Goldring, D. R. Robinson, and S. M. Krane, in *Collagenase in Normal and Pathologic Connective Tissues,* Wiley, New York, 1980, p. 84.
41. G. Vaes, *Biochem. J., 123*:23 (1971).
42. E. Harper, K. J. Bloch, and J. Gross, *Biochemistry, 10*:3035 (1971).
43. S. Abe and Y. Nagai, *J. Biochem., 73*:1007 (1972).
44. C. A. Vater, C. L. Mainardi, and E. D. Harris, Jr., *J. Clin. Invest., 62*:987 (1978).
45. A. Sellers, E. Cartwright, G. Murphy, and J. J. Reynolds, *Biochem. J., 163*:303 (1977).
46. G. P. Stricklin, J. J. Jeffrey, W. T. Roswit, and A. Z. Eisen, *Biochemistry, 22*:61 (1983).
47. H. Nagase, R. C. Jackson, E. E. Brinckerhoff, C. A. Vater, and E. D. Harris, Jr., *J. Biol. Chem., 256*:11951 (1981).
48. C. A. Vater, H. Nagase, and E. D. Harris, Jr., *J. Biol. Chem., 258*:9374 (1983).
49. G. P. Stricklin, A. Z. Eisen, E. A. Bauer, and J. J. Jeffrey, *Biochemistry, 17*:2331 (1977).
50. H. Nagase, C. E. Brinckerhoff, C. A. Vater, and E. D. Harris, Jr., *Biochem. J., 214*: 281 (1983).
51. H. G. Welgus, J. J. Jeffrey, and A. Z. Eisen, *J. Biol. Chem., 256*:9516 (1981).
52. E. D. Harris, Jr., and P. A. McCroskery, *N. Engl. J. Med., 290*:1 (1974).
53. A. Z. Eisen, E. A. Bauer, and J. J. Jeffrey, *Proc. Natl. Acad. Sci. USA, 68*:248 (1971).
54. D. E. Woolley, D. R. Roberts, and J. M. Evanson, *Nature, 261*:325 (1976).
55. K. E. Kuettner, J. Hiti, R. Eisenstein, and E. Harper, *Biochem. Biophys. Res. Commun., 72*:40 (1976).
56. P. J. Roughley, G. Murphy, and A. J. Barrett, *Biochem. J., 169*:721 (1978).
57. G. P. Stricklin and H. G. Welgus, *J. Biol. Chem., 258*:2252 (1983).
58. C. A. Vater, C. L. Mainardi, and E. D. Harris, Jr., *J. Biol. Chem., 254*:3045 (1978).
59. A. Sellers, E. Cartwright, G. Murphy, and J. J. Reynolds, *Biochem. J., 163*:303 (1977).
60. G. Murphy, E. Cartwright, A. Sellers, and J. J. Reynolds, *Biochim. Biophys. Acta, 483*:493 (1977).
61. T. E. Cawston, A. Galloway, E. Mercer, G. Murphy, and J. J. Reynolds, *Biochem. J., 195*:159 (1981).
62. G. Murphy, T. E. Cawston, and J. J. Reynolds, *Biochem. J., 195*:167 (1981).
63. H. W. Macartney and H. Tschesche, *Eur. J. Biochem., 139*:79 (1983).
64. H. Nagase, E. D. Harris, Jr., J. F. Woessner, and K. Brew, *J. Biol. Chem., 258*:7481 (1983).
65. M. S. Hibbs, K. A. Hasty, A. H. Kang, and C. L. Mainardi, *Arthritis Rheum., 26*:s42 (1983).

66. H. W. Macartney and H. Tschesche, *Eur. J. Biochem., 130*:71 (1983).
67. G. Murphy, J. J. Reynolds, U. Bretz, and M. Baggiolini, *Biochem. J., 203*:209 (1982).
68. K. A. Hasty, M. S. Hibbs, A. H. Kang, and C. L. Mainardi, *J. Exp. Med.*, (1984) (in press).
69. D. E. Woolley, M. J. Crossley, and J. M. Evanson, *Eur. J. Biochem., 69*:421 (1976).
70. A. L. Horwitz, A. J. Hance, and R. G. Crystal, *Proc. Natl. Acad. Sci. USA, 74*:897 (1977).
71. P. Gehron-Robey and G. R. Martin, *Collagen Rel. Res., 1*:27 (1981).
72. R. W. Glanville, A. Rauter, and P. Fietzek, *Eur. J. Biochem., 95*:383 (1979).
73. T. Salo, L. A. Liotta, and K. Tryggvason, *J. Biol. Chem., 258*:305 (1983).
74. L. I. Fessler, K. Duncan, J. H. Fessler, T. Salo, and K. Tryggvason, *J. Biol. Chem., 259*:9783 (1984).
75. H. Sage, R. G. Woodbury, and P. Bornstein, *J. Biol. Chem., 254*:9893 (1979).
76. C. L. Mainardi, S. N. Dixit, and A. H. Kang, *J. Biol. Chem., 255*:5435 (1980).
77. K. T. Hughes, G. A. Coles, T. R. Harry, and M. Davies, *Biochim. Biophys. Acta, 662*:111 (1981).
78. G. Murphy, T. E. Cawston, W. A. Galloway, M. J. Barnes, R. A. D. Bunning, E. Mercer, J. J. Reynolds, and R. E. Burgeson, *Biochem. J., 199*:807 (1981).
79. G. Murphy, U. Bretz, M. Baggiolini, and J. J. Reynolds, *Biochem. J., 192*:516 (1980).
80. T. Kalebic, S. Garbisa, B. Glaser, and L. A. Liotta, *Science, 221*:281 (1983).
81. S. Gay, A. Martinez-Hernandez, R. K. Rhodes, and E. J. Miller, *Collagen. Rel. Res., 1*:377 (1981).
82. L. I. Fessler, C. A. Kumamoto, M. E. Meis, and J. H. Fessler, *J. Biol. Chem., 256*: 9640 (1981).
83. D. Broek, E. F. Eikenberry, P. T. Fietzek, and B. Brodsky, in *The Chemistry and Biology of the Mineralized Connective Tissue* (A. Veis, ed.), Elsevier-North Holland, New York, 1981, p. 79.
84. S. F. Elstow and J. B. Weiss, *Collagen Rel. Res., 3*:181 (1983).
85. L. A. Liotta, W. L. Lanzer, and S. Garbisa, *Biochem. Biophys. Res. Commun., 98*: 184 (1981).
86. E. A. Rich, J. M. Seyer, A. H. Kang, and C. L. Mainardi, *Am. Rev. Respir. Dis., 128*:166 (1983).
87. H. Sage, P. Pritzl, and P. Bornstein, *Biochemistry, 20*:3778 (1981).
88. Z. Werb, M. J. Banda, J. H. McKerrow, and R. A. Sandhaus, *J. Invest. Dermatol., 79*:154s (1982).
89. M. Baggiolini, U. Bretz, and B. Dewald, in *Neutral Proteinases of Human Polymorphonuclear Leukocytes* (K. Haveman and A. Janoff, eds.), Urban and Schwartzenberg, Baltimore, 1978, p. 3.
90. J. B. Karlinsky and G. L. Snider, *Am. Rev. Respir. Dis., 117*:1109 (1978).
91. M. J. Banda and Z. Werb, *Biochem. J., 193*:589 (1981).
92. M. J. Banda, E. J. Clark, and Z. Werb, *J. Exp. Med., 152*:1563 (1980).
93. M. C. Bourdillon, D. Brechemeier, N. Blaes, J. C. Derouette, W. Hornebeck, and L. Robert, *Cell. Biol. Int. Rep., 4*:313 (1980).
94. J. Nakao, W-C. Chang, S-I. Murota, and H. Orimo, *Atherosclerosis, 36*:539 (1980).
95. Y. Legrand, J. P. Caen, L. Robert, and J. L. Wautier, *Thromb. Haemostasis, 39*:785 (1975).
96. R. Laub, G. Huybrechts-Gobin, C. Peeters-Joris, and G. Vaes, *Biochim. Biophys. Acta, 721*:425 (1982).
97. B. Johnson-Wint, *Proc. Natl. Acad. Sci. USA, 77*:5331 (1980).
98. C. Biswas, *Biochem. Biophys. Res. Commun., 109*:1026 (1982).

99. M. Kh. Dabbous, M. El-Torky, Sr., B. Brinkley, and N. Sobhy, *Int. J. Cancer, 31*: 357 (1983).
100. N. Henry, A-L. van Lamsweerde, and G. Vaes, *Cancer Res., 43*:5321 (1983).
101. J. Uitto, E. A. Bauer, D. J. Santa Cruz, B. Holtman, and A. Z. Eisen, *J. Invest. Dermatol., 78*:136 (1982).
102. T. M. Hassell, *J. Oral Pathol., 11*:310 (1982).
103. J. Uitto, E. A. Bauer, and A. Z. Eisen, *J. Clin. Invest., 64*:921 (1979).
104. T. I. Morales, J. F. Woessner, J. M. Marsh, and W. J. LeMaire, *Biochim. Biophys. Acta, 756*:119 (1983).
105. J. F. Woessner, *Biochem. J., 180*:95 (1979).
106. J. M. Evanson, J. J. Jeffrey, and S. M. Krane, *Science, 158*:499 (1967).
107. G. S. Lazarus, R. S. Brown, J. R. Daniels, and H. M. Fullmer, *Science, 159*:1482 (1968).
108. E. D. Harris, Jr., D. R. DiBona, and S. M. Krane, *J. Clin. Invest., 48*:2104 (1969).
109. J-M. Dayer, S. M. Krane, R. G. G. Russell, and D. R. Robinson, *Proc. Natl. Acad. Sci. USA, 73*:945 (1976).
110. D. E. Woolley, E. D. Harris, Jr., C. L. Mainardi, and C. E. Brinckerhoff, *Science, 200*:773 (1978).
111. D. E. Woolley, M. J. Crossley, and J. M. Evanson, *Arthritis Rheum., 20*:1231 (1977).
112. H. Nagase, T. E. Cawston, M. DeSilva, and A. J. Barrett, *Biochim. Biophys. Acta, 702*:133 (1982).
113. J. D. Sandy, A. Spiritana, H. L. G. Brown, and D. A. Lowther, *Biochem. J., 193*: 193 (1981).
114. H. Menninger, R. Putzier, W. Mohr, D. Wessinghage, and K. Tillman, *Z. Rheumatol., 39*:145 (1980).
115. M. G. Ehrlich, H. J. Mankin, H. Jones, R. Wright, C. Crisppin, and G. Vigliana, *J. Clin. Invest., 59*:226 (1977).
116. A. I. Sapolsky, H. Keiser, D. S. Howell, and J. F. Woessner, *J. Clin. Invest., 58*: 1030 (1976).
117. K. Deshmukh-Phadke, M. Lawrence, and S. Nanda, *Biochem. Biophys. Res. Commun., 85*:490 (1978).
118. J. T. Dingle, J. Saklatvala, R. Hembry, J. Tyler, H. B. Fell, and R. Jubb, *Biochem. J., 184*:177 (1979).
119. J. Saklatvala, V. A. Curry, and S. J. Sarsfield, *Biochem. J., 215*:385 (1983).
120. J. Saklatvala and S. J. Sarsfield, *Biochem. J. 202*:705, (1982).
121. H. E. Jaslin and J. T. Dingle, *J. Clin. Invest. 68*:571 (1981).
122. J. Saklatvala, S. J. Sarsfield, and L. M. C. Pislworth, *Biochem. J., 209*:337 (1983).
123. M. G. Ehrlich, P. A. Houle, G. Vigliani, and H. J. Mankin, *Arthritis Rheum., 21*: 761 (1978).
124. J. P. Pelletier, J. Martel-Pelletier, D. S. Howell, L. Ghandurmnaymneh, J. E. Enis, and J. F. Woessner, *Arthritis Rheum., 26*:63 (1983).
125. D. J. McCarty, P. B. Halverson, G. F. Carrera, B. J. Brewer, and F. Kozin, *Arthritis Rheum., 24*:464 (1981).
126. H. Birkedal-Hansen, in *Collagenase in Normal and Pathologic Tissue,* Wiley, New York, 1980, p. 127.
127. D. W. Pettigrew, H-M. Wang, J. Sodek, and D. M. Brunette, *J. Periodont. Res., 16*:637 (1981).
128. H. Birdedal-Hansen, C. M. Cobb, R. E. Taylor, and H. M. Fullmer, *Biochim. Biophys. Acta, 429*:229 (1976).
129. H. Birkedal-Hansen, H. Cobb, R. E. Taylor, and H. M. Fullmer, *J. Biol. Chem., 251*:3162 (1976).

130. S. Hurum, J. Sodek, and J. E. Aublin, *Biochem. Biophys. Res. Commun., 107*: 357 (1982).
131. H. Birkedal-Hansen and R. E. Taylor, *Biochem. Biophys. Res. Commun., 107*: 1173 (1982).
132. S. Geiger and E. Harper, *J. Dent. Res., 59*:11 (1980).
133. P. Birek, H-M. Wang, D. M. Brunette, and A. H. Melcher, *Lab. Invest., 43*:61 (1980).
134. M. B. Berman, in *Collagenase in Normal and Pathologic Connective Tissues,* Wiley, New York, 1980, p. 141.
135. M. B. Berman, C. Dohlman, M. Gnadinger, and P. F. Davison, *Exp. Eye Res., 11*: 255 (1971).
136. K. R. Kenyon, M. Berman, J. Rose, and J. Gage, *Invest. Ophthalmol. Visual Sci., 18*:570 (1979).
137. J. M. Gordon, E. A. Bauer, and A. Z. Eisen, *Arch. Ophthalmol., 98*:341 (1980).
138. W. W-Y. Kao, J-P. Vergnes, J. Ebert, C. V. Sundar-Raj, and S. I. Brown, *Biochem. Biophys. Res. Commun., 107*:929 (1982).
139. A. I. Oikarinen, J. J. Zone, A. R. Ahmed, U. Kiistala, and J. Uitto, *J. Invest. Dermatol., 81*:261 (1983).
140. K. H. Singer, N. J. Sawka, H. R. Samowitz, and G. S. Lazarus, *J. Invest. Dermatol., 74*:363 (1980).
141. D. E. Woolley, L. C. Tetlow, C. J. Mooney, and J. M. Evanson, in *Proteinase in Tumor Invasion* (P. Strauli, A. J. Barrett, and A. Baici, eds.), Raven Press, New York, 1980, p. 97.
142. M. Paranjpe, L. Engel, N. Young, and L. A. Liotta, *Life Sci., 26*:1223 (1980).
143. E. A. Bauer, J. M. Gordon, J. M. Reddick, and A. Z. Eisen, *J. Invest. Dermatol., 69*:363 (1977).
144. L. A. Liotta, K. Tryggvason, S. Garbisa, P. Gehron-Robey, and S. Abe, *Nature (London), 284*:67 (1980).
145. T. Salo, L. A. Liotta, J. Keski-Oja, T. Turpeenniemi-Hujanen, and K. Tryggvason, *Int. J. Cancer, 30*:669 (1982).
146. R. H. Kramer and G. L. Nicholson, *Proc. Natl. Acad. Sci. USA, 76*:5704 (1979).
147. R. H. Kramer, K. G. Vogel, and G. L. Nicholson, *J. Biol. Chem., 247*:2678 (1982).
148. I. Ikazaki and K. Maruyama, *J. UOEH, 2*:401 (1980).
149. K. Maruyama, L. Feinman, Z. Fainsilber, M. Nakano, I. Okazaki, and C. S. Lieber, *Life Sci., 30*:1379 (1982).
150. Y. Nagai, H. Hori, R. Hata, H. Konomi, H. Sunada, *Biomed. Res., 3*:345 (1982).
151. K. Fujiwara, T. Saka, T. Oda, and S. Igarashi, *Biochem. Biophys. Res. Commun., 54*:531 (1973).
152. J. E. Gadek, J. A. Kelman, G. Fells, S. E. Weinberger, A. L. Horwitz, H. Y. Reynolds, J. D. Fulmer, and R. G. Crystal, *N. Engl. J. Med., 301*:737 (1979).
153. D. E. Neiwhoehner and J. R. Hoidal, *Science, 217*:359 (1982).
154. C. T. Lee, A. M. Fein, M. Lippman, H. Holtzman, P. Kimbel, and G. Weinbaum, *N. Engl, J. Med., 304*:191 (1981).
155. G. Vaes, in *Collagenase in Normal and Pathologic Connective Tissues,* Wiley, New York, 1980, p. 185.
156. C. Francois-Gillet, J-M. Delaise, Y. Eeckhout, and G. Vaes, *Biochim. Biophys. Acta, 673*:1 (1981).
157. D. E. Woolley, J. S. Tucker, G. Green, and J. M. Evanson, *Biochem. J., 153*:119 (1976).

158. I. Okazaki, K. Maruyama, K. Kashiwazaki, M. Oda, and M. Tsuchiya, in *Collagen Degradation and Mammalian Collagenase,* Excerpta Medica, Amsterdam, 1982, p. 241.
159. Z. Jurukova, *Artery, 8*:275 (1980).
160. Z. Jurukova and C. Milenkov, *Exp. Mol. Pathol., 37*:37 (1982).

Index

A

B

C

D

E

F

G

H

I

J

K

L

M

N

O

P

R

S

T

U

V

W

X

Z

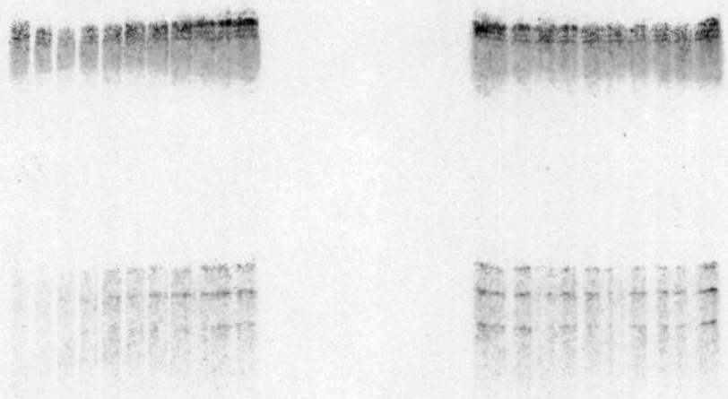

B0[illegible] 11+17+(+1+17

BOOK 10+16+16+16+16=74

G-3807